# Automotive Technician Certification Test Preparation Manual

## Third Edition

## Don Knowles

### Knowles Automotive Training
### Moose Jaw, Saskatchewan
### Canada

DELMAR

THOMSON LEARNING ™

Australia Canada Mexico Singapore Spain United Kingdom United States

**DELMAR**

**THOMSON LEARNING**

Automotive Technician Certification Test Preparation Manual, Third Edition
Don Knowles

**Vice President, Technology Professional Business Unit:**
Gregory L. Clayton

**Product Development Manager:**
Kristen L. Davis

**Product Manager:**
Kimberley Blakey

**Director of Marketing:**
Beth A. Lutz

**Marketing Manager:**
Brian McGrath

**Marketing Coordinator:**
Jennifer Stall

**Production Director:**
Patty Stephan

**Production Manager:**
Andrew Crouth

**Content Project Manager:**
Kara A. DiCaterino

**Art Director:**
Robert Plante

Library of Congress Cataloging-in-Publication Data

Knowles, Don.
   Automotive technician certification test preparation manual / Don Knowles. —3rd ed.
      p. cm.
   Includes index.
   ISBN 1-4180-4926-3
   1. Automobiles—Maintenance and repair—Examinations—Study guides.
2. Automobile mechanics—Certification—United States. I. Title.
   TL152.K575 2006
   629.28'7 076—dc22
                                2006028155

## NOTICE TO THE READER

Publisher does not warrant or guarantee any of the products described herein or perform any independent analysis in connection with any of the product information contained herein. Publisher does not assume, and expressly disclaims, any obligation to obtain and include information other than that provided to it by the manufacturer.

The reader is expressly warned to consider and adopt all safety precautions that might be indicated by the activities herein and to avoid all potential hazards. By following the instructions contained herein, the reader willingly assumes all risks in connection with such instructions.

The publisher makes no representation or warranties of any kind, including but not limited to, the warranties of fitness for particular purpose or merchantability, nor are any such representations implied with respect to the material set forth herein, and the publisher takes no responsibility with respect to such material. The publisher shall not be liable for any special, consequential, or exemplary damages resulting, in whole or part, from the readers' use of, or reliance upon, this material.

# Contents

## 3  Manual Drive Train and Axles

## 4  Suspension and Steering

## 5 Brakes

## 6 Electrical/Electronic Systems

## 7  Heating and Air Conditioning

## 8  Engine Performance

## 9 Advanced Engine Performance

## Glossary ....................................................617

# National Institute for Automotive Service Excellence Certification

## Pretest

The purpose of this pretest is to test your understanding of ASE certification and test questions. If you answer any of the pretest questions incorrectly, complete a careful study of the information in the chapter. Your understanding of ASE test categories will help you in future career planning. A knowledge of ASE test questions is essential before writing any certification tests.

1. ASE was established
   A. at the request of the car manufacturers.
   B. at the request of new car dealers.
   C. at the request of technicians.
   D. as a result of congressional hearings.

2. ASE certification is available in
   A. 8 areas for automobile Master Technician certification.
   B. 6 areas for automobile Master Technician certification.
   C. 2 areas for automobile specialist certification.
   D. 1 area for automobile specialist certification.

3. To maintain valid certification ASE automobile recertification tests must be completed every
   A. 2 years.
   B. 3 years.
   C. 5 years.
   D. 10 years.

4. ASE automobile specialist testing is available in
   A. Engine Machinist, Suspension and Steering, and Alternate Fuels.
   B. Advanced Engine Performance, Electrical, and Brakes.
   C. Engine Machinist, Alternate Fuels, and Advanced Engine Performance.
   D. Alternate Fuels, Engine Performance, and Advanced Engine Performance.

5. While discussing ASE test locations and schedules

   Technician A says a Technician may write the tests at the most convenient location even when this location is in a different state.

   Technician B says ASE tests are held at 450 to 500 locations in the United States.

   Who is correct?
   A. A only
   B. B only
   C. Both A and B
   D. Neither A nor B

6. While discussing the writing of ASE test questions
Technician A says ASE test questions are written and screened by a panel of experts in the automotive service industry.
Technician B says ASE test questions are pretested by automotive instructors selected from across the United States.
Who is correct?
   A. A only
   B. B only
   C. Both A and B
   D. Neither A nor B

7. While discussing the types of ASE test questions
Technician A says some multiple-choice questions require the Technician to select one incorrect response from four responses.
Technician B says some multiple-choice questions contain two statements made by Technicians A and B, and the Technician answering the question must determine if one, both, or neither of the statements are correct.
Who is correct?
   A. A only
   B. B only
   C. Both A and B
   D. Neither A nor B

8. While discussing study procedures in preparation for ASE tests
Technician A says you should study while watching a basketball game on television.
Technician B says you should do all your studying in the last two days before the tests.
Who is correct?
   A. A only
   B. B only
   C. Both A and B
   D. Neither A nor B

9. While discussing the writing of ASE tests
Technician A says you should be careful to align the test book and answer sheet properly.
Technician B says after you have marked the answer to a question you should be sure the question number on the answer sheet is the same as the question number in the test book.
Who is correct?
   A. A only
   B. B only
   C. Both A and B
   D. Neither A nor B

10. The following statements about ASE test questions and results are correct EXCEPT
   A. test results in ASE files are confidential.
   B. test results are mailed to your employer.
   C. some test questions measure the Technician's knowledge of basic theory.
   D. some test questions measure the Technician's knowledge of diagnostic or repair procedures.

## Answers to Pretest

1. D, 2. A, 3. C, 4. C, 5. C, 6. A, 7. C, 8. D, 9. C, 10. B

# Introduction

A careful study of this chapter should help you understand ASE test procedures and certification of Automotive Technicians. This chapter also provides helpful suggestions for ASE test preparation. Your test preparation will play a significant role in your success in passing the tests. Studying the complete book will help you prepare for all the ASE automobile certification tests.

# National Institute for Automotive Service Excellence (ASE)

## Introduction to ASE

During the late 1960s members of the United States Congress received a large volume of complaints about the quality of automotive repair service. As a result of these complaints, congressional hearings were held regarding automotive service. Many experts from various segments of the automotive industry testified at these hearings. From these testimonies it was revealed that the major cause of inaccurate automotive service was technician incompetence. To correct this problem, Congress approved the establishment of the National Institute for Automotive Service Excellence (ASE). Originally, ASE was formed by the National Auto Dealers Association and the Motor Vehicle Manufacturers Association. Shortly thereafter, ASE became an independent nonprofit organization. At the present time in the USA, ASE certified technicians include 293,160 Automobile technicians, 95,481 Master Automobile Technicians and 49,071 Advanced Engine Performance Specialist Technicians.

## ASE Certification

ASE provides voluntary certification tests for Automotive Technicians in the following subject areas:

1. Engine Repair (Test A1)
2. Automatic Transmission/Transaxle (Test A2)
3. Manual Drive Train and Axles (Test A3)
4. Suspension and Steering (Test A4)
5. Brakes (Test A5)
6. Electrical/Electronic Systems (Test A6)
7. Heating and Air-Conditioning (Test A7)
8. Engine Performance (Test A8)

When an Automotive Technician successfully passes all eight ASE automobile certification tests, the Technician is certified as a Master Technician. When a Technician passes an ASE test in one of the eight areas, an Automotive Technician's shoulder patch is issued by ASE. If a Technician passes all eight tests, he or she receives a Master Technician's shoulder patch. Technicians must prove two years' experience in automotive service work prior to certification in any ASE test area. Successful completion of an automotive training program at a recognized institution may be substituted for one year of automotive service work.

Recertification tests are required at 5-year intervals to maintain valid ASE certification. Because the subject areas and tasks are updated periodically, the tests may have been updated several times in a 5-year period. Therefore, the recertification exam tests the Technician's understanding of current technology. Recertification tests contain approximately one-half the questions in each subject area compared to the regular certification tests.

ASE also offers specialist automobile testing with the Engine Machinist, Alternate Fuels, and Advanced Engine Performance Specialist tests. In the Engine Machinist test area Technicians may be certified as a Cylinder Head Specialist, Cylinder Block Specialist, or Assembly Specialist. These three tests are available in the Diesel Engine or Gasoline Engine category.

Certification tests are provided by ASE in the Medium/Heavy Truck technology. Test categories in this subject area include Gasoline Engines, Diesel Engines, Drive Train, Brakes, Suspension and Steering, Electrical Systems, Heating, Ventilation and Air-Conditioning, and Preventative Maintenance Inspection (PMI). In 2000, Technicians who pass the Gasoline Engine or Diesel Engine test plus the Drive Train, Brakes, Suspension and Steering, Electrical Systems, and Preventative Maintenance and Inspection (PMI) tests receive Medium/Heavy Duty Truck Master Technician Certification. In 2001 successful completion of the Heating, Ventilation, and Air-Conditioning test is also required for Master Technician Certification. The Gasoline Engine test will still be available, but it will not be required for Master Technician certification beginning January 1, 2001.

ASE also provides certification tests in Collision Repair and Refinishing tests. These tests include Painting and Refinishing, Nonstructural Analysis and Damage Repair, Structural Analysis and Damage Repair, and Mechanical and Electrical Components. A specialist test is available in Damage Analysis and Estimating.

ASE certification also is available in Parts: Medium/Heavy Truck Parts Specialist, Automobile Parts Specialist, and Medium/Heavy Truck Aftermarket Parts Specialist.

ASE also provides a series of tests to certify Technicians in the School Bus area. School Bus tests include Diesel Engine, Brakes, Suspension and Steering, and Electrical/Electronic Systems. Technicians who pass the Diesel Engine test and the Electrical/Electronic Systems test in the School Bus or Medium/Heavy Truck areas are eligible to write an Electronic Diesel Engine Diagnosis and Specialist test.

A Technician may choose to complete all eight Automobile tests plus the Body Repair and Refinishing tests, and Medium/Heavy Truck tests. Upon the successful completion of all these tests the Technician is certified as a World Class Technician.

# National Automotive Technician's Education Foundation (NATEF) Certification

ASE offers voluntary certification for Automotive and Autobody Technician training programs. This certification of training programs is provided on the recommendations of the National Automotive Technicians Education Foundation (NATEF). NATEF goals include the development, encouragement, and improvement of automotive technical education. This foundation evaluates automotive technical training programs in the following areas:

1. Purpose
2. Administration
3. Learning Resources

4. Finances
5. Student Services
6. Instruction
7. Equipment
8. Facilities
9. Instructional Staff
10. Cooperative Work Agreements

NATEF certification of automotive training programs involves the completion of self-evaluation forms, and a two-day review by an independent inspection team. Training programs must request certification in four of the eight automotive certification areas. These areas include Brakes, Electrical/Electronic Systems, Suspension and Steering, and Engine Performance. NATEF tasks have a priority rating of 1, 2, or 3. An automotive training program must show the ability to deliver 95 percent of the number 1 priority tasks, 80 percent of the number 2 priority tasks, and 50 percent of the number 3 priority tasks. NATEF estimates there are approximately 2,500 secondary and postsecondary Automotive Technology training programs in the United States, and over 900 of these programs are NATEF certified.

## ASE Test Locations and Schedules

ASE tests are given in 450 to 500 locations across the United States and several other countries. Technicians may take the tests of their choice at the most convenient location, even if this location is in a different state. Tests are conducted in May and November each year. An ASE Test Registration Booklet containing a test registration form, a list of test centers, and a test schedule may be obtained by writing to the National Institute for Automotive Service Excellence, 13505 Dulles Technology Drive, Herndon, VA 22071-3415. If there is no test center near your home, arrangements for a special test center may be completed with ASE. The requirements that must be met to establish a special test center include:

1. There must be no test center within 50 miles of your home.
2. A minimum of twenty Technicians must register as a group for each of the test dates requested.
3. Tests at special test centers must be given at the same time as tests at the regular test centers.

During January/February and July/August each year ASE offers computer-based testing at over 200 test centers nationwide. Regular and recertification tests are available in all eight areas of Automobile certification and other test areas. These tests are only available at secure, proctored computer-based testing centers. These tests cannot be taken at home via the internet.

## ASE Tests

ASE tests are written by a committee of experts in the automotive service industry. These committees include automotive instructors, trainers employed by car manufacturers, and test equipment manufacturers. All the test questions are carefully screened by the committee of experts. After this screening the test questions are pretested by Technicians selected from across the United States. The test questions that are not confusing, or easily misinterpreted, become part of a series of questions from which the actual ASE test questions are selected.

ASE test questions are designed to measure the Technician's knowledge of basic theory and diagnostic or repair procedures. Many test questions are based on a specific adjustment, repair, or diagnostic problem. Technicians must have an understanding of the adjustment, repair, or diagnostic problem to answer the question correctly.

ASE test results are confidential. Test results will not be released to anyone without written consent from the Technician.

# Types of ASE Test Questions

Each ASE test contains forty to eighty multiple-choice questions. In this type of question the Technician must read each of the four responses carefully, and then select the response that is the correct answer. In many cases questions are answered incorrectly because the Technician did not take time to thoroughly read the question and all the responses. Always take time to carefully read the complete question and responses!

One type of multiple-choice question has three wrong answers and one correct answer. Some of the responses may be almost correct. Therefore, it is important not to pick one response without reading all the responses. When you are not sure of the answer, eliminate the responses that you know are incorrect to arrive at the correct response. An example of a multiple-choice question with three wrong answers and one correct answer follows.

When a computer has a read-only memory (ROM) chip, the microprocessor can
  A.  read information from the ROM.
  B.  write information into the ROM.
  C.  read information from and write information into the ROM.
  D.  erase information in the ROM.

A careful reading of this question indicates that you are being asked what functions the microprocessor can perform in relation to the ROM chip. Because the information is permanently programmed into a ROM chip, the microprocessor cannot erase information in the ROM. Therefore, response D is wrong. The microprocessor can read information from the ROM, but it cannot write information into the ROM. Therefore, responses B and C are wrong and response A is correct.

Another type of multiple-choice question used on the ASE tests states that all the responses are correct EXCEPT one. Because the correct answer in this type of question is the incorrect response, you must determine which response is wrong. Always read the question and all the responses carefully. An example of a multiple-choice question follows in which you must select the one incorrect response.

The following defines the purpose of engine oil EXCEPT
  A.  provides a sealing action between the piston rings and cylinder walls.
  B.  cleans engine components.
  C.  cools engine components.
  D.  increases the compression ratio.

Engine oil helps to provide a seal between the piston rings and the cylinder walls. Engine components are cooled, cleaned, and lubricated by the engine oil. Therefore, responses A, B, and C are correct. Because compression ratio is determined by the piston stroke and combustion chamber volume, engine oil does not increase the compression ratio. We can conclude that response D is the one incorrect response that must be identified.

Another type of multiple-choice question used in ASE tests involves statements from two Technicians identified as Technician A and Technician B. When answering this type of question you must determine if one, both, or neither of the statements are correct. A multiple-choice question follows with statements from Technician A and Technician B.

Technician A says excessive positive camber on a front wheel may cause excessive wear on the inside edge of the tire tread.

Technician B says excessive toe-in on the front wheels may cause excessive wear on the center of the tire tread.

Who is correct?
A.  A only
B.  B only
C.  Both A and B
D.  Neither A nor B

The first step in answering this question is to determine the effect of excessive positive camber. Because excessive positive camber tilts the wheel outward at the top, this problem causes excessive tread wear on the outside edge of the tire tread. Excessive toe-in caused feathered tread wear across the entire tread, not just on the center of the tire tread. Therefore, neither of the statements by Technician A or Technician B are correct, and the right answer is D.

# ASE Test Preparation

## Studying

Following these suggestions will help you study for the ASE tests:

1.  Leave plenty of time for studying prior to the tests depending on the number of tests you are scheduled to write. Do not attempt to rush through your study in the final days before the tests.

2.  Obtain an ASE test preparation guide.

3.  After studying all of the information regarding the tests you are writing, identify your weak areas and concentrate your study on these areas.

4.  Set a specific study schedule that provides adequate time to complete your studies without interrupting your normal sleep and work schedule.

5.  Select a quiet study location where you will not be interrupted.

6.  Study in a comfortable position. Because drowsiness may reduce your concentration while lying down, avoid studying in this position.

7.  A large meal may cause drowsiness. Do not attempt to study after eating a heavy meal.

8.  Make notes of important facts that you learned during your studies, or highlight these facts in the ASE test preparation guide.

9.  Examine all diagrams carefully in the test preparation guide. Some ASE questions relate to diagrams.

10. Be sure you really understand the theory or service procedure related to each question in the test preparation guide. The questions on the ASE test will not be exactly the same as the ones in the test preparation guide. However, the ASE questions will be related to the same theory or service procedure as the questions in the test preparation guide.

11. Take a short study break every half hour, and complete a few exercises during this break to remain alert.

12. ASE offers online self-assessment quizzes to help technicians prepare for the certification tests. At present these tests are available in Gasoline Engines, Drive Train, Undercar, Electrical/Electronic Systems, and HVAC.

## Preparation Immediately Prior to the Tests

On the day and evening prior to the test your activities may affect your performance during the tests. For example, if you watch a late movie on the night before the test, you may be tired during the test, and this tiredness may reduce your ability to read and think during the test. The following suggestions regarding the day and night before the test may improve your performance during the test.

1. Be sure you know the location of the building and the room where you will write the tests. If you are writing the tests in a city that you are not familiar with, drive to the test location and find the proper visitor's parking area; then find the ASE test room location.

2. Limit your study to reviewing your notes or highlighted areas in the ASE test preparation guide.

3. Maintain your regular schedule for eating and sleeping.

4. Place everything you need for writing the tests in a briefcase or suitable bag. These items include your admission ticket, photo ID, four sharpened Number 2 pencils, and a watch.

5. Do not drink alcoholic beverages.

6. Avoid stressful situations on the day before the test.

7. Maintain mental alertness in the hours before the test. For example, you will probably be more mentally alert if you participated in some moderate exercise compared to your mental alertness if you watched three hours of television.

8. Know the exact test schedule and determine the time you will spend on each test.

9. Arrive at the ASE test site fifteen to twenty minutes prior to the beginning of the tests.

## Taking the Tests

Prior to the tests the monitor will provide you with some verbal instructions. Listen carefully to these instructions, and read the directions carefully in the ASE test booklet. Question the monitor about anything you do not understand before the tests begin.

Align the test book and answer sheet carefully before you begin answering any of the questions. Check this alignment periodically as you answer the questions. Do not place any marks on the answer sheet other than your answers. During the correction process the computer may mistake other marks on the answer sheet for an answer. When you need scratch paper you may write in the test booklet.

Do not rush as you take the tests. Pace yourself according to your planned schedule. If you are uncertain about the answer to a question, place a check mark beside the question in the test booklet and proceed to the next question. Answer the questions that you feel confident about the answers. When you have completed the test, go back and answer the check-marked questions. Follow these suggestions while taking the tests:

1. Read each question completely and carefully so you understand the question.

2. After you have marked the answer to a question, be sure the question number on the answer sheet matches the question number in the test booklet.

3. Do not spend too much time on any question. Maintain your planned schedule to complete the tests.

4. Answer every question. If you are uncertain about the answer to a question, eliminate the wrong answers to arrive at the correct answer, as mentioned previously. Often a careful analysis of the question and possible answers will help you arrive at the correct answer. It is better to provide an answer that you are unsure of rather than leaving a question unanswered.

5. Review your answers when you have completed the test.

6. After you have answered a question do not change your answer unless you are absolutely sure it is wrong.

# About the Author

Don Knowles was employed as an automotive instructor at the Saskatchewan Institute of Applied Science and Technology (SIAST) Palliser Campus from 1961 to 1987. During his tenure at SIAST, he produced over 60 automotive training videos in cooperation with the SIAST audio visual department and wrote a computer-assisted instruction (CAI) program coordinated with each video. Don's specialty was compiling and teaching mobile training programs for experienced technicians.

Don has been writing automotive textbooks and other training materials since 1981. During this time he has written 50 automotive textbooks, student manuals, and training CDs in automotive and medium/heavy truck technology. Don has also written many short articles for service bulletins.

Since 1987 Don has been owner/manager of Knowles Automotive Training. Knowles Automotive Training is an educational member of the Automotive Service Association (ASA). Don is ASE L1 certified and ASE master-certified in automotive and medium/heavy truck. He has been a member of the Society of Automotive Engineers (SAE) for over 25 years and is also a member of the Council of Advanced Automotive Trainers (CAAT) and the North American Council of Automotive Teachers (NACAT).

# Introduction

The purpose of this guide is to help you successfully complete all eight ASE Automotive Certification tests. This guide contains the same types of questions as the ASE tests. Approximately 20 percent of the questions on ASE tests are based on a diagram. Similarly, the same percentage of questions in this guide are based on a diagram. Question answers and analysis are provided at the end of each chapter and on separate answer sheets provided with the guide. A pretest at the beginning of each chapter gives the reader an indication of the amount of study he or she requires to successfully complete the ASE test in that subject area. Three sample questions from the guide follow:

1. A vehicle with a spark control system has an open wire between the knock sensor and the knock sensor module. This may result in higher than specified emissions of

   A. carbon monoxide (CO).

   B. carbon dioxide ($CO_2$).

   C. hydrocarbons (HC).

   D. oxides of nitrogen (NOx).

Answer D

A defective spark control system may allow excessive spark advance, which causes engine detonation. Because detonation causes high cylinder temperature, this action results in NOx emissions. A, B, and C are wrong, and D is correct.

2. A fuel-injected engine has a hard starting problem after it is shut off for several hours. The vehicle driveability and emission levels are normal. The ignition system is always firing during cold cranking.
   Technician A says the fuel pump check valve may be leaking.
   Technician B says the ECT sensor may be the problem.
   Who is correct?

   A. A only

   B. B only

   C. Both A and B

   D. Neither A nor B

Answer A

A leaking fuel pump check valve causes the fuel to drain back out of the fuel system into the tank when the engine is shut off. This action causes hard starting without any other driveability or emission problems. A is correct. A malfunctioning ECT sensor may cause hard starting, but this problem also results in other driveability and emission problems. Therefore, B is wrong.

3. All of the following problems may be caused by an engine thermostat that is stuck open EXCEPT

   A. a rich air–fuel ratio and reduced fuel economy.

   B. an inoperative EGR system.

   C. improper cooling fan operation.

   D. an improper air charge temperature sensor signal.

Answer D

This question asks for the statement that is not true. If the engine thermostat is stuck open, the air–fuel ratio is rich and fuel economy is reduced because the ECT sensor informs the PCM regarding the lower coolant temperature. This problem also affects the EGR and cooling fan operation. Therefore, A, B, and C are true, but none of these is the requested answer.

Because the air charge sensor sends a signal to the PCM in relation to air intake temperature, this signal is not affected by a thermostat that is stuck open. Therefore, D is not true, and this is the requested answer.

# 1 Engine Repair

## Pretest

The purpose of this pretest is to determine the amount of review you may require prior to taking the ASE Engine Repair Test. If you answer all the pretest questions correctly, complete the questions and study the information in this chapter to prepare for the ASE Engine Repair Test.

If two or more of your answers to the pretest questions are incorrect, complete a thorough study and review of the questions and information in this chapter. The pretest answers are located at the end of the pretest.

1. If an engine experiences a sharp, metallic rapping noise only when it is idling, the trouble could be worn
   A. main bearings.
   B. piston pins.
   C. connecting rod bearings.
   D. piston skirts.

2. On a port fuel-injected engine, the cause of excessive black smoke in the exhaust could be
   A. low fuel pressure.
   B. an intake manifold vacuum leak.
   C. high oxygen sensor voltage.
   D. high fuel pressure.

3. During a cylinder power balance test on a four-cylinder port fuel-injected engine, one cylinder has a 75-rpm drop and the other cylinders have a 125-rpm drop.
   Technician A says the cylinder with a 75-rpm drop may have a burned exhaust valve.
   Technician B says the cylinder with a 75-rpm drop may have a bad fuel injector.
   Who is correct?
   A. A only
   B. B only
   C. Both A and B
   D. Neither A nor B

4. While discussing cylinder head service, Technician A says torque-to-yield bolts may be reused if they are not damaged.
   Technician B says torque-to-yield bolts are usually tightened to a specified torque and then rotated a certain number of degrees.
   Who is correct?
   A. A only
   B. B only
   C. Both A and B
   D. Neither A nor B

5. The cause of excessive valve stem height could be
   A. an improperly installed valve guide.
   B. excessive material machined from the head surface.
   C. a worn valve guide and stem.
   D. excessive material removed from the valve seat.

6. While diagnosing the timing belt condition, Technician A says some vehicle manufacturers recommend checking timing belt wear by measuring the installed length of the belt tensioner.
   Technician B says that on some engines the valves will strike the top of the pistons if the timing belt jumps several notches on the camshaft sprocket.
   Who is correct?
   A. A only
   B. B only
   C. Both A and B
   D. Neither A nor B

7. While discussing engine sealants, Technician A says RTV sealer dries in the absence of air.
   Technician B says anaerobic sealer dries in the absence of air.
   Who is correct?
   A. A only
   B. B only
   C. Both A and B
   D. Neither A nor B

8. While measuring cylinder taper and out-of-round, Technician A says the cylinder diameter should be measured at two vertical locations.
   Technician B says cylinder out-of-round is the difference between the cylinder diameter in the thrust and axial directions.
   Who is correct?
   A. A only
   B. B only
   C. Both A and B
   D. Neither A nor B

9. While discussing connecting rod service, Technician A says each connecting rod should be measured for a bent condition.
   Technician B says each connecting rod should be measured for a twisted condition.
   Who is correct?
   A. A only
   B. B only
   C. Both A and B
   D. Neither A nor B

10. A cooling system is pressurized to 15 psi with a pressure tester. After five minutes, the pressure reading on the tester is 3 psi and there are no external leaks.
    Technician A says the automatic transmission cooler may be leaking.
    Technician B says the head gasket may be leaking.
    Who is correct?
    A. A only
    B. B only
    C. Both A and B
    D. Neither A nor B

11. When diagnosing a no-start condition, if a 12 V test light connected from the negative primary coil terminal to ground flashes while cranking the engine, the trouble could be a malfunctioning
    A. pickup coil.
    B. ignition module.
    C. ignition coil.
    D. ignition switch.

12. Technician A says an intake manifold vacuum leak may cause cylinder misfiring at idle speed.
    Technician B says an intake manifold vacuum leak may cause cylinder misfiring at 2,500 rpm.
    Who is correct?
    A. A only
    B. B only
    C. Both A and B
    D. Neither A nor B

13. An accumulation of oil in the air cleaner may be caused by
    A. a partially restricted PCV valve.
    B. a PCV valve that is stuck open.
    C. a leaking rocker arm cover gasket.
    D. a leaking PCV clean air hose.

14. All of the following statements about a battery load test are true EXCEPT that
    A. The battery should be discharged at one-half the cold cranking rating during the test.
    B. The battery temperature should be above 40°F before performing the test.
    C. The battery's specific gravity should be above 1,190 before performing the test.
    D. The battery is satisfactory if the voltage remains above 8.8 V after fifteen seconds with the battery temperature at 70°F.

# Answers to Pretest

**1.  B**    Loose main bearings cause a thumping noise especially when the engine is first started. Worn connecting rod bearings result in a clattering noise that is more pronounced when the engine is accelerated and decelerated. Worn piston skirts cause a slapping noise that is most noticeable when the engine is cold. Worn piston pins result in a sharp, metallic rapping noise during engine idle. B is the correct answer.

**2.  D**    Low fuel pressure, an intake manifold vacuum leak, or high oxygen sensor voltage cause a lean air-fuel ratio that does not produce any black smoke in the exhaust. High fuel pressure results in excessive fuel being injected each time the injectors open. This condition causes a rich air-fuel ration and black smoke in the exhaust. Thus, D is the correct answer.

**3.  C**    Technicians A and B are both correct because either a burned exhaust valve or a defective fuel injector result in partial misfiring of the cylinder and less rpm drop on that cylinder during a cylinder balance test. Therefore, C is the correct answer.

**4.  B**    Technician A is wrong because torque-to-yield bolts must never be reused. Technician B is correct because torque-to-yield bolts are usually tightened to a specified torque and then rotated a certain number of degrees. B is the correct answer.

**5.  D**    An improperly installed valve guide, or excessive material machined from the cylinder head surface do not affect the valve stem height. A worn valve guide and stem allows excessive movement of the stem within the guide, but this condition does not affect valve stem height. Excessive material removed from the valve seat moves the valve stem upward and results in excessive valve stem height. Therefore, D is the correct answer.

**6.  C**    Some vehicle manufacturers recommend checking timing belt tension by measuring the installed length of the belt tensioner. In some engines, the valves strike the top of the pistons if the timing belt jumps several notches. Therefore, Technicians A and B are both right and C is the correct answer.

**7.  B**    RTV sealant dries in the presence of air, and anerobic sealant dries in the absence of air. Therefore, Technician A is wrong and Technician B is right. B is the correct answer.

**8.  B**    Cylinder diameter should be measured at three vertical locations. Therefore, Technician A is wrong. Cylinder out-of-round is the difference between the cylinder diameter in the thrust (crosswise) and axial (lengthwise) directions. Technician B is correct. Thus, B is the right answer.

**9.  C**    Connecting rods should be measured for bent and twisted conditions. Therefore, Technicians A and B are both right, and C is the correct answer.

**10.  C**    A leaking transmission cooler may cause a coolant leak into the transmission fluid. A leaking head gasket may cause a coolant leak into a cylinder. Either defect causes the pressure to drop during a cooling system pressure test. Therefore, Technicians A and B are both right and C is the correct answer.

**11.  C**    When a 12 V test light connected from the negative primary coil terminal to ground flashes while cranking the engine, the ignition switch must be supplying voltage to the primary circuit, and the module and pickup coil must be functioning properly. Therefore A, B, and D are wrong, and C is the correct answer.

**12.  A**    An intake manifold vacuum leak causes cylinder misfiring at idle speed, because vacuum is high under this condition, and the leak leans out the air-fuel ratio. When the engine is operating at a higher rpm, the intake vacuum decreases and the vacuum leak does not affect the air-fuel ratio as much as it does at idle. Therefore, Technician A is right, and Technician B is wrong. A is the correct answer.

**13.  A**    A partially restricted PCV valve causes pressure buildup in the engine, and oil vapor may be forced through the PCV clean air hose into the air cleaner. A PCV valve that is stuck open, a leaking rocker arm cover gasket, or PCV clean air hose do not cause this problem. Therefore, A is the correct answer.

**14.  D**    During a battery load test, the battery should be discharged at one-half the cold cranking rating, the battery temperature should be above 40°F before the test, and the specific gravity should be above 1,190 before performing the test. If the battery is satisfactory, the voltage remains above 9.6 V at 70°F. Therefore, A, B, and C, are wrong, and D is the correct answer.

# General Engine Diagnosis

## ASE Tasks, Questions, and Related Information

In this chapter, each task in the Engine Repair category is followed by a question and some information related to the task. If you answer any question incorrectly, study this information very carefully until you understand the correct answer. Question answers and analysis are provided at the end of this chapter and in the answer sheets provided with this book.

## Task 1  Verify driver's complaint and/or road test vehicle; determine necessary action.

1. While discussing a basic diagnostic procedure, Technician A says the most complicated diagnostic tests should be performed first.

   Technician B says the customer complaint must be identified.

   Who is correct?

   A. A only

   B. B only

   C. Both A and B

   D. Neither A nor B

**Hint**     *The Technician must be familiar with a basic diagnostic procedure such as the following:*
- *Listen carefully to the customer's complaint, question the customer, and check technical service bulletins (TSBs) to obtain more information regarding the problem.*
- *Identify the complaint, road test the vehicle if necessary.*
- *Think of the possible causes of the problem.*
- *Perform diagnostic tests to locate the exact cause of the problem. Always start with the easiest, quickest test.*
- *Be sure the customer's complaint is eliminated. Road test the vehicle if necessary.*

## Task 2  Determine if no-crank, no-start, or hard starting condition is an ignition system, cranking system, fuel system, exhaust system, or engine mechanical problem.

2. A throttle body injected engine has a no-start complaint, and no fuel is discharged from the injectors when the engine is cranking.

   Technician A says to check the voltage supply to the fuel pump.

   Technician B says the fuel filter may be partially restricted.

   Who is correct?

   A. A only

   B. B only

   C. Both A and B

   D. Neither A nor B

**Hint**     *A test spark plug may be connected to several spark plug wires to ground while cranking the engine. If the test spark plug is firing, the ignition system is probably not the cause of the problem.*
*When the cranking system is the cause of one of these problems, the engine will refuse to crank, or it will be cranking slowly. A starter draw test may be performed to indicate the starter condition.*
*On a throttle body injection system, the fuel may be seen discharging from the injectors when the engine is cranked. Always be sure the fuel system has the specified fuel pressure. Noid lights may be connected across the injector terminals on port-injected engines to determine if the injectors are being energized. The noid lights should flash when the engine is cranking. Many engine mechanical problems that cause no-start or hard-start conditions may be diagnosed by performing a compression test and visually inspecting the valve timing.*

## Task 3  Inspect engine assembly for fuel, oil, coolant, and other leaks; determine necessary action.

3. A cooling system is pressurized with a pressure tester to locate a coolant leak. After 15 minutes, the tester gauge has dropped from 15 psi to 5 psi, and there are no visible signs of coolant leaks in the engine compartment.

Technician A says the engine may have a leaking head gasket.

Technician B says the heater core may be leaking.

Who is correct?

A.  A only

B.  B only

C.  Both A and B

D.  Neither A nor B

**Hint**    *Basic fuel, lubricating, and cooling systems and components must be understood. The location of all possible leaks in these systems must be identified. Coolant leaks may be internal or external in relation to the engine.*

## Task 4    Listen to engine noises and vibrations; determine necessary action.

4.  A heavy thumping noise occurs with the engine idling, but the oil pressure is normal.

This noise may be caused by

A.  worn pistons and cylinders.

B.  loose flywheel bolts.

C.  worn main bearings.

D.  loose camshaft bearings.

**Hint**    *You must be familiar with noises caused by bad main bearings, connecting rod bearings, pistons, piston pins, piston rings, and ring ridge in the engine block.*

*Valve train and camshaft noises must be understood. You must be able to recognize combustion chamber noise and flywheel and vibration damper noises. A defective or sticking valve lifter causes a clicking noise that is usually most noticeable when the engine is first started. Worn main bearings cause a thumping noise when the engine is first started, and this problem also results in low oil pressure. Loose connecting rod bearings cause a clattering noise that is most noticeable when the engine is accelerated and decelerated.*

## Task 5    Diagnose the cause of excessive oil consumption, coolant consumption, unusual engine exhaust color, odor, and sound; determine necessary action.

5.  A port fuel-injected engine has a steady "puff" noise in the exhaust with the engine idling. The cause of this problem could be

A.  a burned exhaust valve.

B.  excessive fuel pressure.

C.  a restricted fuel return line.

D.  a sticking fuel pump check valve.

**Hint**    *The causes of excessive oil consumption such as worn rings, scored cylinder walls, worn turbocharger seals, oil leaks, or worn valve guides, seals, and stems must be understood.*

*Unusual exhaust colors and their causes include:*

- *Blue exhaust—Excessive oil consumption, may be more noticeable on acceleration and deceleration*
- *Black exhaust—Rich air-fuel ratio, excessive fuel consumption*
- *Gray exhaust—Coolant leaking into the combustion chambers. This may be more noticeable when the engine is first started. Normal exhaust noise contains steady pulses at the tailpipe. A puff noise in the exhaust at regular intervals usually indicates a cylinder misfire caused by a compression, ignition, or fuel system problem. Erratic exhaust pulses at the tailpipe indicate a rough idle condition caused by ignition or fuel system defects. Excessive exhaust noise indicates a leak in the exhaust system.*

*A high-pitched squealing noise during hard acceleration may be caused by a small leak in the exhaust system, particularly in the exhaust manifolds or exhaust pipe. An intake manifold vacuum leak causes a high-pitched whistle at idle and low speeds. This whistle gradually decreases when the engine is accelerated and the intake vacuum decreases.*

*An excessive sulfur smell in the exhaust indicates a rich air-fuel ratio on vehicles with catalytic converters.*

## Task 6   Perform engine vacuum tests; determine necessary action.

6. With the engine idling, a vacuum gauge connected to the intake manifold fluctuates as illustrated in Figure 1–1. These vacuum gauge fluctuations may be caused by
   A. late ignition timing.
   B. intake manifold vacuum leaks.
   C. a restricted exhaust system.
   D. sticky valves.

**Figure 1–1** Vacuum gauge reading. *(Courtesy of Sun Electric Corporation)*

**Hint**   *The vacuum gauge should be connected directly to the intake manifold to diagnose engine and related system conditions. When a vacuum gauge is connected to the intake manifold, the reading on the gauge should provide a steady reading between 17 and 22 inches of mercury (in. Hg) with the engine idling. Abnormal vacuum gauge readings indicate the following problems:*
   * *A low, steady reading indicates late ignition timing.*
   * *If the vacuum gauge reading is steady and much lower than normal, the intake manifold has a significant leak.*
   * *When the vacuum gauge pointer fluctuates between approximately 11 and 16 in. Hg on a carbureted engine at idle speed, the carburetor idle mixture screws require adjusting. On a fuel-injected engine, the injectors require cleaning or replacing.*
   * *Burned or leaking valves cause a vacuum gauge fluctuation between 12 and 18 in. Hg.*
   * *Weak valve springs result in a vacuum gauge fluctuation between 10 and 25 in. Hg.*
   * *A leaking head gasket may cause a vacuum gauge fluctuation between 7 and 20 in. Hg.*
   * *If the valves are sticking, the vacuum gauge fluctuates between 14 and 18 in. Hg.*
   * *When the engine is accelerated and held at a steady higher rpm, if the vacuum gauge pointer drops to a very low reading, the catalytic converter or other exhaust system components are restricted.*

## Task 7   Perform cylinder power balance tests; determine necessary action.

7. During a cylinder balance test on an engine with electronic fuel injection, cylinder number 3 provides very little rpm drop.
   Technician A says the ignition system may be misfiring on the number 3 cylinder.
   Technician B says the engine may have an intake manifold vacuum leak.
   Who is correct?
   A. A only
   B. B only
   C. Both A and B
   D. Neither A nor B

*Hint*    *If the cylinder is working normally, a noticeable rpm decrease occurs when the cylinder misfires.*

*If there is very little rpm decrease when the analyzer causes a cylinder to misfire, the cylinder is not contributing to engine power. Under this condition, the engine compression, ignition system, and fuel system should be checked to locate the cause of the problem. An intake manifold vacuum leak may cause a cylinder misfire with the engine idling or operating at low speed. If this problem exists, the misfire will disappear at a higher speed when the manifold vacuum decreases. When all the cylinders provide the specified rpm drop, the cylinders are all contributing equally to the engine power.*

## Task 8   Perform cylinder cranking compression tests; determine necessary action.

8. During a compression test, a cylinder has 40 percent of the specified compression reading. When the technician performs a wet test, the compression reading on this cylinder is 75 percent of the specified reading. The cause of the low compression reading could be

   A. a burned exhaust valve.

   B. worn piston rings.

   C. a bent intake valve.

   D. a worn camshaft lobe.

*Hint*    *The ignition and fuel injection system must be disabled before proceeding with the compression test. During the compression test, the engine is cranked through four compression strokes on each cylinder and the compression readings recorded. Lower-than-specified compression readings may be interpreted as follows:*

   • *Low compression readings on one or more cylinders indicate worn rings, valves, a blown head gasket, or a cracked cylinder head. A gradual buildup on the four compression readings on each stroke indicates worn rings, whereas little buildup on the four strokes usually is the result of a burned exhaust valve.*

   • *When the compression readings on all the cylinders are even but lower than the specified compression, worn rings and cylinders are indicated.*

   • *Low compression on two adjacent cylinders is caused by a leaking head gasket or cracked cylinder head.*

   • *Higher-than-specified compression usually indicates carbon deposits in the combustion chamber.*

   • *Zero compression on a cylinder usually is caused by a hole in a piston, or a severely burned exhaust valve. If the zero compression reading is caused by a hole in the piston, the engine will have excessive blowby.*

*When the engine spins freely and compression in all cylinders is low, check the valve timing.*

*If a cylinder compression reading is below specifications, a wet test may be performed to determine if the valves, or the rings, are the cause of the problem. Squirt approximately two or three teaspoons of engine oil through the spark plug opening into the cylinder with the low compression reading.*

*Crank the engine to distribute the oil around the cylinder wall and then retest the compression.*

*If the compression reading improves considerably, the rings (or cylinders) are worn. When there is little change in the compression reading, one of the valves is leaking.*

## Task 9   Perform cylinder leakage tests; determine necessary action.

9. During a leakage test, cylinder number 2 has 50-percent leakage and air is escaping from the PCV valve opening.

   Technician A says the intake valve in the number 2 cylinder may be leaking.

   Technician B says the rings in the number 2 cylinder may be worn.

   Who is correct?

   A. A only

   B. B only

C. Both A and B

D. Neither A nor B

**Hint**   *During the leakage test, a regulated amount of air from the shop air supply is forced into the cylinder with both exhaust and intake valves closed. The gauge on the leakage tester indicates the percentage of leakage in the cylinder. A gauge reading of 0 percent indicates there is no cylinder leakage, and if the reading is 100 percent, the cylinder is not holding any air.*

*If the reading on the leakage tester exceeds 20 percent, check for air escaping from the tailpipe, positive crankcase valve (PCV) opening, and the top of the throttle body or carburetor. Air escaping from the tailpipe indicates an exhaust valve leak. When the air is coming out of the PCV valve opening, the piston rings are leaking. An intake valve is leaking if air is escaping from the top of the throttle body, or carburetor. Remove the radiator cap and check the coolant for bubbles, which indicate a leaking head gasket or cracked head.*

# Cylinder Head and Valve Train Diagnosis and Repair

## ASE Tasks, Questions, and Related Information

**Task 1**   Remove cylinder heads, disassemble, clean, and prepare for inspection.

10. Technician A says a steel scraper should be used to clean the gasket surfaces on aluminum cylinder heads.

   Technician B says some manufacturers provide a detorque sequence for head bolts in aluminum cylinder heads.

   Who is correct?

   A. A only

   B. B only

   C. Both A and B

   D. Neither A nor B

**Hint**   *When removing, cleaning, and inspecting aluminum cylinder heads, some special precautions must be observed. Because aluminum cylinder heads are more subject to warping, some vehicle manufacturers provide a detorque sequence for loosening the head bolts on these heads. Use a plastic scraper to remove gasket material from aluminum cylinder heads. A steel scraper may gouge aluminum cylinder heads and this may result in a coolant leak. Be sure the workbench is clean, and set aluminum cylinder heads on several clean shop towels on the bench. Sharp objects on the bench may scratch or gouge aluminum heads.*

**Task 2**   Visually inspect cylinder heads for cracks, warpage, corrosion, and leakage, and check passage condition; determine needed repairs.

11. The feeler gauge measurement in Figure 1–2 is 0.014 in. (0.35 mm).

   Technician A says the cylinder head should be resurfaced.

   Technician B says block warpage should be measured.

   Who is correct?

   A. A only

   B. B only

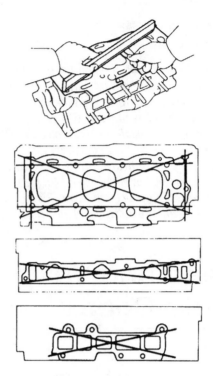

**Figure 1–2** Measuring cylinder head warpage. *(Reprinted with permission)*

C. Both A and B

D. Neither A nor B

*Hint*    *An electromagnetic-type tester and iron filings may be used to check for cracks in cast iron heads. A dye penetrant may be used to locate cracks in aluminum heads.*

*Use a straightedge and a feeler gauge to measure cylinder head warpage. Place the straightedge diagonally across the head-to-block surface at two locations. Use the same method to check warpage on the cylinder head, intake manifold, and exhaust manifold mounting surfaces. The old head gasket may be examined to determine if proper sealing existed between the head and block.*

**Task 3**    **Inspect and verify valve springs for squareness, pressure, and free height comparison; replace as necessary.**

12. While discussing the measurement in Figure 1–3, Technician A says the valve spring is being measured for installed height.

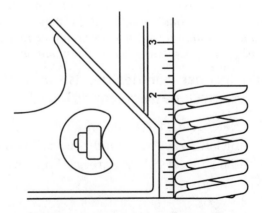

**Figure 1–3** Valve spring measurement.

Technician B says the valve spring must be held in one position.

Who is correct?

A.  A only

B.  B only

C.  Both A and B

D.  Neither A nor B

13.  While discussing valve spring tension measurement (Figure 1–4), Technician A says the valve spring tension should be measured at the installed spring height.

Technician B says if a shim is to be used under the valve spring, the shim should be installed under the spring on the tension gauge.

Who is correct?

A.  A only

B.  B only

C.  Both A and B

D.  Neither A nor B

**Figure 1–4** Valve spring measurement.

*Hint*      *Valve springs must be measured for free length and squareness by placing each spring against a steel square and surface plate on a level surface. Free length is the height of the valve spring measured with no tension on the spring. Squareness is the vertical straightness of the valve spring. The valve spring must be replaced if it does not have the specified free length. When the spring is rotated, a spring height variance of more than 1/16 in. indicates a bent spring that must be replaced.*

*A valve spring tester is used to measure valve spring tension. When a valve spring is placed in the valve spring tension gauge, the spring tension must be measured at the installed valve spring height. If a shim is to be installed under the spring, the shim must be placed under the spring on the tension gauge.*

## Task 4    Inspect valve spring retainers, rotators, locks, and valve lock grooves.

14. Technician A says worn valve lock grooves may cause the valve locks to fly out of place with the engine running, resulting in severe engine damage.
    Technician B says worn valve lock grooves may cause a clicking noise with the engine idling.
    Who is correct?
    A. A only
    B. B only
    C. Both A and B
    D. Neither A nor B

*Hint*    *Valve spring retainers and locks must be checked for wear, scoring, or damage. When any of these conditions are present, replace the components. The valve lock grooves on the valve stems must be inspected for wear, particularly round shoulders. If these shoulders are uneven or rounded, replace the valve.*

## Task 5    Replace valve stem seals.

15. While discussing valve stem seals, Technician A says worn valve stem seals may cause rapid valve stem and guide wear.
    Technician B says worn valve stem seals may cause excessive oil consumption.
    Who is correct?
    A. A only
    B. B only
    C. Both A and B
    D. Neither A nor B

*Hint*    *Prior to valve installation, the valve stem and guide should be lubricated lightly with the manufacturer's specified engine oil. Install the valve spring seat and slide the seal over the top of the valve stem. Be careful not to damage the seal on the valve lock grooves. The seal must be properly positioned on the valve stem. Some seals are pushed down over the guide.*

## Task 6    Inspect valve guides for wear; check valve guide height and stem-to-guide clearance; determine needed repairs.

16. While discussing valve stem-to-guide measurement, Technician A says the valve stems and guides should be measured at three vertical locations.
    Technician B says the valve guide diameter should be measured with a hole, or snap, gauge.
    Who is correct?
    A. A only
    B. B only
    C. Both A and B
    D. Neither A nor B

17. The measurement being performed in Figure 1–5 is
    A. valve concentricity.
    B. valve seat concentricity.
    C. valve face-to-seat contact.
    D. valve stem-to-guide clearance.

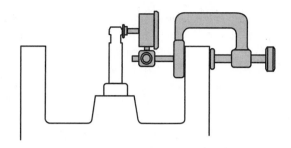

**Figure 1–5** Valve measurement.

**Hint**    *The valve guide should be measured near the top and bottom, and in the center with a hole gauge. Measure the valve stem diameter with a micrometer in the same three positions, and subtract the stem readings from the guide measurements to obtain the clearance. An alternate method for measuring stem-to-guide clearance is to install the valve in the guide with the valve 1/8 in. off the seat. Mount a dial indicator against the valve stem below the lock groove and move the valve stem from side-to-side while observing the clearance reading on the dial indicator.*

*If the valve stem-to-guide clearance is more than specified, the valve guides may be replaced, knurled, or bored out and a thin-wall liner installed. Excessive valve stem-to-guide clearance may result in improper valve seating and lower compression. Increased oil consumption may result from excessive valve stem-to-guide clearance.*

*Valve guide height usually is measured from the top of the spring seat to the top of the guide.*

## Task 7  Inspect valves and valve seats; determine needed repairs.

18. A valve margin of 1/64 in. may cause
    A. a clicking noise at idle speed.
    B. valve overheating and burning.
    C. improper valve seating.
    D. valve seat recession.

19. The valves in Figure 1–6 are from an engine with valve rotators. As indicated in the figure, the valve stem tip wear patterns indicate
    A. valve 1 has a normal wear pattern.
    B. valve 1 has a worn-out rotator.
    C. valve 2 has a worn valve guide.
    D. valve 3 has a weak valve spring.

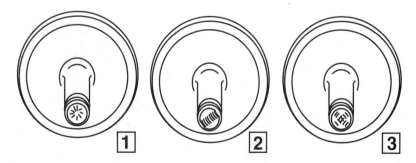

**Figure 1–6** Valve stem tip condition.

20. Valve seats are typically ground to an angle of
    A. 15 or 20 degrees.
    B. 20 or 30 degrees.
    C. 30 or 45 degrees.
    D. 45 or 60 degrees.

*Hint*        *Measure the overall length of the valves. If the length of any valve is not within the manufacturer's specifications, replace the valve. Inspect the surface of the valve tip for wear and score marks. When these conditions are present, resurface the tip on a valve stem grinder, or replace the valve. Do not grind the valve stem tip until the overall valve length is less than specified.*

*Measure the valve margin thickness. When this thickness is less than specified, replace the valve. If the valve margin is less than 1/32 in. after resurfacing, valve overheating and reduced valve life occur. Reface the valve face on a valve grinder. If the specified valve seat angle is 45 degrees, many vehicle manufacturers recommend grinding the valve face to an interference angle of 44.5 degrees.*

*The proper sized pilot is placed in the valve guide prior to valve seat resurfacing. A grinding wheel is placed over this pilot, and the wheel is rotated with an electric drive tool. A grinding stone of the proper size and angle must be installed on the grinding wheel. This grinding wheel must fit on the valve seat without touching any other part of the head surface.*

*Valve seat inserts may be removable or integral with the cylinder head. Removable valve seat inserts may be removed with a special puller or a pry bar. A special driver is used to install the valve seat insert, and the insert should be staked after installation.*

## Task 8    Check valve face-to-seat contact and valve seat concentricity (runout).

21. While discussing a valve seat contact area that is too low on the valve face, Technician A says a 60-degree and 45-degree grinding stone should be used to raise the valve seat contact area.

    Technician B says the low valve seat contact area on the valve face may cause a clicking noise at idle speed.

    Who is correct?
    A. A only
    B. B only
    C. Both A and B
    D. Neither A nor B

*Hint*        *After the valve seats are resurfaced, install blue on the valve face and install the valve against the seat. Remove the valve and observe the seat pattern on the valve face. If the blue appears 360 degrees around the valve face, the valve is concentric. When the blue does not appear 360 degrees around the valve face, replace the valve. A valve seat concentricity tester containing a dial indicator may also be used to measure valve seat concentricity (Figure 1–7).*

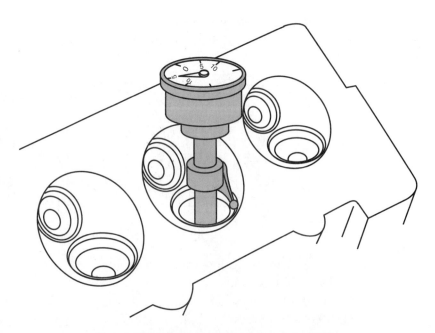

**Figure 1–7** Measuring valve seat concentricity.

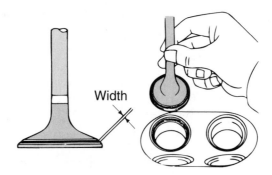

**Figure 1–8** Checking valve seat contact area and width on the valve face. *(Reprinted with permission)*

*The seat contact area on the valve face should be the width specified by the vehicle manufacturer.*

*This contact area must be located in the center of the valve face (Figure 1–8).*

*When the seat contact area is too high on the valve face, use a 30-degree and a 45-degree grinding stone to lower the seat contact area (Figure 1–9). If the seat contact area is too low on the valve face, use a 60-degree and a 45-degree grinding stone to raise the seat area (Figure 1–10).*

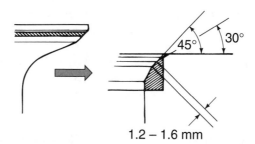

**Figure 1–9** Using a 30-degree and a 45-degree grinding stone to lower the seat contact area on the valve face. *(Reprinted with permission)*

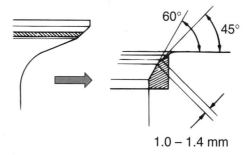

**Figure 1–10** Using a 60-degree and a 45-degree grinding stone to raise the seat contact area on the valve face. *(Reprinted with permission)*

## Task 9  Check valve spring installed (assembled) height and valve stem height; determine needed repairs.

22. Measurement B in Figure 1–11 is more than specified.
    Technician A says this problem may bottom the lifter plunger.
    Technician B says a shim should be installed under the valve spring.

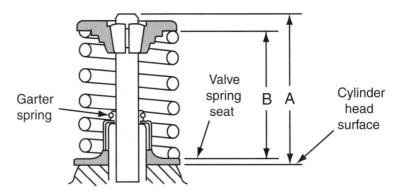

**Figure 1–11** Valve spring measurement.

Who is correct?
A. A only B.
B. B only
C. Both A and B
D. Neither A nor B

**Hint**    *Measure the installed valve stem height from the cylinder head surface on the spring seat to the valve stem tip. If this measurement is more than specified, the valve stem is stretched, or too much material has been removed from the seat. When this measurement is still excessive with a new valve, replace the seat or cylinder head. Excessive valve stem height moves the plunger downward in the valve lifter.*

*Measure the installed valve spring height from the lower edge of the top retainer to the top edge of the spring seat. When this measurement is excessive, install the thickness of shim specified by the manufacturer under the valve spring seat. Excessive installed valve spring height reduces valve spring tension, which may result in valve float and cylinder misfiring at higher speeds.*

## Task 10    Inspect pushrods, rocker arms, rocker arm pivots, and shafts for wear, bending, cracks, looseness, and blocked oil passages; repair or replace as required.

23. While discussing valve train service, Technician A says excessive valve spring tension may cause bent pushrods.
    Technician B says an improper valve timing may cause bent pushrods.
    Who is correct?
    A. A only
    B. B only
    C. Both A and B
    D. Neither A nor B

**Hint**    *Pushrods should be inspected for a bent condition and wear on the ends. Roll the pushrod on a level surface to check for a bent condition. Bent pushrods usually indicate interference in the valve train such as a sticking valve or improper valve adjustment.*

*Check the rocker arm shafts for wear and scoring in the rocker arm contact area. Worn rocker arms, shafts, or pivots cause improper valve adjustment and a clicking noise in the valve train.*

## Task 11    Inspect and replace hydraulic or mechanical lifters/lash adjusters.

24. Technician A says hydraulic valve lifter bottoms should be flat or concave.
    Technician B says a sticking lifter plunger may cause a burned exhaust valve.
    Who is correct?
    A. A only
    B. B only

C. Both A and B

D. Neither A nor B

***Hint***      *When the valve train is serviced, the valve lifters should be removed, cleaned, and tested for leakdown. If the lifter bottoms are pitted or worn, replace the lifters. Lifter bottoms must be convex.*

*Replace lifters with flat or concave bottoms. When the lifter bottoms are scored or concave, replace the camshaft.*

*Sticking lifter plungers cause a clicking noise, especially when the engine is started. Burned valves may be caused by sticking lifter plungers. If the lifters are cleaned and reassembled, they should be tested in a leakdown tester. This tester checks the time required to bottom the plunger with a specific weight applied to the plunger. When the lifters leak down too quickly, a clicking noise may be heard in the valve train with the engine idling. If the camshaft is replaced, new valve lifters must be installed.*

## Task 12      Adjust valves on engines with mechanical or hydraulic lifters.

25. While adjusting mechanical valve lifters, Technician A says that when the valve clearance is checked on a cylinder, the piston in that cylinder should be at TDC on the exhaust stroke.

    Technician B says some mechanical valve lifters have removable shim pads available in various thicknesses to provide the proper valve clearance.

    Who is correct?

    A. A only

    B. B only

    C. Both A and B

    D. Neither A nor B

***Hint***      *In engines with mechanical valve lifters, some rocker arms have an adjustment screw and a lock nut on the valve stem end of the rocker arm. With the piston at TDC on the compression stroke, place a feeler gauge of the specified thickness between the adjusting screw and the valve stem (Figure 1–12). If the valve adjustment is correct, the feeler gauge slides between the adjusting screw and the valve stem with a light push fit. Some engines with mechanical valve lifters have removable metal pads in each lifter or spring retainer. Pads with different thicknesses are available to provide the specified valve clearance.*

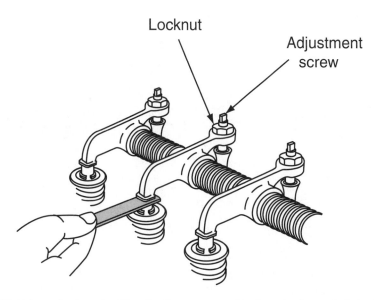

**Figure 1–12** Valve adjustment with adjusting screw and locknut in the end of the rocker arm.

*Some valve trains have hydraulic valve lifters and individual rocker arm pivots retained with self-locking nuts. These valve trains require an initial adjustment of the rocker arm nut to position the lifter plunger. With the valve closed, loosen the rocker arm nut until there is clearance between the end of the rocker arm and the valve stem. Slowly turn the rocker arm nut clockwise while rotating the pushrod. Continue rotating the rocker arm nut until the end of the rocker arm contacts the end of the valve stem, and the pushrod becomes harder to turn. Continue turning the rocker arm nut clockwise the number of turns specified in the service manual. In some engines, this specification is one turn plus or minus one-quarter turn.*

**Task 13    Inspect and replace camshaft(s) (includes checking drive gear wear and backlash, end play, sprocket and chain wear, overhead cam drive sprocket(s), drive belt(s), belt tension, tensioners, and cam sensor components).**

26. When the timing gear teeth are meshed directly with the crankshaft gear teeth, Technician A says the timing gear backlash may be measured with a dial indicator.

    Technician B says that on this type of engine the timing gear backlash may be measured with a micrometer.

    Who is correct?

    A. A only

    B. B only

    C. Both A and B

    D. Neither A nor B

**Hint**    *When the camshaft gear teeth are directly meshed with the crankshaft gear, the camshaft gear backlash may be measured with a dial indicator positioned against one of the camshaft gear teeth.*

*Many timing belts have a hydraulic tensioner that is operated by engine oil pressure. Some manufacturers recommend measuring the installed length of the tensioner to determine the belt wear. If the tensioner length exceeds the manufacturer's specifications, replace the timing belt (Figure 1–13).*

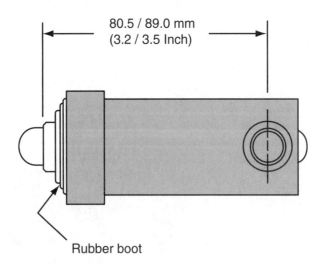

**Figure 1–13** Measuring installed belt tensioner length to determine timing belt wear.

*Some timing belts must be installed with an identification mark facing toward the belt cover. Prior to belt installation, the timing marks on the camshaft and crankshaft sprockets must be properly aligned with the marks on the engine casting (Figure 1–14).*

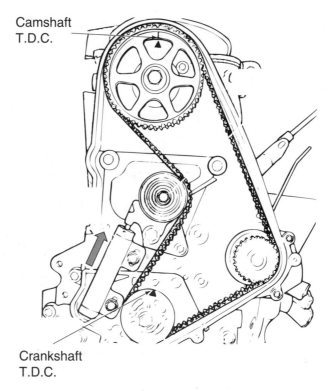

**Figure 1–14** The timing marks on the camshaft and crankshaft sprockets must be properly aligned with the marks on the engine casting. *(Courtesy of Chrysler Corporation)*

*Some engines have steel timing chains. On these engines, the crankshaft sprocket timing mark must be aligned with the timing mark on the camshaft gear prior to the camshaft gear and chain installation. Timing chain stretch and wear may be measured on these engines with a socket and flex handle installed on one of the camshaft sprocket retaining bolts. Rock the camshaft sprocket back and forth without moving the crankshaft gear, and measure the movement on one of the chain links on the camshaft sprocket.*

*On other engines, the timing chains have copper-colored links that must be aligned with the timing marks on the camshaft and crankshaft sprockets when the chain and gears are installed.*

## Task 14   Inspect and measure camshaft journals and lobes; measure camshaft lift.

27. The measurement in Figure 1–15 is checking
    A. the camshaft lobe lift.
    B. the camshaft journal condition.
    C. the pushrod length.
    D. the valve lifter condition.

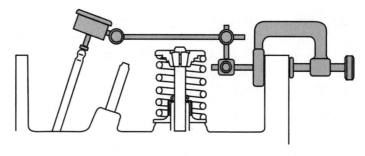

**Figure 1–15** Valve train measurement.

**Hint**    *The outer camshaft journals should be placed in V-blocks to measure the camshaft runout (Figure 1–16). To measure the camshaft runout, position a dial indicator against the other camshaft journals and rotate the camshaft.*

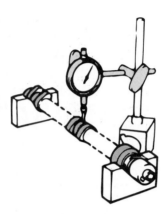

**Figure 1–16** Measuring camshaft runout. *(Reprinted with permission)*

*A micrometer may be placed from the highest point on the camshaft lobe to the side opposite the lobe to measure the lobe wear (Figure 1–17). If the wear on any lobe exceeds specifications, replace or regrind the camshaft. Most manufacturers do not recommend camshaft regrinding.*

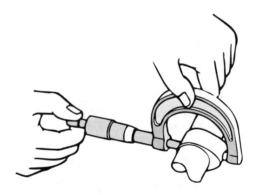

**Figure 1–17** Measuring camshaft lobe height. *(Reprinted with permission)*

*Use a micrometer to measure the diameter of each camshaft journal. Measure this diameter in several locations. If the diameter is less than specified, replace the camshaft.*

## Task 15    Inspect and measure camshaft bore for wear, damage, out-of-round, and alignment; determine needed repairs.

28. When discussing camshaft bearing clearance, Technician A says excessive camshaft bearing clearance may result in lower-than-specified oil pressure.

    Technician B says excessive camshaft bearing clearance may cause a thumping noise when the engine is idling.

    Who is correct?

    A.  A only

    B.  B only

    C.  Both A and B

    D.  Neither A nor B

**Hint**     *The camshaft bearing out-of-round should be measured at several locations around the bearing with a telescoping gauge. When the out-of-round exceeds specifications, bearing replacement is necessary. On overhead cam engines with removable camshaft bearing caps, these caps must be torqued to specifications before measuring the bearing out-of-round.*

*When the lower half of the camshaft bearings are positioned on top of the cylinder head, a straightedge may be placed on these bearing surfaces with the bearing caps removed to measure bearing alignment. Measure the clearance between the straightedge and each bearing bore to determine the bore alignment. When the camshaft bearing bores are improperly aligned, replace the cylinder head.*

*On overhead cam engines with removable camshaft bearing caps, Plastigage may be used to measure the bearing clearance (Figure 1–18).*

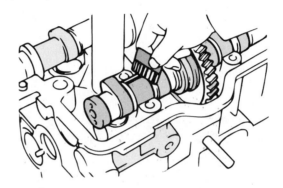

**Figure 1–18** Measuring camshaft bearing clearance with Plastigage. *(Reprinted with permission)*

## Task 16   Time camshaft(s) to crankshaft.

29. Technician A says improper valve timing may cause reduced engine power.
    Technician B says improper valve timing may cause bent valves in some engines.
    Who is correct?
    A. A only
    B. B only
    C. Both A and B
    D. Neither A nor B

**Hint**     *With the timing gear cover removed, the camshaft timing may be measured by checking the position of the marks on the camshaft and crankshaft sprockets. These marks must be aligned as indicated in the vehicle manufacturer's service manual (Figure 1–19).*

*Valve timing may be checked by observing the valve position in relation to the piston position. With any piston at TDC on the exhaust stroke, the intake valve should be opening, and the exhaust valve should be closing. This valve position with the piston at TDC on the exhaust stroke is called valve overlap. If the valves do not open properly in relation to the crankshaft position, the valve timing is not correct. Under this condition, the timing chain (or belt) cover should be removed to check the position of the camshaft sprocket in relation to the crankshaft sprocket position. When the valve timing is incorrect, the timing chain (or belt) and/or sprockets must be replaced.*

## Task 17   Inspect cylinder head mating surface condition and finish, reassemble and install gasket(s) and cylinder head(s); replace and tighten fasteners according to manufacturers' procedures.

30. While discussing torque-to-yield head bolts, Technician A says that compared to conventional head bolts, torque-to-yield bolts provide more uniform clamping force.

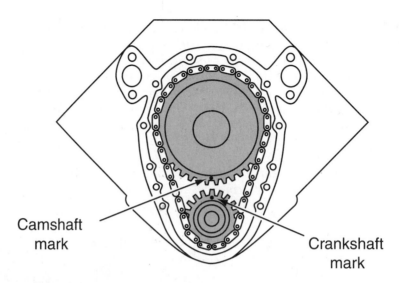

**Figure 1–19** Alignment of timing marks on crankshaft and camshaft sprockets.

Technician B says torque-to-yield bolts are tightened to a specific torque, and then rotated a certain number of degrees.

Who is correct?

A. A only

B. B only

C. Both A and B

D. Neither A nor B

*Hint*    *All head bolt openings in the block must be checked for the presence of oil or foreign material, which may cause false torque readings or a cracked block. Use compressed air to remove any foreign material from head bolt openings.*

*Many engines now have torque-to-yield head bolts that must be replaced each time they are removed. Head bolts must be tightened to the specific torque in the proper sequence. Torque-to-yield head bolts usually are tightened to a specific torque and then rotated the specified number of degrees.*

# Engine Block Diagnosis and Repair

## ASE Tasks, Questions, and Related Information

**Task 1**    **Disassemble engine block; clean and prepare components for inspection and reassembly.**

31.  All of the following statements about block disassembly are true EXCEPT:

A. The ring ridge should be removed before removing the pistons and connecting rods.

B. The connecting rods must be reinstalled with their markings facing in the specified direction.

C. The main bearing caps must be installed in their original position.

D. After ridge reaming, use compressed air to blow metal filings from the cylinders.

**Hint**    *When disassembling an engine block, the oil pan and oil pump should be removed first, followed by the timing gear cover and valve train components. A ridge reamer must be used to remove the ring ridge from the top of the cylinder walls prior to piston and connecting rod removal. Use an oily rag to remove all the metal filings from the cylinder. Observe the markings on the connecting rods before removing the pistons and connecting rods so these components can be installed in the original position. Remove the flex plate or flywheel. If the main bearing caps are not marked, use a center punch to place identification marks on these caps and on the block.*

*These caps must be reinstalled in their original position. Remove the main bearing caps and crankshaft. Remove the core plugs and oil gallery plugs, and clean the block in a cleaning tank.*

## Task 2    Visually inspect engine block for cracks, corrosion, passage condition, core and gallery plug hole condition, surface warpage, and surface finish and condition; determine necessary action.

32. Technician A says a warped cylinder head mounting surface on an engine block may cause valve seat distortion.

    Technician B says a warped cylinder head mounting surface on an engine block may cause coolant and combustion leaks.

    Who is correct?

    A. A only

    B. B only

    C. Both A and B

    D. Neither A nor B

**Hint**    *A cast iron block may be inspected for cracks with an electromagnetic crack detector. A dye penetrant may be used to check for cracks in an aluminum block. The block may be pressurized with a pressure tester to check for cracks. During an engine rebuild procedure, the core plugs and oil gallery plugs should be replaced.*

*Inspect the core plug bores before installing the new plugs. If the bore is damaged, it may be repaired by boring it to the next specified oversize plug. Oversize core plugs are stamped with the letters OS. Before installing these plugs, coat the sealing edge with a nonhardening water-resistant sealer. Oil gallery plugs should be coated with an oil-resistant sealer. A block may have dish-type, cup-type, or expansion-type core plugs. Install the core plugs with the proper special driving tool.*

*The cylinder head mounting surfaces on the block must be checked for warpage with a straightedge and a feeler gauge (Figure 1–20). If the warpage on these surfaces exceeds the manufacturer's specifications, the block surfaces must be resurfaced.*

## Task 3    Inspect and repair damaged threads where allowed; install core and gallery plugs.

33. While discussing heli-coil installation, Technician A says the first step is to use a tap, and thread the opening to match the external threads on the heli-coil.

    Technician B says the heli-coil should be installed with the proper-sized drill bit.

    Who is correct?

    A. A only

    B. B only

    C. Both A and B

    D. Neither A nor B

**Hint**    *If threads are damaged, the opening may be drilled and threaded, and then a heli-coil may be installed to provide a thread that's the same as the original (Figure 1–21).*

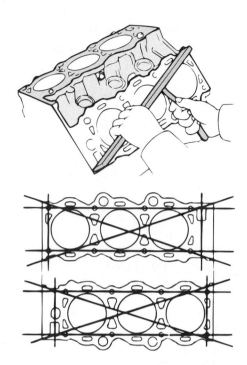

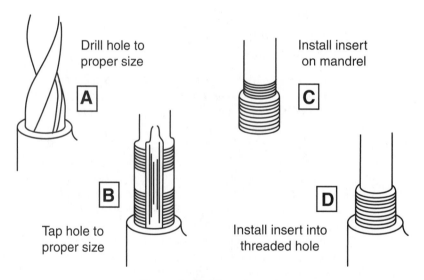

**Figure 1–20** Measuring warpage on the top surfaces of the block. *(Reprinted with permission)*

Drill hole to
proper size

A

Install insert
on mandrel

C

Tap hole to
proper size

B

Install insert into
threaded hole

D

**Figure 1–21** Thread repair.

## Task 4    Inspect and measure cylinder walls; remove cylinder wall ridges; hone and clean cylinder walls; determine need for further action.

34. If new rings are installed without removing the ring ridge, the following may result:
    A. the piston skirt may be damaged.
    B. the piston pin may be broken.
    C. the connecting rod bearings may be damaged.
    D. the piston ring lands may be broken.

35. While discussing cylinder measurement, Technician A says the cylinder taper is the difference between the cylinder diameter at the top of the ring travel compared to the cylinder diameter at the center of the ring travel.

Technician B says cylinder out-of-round is the difference between the axial cylinder bore diameter at the top of the ring travel compared to the thrust cylinder bore diameter at the bottom of the ring travel.

Who is correct?

A. A only

B. B only

C. Both A and B

D. Neither A nor B

36. While deglazing and cleaning a cylinder
    A. 120-grit stones may be used on the cylinder hone.
    B. wash the cylinder with soapy water after glazing.
    C. wash the cylinder with an oil-based solvent after deglazing.
    D. 400-grit emery paper and hand pressure may be used.

**Hint**     *If the amount of cylinder wear does not require cylinder reboring, remove the ring ridge at the top of each cylinder with a ridge reamer (Figure 1–22). Always consult the engine manufacturer's service manual regarding ring ridge removal. While using the ridge reamer, be careful not to mark the cylinder wall below the ring ridge. Remove the ring ridge only. Do not remove any metal from the cylinder wall below the ring ridge.*

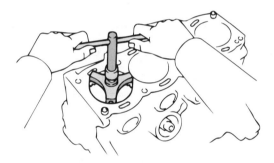

**Figure 1–22** Ring ridge removal. *(Reprinted with permission)*

*Use a dial bore gauge to measure the cylinder diameter in three vertical locations. These locations are just below the ring ridge at the top of the cylinder, in the center of the ring travel, and just above the lowest part of the ring travel. Cylinder taper is the difference in the cylinder diameter at the top of the ring travel compared to the diameter at the bottom of the ring travel.*

*In each of the three cylinder vertical measurement locations, measure the cylinder diameter in the thrust direction and in the axial direction (Figure 1–23). Cylinder out-of-round is the difference between the cylinder diameter in the thrust and axial directions. If the cylinder out-of-round exceeds specifications, rebore the cylinder.*

*If cylinder wear, out-of-round, and taper do not exceed specifications, the cylinders may be deglazed. Cylinders may be deglazed with 220- or 280-grit stones installed on a cylinder hone, or a hone-type brush. After deglazing, the cylinder should be cleaned with hot, soapy water and a stiff-bristle brush. Clean the residue from the cylinder with a soft lint-free cloth. Rinse the block and dry it thoroughly. Coat all machined surfaces in the block with a light coating of the manufacturer's recommended engine oil.*

*If one cylinder requires reboring, most manufacturers recommend reboring all the cylinders to the same size. Cylinder reboring usually is done with a honing machine or boring bar. The proper stones must be used on the honing machine depending on the block material. Honing stones have an identification number. Stones with a higher number have a finer grit. When the honing operation is completed, the cylinders should have a 60-degree crosshatch pattern. After cylinder honing, the same procedure for block cleaning should be followed as discussed previously in cylinder deglazing.*

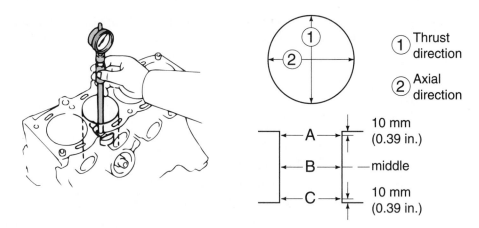

**Figure 1–23** Cylinder measurement. *(Reprinted with permission)*

## Task 5
**Inspect crankshaft for end play, straightness, journal damage, keyway damage , thrust flange and sealing surface condition, and visual surface cracks; check oil passage condition; measure journal wear; check crankshaft sensor reluctor ring (where applicable); determine necessary action.**

37. When measuring the crankshaft journal in Figure 1–24, the difference between measurements
    A. A and B indicates horizontal taper.
    B. C and D indicates vertical taper.
    C. A and C indicates out-of-round.
    D. A and D indicates vertical taper.

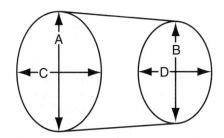

**Figure 1–24** Crankshaft journal measurement.

*Hint*    *Use a micrometer to measure each crankshaft journal for vertical taper, horizontal taper, and out-of-round. The out-of-round should be measured at two locations on each side of the journal.*

*When the journal out-of-round or taper exceeds specifications, or journal scoring is evident, journal grinding is required.*

## Task 6    **Inspect and measure main bearing bores and cap alignment and fit.**

38. Main bearing bore alignment may be measured with
    A. a dial indicator.
    B. a micrometer.
    C. a telescoping gauge.
    D. a straightedge and a feeler gauge.

*Hint*    *Main bearing bore alignment may be measured with a specially machined arbor. This arbor is installed in the main bearing saddles with the bearing inserts removed. With the main bearing caps tightened to the specific torque, the arbor should turn when rotated with a 1 ft. (3.04 cm) bar. If the arbor does not turn, one or more of the main bearing saddles is out of alignment.*

*Main bearing bore alignment may also be measured with a feeler gauge and a straightedge. With the bearing caps removed, the straightedge is placed on the main bearing saddles. If a feeler gauge with a thickness one-half the specified bearing clearance can be slid between the straightedge and any bearing saddle, the main bearing bore is out of alignment. The main bearing bores may be line bored to correct bore misalignment. Main bearing bore out-of-round may be measured with a dial bore gauge.*

**Task 7**    **Install main bearings and crankshaft; check bearing clearances and end play; replace/retorque bolts according to manufacturers' procedures.**

39. A crankshaft has more than the specified end play.
    Technician A says the thrust surfaces on the sides of one main bearing may be worn.
    Technician B says a thicker shim should be installed between the rear edge of the rear main bearing and the rear crankshaft flange.
    Who is correct?
    A. A only
    B. B only
    C. Both A and B
    D. Neither A nor B

*Hint*    *Main bearing clearance may be measured with a strip of Plastigage placed between the bearing and the crankshaft. The bearing cap must be torqued to specifications, followed by cap removal. The width of the Plastigage strip is then measured with a gauge scale on the Plastigage package to determine the bearing clearance. The crankshaft end play may be measured with a dial indicator positioned against the flywheel mounting surface on the rear of the crankshaft.*

*When the crankshaft is moved forward or rearward, the end play is indicated on the dial indicator.*

*The crankshaft end play is determined by the clearance between the friction surfaces on the sides of the thrust-type main bearing insert and the matching thrust surfaces on the crankshaft.*

**Task 8**    **Inspect camshaft bearings for excessive wear and alignment; install camshaft, timing chain, and gears; check end play.**

40. The tool shown in Figure 1–25 is used to
    A. remove camshaft bearings.
    B. install camshaft bearings.
    C. remove and install camshaft bearings.
    D. measure camshaft bearing alignment.

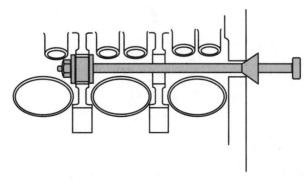

**Figure 1–25** Engine block service tool.

*Hint*    *If the camshaft bearings are mounted in the block, the oil hole in the bearing must be aligned with the oil hole in the block. Many overhead cam engines do not have removable camshaft bearings.*

*Inspect the camshaft bearings or bearing bores for scoring, roughness, and wear. Camshaft bearings or bearing bores should be measured at two different locations with a telescoping gauge.*

*Measure the camshaft journals with a micrometer, and subtract the journal diameter from the bearing diameter to obtain the clearance. If the wear exceeds specifications, replace the bearings, or cylinder head.*

*When the camshaft bearing bores are in the cylinder head, the bearing caps should be removed and a straightedge positioned across the bearing bores in the cylinder head to measure bearing alignment. Insert a feeler gauge between the straightedge and each bearing bore to measure any misalignment.*

## Task 9    Inspect auxiliary (balance, intermediate, idler, counterbalance, or silencer) shaft(s), drive(s), and support bearings for damage and wear; determine necessary action.

41. Technician A says improper balance shaft timing causes severe engine vibrations. Technician B says the balance shafts are timed in relation to the camshaft. Who is correct?
    A. A only
    B. B only
    C. Both A and B
    D. Neither A nor B

*Hint*    *The balance shafts should be checked for runout with the same procedure used for measuring camshaft runout. The balance shaft journals should be measured for taper with the same procedure for measuring crankshaft journal taper. When the balance shafts are installed, they must be properly timed to the crankshaft.*

## Task 10    Inspect, measure, service, repair, or replace pistons and piston pins; identify piston and bearing wear patterns that indicate connecting rod alignment problems; determine necessary action.

42. The tool in Figure 1–26 is used to
    A. widen the piston ring grooves.
    B. deepen the piston ring grooves.
    C. remove and replace piston rings.
    D. remove carbon from the ring grooves.

43. A bent connecting rod may cause
    A. uneven connecting rod bearing wear.
    B. uneven main bearing wear.
    C. uneven piston pin wear.
    D. excessive cylinder wall wear.

*Hint*    *Piston ring grooves should be cleaned using a ring groove cleaning tool. Use a micrometer to measure the piston diameter at right angles to the piston pin bores and 1 in. below the bottom edge of the lowest ring groove (Figure 1–27).*

*Uneven wear on the edges of the piston skirt next to the pin hole may be caused by connecting rod misalignment. Connecting rod misalignment may also cause V-shaped connecting rod bearing wear.*

## Task 11    Inspect connecting rods for damage, bore condition, and pin fit; determine necessary action.

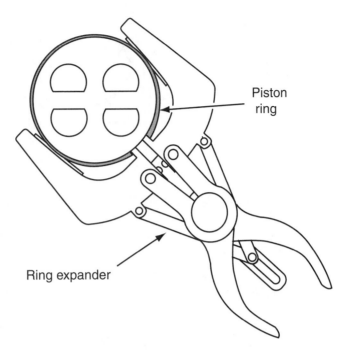

**Figure 1–26** Piston service tool.

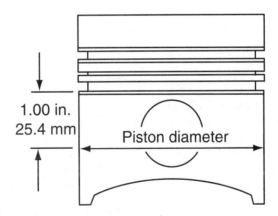

**Figure 1–27** Measuring piston diameter.

44. Technician A says the curvature of connecting rod bearings is slightly larger than the curvature of the bearing bores, and this feature is called bearing spread.

    Technician B says that when a connecting rod bearing half is installed, the bearing edges extend slightly from the mounting area, and this feature is called bearing crush.

    Who is correct?
    A. A only
    B. B only
    C. Both A and B
    D. Neither A nor B

**Hint**    *Measure the connecting rod bore for out-of-round and proper bore size. Assemble the connecting rod on the rod alignment measuring tool. Place a feeler gauge between the upper projection on the fixture and the machined tool surface to measure the rod bend (Figure 1–28). Install a feeler gauge between the side projections on the fixture and the machined tool surface to measure rod twist (Figure 1–29). If rod bend or twist exceeds specifications, replace the connecting rod.*

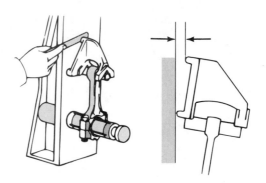

**Figure 1–28** Measuring connecting rod bend. *(Reprinted with permission)*

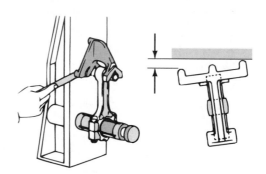

**Figure 1–29** Measuring connecting rod twist. *(Reprinted with permission)*

*Some connecting rods have a bushing in the rod opening. The oil hole in the bushing must be aligned with the oil hole in the connecting rod. The bushing may be honed in a pin hole machine to obtain the specified pin-to-bushing clearance. When a bushing is positioned in the upper rod opening, some manufacturers recommend heating the pistons in a piston heater prior to piston pin installation.*

*In some engines, the piston pins are pressed into the upper rod opening. The pin and connecting rod must be centered in the piston bores, and the marks on the piston and connecting rod must be properly positioned before the connecting rod and piston are assembled.*

**Task 12    Inspect, measure, and install or replace piston rings; assemble piston and connecting rod; install piston/rod assembly; check bearing clearance and sideplay; replace/retorque fasteners according to manufacturers' procedures.**

45. While discussing piston ring service, Technician A says the ring gap should be measured with the ring positioned at the top of the ring travel in the cylinder.

    Technician B says the two compression rings are interchangeable on most pistons.

    Who is correct?

    A.  A only

    B.  B only

    C.  Both A and B

    D.  Neither A nor B

**Hint**    *Position the piston ring squarely in the cylinder at the bottom of the ring travel, and measure the ring gap with a feeler gauge (Figure 1–30). If the gap is less than specified, file a small amount of metal from the end of the ring. When the gap is more than specified, the ring or cylinder bore is worn, or the ring is the wrong size.*

*Piston rings are removed and installed with a ring expander, and must be installed in the proper vertical direction on the piston (Figure 1–31). Piston ring gaps must be positioned*

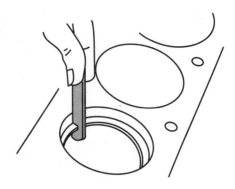

**Figure 1–30**  Measuring ring gap.

*properly around the piston to minimize blowby. The rings and piston should be coated with the specified engine oil; a ring compressor is then tightened on the piston to compress the rings.*

*Protecting sleeves must be installed on the connecting rod bolts prior to piston installation. Be sure the piston and connecting rod marks are properly positioned and then tap the piston gently into the cylinder.*

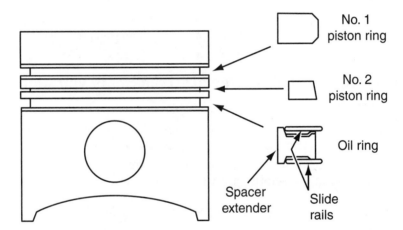

**Figure 1–31**  Piston rings must be installed in the proper vertical direction on the piston.

*To measure connecting rod bearing clearance, install a strip of Plastigage across the journal and then tighten the bearing cap to the specified torque. Remove the bearing cap and measure the width of the Plastigage on the journal (use the scale provided on the Plastigage package) to determine the bearing clearance (Figure 1–32).*

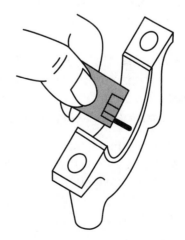

**Figure 1–32**  Bearing clearance measurement.

*A feeler gauge should be inserted between the side of the connecting rod and the edge of the crankshaft journal to measure the side clearance (Figure 1–33). If this clearance exceeds specifications, the sides of the connecting rod or the crankshaft journal are worn.*

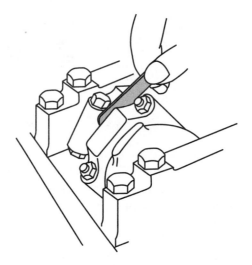

**Figure 1–33** Measuring connecting rod side clearance.

## Task 13   Inspect, reinstall, or replace crankshaft vibration damper (harmonic balancer).

46. Technician A says the vibration damper counterbalances the back-and-forth twisting motion of the crankshaft each time a cylinder fires.

   Technician B says if the seal contact area on the vibration damper hub is scored, the damper assembly must be replaced.

   Who is correct?

   A. A only

   B. B only

   C. Both A and B

   D. Neither A nor B

**Hint**   *Inspect the rubber between the inner hub and outer inertia ring on the vibration damper. If this rubber is cracked, oil-soaked, swollen, deteriorated, or loose, replace the damper. Inspect the seal contact area on the vibration damper hub for ridging and scoring. When these conditions are present, replace the damper, or machine the damper hub and install the proper sleeve on the hub to provide a new seal contact area. A special puller and installer tool is required to remove and install the vibration damper.*

## Task 14   Inspect crankshaft flange and flywheel mating surfaces; inspect and replace crankshaft pilot bearing/bushing (if applicable); inspect flywheel/flexplate for cracks and wear (includes flywheel ring gear); measure flywheel runout; determine necessary action.

47. Technician A says metal burrs on the crankshaft flange may cause excessive wear on the ring gear and starter drive gear teeth.

   Technician B says metal burrs on the crankshaft flange may cause improper torque-converter-to-transmission alignment.

   Who is correct?

   A. A only

   B. B only

C. Both A and B

D. Neither A nor B

48. Technician A says excessive flywheel runout may cause grabbing, erratic clutch operation.

Technician B says the pressure plate should always be reinstalled in the original position on the flywheel.

Who is correct?

A. A only

B. B only

C. Both A and B

D. Neither A nor B

49. A worn pilot bearing may cause a growling, rattling noise

A. while driving at a steady 20 mph.

B. in reverse with the clutch engaged.

C. while accelerating in low gear.

D. with the clutch pedal pressed.

**Hint**    *Inspect the crankshaft flange and the flywheel-to-crankshaft mating surface for metal burrs (Figure 1–34). Remove any metal burrs with fine emery paper. Be sure the threads in the crankshaft flange are in satisfactory condition. Replace the flywheel bolts and retainer if any damage is visible on these components. Install the flywheel, retainer, and bolts, and tighten the bolts in sequence to the manufacturer's specified torque.*

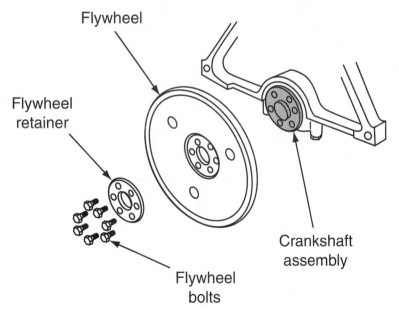

Flywheel

Flywheel retainer

Crankshaft assembly

Flywheel bolts

**Figure 1–34** Flywheel and crankshaft mounting flange.

*Inspect the flywheel for scoring and cracks in the clutch contact area. Minor score marks and ridges may be removed by resurfacing the flywheel.*

*Mount a dial indicator on the engine flywheel housing, and position the dial indicator stem against the clutch contact area on the flywheel (Figure 1–35). Rotate the flywheel to measure the flywheel runout. If the flywheel runout exceeds specifications, replace the flywheel.*

*Place your finger in the inner pilot bearing race and rotate the race. If the bearing feels rough or loose, replace the bearing. A new transmission mainshaft may be positioned in the pilot bushing.*

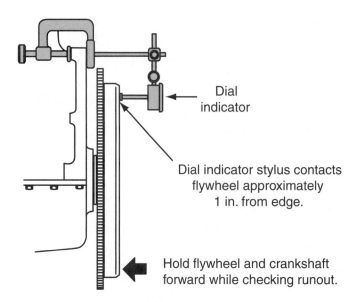

**Figure 1–35** Measuring flywheel runout. *(Reprinted with permission)*

*If the mainshaft has excessive movement in the bushing, replace the bushing. The proper driver must be used to install the pilot bearing or bushing.*

## Task 15    Inspect and replace pans and covers.

50. Technician A says a springless seal may be used in the timing gear cover.
    Technician B says that when a seal is installed the seal lip must face in the direction of oil flow.
    Who is correct?
    A. A only
    B. B only
    C. Both A and B
    D. Neither A nor B

**Hint**    *Gaskets are used to seal minor variations between two flat surfaces. Oil pan gaskets, or rocker arm cover gaskets usually are manufactured from cork, rubber, or a combination of rubber and cork or rubber and silicone. Always inspect the gasket mounting surfaces on pans and rocker arm covers for warpage. Warped mounting surfaces must be straightened.*

*Seals may be classified as springless or spring loaded. Springless seals are used in front wheel hubs where they seal a heavy lubricant. The garter spring in a spring-loaded seal provides extra lip sealing force to compensate for lip wear, shaft runout, and bore eccentricity. A fluted lip seal may be used to direct oil back into a housing. A sealer is painted on the case of some seals to prevent leaks between the seal case and the bore. When a seal is installed, the garter spring must face toward the fluid flow. Split rubber seals, or rope seals, may be used in rear main bearings. Prior to seal installation, the shaft and seal bore must be inspected for scratches and roughness. Seal lips should be lubricated before installation, and the proper seal driver must be used to install the seal.*

## Task 16    Assemble the engine using gaskets, seals, and formed-in-place (tube-applied) sealants, thread sealers, etc. according to manufacturers' specifications.

51. When installing RTV sealer
    A. the components to be sealed should be washed with an oil-based solvent.
    B. an RTV bead 1/8-in. wide should be placed in the center of the sealing surface.

C. the RTV bead should be placed on one side of any bolt holes.

D. the RTV bead should be allowed to dry for ten minutes before component installation.

*Hint*     *Room temperature vulcanizing (RTV) sealer may be used in place of conventional gaskets on oil pans and rocker arm covers. RTV sealer dries in the presence of air by absorbing moisture from the air. All the old material must be removed from the surface area before the RTV sealer is applied. A chlorinated solvent must be used to clean the RTV sealer mounting area. If oil-based solvents are used to clean this mounting area, an oily residue is left on the area that prevents RTV sealer adhesion. Apply a 1/8-in. diameter bead of RTV sealer to the center of one surface to be sealed. This bead must surround bolt holes in the mounting area. Do not apply excessive amounts of RTV sealer, and always assemble the components within five minutes after the RTV application; otherwise, curing will occur.*

*Anaerobic sealer may be used in place of a gasket between two machined surfaces. This sealer dries in the absence of air, and the same cleaning procedure must be used for RTV and anaerobic sealer.*

# Lubrication and Cooling Systems Diagnosis and Repair

## ASE Tasks, Questions, and Related Information

**Task 1**   **Diagnose engine lubrication system problems; perform oil pressure tests; determine necessary action.**

52. All of the following are causes of low engine oil pressure EXCEPT
    A. worn camshaft bearings.
    B. worn crankshaft bearings.
    C. weak oil pressure regulator spring tension.
    D. restricted pushrod oil passages.

*Hint*     *When testing engine oil pressure, the oil sending unit usually is removed and an oil pressure gauge is connected to the sending unit opening. The engine should be at normal operating temperature, and the oil pressure test usually is performed at idle speed and a higher speed such as 2,500 rpm.*

**Task 2**   **Disassemble, inspect and measure oil pump (includes gears, rotors, housing, and pick-up assembly), pressure relief devices, and pump drive; determine necessary action; replace oil filter.**

53. The following are normal oil pump component measurements EXCEPT
    A. inner rotor diameter.
    B. clearance between the rotors.
    C. inner and outer rotor thickness.
    D. outer rotor to housing clearance.

*Hint*     *Inspect the oil pump pressure relief valve for sticking and wear. If this valve sticks in the closed position, oil pressure is excessive. A pressure relief valve stuck in the open position results in low oil pressure.*

*Measure the thickness of the inner and outer rotors with a micrometer. When this thickness is less than specified on either rotor, replace the rotors or the oil pump. The following oil pump measurements should be performed with a feeler gauge:*
- *Measure pump cover flatness with a feeler gauge positioned between a straightedge and the cover.*
- *Measure the clearance between the outer rotor and the housing.*
- *Measure the clearance between the inner and outer rotors with the rotors installed.*
- *Measure the clearance between the top of the rotors and a straightedge positioned across the top of the oil pump.*

## Task 3   Perform cooling system tests; determine necessary action.

54. The tester in Figure 1–36 may be used to test the following items EXCEPT
    A. cooling system leaks.
    B. radiator cap pressure relief valve.
    C. the coolant's specific gravity.
    D. heater core leaks.

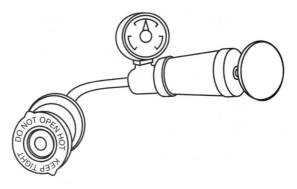

**Figure 1–36**  Pressure tester.

**Hint**    *A pressure tester may be connected to the radiator filler neck to check for cooling system leaks.*
   *Operate the tester pump and apply 15 psi to the cooling system. Inspect the cooling system for external leaks with the system pressurized. If the gauge pressure drops more than specified by the vehicle manufacturer, the cooling system has a leak. If there are no visible external leaks, check the front floor mat for coolant dripping out of the heater core. When there are no external leaks, check the engine for combustion chamber leaks.*
   *The radiator pressure cap may be tested with the pressure tester. When the tester pump is operated, the cap should hold the rated pressure. Always relieve the pressure before removing the tester.*

## Task 4   Inspect, replace, and adjust drive belt(s), tensioner(s), and pulleys.

55. In a drive belt system in which one belt drives only the alternator, a loose alternator belt may cause
    A. a discharged battery.
    B. a squealing noise while decelerating.
    C. a damaged alternator bearing.
    D. engine overheating.

**Hint**    *Because the friction surfaces are the sides of a V-belt, the belt must be replaced if the sides are worn and the belt is contacting the bottom of the pulley.*
   *The belt tension may be checked with the engine shut off, and a belt tension gauge placed over the belt at the center of the belt span (Figure 1–37). A loose or worn belt may cause a squealing noise when the engine is accelerated.*

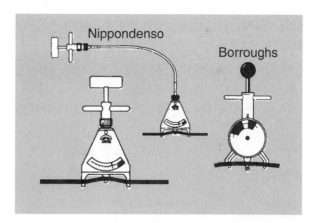

**Figure 1–37** Measuring belt tension. *(Reprinted with permission)*

*The belt tension also may be checked by measuring the amount of belt deflection with the engine shut off. Use your thumb to depress the belt at the center of the belt span. If the belt tension is correct, the belt should have 1/2-in. deflection per foot of belt span.*

*Ribbed V-belts usually have a spring-loaded belt tensioner, with a belt wear indicator scale on the tensioner housing. If a power steering pump belt requires tightening, always pry on the pump ear, not on the housing.*

## Task 5   Inspect and replace engine cooling and heater system hoses and fittings.

56. Technician A says a collapsed upper radiator hose may be caused by an inoperative pressure release valve in the radiator cap.

    Technician B says a collapsed upper radiator hose may be caused by a plugged hose between the radiator filler neck and the recovery reservoir.

    Who is correct?

    A. A only

    B. B only

    C. Both A and B

    D. Neither A nor B

**Hint**   *All cooling system hoses should be inspected for soft spots, swelling, hardening, chafing, leaks, and collapsing. If any of these conditions are present, hose replacement is necessary. Hose clamps should be inspected to make sure they are tight. Some radiator hoses contain a wire coil inside them to prevent hose collapse as the coolant temperature decreases. Remember to include heater hoses and the bypass hose in the hose inspection. Prior to hose removal, the coolant must be drained from the radiator.*

## Task 6   Inspect, test, and replace thermostat, bypass, and housing.

57. The thermostat is stuck open on a port fuel-injected engine. This problem may cause

    A. a rich air-fuel ratio.

    B. a lean air-fuel ratio.

    C. excessive fuel pressure.

    D. engine overheating.

**Hint**   *The thermostat may be suspended with a thermometer in a container filled with water. Do not allow the thermostat to contact the bottom of the container. Heat the water while observing the thermostat valve and the thermometer. The thermostat valve should begin to open when the temperature on the thermometer is equal to the rated temperature stamped on the thermostat. Replace the thermostat if it does not open at the rated temperature.*

*Many thermostats are marked for installation in the proper direction. Inspect the bypass hose for cracks, deterioration, and restrictions; replace the hose if these conditions are present.*

**Task 7**   Inspect coolant; drain, flush, and refill cooling system with recommended coolant; bleed air as required.

58. While discussing the cooling system service, Technician A says if the cooling system pressure is reduced, the coolant boiling point is increased.

Technician B says when more antifreeze is added to the coolant the coolant boiling point is increased.

Who is correct?

A. A only

B. B only

C. Both A and B

D. Neither A nor B

**Hint**   *If the radiator tubes and coolant passages in the block and cylinder head are restricted with rust and other contaminants, these components may be flushed. Cooling system flushing equipment is available for this purpose. Always operate the flushing equipment according to the equipment manufacturer's directions, and be sure that your service procedure conforms to pollution laws in your state. Engine coolant must be recycled or handled as a hazardous waste material.*

*Coolant reconditioning machines are available to remove harmful particles and restore corrosion additives so the coolant can be returned to the cooling system.*

**Task 8**   Inspect and replace water pump.

59. Technician A says a damaged water pump bearing may cause a growling noise when the engine is idling.

Technician B says the water pump bearing may be ruined by coolant leaking past the pump seal.

Who is correct?

A. A only

B. B only

C. Both A and B

D. Neither A nor B

**Hint**   *With the engine shut off, grasp the fan blades, or the water pump hub, and try to move the blades from side to side. This action checks for looseness in the water pump bearing. If there is any side-to-side movement in the bearing, water pump replacement is required.*

*Check for coolant leaks, rust, or residue at the water pump drain hole in the bottom of the pump and at the inlet hose connected to the pump. When coolant is dripping from the pump drain hole, replace the pump. The water pump may be tested with the pressure tester connected to the radiator filler neck.*

**Task 9**   Inspect, test, and replace radiator, heater core, pressure cap, and coolant recovery system.

60. An excessively high coolant level in the recovery reservoir may be caused by any of these problems EXCEPT

A. restricted radiator tubes.

B. a thermostat that is stuck open.

C. a loose water pump impeller.

D. an inoperative electric-drive cooling fan.

**Hint**   *The radiator cap should be inspected for a damaged sealing gasket or vacuum valve. If the pressure cap, sealing gasket, or seat is damaged, the engine will overheat, and the coolant will be lost to the coolant recovery system. Under this condition, the coolant recovery container becomes overfilled with coolant.*

*If the cap vacuum valve is sticking, a vacuum may occur in the cooling system after the engine is shut off and the coolant temperature decreases. This vacuum may cause collapsed cooling system hoses. A pressure tester may be used to test the pressure cap and pressure test the entire cooling system.*

*The coolant level should be at the appropriate mark on the recovery container, depending on engine temperature.*

## Task 10   Clean, inspect, test, and replace fan (both electrical and mechanical), fan clutch, fan shroud, air dams, and cooling related temperature sensors.

61. In Figure 1–38, an open ground circuit on the engine temperature switch may cause
    A. continual cooling fan motor operation.
    B. a completely inoperative cooling fan motor.
    C. a burned-out cooling fan motor.
    D. engine overheating.

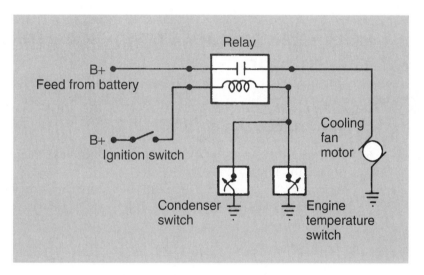

**Figure 1–38** Electric-drive cooling fan circuit. *(Reprinted with permission from SAE Paper 790722 ©1979 Society of Automotive Engineers, Inc.)*

**Hint**     *If the radiator shroud is loose, improperly positioned, or broken, airflow through the radiator is reduced, and engine overheating may result.*

*The viscous-drive fan clutch should be visually inspected for leaks. If there are oily streaks radiating outward from the hub shaft, the fluid has leaked out of the clutch.*

*With the engine shut off, rotate the cooling fan by hand. If the viscous clutch allows the fan blades to rotate easily both hot and cold, the clutch should be replaced. A slipping viscous clutch results in engine overheating. If there is any looseness between the viscous clutch and the shaft, replace the viscous clutch.*

*If the electric-drive cooling fan does not operate at the coolant temperature specified by the vehicle manufacturer, engine overheating will result, especially at idle and lower speeds when air-flow through the radiator is reduced.*

## Task 11   Inspect, test, and replace internal and external oil coolers.

62. An engine oil cooler helps to prevent
    A. oxidation of the engine oil.
    B. excessive oil pressure.
    C. oil pump wear.
    D. main bearing wear.

*Hint*    *Some engines, such as diesels and turbocharged engines, have an oil cooler. This oil cooler may be contained in one of the radiator tanks, or mounted separately near the front of the engine. External oil coolers should be inspected for leaks and restricted airflow passages.*

*The maximum oil operating temperature is approximately 250°F (121°C). Hot oil combines with oxygen in the air to form carbon and a sticky varnish. This process is called oil oxidation.*

*An engine oil cooler reduces oil temperature and helps prevent oil oxidation, especially on engines operating under heavy load.*

# Fuel, Electrical, Ignition, and Exhaust Systems Inspection and Service

## ASE Tasks, Questions, and Related Information

**Task 1**  **Inspect, clean, or replace fuel and air induction system components, intake manifold, and gaskets.**

63. Technician A says an intake manifold vacuum leak may cause a cylinder misfire with the engine idling.

Technician B says an intake manifold vacuum leak may cause a cylinder misfire during hard acceleration.

Who is correct?

A. A only

B. B only

C. Both A and B

D. Neither A nor B

*Hint*    *The intake manifold should be inspected for cracks, and the surface of the manifold that fits against the cylinder head should be checked for warpage with a straightedge and a feeler gauge.*

*Many intake manifolds are made from a plastic compound. These manifolds transfer less heat to the air-fuel mixture compared to aluminum or cast-iron intake manifolds. Always observe the specified torque specifications when servicing plastic manifolds. Special equipment is now available to blow smoke into the intake manifold with the engine not running to detect manifold leaks. Since manifold vacuum is high at idle speed, a manifold vacuum leak causes more air to enter the intake at this speed. Therefore, a manifold vacuum leak may cause a cylinder misfire at idle, and the misfire disappears as the engine speed is increased.*

**Task 2**  **Inspect, service or replace air filters, filter housings, and intake ductwork.**

64. While cleaning a pleated paper-type air filter element, the air gun should be held

A. 6 in. from the outside of the air filter element.

B. directly against the outside of the air filter element.

C. directly against the inside of the air filter element.

D. 6 in. from the inside of the air filter element.

*Hint*    *Air filter elements should be changed at the manufacturer's recommended service intervals. When the vehicle is operating in extremely dusty conditions, the air filter should be changed at more frequent intervals. If the air filter element contains small holes, dust enters the engine and causes very fast cylinder wall and piston wear. A restricted air filter element reduces airflow into the engine and causes a rich air-fuel ratio.*

## Task 3  Inspect turbocharger/supercharger; determine necessary action.

65. Reduced turbocharger boost pressure may be caused by
   A. a wastegate valve stuck closed.
   B. a wastegate valve stuck open.
   C. a leaking wastegate diaphragm.
   D. a disconnected wastegate linkage.

**Hint**  *Excessive smoke in the exhaust may indicate worn turbocharger seals. The technician must remember that worn valve guide seals or piston rings also cause oil consumption and blue smoke in the exhaust.*

*Check for intake system leaks. If there is a leak in the intake system before the compressor housing, dirt may enter the turbocharger and damage the compressor or turbine wheel blades.*

*When a leak is present in the intake system between the compressor wheel housing and the cylinders, turbocharger pressure is reduced.*

*Turbocharger boost pressure may be tested with a pressure gauge connected to the intake manifold.*

*The boost pressure should be tested during hard acceleration while driving the vehicle.*

## Task 4  Test engine cranking system; determine needed repairs.

66. During the battery test in Figure 1–39
   A. the battery should be discharged at two-thirds of the cold cranking rating.
   B. the battery should be discharged at one-half of the amp-hour rating.
   C. the voltage should remain above 9.6 V at 70°F battery temperature.
   D. the load should be applied to the battery for 20 seconds.

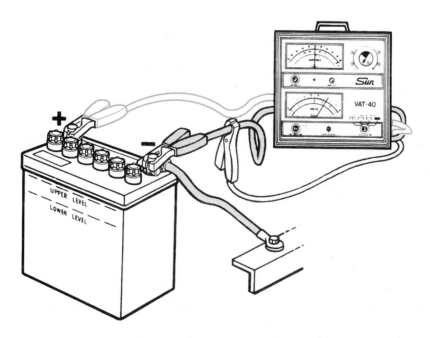

**Figure 1–39** Battery test.

**Hint**  *The hydrometer tests the specific gravity of the battery electrolyte. A fully charged battery should have a specific gravity of 1.265. A voltmeter may be connected across the battery terminals to test the open circuit battery voltage. The specific gravity and the open circuit voltage indicate the battery state of charge (Figure 1–40).*

| State of charge | Specific gravity | Open-circuit voltage |
|---|---|---|
| 100% | 1.265 | 12.6 |
| 75% | 1.225 | 12.4 |
| 50% | 1.190 | 12.2 |
| 25% | 1.155 | 12.0 |
| Dead | 1.120 | 11.9 |

**Figure 1–40**  Battery specific gravity and open circuit voltage. *(Reprinted with permission)*

*A battery load or capacity tester is used to test the battery capacity. During this test, the battery is discharged at one-half the cold cranking amperes for 15 seconds, and the voltage is recorded at the end of this time. A satisfactory battery at 70°F has 9.6 V at the end of the capacity test.*

67. Technician A says the negative battery cable must be removed prior to starter removal.

   Technician B says if there are shims between the starter mounting flange and the flywheel housing, these shims should be discarded when the starter is reinstalled. Who is correct?

   A.  A only

   B.  B only

   C.  Both A and B

   D.  Neither A nor B

**Hint**    *Always remove the negative battery cable before removing the starter. If the vehicle is air bag–equipped, wait the specified length of time after the negative battery cable is removed before working on the vehicle. If there are shims between the starter motor mounting flange and the flywheel housing, these shims must be installed in their original position when the starter is reinstalled.*

   *The starter mounting flange and the flywheel housing mounting surface must be free from dirt, paint, and other debris.*

## Task 5  Inspect and replace positive crankcase ventilation (PCV) system components.

68. In Figure 1–41, the hose from the PCV valve to the intake manifold is restricted. This problem could result in

   A.  an acceleration stumble.

   B.  oil accumulation in the air cleaner.

   C.  engine surging at low speed.

   D.  engine detonation during acceleration.

**Hint**    *If the PCV valve is stuck in the open position, excessive airflow through the valve causes a lean air-fuel ratio and possible rough idle operation or engine stalling. When the PCV valve, or hose, is restricted, excessive crankcase pressure forces blowby gases through the clean air hose and filter in the air cleaner. Worn rings or cylinders cause excessive blowby gases and increased crankcase pressure, which forces blowby gases through the clean air hose and filter in the air cleaner. A restricted PCV valve, or hose, may result in the accumulation of moisture and sludge in the engine and engine oil.*

   *Some vehicle manufacturers recommend removing the PCV valve from the rocker arm cover, and placing your finger over the valve with the engine idling. When there is no vacuum*

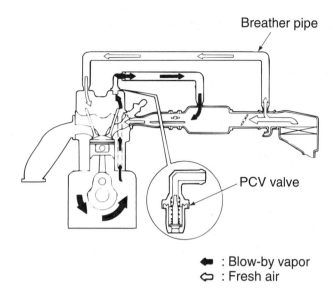

**Figure 1–41** PCV system. *(Courtesy of American Honda Motor Co., Inc.)*

*at the PCV valve, the valve, or manifold inlet are restricted. Replace the restricted component or components.*

## Task 6 Visually inspect and reinstall primary and secondary ignition system components; time distributor.

69. The ohmmeter in Figure 1–42 is connected to test
    A. the primary winding for shorts.
    B. the primary winding for grounds.
    C. the secondary winding for shorts.
    D. the secondary winding for grounds.

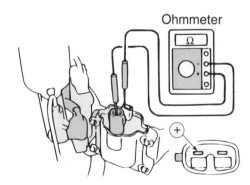

**Figure 1–42** Testing ignition coil. *(Reprinted with permission)*

**Hint**    *The coil secondary winding should be tested for shorts, grounds, and open circuits with an ohmmeter. Always inspect the coil tower for cracks. Secondary coil voltage should be tested with a test spark plug or an oscilloscope. When timing the distributor to the engine, the number 1 piston should be positioned at top dead center (TDC) on the compression stroke, and the rotor must be pointing toward the number 1 spark plug wire terminal in the distributor cap. The high points on the pickup coil and the timer core must be aligned.*

## Task 7   Inspect and diagnose exhaust system; determine needed repairs.

70. All of the following statements regarding manifold heat control valves are true EXCEPT
    A. A manifold heat control valve improves fuel vaporization in the intake manifold, especially when the engine is cold.
    B. A manifold heat control valve stuck in the closed position causes a loss of engine power.
    C. A manifold heat control valve stuck in the open position may cause an acceleration stumble.
    D. A manifold heat control valve stuck in the closed position reduces the intake manifold temperature.

**Hint**    *The exhaust manifold should be inspected for cracks and damage to the flange. All foreign material should be cleaned from the exhaust manifold and cylinder head mating surfaces. These surfaces should be checked for warpage with a straightedge and a feeler gauge. If the exhaust manifold contains a heat control valve, this valve should be inspected for erosion and free movement.*

# Post Test

1. An engine has a clattering noise when the engine is accelerated and decelerated. The most likely cause of this problem is
   A. worn main bearings.
   B. loose connecting rod bearings.
   C. sticking valve lifters.
   D. worn push rods.

2. An engine has gray smoke in the exhaust when it is first started.
   Technician A says the piston rings are worn.
   Technician B says the valve guides and seals are worn.
   Who is correct?
   A. A only
   B. B only
   C. Both A and B
   D. Neither A nor B

3. A longitudinally mounted V6 engine has cylinders 1, 3, and 5 on the side of the engine next to the firewall, and cylinders 2, 4, and 6 on the side next to the radiator. A compression test indicates cylinders 3 and 5 have 55 psi compression and all the other cylinders have 140 psi compression. The most likely cause of this problem is
   A. burned exhaust valves in cylinders 3 and 5.
   B. a leaking head gasket between cylinders 3 and 5.
   C. worn rings on cylinders 3 and 5.
   D. worn valve guides and seals in cylinders 3 and 5.

4. An engine has excessive oil consumption and the exhaust contains blue smoke especially during deceleration.
   Technician A says to check the valve stem-to-guide clearance and valve stem seals.
   Technician B says to pressure test the cooling system.
   Who is correct?
   A. A only
   B. B only

C. Both A and B

D. Neither A nor B

5. An engine has an excessively low oil pressure problem. When the engine is disassembled, the oil pump, main bearing clearance, and connecting rod bearing clearance are within specifications.

   Technician A says the camshaft bearings may be worn.

   Technician B says the push rods and rocker arms may be worn.

   Who is correct?

   A. A only

   B. B only

   C. Both A and B

   D. Neither A nor B

6. The bearing inserts in a connecting rod have a V-shaped wear pattern, and the piston attached to this connecting rod has uneven wear on the piston skirt. The most likely cause of this problem is

   A. a worn connecting rod bearing.

   B. excessive piston clearance.

   C. worn piston rings.

   D. A connecting rod misalignment.

7. When measuring piston ring gap clearance and installing piston rings, Technician A says the ring gaps should be positioned directly below each other on the piston.

   Technician B says if the ring gap is excessive, and that the cylinder may have excessive wear.

   Who is correct?

   A. A only

   B. B only

   C. Both A and B

   D. Neither A nor B

8. All of the following defects will cause the pressure to decrease during a cooling system pressure test EXCEPT

   A. a leaking heater core.

   B. leaking core plugs in the engine block.

   C. plugged air passages in the radiator.

   D. a leaking transmission cooler.

9. A customer complains about lack of heat in the vehicle interior. The coolant level and thermostat are proven to be satisfactory.

   Technician A says there may be air in the cooling system and a cooling system bleeding procedure should be performed.

   Technician B says the head gasket may be leaking.

   Who is correct?

   A. A only

   B. B only

   C. Both A and B

   D. Neither A nor B

10. An electric cooling fan is controlled by the powertrain control module (PCM). The cooling fan does not come on when the engine is hot, and the customer complains about overheating.

    Technician A says to use a scan tool to check the cooling fan motor and circuit by commanding the PCM to operate the cooling fan.

    Technician B says to use a scan tool to check the engine coolant sensor signal to the PCM.

Who is correct?
A. A only
B. B only
C. Both A and B
D. Neither A nor B

11. When a turbocharger is disassembled, it's discovered the compressor wheel vanes are badly pitted and damaged. The most likely cause of this problem is
   A. a sticking turbocharger wastegate and excessive boost pressure.
   B. a leak in the air intake system between the turbocharger and the air cleaner.
   C. a leak in the air intake system between the turbocharger and the engine.
   D. worn turbocharger bearings.

12. A starter motor draws 350 amperes and the engine has a slow cranking complaint. The specified starter draw is 200 amperes. All of the following defects may be the cause of the problem EXCEPT
   A. worn starting motor bushings.
   B. worn starting motor brushes.
   C. a defective starting motor armature.
   D. a battery that indicates 9.6 V on a load test at 70°F.

13. An engine has an accumulation of oil in the air cleaner and air filter.
   Technician A says the PCV valve or hose may be restricted.
   Technician B says the engine may have badly worn piston rings.
   Who is correct?
   A. A only
   B. B only
   C. Both A and B
   D. Neither A nor B

14. An ignition system experiences repeated module failure. The most likely cause of this problem is
   A. a shorted primary winding.
   B. a shorted secondary winding,
   C. an open pickup coil.
   D. a leaking distributor cap and rotor.

15. When measuring an engine cylinder for wear with a bore gauge, Technician A says the cylinder should be measured at two vertical locations.
   Technician B says the cylinder should be measured in the thrust and axial directions to determine the cylinder out-of-round.
   Who is correct?
   A. A only
   B. B only
   C. Both A and B
   D. Neither A nor B

16. An engine with a coil-near-spark plug ignition system has a no-start condition, and there is no spark at any of the spark plugs.
   Technician A says to test the crankshaft sensor and circuit.
   Technician B says to test for voltage at each ignition coil with the ignition switch on.
   Who is correct?
   A. A only
   B. B only

C. Both A and B

D. Neither A nor B

17. An engine experiences a high-pitched squealing noise only when the engine is idling. The most likely cause of this problem is
    A. a leak in the intake manifold.
    B. a leak in the exhaust manifold.
    C. a loose alternator and water pump drive belt.
    D. a loose camshaft belt.

18. One exhaust valve in a cylinder head has excessive valve stem installed height. Technician A says too much material may have been machined from the valve seat.
    Technician B says this condition moves the plunger upward in the valve lifter.
    Who is correct?
    A. A only
    B. B only
    C. Both A and B
    D. Neither A nor B

19. Technician A says shims may be installed under the valve springs to correct excessive valve spring installed height.
    Technician B says excessive valve spring installed height causes reduced valve spring tension and valve float at high speed.
    Who is correct?
    A. A only
    B. B only
    C. Both A and B
    D. Neither A nor B

20. When a piston is positioned at TDC on the exhaust stroke and the valve timing is correct
    A. the intake valve should be wide open.
    B. the exhaust valve should be opening.
    D. the intake valve should be opening.
    D. the exhaust valve should be closed

# Answers and Analysis

**1. B** The easiest, quickest test should be performed first. Therefore, A is wrong. The customer complaint must be identified during a basic diagnostic procedure. Thus, B is the correct answer.

**2. A** Technician B is wrong because with a partially restricted fuel filter, some fuel is still discharged from the injectors. Technician A is right because if there is no voltage supplied to the fuel pump, the fuel pressure is zero, and no fuel is discharged from the injectors. Therefore, A is the right answer.

**3. C** Technicians A and B are both correct because neither a leaking head gasket nor heater core will cause leaks in the engine compartment, but they will reduce the test pressure. C is the right answer.

**4. B** Worn pistons and cylinders would cause a rapping noise while accelerating, and worn main bearings would cause a thumping noise when the engine is started. Therefore, A and C are wrong.

Loose camshaft bearings usually do not cause a noise unless severely worn, so D is wrong. Loose flywheel bolts may cause a thumping noise at idle. Thus, B is the correct answer.

**5.  A**     Excessive fuel pressure or a restricted return fuel line causes a rich air-fuel ratio. Therefore, B and C are wrong.

A sticking fuel pump check valve may cause hard starting, so D is wrong.

A burned exhaust valve causes a "puff" noise in the exhaust. Thus, A is right.

**6.  D**     Late ignition timing, or an intake manifold vacuum leak, causes a low steady vacuum gauge reading. Therefore, A and B are wrong.

A restricted exhaust system causes a low vacuum reading at high speeds, so C is wrong.

Sticking valves cause a vacuum gauge fluctuation between 12 in. and 18 in. Hg at idle. Thus, D is the correct answer.

**7.  C**     A cylinder with very little rpm drop indicates cylinder misfiring, which may be caused by the ignition system or an intake manifold vacuum leak. Therefore, both A and B are right, and C is the correct answer.

**8.  B**     If the compression increases significantly during a wet test, worn rings and cylinders are indicated. Therefore, A, C, and D are wrong; B is the right answer.

**9.  B**     A leaking intake valve causes air to escape from the air intake. Therefore, A is wrong.

Worn rings cause air to escape from the PCV valve opening during the cylinder leakage test. Thus, B is the right answer.

**10.  B**    Technician A is wrong because a plastic scraper should be used to clean gasket surfaces on aluminum heads. A steel scraper gouges these heads. Technician B is right because some manufacturers provide a detorque sequence for aluminum heads. Therefore, B is the right answer.

**11.  C**    If the cylinder head warpage is 0.014 in., the head must be resurfaced. Block warpage may have caused the head distortion. Therefore, A and B are correct, and C is the right answer.

**12.  D**    The valve spring must be rotated while measuring height and squareness. Neither A nor B are correct; thus, D is the right answer.

**13.  C**    Technician A is correct because the valve spring tension must be measured at the installed spring height. Technician B is also correct because if a shim is required, the shim must be installed under the spring on the tension gauge. Because both technicians are correct, C is the right answer.

**14.  A**    Worn valve spring locks will not cause a clicking noise because they are continually loaded by the valve spring tension. Therefore, B is wrong.

Worn valve lock grooves may cause the locks to fly out of place with the engine running causing severe engine damage. Thus, A is right.

**15.  B**    Worn valve stem seals allow oil to run down the stems into the cylinder causing excessive oil consumption, but worn seals will not increase stem and guide wear. Therefore, A is wrong, and B is the right answer.

**16.  C**    Valve stems should be measured at three locations, and valve guides are measured with a snap or hole gauge. Both A and B are correct, and C is the right answer.

**17.  D**    Valve stem-to-guide clearance is measured in Figure 1–5. Therefore, A, B, and C are wrong, and D is the right answer.

**18.  B**    A reduced valve margin causes valve overheating and burning. This problem does not cause improper valve seating, valve seat recession, or a clicking noise at idle speed. Therefore, A, C, and D are wrong, and B is the right answer.

**19.  A**    Valve 1 indicates normal wear, valve 2 indicates partial valve rotation, and valve 3 indicates complete lack of valve rotation. Therefore, B, C, and D are wrong, and A is the right answer.

**20.  C**    Most valve seats have an angle of 30 degrees or 45 degrees. Therefore, A, B, and D are wrong, and C is the right answer.

**21.  A**   Improper valve seat contact area on the valve face does not cause a clicking noise at idle speed. B is wrong. When the valve seat contact area is too low on the valve face, 60-degree and 45-degree stones are used to move the seat contact area upward. Thus, A is the right answer.

**22.  B**   Excessive installed valve spring height would not bottom the lifter unless the installed valve stem height is also more than specified. Therefore, A is wrong.

   If the installed valve spring height is more than specified, a shim may be installed under the spring to correct this problem. Thus, B is right.

**23.  C**   Bent pushrods may be caused by sticking valves or excessive valve spring tension. Improper valve timing may cause the valves to strike the pistons in some engines, resulting in bent pushrods. Therefore, both A and B are correct, and C is the right answer.

**24.  B**   Lifter bottoms must be convex. Therefore, A is wrong.

   A sticking lifter plunger may hold the valve open, resulting in a burned valve. B is the right answer.

**25.  B**   When measuring valve clearance, the piston must be TDC compression with both valves closed. Therefore, A is wrong. Some mechanical lifters have removable shim pads. Thus, B is the right answer.

**26.  A**   When the timing gear teeth are meshed with the crankshaft gear teeth, the gear backlash is measured with a dial indicator. Therefore, B is wrong, and A is the right answer.

**27.  A**   On some engines, the cam lobe lift is measured with a dial indicator positioned against the top of the pushrod. Therefore, B, C, and D are wrong, and A is right.

**28.  A**   Excessive camshaft bearing clearance causes low oil pressure, but this problem does not cause a thumping noise with the engine idling. Therefore, B is wrong, and A is the right answer.

**29.  C**   Improper valve timing may cause the valves to strike the pistons in some engines, causing bent valves. This problem also reduces engine power. Thus, both A and B are correct, and C is the right answer.

**30.  C**   Torque-to-yield bolts do provide more clamping force, because they are tightened to a specific torque and then rotated a certain number of degrees. A and B are both correct, and C is the right answer.

**31.  D**   The ring ridge should be removed before extracting the pistons and connecting rods. Therefore, statement A is correct.

   The connecting rods must be reinstalled with their markings facing in the specified direction.

   Therefore, statement B is correct.

   The main bearings must be installed in their original position, and so statement C is correct.

   After ridge reaming, an oily rag should be used to remove metal filings from the cylinder. Do not blow metal filings with compressed air! Therefore, statement D is incorrect and this is the right answer.

**32.  C**   A warped head mounting surface on the block may cause valve seat distortion when the head bolts are torqued, and this problem may also cause coolant and combustion leaks.

   Therefore, both A and B are correct, and C is the right answer.

**33.  D**   The first step in heli-coil installation is to drill the opening, and a special tool is used to install the heli-coil. Therefore, both A and B are wrong, and D is the right answer.

**34.  D**   Installing new rings without removing the ring ridge may cause broken ring lands.

   Therefore, A, B, and C are wrong, and D is the right answer.

**35.  D**   Cylinder taper is the difference between the cylinder diameter at the top of the ring travel compared to the diameter at the bottom of the ring travel. Therefore, A is wrong.

   Cylinder out-of-round is the difference between the axial bore diameter and the thrust bore diameter at the same cylinder position. Therefore, B is also wrong, and D is the right answer.

**36. B**    Honing stones with a lower number have a coarser grit. Cylinders may be deglazed with 220-to 280-grit stones, and the cylinder should be washed with soapy water after deglazing.

Therefore, A, C, and D are wrong, and B is the right answer.

**37. C**    A and B measure vertical taper, C and D measure horizontal taper, A and C measure out-of-round, and A and D is not a valid measurement. Therefore, A, B, and D are wrong, and C is the right answer.

**38. D**    Main bearing bore alignment may be measured with a straightedge and a feeler gauge.

Therefore, A, B, and C are wrong, and D is the right answer.

**39. A**    Technician A is correct because worn thrust surfaces on the sides of one main bearing will cause excessive crankshaft end play.

Technician B is wrong because a shim is not installed between the rear edge of the rear main bearing and the rear of the crankshaft flange to correct excessive crankshaft end play.

Therefore, A is the right answer.

**40. C**    The tool in Figure 1–25 is used to remove and install camshaft bearings. Therefore, A, B, and D are wrong, and C is the right answer.

**41. A**    The balance shafts are timed in relation to the crankshaft, and improper balance shaft timing results in severe engine vibrations. Therefore, B is wrong and A is the right answer.

**42. C**    The tool in Figure 1–26 is a piston ring removal and replacement tool. Therefore, A, B, and D are wrong, and C is the right answer.

**43. A**    A bent connecting rod causes uneven connecting rod bearing wear, but this problem does not influence main bearing, piston pin, or cylinder wall wear. Therefore, B, C, and D are wrong, and A is the right answer.

**44. C**    Bearing spread is the slightly larger curvature of the bearing insert compared to the bearing bore. Bearing crush is the bearing design that allows the bearing edges to be extended slightly above the mounting area when the bearing is installed. Therefore, both A and B are correct, and C is the right answer.

**45. D**    Ring gap should be measured with the ring at the bottom of the ring travel, and compression rings must not be interchanged on most pistons. Therefore, both A and B are wrong, and D is the right answer.

**46. A**    If the damper seal contact area is scored, some damper hubs may be machined and a sleeve installed to provide a new seal contact area. Therefore, B is wrong.

The vibration damper counterbalances the back-and-forth twisting crankshaft motion. A is the right answer.

**47. C**    Metal burrs on the crankshaft flange cause flywheel and torque converter misalignment, and this flywheel misalignment may cause excessive ring gear and starter drive gear wear.

Therefore, both A and B are correct, and C is the right answer.

**48. C**    Flywheel runout causes an uneven clutch plate mating surface, resulting in clutch grabbing, and the pressure plate must be installed in the original position to maintain proper balance.

Therefore, both A and B are correct, and C is the right answer.

**49. D**    With the clutch engaged, the clutch plate is held firmly between the flywheel and pressure plate, and the crankshaft and complete clutch assembly are turning together. Under this condition, the pilot bearing inner and outer races are turning at the same speed, thus preventing bearing noise under this condition. Therefore, A, B, and C are wrong.

With the clutch pedal depressed, the clutch disc and input shaft are free to move up and down in the worn pilot bearing, resulting in a rattling noise. D is the right answer.

**50. D**    The timing gear cover seal contains a garter spring, and the seal lip must face toward the oil flow. Therefore, A and B are both wrong, and D is the right answer.

**51. B**    RTV sealer cures in five minutes so components using this sealer must be assembled quickly.

Chlorinated solvent must be used to clean RTV sealed components. The RTV bead must surround any bolt holes. Therefore, A, C, and D are wrong.

The RTV bead that's 1/8-in. wide should be placed in the center of the sealing surface. B is the right answer.

**52.  D**  Worn camshaft bearings, crankshaft bearings, or a weak pressure regulator spring would cause low oil pressure. Therefore, A, B, and C are wrong.

Restricted pushrod passages will not cause low oil pressure; therefore, response D is the right answer.

**53.  A**  B, C, and D are valid oil pump measurements, and so these responses are wrong.

If the clearance between the rotors is normal, there is no need to measure the inner rotor diameter. Therefore, A is not a normal measurement, and this is the correct answer.

**54.  C**  A, B, and D are valid cooling system tests with the pressure tester. Therefore, these are not the requested answer.

The cooling system pressure tester does not measure coolant specific gravity, making C the requested answer.

**55.  A**  A squealing on acceleration is also a result of a loose alternator belt. Therefore, B is wrong.

A loose alternator belt would not damage the alternator bearing, and so C is wrong.

The alternator belt may not drive the water pump, so a loose alternator belt may not cause engine overheating. D is wrong.

A loose alternator belt reduces alternator output and causes a discharged battery. A is the right answer.

**56.  B**  A defective pressure release valve in the radiator cap allows coolant to escape from the radiator to the recovery reservoir. Therefore, A is wrong.

A plugged recovery reservoir hose prevents coolant flow from the reservoir into the radiator as the engine cools down, resulting in a collapsed upper hose. B is right.

**57.  A**  If the thermostat is stuck open, the coolant temperature remains lower than normal and the coolant temperature sensor informs the computer regarding this low coolant temperature.

The computer provides a richer air-fuel ratio in relation to the low coolant temperature, resulting in reduced fuel economy. Therefore, B, C, and D are wrong, and A is the right answer.

**58.  B**  When the cooling system pressure is reduced, the coolant boiling point is decreased. Therefore, A is wrong.

Adding more antifreeze to the coolant solution increases the boiling point. Thus, B is the correct answer.

**59.  C**  A defective water pump bearing may cause a growling noise at idle speed, and the bearing may have been contaminated by coolant leaking past the pump seal. Therefore, both A and B are correct, and C is the right answer.

**60.  B**  Responses A, C, or D cause engine overheating and excessive coolant in the recovery reservoir.

Therefore, these responses are not the requested answer.

A thermostat that is stuck open reduces coolant temperature and does not cause a high coolant level in the recovery reservoir. Since B is not a cause of the problem, this is the right answer.

**61.  D**  If the temperature switch ground circuit is open and the A/C is off, the circuit from the fan relay winding to ground cannot be completed, resulting in an inoperative fan motor and engine overheating. When the A/C is on, the fan relay winding may be grounded through the condenser switch to provide cooling fan operation and some reduction in engine temperature.

Therefore, A, B, and C are wrong, and D is right.

**62.  A**  An engine oil cooler reduces oil temperature, which prevents oil oxidation. Therefore, B, C, and D are wrong, and A is the right answer.

**63.  A**    An intake manifold vacuum leak may cause a lean air-fuel ratio and a cylinder misfire with the engine idling. Since manifold vacuum decreases with throttle opening, an intake manifold vacuum leak does not allow as much air into the intake during hard acceleration and the idle misfire may disappear. Therefore, A is right and B is wrong, and A is the right answer.

**64.  D**    While cleaning most paper air-cleaner elements, the air gun should be held 6 in. from the inside of the element because air normally flows from the outside to the inside of the element.

Therefore, A, B, and C are wrong, and D is the right answer.

**65.  B**    A wastegate valve stuck closed, a leaking wastegate diaphragm, or a disconnected wastegate linkage can all cause excessive boost pressure. Therefore, A, C, and D are wrong. A wastegate valve stuck open reduces boost pressure; thus, B is the right answer.

**66.  C**    During the battery load test, the battery should be discharged at one-half the cold cranking rating for 15 seconds. Therefore, A, B, and D are wrong. During the load test, the battery voltage should remain above 9.6 V with the battery temperature at 70°F. Therefore, C is the right answer.

**67.  A**    If there are shims between the starter mounting flange and the flywheel housing, these shims must be reinstalled in their original position. Therefore, B is wrong. The negative battery cable must be removed prior to starter removal. Therefore, A is the right answer.

**68.  B**    If the hose from the PCV valve to the intake manifold is restricted, excessive crankcase pressure forces oil vapors through the clean air hose from the rocker arm cover into the air cleaner. Therefore, B is the right answer.

**69.  C**    The ohmmeter is connected to test the secondary winding for shorts. Therefore, C is the right answer.

**70.  D**    A manifold heat control valve improves fuel vaporization in the intake manifold; therefore, A is not the requested answer.

A manifold heat control valve stuck in the closed position causes a loss of engine power; therefore, B is not the requested answer.

A manifold heat control valve stuck in the open position may cause an acceleration stumble; therefore, C is not the requested answer.

A manifold heat control valve stuck in the closed position increases the intake manifold temperature; therefore, D is wrong, and this is the requested answer.

# Answers to Post Test

**1.  B**    Worn main bearings cause a thumping noise when the engine is first started. Sticking valve lifters cause a clicking noise when the engine is first started. Worn pushrods may cause a clicking noise during engine idle. Therefore, A, C, and D are wrong. Loose connecting rod bearings cause a clattering noise when the engine is accelerated and decelerated. Therefore, B is the right answer.

**2.  D**    Gray smoke in the exhaust when the engine is first started is caused by a coolant leak into the combustion chamber(s). Therefore, Technicians A and B are both wrong, and D is the right answer.

**3.  B**    The most likely cause of low compression on two adjacent cylinders is a leaking head gasket. Therefore, A, C, and D are wrong, and B is the right answer.

**4.  A**    A coolant leak into the combustion chambers causes gray smoke in the exhaust. Therefore, Technician B is wrong. Blue smoke in the exhaust during deceleration may be caused by worn valve guides and valve stem seals. Therefore, Technician A is correct, making A the right answer.

**5. A**   Worn pushrods and rocker arms do not affect engine oil pressure. Therefore, Technician B is wrong. Worn camshaft bearings cause low oil pressure. Therefore, Technician A is correct, and A is the right answer.

**6. D**   Connecting rod misalignment results in a V-shaped wear pattern on the connecting rod bearings, and uneven wear on the piston skirt. Therefore, A, B, and C are wrong, and D is the right answer.

**7. B**   When the piston rings are installed, the ring gaps should be staggered around the piston. Therefore, Technician A is wrong. Excessive ring gap may be caused by a worn cylinder. Therefore, Technician B is correct, and B is the right answer.

**8. C**   A leaking heater core, leaking core plugs in the block, or a leaking transmission cooler will cause the pressure to decrease during a cooling system pressure test. Therefore, A, B, and D are not the requested answer. Plugged air passages in the radiator have no effect on the pressure during a cooling system pressure test. Therefore, C is the requested answer.

**9. A**   A leaking head gasket may cause a lack of heat in the vehicle interior, but this defect also causes a coolant leak and low coolant level. Therefore, Technician B is wrong. Air in the cooling system may cause a lack of heat in the vehicle interior. Therefore, Technician A is correct, and A is the right answer.

**10. C**   An inoperative cooling fan may be caused by a defect in the cooling fan motor and related circuit. The cooling fan motor and circuit may be tested by commanding the PCM to operate the cooling fan motor. Therefore, Technician A is correct. A defective engine coolant sensor signal may cause improper cooling fan motor operation. Therefore, Technician B is also correct, and C is the right answer.

**11. B**   A sticking turbocharger wastegate and excessive boost pressure does not cause a pitted compressor wheel. A leak in the air intake system between the turbocharger and the engine causes dirt particles to enter the engine, but these particles do not enter the turbocharger. Worn turbocharger bearings may cause excessive end play on the turbocharger shaft and compressor wheel damage. However, this problem does not cause a pitted compressor wheel. Therefore, A, C, and D are wrong. A leak in the air intake system between the air cleaner and the turbocharger causes dirt particles to flow past the compressor wheel, resulting in pitted compressor wheel vanes. Therefore, B is the right answer.

**12. D**   Worn starting motor bushings, or brushes, or a defective starting motor armature may cause excessive starting motor current draw. Therefore, A, B, and C are not the requested answer. A battery that indicates 9.6 V on a load test at 70°F is in satisfactory condition. Therefore, the battery does not cause excessive starting motor draw, and D is the requested answer.

**13. C**   An accumulation of oil in the air cleaner and air filter may be caused by a restricted PCV valve or hose, or excessive pressure in the engine resulting from worn piston rings. Therefore, Technicians A and B are both correct, and C is the right answer.

**14. A**   A shorted secondary winding, an open pickup coil, or a leaking cap and rotor do not affect the module. Therefore, B, C, and D are wrong. A shorted primary winding results in excessive primary current and possible module damage. Therefore, A is the right answer.

**15. B**   The cylinders should be measured at three vertical locations. Therefore, Technician A is wrong. The cylinder should be measured in the thrust and axial directions to determine the cylinder out-of-round. Therefore, Technician B is correct, and B is the right answer.

**16. C**   When there is no spark at any of the spark plugs on a coil-near-spark-plug ignition system, the crankshaft sensor or circuit may be defective, or there may be no voltage supplied to any of the coils. Therefore, Technicians A and B are both correct, and C is the right answer.

**17. A**   A leak in the exhaust manifold may cause a squealing noise during acceleration. A loose alternator and water pump drive belt may also cause a squealing noise during acceleration. A loose camshaft belt may cause the belt to jump several notches on the camshaft gear resulting in improper valve timing. Therefore, B, C, and D are wrong. A leak in the intake manifold may cause a high-pitched squeal with the engine idling. Therefore, A is the right answer.

**18.  A**   Excessive valve stem installed height causes the plunger to move downward in the lifter. Therefore, Technician B is wrong. Excessive valve stem installed height may be caused by machining too much material from the valve seat. Therefore, Technician A is correct, and A is the right answer.

**19.  C**   Shims may be installed under the valve springs to correct excessive valve spring installed height, and excessive valve spring installed height causes reduced valve spring tension and valve float at high speed. Therefore, both Technicians A and B are correct, and C is the right answer.

**20.  C**   When a piston is at TDC on the exhaust stroke, the intake valve should be starting to open, and the exhaust valve should be closing. Therefore, A, B, and D are wrong, and C is the right answer.

# 2 Automatic Transmission/ Transaxle

## Pretest

The purpose of this pretest is to determine the amount of review that you may require prior to taking the ASE Automatic Transmission and Transaxle Test. If you answer all the pretest questions correctly, complete the questions and study the information in this chapter to prepare for the ASE Automatic Transmission and Transaxle Test. If two or more of your answers to the pretest questions are incorrect, complete a thorough study of the questions and information in this chapter.

The pretest answers are located at the end of the pretest, and are also in the answer sheets supplied with this book.

1. An automatic transaxle has a loss of automatic transmission fluid (ATF) and there are no visible external leaks.
   Technician A says the vacuum modulator diaphragm may be leaking.
   Technician B says the transaxle cooler may be leaking.
   Who is correct?
   A. A only
   B. B only
   C. Both A and B
   D. Neither A nor B

2. The ATF in a transmission is milky in color. The cause of this problem could be
   A. burned clutch discs.
   B. burned bands.
   C. a worn case.
   D. a leaking transmission cooler.

3. A non-computer-controlled transaxle has higher-than-specified fluid pressure in all gear selector positions.
   Technician A says the pump may be damaged.
   Technician B says the pressure regulator valve may be stuck.
   Who is correct?
   A. A only
   B. B only
   C. Both A and B
   D. Neither A nor B

4. A computer-controlled transaxle provides torque converter clutch (TCC) lockup at 25 mph. The specified lockup vehicle speed is 46 mph.
   Technician A says the TCC lockup solenoid may be malfunctioning.
   Technician B says the vehicle speed sensor (VSS) may be malfunctioning.
   Who is correct?
   A. A only
   B. B only

C. Both A and B

D. Neither A nor B

5. If an electronic problem occurs in a four-speed computer-controlled transaxle, the transaxle will

A. operate only in reverse.

B. provide 1 - 2 and 2 - 3 upshifts.

C. provide only one forward gear.

D. provide only first gear.

6. When a vacuum gauge is connected into the transmission modulator vacuum hose with a T-fitting, the vacuum is a steady 12 in. Hg with the engine idling.

Technician A says the ignition timing may be later than specified.

Technician B says the engine may have a sticking valve.

Who is correct?

A. A only

B. B only

C. Both A and B

D. Neither A nor B

7. During an air pressure test on the forward clutch, the clutch application is not heard, but there is no air escaping. The cause of the problem could be

A. damaged forward clutch piston seals.

B. a cracked forward clutch drum.

C. a plugged forward clutch fluid passage.

D. damaged forward clutch hub steel rings.

8. All the upshifts occur at a higher speed than specified in a non-computer-controlled transaxle.

Technician A says the cause of this problem may be high governor pressure.

Technician B says the cause of this problem may be low throttle pressure.

Who is correct?

A. A only

B. B only

C. Both A and B

D. Neither A nor B

9. A vehicle has a computer-controlled transaxle, cruise control, and an electronic instrument panel. The cruise control module is separate from the powertrain control module. The torque converter clutch, cruise control, and speedometer are inoperative.

The cause of this problem could be a malfunctioning

A. vehicle speed sensor.

B. powertrain control module.

C. engine coolant temperature sensor.

D. crankshaft sensor.

10. A four-speed automatic transmission slips consistently in second and fourth gear.

Technician A says the 2 - 4 band adjustment may be too loose.

Technician B says the 2 - 4 servo piston seal may be leaking.

Who is correct?

A. A only

B. B only

C. Both A and B

D. Neither A nor B

11. A computer-controlled transmission suddenly shifts into neutral while driving with the gear selector in the overdrive position. The cause of this problem could be a malfunctioning
    A. manual lever position sensor (MLPS).
    B. shift solenoid.
    C. transmission oil temperature (TOT) sensor.
    D. output speed sensor (OSS).

12. During a transmission stall test, the engine's rpm is higher than specified. All of the following could be the cause of the problem EXCEPT
    A. a seized one-way stator clutch in the torque converter.
    B. slipping clutch discs.
    C. slipping band.
    D. lower-than-specified fluid pressure.

13. The input shaft end play in an automatic transaxle is less than specified. The cause of this problem could be
    A. a worn pump.
    B. a worn transaxle case.
    C. improper selective washer thickness.
    D. improper forward clutch clearance.

14. A fuel-injected engine with a four-speed automatic transaxle has a loss of power at high speed. All of the following could be the cause of the problem EXCEPT
    A. a seized one-way stator clutch in the torque converter.
    B. low fuel system pressure.
    C. a slipping one-way stator clutch in the torque converter.
    D. a restricted exhaust system.

# Answers to Pretest

**1.  C**   A leaking modulator diaphragm or a leaking transmission cooler may cause a loss of transmission fluid without any visible external leaks. Therefore, Technicians A and B are both right, and C is the correct answer.

**2.  D**   Burned clutch discs or bands may cause the transmission fluid to be darkened and have a burned odor. A worn case may cause aluminum filings in the transmission fluid. A leaking transmission cooler contaminates the transmission fluid with engine coolant, and this causes the fluid to have a milky color. Therefore, D is the correct answer.

**3.  B**   A damaged transmission pump may cause low fluid pressure, but a sticking pressure regulator valve may cause higher than normal transmission fluid pressure in all gear selector positions. B is the correct answer.

**4.  B**   A malfunctioning TCC solenoid would not cause low TCC engagement speed. The PCM operates the TCC in response to the VSS signal. Therefore, a malfunctioning VSS sensor may cause TCC engagement at a lower-than-specified vehicle speed. Therefore, Technician A is wrong and Technician B is right, making B the correct answer.

**5.  C**   If an electronic defect occurs in a computer-controlled four-speed transaxle, the transaxle operates only in one forward gear and reverse. Therefore, C is the correct answer.

**6.  A**   Late ignition timing may cause a low steady engine vacuum reading. A sticking engine valve causes an erratic vacuum reading. Therefore, Technician B is wrong, and Technician A is right, so A is the correct answer.

**7.  C**   During an air pressure test on the forward clutch drum, a leaking forward clutch seals, a cracked forward clutch drum, or damaged forward clutch hub steel rings cause air to escape.

Therefore, these answers are wrong. A plugged forward clutch passage causes failure of the clutch to apply and no air to escape. Therefore, C is the correct answer.

**8.  D**  In a non-computer-controlled transmission, upshifts that occur at a higher-than-specified speed may be caused by high throttle pressure or low governor pressure. Therefore, both Technicians A and B are both wrong, and D is the correct answer.

**9.  A**  In a computer-controlled transaxle, proper operation of the cruise control, torque converter clutch, and speedometer all depend on a satisfactory signal from the vehicle speed sensor. Therefore, a malfunctioning vehicle speed sensor signal may affect the operation of these three systems. A is the correct answer.

**10.  A**  A leaking servo piston seal causes momentary slipping or prolonged shifting. Therefore, Technician B is wrong. A loose 2 - 4 band adjustment causes consistent transmission slipping in second and fourth gears. Therefore, Technician A is right, and A is the correct answer.

**11.  A**  A malfunctioning manual lever position sensor (MLPS) may send an erroneous signal to the transmission computer indicating the MLPS has been moved to the neutral position, when the driver did not shift the transmission lever into this position. Under this condition, the transmission shifts into neutral. Therefore, A is the correct answer.

**12.  A**  Slipping clutch discs, a slipping band, or lower-than-specified transmission pressure may cause higher-than-specified rpm during a transmission stall test. Therefore, B, C, and D are not the requested answers. The one-way stator clutch in the torque converter is normally holding during a stall test. Therefore, a seized one-way clutch does not affect the stall speed, and A is the requested answer.

**13.  C**  The input shaft end play is determined by selective washer thickness. Therefore, C is the correct answer.

**14.  C**  Low fuel system pressure or a restricted exhaust system may cause a loss of power at high speed, and thus B or D are not the requested answer. A seized one-way stator clutch in the torque converter may cause a loss of power at high speed, because this clutch must free wheel at higher speeds. Therefore, A is not the requested answer. A slipping one-way clutch in the torque converter may cause a loss of power at low speed, but this defect does not cause a loss of power at high speed. C is the requested answer.

# General Transmission/Transaxle Diagnosis, Mechanical/Hydraulic Systems

## ASE Tasks, Questions, and Related Information

In this chapter, each task in the Automatic Transmission and Transaxle category is followed by a question and some information related to the task. If you answer any question incorrectly, study this information very carefully until you understand the correct answer. Question answers and analysis are provided at the end of this chapter and in the answer sheets provided with this book.

**Task 1  Road test the vehicle to verify mechanical/hydraulic system problems based on driver's concern; determine necessary action.**

1. The customer complains about fluid usage in an automatic transmission, and there are no visible signs of fluid leaks.

   Technician A says the transmission vent may be plugged or restricted.

   Technician B says the vacuum modulator diaphragm may be leaking.

   Who is correct?

   A. A only

   B. B only

   C. Both A and B

   D. Neither A nor B

**Hint**     *(Refer to the general diagnostic procedure provided in Chapter 1.) Prior to the road test, check the transmission fluid level and condition. Some transmission problems are caused by improper fluid level or contaminated fluid. Because transmission problems may be caused by improper engine performance, correct any engine performance problems prior to the road test. The engine and transmission should be at normal operating temperature prior to the road test. During the road test, operate the transmission in all gear selector positions, and run the engine under various conditions. Operate the vehicle under the conditions where the driver complaint occurred. Record all abnormal transmission operation noises and vibrations.*

**Task 2**  **Diagnose noise, vibration, harshness, and shift quality problems; determine necessary action.**

2. An automatic transmission has a whining noise that occurs in all gears while driving the vehicle. This noise is also present with the engine running and the vehicle stopped.

   Technician A says the rear planetary gear set may be damaged.

   Technician B says the oil pump may be damaged.

   Who is correct?

   A. A only

   B. B only

   C. Both A and B

   D. Neither A nor B

3. A rear-wheel drive vehicle has a vibration that increases in relation to vehicle speed. This vibration also is present when the engine is accelerated with the vehicle stopped and the gear selector in neutral or park. The cause of this problem could be

   A. improper torque converter balance.

   B. improper drive shaft balance.

   C. improper drive shaft angles.

   D. worn engine mounts.

**Hint**     *When diagnosing noise and vibration problems pay careful attention to the exact conditions when the noise or vibration occurs. For example, if a vibration occurs while driving the vehicle and with the engine running and the vehicle not moving, the cause of the vibration is likely in the engine or torque converter. When the vibration changes with an alteration in engine speed, the problem may be in the torque converter. If the vibration changes with an alteration in vehicle speed, the problem probably is in the driveline or transmission output shaft.*

**Task 3**  **Diagnose unusual fluid usage, type, level, and condition problems; determine necessary action.**

4. Fluid sometimes escapes from the dipstick tube on an automatic transaxle.

   Technician A says the transaxle fluid may be contaminated.

Technician B says the transaxle cooler may be the problem.
Who is correct?
A. A only
B. B only
C. Both A and B
D. Neither A nor B

5. Lubricant is leaking from the torque converter access cover. When this cover is removed, the shell of the converter is wet with fluid, but the front of the converter is dry. The cause of the problem could be
A. a leaking transmission oil pump seal.
B. a leaking rear main bearing.
C. a loose rear main bearing.
D. a leaking converter drain plug.

6. The fluid in an automatic transaxle is a dark brown color and smells burned.
Technician A says this problem may be caused by a worn front planetary sun gear.
Technician B says this problem may be caused by worn friction-type clutch plates.
Who is correct?
A. A only
B. B only
C. Both A and B
D. Neither A nor B

**Hint**    *Normal automatic transmission fluid is pink or red. If the fluid is dark brown or blackish and has a burned odor, the fluid has been overheated, possibly from burned clutches or bands. Milky colored fluid likely is caused by coolant contamination from a leaking transmission cooler.*
*Silvery metal particles in the fluid indicate damaged metal transmission components. If the dipstick feels sticky and is difficult to wipe clean, the fluid contains varnish. This varnish formation indicates transmission fluid and filter changes have been neglected.*

## Task 4  Perform pressure tests; determine necessary action.

7. An automatic transaxle has low pressure in third gear only.
Technician A says the transaxle may have an internal leak.
Technician B says the TV cable may be misadjusted.
Who is correct?
A. A only
B. B only
C. Both A and B
D. Neither A nor B

8. All the transmission pressures are normal at idle speed but low at wide open throttle.
Technician A says the TV cable may need adjusting.
Technician B says the vacuum modulator may be the problem.
Who is correct?
A. A only
B. B only
C. Both A and B
D. Neither A nor B

9. During a transmission pressure test, the pressure gradually decreases at higher engine speeds. The cause of this problem could be
A. a worn oil pump.
B. a restricted oil filter.

C. a stuck pressure regulator.

D. a plugged modulator hose.

**Hint**    *Transmission or transaxle pressure tests may be performed to diagnose internal problems such as rough or improperly timed shifts. These two problems may be caused by excessive line pressure, which may be tested with a pressure test. To perform the pressure test, the technician requires two pressure gauges, a tachometer, and specifications for the transmission being tested.*

*If the transmission has a vacuum modulator, then a vacuum gauge and a hand vacuum pump also are required. Various plugs in the transaxle case must be removed to install the pressure gauges (Figure 2–1).*

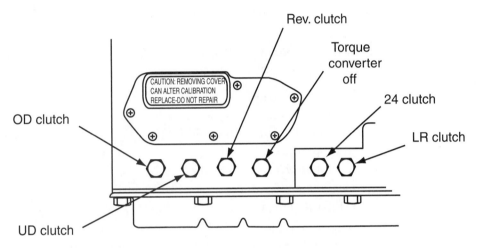

**Figure 2–1** Pressure test plug locations.

*If the transaxle pressures are low at slow idle, the problem may be in the pump, filter, fluid level, or pressure regulator. An internal leak may also be the cause of this problem.*

*When the pressure is low only in a specific gear, the problem is likely an internal leak. A partially plugged transaxle oil filter may cause a gradual reduction in pressure at higher engine speeds.*

*If all the pressures are low at wide open throttle (WOT), pull on the TV cable or disconnect the vacuum modulator hose and test the pressures. When the pressures increase, check the TV cable adjustment, cable condition, vacuum modulator and hose.*

## Task 5    Perform stall tests; determine necessary action.

10. During a stall test, the engine rpm is less than specified.
    Technician A says the turbine in the torque converter may be the problem.
    Technician B says some of the transmission clutches may be slipping.
    Who is correct?
    A. A only
    B. B only
    C. Both A and B
    D. Neither A nor B

11. During a stall test, the stall speed is above specifications.
    Technician A says the exhaust system may be restricted.
    Technician B says the engine may have low compression.
    Who is correct?
    A. A only
    B. B only

C. Both A and B

D. Neither A nor B

**Hint**    *During a stall test, a tachometer is connected to the ignition system, and this meter must be located where it can be seen by the driver. Apply the parking brake and place blocks in front of the vehicle's tires. Press and hold the brake pedal and place the gear selector in drive. Press the accelerator pedal to the WOT position, and note the rpm on the tachometer at which the engine stalled.*

*When the stall speed is below specifications, the torque converter stator clutch may be slipping, or the exhaust system may be restricted. An engine that is improperly tuned or has low compression also reduces stall speed.*

*If the stall speed is above specifications, the clutches and bands in the transmission may be slipping.*

## Task 6   Perform torque converter clutch (lock-up converter) mechanical/hydraulic system tests; determine necessary action.

12. A torque converter clutch does not lock up at any vehicle speed or engine temperature.

    The cause of the problem could be

    A. an inoperative TCC solenoid.

    B. an inoperative stator clutch.

    C. a leaking second speed servo piston seal.

    D. a sticking pressure regulator valve.

13. The engine shudders immediately after TCC lockup.

    Technician A says the engine may have an ignition problem.

    Technician B says the fuel injection system may have a lean condition.

    Who is correct?

    A. A only

    B. B only

    C. Both A and B

    D. Neither A nor B

**Hint**    *In a lockup torque converter, a lockup plate is mounted between the turbine and the front cover.*

*This lockup plate is splined to an extension on the front of the turbine. A friction material about 1 in. wide is mounted on the front of the lockup plate near the outer edge of the plate.*

*In the unlocked mode, a TCC solenoid directs oil through the hollow transmission input shaft and out between the lockup plate and the front of the converter. This action keeps the lockup plate away from the front of the converter so the friction material on the lockup plate does not contact the front of the converter.*

*When the TCC is in the locked mode, the TCC solenoid plunger is moved so oil is directed through a passage in the hub area of the converter. This oil flow forces the lockup plate against the front of the converter so the friction material on this plate contacts the front of the converter.*

*Under this condition, engine torque is transmitted directly from the flex plate through the front of the converter and the lockup plate to the input shaft. This action prevents slippage in the converter and improves fuel economy. In many four-speed automatic transmissions, the torque converter is usually locked up in third or fourth gear above a specific vehicle speed and at a certain throttle opening. The torque converter clutch is usually not locked up below a specific engine or transmission temperature.*

*An engine shudder after the TCC locks up may be caused by an ignition problem or a lean air-fuel ratio that results in cylinder misfiring. This problem may also be caused by damaged engine or transmission mounts.*

## Task 7  Diagnose mechanical and vacuum control systems; determine necessary action.

14. The vacuum gauge is connected to the modulator system on a noncomputer-controlled transmission with a T-fitting (Figure 2–2). With the engine idling, the vacuum is 18 in. Hg. With the engine running at 2,000 rpm with a steady throttle, and the vehicle speed at 55 mph, the vacuum is 2 in. Hg. The shifts occur at a higher speed than specified, and the stall test rpm is lower than specified.

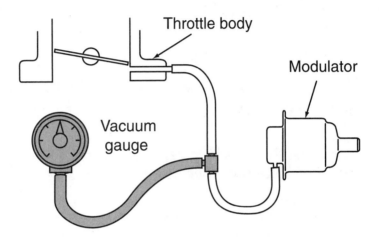

**Figure 2–2** Testing vacuum modulator vacuum system.

Technician A says the ignition timing may be later than specified.
Technician B says the exhaust system may be restricted.
Who is correct?
A. A only
B. B only
C. Both A and B
D. Neither A nor B

15. During an air pressure test on the reverse clutch, the clutch application is not heard, and a hissing noise is evident.
Technician A says the transmission case may be cracked.
Technician B says the reverse clutch drum may be cracked.
Who is right?
A. A only
B. B only
C. Both A and B
D. Neither A nor B

**Hint**    *When pressure tests and a road test indicate the problem is in one of the transaxle apply components, air pressure tests may be performed to test these components. With the valve body removed, apply clean moisture-free air at 40 psi to the appropriate passage in the transaxle housing.*

*These passages are identified in the service manual. When air pressure is applied to a component such as the forward clutch, the technician should be able to hear the clutch piston application.*

*If excessive air can be heard escaping, there is an internal leak, which is probably located at the forward clutch piston seal.*

*The transmission vacuum modulator depends on normal engine vacuum for proper operation. Intake manifold vacuum leaks, late ignition timing, or a restricted exhaust system affect the intake manifold vacuum and cause improper modulator operation and transmission shifting.*

# General Transmission/Transaxle Diagnosis, Electronic Systems

## ASE Tasks, Questions, and Related Information

**Task 1**   Road test the vehicle to verify electronic system problems based on driver's concern; determine necessary action.

16. An electronically controlled transaxle remains in second gear at all forward vehicle speeds. The cause of the problem may be
    A. a worn oil pump.
    B. a restricted filter.
    C. an inoperative shift solenoid.
    D. an improper linkage adjustment.

17. The transmission control indicator light (TCIL) on a vehicle is flashing, but the MIL is not illuminated.
    Technician A says there may be a mechanical problem in the transmission.
    Technician B says there should be a diagnostic trouble code (DTC) stored in the PCM.
    Who is correct?
    A. A only
    B. B only
    C. Both A and B
    D. Neither A nor B

**Hint**   *On-board diagnostic II (OBD II) systems are required on all cars and light trucks manufactured since 1996. In these systems, the PCM monitors emission control systems as well as electronic engine and transmission functions. On many vehicles, the computer that controls transmission functions is contained in the PCM. Some vehicles do have a transmission computer that is separate from the PCM. In our discussions, we will assume that the PCM controls transmission functions.*

*If a problem occurs and exhaust emissions increase to 1.5-percent higher than the maximum allowable emissions for that model year, the PCM sets a diagnostic trouble code (DTC). In most cases, if the defect occurs in two consecutive drive cycles, the PCM illuminates the MIL. In OBD II systems, the PCM continually monitors engine and transmission inputs and outputs, and if a problem occurs in any of these components, the PCM illuminates the MIL immediately.*

*A scan tool should be connected to the data link connector (DLC) under the dash during a road test. The scan tool displays many transmission functions, such as gear ratios and input sensor signals, and indicates the status of all the transmission shift solenoids and the TCC solenoid.*

*If a problem occurs in the transmission electronic system, the PCM illuminates the MIL.*

*Some transmissions in the 2000 model year are capable of not commanding a mechanically failed component again during that ignition on cycle; a DTC is stored in the PCM, but the MIL is not illuminated. Under this condition, if the vehicle has a TCIL, this light begins flashing. For example, if a set of clutches that provides third gear is slipping, the third gear ratio indicated on the scan tool is incorrect. Under this condition, a DTC is set in the PCM, but the MIL is not illuminated and the TCIL begins flashing. The PCM no longer commands third gear during that ignition on cycle, but other gears are provided normally.*

**Task 2**  **Perform pressure tests on transmissions equipped with electronic pressure control; determine necessary action.**

18.  The solenoid in Figure 2–3 is

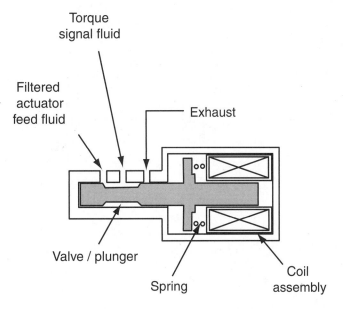

**Figure 2–3** Solenoid design.

    A.  a shift solenoid.

    B.  an electronic pressure control (EPC) solenoid.

    C.  a torque converter clutch (TCC) solenoid.

    D.  an overdrive solenoid.

19.  While discussing electronically controlled transmission oil pressure, Technician A says the PCM commands the EPC solenoid to provide higher transmission pressure during hard acceleration.

Technician B says an oil leak at the manual valve may reduce transmission pressure.

Who is correct?

    A.  A only

    B.  B only

    C.  Both A and B

    D.  Neither A nor B

**Hint**  *In many electronically controlled transmissions and transaxles, the pressure is electronically controlled by an electronic pressure control (EPC) solenoid. The PCM operates the EPC solenoid to provide the required pressure under all engine and transmission operating conditions. In an electronically controlled transmission or transaxle, low oil pressure may be caused by a mechanical* *problem such as a worn oil pump or an internal oil leak. Low oil pressure may also be caused by an electrical problem in the EPC solenoid and related circuit. If there is an electrical problem in this system, there should be a DTC in the PCM memory and the MIL will be illuminated. The PCM and the EPC solenoid replace the throttle valve (TV) cable or vacuum modulator in previous hydraulically controlled automatic transmissions.*

### Task 3  Perform torque converter clutch (lock-up converter) electronic system tests; determine necessary action.

20. The TCC system in Figure 2–4 locks up at 28 mph rather than the specified lock-up speed of 42 mph. The cause of this problem could be a malfunctioning

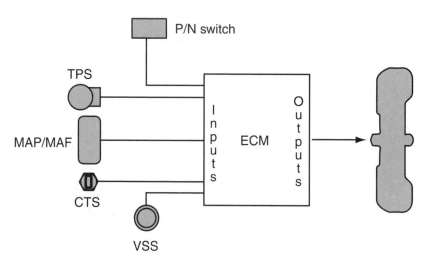

**Figure 2–4** Typical ECM inputs for torque converter clutch control.

    A. MAP sensor.
    B. park/neutral switch.
    C. ECT sensor.
    D. VSS sensor.

21. A vehicle with an electronically controlled transmission and a lock-up torque converter experiences engine stalling when the brakes are applied and the engine is decelerated from 55 mph.
Technician A says the TPS sensor may be the problem.
Technician B says the brake switch may be the problem.
Who is correct?
    A. A only
    B. B only
    C. Both A and B
    D. Neither A nor B

**Hint**  *In an electronically controlled transmission, the TCC lockup is controlled by the PCM. The PCM operates a TCC solenoid in the transmission. This solenoid supplies, or does not supply, oil pressure to a valve in the valve body. This oil pressure determines the valve position, which in turn supplies oil to the appropriate input location in the torque converter to lock or unlock the TCC.*

*The PCM uses input information from the engine coolant temperature (ECT) sensor, throttle position sensor (TPS), vehicle speed sensor (VSS), manifold absolute pressure (MAP) sensor, park/neutral switch, and brake pedal switch to control the TCC. The PCM does not lock the torque converter if the engine coolant is below a specific temperature, the throttle is wide open, or the vehicle speed is below a certain value. When the brake pedal is depressed, the brake pedal switch signals the PCM to release the TCC.*

## Task 4  Diagnose electronic transmission control systems using appropriate test equipment and service information; determine necessary action.

22. When driving in fourth gear at 60 mph with a constant throttle opening on a level road with a computer-controlled transmission, the transmission suddenly shifts into third gear without moving the gear selector. The MOST likely cause of this problem is
    A. an inoperative manual lever position sensor (MLPS).
    B. an inoperative engine coolant temperature (ECT) sensor.
    C. an inoperative TCC solenoid.
    D. an inoperative clutch set.

23. While discussing electronically controlled transmissions, Technician A says that on some electronically controlled transmissions the PCM can detect clutch slippage by comparing the transmission input speed sensor (TISS) and transmission output speed sensor (TOSS) input signals.
    Technician B says the transmission oil temperature (TOT) sensor input does not affect transmission shift timing.
    Who is correct?
    A. A only
    B. B only
    C. Both A and B
    D. Neither A nor B

24. In a computer-controlled four-speed automatic transmission, the specified second gear ratio is 1.48:1. The second gear ratio indicated on the scan tool is 1.61:1. The cause of this problem could be
    A. an inoperative shift solenoid.
    B. a slipping clutch set or band.
    C. a malfunctioning TOSS sensor.
    D. a malfunctioning ECT sensor.

25. A vehicle is equipped with an OBD II system and an electronically controlled four-speed automatic transmission with an electronic pressure control (EPC) solenoid.
    The engine has a severe lack of power, and the engine rpm is higher than normal during acceleration.
    Technician A says to check the EPC solenoid and related electrical circuit.
    Technician B says to check the TCC solenoid and related electrical circuit.
    Who is correct?
    A. A only
    B. B only
    C. Both A and B
    D. Neither A nor B

**Hint**    *When diagnosing transmission electronic problems, a scan tool is connected to the data link connector (DLC) under the dash. On OBD II vehicles, a sixteen-terminal DLC is located under the dash near the steering column. Data regarding transmission inputs and outputs and other transmission functions such as gear ratios may be displayed on the scan tool. Gear ratios are determined by transmission design and proper operation of mechanical components such as clutches and bands. If an improper gear ratio is displayed on the scan tool, a clutch set or band are usually slipping. Transmission inputs may include the transmission input shaft speed*

*(TISS) sensor, transmission output shaft speed (TOSS) sensor, and/or vehicle speed sensor (VSS), transmission oil temperature (TOT) sensor, and the manual lever position sensor (MLPS). In some OBD II vehicles, the MLPS sensor may be referred to as a digital transmission range (DTR) sensor.*

*When the TOT sensor input indicates that the transmission oil is cold, the PCM slightly delays the transmission shifts. The TISS, TOSS, and VSS signals are used to provide proper transmission shifting. In some transmissions, the PCM compares the TISS and TOSS signals to detect clutch slipping. The MLPS is one of the most important sensors because it signals the PCM regarding the gearshift selector position, and the PCM then operates the shift solenoids to supply the proper transmission gear. If the MLPS is a problem, the PCM may shift the transmission into a different gear than the gear selected by the gearshift position. Some electronically controlled transmissions have a group of hydraulic pressure switches in place of the MLPS. These switches send input signals to the PCM in relation to the gear provided by the transmission, which should be the same as the gear selected by the driver. For proper transmission control, the PCM uses some inputs that are also used to control engine functions. These inputs may include the TPS, engine coolant temperature (ECT) sensor, engine rpm, system voltage, and brake switch.*

*Transmission outputs may include two or more shift solenoids, a TCC solenoid, and the electronic pressure control (EPC) solenoid. In some OBD II vehicles, a feedback voltage signal is sent from the EPC solenoid to the PCM. If the PCM senses higher than normal EPC current flow and low transmission oil pressure, the PCM may operate the engine output functions to reduce engine torque and power and prevent transmission damage.*

*In an OBD II vehicle, the PCM continually monitors the transmission inputs and outputs. If an electrical problem occurs in the transmission, a DTC is set in the PCM memory, and the MIL is usually illuminated. A DTC indicates a problem in a specific area. For example, the scan tool may display a DTC representing a shift solenoid. This DTC may be caused by an electrical problem in the shift solenoid winding or in the wires from the PCM through the wiring harness and transmission connector to the shift solenoid. An ohmmeter and voltmeter may be used to test the shift solenoid winding and connecting wires. When certain electrical problems are sensed by the PCM that could result in transmission damage or hazardous conditions, the PCM enters a limp-in mode and provides only one forward gear, usually second gear. Because reverse gear is determined by manual valve position, this gear is also available. Under this condition, the vehicle may be driven at a reduced speed to a service facility.*

## Task 5   Verify proper operation of charging system; check battery, connections, vehicle power, and grounds.

26. A vehicle with an electronically controlled four-speed automatic transmission experiences repeated PCM failure and replacement, but there are no other problems with the vehicle. The battery remains fully charged, and the battery electrolyte level is always normal.

    Technician A says there may be an intermittent open circuit in the wire connected from the alternator battery terminal to the positive battery terminal.

    Technician B says the voltage regulator may be allowing continual high charging system voltage.

    Who is correct?

    A. A only

    B. B only

    C. Both A and B

    D. Neither A nor B

**Hint**     *On a vehicle with an electronically controlled transmission, problems in the starting and charging systems may result in improper transmission shifting and PCM damage. For example, loose electrical connections on the battery or in the charging system may cause momentary open circuits and voltage spikes, which cause computer damage. If the voltage regulator allows a* *continual high charging system voltage, the PCM may be damaged, but other electrical/electronic equipment on the vehicle will likely be damaged as well. Excessive charging system voltage also results in excessive water being gassed from the battery.*

*In most of these systems, the PCM receives inputs from system voltage and cranking voltage. If these input signals are out of range, the PCM may not provide proper transmission shifting.*

*High resistance in vehicle ground circuits may cause unusual computer operation. Ground circuits may be tested for excessive resistance by measuring the voltage drop across the circuit with a normal current flow in the circuit. Problems in the alternator, such as open or shorted diodes or stator, may cause erratic alternator voltage and voltage spikes that can result in unusual PCM operation such as improper transmission shifting and the unusual operation of engine output functions.*

*The normal charging system voltage may be checked with a voltmeter connected across the battery terminals and the engine speed at 2,000 rpm. An alternator output test may be performed to determine the electrical condition of the alternator, and a starter draw test will indicate the electrical condition of the starter.*

## Task 6   Differentiate between engine performance and transmission/transaxle related problems; determine necessary action.

27. An engine with a computer-controlled transmission experiences a shudder problem momentarily when the torque converter clutch locks up at 42 mph. There are no other drivability problems.

    Technician A says to use an exhaust gas analyzer to test the air-fuel ratio.

    Technician B says to use an ignition scope to check the spark plugs and wires.

    Who is correct?

    A. A only

    B. B only

    C. Both A and B

    D. Neither A nor B

**Hint**   *Certain engine and transmission defects may cause similar drivability problems. For example, a seized one-way clutch in the torque converter causes a loss of power at higher speeds. A restricted exhaust system also causes this problem. However, a restricted exhaust system also causes very low intake manifold vacuum at higher speeds. In many diagnostic situations, accurate testing usually pinpoints the root cause of the problem.*

## Task 7   Diagnose poor shift quality resulting from problems in the electronic transmission control system; determine necessary action.

28. A customer complains about reduced fuel mileage and late transmission shifting. The most likely cause of these complaints is

    A. a defective vehicle speed sensor (VSS).

    B. a defective transmission input speed sensor (TISS).

    C. a defective engine coolant temperature (ECT) sensor.

    D. a defective transmission output speed sensor (TOSS).

**Hint**   *The transmission computer depends on accurate input signals from all the input sensors to provide proper transmission shift quality and timing. Defects in any of the input sensors or sensor wiring may cause improper transmission shifting. Some sensor signals such as the engine coolant sensor signal may be sent to both the engine and transmission computers, and a defective signal from this sensor may affect both engine and transmission functions. When diagnosing transmission shift problems, the first step is to verify the customer's complaint. One of the next steps is to visually inspect the condition of the input sensors and sensor wiring. A scan tool may be connected to the data link connector (DLC) to test the sensor data.*

# Transmission/Transaxle Maintenance and Adjustment

## ASE Tasks, Questions, and Related Information

**Task 1**  Inspect, adjust, and replace manual valve shift linkage, transmission range sensor/switch and park/neutral position switch (inhibitor/neutral safety switch).

29. Technician A says an improper shift linkage adjustment may cause premature transmission clutch failure.

   Technician B says an improper shift linkage adjustment may cause higher than normal fluid pressure.

   Who is correct?

   A. A only

   B. B only

   C. Both A and B

   D. Neither A nor B

**Hint**    *Most automatic transmissions or transaxles have a cable or rod-type shift linkage connected from the gear selector lever to the transmission lever. If the shift linkage adjustment is not correct, the manual valve is improperly positioned (Figure 2–5). Improper manual valve position may result in low fluid pressure, improper clutch application, and excessive clutch wear. To check the shift linkage adjustment, remove the shift linkage or cable from the transmission lever. Place the gear selector in the specified position, which often is the park position. Move the transmission lever to the park position and install and tighten the linkage on the transmission lever. Move the gear selector through all the gear positions, but be sure there is a detent in each position.*

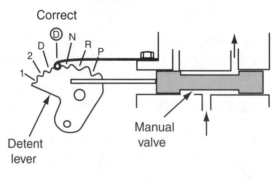

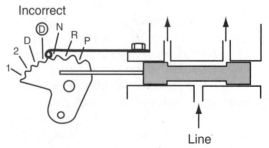

**Figure 2–5** Wrong manual valve position caused by improper shift linkage adjustment.

## Task 2  Inspect, adjust, and replace cables or linkages for throttle valve (TV), kickdown, and accelerator pedal.

30. When the throttle valve cable is improperly adjusted so throttle pressure is higher than normal, the transmission shifts occur
    A. at a lower vehicle speed than specified.
    B. at the specified vehicle speed.
    C. at the same vehicle speed.
    D. at a higher vehicle speed than specified.

**Hint**    *The TV cable connects the accelerator pedal to the throttle valve in the transmission valve body. In some transmissions, the TV cable movement controls the throttle valve and the downshift valve. In other applications, the vacuum modulator controls the throttle valve and the TV cable controls the downshift valve.*

*The throttle valve position controls throttle pressure. Upshifts occur when governor pressure overcomes throttle pressure on a specific shift valve. Downshifts occur when throttle pressure overcomes governor pressure and moves a certain shift valve.*

*An improperly adjusted TV cable may cause lower than normal throttle pressure and early upshifts. This misadjustment may also result in high throttle pressure and delayed, harsh upshifts.*

*A typical TV cable adjustment involves releasing the cable lock tab, and pulling the cable fully in the readjust direction (Figure 2–6). Hold the throttle lever fully clockwise against its stop and press the lock tab into the locked position.*

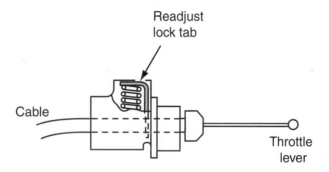

**Figure 2–6** Throttle valve cable adjustment.

*Some transmissions have a kickdown switch in place of a downshift linkage and valve. When the throttle is wide open, the throttle linkage closes the kickdown switch and operates a downshift solenoid in the transmission to provide the necessary downshift (Figure 2–7).*

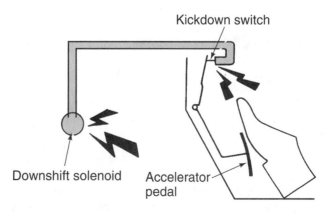

**Figure 2–7** Kickdown switch. *(Used with permission from Nissan North America, Inc.)*

*In computer-controlled transmissions and transaxles, the computer operates shift solenoids to control all upshifts and downshifts in response to sensor input signals, and a throttle valve and downshift valve are not required.*

**Task 3**    **Replace fluid and filter(s); verify proper fluid level.**

31. Technician A says that transmission filters may be cleaned and reused.
Technician B says that transmission fluid oxidizes faster at lower temperatures.
Who is correct?
A. A only
B. B only
C. Both A and B
D. Neither A nor B

**Hint**    *The interval for transmission and transaxle fluid changes depends on the type of transmission and the type of service. Severe service such as trailer towing, or commercial vehicle operation, requires more frequent fluid changes, because this type of service involves higher transmission temperatures. In many transmissions and transaxles, the oil pan must be removed to drain the fluid because there is no drain plug in the oil pan. The filter usually is bolted to the bottom of the valve body and the filter should be replaced, not cleaned. A gasket or O-ring between the filter and the valve body or case must be replaced. Fiber material in the oil pan indicates worn bands or clutches. Steel particles in the oil pan indicate damaged components such as gear sets or the oil pump. Aluminum particles in the oil pan indicate a damaged case.*

# In-Vehicle Transmission/Transaxle Repair

## ASE Tasks, Questions, and Related Information

**Task 1**    **Inspect, adjust, and replace vacuum modulator, valve, lines, and hoses.**

32. A vehicle is operating where the temperature is 0°F, and the transmission has a vacuum modulator. The transmission experiences repeated clutch piston seal failures, and the driver complains about harsh, late shifting.
Technician A says moisture may be freezing in the vacuum modulator diaphragm chamber.
Technician B says the manual valve shift linkage may require adjusting.
Who is correct?
A. A only
B. B only
C. Both A and B
D. Neither A nor B

**Hint**    *The vacuum hose and vacuum modulator may be tested by connecting a vacuum gauge to the modulator hose with a T-fitting. If the vacuum is less than specified, check the engine vacuum, connecting hose, and modulator. When a hand-operated vacuum pump is connected to the modulator, it should hold 18 in. Hg of vacuum. Some vacuum modulators may be adjusted with a hand-operated vacuum pump and the proper length of gauge pins.*

**Task 2** **Inspect, adjust, repair, and replace governor cover, seals, sleeve/bore, valve, weight, springs, retainers, and gear.**

33. The shifts in an automatic transaxle occur at a higher speed than specified. All of these items could be the cause of the problem EXCEPT
    A. a sticking governor valve.
    B. excessive governor spring tension.
    C. worn governor weights and pins.
    D. weak governor spring tension.

**Hint**    *When a road test and transmission pressure tests indicate a governor problem, the governor should be removed, cleaned, and inspected. Improper shift points is one of the most common complaints related to governor operation. The transmission must be disassembled to access some governors. Other transmissions have a governor that is removable from the outside of the transmission (Figure 2–8).*

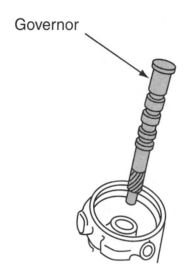

Governor

**Figure 2–8** Externally removeable governor.

*Many governors contain primary and secondary valves. The primary governor valve must be removed first, followed by the secondary valve retaining pin, spring, and secondary valve. All worn components must be replaced. After cleaning and lubricating with ATF, these valves must move freely in their bores. Some governors are mounted on the transmission output shaft; a drive ball in this shaft forces the governor to rotate with the shaft (Figure 2–9).*

**Task 3** **Inspect and replace external seals and gaskets.**

34. A transaxle experiences repeated pump seal failure.
    Technician A says the pump body bushing may be worn.
    Technician B says the governor pressure is higher than specified.
    Who is correct?
    A. A only
    B. B only
    C. Both A and B
    D. Neither A nor B

**Hint**    *Seal mounting bores must be inspected for scratches, metal burrs, and cracks. All seals must be removed with the proper puller, and installed with the appropriate seal driver. Because a plugged vent may cause seal leakage, always inspect the vent for restrictions.*

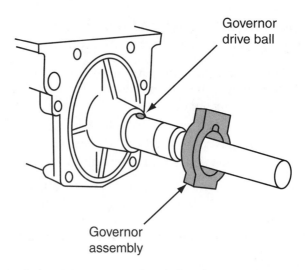

**Figure 2–9** Governor drive ball in the output shaft.

## Task 4    Inspect and replace drive shaft yoke and drive axle bushings and seals.

35. An extension housing bushing and the bushing contact area on the drive shaft slip yoke are severely pitted and scored, and the housing seal is leaking. There are no other transmission problems or complaints.

Technician A says there may be excessive resistance in the battery ground between the chassis and the engine.

Technician B says the transmission fluid may be contaminated because fluid and filter changes have not been performed.

Who is correct?
A.  A only
B.  B only
C.  Both A and B
D.  Neither A nor B

*Hint*    *Inspect the extension housing for cracks and metal burrs in the seal bore and transmission case mating surfaces. Replace the extension housing bushing if it is worn. Inspect the speedometer drive assembly for worn components.*

## Task 5    Check condition of engine cooling system; inspect, test, and flush or replace transmission cooler, lines, and fittings.

36. With the engine idling, the ATF flow through a transmission cooler should be
A.  one quart in 60 seconds.
B.  one pint in 60 seconds.
C.  one quart in 40 seconds.
D.  one quart in 20 seconds.

*Hint*    *Transmissions may have a cooler in one of the radiator tanks, an external cooler, or both. If the internal cooler develops a leak, the ATF becomes contaminated with coolant, and the coolant is contaminated with ATF. When ATF is contaminated with coolant, the ATF has a milky color, and the coolant may be visible as bubbles on the transmission dipstick. If ATF enters the coolant, the ATF may be seen floating on top of the coolant when the radiator cap is removed. A leaking internal transmission cooler causes a loss of ATF without any visible signs of external leaks.*

*A transmission cooler may be tested for leaks by plugging one cooler line and supplying regulated air pressure up to 75 psi air pressure to the other line. Bubbles and/or air escaping indicate the leak's location.*

*Restricted transmission coolers may cause overheating of the transmission, resulting in severe damage to clutches, bands, and seals. To test the transmission cooler for restriction, disconnect the outlet line and connect a hose from the outlet into an empty container. Start the engine and allow it to run for 20 seconds. One quart of ATF should flow through the cooler into the container.*

*If the cooler flow is less than one quart, the cooler or inlet line is restricted. Transmission coolers may be flushed with compressed air or an approved cleaning solution and compressed air. Coolers should always be flushed in both directions when a transmission is overhauled.*

*Compressed air should be used to remove debris from external cooler air passages. All cooler lines should be inspected for leaks and restrictions.*

### Task 6   Inspect and replace speedometer/speed sensor drive gear, driven gear, and retainers.

37. An erratic speedometer could be caused by all of these problems EXCEPT
    A. a missing drive gear retaining clip.
    B. a dry speedometer cable.
    C. worn speedometer gears.
    D. worn driven gear retaining bushing.

**Hint**    *The speedometer drive and driven gears should be inspected for worn or damaged teeth. In some transmissions, the speedometer drive gear is machined into the output shaft, whereas in other transmissions this drive gear is splined onto the output shaft or held in place with a clip (Figure 2–10). If the speedometer drive gear is a machined part of the output shaft, the complete shaft must be replaced if this gear is damaged. Inspect all gear retaining clips for looseness or damage. The speedometer driven gear and bushing assembly is retained in the transaxle housing with a bolt (Figure 2–11). If the speedometer driven gear bushing assembly is loose, it should be replaced.*

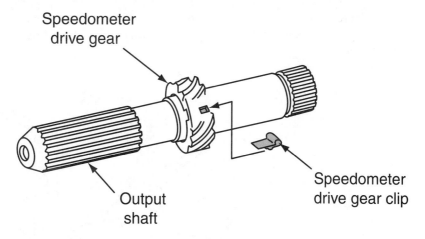

**Figure 2–10** Speedometer drive gear held on the output shaft with a clip.

### Task 7   Inspect valve body mating surfaces, bores, valves, springs, sleeves, retainers, brackets, check balls, screens, spacers, and gaskets; replace as necessary.

38. The clearance between the valves and matching valve body bores should not exceed
    A. 0.001 in.
    B. 0.003 in.

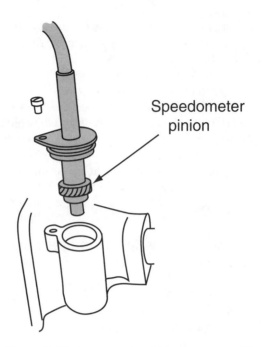

**Figure 2–11** Speedometer driven gear assembly.

C. 0.005 in.

D. 0.008 in.

*Hint*    *When pressure tests indicate there may be a problem in the valve body, this component should be removed, cleaned, and inspected. The valve body may be removed with the transmission in the vehicle, or after transmission removal for an overhaul. When the valve body is removed, be careful not to lose the steel balls (Figure 2–12). The appropriate service manual usually provides a diagram showing the proper ball positions.*

*When the valves, springs, pins, and balls are removed from the valve body, note the position of each component, and lay these components in the proper order on clean, lint-free shop towels.*

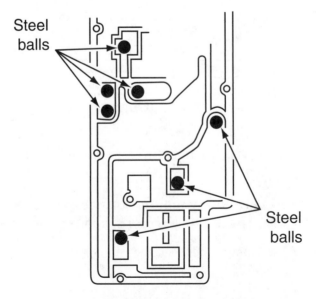

**Figure 2–12** Valve body and steel balls.

*Clean the valve body in an approved cleaning solution. Check all valves and bores for scratches and metal burrs. Minor scratches or metal burrs may be removed from valves with crocus cloth. If a valve does not move freely in its bore after cleaning, replace the valve. Check for wear on all valves and bores. The clearance between each valve and the matching bore should not exceed 0.001 in. If a bore is worn, replace the valve body.*

*Always install new valve body gaskets, and place the gasket, or gaskets, on the valve body to be sure they fit properly and do not cover any openings. Valve body and transmission case mating surfaces should be checked for warpage with a straightedge.*

## Task 8  Check and adjust valve body bolt torque.

39. After a noncomputer-controlled transaxle and valve body overhaul, the transaxle does not complete a 1 - 2 upshift, and shifts from first gear to third gear. All other shifts are normal, and this problem was not present before the overhaul.

    Technician A says the valve body torque may be excessive.

    Technician B says the governor pressure may be too low.

    Who is correct?

    A. A only
    B. B only
    C. Both A and B
    D. Neither A nor B

**Hint**  *Be sure each valve body bolt is installed in the proper position. These bolts usually are different lengths. Installing these bolts in the wrong holes may cause improper torque or transmission case damage. Be sure the gaskets and spacer are properly positioned prior to valve body installation.*

*Clean petroleum jelly may be used to hold the gaskets in place during valve body installation.*

*Valve body bolts must be tightened to the specified torque in the proper sequence. If this bolt torque is excessive, valve body warping and valve sticking may occur.*

## Task 9  Inspect servo bore, piston, seals, pin, spring, and retainers; repair or replace as necessary and adjust bands.

40. The procedure shown in Figure 2–13 is to determine the proper
    A. servo piston thickness.
    B. band anchor length.

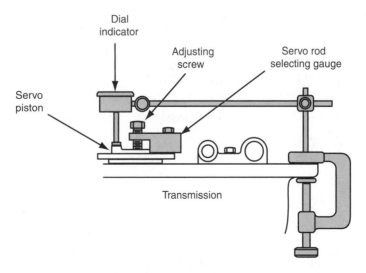

**Figure 2–13** Servo measurement.

C. servo piston selective pin.

D. servo piston adjustment.

**Hint**    *A servo is a hydraulically operated piston and pin that is used to apply a band. Servo pistons and bores must be cleaned and inspected for cracks, scratches, metal burrs, and wear. Servo piston-to-bore clearance usually is 0.003 to 0.005 in. Servo piston seals may be made of Teflon or cast iron with hooked ends. Some servo pistons have molded rubber seals. All piston seals and bores must be lubricated with the proper ATF prior to installation. Transmissions without band adjustment screws have selective servo apply pins, and the Technician must determine the proper pin length to provide proper band application.*

## Task 10    Inspect accumulator bore, piston, seals, spring, and retainers; repair or replace as necessary.

41. An automatic transaxle has a complaint of harsh 3 - 4 upshifts. All the other shifts are normal.

    Technician A says the fourth accumulator piston may be stuck.

    Technician B says the pressure regulator valve is sticking.

    Who is correct?

    A. A only

    B. B only

    C. Both A and B

    D. Neither A nor B

**Hint**    *Accumulators are spring-loaded pistons that provide a hydraulic cushion to reduce shift harshness.*

*Accumulator piston, seal, and spring service is similar to servo service. Some accumulator pistons have O-ring seals, which must be replaced during an overhaul (Figure 2–14).*

## Task 11    Inspect parking gear, inspect and replace parking pawl, shaft, spring, and retainer.

42. All of these statements about a parking pawl are true EXCEPT

    A. the parking pawl locks the input shaft.

    B. the parking pawl is mechanically operated.

    C. the parking pawl's projection engages in a notched drum.

    D. the parking pawl is retained on a pivot pin.

**Hint**    *The parking pawl opening and pivot pin should be inspected for wear. Check the pawl spring for proper tension. The projection on the pawl should be checked for wear. Inspect the linkage from the pawl to the shift lever for proper operation and wear.*

## Task 12    Inspect, test, adjust, repair, or replace electrical/electronic components and circuits including computers, solenoids, sensors, relays, terminals, connectors, switches, and harnesses.

43. A computer-controlled transaxle has a relay that supplies voltage to the solenoids and switches in the transaxle when the ignition switch is turned on. The computer senses a problem in the input speed sensor, and does not close the relay. Under this condition, the transaxle operates in

    A. first gear and reverse.

    B. second gear and reverse.

    C. first and second gear.

    D. second and third gear.

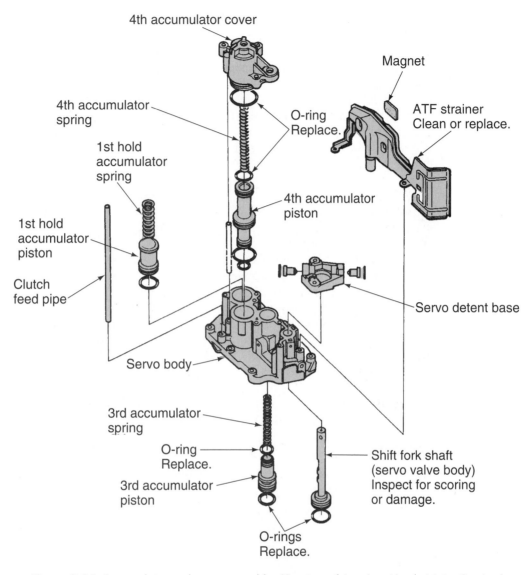

**Figure 2–14** Accumulator and servo assembly. *(Courtesy of American Honda Motor Co., Inc.)*

44. When diagnosing a computer-controlled transmission or transaxle, the first sensor to be verified should be
   A. the manual lever position sensor (MLPS).
   B. the transaxle input speed sensor (TISS).
   C. the transaxle output speed sensor (TOSS).
   D. the transaxle oil temperature sensor (TOT).

*Hint*    *Solenoids, such as the TCC or shift solenoids, may be tested with an ohmmeter. When the ohmmeter leads are connected to the solenoid terminals, a reading below the specified value indicates a shorted solenoid winding. If an infinity ohmmeter reading is obtained, the solenoid winding is open. When the ohmmeter leads are connected from one of the solenoid terminals to ground, a low reading indicates a grounded solenoid winding, whereas an infinity reading indicates the winding is not grounded.*

*Some transmissions contain normally open, or normally closed, oil pressure switches. These switches usually inform the computer regarding the present operating gear in the transmission.*

*These switches may be closed with air pressure for test purposes. When air pressure is applied to a normally open switch, an ohmmeter connected to the switch terminals should indicate a very low reading. With no pressure applied to the switch, an infinity ohmmeter reading should be obtained.*

*When diagnosing a computer-controlled transmission or transaxle, one of the first tests should be to verify the adjustment and operation of the manual lever position sensor (MLPS).*

*The MLPS sensor is operated by the manual valve shift linkage, and this sensor informs the computer regarding the gear selector position. The computer then provides the selected gear. As the gear selector is moved, various resistors inside the MLPS sensor are connected through the sensor terminals to the computer. A malfunctioning MLPS sensor, or an improper sensor adjustment, may cause improper transmission shifting.*

*Many transmission sensors are permanent magnet generators that produce an AC voltage signal proportional to rotational speed in the transmission. An ohmmeter may be connected to these sensor terminals to check the sensor winding for open circuits, grounds, and shorts. An AC voltmeter may be connected to the sensor terminals to measure the AC voltage signal with the transmission operating normally.*

*Some electronically controlled transaxles have a computer-controlled relay that supplies voltage to the transmission solenoids and switches when the ignition switch is turned on. If the computer senses a serious electronic problem, it does not energize the relay, and the voltage supply to the solenoids and switches is shut off. Under this condition, the transaxle operates in only second gear while moving forward. Because reverse is a function of manual valve position, this gear is still available. The relay may be tested by energizing the relay winding with a 12 V battery and connecting an ohmmeter to the relay contact terminals. The ohmmeter should indicate very low resistance across the relay contacts when the winding is energized.*

*A scan tool may by connected to the data link connector (DLC) under the dash to obtain data and diagnostic trouble codes (DTCs) from the transmission computer.*

## Task 13   Inspect, replace, and align power train mounts.

45. A front-wheel-drive vehicle experiences intermittent shifting. Sometimes the transaxle shifts normally, and occasionally it misses a shift.

Technician A says the manual valve shift linkage may need adjusting.

Technician B says the engine or transaxle mounts may be broken.

Who is correct?

A. A only
B. B only
C. Both A and B
D. Neither A nor B

**Hint**   *Worn or broken engine or transmission mounts may cause excessive engine and transaxle or transmission movement during acceleration. This excessive movement changes the effective length of the shift and throttle cables or linkages, which may result in erratic transaxle shifting.*

*On a front-wheel-drive (FWD) vehicle, worn or improperly positioned engine cradle mounts may cause improper transaxle shifting, drive axle vibration, and incorrect front suspension angles.*

*Some engine cradles have an alignment hole to check cradle alignment with the chassis (Figure 2–15).*

*Worn or broken engine or transmission mounts on a rear-wheel-drive (RWD) vehicle may cause improper shifting and incorrect drive shaft angles, resulting in vibration.*

*To check the mounts lift up on the engine, transmission, or transaxle and check for excessive movement in the mounts. The torque reaction of the mounts also should be checked. Set the parking brake and apply the foot brake firmly. Start the engine and place the gear selector in drive.*

*Accelerate the engine to 2,000 rpm and observe the engine movement. Excessive engine and transmission, or transaxle, movement indicates weak or broken mounts.*

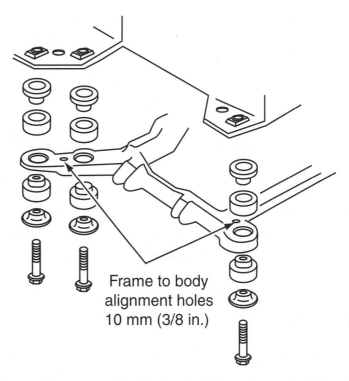

**Figure 2–15** Engine cradle, mounts, and alignment hole from a front-wheel-drive (FWD) vehicle.

# Off-Vehicle Transmission/Transaxle Repair, Removal, Disassembly, and Assembly

## ASE Tasks, Questions, and Related Information

**Task 1**  Remove and replace transmission/transaxle; inspect engine core plugs, rear crankshaft seal, transmission dowel pins, and dowel pin holes.

46. All of the following statements about transmission/transaxle removal and replacement are true EXCEPT
    A. the negative battery cable should be disconnected prior to transmission removal.
    B. the drive shaft should be marked in relation to the differential flange.
    C. the front drive axles should be marked in relation to the front hubs.
    D. the engine support fixture should be installed before loosening the transaxle-to-engine bolts.

*Hint*     *Always disconnect the battery ground cable prior to transaxle or transmission removal. The transaxle fluid should be drained prior to removal. Raise the vehicle on a lift, and place a large drain pan under the oil pan. If the transaxle oil pan does not have a drain plug, remove the bolts on one side of the oil pan and loosen the bolts on the other side to allow the pan to drop partially downward to drain the fluid.*

*On a front-wheel-drive vehicle, the drive axles have to be removed prior to transaxle removal.*

*Most manufacturers recommend loosening the outer drive axle retaining nuts with the drive wheel contacting the floor and the brakes applied. On rear-wheel-drive vehicles, the drive shaft must be removed prior to transmission removal. Always mark the drive shaft in relation to the differential flange.*

*Always install an engine support fixture before loosening the transaxle-to-engine retaining bolts. Support the transaxle or transmission securely on a transmission jack during removal and installation.*

*After the transmission is removed, the transmission dowel pins and dowel pin holes should be inspected for wear. Worn dowel pins or dowel pin openings may cause transmission misalignment and vibration problems.*

## Task 2    Disassemble, clean, and inspect transmission case, sub-assemblies, and mating surfaces.

47. If the input shaft end play in Figure 2–16 is more than specified
    A. the input shaft must be replaced.
    B. the pump must be replaced.
    C. the transmission case is worn.
    D. a thicker selective washer is required.

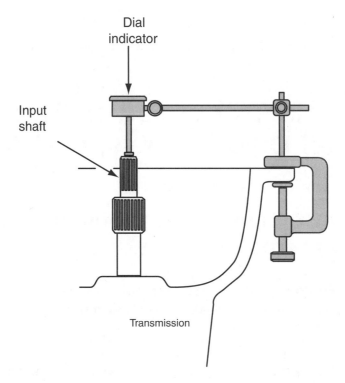

**Figure 2–16** Typical transmission input shaft end play measurement.

*Hint*    *Before the transmission or transaxle is disassembled, end play should be measured on various components as recommended in the vehicle manufacturer's service manual. Components usually are removed in this order: external components such as the modulator and bell housing, valve body, servos, accumulators, pump, input shaft and front clutch and gear set, rear clutch and gear set. Visually inspect all components during disassembly. Clean all components with an approved cleaning solution, and inspect all parts for wear and damage. Always inspect the oil pan for band and clutch material and aluminum, steel, or brass cuttings.*

## Task 3    Assemble after repair.

48. While performing the measurement in Figure 2–17, the dial indicator reading is more than specified. To correct this problem install
    A. a thicker selective thrust washer.
    B. new friction and steel clutch plates.
    C. a thicker reaction plate.
    D. a new clutch drum.

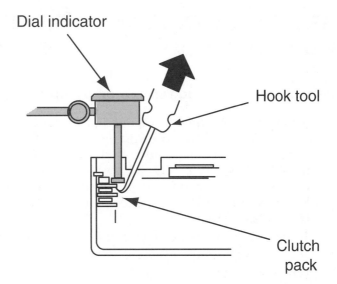

**Figure 2–17** Clutch plate clearance measurement.

**Hint**    *During transmission or transaxle assembly, replace all seals, gaskets, and O-rings. Lubricate all seal lips, O-rings, and clutch plates with the specified type of ATF. All seals must be installed in the proper direction. Be sure all components are installed in their original location. Thrust washers may be held in place with clean petroleum jelly. Perform all end play measurements and band adjustments recommended in the vehicle manufacturer's service manual. A general assembly procedure includes the following: rear gear set clutches and bands, front gear set clutches and bands, pump, valve body, servos, accumulators, and external components.*

## Task 4    Inspect converter flex (drive) plate, converter attaching bolts, converter pilot, converter pump drive surfaces, converter end play, and crankshaft bore.

49. An engine has a clunking noise that usually occurs during deceleration and sometimes with the engine idling. The cause of this noise could be loose
    A. main bearings.
    B. converter-to-flex plate bolts.
    C. connecting rod bearings.
    D. piston pins.

**Hint**    *The converter flex plate should be inspected for cracks and warping. Inspect the ring gear for broken, damaged teeth. Check the torque on the flex plate bolts, and inspect the flex plate bolt holes for wear. Check the converter attaching bolts for wear and proper torque. The converter hub must be smooth in the pump seal contact area. Minor scratches in this area may be removed with fine crocus cloth. If the converter hub is scored or scratched in the seal contact area, replace the converter. Inspect the converter drive lugs for wear.*

# Off-Vehicle Transmission/Transaxle Repair, Gear Train, Shafts, Bushings, Oil Pump, and Case

## ASE Tasks, Questions, and Related Information

**Task 1**   **Inspect, measure, and replace oil pump components.**

50. A transmission experiences a pump seal failure, and during this failure the transmission was severely overheated. The pump and seal were replaced during a transmission overhaul, and all the converter tests were satisfactory. The transmission now has a high-pitched whining noise that increases in relation to engine speed. This noise is present with the engine running in park or neutral, and while driving the car. The engine has normal power, and there are no other transmission complaints.

Technician A says the converter hub may be misaligned.

Technician B says the converter one-way clutch may be slipping.

Who is correct?

A. A only

B. B only

C. Both A and B

D. Neither A nor B

**Hint**   *Before removing the oil pump gears, the gears should be marked in relation to each other so they may be assembled in the original position. Inspect the gears and pump surfaces for wear, scoring, and damage. Inspect the pump bushing for wear. Check the stator shaft for looseness in the pump cover.*

*Use a feeler gauge to measure the clearance between the outer gear and the housing. Measure the clearance between the outer gear teeth and the crescent. Use a feeler gauge and a straightedge to measure between the top of the gears and the pump cover (Figure 2–18). On a vane-type pump, measure the thickness of the rotor, vanes, and slide with a micrometer. If any of these clearances and measurements are not within specifications, replace the pump.*

*Replace the pump O-ring, gasket, seal rings, seal, and thrust washer. Clean and inspect the pressure regulator valve, guide, and spring.*

**Figure 2–18** Clearance measurement oil pump gears to pump cover.

## Task 2  Check bearing preload; determine needed service.

51. The output shaft turning torque is 8 in.-lbs., and the specified turning torque is 4 in.-lbs. A 5.44 mm selective shim is installed on the output shaft.

    Technician A says to install a 5.50 mm selective shim.

    Technician B says to install a 5.35 mm shim.

    Who is correct?

    A. A only

    B. B only

    C. Both A and B

    D. Neither A nor B

**Hint**  *Bearing preload adjustments are necessary on some transaxle bearings such as the output shaft bearings. Some transaxles have a selective shim positioned between two tapered roller bearings that support the output shaft in the transaxle case (Figure 2-19). Use the vehicle manufacturer's special tools to install the output shaft assembly in the output shaft bearings. Install a new output shaft nut, and tighten this nut to the specified torque. A special tool and torque wrench are used to measure the output shaft turning torque (Figure 2-20). If the turning torque is more than specified, install a thicker selective shim. When the turning torque is less than specified, a thinner selective shim is required. After the specified turning torque is obtained, use a special tool to stake the output shaft nut.*

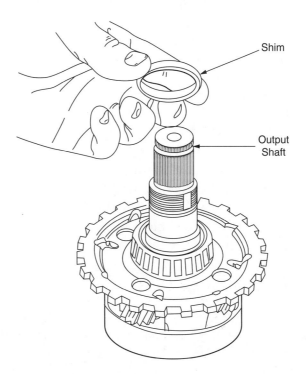

Shim

Output Shaft

**Figure 2–19** Installing selective shim.

## Task 3  Check end play; inspect, measure, and replace thrust washers and bearings as needed.

52. A transmission experiences repeated pump seal failure, and there are no other transmission complaints.

    Technician A says the pressure regulator valve may be sticking.

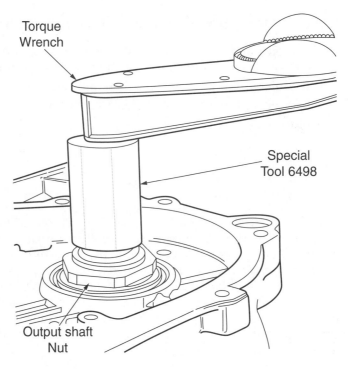

**Figure 2–20**  Measuring output shaft turning torque.

Technician B says the drainback hole behind the seal may be plugged.

Who is correct?

A.  A only

B.  B only

C.  Both A and B

D.  Neither A nor B

53.  The purpose of the measurement in Figure 2–21 is to determine the proper

A.  clutch pack retaining ring thickness.

B.  clutch pack reaction plate thickness.

C.  selective washer thickness.

D.  steel clutch plate thickness.

**Hint**    *Prior to transaxle disassembly, measure the input shaft end play with a dial indicator (Figure 2–22). With the dial indicator stem positioned against the input shaft, pull this shaft in and out to measure the end play. If this measurement is not within specifications, install a number four thrust plate with the proper thickness to provide the specified end play during transaxle reassembly. The number four thrust plate is located behind the overdrive clutch hub and in front of the front sun gear (Figure 2–23). End play measurements vary depending on the transmission or transaxle. Always perform the end play measurements outlined in the vehicle manufacturer's service manual.*

*All thrust washers should be inspected for wear, scoring, flaking, and damaged tabs. Some thrust washers are available in various selective fit thicknesses to provide the proper end play between components.*

*When thrust washers are worn, component clearance is reduced, which may cause clunking and a rubbing noise. Thrust washer thickness may be measured with a micrometer. All bearings should be inspected for looseness and roughness.*

## Task 4  Inspect and replace shafts.

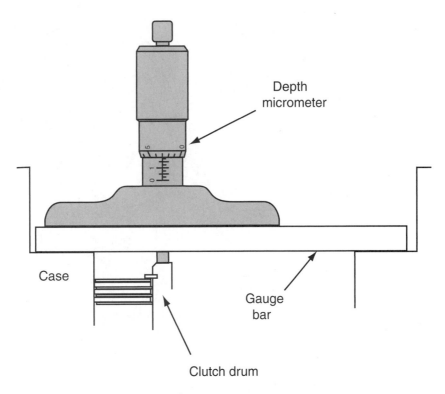

**Figure 2–21** Measurement at reverse clutch drum face.

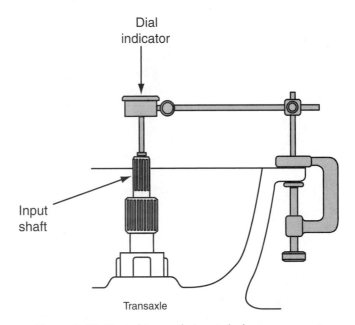

**Figure 2–22** Typical transaxle input shaft measurement.

54. The vehicle owner complains about harsh torque converter clutch operation in the four-speed transmission partially shown in Figure 2–24. Converter clutch lockup does occur at the specified speed.

Technician A says the vehicle speed sensor may be malfunctioning.

Technician B says the number nine check ball in the turbine shaft is sticking.

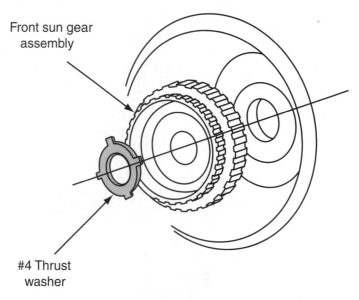

**Figure 2–23**  Number 4 thrust washer.

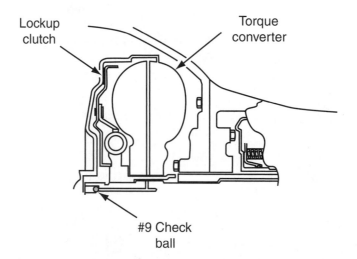

**Figure 2–24**  Typical four-speed automatic transmission with torque converter clutch lockup.

Who is correct?
A.  A only
B.  B only
C.  Both A and B
D.  Neither A nor B

**Hint**  *The bushing contact area on all shafts must be inspected for scoring and wear. During a transmission overhaul, replace all shaft O-rings, and Teflon or cast iron sealing rings. Most Teflon sealing rings must be cut from the shaft. Some shafts contain fluid passages and a check ball. Blow out these passages with compressed air, and be sure the check ball moves freely and seats properly.*

**Task 5**  **Inspect oil delivery circuit, including seal rings, ring grooves, sealing surface areas, feed pipes, orifices, and encapsulated check valves (balls).**

55. The tool in Figure 2–25 is used to
   A. size Teflon oil seal rings.
   B. align the oil pump halves.
   C. check for gear interference.
   D. install Teflon oil seal rings.

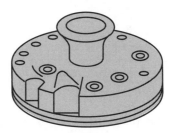

**Figure 2–25** Oil pump service tool.

*Hint*     *All sealing rings should be inspected for wear, scoring, scratches, and nicks. Check all ring grooves for wear. Inspect the sealing surfaces contacted by the sealing rings for wear, scoring, and ridges.*

## Task 6   Inspect and replace bushings.

56. All of these statements about bushing inspection and measurement are true EXCEPT
   A. a scored shaft in the bushing contact area may indicate a lack of lubrication.
   B. shaft to bushing clearance may be measured with a wire-type feeler gauge.
   C. shaft to bushing clearance may be measured with a vernier caliper and a micrometer.
   D. normal bushing clearance is 0.015 to 0.025 in.

*Hint*     *All bushings should be inspected for wear, scoring, and burning. Some shafts have internal bushings on which other shafts are supported. These internal bushings must always be checked for wear. Because the shaft contains much harder material compared to a bushing, a scored shaft usually indicates lack of lubrication from a plugged passage, or a bushing in which the oil feed hole is not aligned with the oil hole in the case.*

*A wire-type feeler gauge may be inserted between the bushing and the shaft to measure bushing wear. If this type of measurement is not possible, the inside diameter of the bushing and the outside diameter of the shaft may be measured with a vernier caliper and micrometer to determine the bushing clearance. Normal shaft to bushing clearance is 0.0005 to 0.015 in. One of the most critical bushing fits is the converter hub to pump bushing clearance, which is 0.0005 in.*

## Task 7   Inspect and measure planetary gear assembly; replace parts as necessary.

57. To provide reverse gear with a planetary gear set
   A. the carrier is held and the annulus gear drives the sun gear.
   B. the sun gear is held and the annulus gear drives the carrier.
   C. the carrier is held and the sun gear drives the annulus gear.
   D. the annulus gear is held and the sun gear drives the carrier.

*Hint*     *All planetary gear teeth should be checked for wear, chips, and damage. Inspect all splines in the gear sets and drums for wear and damage. Measure the end play between all planetary pinion gears and the planetary carrier. Be sure the planetary pinions rotate freely and smoothly. Inspect all thrust washers and thrust bearings for wear.*

## Task 8   Inspect case bores, passages, bushings, vents, mating surfaces, thread condition, and dowel pins; repair or replace as necessary.

58. Technician A says case porosity between two fluid passages could result in pressure bleed off and clutch burnout.

    Technician B says case porosity between two fluid passages could result in the transmission being in two gears at once, resulting in jam-up.

    Who is correct?

    A. A only

    B. B only

    C. Both A and B

    D. Neither A nor B

**Hint**    *The transmission or transaxle case must be thoroughly cleaned with an approved cleaning solution. Blow out all passages with compressed air. Be sure to remove and clean all filter screens in case passages. Inspect and blow out the vent. Blow out all threaded openings with compressed air, and inspect the thread condition in all threaded openings.*

*Inspect the case for porosity and cracks. If porosity is suspected in an oil passage, plug one end of the passage and supply air pressure to the other end. When the passage is porous, air can be heard escaping, or ATF bubbles when placed on the porous area.*

*Inspect all bores such as servo, seal, and bearing bores for scratches, scoring, and ridges.*

*Minor scratches may be removed with crocus cloth. Deep scratches, scoring, and ridges require case replacement. Inspect all clutch plate splines in the case for wear and damage. Use a straight-edge to check the valve body mating surface for warpage.*

## Task 9    Inspect and replace transaxle drive chains, sprockets, gears, bearings, and bushings.

59. The tool in Figure 2–26 is used to

    A. remove and install the drive chain and sprockets.

    B. measure drive chain and sprocket wear.

    C. measure wear in sprocket support bearing wear.

    D. install a sprocket chain tightener.

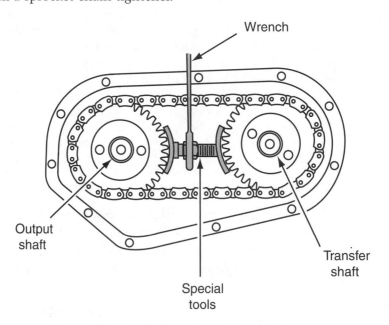

**Figure 2–26** Drive chain service tool.

**Hint**    *Inspect the drive chain and sprockets for wear and damage. The sprocket support bushings or bearings should be checked for wear, scoring, and flaking. Chain condition may be measured by deflecting the chain in each direction with a screwdriver. Measure the amount of chain deflection,* *and compare this measurement to specifications. If the chain deflection is excessive, chain and/or sprocket replacement is necessary.*

## Task 10  Inspect, measure, repair, adjust or replace transaxle final drive components.

60. The turning torque in Figure 2–27 is less than specified. To correct this problem
    A. thicker spacers are required behind both differential side gears.
    B. a thicker spacer is required behind one side gear bearing cup.
    C. thicker spacers are required behind both side bearing cups.
    D. thicker spacers are required behind both side bearings.

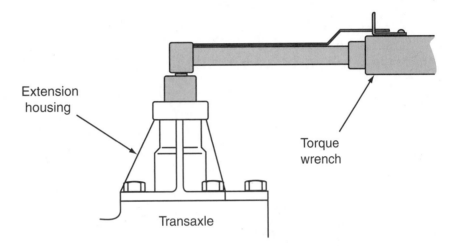

**Figure 2–27** Differential turning torque measurement.

**Hint**     *Inspect the differential ring gear and pinion gear for chipped and damaged teeth. The differential side bearings and bearing cups should be inspected for scoring, pitting, and damage. Check the side gears, pinion gears, and spacers for chipped and worn teeth, and scoring or wear on the spacer surfaces. Inspect the side gear and pinion gear contact areas in the case for scoring or wear.*

*Typical differential adjustments are side gear end play, differential end play, and turning torque. End play must be measured on each side gear with a special tool and a dial indicator placed against the side gear. Move the side gear up and down to measure the end play (Figure 2–28). Side gear spacers are available in various thicknesses to obtain the specified end play.*

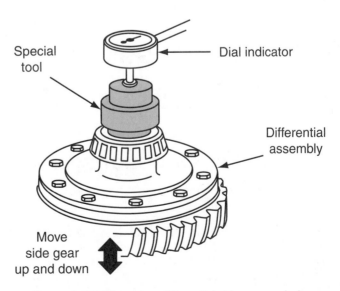

**Figure 2–28** Measuring differential side gear end play.

*The differential end play is measured with the differential assembled in the case and a gauging shim installed behind the bearing cup in the differential bearing retainer. Pry the differential upward with a dial indicator and special tool installed against the upper side of the differential (Figure 2–29). The specified shim thickness is equal to the end play plus a specific thickness for side bearing preload. After the differential is assembled with the proper shim thickness for differential end play and side bearing preload, the differential turning torque must be measured.*

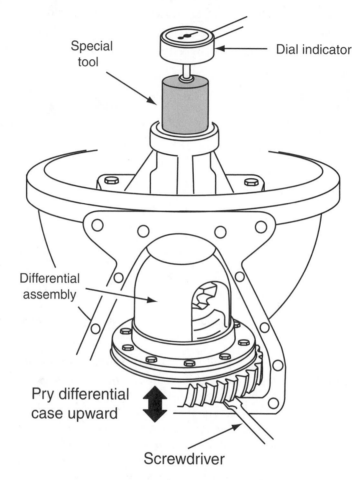

**Figure 2–29** Measuring differential end play.

# Off-Vehicle Transmission/Transaxle Repair, Friction and Reaction Units

## ASE Tasks, Questions, and Related Information

### Task 1    Inspect hydraulic clutch pack assembly; replace parts as necessary.

61. The tool in Figure 2–30 is used to install clutch
    A.  pistons and oil seals.
    B.  hub Teflon rings.

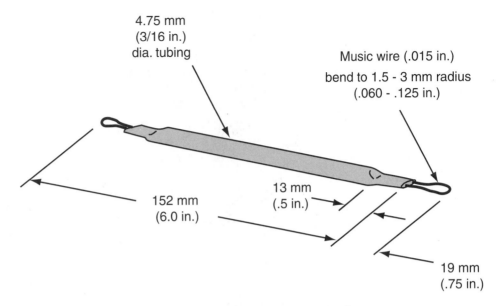

4.75 mm
(3/16 in.)
dia. tubing

Music wire (.015 in.)
bend to 1.5 - 3 mm radius
(.060 - .125 in.)

152 mm
(6.0 in.)

13 mm
(.5 in.)

19 mm
(.75 in.)

**Figure 2–30** Clutch pack service tool.

      C. hub cast iron rings.

      D. drum snap rings.

**Hint**    *A spring compressor must be installed on some clutch packs prior to removing the retainer. The snap ring may be removed from other clutch packs without using a spring compressor. Always refer to the vehicle manufacturer's service manual. Some clutch packs are retained in the transaxle or transmission case with a snap ring.*

    *Inspect all the friction clutch plates for wear, flaking, damage, overheating, glazing, and warping. Inspect the steel plates for wear, scoring, damage, and overheating. Use a straightedge to measure steel plate warpage. Check all clutch plate splines for damage. Inspect wave, or reaction, plates for damage. Be sure the check balls in the drum and piston are free and seat properly.*

    *Measure the clutch piston spring height.*

    *Do not interchange clutch pack components. Prior to reassembly, soak all clutch discs for thirty minutes in the specified ATF. Replace all clutch piston and other seals. Be sure these seals fit properly; coat all seals and seal contact areas with the specified ATF. Seal lips must face in the direction that fluid pressure is applied against the seal.*

## Task 2   Measure and adjust clutch pack clearance.

    62. All of the following problems could result in clutch disc burning EXCEPT

      A. a sticking clutch drum check ball.

      B. reduced clutch pack clearance.

      C. a damaged clutch piston seal.

      D. higher-than-specified line pressure.

**Hint**    *After the clutch pack is assembled, the clutch pack clearance must be measured with a feeler gauge or dial indicator. The feeler gauge usually is inserted between the backing plate and the upper clutch disc. On some clutch packs, backing plates of various thicknesses are available to adjust the clearance, whereas in other clutch packs this clearance is adjusted with snap rings of different thicknesses.*

## Task 3   Air test the operation of clutch and servo assemblies.

    63. The tool in Figure 2–31 is used for

      A. oil passage location.

      B. check ball location.

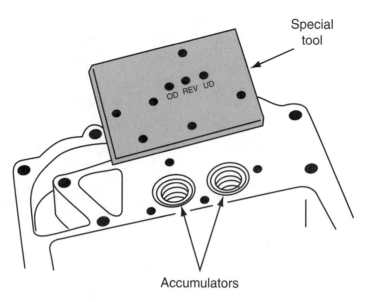

**Figure 2–31** Transaxle service tool.

      C. clutch pack air pressure testing.

      D. valve body positioning.

**Hint**    *Some vehicle manufacturers recommend assembling the pump and front clutch assembly.*

    *Apply regulated air pressure at 35 psi to the clutch apply port, and listen to the clutch apply and release. The clutch should apply with a noticeable thud and release quickly. If air can be heard escaping at the clutch piston, the piston seals are leaking. Some air may escape at the metal or Teflon rings in the clutch drum hub.*

    *Other vehicle manufacturers recommend assembling the clutch packs, gear sets, and pump, and then applying air pressure to the appropriate apply passages in the transmission case. Servos may be air pressure tested in the same way as clutch packs.*

## Task 4  Inspect one-way clutch assemblies; replace parts as necessary.

    64. A four-speed automatic transmission has a buzzing noise in second, third, and fourth gear. This noise does not occur in first gear.

        Technician A says the low/roller clutch may be severely brinnelled.

        Technician B says the forward clutch discs may be badly worn.

        Who is correct?

        A. A only

        B. B only

        C. Both A and B

        D. Neither A nor B

**Hint**    *After disassembly, the rollers in a roller clutch should be inspected for smooth finishes with no evidence of flatness. If any rollers are scored, pitted, or have flat surfaces, replace the roller clutch. When the roller surfaces indicate brinnelling, the roller clutch has been subjected to severe loads from hard driving or transmission abuse. Inspect the roller clutch races and cams for wear, scoring, pitting, and damage.*

    *In a sprag clutch inspect, the sprag faces and races for wear, scoring, and torn face surfaces.*

    *When a roller clutch or sprag is worn or damaged, always inspect the fluid lubrication hole to the unit to ensure there is adequate lubrication.*

    *After assembly, be sure the sprag or roller clutch allows the two components to freewheel in one direction and hold in the opposite direction (Figure 2–32). Be sure the holding and free-wheeling action is in the proper direction.*

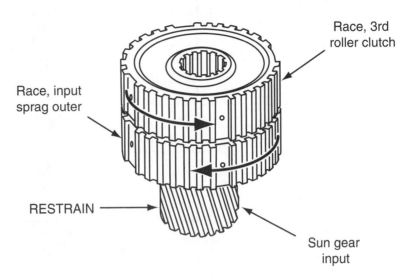

Must freewheel in direction of
arrows and hold in opposite direction.

Race, 3rd
roller clutch

Race, input
sprag outer

RESTRAIN

Sun gear
input

Figure 2–32 Checking sprag or roller clutch operation.

## Task 5  Inspect bands and drums (housings/cylinders); replace and/or adjust as necessary.

65. A transaxle experiences premature band failure, and the band is worn severely on the outer edges but not in the center.
    Technician A says the band strut may be worn unevenly, resulting in band misalignment.
    Technician B says the band contact area on the clutch drum is dished.
    Who is correct?
    A. A only
    B. B only
    C. Both A and B
    D. Neither A nor B

66. A four-speed automatic transmission slips in first gear or manual first gear, and there is no engine braking when coasting in first gear (Figure 2–33). The cause of these problems could be
    A. a malfunctioning O/D one-way clutch.
    B. a malfunctioning forward clutch.
    C. a malfunctioning intermediate band.
    D. a malfunctioning rear one-way clutch.

67. An improper band adjustment may cause
    A. shifts at a lower vehicle speed than specified.
    B. transmission slipping in some gears.
    C. shifts at a higher vehicle speed than specified.
    D. transmission slipping in all gears.

**Hint**    *Bands should be inspected for chips, cracks, burning, glazing, and uneven wear patterns. Be sure the band material is not worn excessively. Inspect all the band anchors, struts, levers, and pivots. The clutch drum surface on which the band makes contact should be inspected for*

| | | Park | Rev. | Neut. | 1st | 2nd | 3rd | 4th | Man. 1st | Man. 2nd |
|---|---|---|---|---|---|---|---|---|---|---|
| Overdrive band | | | | | | | | A | | |
| Intermediate band | | | | | | A | | | | A |
| Low/Reverse band | | | A | | | | | | A | |
| Forward clutch | | | | | A | A | A | A | A | A |
| Direct/Reverse clutch | | | A | | | | A | A | | |
| Coast clutch | | | AW | | AW | AW | AW | | A | A |
| Drive | O/D one-way clutch | | H | | H | H | H | | H | H |
| Coast | | | OR | | OR | OR | OR | OR | OR | OR |
| Drive | Rear one-way clutch | | | | H | OR | OR | OR | H | |
| Coast | | | | | OR | OR | OR | OR | H | |
| Shift solenoid 1 | | ON | ON | ON | ON | ON | OFF | OFF | ON | ON |
| Shift solenoid 2 | | OFF | OFF | OFF | OFF | ON | OFF | OFF | OFF | ON |
| Shift solenoid 3 | | OFF | OFF | OFF | OFF | OFF | OFF | ON | OFF | OFF |
| Coast clutch solenoid | | OFF | OFF | OFF | OFF | OFF | O/W | | ON | ON |
| Engine braking | | NO | O/W | NO | NO | YES | YES | NO | YES | YES |

A = Applied      H = Hold      OR = Overrunning
AW = Applied with "TCS" on - overdrive cancelled      OW = On with "TCS" on - CCS applied

**Figure 2–33** Clutch, band, one-way clutch, and solenoid application.

*burning scoring, glazing, and distortion. Use a straightedge to check the drum for flatness in the band contact area. If the clutch drum is dished in the band contact area, band distortion and uneven, rapid band wear occurs.*

*Many band adjustments may be performed externally, whereas in other transmissions the oil pan must be removed to complete this adjustment. Improper band adjustment may cause harsh shifting, slipping, band burning, and premature failure.*

*To complete the band adjustment, loosen the adjuster locknut, and tighten the adjuster to the specified torque to seat the band. Loosen the band adjuster the specified number of turns, and hold the adjuster in this position while tightening the locknut to the specified torque.*

# Post Test

1. When diagnosing a noncomputer-controlled automatic transmission, Technician A says an intake manifold vacuum leak may affect transmission operation.
   Technician B says a thermostat stuck open may affect transmission operation.
   Who is correct?
   A. A only
   B. B only
   C. Both and B
   D. Neither A nor B

2. A rear wheel drive vehicle has a vibration that occurs above 50 mph. The vibration does not occur at lower speeds or when the engine is accelerated with the transmission in neutral or park. The cause of this problem could be
   A. improper torque converter balance.
   B. improper drive shaft balance.
   C. improper flexplate balance.
   D. loose flexplate bolts.

3. The fluid in an automatic transaxle is a dark brown color and smells burned. Technician A says this problem may be caused by a worn front planetary gear set. Technician B says this problem may be caused by worn friction-type clutch plates. Who is correct?
   A. A only
   B. B only
   C. Both A and B
   D. Neither A nor B

4. When testing transaxle pressure, the pressure is lower than specified at the reverse clutch plug, and the specified pressure is obtained at all the other test plugs. The cause of this problem is
   A. a worn transmission oil pump.
   B. leaking reverse clutch drum seals.
   C. a sticking pressure regulator valve.
   D. worn torque converter hub projections.

5. A computer-controlled transmission has the specified fluid pressure with the engine idling in neutral or park. The transmission pressure is less than specified at higher speeds and loads in all gears. The cause of this problem could be
   A. a sticking electronic pressure control (EPC) solenoid.
   B. a sticking shift solenoid.
   C. an open TCC solenoid winding.
   D. a sticking governor.

6. In many rear wheel drive vehicles, the highest transmission pressure is required at wide open throttle in
   A. first gear.
   B. second gear.
   C. reverse.
   D. fourth gear.

7. When the torque converter clutch (TCC) is released
   A. the TCC solenoid is energized.
   B. the fluid is directed into the hub of the converter.
   C. the fluid is directed over the top of the turbine.
   D. the fluid is directed through the input shaft.

8. When testing the vacuum modulator vacuum supply system, with the engine idling the vacuum is 16 in. Hg. at the throttle body port. With the hose removed from the vacuum modulator, the vacuum is 10 in. Hg at the end of the hose. When the vacuum gauge is connected into the modulator hose with a T-fitting, the vacuum is 10 in. Hg.
   Technician A says the modulator vacuum hose may be leaking.
   Technician B says the modulator diaphragm may be leaking.
   Who is correct?
   A. A only
   B. B only

C. Both A and B

D. Neither A nor B

9. A computer-controlled transmission does not provide a 3 - 4 upshift. With a transmission tester connected to the transmission, the 3 - 4 upshift is still not provided.

Technician A says the problem may be in the transmission computer.

Technician B says the problem may be in the wiring between the computer and the transmission.

Who is correct?

A. A only

B. B only

C. Both A and B

D. Neither A nor B

10. When diagnosing a computer-controlled transaxle, a diagnostic trouble code (DTC) representing the TCC solenoid is obtained on a scan tool.

Technician A says one of the TCC solenoid wires may be grounded inside the transaxle.

Technician B says there may be an open wire in the transaxle electrical connector.

Who is correct?

A. A only

B. B only

C. Both A and B

D. Neither A nor B

11. A computer-controlled transaxle does not shift into reverse, but all the other shifts occur normally. The most likely cause of this problem is

A. a defective shift solenoid.

B. a defective transmission computer.

C. a defective manual lever position sensor (MLPS).

D. an improper shift linkage adjustment.

12. All the upshifts occur at a higher vehicle speed than specified in a computer-controlled transaxle. The most likely cause of this problem is

A. a defective shift solenoid.

B. a defective governor.

C. an improper throttle valve (TV) adjustment.

D. a defective throttle position sensor (TPS).

13. An improperly adjusted throttle valve (TV) cable may cause any of the following problems EXCEPT

A. early upshifts.

B. late upshifts.

C. harsh shifts.

D. a buzzing noise.

14. During a fluid change on a four-speed transaxle, steel filings and particles are found in the oil pan.

Technician A says the planetary gear set may be defective.

Technician B says the transaxle case may be worn.

Who is correct?

A. A only

B. B only

C. Both and B

D. Neither A nor B

15. An improper vacuum modulator operation may be caused by all of the following problems EXCEPT
    A. late basic ignition timing.
    B. a rich air-fuel ratio.
    C. a leak in the vacuum brake booster.
    D. improper modulator adjustment.

16. The transmission extension housing seal is leaking and the extension housing bushing is worn on one side. The most likely cause of this problem is
    A. metal burrs between the extension housing and the transmission case.
    B. a worn rear transmission mount and improper drive shaft angles.
    C. a worn outer surface on the slip yoke in the seal contact area.
    D. a missing balance weight on the drive shaft.

17. While discussing the transmission cooler service, Technician A says the cooler may be leak tested with air pressure at 150 psi (1034 kPa).
    Technician B says that when a transmission is overhauled the cooler should be flushed.
    Who is correct?
    A. A only
    B. B only
    C. Both A and B
    D. Neither A nor B

18. When installing transmission/transaxle valve bodies
    A. the valve body bolts are equal in length.
    B. gaskets may be held in place with gasket shellac.
    C. excessive valve body bolt torque may cause a sticking valve.
    D. excessive pressure may be caused by a missing steel ball.

19. While discussing computer-controlled transmissions. Technician A says the manual lever position sensor (MLPS) is operated by transmission fluid pressure.
    Technician B says a defective MLPS may cause late upshifts.
    Who is correct?
    A. A only
    B. B only
    C. Both A and B
    D. Neither A nor B

20. On a front-wheel-drive vehicle, the front cradle alignment hole is severely misaligned with the matching hole in the chassis. This defect may cause all of the following problems EXCEPT
    A. improper transaxle shifting.
    B. drive axle vibration.
    C. incorrect front wheel alignment.
    D. a whining noise with the engine idling.

# Answers and Analysis

**1.  B**  A plugged or restricted transmission vent may cause pressure buildup in the transmission and loss of fluid from some of the seals, but these leaks would be visible. Therefore, A is wrong.

A leaking modulator diaphragm would allow the intake manifold vacuum to pull transmission fluid through the diaphragm into the intake manifold. This leak is not externally visible. Thus, B is right.

**2.  B**    The rear planetary gear set is not turning with the engine running and the vehicle stopped, so an inoperative rear planetary gear set would not cause this noise. Therefore, A is wrong.

Because the oil pump is turning continually with the engine running, this component may be the cause of the whining noise. Thus, B is right.

**3.  A**    Improper drive shaft balance, drive shaft angles, or worn engine mounts would not cause a vibration with the engine running and the vehicle stopped. Therefore, B, C, and D are wrong.

Improper torque converter balance would result in a vibration with the engine running and the vehicle stopped, or while driving the vehicle. A is the right answer.

**4.  C**    Contaminated transaxle fluid may cause excessive fluid foaming, and fluid may be forced from the dipstick tube, so A is right.

An inoperative transaxle cooler may cause transaxle overheating, which results in fluid being forced from the dipstick tube. A leaking transaxle cooler may contaminate the transaxle fluid with coolant, resulting in fluid foaming and fluid escaping from the dipstick tube. Therefore, both A and B are right, and the correct answer is C.

**5.  A**    A leaking rear main bearing seal or loose rear main bearing may cause a leak from the converter access cover but the front of the converter would be wet with this problem.
Therefore, B and C are wrong.

If the converter has a drain plug, it usually is located in the front of the converter, so a leak at this location would also result in fluid on the front of the converter.

A leak at the transmission oil pump would cause the shell of the converter to be wet, so A is right.

**6.  B**    Dark brown ATF usually is caused by overheating resulting from burned clutches or bands. B is correct. A worn front planetary sun gear would not cause this problem. A is wrong.

**7.  A**    An improperly adjusted TV cable may cause low pressure at WOT, but this problem would not result in low pressure in third gear only. Therefore, B is wrong.

A leak at an internal location such as a clutch piston may cause low pressure in one gear only. Thus, A is correct.

**8.  C**    An improperly adjusted TV cable, or a malfunctioning vacuum modulator may cause normal pressure at idle speed, but low pressure at WOT. Therefore, both A and B are right, and C is the correct answer.

**9.  B**    An inoperative oil pump would result in low pressures at idle speed, and so A is wrong. If the pressure regulator is stuck, the pressures will not increase during wide throttle openings.
Therefore, C is wrong.

A plugged modulator hose may cause high pressures at low speed; thus, D is wrong.

A restricted oil filter may cause the normal transmission pressure to gradually decrease at higher engine speeds.

**10.  D**    An inoperative turbine in the converter, or slipping clutches in the transmission, would cause higher-than-specified stall speed. Therefore, both A and B are wrong, and the right answer is D.

A slipping converter stator clutch reduces torque multiplication in the converter, and causes the fluid in the converter to be directed against the rotation of the impeller pump and engine. This reduces stall speed.

**11.  D**    Low engine compression or a restricted exhaust system would result in a lower-than-specified stall speed. Higher-than-specified stall speed usually is caused by slipping clutches and bands. Therefore, both A and B are wrong, and D is the right answer.

**12.  A**    An inoperative stator clutch may cause a loss of engine power, and lower stall speed, but this problem does not affect TCC lockup. Therefore, B is wrong.

A leaking second speed servo piston seal could result in improper intermediate band application and slipping in second gear. Because TCC lockup only occurs in third and fourth gear, this problem should not affect TCC lockup. Therefore, C is wrong.

A sticking pressure regulator may reduce pressure at wide throttle opening, but it should not affect TCC lockup, so D is wrong.

An inoperative TCC solenoid could prevent TCC lockup. Thus, A is right.

**13. C**  An ignition problem that causes misfiring or a fuel system problem (such as partially restricted injectors that cause a lean condition), may result in shudder after TCC lockup.

Both A and B are right, and C is the correct answer.

**14. B**  With late ignition timing, the vacuum would be less than 18 in. Hg at idle. Therefore, A is wrong.

Restricted exhaust could result in normal vacuum at idle, very low vacuum at part throttle, low stall speed, and delayed upshifts. B is right.

**15. C**  During an air pressure test, a cracked transmission case surrounding the reverse clutch fluid passage or a cracked reverse clutch drum may result in loss of clutch application and a hissing noise from the escaping air. Both A and B are right, and C is the correct answer.

**16. C**  When the computer detects an electronic problem, the transaxle remains in second gear in all forward vehicle speeds. Therefore, C is right, and A, B, and D are wrong because they are mechanical problems.

**17. C**  If the TCIL is flashing, the computer has detected an improper gear ratio likely caused by a slipping band or clutch, and a DTC should be set in the computer memory. Therefore, both Technicians A and B are right, and C is the correct answer.

**18. B**  The solenoid in Figure 2–3 is an electronic pressure control (EPC) solenoid. Therefore, A, C, and D are wrong, and B is right.

**19. C**  In an electronically controlled transmission, the PCM commands the EPC solenoid to supply higher pressure during hard acceleration, and an oil leak at the manual valve would reduce transmission pressure. Therefore, both Technicians A and B are right, and C is the correct answer.

**20. D**  If a TCC does not lock up at the specified speed, the VSS sensor is the most likely problem, because this signal informs the PCM to lock up the converter at the specified speed.

Therefore, A, B, and C are wrong, and D is right.

**21. B**  If the engine is stalling when decelerating from 55 mph, the TCC may not be releasing properly. Because the TPS does not affect TCC release, a problem in this component would not cause the problem. Therefore, A is wrong.

The PCM releases the TCC when it receives a signal from the brake switch. Therefore, an inoperative brake switch may cause improper TCC release and engine stalling. B is the right answer.

**22. A**  If the transmission downshifts when driving at a constant throttle opening on a level road, the most likely problem is a wrong signal from the MLPS sensor, which informs the PCM that the driver has selected another gear. Therefore, B, C, and D are wrong, and A is the right answer.

**23. A**  In an electronically controlled transmission, the TOT sensor signal informs the PCM to delay the upshifts to a slightly higher rpm when the engine is cold. Therefore, B is wrong.

In some computer-controlled transmissions, the PCM detects clutch slipping by comparing the TISS and TOSS signals. Therefore, A is the right answer.

**24. B**  The components listed in A, C, and D do not affect the gear ratio indicated on the scan tool; therefore, these responses are wrong. If the indicated gear ratio on the scan tool is different from the specified gear ratio, a clutch set may be slipping. Therefore, B is the right answer.

**25. A**  A problem in the TCC solenoid or related circuit does not cause a severe lack of engine power. Therefore, B is wrong.

In some OBDII systems with electronically controlled transmissions, the PCM increases EPC solenoid current to lower transmission oil pressure. If the PCM detects higher than normal EPC solenoid current, which supplies lower than normal transmission oil pressure, the PCM retards the ignition timing and leans the air-fuel ratio to lower engine power and torque to prevent transmission damage. Therefore, the EPC solenoid and related circuit should be checked if the engine has a severe lack of power. Thus, A is the right answer.

26.    A    If the voltage regulator allows continually high-charging system voltage, other electrical components would be damaged besides the PCM. Therefore, B is wrong.

An intermittent open circuit between the alternator battery terminal and the positive battery terminal causes high-voltage spikes and damage to computers such as the PCM.

Therefore, A is the right answer.

27.    C    A lean air-fuel ratio may cause a momentary lean misfire condition when the torque converter locks up. High resistance in the spark plug wire or spark plugs may also cause momentary misfiring when the torque converter locks up. Therefore, Technicians A and B are both correct, and C is the right answer.

28.    C    A defective vehicle speed sensor (VSS), transmission input speed sensor (TISS), or transmission output speed sensor (TOSS) may affect transmission shifting, but defects in these sensors do not cause reduced fuel mileage. Therefore, A, B, and D are wrong. A defective engine coolant temperature (ECT) sensor that indicates low engine temperature reduces fuel economy and causes late transmission shifting.

29.    A    A misadjusted manual valve shift linkage may cause low fluid pressure. Therefore, B is wrong.

This low fluid pressure may cause clutch slipping and premature failure. Thus, A is right.

30.    D    If the throttle valve cable is misadjusted so throttle pressure is higher than specified, the transmission shifts occur at a higher-than-specified vehicle speed, because the vehicle speed must be higher for governor pressure to overcome throttle pressure and cause the upshifts to occur. Therefore, A, B, and C are wrong, and D is the right answer.

31.    D    Transmission filters should be replaced, not cleaned. Therefore, A is wrong.

Transmission fluid oxidizes faster at higher temperatures, and so B is also wrong. Thus, the correct answer is D.

32.    A    An improper manual valve shift linkage adjustment may cause reduced pressure, transmission slipping, and premature clutch failure. Therefore, B is wrong.

If moisture is freezing in the modulator diaphragm, the vacuum is not applied to the diaphragm. Under this condition, the pressure is high, and repeated clutch piston failure plus harsh, late shifting may occur. Thus, A is the right answer.

33.    D    This question asks us to select the component that is not the cause of the problem. A sticking governor valve, excessive governor spring tension, or worn weights and pins could result in reduced governor valve movement in relation to speed. This action causes the shifts to occur at a higher speed. Therefore, A, B, or C could be the cause of the problem, but they are not the requested answer.

A weak governor spring tension would cause excessive governor valve movement in relation to vehicle speed, resulting in earlier shifts. D would not cause the problem, so it is the right answer.

34.    A    High governor pressure affects shift timing and quality, but this problem does not cause high pressure in the pump. Pump pressure is determined by the pressure regulator valve. Therefore, B is wrong.

A worn pump body bushing allows excessive torque converter hub movement in this bushing, and this movement damages the pump seal. Thus, A is right.

35.    A    If the transmission fluid is contaminated from the lack of fluid and filter changes, the extension housing bushing and drive shaft yoke may be scored but not pitted. If the fluid is contaminated, other problems would occur such as sticking valves in the valve body and improper shifting. Therefore, B is wrong.

High resistance in a battery ground between the chassis and the engine may cause the current to flow from the chassis through some of the powertrain components to the battery ground cable attached to the engine. This may cause arcing and pitting of powertrain components such as the extension housing bushing and drive shaft yoke. Thus, A is the right answer.

**36. D** With the engine idling, the ATF flow through a transmission cooler should be one quart in 20 seconds. Therefore, A, B, and C are wrong, and D is right.

**37. A** This question asks for the response that is not the cause of the problem. A dry cable, worn gears, or worn drive gear retaining bushing could result in erratic speedometer operation. Therefore, B, C, and D are not the requested answer.

A missing drive gear retaining clip would cause the output shaft to rotate inside this gear without turning the gear. This problem would cause an inoperative speedometer. Thus, A is the right answer.

**38. A** The clearance between the valves and valve body bores should not exceed 0.001 in. Therefore, B, C, and D are wrong, and A is right.

**39. A** Low governor pressure would affect the other shifts, not just the 1 - 2 upshift. Therefore, B is wrong.

Excessive valve body torque may have warped the valve body causing the 1 - 2 shift valve to stick and preventing the 1 - 2 upshift. Thus, A is the right answer.

**40. C** The procedure shown in Figure 2–13 is to determine the proper servo piston selective pin. Therefore, A, B, and D are wrong, and C is the correct answer.

**41. A** A sticking pressure regulator valve could result in high transaxle pressures and harsh shifting in all gears, not just in fourth gear. Therefore, B is wrong.

A sticking fourth gear accumulator piston may cause harsh shifting only in fourth gear.

**42. A** This question asks for the response that is not correct. The parking pawl is mechanically operated by the shift lever, and the pawl projection engages in a notched drum. Therefore, statements B and C are right, but they are not the requested answer. The pawl is retained on a pivot pin, so D is not the requested answer.

When the pawl is engaged, it locks the output shaft. Therefore, statement A is not correct, and this is the right answer.

**43. B** When the computer senses an electronic problem and fails to close the transaxle relay, the transaxle operates in second gear and reverse. Therefore, A, C, and D are wrong and B is the right answer.

**44. A** When diagnosing a computer-controlled transmission or transaxle for improper shifting, the first sensor to verify is the manual lever position sensor (MLPS), because this sensor informs the computer regarding the gear selector position. The computer supplies the proper gear in relation to the input. Therefore, B, C, and D are wrong, and A is right.

**45. B** An improperly adjusted manual valve shift linkage may result in transaxle slipping, but the condition should be constant. Therefore, A is wrong.

Broken engine or transaxle mounts cause excessive engine and transaxle movement, which changes the effective length of the throttle valve cable and results in erratic shifting. Thus, B is the right answer.

**46. C** This question asks for the statement that is not correct. Prior to transmission/transaxle removal, the negative battery terminal should be removed, the drive shaft should be marked in relation to the differential flange, and the engine support fixture must be installed.

Therefore, A, B, and D are correct statements, but they are not the requested answer.

The front drive axles do not require marking in relation to the front hubs. Therefore, statement C is incorrect, and this is the requested answer.

**47. D** If the input shaft end play is excessive, a thicker selective washer is required on the inner end of the input shaft. Therefore, A, B, and C are wrong, and D is the right answer.

**48.  C**  If the clutch plate clearance is more than specified, a thicker reaction plate must be installed in the clutch pack. Therefore, A, B, and D are wrong, and C is the right answer.

**49.  B**  Loose main bearings cause a thumping noise when the engine is first started, so A is wrong.

Loose connecting rod bearings cause a knocking noise that is most noticeable during acceleration, and worn piston pins cause a constant metallic knocking noise at idle. Therefore, C and D are wrong.

Loose flex plate-to-converter bolts may cause a clunking noise during deceleration or possibly at idle. Therefore, B is the right answer.

**50.  A**  If the converter one-way clutch was slipping, the engine would have a power loss at lower speeds, and the question informs us that all converter tests were satisfactory. This would include a one-way clutch test. Therefore, B is wrong.

When the transmission was overheated, the converter hub was misaligned and the hub was rubbing on the pump bushing. With the transmission removed and the converter bolted to the flex plate, a converter hub misalignment measurement with a dial indicator proved the hub was misaligned. Thus, A is the right answer.

**51.  A**  A thicker shim is required to reduce the turning torque. Therefore, Technician A is right, and Technician B is wrong, making A the correct answer.

**52.  B**  A sticking pressure regulator valve may cause high pressure and repeated pump seal failure, but this problem would also result in other complaints such as hard shifting. Therefore, A is wrong.

A plugged drainback hole behind the pump seal may cause excessive pressure on the seal and repeated seal failure. Thus, A is the right answer.

**53.  C**  The purpose of the measurement in Figure 2–21 is to determine the proper selective washer thickness. Therefore, A, B, and D are wrong, and C is right.

**54.  B**  An inoperative vehicle speed sensor could result in torque converter clutch (TCC) lockup at the wrong speed. Because the question indicates lockup speed is normal, A is wrong.

The number nine check ball in the turbine shaft seats against an orificed seat when the TCC is applied. This check ball action controls the exhausting of fluid from the front of the lockup plate, which controls TCC application. A stuck check ball may allow faster exhausting of the fluid in front of the lockup plate and harsh TCC application. Therefore, B is the right answer.

**55.  A**  The tool in Figure 2–23 is used to size Teflon sealing rings. Therefore, B, C, and D are wrong, and A is the right answer.

**56.  D**  This question asks for the statement that is not correct. A scored shaft in the bushing contact area may indicate a lack of lubrication, and bushing to shaft clearance may be measured with a wire-type feeler gauge or a vernier caliper and micrometer. Therefore, A, B, and C are correct, but they are not the requested answer.

Normal bushing clearance is 0.0005 to 0.015 in., so D is wrong, making it the requested answer.

**57.  C**  If the carrier is held and the annulus gear drives the sun gear, a reverse overdrive is provided, which is not practical, so A is wrong.

When the sun gear is held and the annulus gear drives the carrier, a forward overdrive is provided. Therefore, B is wrong.

When the annulus gear is held and the sun gear drives the carrier, a forward overdrive is provided, so D is wrong.

If the carrier is held and the sun gear drives the annulus gear, a reverse reduction is provided. Therefore, C is the right answer.

**58.  C**  Transmission or transaxle case porosity between two fluid passages may cause pressure bleed off and clutch burnout, so A is right. This problem also may cause the transmission to be in two gears at once, causing jam-up. Therefore, both A and B are right, and C is the correct answer.

**59.  A**   The tool shown in Figure 2–24 is used to remove and install the drive chain and sprockets. Therefore, B, C, and D are wrong, and A is the right answer.

**60.  B**   To increase the turning torque, a thicker shim is required behind one side bearing cup. Therefore, A, C, and D are wrong, and B is the right answer.

**61.  A**   The tool in Figure 2–28 is used to install clutch pistons and seals. Therefore, B, C, and D are wrong, and A is the right answer.

**62.  D**   This question asks for the response that is not correct. A sticking drum check ball, reduced clutch pack clearance, or a damaged clutch piston seal may result in clutch disc burning.

Therefore, statements A, B, and C are correct, but they are not the requested answer.

High line pressure may cause harsh shifting and clutch piston seal damage, but it does not result in clutch disc burning. Statement D is incorrect, so this is the requested answer.

**63.  C**   The tool in Figure 2–29 is used for air pressure testing the clutch packs. Therefore, A, B, and D are wrong, and C is the right answer.

**64.  A**   Because the forward clutch is applied in all forward gears, worn forward clutch discs could result in slipping in all forward gears. However, this problem should not result in a buzzing noise. Therefore, B is wrong.

Because the low/roller clutch is freewheeling in second, third, and fourth gear, and holding in first gear, severe brinnelling of this roller clutch may cause a buzzing noise in second, third, and fourth gears. Thus, A is the right answer.

**65.  B**   Uneven wear on a band strut may cause band misalignment, and the resulting wear pattern on the band may be curved across the band rather than circular with the band. However, this problem would not cause severe wear on the outer band edges, so A is wrong.

A dished band contact area on the drum would result in excessive wear on the outer band edges. Therefore, B is the right answer.

**66.  D**   An inoperative O/D one-way clutch causes slipping in reverse, first, second, and third gears, so A is wrong.

An inoperative forward clutch causes slipping in all forward gears, thus B is wrong.

If the intermediate band is inoperative, slipping occurs in second gear, so C is wrong.

An inoperative rear one-way clutch causes slipping in first gear or manual first gear with no engine braking while coasting in first gear. Therefore, D is correct.

**67.  B**   Because shift timing in relation to vehicle speed is a function of throttle pressure and governor pressure, band adjustment does not affect shift timing. Therefore, A and C are wrong. Because the band, or bands, are not applied in all gears, improper band adjustment does not cause slipping in all gears. Therefore, D is wrong and B is the correct answer.

# Answers to Post Test

**1.  A**   With a noncomputer-controlled transmission, an engine thermostat stuck open does not affect transmission operation. Therefore, Technician B is wrong. An intake manifold vacuum leak affects transmission shifting especially on models with a vacuum modulator. Technician A is right, so A is the correct answer.

**2.  B**   Improper torque converter or flexplate balance causes a vibration when the engine is accelerated with the transmission in neutral or park, and thus A and C are incorrect. Loose flexplate bolts may cause a thumping noise with the engine idling, and so D is incorrect. Improper drive shaft balance may cause a vibration above a certain vehicle speed. Therefore, B is correct.

**3.  B**   A worn front planetary gear set does not cause dark brown transmission fluid with a burned smell. Therefore, A is incorrect. If the transmission is overheated, the bands and clutches may be worn and burned and this condition causes dark brown fluid with a burned smell. Thus, B is correct.

**4.  B**    A worn transmission pump reduces the pressure at all the test plugs, and so A is incorrect. A sticking pressure regulator valve may reduce the pressure at all the test plugs; thus, C is incorrect. Worn torque converter hub projections may cause improper or no oil pump drive, and this reduces the pressure at all the test plugs. D is incorrect. Leaking reverse clutch drum seals reduce the fluid pressure only in reverse. Therefore, B is correct.

**5.  A**    A sticking shift solenoid has no affect on transmission pressure, and so B is incorrect. An open TCC solenoid winding has no affect on transmission pressure; thus, C is incorrect. A governor is not required in a computer-controlled transmission. Therefore, D is incorrect. The electronic pressure control solenoid is operated by the computer to control transmission pressure. A sticking EPC solenoid may cause lower transmission pressure at higher engine speed and load. Thus, A is correct.

**6.  C**    In many rear-wheel-drive vehicles, the highest transmission pressure is required at wide open throttle in reverse gear. Therefore, A, B, and D are incorrect, and C is the correct answer.

**7.  D**    When the TCC is released, the TCC solenoid is de-energized, and so A is incorrect. When the TCC is released, fluid is directed through the input shaft and out in front of the lockup plate. Therefore, B and C are wrong, and D is correct.

**8.  A**    The same vacuum is available at the end of the hose with the hose removed or connected to the vacuum modulator. Therefore, B is incorrect. When the vacuum is less at the end of the modulator hose compared to the vacuum at the throttle body port, the modulator hose is leaking. Thus, A is correct.

**9.  D**    The transmission tester eliminates the computer and the wiring from the computer to the transmission. If the problem is still present with the tester connected, the defect is in the transmission. Therefore, Technicians A and B are both wrong, and D is the correct answer.

**10.  C**    A grounded TCC solenoid wire inside the transaxle may cause a DTC related to the TCC solenoid. An open TCC solenoid wire in the transaxle electrical connector may also cause a DTC related to the TCC solenoid. Therefore, Technicians A and B are both right, and C is the correct answer.

**11.  D**    In reverse, fluid is directed to the reverse clutch or band by manual valve position. Therefore, A, B, and C are wrong, and D is correct.

**12.  D**    A defective shift solenoid only affects one or two of the transaxle shifts, and so A is incorrect. Computer-controlled transaxles do not have a governor or a throttle valve. Therefore, B and C are wrong. A defective TPS may cause improper transaxle shifting. D is correct.

**13.  D**    An improperly adjusted TV cable may cause early upshifts, late upshifts, or harsh shifts. Therefore, A, B, and C are not the requested answer. An improperly adjusted TV cable does not cause a buzzing noise. Therefore, D is the requested answer.

**14.  A**    A worn transaxle case causes aluminum filings in the oil pan. Therefore, Technician B is wrong. Steel particles and filings may come from a defective planetary gear set. Therefore, Technician A is right, making A the correct answer.

**15.  B**    Improper modulator operation may be caused by late ignition timing, a leak in the vacuum brake booster, or improper modulator adjustment. Therefore, A, C, and D are not the requested answer. A rich air-fuel ratio does not cause improper vacuum modulator operation. Therefore, B is the requested answer.

**16.  A**    Improper drive shaft angles or a mission balance weight on the drive shaft cause a vibration problem. Therefore, B and D are wrong. A worn outer surface on the slip yoke in the seal contact area may cause seal leakage, but this problem does not cause wear on one side of the bushing. Thus, C is incorrect. Metal burrs between the extension housing and the transmission may cause improper extension housing alignment, seal leakage, and wear on one side of the extension housing bushing. Therefore, A is correct.

**17.  B**    The cooler should be leak tested with regulated air pressure, and the pressure must not exceed 75 psi (517 kPa). Therefore, Technician A is wrong. When a transmission is overhauled, the cooler should be flushed. Thus, Technician B is right, making B the correct answer.

**18.  C**   Most valve body bolts have different lengths. Thus, A is incorrect. Gaskets should be held in place with clean petroleum jelly, making answer B incorrect. A missing steel ball usually causes improper shifting, so D is incorrect. Excessive valve body bolt torque may warp the valve body and cause a sticking valve. Therefore, C is correct.

**19.  D**   The MLPS is operated by the transmission shift linkage. Therefore, Technician A is wrong. A defective MLPS does not cause late upshifts, and so Technician B is wrong. Since Technicians A and B are both wrong, D is the correct answer.

**20.  D**   Improper cradle alignment on a front-wheel-drive vehicle may cause improper transaxle shifting, drive axle vibration, and incorrect front wheel alignment. Therefore, A, B, and C are not the requested answer. Improper cradle alignment does not cause a whining noise with the engine idling. Thus, D is the requested answer.

# 3 Manual Drive Train and Axles

## Pretest

The purpose of this pretest is to determine the amount of review that you may require prior to taking the ASE Manual Drive Train and Axles Test. If you answer all the pretest questions correctly, complete the questions and study the information in this chapter to prepare for the ASE Manual Drive Train and Axles Test.

If two or more of your answers to the pretest questions are incorrect, complete a thorough study of the questions and information in this chapter. The pretest answers are located at the end of the pretest, and are also in the answer sheets supplied with this book.

1. A vehicle has a final drive ratio of 3.42:1, and a first gear ratio of 3.65:1. The overall gear ratio in first gear is
   A. 7.07:1.
   B. 10.56:1.
   C. 12.48:1.
   D. 13.76:1.

2. The rear axle ratio is changed from 3.08:1 to 3.73:1.
   Technician A says the engine rpm will be higher at the same vehicle speed in any gear.
   Technician B says the engine will have less torque at the same vehicle speed in any gear.
   Who is correct?
   A. A only
   B. B only
   C. Both A and B
   D. Neither A nor B

3. A typical overdrive transmission gear ratio would be
   A. 0.46:1.
   B. 0.70:1.
   C. 0.92:1.
   D. 0.96:1.

4. The differential case and ring gear are rotating at 100 rpm, and one rear drive wheel is not turning. The opposite rotating rear drive wheel is turning at
   A. 100 rpm.
   B. 200 rpm.
   C. 300 rpm.
   D. 400 rpm.

5. Hard transaxle shifting may be caused by
   A. a bent shift rail.
   B. rough input shaft bearing.

    C. excessive output shaft end play.

    D. worn clutch release bearing.

6. A hydraulic clutch does not disengage properly, resulting in hard shifting.
   Technician A says there may be air in the hydraulic clutch system.
   Technician B says there may be too much clutch pedal free play.
   Who is correct?
   A. A only
   B. B only
   C. Both A and B
   D. Neither A nor B

7. A clutch grabs and chatters while engaging.
   Technician A says the torsion springs in the clutch plate may be weak or broken.
   Technician B says the clutch release bearing may be rough.
   Who is correct?
   A. A only
   B. B only
   C. Both A and B
   D. Neither A nor B

8. When a flywheel requires resurfacing, the amount of metal machined from the flywheel surface should be
   A. 0.005 to 0.010 in.
   B. 0.015 to 0.025 in.
   C. 0.010 to 0.040 in.
   D. 0.010 to 0.090 in.

9. A bell housing bore has excessive runout.
   Technician A says the bell housing bore may be machined to correct the problem.
   Technician B says the offset dowels in the back of the engine block may be rotated to correct this excessive runout.
   Who is correct?
   A. A only
   B. B only
   C. Both A and B
   D. Neither A nor B

10. All of the following statements about synchronizer assemblies are true EXCEPT
    A. Synchronizer hubs and sleeves should be marked before disassembly so they can be reassembled in the same position.
    B. Synchronizer hubs are reversible.
    C. Worn blocking rings may cause hard shifting.
    D. Threads in the cone area of the blocking ring must be sharp.

11. A vehicle has a clicking noise in the differential while driving straight ahead. The cause of the problem could be damaged
    A. side gear teeth.
    B. pinion/spicer gear teeth.
    C. pinion shaft and case.
    D. ring gear and pinion teeth.

12. Technician A says in some transaxles the differential side bearing preload may be adjusted with adjuster nuts behind the side bearing cups.

Technician B says in some transaxles the differential side bearing preload may be adjusted by changing the shim thickness behind one of the side bearings.

Who is correct?

A.  A only

B.  B only

C.  Both A and B

D.  Neither A nor B

13.  The differential turning torque in a transaxle is less than specified.

Technician A says a thinner thrust washer may be installed behind both differential side gears.

Technician B says a thicker thrust washer may be installed behind one of the differential side gears.

Who is correct?

A.  A only

B.  B only

C.  Both A and B

D.  Neither A nor B

14.  The transfer case on a four-wheel-drive (4WD) vehicle contains a planetary gear set. Technician A says in 4WD low the annulus gear is locked to the case.

Technician B says in 4WD low the input shaft is driving the sun gear and the planetary carrier is the output to the front and rear drive shafts.

Who is correct?

A.  A only

B.  B only

C.  Both A and B

D.  Neither A nor B

# Answers to Pretest

**1.  C**    The first gear ratio is multiplied by the final drive gear ratio to obtain the overall gear ratio. Therefore, $3.42 \times 3.65 = 12.48$. C is the correct answer.

**2.  A**    When the rear axle ratio is changed from 3.08 to 3.73, the engine rpm is higher at the same vehicle speed in any gear. Therefore, Technician A is right, and A is the correct answer. Because the engine rpm is higher in any gear, the engine torque is increased. Therefore, Technician B is wrong.

**3.  B**    A typical overdrive transmission gear ratio is 0.70:1. Therefore, A, C, and D are wrong, and B is the correct answer.

**4.  B**    If the differential case and ring gear are rotating at 100 rpm with one drive wheel held, the opposite drive wheel is rotating at 200 rpm. Therefore, A, C, and D are wrong, and B is the correct answer.

**5.  A**    A rough input shaft bearing causes a growling noise with the engine idling in neutral and the clutch engaged. Thus, B is wrong. Excessive output shaft end play may cause clunking noises and jumping out of gear. Therefore, C is wrong. A worn clutch release bearing may cause a growling noise with the clutch pedal depressed, and so D is wrong. Hard transaxle shifting may be caused by a bent shift rail, making answer A correct.

**6.  C**    Air in the hydraulic system or too much clutch pedal free play may cause improper disengagement of a hydraulic clutch. Therefore, Technicians A and B are both right, and C is the correct answer.

**7.  A**    Weak or broken torsion springs in a clutch disc may cause clutch grabbing and chattering. Technician A is right. A rough clutch release bearing does not cause clutch grabbing and chattering. Technician B is wrong. Thus, A is the correct answer.

**8.  C**    When a flywheel requires resurfacing, the amount of metal machined from the flywheel surface is 0.010 to 0.040 in. Therefore, A, B, and D are wrong, and C is the correct answer.

**9.  B**    When a bell housing bore has excessive runout, the bore may not be machined to correct the problem. Thus, Technician A is wrong. Offset dowels in the back of the engine block may be rotated to correct bell housing bore runout. Therefore, Technician B is right, and B is the correct answer.

**10.  B**    Synchronizer hubs and sleeves should be marked before disassembly. A is correct but this is not the requested answer. Worn blocking rings may cause hard shifting, and so C is right, but this is not the requested answer. Sharp threads in the cone area of the blocking ring must be sharp, and thus D is right but this is not the requested answer. Synchronizer hubs are not reversible. Therefore, B is wrong, making it the requested answer.

**11.  D**    Damaged side gear or pinion gear teeth, or a damaged pinion shaft and case may cause noise when turning a corner. Therefore, A, B, and C are wrong. Damaged ring gear and pinion teeth may cause a clicking noise when driving straight ahead. Thus, D is the correct answer.

**12.  D**    In transaxles, the differential side bearing preload is not adjusted by adjuster nuts behind the side bearing cups, or shims behind the side bearings. Technicians A and B are both wrong, and D is the correct answer.

**13.  D**    To increase the differential turning torque in a transaxle, a thicker shim must be installed behind one of the differential side bearing cups. Therefore, both Technicians A and B are wrong, and D is the correct answer.

**14.  C**    In 4WD low, the annulus gear is locked to the case and the input shaft drives the sun gear while the planetary carrier is the output to the front and rear drive shafts. Therefore, Technicians A and B are both right, and C is the correct answer.

# Clutch Diagnosis and Repair

## ASE Tasks, Questions, and Related Information

In this chapter, each task in the Manual Transmission and Transaxle category is provided followed by a question and some information related to the task. If you answer any question incorrectly, study this information very carefully until you understand the correct answer. Question answers and analysis are provided at the end of this chapter and in the answer sheets provided with this book.

**Task 1**  **Diagnose clutch noise, binding, slippage, pulsation, chatter, pedal feel/effort, and release problems; determine needed repairs.**

   1. Clutch chatter may be caused by
      A. a worn, rough clutch release bearing.
      B. a worn, rough pilot bearing.
      C. excessive input shaft end play.
      D. weak clutch plate torsional springs.

**Hint**    *A growling noise that occurs only when the clutch pedal is depressed is likely caused by a worn clutch release bearing. A rattling noise with the clutch pedal depressed may be caused by a worn* *pilot bearing or bushing. If a growling noise occurs with the transmission in neutral, the engine idling, and the clutch pedal released, the input shaft bearing is likely defective. Clutch slippage may be caused by worn or contaminated clutch facings or improper clutch linkage adjustment. Clutch pedal pulsation may be caused by uneven pressure plate lever height. Clutch chatter may be caused by weak clutch plate torsional springs, improper clutch housing alignment, contaminated clutch facings, or worn power train mounts. Improper clutch release may be caused by improper clutch linkage adjustment or clutch disc splines sticking on the input shaft splines.*

**Task 2**   **Inspect, adjust, and replace clutch pedal linkage, cables and automatic adjuster mechanisms, brackets, bushings, pivots, springs, and electrical switches.**

2. Lack of clutch pedal free play may cause
   A.  hard shifting.
   B.  improper clutch release.
   C.  transaxle gear damage.
   D.  clutch slipping.

3. A clutch with a self-adjusting cable has
   A.  one inch of clutch pedal free play.
   B.  two inches of clutch pedal free play.
   C.  a constant running release bearing.
   D.  an overcenter assist spring.

**Hint**    *Some vehicles have linkages connected from the clutch pedal to the clutch release fork (Figure 3–1). In other clutch systems, the clutch pedal is connected to the release fork by a cable. Many clutch linkages or cables have an adjustment to set the clutch pedal free play. Clutch pedal free play is the amount of pedal movement before the release bearing contacts the pressure plate release fingers or diaphragm. Many late-model vehicles have a self-adjusting clutch cable. In these clutches, the cable is wrapped around and attached to a toothed wheel, and a ratcheting spring-loaded pawl is engaged with the toothed wheel (Figure 3–2). Each time the clutch pedal is released, the pawl removes any slack from the cable by engaging the next tooth on the wheel.*

*A self-adjusting clutch has no built-in free play, and a constant running release bearing.*

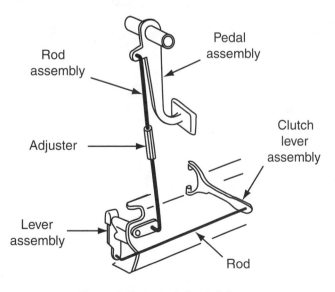

**Figure 3–1** Typical clutch linkage.

**Task 3**   **Inspect, adjust, replace, and bleed hydraulic clutch slave cylinder/actuator, master cylinder, lines, and hoses; clean and flush hydraulic system; refill with proper fluid.**

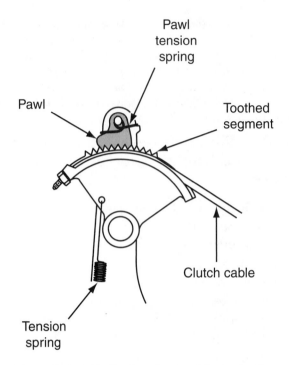

**Figure 3–2** Self-adjusting clutch cable mechanism.

4. The clutch in the hydraulic clutch system in Figure 3–3 fails to disengage properly when the clutch pedal is fully depressed. The cause of this problem could be
   A. less-than-specified clutch pedal free play.
   B. air in the clutch hydraulic system.
   C. worn clutch facings.
   D. a scored pressure plate.

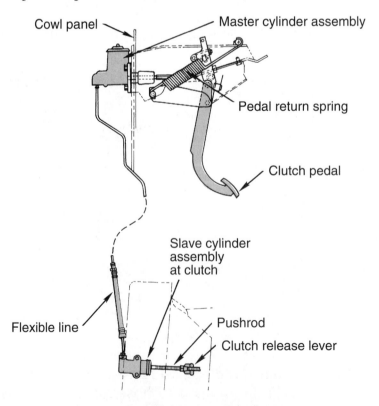

**Figure 3–3** Typical clutch hydraulic system.

*Hint*     *In a hydraulic clutch system, the clutch pedal is connected to the clutch master cylinder pushrod, and a hydraulic line is connected from the clutch master cylinder to the slave cylinder.*

## Task 4     Inspect, adjust, and replace release (throw-out) bearing, bearing retainer, lever, and pivot.

5. While discussing a clutch with an adjustable linkage, Technician A says the clutch pedal free play adjustment sets the distance between the release bearing and the pressure plate fingers.

   Technician B says a worn release bearing is noisy with the clutch pedal released.

   Who is correct?
   A. A only
   B. B only
   C. Both A and B
   D. Neither A nor B

*Hint*     *The clutch release bearing is connected to the release fork, and the center opening in this bearing is mounted on a machined hub that is bolted to the front of the transmission (Figure 3–4).*

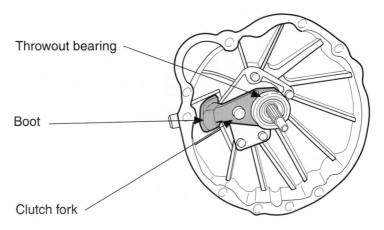

Throwout bearing

Boot

Clutch fork

**Figure 3–4** Clutch release bearing and fork. *(Used with permission from Nissan North America, Inc.)*

   *When the clutch pedal is depressed, the release bearing pushes on the pressure plate release fingers or thrust pad. This action moves the pressure plate fingers or levers against the pressure plate spring pressure. Movement of the pressure plate fingers forces the pressure plate away from the clutch disc to release the clutch, thus interrupting power flow from the engine to the transmission.*

## Task 5     Inspect and replace clutch disc and pressure plate assembly; inspect input shaft splines.

6. Clutch chatter may be caused by
   A. excessive crankshaft end play.
   B. loose engine main bearings.
   C. a badly scored pressure plate.
   D. improper pressure plate to flywheel position.

*Hint*     *A straightedge and a feeler gauge should be used to measure pressure plate warpage (Figure 3–5). If the flywheel does not have locating dowels, always punch mark the pressure plate and flywheel so the pressure plate is reinstalled in the original position.*

**Figure 3–5** Measuring pressure plate warpage.

*Clutch facing wear should be checked by measuring the distance from the facing surface to the rivet heads with a depth gauge or vernier caliper (Figure 3–6). When the clutch disc and pressure plate are installed on the flywheel, the clutch disc must be installed in the proper direction.*

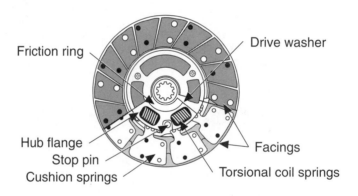

**Figure 3–6** Clutch disc components.

*An alignment tool must be installed through the clutch disc hub into the pilot bearing to align the disc properly while the pressure plate bolts are tightened.*

7. The distance from the rivet heads in the clutch facing surface should be at least
   A. 0.005 in.
   B. 0.008 in.
   C. 0.012 in.
   D. 0.025 in.

## Task 6  Inspect and replace pilot bearing/bushing; inspect pilot bearing/bushing mating surfaces.

8. A worn pilot bearing may cause a rattling, growling noise while
   A. the engine is idling and the clutch pedal is fully depressed.
   B. decelerating in high gear with the clutch pedal released.
   C. accelerating in low gear with the clutch pedal released.
   D. the engine is idling in neutral with the clutch pedal released.

**Hint**    *The pilot bearing or bushing is mounted in the center of the flywheel. This bearing supports the front of the transmission input shaft, and the clutch disc is splined to this shaft. The pilot*

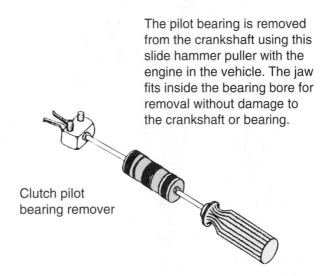

The pilot bearing is removed from the crankshaft using this slide hammer puller with the engine in the vehicle. The jaw fits inside the bearing bore for removal without damage to the crankshaft or bearing.

Clutch pilot bearing remover

**Figure 3–7** Pilot bearing puller. *(Courtesy of Kent-Moore Division, SPX Corp.)*

*bearing may be a ball-type or roller-type bearing, or a bushing. The proper puller must be used to remove the pilot bearing (Figure 3–7). The pilot bearing must be installed with the proper driving tool.*

## Task 7   Inspect, and measure flywheel and ring gear; inspect dual-mass flywheel damper where required; repair or replace as necessary.

9. Technician A says if too much material is removed during flywheel resurfacing, the torsion springs on the clutch disc may contact the flywheel bolts, resulting in noise while engaging and disengaging.

   Technician B says if excessive material is removed when the flywheel is resurfaced, the slave cylinder may not have enough travel to release the clutch properly.

   Who is correct?

   A. A only

   B. B only

   C. Both A and B

   D. Neither A nor B

**Hint**   *If the flywheel surface is scored, burned, or worn it should be resurfaced or replaced. During the resurfacing operation 0.010 to 0.040 in. of metal may be removed from the flywheel surface. Do not remove excessive material from the flywheel surface.*

   *Worn or chipped flywheel ring gear teeth may cause improper flywheel balance and engine vibrations. Improper starting motor engagement also results from worn or chipped flywheel ring gear teeth. When the flywheel is removed from the crankshaft, punch or index marks must be placed on the flywheel and crankshaft so the flywheel is reinstalled in the original position to maintain proper flywheel balance.*

## Task 8   Inspect engine block, clutch (bell) housing, transmission case mating surfaces, and alignment dowels; determine needed repairs.

10. Excessive misalignment between the bell housing and the engine block may cause

    A. reduced clutch pedal free play.

    B. a growling noise with the clutch pedal depressed.

    C. a vibration at higher speeds.

    D. clutch grabbing and chatter.

11. Technician A says shims may be installed between the bell housing and engine block mating surfaces to correct bell housing face runout.

Technician B says excessive bell housing face runout may be caused by overheated clutch disc facings.

Who is correct?

A. A only

B. B only

C. Both A and B

D. Neither A nor B

12. To correct excessive runout on the dial indicator in Figure 3–8

A. replace the engine mounts.

B. adjust eccentric bell housing dowels.

C. install bell housing shims.

D. replace the clutch disc.

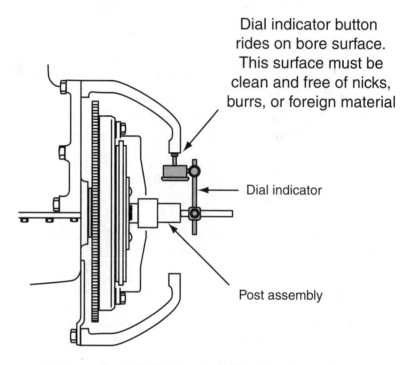

Dial indicator button rides on bore surface. This surface must be clean and free of nicks, burrs, or foreign material

Dial indicator

Post assembly

**Figure 3–8** Measuring bell housing bore.

**Hint**    *The mating surfaces of the engine block and bell housing should be inspected for metal burrs and accumulation of foreign material. Clean these mating surfaces with an approved cleaning solution. Remove any metal burrs with a fine-toothed file. Follow the same procedure to clean and inspect the mating surfaces of the bell housing and transmission, including the bell housing bore.*

*When bell housing face squareness is measured, a dial indicator is attached to a steel post mounted in the clutch disc splined opening, and the dial indicator stem contacts the bell housing face (Figure 3–9). If the bell housing face runout is excessive, shims may be installed between the bell housing and the engine block to correct this misalignment.*

*When performing the bell housing bore runout measurement, a special tool is connected between the dial indicator stem and the bell housing bore. Some engines have offset dowel pins in the block surface that mates with the bell housing surface. These offset dowel pins may be rotated to correct the bell housing bore runout. If the offset dowel adjustment is not enough to correct the bell housing bore runout, replace the bell housing.*

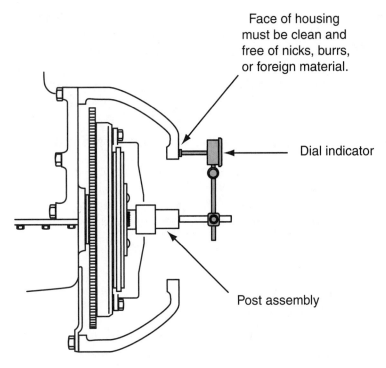

**Figure 3–9** Measuring bell housing face.

## Task 9 Measure flywheel surface runout and crankshaft end play; determine needed repairs.

13. With the dial indicator positioned as shown in Figure 3–10, the measurement being performed is
    A. crankshaft end play.
    B. crankshaft warpage.
    C. rear main bearing wear.
    D. rear engine block alignment.

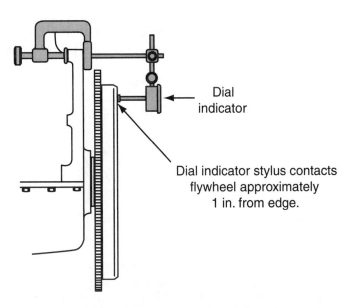

**Figure 3–10** Dial indicator positioned against flywheel surface.

*Hint*    *A dial indicator may be positioned against the clutch facing contact area on the flywheel while rotating the flywheel to measure flywheel runout. Hold the crankshaft forward or backward while measuring flywheel runout. If this runout is excessive, resurface or replace the flywheel.*

*With the dial indicator positioned against the flywheel surface, the crankshaft end play may be measured by forcing the flywheel and crankshaft forward and rearward while observing the dial indicator reading.*

## Task 10    Inspect, replace, and align powertrain mounts.

14. Technician A says sagged transmission mounts may cause improper drive shaft angles on a rear-wheel-drive car.

    Technician B says improper drive shaft angles may cause a vibration that is constant when the vehicle is accelerated and decelerated.

    Who is correct?

    A. A only

    B. B only

    C. Both A and B

    D. Neither A nor B

*Hint*    *Engine and transmission mounts should be inspected for broken, sagged, oil-soaked, or deteriorated conditions. Any of these mount conditions may cause a grabbing, binding clutch. On a rear-wheel-drive car, defective engine or transmission mounts may cause improper drive shaft angles, which result in a vibration that changes in intensity when the vehicle is accelerated and decelerated.*

# Transmission Diagnosis and Repair

## ASE Tasks, Questions, and Related Information

## Task 1    Diagnose transmission noise, hard shifting, gear clash, jumping out of gear, fluid condition and type, and fluid leakage problems; determine needed repairs.

15. While discussing a manual transmission that jumps out of second gear, Technician A says there is excessive end play between the second speed gear and its matching synchronizer.

    Technician B says the detent springs on the shift rail may be weak.

    Who is correct?

    A. A only

    B. B only

    C. Both A and B

    D. Neither A nor B

*Hint*    *A growling noise is usually caused by worn transaxle bearings. Worn or damaged gears may cause a clicking or grinding noise. If this type of noise only occurs on one gear, the gears meshed in that gear are damaged. Hard shifting may be caused by a bent shift rail, a damaged or worn synchronizer, misadjusted shift linkage, improper fluid type, or improper clutch adjustment. Gear clash may also be caused by improper clutch adjustment, or worn synchronizers. Jumping out of gear may be caused by excessive gear end play and worn gear or synchronizer teeth.*

### Task 2   Inspect, adjust, and replace transmission external shifter assembly, shift linkages, brackets, bushings/grommets, pivots, and levers.

16. The shift lever adjustment usually is performed with the transmission in
    A. neutral.
    B. first gear.
    C. second gear.
    D. reverse gear.

**Hint**   *Most external shift linkages and cables require adjusting, and a similar adjustment procedure is used on some vehicles. Raise the vehicle on a lift and place the shift lever in neutral to begin the shift linkage adjustment. With a lever-type shift linkage, install a 1/4-in. rod in the adjustment hole in the shifter assembly (Figure 3–11). Adjust the shift linkages by loosening the rodretaining locknuts and moving the levers until the 1/4-in. rod fully enters the alignment holes. Tighten the locknuts and check the shift operation in all gears.*

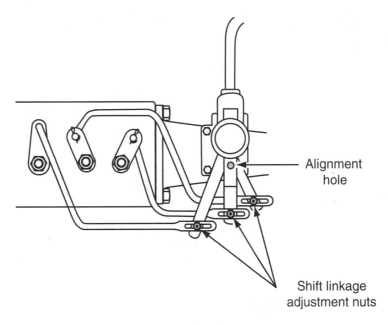

Alignment hole

Shift linkage adjustment nuts

**Figure 3–11** Typical adjustment of floor-mounted shift levers.

*Rod-type shift linkages are adjusted with basically the same procedure as lever-type linkages. When the alignment pin is in place, adjust the shift rod so the pin slides freely in and out of the alignment hole.*

### Task 3   Inspect and replace transmission gaskets, sealants, seals, and fasteners; inspect sealing surfaces.

17. Repeated extension housing seal failure may be caused by
    A. a scored drive shaft yoke.
    B. excessive output shaft end play.
    C. excessive input shaft end play.
    D. a worn output shaft bearing.

**Hint**   *Fluid leaks in a manual transmission may occur at the extension housing seal, shift cover gasket, vent, backup light switch, drain plug, speedometer drive, or front bearing retainer gasket. If the extension housing seal is leaking, it may be replaced with the transmission in the vehicle.*

*When this seal is replaced, always check the fit of the front drive shaft yoke in the extension housing bushing and inspect this yoke for scoring in the seal contact area. A worn extension housing bushing or a scored front drive shaft yoke will cause repeated extension housing seal failure.*

## Task 4  Remove and replace transmission; inspect transmission mounts.

18. During manual transmission removal and replacement
    A. the drive shaft may be installed in any position on the differential pinion gear flange.
    B. the transmission weight may be supported by the input shaft in the clutch disc hub.
    C. the engine support fixture should be installed after the transmission-to-engine bolts are loosened.
    D. the clutch disc must be aligned with an aligning tool prior to transmission installation.

**Hint**    *Prior to transmission or transaxle removal, the battery ground cable, shift linkages or cables, and speedometer cable must be removed. On a rear-wheel-drive car, the drive shaft must be removed. Mark the rear drive shaft flange in relation to the yoke before drive shaft removal to ensure drive shaft installation in the original position. Drain the transmission lubricant. Before the transmission retaining bolts are loosened, an engine support fixture must be installed to support the weight of the engine. Use a transmission jack to support the weight of the transmission during the removal process.*

*Install a clutch disc alignment tool through the clutch disc hub into the pilot bearing to be sure the clutch disc is properly aligned before attempting to install the transmission (Figure 3–12).*

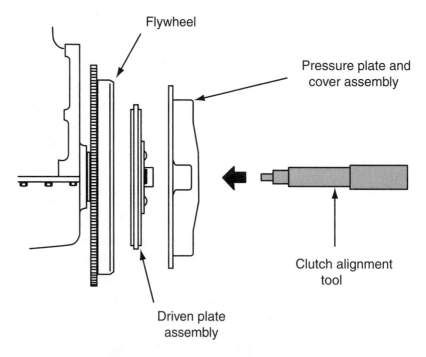

**Figure 3–12** Clutch disc alignment tool.

## Task 5  Disassemble and clean transmission components; reassemble transmission.

19. Technician A says if the synchronizer sleeve does not slide smoothly over the blocker ring and gear teeth, hard shifting will occur.
    Technician B says the synchronizer hub and sleeve must be marked in relation to each other prior to disassembly.

Who is correct?

A. A only

B. B only

C. Both A and B

D. Neither A nor B

**Hint**    *Some transmissions have a seal and shim between the bell housing and the transmission case. This shim helps to control end play. Remove the shift-lever cover, shift forks, and the related mechanism.*

     *Remove the rear extension housing and front bearing retainer. Some front bearing retainers have a seal and shim behind the retainer. Remove the input gear and shaft, and the output shaft assembly. Remove the counter shaft and counter gear assembly followed by the reverse idle shaft and reverse idler gear.*

     *Before a synchronizer assembly is removed from a shaft, mark the alignment of the collar and hub so these components may be reassembled in the original position (Figure 3–13).*

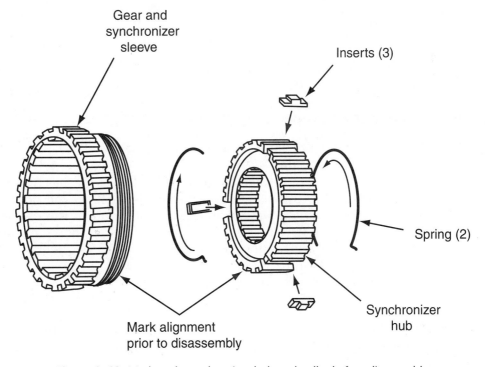

**Figure 3–13** Mark each synchronizer hub and collar before disassembly.

## Task 6   Inspect, repair, and/or replace transmission shift cover and internal shift forks, bushings, bearings, levers, shafts, sleeves, detent mechanisms, interlocks, and springs.

20. A bent shift rail may cause

     A. the transmission to jump out of gear.

     B. hard shifting in some gears.

     C. gear clash during some shifts.

     D. gear noise in some gears.

**Hint**    *Shift rails should be inspected to be sure they are not bent, broken, or worn. Inspect all the holes and notches in the shift rails for wear or damage (Figure 3–14). When each shift rail is installed in the appropriate bore in the case of shift cover, check for excessive movement between the rail and the bore. Inspect all detent springs to be sure they are not worn, bent, or weak.*

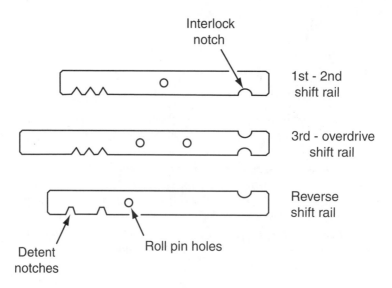

**Figure 3–14** Shift rail inspection.

*Check the interlock plates for flatness with a straightedge. If the interlock plates are worn on the surface that contacts the shift rails, replace the plates.*

## Task 7    Inspect and replace input (clutch) shaft, bearings, and retainers.

Gear and synchronizer sleeve Inserts (3) Mark alignment prior to disassembly Synchronizer hub Spring (2)

21. A manual transmission has a growling noise with the engine idling and the clutch pedal released with the transmission in neutral. The noise disappears when the clutch pedal is depressed and the transmission is placed in gear. The cause of this noise could be a worn
    A. pilot bearing in the crankshaft.
    B. input shaft pilot bearing contact area.
    C. input shaft needle bearings.
    D. main shaft ball bearing.

**Hint**    *The tip of the input shaft that rides in the pilot bearing must be smooth. Worn input shaft splines may cause the clutch disc to stick on these splines, resulting in improper clutch release.*

*Be sure the input shaft teeth that mesh with the synchronizer collar are not worn or chipped.*

*Remove the roller bearings from the inner end of the input shaft. These bearings and the bearing contact area in the input shaft must be inspected for wear, pitting, and roughness. Check the input shaft bearing for smooth rotation and looseness.*

## Task 8    Inspect and replace main shaft, gears, thrust washers, bearings, and retainers/snap rings; measure gear clearance/end play.

22. The transmission in Figure 3–15 has a growling noise in first gear only. The cause of this problem could be
    A. a worn first gear blocking ring.
    B. a worn first gear synchronizer sleeve.
    C. a chipped and worn first speed gear teeth.
    D. a worn, rough main shaft bearing.

Neutral

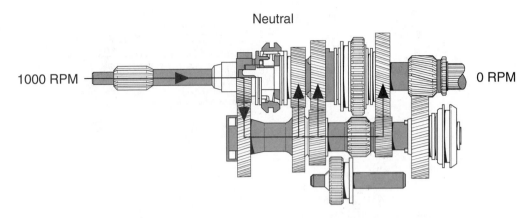

1000 RPM                                                                    0 RPM

**Figure 3–15** Five-speed manual transmission.

**Hint**    *Inspect all the gear teeth on the main shaft gears for chips, pitting, cracks, wear, and discoloration because of overheating. A normal gear tooth wear pattern appears as a polished finish with little wear on the gear face. Inspect the bearing surfaces in the gear bores for roughness, pitting, and scoring.*

*After the gears are removed from the main shaft, inspect all the bearing surfaces on this shaft for roughness, pitting, and scoring (Figure 3–16). Inspect the gear journal areas on the main shaft for the same conditions.*

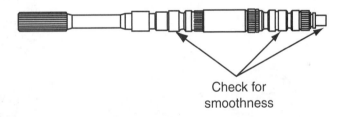

Check for
smoothness

**Figure 3–16**  Bearing contact surfaces on the main shaft must be smooth.

## Task 9  Inspect and replace synchronizer hub, sleeve, keys (inserts), springs, and blocking (synchronizing) rings/mechanisms; measure blocking ring clearance.

23. While inspecting synchronizer assemblies
    A. the dog teeth tips on the blocking rings should be flat with smooth surfaces.
    B. the threads in the cone area of the blocking rings should be sharp and not dulled.
    C. the clearance is not important between the blocking ring and the matching gear's dog teeth.
    D. the sleeve should fit snugly on the hub and offer a certain amount of resistance to movement.

24. The clearance on the fourth speed gear shown in Figure 3–17 is less than specified.
    Technician A says this may result in noise while driving in fourth gear.
    Technician B says this problem may cause hard shifting into fourth gear.
    Who is correct?
    A. A only
    B. B only

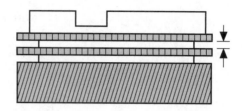

**Figure 3–17** Use a feeler gauge to measure the distance between the blocking ring and the dog teeth on the matching gear.

    C. Both A and B
    D. Neither A nor B

**Hint**    *Be sure the sleeve moves freely on the hub. Inspect the insert springs to be sure they are not bent, distorted, or broken (Figure 3–18). Inspect the blocking rings for cracks, breaks, and flatness.*

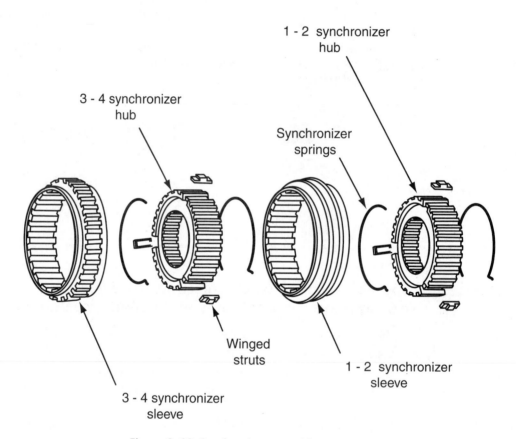

**Figure 3–18** Synchronizer assembly components.

    *The dog teeth on the blocking rings must be pointed with smooth surfaces. Threads in the cone area of the blocking rings must be sharp and not dulled. With the blocking ring positioned on its matching gear, use a feeler gauge to measure the distance between the blocking ring and the matching gear's dog teeth.*

**Task 10**  **Inspect and replace countershaft, counter (cluster) gear, bearings, thrust washers, and retainers/snap rings.**

25. The counter gear shaft and needle bearings are pitted and scored.

Technician A says the transmission may have a growling noise with the engine idling, the transmission in neutral, and the clutch pedal released.

Technician B says the transmission may have a growling noise while driving in any gear.

Who is correct?

A. A only

B. B only

C. Both A and B

D. Neither A nor B

*Hint*     *Inspect all the gear teeth on the counter gear for chips and cracks. The wear pattern on all the gear teeth should indicate a polished finish with very little wear. Inspect the bearing surfaces in the counter gear bore for pitting, scoring, or overheating. Inspect the counter gear shaft for wear, pitting, and roughness in the bearing contact areas. All counter gear needle or ball bearings should be inspected for roughness and looseness. Replace the bearings if these conditions are present. Inspect all counter gear thrust washers and retainers for wear.*

## Task 11   Inspect and replace reverse idler gear, shaft, bearings, thrust washers, and retainers/snap rings.

26. In the transmission shown in Figure 3–19, the reverse idler gear is rotating in

A. first gear.

B. fifth gear.

C. third and fourth gear.

D. reverse gear.

*Hint*     *The reverse idler gear teeth should be inspected for chips, pits, and cracks. Check the gear bore for roughness and scoring. The needle bearings and shaft must be inspected for roughness, scoring, and pitting.*

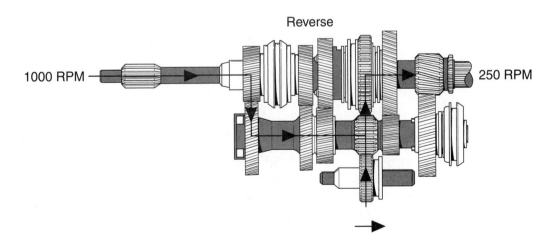

Reverse

1000 RPM     250 RPM

**Figure 3–19** Five-speed manual transmission

## Task 12   Measure and adjust shaft/gear, and synchronizer end play.

27. Excessive input shaft end play in the transmission in Figure 3–19 may cause the transmission to jump out of

A. first gear.

B. second gear.

C. fourth gear.

D. All of the above

**Hint**    *Some vehicle manufacturers provide service procedures and specifications for measuring the end play on various transmission gears. In many end play adjustments, a dial indicator is mounted on the transmission case, and the dial indicator stem is positioned on the gear. Push the gear back and forth and observe the end play reading on the dial indicator. Excessive counter gear end play usually is caused by worn thrust washers.*

## Task 13  Measure and adjust bearing preload or end play.

28. A manual transmission has a clunking noise during acceleration and deceleration.

Technician A says the main shaft may have excessive end play.

Technician B says the clutch facings and pressure plate friction surface may be worn.

Who is correct?

A. A only

B. B only

C. Both A and B

D. Neither A nor B

**Hint**    *Bearing preload is determined by selective shim thickness in some manual transmissions. To determine the proper selective shim thickness for some main shaft bearings, use a depth micrometer to measure from the transmission case to the top of the main shaft output bearing (measurement A). Be sure this bearing is properly seated. Measure from the support ring to the transmission case with a new gasket installed on the case (measurement B) (Figure 3–20). The proper selective shim thickness is calculated by subtracting measurement B from measurement A.*

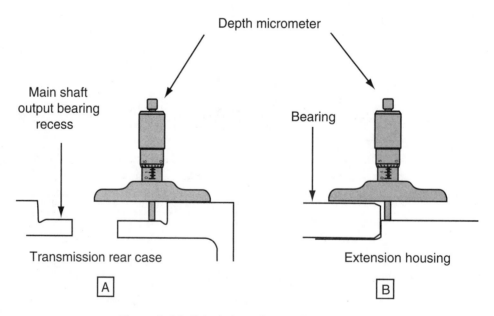

**Figure 3–20** Calculating selective shim thickness.

*Then subtract 0.003 in. from the calculated shim thickness to compensate for the compression of the housing gasket when the extension housing is bolted to the case. Reduced bearing preload may cause excessive end play and a clunking noise in the transmission. Excessive bearing preload may result in overheated transmission components.*

## Task 14   Inspect, repair, and replace extension housing and transmission case mating surfaces, bores, bushings, and vents.

29. Premature wear on the extension housing bushing may be caused by
    A. worn speedometer drive and driven gears.
    B. metal burrs on the rear transmission mating surface.
    C. a plugged transmission vent opening.
    D. excessive transmission main shaft end play.

**Hint**   *Inspect the extension housing for cracks, and check the mating surfaces of this housing and the transmission case metal burrs and gouges. Remove any metal burrs or gouges with a fine-toothed file. Replace the extension housing seal and check the bushing in this housing for wear or damage.*

## Task 15   Inspect and replace speedometer drive gear, driven gear, and retainers.

30. Erratic speedometer operation with the speedometer drive in Figure 3–21 may be caused by
    A. a worn adapter bushing and distance sensor.
    B. a worn extension housing bushing.
    C. excessive transmission main shaft end play.
    D. a distance sensor electrical defect.

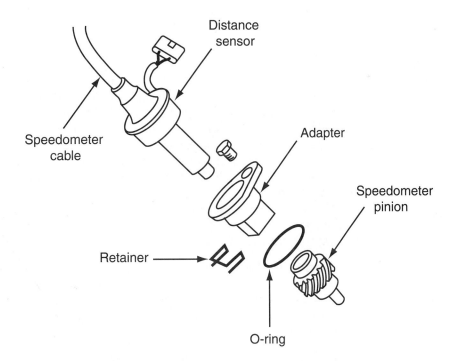

**Figure 3–21** Speedometer drive and related components.

**Hint**   *The speedometer driven gear should be inspected for worn teeth and looseness. Be sure the adapter is not loose on the cable extension. A worn adapter may cause erratic speedometer operation.*

*During a transmission overhaul, replace the speedometer drive O-ring. Inspect the speedometer drive gear for worn or damaged teeth and looseness on the main shaft. Be sure the speedometer drive gear is securely retained on the main shaft.*

## Task 16    Inspect, test, and replace transmission sensors and switches.

31. Component 19 in Figure 3–22 is
    A. a vehicle speed sensor.
    B. a computer-aided gear select solenoid.
    C. a backup light switch.
    D. a shift solenoid.

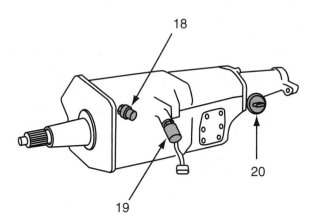

**Figure 3–22** Externally mounted components, manual 6-speed transmission.

**Hint**    *Manual transmissions may have various sensors, switches, and solenoids. These components may include a vehicle speed sensor (VSS), a backup light switch, and a computer-aided gear select solenoid. The VSS sends a voltage signal to the PCM in relation to vehicle speed. The PCM uses this signal to operate outputs such as the cruise control. The backup light switch is operated by linkage inside the transmission to operate the backup lights when the transmission is placed in reverse. The computer-aided gear select solenoid ensures good fuel economy and compliance with federal fuel economy standards by inhibiting second and third gears when shifting out of first gear under these conditions:*
  • *Coolant temperature is above 122°F.*
  • *Vehicle speed is between 12 and 19 mph.*
  • *Throttle opening is less than 35 percent.*

## Task 17    Inspect lubrication systems.

**Hint**    *If manual transmission components are not properly lubricated, rapid component wear occurs. Therefore, all lubrication system components must be carefully inspected and worn or damaged components must be replaced.*

## Task 18    Check fluid level, and refill with proper fluid.

32. While discussing manual transmission and transaxle lubricants, Technician A says that compared to a 90W gear oil, ATF reduces friction and improves horsepower and fuel economy.
    Technician B says thicker gear oils have lower classification numbers.
    Who is correct?
    A. A only
    B. B only

C. Both A and B

D. Neither A nor B

**Hint**     *Some vehicle manufacturers recommend a mineral oil with an extreme pressure (EP) additive for manual transmissions. The most common gear oil classifications are SAE 75W, 75W-80, 80W-90, 85W-90, 90, or 140. Thicker gear oils have higher classification numbers (Figure 3–23). Some manual transmissions require engine oil or automatic transmission fluid (ATF).*

*Many manual transaxles require 30W engine oil, 90W gear oil, or ATF.*

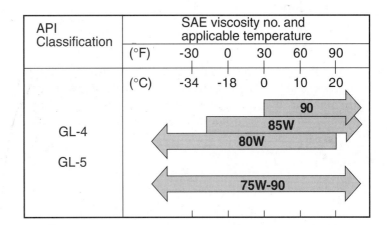

**Figure 3–23** Transmission and transaxle lubricant ratings. *(Provided by Subaru of America, Inc.)*

# Transaxle Diagnosis and Repair

## ASE Tasks, Questions, and Related Information

**Task 1**  **Diagnose transaxle noise, hard shifting, gear clash, jumping out of gear, fluid condition and type, and fluid leakage problems; determine needed repairs.**

33. A fully synchronized four-speed manual transaxle experiences gear clash in all forward gears and reverse. The cause of this problem could be

A. a worn fourth-speed synchronizer.

B. the clutch disc sticking on the input shaft.

C. excessive main shaft end play.

D. a worn 3 - 4 shifter fork.

**Hint**     *A growling noise is usually caused by worn transaxle bearings. Worn or damaged gears may cause a clicking or grinding noise. If this type of noise only occurs on one gear, the gears meshed in that gear are damaged. Hard shifting may be caused by a bent shift rail, a damaged or worn synchronizer, a misadjusted shift linkage, or an improper clutch adjustment. Gear clash may also be caused by improper clutch adjustment, or worn synchronizers. Jumping out of gear may be caused by excessive gear end play and worn gear or synchronizer teeth.*

## Task 2    Inspect, adjust, lubricate, and replace transaxle external shift assemblies, linkages, brackets, bushings/grommets, cables, pivots, and levers.

34. Technician A says an improper shift linkage adjustment may cause hard transaxle shifting.

    Technician B says an improper shift linkage adjustment may result in the transaxle sticking in gear.

    Who is correct?

    A. A only

    B. B only

    C. Both A and B

    D. Neither A nor B

**Hint**    *Many shift linkages and cables require adjusting, and a similar adjustment procedure is used on some vehicles. Loosen the adjusting screw on the transaxle crossover cable (Figure 3–24). Slide a 1/4-in. rod through the alignment hole in the transmission lever and case to hold the transaxle in the neutral position, and tighten the adjusting screw to the specified torque (Figure 3–25).*

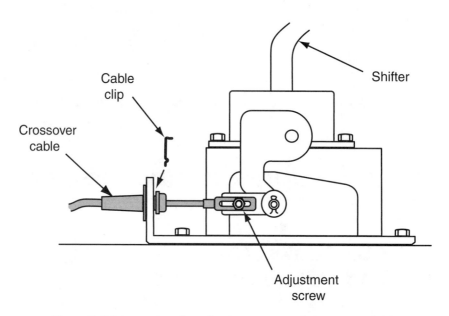

**Figure 3–24**  Loosening the adjusting screw on the crossover cable.

## Task 3    Inspect and replace transaxle gaskets, sealants, seals, and fasteners; inspect sealing surfaces.

35. While discussing repeated transaxle drive axle seal leakage and replacement, Technician A says this problem may be caused by a plugged transaxle vent.

    Technician B says this problem may be caused by a worn outer drive axle joint.

    Who is correct?

    A. A only

    B. B only

    C. Both A and B

    D. Neither A nor B

**Hint**    *Transaxle leaks may occur at the drive axle seals, vent, sealing surfaces between the case sections, or the drain plug. During a transaxle overhaul, all seals and gaskets should be replaced. Seal bores should be inspected for metal burrs and scratches. Inspect all seal lip contact areas for*

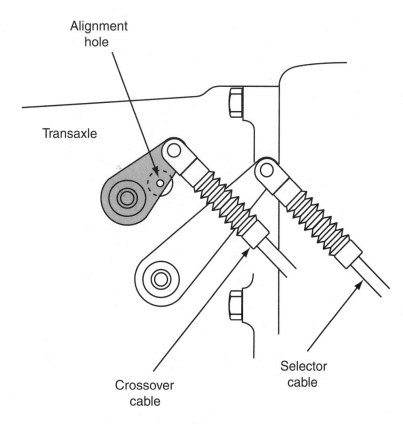

Alignment
hole

Transaxle

Crossover
cable

Selector
cable

**Figure 3–25** Installing a 1/4 in. drill bit in the transaxle lever and case to hold the transaxle in neutral.

*roughness or scoring. Coat the outside diameter of each seal case with an appropriate sealer prior to installation. Always use the proper driver to install each seal. Coat the seal lips and lip contact area with the manufacturer's specified transaxle lubricant.*

## Task 4    Remove and replace transaxle; inspect, replace, and align transaxle mounts and subframe/cradle assembly.

36. Technician A says a misaligned engine and transaxle cradle may cause drive axle vibrations.

    Technician B says a misaligned engine and transaxle cradle may cause improper front suspension angles.

    Who is correct?

    A.  A only

    B.  B only

    C.  Both A and B

    D.  Neither A nor B

**Hint**    *Prior to transaxle removal, the battery ground cable, shift linkages or cables, speedometer cable, vehicle speed sensor, and all electrical connections must be disconnected. Drain the transaxle lubricant. On front-wheel-drive vehicles, the front drive axles must be removed from the transaxle prior to transaxle removal. Before the transaxle retaining bolts are loosened, an engine support fixture must be installed to support the weight of the engine. On some front-wheel-drive vehicles, the engine cradle, or part of the cradle, must be removed prior to transaxle removal. Use a transmission jack to support the weight of the transaxle during the removal process.*

*Install a clutch disc alignment tool through the clutch disc hub into the pilot bearing to be sure the clutch disc is properly aligned before attempting to install the transaxle.*

## Task 5    Disassemble and clean transaxle components; reassemble transaxle.

37. Refer to Figures 3–26 to 3–31. While disassembling the transaxle

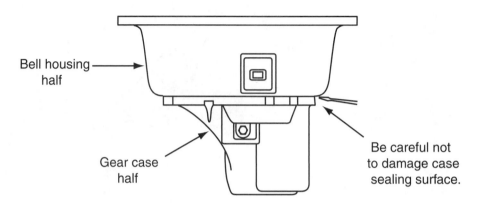

**Figure 3–26** Separating the transaxle halves.

A. the input and output shafts must be removed before the differential.
B. the input and output shafts are retained with snap rings under the transaxle cover.
C. the transaxle should be supported on the bench for input and output shaft removal.
D. the input and output shafts may be pulled by hand from the transaxle bearings.

**Hint**    *Remove the retaining bolts between the two transaxle halves, and separate these halves (Figure 3–26). Be careful not to mark the transaxle case. Remove the differential assembly from the case. Remove the reverse idler shaft bolt, reverse idler gear shaft, and reverse fork bracket (Figure 3–27). Remove the selector shaft spacer and selector shaft. Remove the transaxle cover and use a pair of snap ring pliers to remove the snap rings in the input and output shafts (Figure 3–28). Support the transaxle on the proper bench fixture and shims (Figure 3–29). Use the proper driving fixture to press the input and output shaft assemblies out of the case (Figure 3–30).*

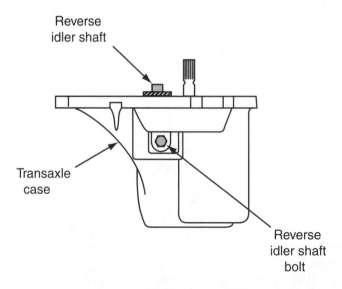

**Figure 3–27** Removing the reverse idler shaft bolt, shaft and gear.

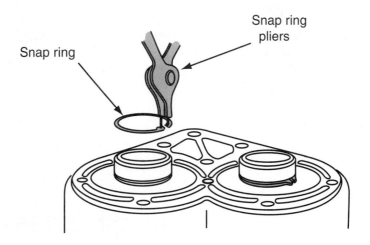

**Figure 3–28** Removing the snap rings on the input and output shafts.

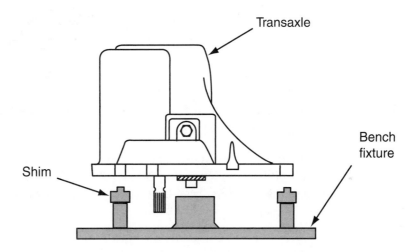

**Figure 3–29** Supporting the transaxle case on the proper fixture and shims.

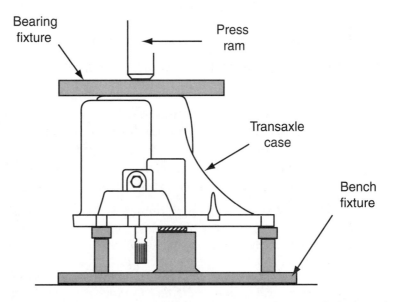

**Figure 3–30** Using the proper fixture to drive the input and output shafts from the case.

*Remove the reverse brake shim, friction cone, blocking ring, and needle bearing (Figure 3–31). Wash all the transaxle components in an approved cleaning solution and blow-dry these components with compressed air.*

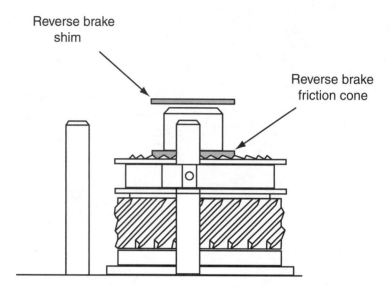

**Figure 3–31** Removing the reverse brake shim and friction cone.

## Task 6   Inspect, repair, and/or replace transaxle shift cover and internal shift forks, levers, bushings, shafts, sleeves, detent mechanisms, interlocks, and springs.

38. While discussing a four-speed manual transaxle that jumps out of third gear, Technician A says the shift rail detent spring tension on the 3 - 4 shift rail may be weak.

    Technician B says there may be excessive wear on the fourth speed gear dog teeth.

    Who is correct?

    A. A only

    B. B only

    C. Both A and B

    D. Neither A nor B

*Hint*   *Remove the shift blocker assembly and shift forks (Figure 3–32). The same basic inspection may be performed on transmission and transaxle shifting mechanisms. Inspection of transmission shifting mechanisms was explained previously in this chapter.*

## Task 7   Inspect and replace input shaft, gears, bearings, and retainers/snap rings.

39. A five-speed manual transaxle has a growling and rattling noise in third gear only. The cause of this noise could be

    A. worn, chipped teeth on the third-speed gear on the input shaft.

    B. worn dog teeth on the third-speed gear on the input shaft.

    C. worn dog teeth on the third-speed synchronizer blocking ring.

    D. worn threads in the cone area of the third-speed blocking ring.

*Hint*   *Worn input shaft splines may cause the clutch disc to stick on these splines, resulting in improper clutch release. The gears on the input shaft should be inspected for cracks and pitted, worn, or broken teeth (Figure 3–33). The bore in each gear and the matching surface on the input shaft should be inspected for roughness, pits, and scoring. Inspect the needle bearings mounted between the gears and the input shaft for roughness and looseness.*

*Use a feeler gauge to measure the clearance between the dog teeth on the third, fourth, and fifth speed gear and the matching blocking ring.*

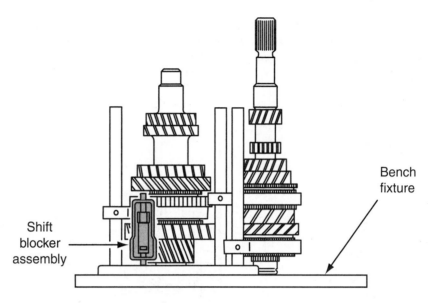

**Figure 3–32** Removing shift blocker assembly and shift forks.

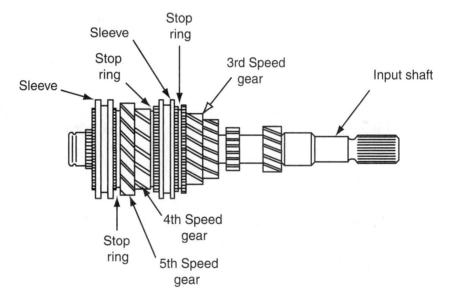

**Figure 3–33** Transaxle input shaft assembly.

**Task 8**  **Inspect and replace output shaft, gears, thrust washers, bearings, and retainers/snap rings.**

40.  The second-speed gear dog teeth and blocking ring teeth are badly worn. This problem may cause
   A.  a growling noise while driving in second gear.
   B.  a vibration while accelerating in second gear.
   C.  hard shifting in second and third gear.
   D.  the transaxle to jump out of second gear.

*Hint*      *The inspection performed on the input shaft and gears should be repeated on the output shaft and gears. In some transaxles, the output shaft assembly is serviced as a complete unit. If any gear or component on the output shaft is worn, the complete assembly must be replaced.*

*Use a feeler gauge to measure the clearance between the dog teeth on the first- and second-speed gear and the matching blocking ring (Figure 3–34). Inspect the output shaft bearings in the transaxle case for roughness or looseness.*

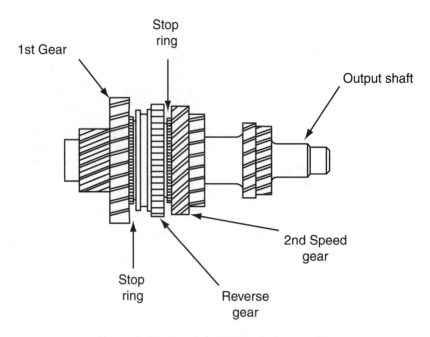

**Figure 3–34** Transaxle output shaft assembly.

## Task 9   Inspect and replace synchronizer hub, sleeve, keys (inserts), springs, and blocking (synchronizing) rings; measure blocking ring clearance.

41. Technician A says synchronizer hubs are reversible on the shaft on which they are mounted.

    Technician B says synchronizer sleeves are reversible on their matching hub.

    Who is correct?

    A. A only

    B. B only

    C. Both A and B

    D. Neither A nor B

**Hint**    *Synchronizer inspection and replacement is basically the same for transmissions and transaxles. Refer to the transmission synchronizer inspection explained previously in this chapter.*

*Synchronizer hubs and sleeves are directional. Always assemble these components in their original position. Prior to disassembly, always mark the synchronizer sleeve and hub so these components are assembled in their original location in relation to each other.*

## Task 10   Inspect and replace reverse idler gear, shaft, bearings, thrust washers, and retainers/snap rings.

42. A transaxle shifts normally into all forward gears, but it will not shift into reverse gear, and there is no evidence of noise while attempting this shift.

    Technician A says the reverse shifter fork may be broken.

    Technician B says the reverse idler gear teeth may be worn.

    Who is correct?

    A. A only

    B. B only

C. Both A and B

D. Neither A nor B

**Hint**    *The reverse idler gear teeth should be inspected for chips, pits, and cracks. Check the gear bore for roughness and scoring. The reverse idler shaft and bearings must be inspected for roughness, scoring, and pitting (Figure 3–35).*

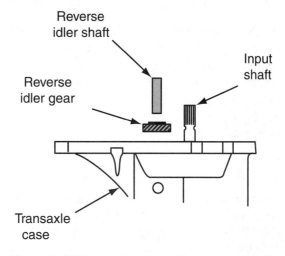

**Figure 3–35** Transaxle reverse idler gear and shaft.

## Task 11   Inspect, repair, and/or replace transaxle case mating surfaces, bores, dowels, bushings, bearings, and vents.

43. Technician A says an epoxy-based sealer may be used to repair a crack in some transaxle cases.

    Technician B says some cracks in transaxle cases may be repaired with Loctite PST.

    Who is correct?

    A. A only

    B. B only

    C. Both A and B

    D. Neither A nor B

**Hint**    *Inspect the mating surfaces of the two transaxle halves for metal burrs and scratches. Inspect the transaxle case for cracks. All bearing bores should be inspected for cracks, scoring, or gouges.*

*Inspect all seal mounting areas for metal burrs and gouges. Be sure the vent is not restricted. A plugged vent may result in pressure buildup in the transaxle, resulting in leaks from the seals.*

*Inspect all shift rail bushings and replace all shift rail seals.*

*Transaxle case replacement often is required if the case is cracked. However, some vehicle manufacturers recommend crack repair with an epoxy-based sealer, depending on the location of the crack. Always refer to the vehicle manufacturer's recommended crack repair procedure in the appropriate service manual. Damaged threads in the case may be repaired with a thread repair kit.*

## Task 12   Inspect and replace speedometer drive gear, driven gear, and retainers.

44. In some transaxles, the speedometer drive gear is mounted on

    A. the input shaft.

    B. the transfer gear.

    C. the differential case.

    D. the drive axle inner hub.

**Hint**     *Some manual transaxles have a combined speedometer drive and speed-distance sensor. The speedometer driven gear should be inspected for worn teeth and looseness. Be sure the adapter is not loose on the speed-distance sensor extension. A worn adapter may cause erratic speedometer operation. During a transmission overhaul, replace the speedometer drive O-ring. Inspect the speedometer drive gear for worn or damaged teeth and looseness on the differential case.*

## Task 13   Inspect, test, and replace transaxle sensors and switches.

45. While discussing a reverse lockout system in a manual transaxle, Technician A says the reverse lockout system is activated by a signal from the neutral position switch.

    Technician B says the reverse lockout system prevents the driver from shifting into reverse at vehicle speeds above 3 mph.

    Who is correct?

    A. A only

    B. B only

    C. Both A and B

    D. Neither A nor B

**Hint**     *Some manual transaxles have a backup light switch, a neutral position switch, a reverse lockout solenoid, a vehicle speed sensor (VSS), and a differential speed sensor. At vehicle speeds below 16 mph, the signal from the VSS informs the PCM to turn off the reverse lockout solenoid. Under this condition, the driver may shift the transaxle into reverse gear. If the vehicle speed exceeds 19 mph, the VSS signal informs the PCM to energize the reverse lockout solenoid. Under this condition, an interlock shift piece is moved against the reverse shift piece so the driver cannot shift into reverse (Figure 3–36). This action helps prevent transaxle damage.*

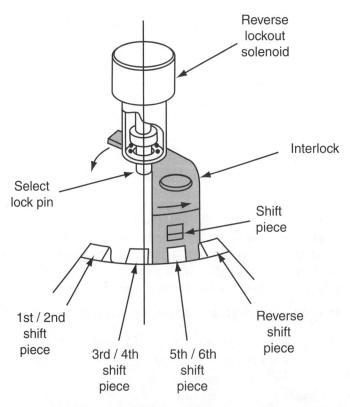

**Figure 3–36** Reverse lockout solenoid and related mechanism.

## Task 14   Diagnose differential assembly noise and vibration problems; determine needed repairs.

46. A manual transaxle chatters while driving straight ahead.

    Technician A says the ring gear and pinion gear teeth may be worn and chipped.

    Technician B says there may be improper preload on the differential components.

    Who is correct?

    A. A only

    B. B only

    C. Both A and B

    D. Neither A nor B

*Hint*     *A clicking noise in all gears when the vehicle is driven straight ahead may be caused by damaged differential ring gear or drive pinion teeth. Loose ring gear bolts may cause a knocking noise when the vehicle is driven straight ahead. Differential chatter when the vehicle is driven straight ahead may result from incorrect preload on differential components. Differential gear chuckle may be caused by worn or damaged differential thrust washers, loose side gears in the case, damaged ring gear teeth, or loose ring gear bolts.*

*A whining noise that changes with acceleration, deceleration, and steady throttle may be caused by worn differential ring gear and drive pinion teeth. Worn differential bearings may cause a growling noise when the vehicle is driven. If a clicking noise occurs while cornering, the differential side gears and pinion gears may have damaged teeth.*

*Differential vibration may be caused by worn or broken transaxle or engine mounts. Damaged or galled differential bearings, or damaged ring gear teeth, may be the cause of differential vibration.*

*These problems also may cause the transaxle noises mentioned in the preceding paragraph.*

## Task 15   Remove and replace differential final drive assembly.

47. While discussing differential side bearing preload in the transaxle case in Figure 3–37, Technician A says differential bearing preload is adjusted by rotating a threaded adjuster on each side of the differential bearings.

    Technician B says the differential bearing preload is automatically adjusted when the case halves are reassembled.

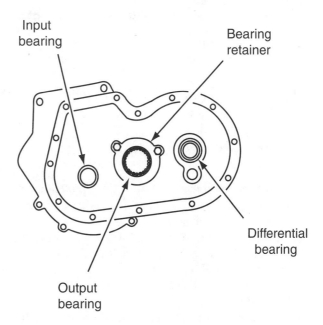

Input
bearing

Bearing
retainer

Differential
bearing

Output
bearing

Figure 3–37 Differential bearing cup in the clutch bell housing side of the transaxle case.

Who is correct?

A. A only

B. B only

C. Both A and B

D. Neither A nor B

**Hint**    *In many transaxles, after the halves of the case have been separated, the differential assembly may be lifted from the case (Figure 3–38). A differential bearing cup is pressed into each half of the case, and in some transaxles a shim for bearing preload adjustment is positioned behind the bearing cup in the clutch bell housing side of the case.*

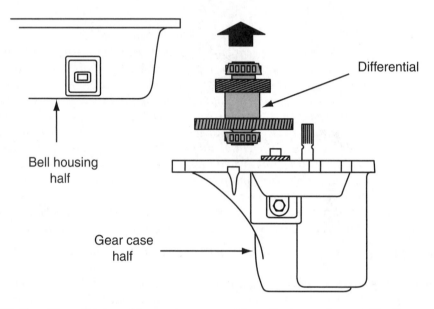

Bell housing half

Differential

Gear case half

**Figure 3–38** The differential assembly may be removed from the transaxle case after the case halves are separated.

## Task 16   Inspect, measure, adjust, and replace differential pinion gears (spiders), shaft, side gears, thrust washers, and case.

48. While determining the proper differential side gear thrust washer thickness

A. measure the end play on one side gear to calculate the side gear spacer washer thickness.

B. the side gear end play is measured with the thrust washers behind the gears.

C. the correct thickness of the side gear thrust washer provides the specified side gear end play.

D. the correct thickness of the side gear thrust washer provides a slight side gear preload.

**Hint**    *Use a hammer and punch to drive the pinion shaft roll pin from the differential case. Remove the pinion shaft, gears, side gears, and thrust washers.*

*Reassemble the differential side gears, pinion gears, pinion gear thrust washers, and shaft. Do not install the side gear thrust washers. Install the roll pin to retain this shaft in the differential case. Rotate the side gears two revolutions in each direction. Install a special tool through one of the axle openings against the side gear and assemble a dial indicator so the stem rests against the tool (Figure 3–39). Move the side gear up and down and record the end play on the dial indicator.*

*Rotate the side gear 90 degrees and record another end play reading. Turn the side gear another 90 degrees and record a third end play reading. Use the smallest end play recorded to*

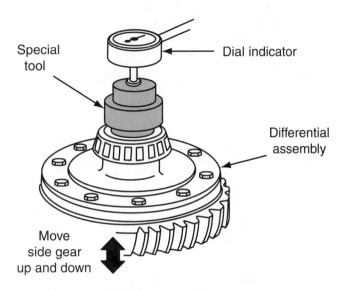

**Figure 3–39** Side gear end play measurement.

*calculate the required side gear washer thickness. The side gear end play must be 0.001 in. to 0.013 in. If the end play recorded was 0.050 in., install a 0.042 in. washer, which provides an end play of 0.008 in.*

*Repeat the side gear end play measurement on the opposite side gear. Disassemble the side gears and pinion gears and reassemble these components with the required side gear thrust washers.*

## Task 17  Inspect and replace differential side bearings; inspect case.

49. Technician A says a preload adjustment shim is positioned between both differential bearings and the differential case.

Technician B says one of the differential bearings must be removed before the speedometer drive gear.

Who is correct?

A. A only

B. B only

C. Both A and B

D. Neither A nor B

**Hint**  *On some transaxles, a special puller is used to remove the bearings from the differential case.*

*These bearings must be installed on the differential case with a special driving tool. The differential bearing cups are removed from the case with special tool (Figure 3–40). A preload shim is located behind one of these bearing cups.*

## Task 18  Measure shaft preload/end play (shim/spacer selection procedure).

50. While discussing input and output shaft preload in manual transaxles, Technician A says in some transaxles selective shims are positioned behind the tapered roller bearings on the input and output shafts.

Technician B says that in some transaxles selective shims are positioned behind the outer bearing races on the input and output shafts to obtain the correct preload.

Who is correct?

A. A only

B. B only

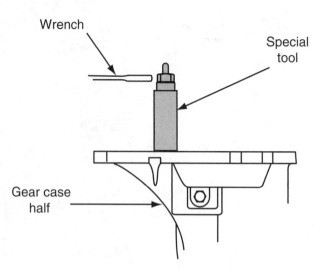

**Figure 3–40** Using a special puller to remove a differential bearing cup from the case.

C. Both A and B

D. Neither A nor B

**Hint**    *In some manual transaxles, shims are positioned behind the bearing races on the input shaft, output shaft, and differential. When assembling the transaxle, the proper shim thickness must be selected. After the input shaft, output shaft, and differential are assembled, special tools are placed over the input shaft, output shaft, and differential outer bearing races. The transaxle halves are assembled with special spacers positioned between the two halves. Special retaining bolts are installed to retain the transaxle halves, and these bolts are tightened to the specified torque. Various selective shim sizes are used to measure the distance between the outer sleeve and the base pad on each special tool (Figure 3–41). The input shaft shim should be two sizes smaller than the largest shim that fits in the tool gap. The differential shim should be three sizes larger than the shim that fits smoothly in the tool gap. The output shaft requires the thickest shim that fits in the tool gap. After the shim selection, the two transaxle halves are separated and the spacers and tools removed. Place the selected shims behind the appropriate bearing race bores in the transaxle case and reassemble the transaxle.*

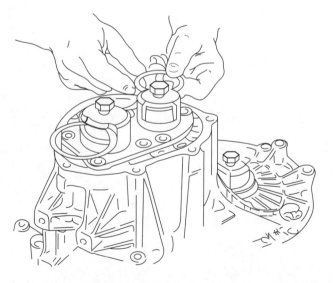

**Figure 3–41** Selecting shims for input shaft, output shaft, and differential.

*After the transaxle is assembled, the input shaft must have the specified turning torque if the shim selections are correct.*

## Task 19   Inspect lubrication systems.

**Hint**     *Some manual transaxles have an oil feed trough that maintains an adequate oil supply to the end bearings (Figure 3–42). Inspect the oil feed trough for damage or a bent condition. A damaged or bent oil feed trough may reduce lubricant flow to the end bearings and cause premature bearing failure. Some output shaft front bearings have an oil feeder behind the bearing to maintain the proper supply of lubricant to the bearing and the shaft supported by the bearing (Figure 3–43).*

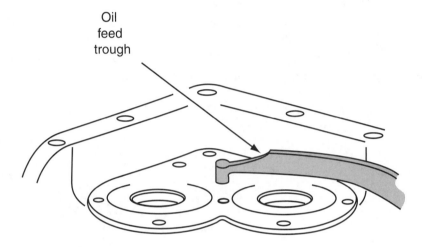

**Figure 3–42** Lubricant oil trough.

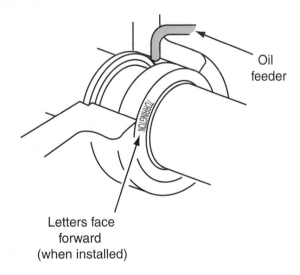

**Figure 3–43** Bearing oil feeder.

## Task 20   Check fluid level, and refill with proper fluid.

51. The needle bearings between the output shaft and output shaft gears are scored and blue from overheating.
    Technician A says the transaxle may have been filled with the wrong lubricant.
    Technician B says the projection is broken off on the oil feeder behind the front output shaft bearing.

Who is correct?
A.  A only
B.  B only
C.  Both A and B
D.  Neither A nor B

**Hint**    *Some manual transaxles require 30 W engine ATF or 90 W gear oil. Some manual transaxles have a dipstick to indicate the proper lubricant level. On other manual transaxles, the lubricant should be level with the bottom of the filler plug hole.*

## Task 21    Measure and adjust differential bearing preload/end play.

52.  During the measurement in Figure 3–44
   A.  a new bearing cup is installed with a shim in the clutch bell housing side of the transaxle case.
   B.  a medium load should be applied to the differential in the upward direction.
   C.  the proper shim thickness is equal to the differential end play recorded on the dial indicator.
   D.  the bolts between the transaxle case halves must be tightened to one-half the specified torque.

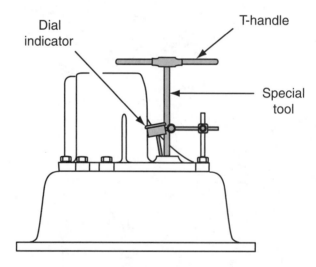

**Figure 3–44** Measuring differential end play.

**Hint**    *After the differential bearing cup and shim have been removed from the clutch bell housing half of the transaxle case, use the proper driving tool to install a new bearing cup without the shim. Install a new bearing cup in the gear case half of the case with the proper driver. Lubricate the differential bearings with the manufacturer's specified transaxle lubricant. Install the differential assembly and reassemble the case halves. Tighten the case bolts to the specified torque.*

*Install the special tool into the differential side gears and install a T-handle on top of this tool. Apply downward pressure to the T-handle and rotate this handle several times in both directions to seat the side bearings. Zero the dial indicator installed against the upper side of the differential case. Apply a medium load to the differential in the upward direction through the opposite drive axle opening. Rotate the differential back and forth with the T-handle while maintaining this upward pressure. Record the end play indicated on the dial indicator.*

*The proper preload shim thickness is equal to the end play plus 0.007 in. (0.18 mm) to provide side bearing preload.*

*After the case is assembled install the special turning tool in the differential in place of the T-handle used for the end play measurement. Install an inch-pound torque wrench on this tool and rotate the differential back and forth while observing the turning torque on the torque wrench. If the turning torque is more than specified, reduce shim thickness by 0.002 in. (0.5 mm). When the turning torque is less than specified, increase the shim thickness by 0.002 in. (0.5 mm).*

# Drive Shaft/Half-Shaft and Universal Joint/Constant Velocity (CV) Joint Diagnosis and Repair (Front and Rear Wheel Drive)

## ASE Tasks, Questions, and Related Information

**Task 1** **Diagnose shaft and universal/CV joint noise and vibration problems; determine needed repairs.**

53. While discussing drive shaft and universal joint diagnosis in rear-wheel-drive vehicles, Technician A says a worn universal joint may cause a squeaking noise that decreases in relation to vehicle acceleration.

    Technician B says a heavy vibration that only occurs during acceleration may be caused by a worn centering ball and socket on a double Cardan U-joint.

    Who is correct?

    A. A only

    B. B only

    C. Both A and B

    D. Neither A nor B

*Hint*    *A clicking noise while cornering at low speed may be caused by a worn outer CV joint. A clunking noise during acceleration or deceleration may be caused by a defective inboard CV joint. When a growling noise or clicking noise is coming from one front wheel, check the bearing and wheel rim in that wheel. A vibration during acceleration may be caused by a worn inboard CV joint. Vibration at highway speeds may be caused by improper wheel balance or worn CV joints.*

**Task 2** **Inspect, service, and replace shafts, yokes, boots, and universal/CV joints; verify proper phasing.**

54. A front-wheel-drive car has a clunking noise while decelerating.

    Technician A says this noise may be caused by a worn inner drive axle joint.

    Technician B says this noise may be caused by a worn front-wheel bearing.

    Who is correct?

    A. A only

    B. B only

    C. Both A and B

    D. Neither A nor B

*Hint*    *Most CV joints are replaced as a complete assembly. Prior to removing the CV joint boot, mark the inner end of the boot in relation to the drive axle so the boot may be installed in the original position. When reassembling the joint, always install all the grease in the joint that is provided in the repair kit.*

*Prior to drive shaft removal, always mark the drive shaft in relation to the differential flange so this shaft may be installed in the original position. After the universal joint retaining clips are removed from the drive shaft, a vice and the proper size of socket may be used to remove the spider from the yoke.*

*In a double Cardan U-joint, the center yoke should be marked in relation to the ball tube yokes prior to disassembly so these components may be assembled in their original location (Figure 3–45). The bearing caps in this type of joint should be removed in the proper sequence using the same procedure as followed for a single U-joint. The centering ball and ball seats must be replaced if they are worn or scored.*

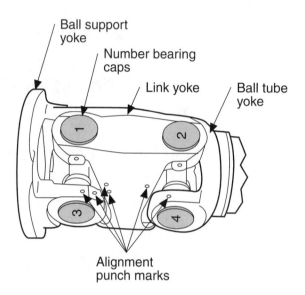

**Figure 3–45** Marking the center yoke in relation to the ball tube yokes in a double Cardan U-joint.

## Task 3   Inspect, service, and replace shaft center support bearings.

55. A light-duty rear-wheel-drive truck has a growling noise that is not influenced by acceleration and deceleration.

    Technician A says an outer rear axle wheel bearing may be rough and worn.

    Technician B says the drive shaft center support bearing may be rough and worn.

    Who is correct?

    A. A only

    B. B only

    C. Both A and B

    D. Neither A nor B

**Hint**   *When the center support bearing is removed from the drive shaft, it should be inspected for looseness and roughness. Inspect the center support bearing cushion for wear, oil soaking, and deterioration. Check for damaged or bent upper and lower center support bearing brackets. Some center support bearings have an F marked on the front side of the bearing.*

## Task 4   Check and correct drive/propeller shaft balance.

56. The markings in Figure 3–46 are required for

    A. drive shaft removal and replacement.

    B. universal joint removal and replacement.

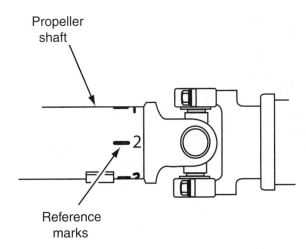

Propeller
shaft

Reference
marks

**Figure 3–46** Chalk mark the drive shaft at four locations spaced 90 degrees apart near the drive shaft balance weight.

    C. drive shaft runout measurement.

    D. drive shaft balance testing.

*Hint*    *Prior to drive shaft balancing, always inspect this shaft for damage, a missing balance weight, or an accumulation of dirt or undercoating. Remove both rear wheel assemblies and install the wheel nuts on the wheel studs with the flat side toward the brake drums. Chalk mark the drive shaft at four locations 90 degrees apart just forward of the drive shaft balance weight. Fabricate a tool to hold the strobe light pickup against the rear axle housing just behind the pinion yoke.*

    *Run the vehicle in gear until the drive shaft vibration is most severe. Because 55 mph indicated on the speedometer has a possible tire speed of 110 mph, do not exceed 55 mph. Point the strobe light at the chalk marks on the drive shaft and note the position of one reference mark.*

    *Gently apply the brakes and shut off the engine. Rotate the drive shaft until the chalk mark is in the same position as it appeared under the strobe light. Install two screw-type hose clamps on the drive shaft near the rear of the shaft. Position these clamps so both heads are beside each other and directly opposite the number that appeared under the strobe light.*

## Task 5   Measure drive shaft runout.

57. Technician A says the dial indicator should be positioned near the front of the drive shaft to measure drive shaft runout.

    Technician B says if the drive shaft runout is excessive, the drive shaft may be straightened in a hydraulic press.

    Who is correct?

    A. A only

    B. B only

    C. Both A and B

    D. Neither A nor B

*Hint*    *A dial indicator must be mounted to the vehicle underbody near the center of the drive shaft to measure drive shaft runout. Position the dial indicator stem against the surface of the drive shaft. Be sure the drive shaft surface is clean and undamaged. Rotate the drive shaft one revolution to measure the runout. When the runout exceeds specifications, replace the drive shaft and recheck the runout. If the runout is still excessive, check for a bent U-joint flange or slip yoke.*

## Task 6   Measure and adjust drive shaft working angles.

58. A rear-wheel-drive vehicle has a vibration that increases in relation to vehicle speed.

    Technician A says the balance pad may have fallen off the drive shaft.

    Technician B says some of the wheels may be out of balance.

Who is correct?
A. A only
B. B only
C. Both A and B
D. Neither A nor B

**Hint**     *Raise the vehicle on a lift so the rear wheels are free to rotate. Clean the outer surface of the bearing caps in the front and rear U-joints. Be sure one of the rear U-joint bearing caps is facing straight down. Install the magnetic end of the inclinometer on the U-joint bearing cap facing downward, and rotate the inclinometer adjusting knob until the weighted cord is centered on the scale (Figure 3–47). Remove the inclinometer and turn the drive shaft 90 degrees. Install the inclinometer on the U-joint bearing cap facing downward, and record the degree reading where the weighted cord appears on the scale. The drive shaft angle is the difference between the zero reading and the reading when the shaft is rotated 90 degrees.*

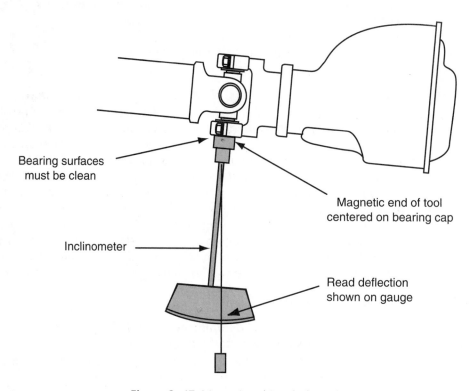

Bearing surfaces
must be clean

Magnetic end of tool
centered on bearing cap

Inclinometer

Read deflection
shown on gauge

**Figure 3–47** Measuring drive shaft angle.

*Repeat the procedure on the front U-joint to obtain the front drive shaft angle. If the drive shaft angles are not within specifications, inspect the engine and transmission mounts for proper position, wear, looseness, breaks, and deterioration. Inspect the rear suspension mounting bushings and arms for wear, looseness, or a bent condition.*

**Task 7    Inspect, service, and replace front wheel bearings, seals, and hubs.**

59. After servicing and lubricating the rear-wheel bearings in a front-wheel-drive car, the bearing adjusting nut is tightened to 20 ft.-lb., and loosened one-half turn. The next step in the bearing adjusting procedure is to
   A. install the nut retainer and cotter key.
   B. tighten the adjusting nut so the cotter key hole lines up properly.
   C. tighten the adjusting nut to 10 to 15 in.-lbs.
   D. check the wheel bearing for free play.

*Hint*     *On some front-wheel-drive vehicles, the front- and rear-wheel bearing hubs are serviced only as a complete assembly. The end play on these wheel bearing hubs may be measured with a dial indicator. If the end play exceeds specifications, the hub must be replaced. Loose front wheel bearings may cause steering wander. The front wheel hubs on many rear-wheel-drive cars, and the rear-wheel hubs on some front-wheel-drive cars contain two tapered roller bearings. These wheel bearings may be removed, cleaned, inspected, repacked with lubricant, and adjusted. When two tapered roller bearings are mounted in the front- or rear-wheel hub, a typical bearing adjustment procedure consists of the following:*
  1. *Tighten the bearing adjustment nut to 17 to 25 ft.-lbs.*
  2. *Back off the adjustment nut one-half turn.*
  3. *Tighten the bearing adjustment nut to 10 to 15 in.-lbs.*
  4. *Install the nut retainer and cotter key.*

# Rear Wheel Drive Axle Diagnosis and Repair, Ring and Pinion Gears

## ASE Tasks, Questions, and Related Information

**Task 1**  **Diagnose noise, vibration, and fluid leakage problems; determine needed repairs.**

60. While driving a vehicle straight ahead, the differential produces a whining noise.
    Technician A says the differential side gears are damaged.
    Technician B says the ring gear and pinion adjustments may be incorrect.
    Who is correct?
    A. A only
    B. B only
    C. Both A and B
    D. Neither A nor B

*Hint*     *A whining noise that changes with acceleration, deceleration, or steady throttle may be caused by defective ring and drive pinion gears. A growling noise from one rear wheel usually indicates a defective rear wheel bearing. A ticking noise from the differential while driving the vehicle may indicate damaged ring and drive pinion gears. Differential chatter may be caused by improper side bearing preload. A clicking or clunking noise while cornering may be caused by damaged side gears and pinion gears. Loose ring gear bolts cause a knocking with the vehicle in motion. Gear chuckle may be caused by loose ring gear bolts, lack of lubricant, or damaged ring gear and drive pinion teeth. Vibration that changes with acceleration and deceleration may be caused by improper drive shaft angles or drive shaft imbalance.*

**Task 2**  **Inspect and replace companion flange and pinion seal; measure companion flange runout.**

61. Technician A says that insufficient pinion nut torque may cause a clunking noise during acceleration or deceleration.
    Technician B says that insufficient pinion nut torque may cause a growling noise with the vehicle in motion.

Who is correct?
A. A only
B. B only
C. Both A and B
D. Neither A nor B

**Hint**   *If there is evidence of fluid leakage in the pinion seal area, the pinion seal must be replaced. A special holding tool is required to hold the differential flange while loosening the pinion nut (Figure 3–48).*

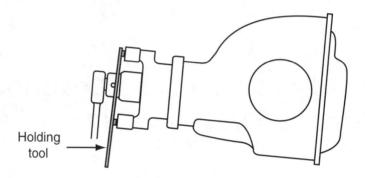

Holding
tool

**Figure 3–48** Differential flange holding tool.

*Coat the outside diameter of the new seal with gasket sealer, and lubricate the seal lips with the manufacturer's specified differential lubricant. Use the proper seal driver to install the pinion seal. Install the differential flange and a new pinion nut. Since the pinion nut torque determines the pinion bearing preload, the torque on this nut is critical. Tighten the pinion nut until the manufacturer's specified turning torque is obtained with an inch-pound torque wrench and socket installed on the pinion nut. When the pinion nut is tightened to obtain the proper preload and turning torque with the original collapsible spacer, a typical turning torque is 6 in.-lbs more than the specified turning torque.*

*Mount a dial indicator on the vehicle chassis, and position the indicator stem against the differential flange. Rotate the flange and observe the flange runout on the dial indicator. A damaged flange with excessive runout may cause a vibration while driving at a constant speed because this problem causes incorrect drive shaft angles.*

## Task 3  Measure ring gear runout; determine needed repairs.

62. Excessive runout on the dial indicator in Figure 3–49 may be caused by excessive
A. differential case runout.
B. side bearing preload.
C. side gear end play.
D. ring gear bolt torque.

**Hint**   *The ring gear runout should be measured prior to disassembling the differential. Mount the dial indicator assembly, and position the dial indicator stem at a 90-degree angle against the back of the ring gear. Turn the dial indicator to zero and rotate the ring gear one revolution. The difference between the highest and lowest dial indicator reading is the ring gear runout.*

*To determine if excessive ring gear runout is caused by the ring gear or the case, remove the ring gear and case and remove the ring gear from the case. Install the case assembly without the ring gear, and be sure the side bearings are in good condition and properly torqued. Position the dial indicator against the back side of the case, and rotate the case one revolution to measure the case runout. If the case runout is normal but the ring gear runout is excessive, replace the ring gear and pinion. When the case runout is excessive, replace the case.*

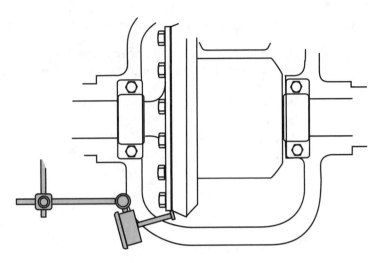

**Figure 3–49** Measuring ring gear runout.

## Task 4   Inspect and replace ring and pinion gear set, collapsible spacers, sleeves (shims), and bearings.

63. All of these statements about ring and pinion gears are true EXCEPT
    A. Hunting-type ring and pinion gear sets must be timed.
    B. Loose ring gear bolts may cause gear chuckle or knocking noise while the vehicle is in motion.
    C. Damaged ring gear and pinion gear teeth may cause a ticking noise while driving the vehicle.
    D. The grooved, painted tooth on the pinion gear must be meshed with the painted notched ring gear teeth on some gear sets.

**Hint**    *If the ring gear teeth are damaged, the ring gear and pinion gear must be replaced as a set. When installing a new ring gear on the case, be sure the bolt holes are aligned in these components.*

*Pilot studs may be used to ensure bolt hole alignment.*

*Many ring gear and pinion gear sets have timing marks that must be aligned when assembling the differential. On some gear sets, one pinion gear tooth is grooved and painted, and the ring gear has a notch between two painted teeth. When the ring and pinion gears are meshed together, the grooved and painted pinion gear tooth must fit between the notched and painted ring gear teeth. Some ring gear and pinion gear sets do not have timing marks. These gear sets are referred to as hunting gears.*

*If the pinion gear requires replacement, always replace the pinion gear and ring gear as a set.*

*When the pinion bearing must be replaced, use a press plate and press to remove the rear pinion bearing. A toothed ring for the ABS is pressed onto the shaft in some differentials. This toothed ring must be removed with a press plate and press prior to removing the rear pinion bearing.*

*Because removal of this toothed ring may destroy its friction fit on the pinion shaft, most manufacturers recommend replacement of this ring if it is removed.*

*Before pressing the new rear pinion bearing onto the pinion shaft, be sure the proper spacer washer is installed behind the pinion bearing. The shim determines the pinion depth. Press a new ABS toothed ring onto the pinion shaft and install a new collapsible sleeve on the shaft.*

*When the differential has been disassembled and overhauled, a new collapsible spacer and pinion nut must be installed. Tighten the pinion nut a small amount at a time until the specified turning torque is obtained with an inch-pound torque wrench and socket installed on the pinion nut. Never loosen the pinion nut to obtain the correct turning torque.*

## Task 5   Measure and adjust drive pinion depth.

64. In Figure 3–50 a 0.063 in. shim that fits between the gauge block and the gauge tube with a light drag and the pinion gear is marked +3. The proper pinion depth shim is
   A. 0.060 in.
   B. 0.063 in.
   C. 0.066 in.
   D. 0.069 in.

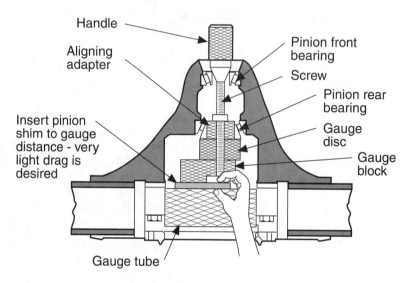

**Figure 3–50** Selecting proper pinion depth shim with an arbor and gauge tube.

65. In Figure 3–51, after the dial indicator is rotated to the zero position with the stem on the gauge plate, when the dial indicator stem is moved off the gauge plate, the dial indicator pointer moves 0.057 in. counterclockwise and the pinion gear is marked minus 4. The proper pinion depth shim is
   A. 0.039 in.
   B. 0.041 in.
   C. 0.042 in.
   D. 0.043 in.

**Hint**     *Pinion depth is the distance from the nose of the pinion gear to the centerline of the axles or differential case (Figure 3–52). This depth normally is adjusted with the shim thickness behind the rear pinion bearing. Some vehicle manufacturers recommend measuring the pinion depth with an arbor and gauge tube.*

*Other manufacturers recommend measuring the pinion depth with a gauge set and dial indicator. The gauge set is installed in the pinion bearings, and a gauge shaft and discs are mounted in the side bearing openings. After the nominal pinion depth shim is determined, check the markings on the rear or side of the pinion shaft ahead of the rear pinion bearing. If the pinion is marked +3, add 0.003 in. (0.025 mm) to the nominal shim thickness. When the pinion is marked with a minus value, subtract this amount from the nominal shim thickness.*

## Task 6   Measure and adjust drive pinion bearing preload (collapsible spacer or shim type).

66. Technician A says the pinion bearings should be lubricated when the pinion turning torque is measured.

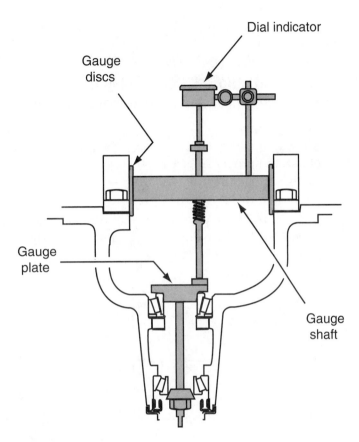

**Figure 3–51** Selecting proper pinion depth shim with a gauge set and dial indicator.

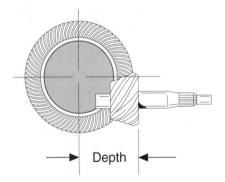

**Figure 3–52** Pinion depth. *(Used with permission from Nissan North America, Inc.)*

Technician B says the pinion nut may be loosened to obtain the specified turning torque.

Who is correct?

A.  A only

B.  B only

C.  Both A and B

D.  Neither A nor B

**Hint**    *The pinion bearing preload is measured with an inch-pound torque wrench and socket installed on the pinion nut. Prior to measuring the pinion turning torque with an inch-pound torque wrench, the pinion shaft assembly should be installed with the bearings lubricated and a new collapsible spacer Axles the proper pinion depth shim. A new pinion shaft nut should be installed. Tighten the pinion nut gradually and keep measuring the turning torque. When the* *specified turning torque is obtained, the pinion bearing preload is correct. Never loosen the pinion nut to obtain the proper turning torque. If the pinion nut is overtightened and the turning torque is excessive, install a new collapsible spacer and repeat the procedure.*

**Task 7**   **Measure and adjust differential (side) bearing preload, and ring and pinion backlash (threaded adjuster or shim type).**

67. In Figure 3–53, the ring gear backlash and side play are zero. Right- and left-side-bearing adjusting nuts are determined while facing the differential from the rear. To obtain the proper ring gear backlash
    A. tighten the right- and left-side-bearing adjusters.
    B. loosen the left-side-bearing adjuster.
    C. loosen the left-side-bearing adjuster and tighten the right-side-bearing adjuster.
    D. loosen the right-side-bearing adjuster and tighten the left-side-bearing adjuster.

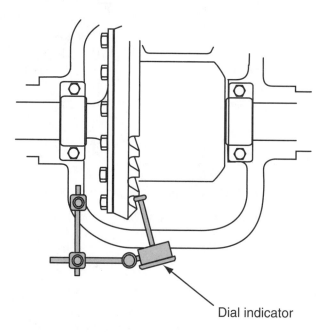

Dial indicator

**Figure 3–53** Measuring ring gear backlash.

*Hint*     *The backlash between the ring gear and pinion gear is adjusted after the differential is assembled.*

*Mount a dial indicator on the differential housing, and position the dial indicator stem against one of the ring gear teeth. Turn the dial indicator to zero and rock the ring gear back and forth against the pinion gear teeth. The ring gear backlash is indicated on the dial indicator.*

*Vehicle manufacturers usually recommend measuring the backlash at several locations around the ring gear.*

*Side-bearing preload limits the amount of lateral differential case movement in the axle housing or carrier. When the differential has threaded adjusters on the outside of the side bearings, loosen the right adjuster and tighten the left adjuster to obtain zero backlash. Turn the right adjuster the specified amount to obtain the proper preload. Then rotate each adjuster the same amount in opposite directions to obtain the specified backlash.*

*When shims are positioned behind the side bearings to adjust backlash and preload, the differential case is pried to one side and the movement is recorded with a dial indicator or feeler gauge. Service spacers, shims, and feeler gauges are installed on each side of the side bearings to obtain zero side play and zero backlash. Then calculate the proper shim thickness to provide the specified backlash and side-bearing preload (Figure 3–54). On some differentials,*

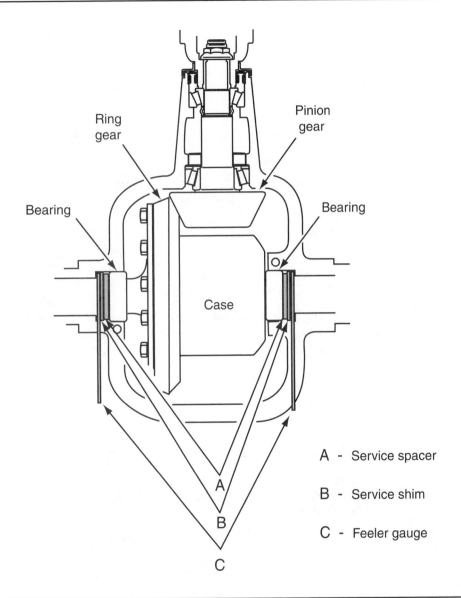

A - Service spacer

B - Service shim

C - Feeler gauge

| EXAMPLE | | |
|---|---|---|
| 0.250" | **Ring gear side** Combined total of: Service spacer (A) Service shim (B) Feeler gauge (C) | **Opposite side** Combined total of: Service spacer (A) Service shim (B) Feeler gauge (C) 0.265" |
| −0.010" / 0.240" | To maintain proper backlash (0.005" - 0.008") ring gear is moved away from the pinion by subtracting 0.010" shim from ring gear side and adding 0.010" to the other side | +0.010" / 0.275" |
| +0.004" | To obtain proper preload on side bearings, add 0.004" shim to each side | +0.004" |
| 0.244" | Shim dimension required for ring gear side | Shim dimension required for opposite side 0.279" |

**Figure 3–54** Adjusting ring gear backlash and side bearing preload with shims.

*the proper shims have to be driven into place behind the side bearings with a special tool and a soft hammer.*

## Task 8    Perform ring and pinion tooth contact pattern checks; determine needed adjustments.

68. Technician A says if the ring tooth contact pattern indicates pinion tooth contact on the toe of the pinion gear, the pinion gear should be moved toward the ring gear.

    Technician B says if the pinion gear teeth have low flank contact on the ring gear teeth, the pinion gear should be moved toward the ring gear.

    Who is correct?
    A.  A only
    B.  B only
    C.  Both A and B
    D.  Neither A nor B

**Hint**    *Observe the tooth contact pattern on the ring gear. This pattern should be centered on the drive side of the ring gear teeth. If the ring gear tooth contact pattern is not correct, the drive pinion has to be moved in relation to the ring gear, or the backlash adjusted to provide the correct pattern (Figure 3–55).*

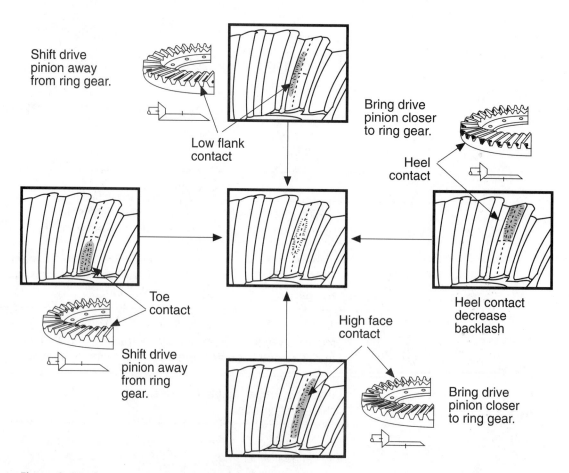

**Figure 3–55** Correct and incorrect ring gear tooth contact patterns. *(Used with permission from Nissan North America, Inc.)*

# Rear Wheel Drive Axle Diagnosis and Repair, Differential Case/Carrier Assembly

## ASE Tasks, Questions, and Related Information

**Task 1**  Diagnose differential assembly noise and vibration problems; determine needed repairs.

69. When diagnosing a rear-wheel-drive differential that vibrates only while turning a corner Technician A says the pinion bearings may be worn and the pinion bearing preload is less than specified.

    Technician B says the bearing surfaces between the side gears or bevel pinions and the differential case may be damaged.

    Who is correct?

    A. A only

    B. B only

    C. Both A and B

    D. Neither A nor B

    Shift drive pinion away from ring gear.

    Toe contact Low flank contact High face contact Bring drive pinion closer to ring gear.

    Heel contact decrease backlash Heel contact Bring drive pinion closer to ring gear.

    Shift drive pinion away from ring gear.

**Hint**  *A whining noise that changes with acceleration, deceleration, or steady throttle may be caused by defective ring and drive pinion gears. A growling noise from one rear wheel usually indicates a defective rear wheel bearing. A ticking noise from the differential while driving the vehicle may indicate damaged ring and drive pinion gears. Differential chatter may be caused by improper side bearing preload. A clicking or clunking noise while cornering may be caused by damaged side gears and pinion gears. Loose ring gear bolts cause a knocking with the vehicle in motion. Gear chuckle may be caused by loose ring gear bolts, lack of lubricant, or damaged ring gear and drive pinion teeth. Vibration that changes with acceleration and deceleration may be caused by improper drive shaft angles or drive shaft imbalance.*

**Task 2**  Remove and replace differential assembly.

70. All of the following statements about differential case and ring gear assembly removal and replacement are true EXCEPT

    A. The ring gear runout should be measured before removal of the case and ring gear assembly.

    B. The case side play should be measured before removal of the case and ring gear assembly.

    C. The side bearing caps should be marked in relation to the housing before removal of the case and ring gear assembly.

    D. The side bearings should be clean and dry before installation of the case and ring gear assembly.

**Hint**  *Measure the ring gear runout and differential case side play before removing the differential case and ring gear. Be sure to mark the side bearing caps in relation to the housing prior to removal. Remove the side-bearing caps and lift the case and ring gear from the housing.*

*Prior to installing the case and ring gear assembly, the side bearings and bearing cups must be lubricated with the vehicle manufacturer's recommended differential lubricant. Install the case and ring gear assembly with the bearing cups and shims or adjuster nuts. Be sure the bearing cap threads are properly seated in the adjuster nut threads when the bearing caps are installed. Tighten the bearing cap bolts to the specified torque.*

**Task 3**  **Inspect, measure, adjust and replace differential pinion gears (spiders), shaft, side gears, thrust washers, and case/carrier.**

*Hint*    *Inspection and side-play measurement of the side gears, pinion gears, thrust washers, and case in a rear-wheel-drive differential is basically the same as the inspection of these components in a transaxle differential. Refer to Task 16 in Transaxle Diagnosis and Repair for this inspection.*

**Task 4**  **Inspect and replace differential side bearings, inspect case/carrier.**

*Hint*    *The inspection and replacement procedure for differential side bearings in a rear-wheel-drive differential is basically the same as the side-bearing inspection and replacement in a transaxle.*
*This inspection and replacement procedure is explained in Task 17 of Transaxle Diagnosis and Repair.*

**Task 5**  **Measure differential case/carrier runout; determine needed repairs.**

71. Technician A says an accurate differential case runout measurement may be performed with scored side bearings.
Technician B says the ring gear runout should be measured before the case runout.
Who is correct?
A. A only
B. B only
C. Both A and B
D. Neither A nor B

*Hint*    *As mentioned previously, the ring gear runout should be measured prior to removing the differential case and ring gear assembly. This procedure is explained in Task 3 of Ring and Pinion Gear Diagnosis. If the ring runout is excessive, the case runout should be measured.*

**Task 6**  **Inspect axle housing and vent.**

*Hint*    *The rear axle housing vent should be inspected to be sure it is not plugged or restricted. A plugged vent may cause pressure buildup in the axle housing, as well as seal or gasket leakage. The axle housing should be inspected for cracks, damage, warpage, and proper mounting.*

# Rear Wheel Drive Axle Diagnosis and Repair, Limited Slip/Locking Differential

## ASE Tasks, Questions, and Related Information

**Task 1**  **Diagnose limited slip differential noise, slippage, and chatter problems; determine needed repairs.**

72. All of these statements about limited slip differentials are true EXCEPT
   A. Friction plates are splined to the rear axle hub.
   B. Steel plates are splined to the differential case.
   C. Each clutch set contains a preload spring.
   D. A special lubricant is required.

*Limited slip differentials have a set of multiple disc clutches behind each side gear to control differential action. The steel plates in each clutch set are splined to the case, and the friction plates between the steel plates are splined to the side gear clutch hub. Each clutch set has a preload spring that applies initial force to the clutch packs. A steel shim in each clutch set controls preload (Figure 3–56). A friction plate always is placed next to the hub.*

*Limited slip differentials require a special lubricant specified by the vehicle manufacturer. The use of improper lubricant in a limited slip differential may result in differential noise such as chattering while cornering. Clicking while turning a corner may be caused by worn limited slip components such as clutches, preload springs, and shims.*

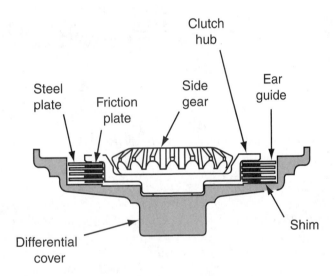

**Figure 3–56** Limited slip differential clutch set.

## Task 2   Inspect, drain, and refill with correct lubricant.

73. A limited slip differential chatters while cornering.
   Technician A says the differential may be filled with the wrong lubricant.
   Technician B says friction and steel plates may be worn and burned.
   Who is correct?
   A. A only
   B. B only
   C. Both A and B
   D. Neither A nor B

**Hint**   *Limited slip differentials should be inspected for fiber and metal cuttings in the bottom of the housing. With the differential components removed, these cuttings may be flushed out of the housing with a solvent gun and an approved cleaning solution. Wipe the bottom of the housing out with a clean shop towel. Always use new gaskets when the differential and the cover are installed in the housing. Fill the differential to the bottom of the filler plug opening with the vehicle manufacturer's specified limited slip differential lubricant.*

## Task 3   Inspect, adjust, and replace clutch (cone/plate) pack or locking assembly.

74. The measurement in Figure 3–57 determines the proper
    A. friction plate thickness.
    B. steel plate thickness.
    C. shim thickness.
    D. preload spring tension.

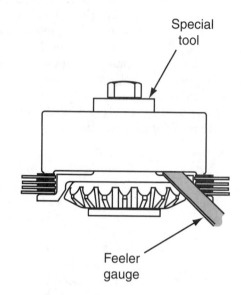

**Figure 3–57** Measuring limited slip differential clutch pack to determine proper preload shim selection.

**Hint**    *Limited slip differential clutch packs should be soaked in the specified lubricant for 30 minutes prior to installation. After each clutch pack is assembled, use the proper gauge and a feeler gauge to determine the correct shim thickness for proper preload.*

*After the clutch packs are assembled in the housing, install a special tool, socket, and torque wrench in one side gear. Hold the opposite side gear and observe the torque required to rotate the side gear. If the turning torque is not within specifications, the problem is in the clutch plates, shim thickness, or preload spring.*

# Rear Wheel Drive Axle Diagnosis and Repair, Axle Shafts

## ASE Tasks, Questions, and Related Information

**Task 1**    **Diagnose rear axle shaft noise, vibration, and fluid leakage problems; determine needed repairs.**

75. Technician A says an axle shaft with excessive runout may cause a vibration while driving at a constant speed of 60 mph (96 kmh).

Technician B says an axle shaft with excessive runout may cause a vibration while accelerating at low speed.

Who is correct?

A. A only

B. B only

C. Both A and B

D. Neither A nor B

**Hint** *Excessive axle shaft runout may cause a constant vibration in a certain speed range. A defective rear axle bearing may cause a growling noise that is not affected by acceleration and deceleration. This noise is most noticeable at low speeds.*

## Task 2   Inspect and replace rear axle shaft wheel studs.

**Hint** *If the threads on the axle studs are damaged, run a die over the threads. When these studs are damaged or bent, they should be replaced. Remove the axle and use a hydraulic press to remove and replace the studs.*

## Task 3   Remove, inspect, and/or replace rear axle shafts, splines, seals, bearings, and retainers.

76. All of the following statements about rear axle shaft and bearing service are true EXCEPT

    A. The retainer should be loosened by striking it with a hammer and chisel.

    B. The bearing should be pressed off the shaft with a hydraulic press.

    C. After the bearing is pressed off the axle, the bearing may be reused.

    D. Press the bearing and retainer onto the axle with a hydraulic press.

**Hint** *After the axle is removed, use a slide hammer-type puller to remove the axle seal from the housing. If the bearing is held on the axle shaft with a retainer, notch the retainer in several places with a hammer and chisel to loosen the retainer. Never heat the retainer during the removal or installation process. Never strike the bearing with a hammer. When the retainer is loosened, the bearing must be pressed off the axle shaft with the proper press plate and a hydraulic press. Press the new bearing and retainer onto the axle shaft.*

*The axle shaft runout may be measured by placing the axle shaft in a pair of V-blocks and positioning a dial indicator against the center of the shaft. Rotate the shaft and observe the runout on the dial indicator.*

*If the axle bearing is pressed into the axle housing, a slide hammer–type puller must be used to remove the bearing.*

## Task 4   Measure rear axle flange runout and shaft end play; determine needed repairs.

77. The cause of excessive runout on the dial indicator in Figure 3–58 could be

    A. a worn axle C-lock.

    B. a bent axle shaft.

    C. a bent differential housing.

    D. a worn rear axle bearing.

**Hint** *Place a dial indicator stem against the axle flange to measure flange runout and end play.*

*Maintain a slight inward pressure on the axle flange and turn the dial indicator to zero. Observe the flange runout while rotating the axle shaft one revolution. If the runout is more than specified, replace the axle shaft.*

*Move the axle inward and outward and observe the dial indicator reading to measure the axle shaft end play. If the end play exceeds specifications, inspect the bearing, retainer plate, or Clock and C-lock groove in the axle.*

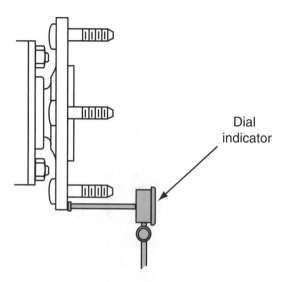

**Figure 3–58** Measuring axle flange runout.

# Four-Wheel Drive/All-Wheel Drive Component Diagnosis and Repair

## ASE Tasks, Questions, and Related Information

**Task 1**  Diagnose drive assembly noise, vibration, shifting, leakage and steering problems; determine needed repairs.

78. While discussing a four-wheel-drive vehicle with a vibration problem that is more noticeable while cornering, Technician A says the U-joints may be worn.
Technician B says the outboard front axle joints may be worn.
Who is correct?
A. A only
B. B only
C. Both A and B
D. Neither A nor B

**Hint**  *A vibration that is most noticeable when changing throttle position may be caused by worn engine mounts, worn U-joints, or improper drive shaft angles. Shifting problems, such as poor engagement, may be caused by defective shift controls, low fluid level, low engine vacuum, or worn shift linkages. If the transfer case jumps out of gear, the shift linkage may be improperly adjusted or the sliding clutch hub teeth may be damaged. Noise from the transfer case may be caused by worn case bearings, damaged gears, low fluid level, or the wrong type of fluid.*

**Task 2**  Inspect, adjust, and repair transfer case manual shifting mechanisms, bushings, mounts, levers, and brackets.

79. A vacuum-shifted 4WD system does not shift into 4WD.
Technician A says the engine vacuum may be low.
Technician B says the vacuum motor at the front axle may be the problem.

Who is correct?

A. A only

B. B only

C. Both A and B

D. Neither A nor B

**Hint**     *A range control linkage adjustment is required on some transfer cases. Place the range control lever in the 2WD position, and place the specified spacer between the gate in the console and the lever. Place the outer lever on the transfer case in the 2WD position. Adjust the linkage so it is in the specified location in relation to the outer lever.*

*The main components in the electric control shift system are the electronic control module, shift motor, panel-mounted switch, speed sensor, shift position sensor, and electromagnetic clutch (Figures 3–59 and 3–60). Use a multimeter to diagnose the system using the vehicle manufacturer's recommended procedure.*

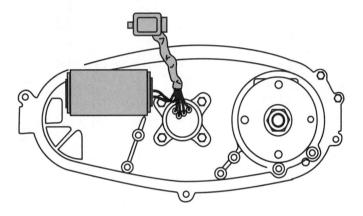

**Figure 3–59** Electric transfer shift control motor and wiring harness.

*Some transfer cases are shifted by engine vacuum. A vacuum motor shifts the transfer case into 4WD, and the vacuum is then diverted to the front axle vacuum motor to lock the front axle into 4WD (Figure 3–61). In these vacuum-operated systems, always be sure the engine vacuum is satisfactory before proceeding with the diagnosis. Check all the vacuum hoses for leaks, kinks, and loose-fitting hoses. When the engine vacuum and hoses are satisfactory, test the vacuum motors at the transfer case and front axle.*

*Inspect all transfer case and engine mounts and brackets for wear, looseness, breaks, oil-soaked conditions, and deterioration. Worn or broken mounts may cause transfer case vibration.*

## Task 3   Remove and replace transfer case.

80. While discussing transfer case removal, Technician A says if the vehicle has a torsion bar front suspension, the torsion bars and rear torsion bar support may have to be removed prior to transfer case removal.

    Technician B says the transfer case lubricant should be drained after the transfer case is removed from the vehicle.

    Who is correct?

    A. A only

    B. B only

    C. Both A and B

    D. Neither A nor B

**Hint**     *Disconnect the negative battery cable. If the vehicle is equipped with an air bag, wait for the time period specified by the vehicle manufacturer before working on the vehicle. The vehicle should be raised on a lift prior to transfer case removal. All safety precautions must be observed regarding lift operation and proper vehicle positioning on the lift. The transfer case drain plug*

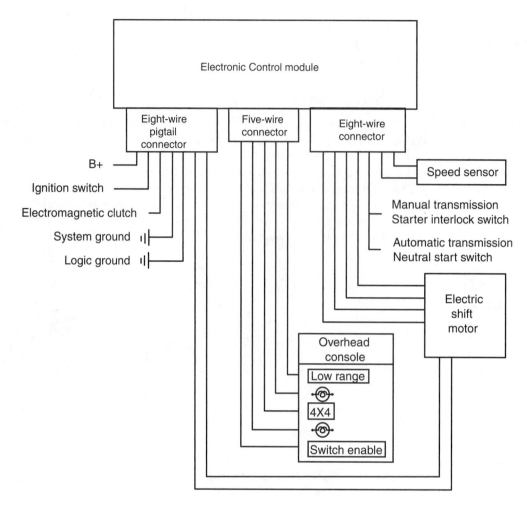

**Figure 3–60** Electric transfer case shift control system.

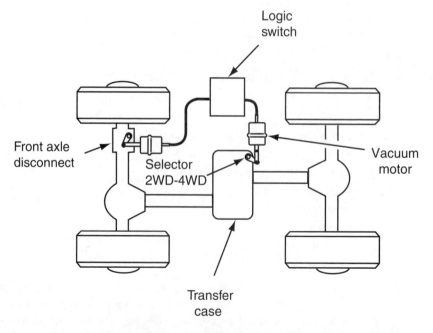

**Figure 3–61** Vacuum-operated transfer case shift control system.

*should be removed and the lubricant drained into a suitable container. The front and rear drive shafts must be marked in relation to their matching flanges, followed by drive shaft removal. If the vehicle has a torsion bar front suspension, the torsion bars and rear torsion bar support may have to be removed to allow transfer case removal. Observe all safety precautions and proper service procedures regarding torsion bar removal. Remove the transfer case skid plate, shift linkage, and all electrical connectors. Support the transfer case on a high-lift jack, and use a safety strap to secure the transfer case on the jack. Remove the transfer case-to-transmission mounting bolts, and lower the transfer case with the jack.*

## Task 4  Disassemble transfer case; clean and inspect internal transfer case components; determine needed repairs.

81. All of the following statements about transfer case service are true EXCEPT
    A. The oil pump may be driven by the output shaft.
    B. Excessive amounts of silicone sealant on the case halves may plug the oil pickup screen.
    C. Compressed air should be used to blow out all lubricant passages.
    D. Transfer case shifting is not affected by the use of an improper transfer case fluid.

**Hint**    *The transfer case should be cleaned externally with a suitable cleaning solution and a brush prior to disassembly. The transfer case must be properly secured in a bench-mounted holding fixture.*
*The disassembly and reassembly procedures vary, depending on the transfer case manufacturer. Always follow the procedure in the vehicle manufacturer's service manual. The following is a general disassembly procedure:*
1. *Remove all electrical sensors, switches, and motors from the transfer case.*
2. *Remove the rear extension housing, followed by the speedometer spacer, drive gear, and steel ball.*
3. *Remove the rear output shaft snap ring, and the transfer case to cover bolts.*
4. *Separate the transfer case cover from the case.*
5. *Remove, clean, and inspect all transfer case components (Figures 3–62 and 3–63).*

*Use a clean shop towel to clean the magnet in the transfer case. An oil pump is mounted in many transfer cases to circulate lubricant to all the case components. The oil pump is sometimes driven by the output shaft. In some transfer cases, the oil pump is not repairable. Use the proper seal drivers to replace all transfer case seals. Replace all worn or damaged transfer case components.*

## Task 5  Reassemble transfer case; refill with proper fluid.

82. Technician A says that in 4WD low, the powerflow in the transfer case is from the input shaft through the sun gear and planetary carrier to provide a gear reduction.
    Technician B says that in 4WD low, the annulus gear in the planetary gear is rotating counterclockwise.
    Who is correct?
    A. A only
    B. B only
    C. Both A and B
    D. Neither A nor B

**Hint**    *Use compressed air to blow out all oil passages in the case or case components. When reassembling the transfer case, some manufacturers recommend sealing the case to the cover with a small bead of black silicone rubber sealant. An excessive amount of sealant may contaminate the transfer case fluid and plug the oil pump pickup screen, resulting in case component failure.*
*After the transfer case is properly assembled and installed, it must be filled to the required level with the manufacturer's specified lubricant. If the wrong fluid is placed in the transfer case, hard shifting and component damage may occur.*

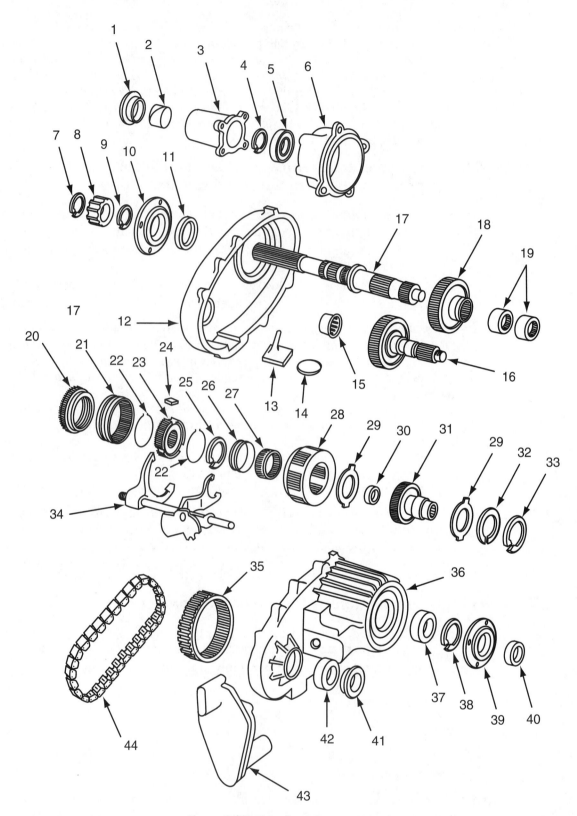

**Figure 3–62** Transfer case components.

| | |
|---|---|
| 1. Rear output shaft seal | 23. Synchronizer hub |
| 2. Extension housing bushing | 24. Synchronizer struts (3) |
| 3. Extension housing | 25. Snap ring |
| 4. Retainer | 26. Range shift hub |
| 5. Rear output shaft bearing | 27. Range gear |
| 6. Pump retainer housing | 28. Planetary carrier |
| 7. Tone wheel retainer | 29. Thrust washer (2) |
| 8. Tone wheel | 30. Main shaft pilot bearing |
| 9. Tone wheel retainer | 31. Input gear |
| 10. Oil pump | 32. Carrier lock ring |
| 11. Oil pump seal | 33. Snap ring |
| 12. Rear case half | 34. Shifting fork mechanism |
| 13. Pump pick-up screen | 35. Annulus gear |
| 14. Magnet | 36. Front case half |
| 15. Front output rear bearing | 37. Input bearing |
| 16. Front output (driven) shaft | 38. Snap ring |
| 17. Main shaft | 39. Input bearing retainer |
| 18. Drive sprocket | 40. Input bearing seal |
| 19. Drive sprocket bearings | 41. Front output shaft seal |
| 20. Main drive synchronizer ring | 42. Front output shaft bearing |
| 21. Synchronizer sleeve | 43. Encoder motor |
| 22. Synchronizer strut spring (2) | 44. Drive chain |

**Figure 3–63** Identification of transfer case components.

## Task 6 Check transfer case fluid, level, condition and type.

*Hint*    *The transfer case should be filled to the bottom edge of the filler plug opening with the vehicle on a level surface. Some manufacturers now specify automatic transmission fluid for the transfer case lubricant.*

## Task 7 Inspect, service, and replace front drive/propeller shaft and universal/CV joints.

83. All of the following problems may cause a vibration on a 4WD vehicle that is more noticeable when changing the throttle position EXCEPT
    A. worn U-joints.
    B. worn front drive axle joints.
    C. incorrect drive shaft angles.
    D. worn drive shaft slip joints.

*Hint*    *Rotate the drive shaft and listen for a squeaking noise from the U-joints. A squeaking noise indicates a dry, worn U-joint. Check for movement in the slip joints when up and down force is applied on the U-joint behind the slip joint. Inspect the drive shafts for impact damage.*
*If the drive shaft has a center bearing, inspect this bearing for looseness and roughness. Measure the drive shaft angles with an inclinometer as explained previously in this chapter. Typical satisfactory drive shaft angles are provided in Figure 3–64.*

## Task 8 Inspect, service, and replace front drive axle universal/CV joints and drive/half shafts.

84. All of the following statements regarding 4WD front-drive axles and joints are true EXCEPT

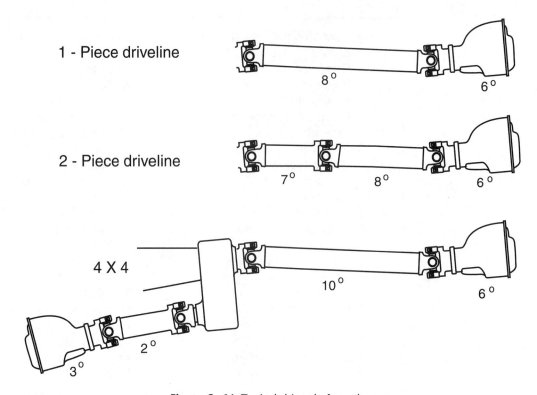

1 - Piece driveline

2 - Piece driveline

4 X 4

**Figure 3–64** Typical drive shaft angles.

A.  The inner tripod joint is held on the axle shaft with a snap ring.
B.  A special swaging tool may be required to tighten the outer boot clamps.
C.  Coat the new joint with the grease supplied with the joint, and discard the remaining grease.
D.  A worn outer CV joint may cause a clicking noise while cornering.

**Hint**     *Many 4WD front-drive axles have an outer CV joint and an inner tripod joint. Remove the outer boot clamps and slide the boot toward the center of the axle. Spread the retaining ring with a pair of snap ring pliers, and then slide the outer CV joint from the axle shaft (Figure 3–65).*
*Many 4WD front axle joints have an inner tripod joint. A snap ring holds the inner spider assembly onto the shaft in this type of joint (Figure 3–66). Inspect the inner and outer joint*

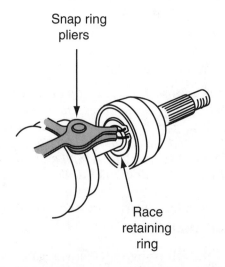

Snap ring
pliers

Race
retaining
ring

**Figure 3–65** Removing outer CV joint from the axle shaft.

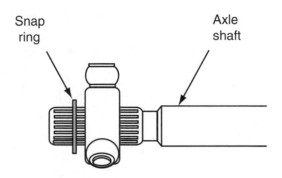

**Figure 3–66** Inner tripod joint, front drive axle 4WD.

*components for looseness, wear, scoring, and damage. Check the axle boots for cracks, oil soak-ing, and deterioration. Replace the boots and axle joints as required. Always install all the grease supplied with the replacement joints in the joint. Always install the boot clamps using the man-ufacturer's recommended procedure. Some outer boot clamps must be installed with a special swaging tool (Figure 3–67).*

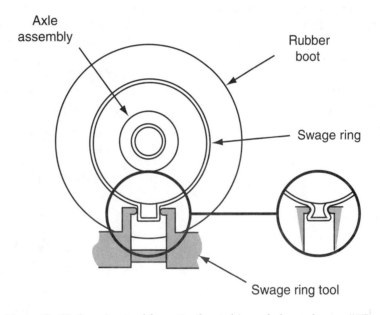

**Figure 3–67** Swaging tool for outer front drive axle boot clamps 4WD.

## Task 9    Inspect, service, and replace front wheel bearings, seals, and hubs.

85. Technician A says automatic locking hubs should be packed with grease. Technician B says the cap on automatic locking hubs should be packed with grease.
    Who is correct?
    A. A only
    B. B only
    C. Both A and B
    D. Neither A nor B

**Hint**    *In a 4WD vehicle, the front-wheel bearings are serviced using the same procedure as 2WD vehicles. Because there is more load on the front-wheel bearings in a 4WD vehicle, front-wheel bearing adjustment is critical. After the front-wheel bearings are lubricated and installed, the inner locknut is tightened to the specified torque (Figure 3–68). Some manufacturers recommend*

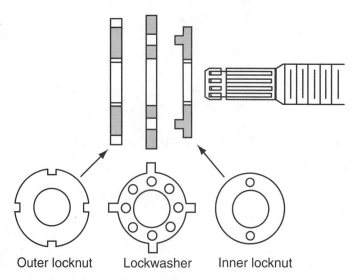

Outer locknut    Lockwasher    Inner locknut

**Figure 3–68** Installing and adjusting wheel bearing locknuts.

*tightening this locknut until the specified front-wheel turning torque is obtained with a spring scale connected to one of the wheel studs. Rotate the hub and then back off the inner locknut the specified amount, typically one-quarter turn. Install the lockwasher and tighten the outer lock-nut to the specified torque.*

*Four-wheel-drive vehicles may have manual or automatic locking hubs (Figure 3–69 and Figure 3–70). After the wheel bearings have been adjusted and the locking screws or nut*

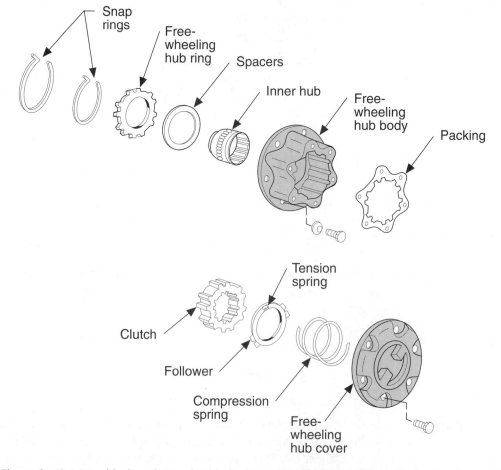

**Figure 3–69** Manual locking front wheel hub. *(Courtesy of Mitsubishi Motor Sales of America, Inc.)*

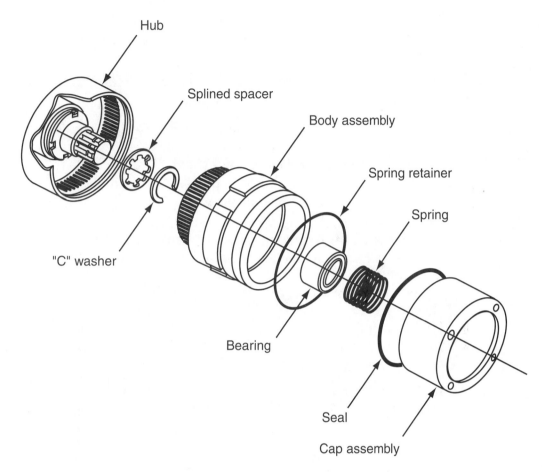

**Figure 3–70** Automatic locking front wheel hub.

*tightened to the specified torque, place some multipurpose grease on the hub inner spines. Do not pack the hub with grease. Slide the locking hub assembly into the wheel hub until it seats. Install the lock ring in the wheel hub groove. Install the lock washer and axle shaft stop on the axle bolt, and then install the bolt and tighten it to the specified torque. Place a small amount of lubricant on the cap seal, and install the cap over the body and into the wheel hub. Install the attaching bolts and tighten them to the specified torque. Rotate the locking hub control from stop to stop to make sure it operates freely. Set both controls to the same auto or lock position.*

## Task 10  Check transfer case and front axle seals and all vents.

86. Technician A says a plugged transfer case vent may cause seal leakage.

    Technician B says the remote transfer case vent helps to prevent moisture from entering the transfer case when driving through water.

    Who is correct?

    A.  A only

    B.  B only

    C.  Both A and B

    D.  Neither A nor B

*Hint      Inspect the front and rear output shaft flanges in the transfer case for evidence of leakage in the seal area. If there is any sign of leakage at the drive shaft flange seals, these seals must be replaced. When the flanges and seals are removed, always inspect the seal contact area on the flanges for roughness and scoring. Replace the flanges if these conditions are present. Inspect the seal contact area in the case for metal burrs and gouges. Remove metal burrs with a fine-toothed file. Prior to seal installation, lube the seal lips and apply sealer to the outer diameter of the seal case. Inspect the remote venting system on the transfer case and differentials for restrictions and kinks.*

## Task 11   Diagnose, test, adjust, and replace electrical/electronic components of four-wheel/all-wheel drive systems.

87. In an electronically shifted transfer case, the shift motor is operated by
   A. the powertrain control module (PCM).
   B. the generic electronic module (GEM).
   C. the transmission control module (TCM).
   D. the body control module (BCM).

88. When discussing an electronically shifted transfer case coupled to an automatic transmission, Technician A says the transfer case will shift into 4L at any speed. Technician B says the transfer case will shift into 4L with the transmission in any gear.
   Who is correct?
   A. A only
   B. B only
   C. Both A and B
   D. Neither A nor B

**Hint**    *Some 4WD vehicles have an electronically shifted transfer case that contains an electric shift motor, a magnetic clutch, a shift position sensor, and a vehicle speed sensor (VSS). (See Figure 3–71.)*

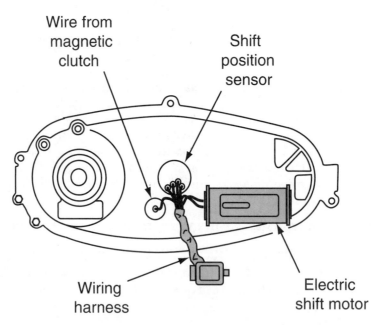

**Figure 3–71** Electronically shifted transfer case.

*This electronic shift system allows shifting between 2H and 4H at any vehicle speed. Therefore, these systems may be referred to as shift-on-the-fly systems. Shifts between 4H and 4L require the vehicle to be stopped, or moving below 3 mph, and the automatic transmission must be shifted into neutral. If the vehicle has a manual transmission, the clutch must be disengaged to allow 4H to 4L shifts. This transfer case system uses an automatic front hub lock/part-time system, which allows all front-drive components to be stationary in 2WD.*

*If the vehicle is in 2WD and the driver selects 4H, the magnetic clutch in the transfer case is energized and immediately spins up the front-drive system from zero to vehicle speed in milliseconds.*

*This action engages the automatic front axle hubs. When the transfer case front and rear output shafts reach the same speed, the spring-loaded lockup collar mechanically engages the main shaft hub to the output shaft drive sprocket to provide 4H. Under this condition, the magnetic clutch is de-energized.*

*The generic electronic module (GEM) analyzes input information to determine if the proper design conditions are present to provide the driver-requested shift. These inputs include the transfer case motor sense plate, transmission range (TR) sensor or neutral safety switch, and vehicle speed sensor (VSS). (See Figure 3–72.) The transfer case motor sense plate sends a signal to the GEM in relation to the transfer case shift motor position. The TR sensor informs the GEM if the transmission is in neutral or some other gear. The VSS signal informs the GEM regarding vehicle speed. When the inputs indicate to the GEM that all the design conditions are met, the GEM operates the transfer case shift motor to provide the driver requested shift (Figure 3–73). The transfer case shift motor drives a rotary shift cam, which moves the mode shift fork or range shift fork to the selected position. After a transfer case shift is completed, the shift position sensor sends a signal to the GEM regarding the shift motor position.*

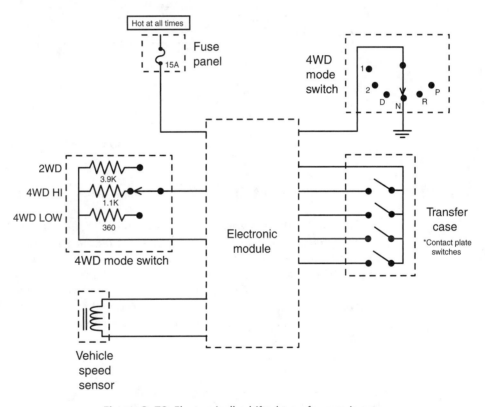

**Figure 3–72** Electronically shifted transfer case inputs.

## Task 12  Test, diagnose, and replace axle actuation and engagement systems (including: viscous, hydraulic, magnetic and mechanical).

89.  When an automatic four-wheel-drive (A4WD) transfer case is operating in A4WD and the front drive shaft begins to turn faster than the rear drive shaft
   A. the computer energizes the clutch coil continually.
   B. the computer reduces the torque to the rear wheels.
   C. the computer increases the clutch coil duty cycle.
   D. the computer reduces the throttle opening.

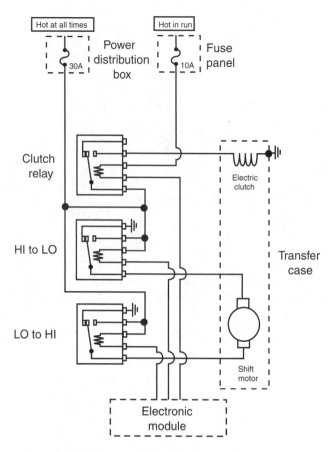

**Figure 3–73** Electronically shifted transfer case outputs.

**Hint**     *Some electronically shifted transfer cases have an automatic four-wheel-drive (A4WD) function.*

*In these systems, the driver may select (A4WD) 4WD High or 4WD Low (Figure 3–74).*

*Hall-effect sensors monitor both input shaft and output shaft speed. The GEM operates the transfer case clutch with a variable duty cycle. During normal cruise driving with the selector switch in A4WD, the GEM operates the transfer case clutch with a minimum duty cycle. This action allows for a slight difference in speed between the front and rear differentials. If the Hall Effect input sensors indicate a specific difference between input and output shaft speed, the GEM increases the clutch duty cycle to eliminate the wheel slip that is causing the difference in speed between the input and output shafts.*

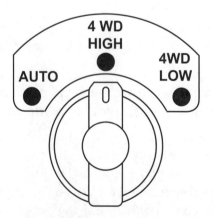

**Figure 3–74** Selector switch for electronically shifted transfer case with automatic four-wheel drive.

*The clutch hub (A in Figure 3–75) is splined to the rear output shaft and internally splined to the clutch discs (B). The clutch housing (C) is splined to the drive sprocket (D) and externally splined to the clutch discs. Therefore, the clutch housing rotates at the front output shaft speed.*

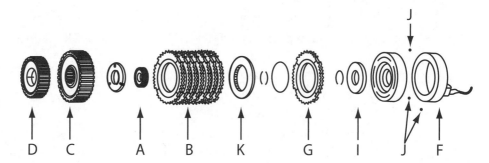

**Figure 3–75** Clutch set and related components in electronically shifted transfer case with automatic four-wheel drive.

*When the clutches are squeezed together, the clutch hub connected to the output shaft drives the drive sprocket through the clutches to provide 4WD. When the GEM increases the transfer case coil (F) duty cycle, the armature (G) is pulled toward the cam/coil housing assembly. Because the armature is splined to the clutch housing (C), the clutch housing is turning at front output shaft speed, and the cam/coil housing also begins to rotate at the same speed. Half of a ball and cam assembly is mounted inside the cam/coil housing. The other half of the ball and cam assembly (I) is splined to the rear output shaft, and the balls (J) are seated between the two cams. Because the two cams are turning at different speeds, the balls climb to the higher part of the cams. This action pushes the cam against the pressure plate (K) and squeezes the clutch discs together.*

*Torque is now transferred from the rear output shaft through the clutch discs to the clutch housing and drive sprocket to the front output shaft to prevent wheel slip.*

*If the driver selects 4WD High or 4WD Low, these shifts are performed in basically the same way as previously explained.*

# Post Test

1. A vehicle experiences clutch pedal pulsations.
   Technician A says this problem may be caused by a defective pressure plate.
   Technician B says this problem may be caused by a worn pilot bushing.
   Who is correct?
   A. A only
   B. B only
   C. Both A and B
   D. Neither A nor B

2. Clutch slipping may be caused by
   A. a worn, rough clutch release bearing.
   B. a worn, rough pilot bearing.
   C. excessive input shaft end play.
   D. oil-soaked clutch facings.

3. In a hydraulic clutch system, grinding occurs while changing gears.
   Technician A says this problem may be caused by low fluid level in the clutch master cylinder.
   Technician B says this problem may be caused by air in the hydraulic clutch system.

Who is correct?

A.  A only

B.  B only

C.  Both A and B

D.  Neither A nor B

4.  A vehicle has repeated clutch release bearing failure. The cause of this problem could be

   A.  lack of clutch pedal free-play.

   B.  a worn clutch pedal linkage.

   C.  a worn pilot bearing.

   D.  clutch housing misalignment.

5.  When servicing clutches and pressure plates

   A.  clutch facing wear is measured from the lining surface to the cushion springs between the linings.

   B.  clutch facing wear is measured with a micrometer.

   C.  a clutch disc installed backwards causes excessive noise during clutch operation.

   D.  a screwdriver is used to align the clutch disc hub with the pilot bearing.

6.  Metal burrs between the clutch housing and the engine block may cause

   A.  a growling noise when the clutch pedal is released.

   B.  a rattling noise with the clutch pedal depressed.

   C.  a vibration when the vehicle speed is above 55 mph.

   D.  clutch chatter and grabbing during clutch engagement.

7.  Technician A says excessive crankshaft end play may cause improper clutch release. Technician B says excessive crankshaft end play may cause clutch slipping.

   Who is correct?

   A.  A only

   B.  B only

   C.  Both A and B

   D.  Neither A nor B

8.  While discussing manual transmission shift linkage adjustments, Technician A says an improper shift linkage adjustment may cause the transmission to jump out of gear. Technician B says the shift linkage is usually adjusted until a specific clearance is obtained between the linkage and the linkage slot.

   Who is correct?

   A.  A only

   B.  B only

   C.  Both A and B

   D.  Neither A nor B

9.  A transmission jumps out of second gear. The most likely cause of this problem is

   A.  a weak detent spring on the shift rail.

   B.  improper transmission lubricant.

   C.  a bent interlock plate.

   D.  a broken shift fork.

10.  The input shaft splines are badly worn and rusted.

   Technician A says this problem may cause a rattling noise in any forward gear.

   Technician B says this problem may cause gear clashing when shifting.

   Who is correct?

   A.  A only

   B.  B only

C. Both A and B

D. Neither A nor B

11. While inspecting manual transmission components
    A. normal gear tooth wear appears as a dull finish with a sharp point on top of the gear teeth.
    B. no measurements are required on the synchronizers, blocking rings, and matching gears.
    C. Normal gear tooth wear is a polished finish with very little wear on the face of the gear teeth.
    D. end play measurements are required on all synchronizer hubs.

12. A four-speed manual transaxle has a hard shifting problem only in third gear. The cause of this problem could be
    A. the transaxle is filled with the wrong lubricant.
    B. the clutch disc is sticking on the input shaft.
    C. a worn third-speed blocking ring and worn third-speed gear dog teeth.
    D. misalignment between the clutch housing and the engine block.

13. Technician A says improper shift linkage adjustment may cause hard transaxle shifting.
    Technician B says an improper shift linkage adjustment may result in the transaxle sticking in gear.
    Who is correct?
    A. A only
    B. B only
    C. Both A and B
    D. Neither A nor B

14. A five-speed manual transaxle has a growling and rattling noise only in third gear. The cause of this noise could be
    A. worn, chipped teeth on the third speed gear on the input shaft.
    B. worn dog teeth on the third speed gear on the input shaft.
    C. worn dog teeth on the third-speed synchronizer blocker ring.
    D. worn threads in the cone area of the third-speed blocking ring.

15. While discussing transaxle case service, Technician A says a scored and worn bearing bore in the transaxle case may be repaired by reboring and installing a sleeve.
    Technician B says cracks in a fluid passage in a transaxle case may be repaired.
    Who is correct?
    A. A only
    B. B only
    C. Both A and B
    D. Neither A nor B

16. All of the following statements about measuring differential end play and turning torque are true EXCEPT
    A. Differential end play is measured with a dial indicator.
    B. The differential side bearing preload is equal to the differential end play plus 0.012 in.
    C. To obtain the proper differential side bearing preload, a selective shim is installed behind one of the side bearing cups in the transaxle case.
    D. If the differential turning torque is more than specified, a thinner selective shim must be installed.

17. A front wheel drive car has a clicking noise while cornering at low speeds. The most likely cause of this problem is
    A. a worn outer CV joint.
    B. a worn wheel bearing.
    C. a worn inner CV joint.
    D. an out-of-balance wheel.

18. While driving straight ahead, a differential in a rear-wheel-drive vehicle produces a clicking noise. The most likely cause of this problem is
    A. damaged differential side gears.
    B. damaged ring gear and drive pinion teeth.
    C. reduced differential side bearing preload.
    D. loose ring gear mounting bolts.

19. All of the following statements about a vacuum-shifted 4WD system are true EXCEPT
    A. The transfer case is shifted into 4WD by a vacuum motor.
    B. The front axle is shifted into 4WD by a vacuum motor.
    C. A vacuum motor and related shift mechanism is located on each front axle.
    D. A vacuum leak may affect transfer case and front axle shifting.

20. While discussing shift control in an electronically shifted 4WD system with an automatic transmission, Technician A says the vehicle speed must be below 8 mph when shifting from 4H to 4L.
    Technician B says the automatic transmission must be in O/D when shifting from 4H to 4L.
    Who is correct?
    A. A only
    B. B only
    C. Both A and B
    D. Neither A nor B

# Answers and Analysis

**1.  D**    Weak clutch plate torsional springs cause clutch chatter, but a worn clutch release bearing, pilot bearing, or excessive input shaft end play does not result in this problem. Thus, D is right.

**2.  D**    When there is excessive free play, the clutch is not released when the clutch pedal is fully depressed, resulting in clutch dragging and hard shifting. If there is no clutch pedal free play, the clutch release bearing maintains pressure on the pressure plate so the clutch is not fully applied. This action results in clutch slipping. Therefore, D is right.

**3.  C**    Many self-adjusting clutch cables have no free play, no overcenter spring, and a constant running release bearing. Thus, C is right.

**4.  B**    If the clutch pedal free play is less than specified, the clutch may not be fully engaged with the pedal released. Therefore, A is wrong.

Worn clutch facings, or a scored pressure plate may cause a slipping clutch, but these problems will not cause improper clutch disengagement. Therefore, C and D are wrong.

If there is air in the clutch hydraulic system, the slave cylinder may not push the clutch release lever enough to disengage the clutch. Thus, B is right.

**5.  A**    A worn release bearing is noisy with the clutch pedal depressed because the release bearing is in contact with the pressure plate. Therefore, B is wrong.

The clutch pedal free-play adjustment sets the distance between the release bearing and the pressure plate fingers. Thus, A is right.

**6. C**   Excessive crankshaft end play causes the pressure plate to move away from the clutch release bearing, which may result in improper clutch release. Loose engine main bearings may cause an oil leak at the rear main bearing, which contaminates the clutch facings with oil, resulting in clutch slipping. Improper pressure plate to flywheel position causes engine vibrations. Therefore, A, B, and D are wrong.

A badly scored pressure plate may cause clutch chatter. Thus, C is right.

**7. C**   Clutch slipping may occur if the rivet heads are 0.012 in. or less below the facing surface. Therefore, C is right, and A, B, and D are wrong.

**8. A**   When the clutch pedal is released, the clutch plate is held firmly between the flywheel and pressure plate. Under this condition the input shaft cannot rattle in a worn pilot bearing. Therefore, B, C, and D are wrong. A worn pilot bearing causes a rattling noise at low speed with the clutch pedal depressed because the clutch plate and input shaft are free to move between the pressure plate and the flywheel. Thus, A is right.

**9. C**   If excessive material is removed from the flywheel the torsion springs on the clutch plate are moved closer to the flywheel, and these springs may contact the flywheel. Removing excessive material from the flywheel moves the pressure plate forward away from the release bearing. This action increases free play so the slave cylinder rod may not move far enough to release the clutch. Technicians A and B are both right, making C is the correct answer.

**10. D**   Excessive bell housing misalignment does not affect clutch pedal free play, noise, or vibration at high speeds. Therefore, A, B, and C are wrong.

Misalignment between the bell housing and the engine block may cause clutch chatter because the clutch disc mounted on the input shaft is not aligned properly with the pressure plate and flywheel. Thus, D is right.

**11. A**   Overheated clutch facings would not result in bell housing face runout. Therefore, B is wrong.

Shims may be placed between the bell housing and the engine block to correct bell housing face runout. Thus, A is right.

**12. B**   The dial indicator in Figure 3–8 is positioned to measure bell housing bore alignment. Bell housing bore misalignment may be corrected by turning eccentric dowels in the engine block to the bell housing mounting surface. Thus, B is right.

**13. A**   The measurement shown in Figure 3–10 is for flywheel runout and crankshaft end play.

Because flywheel runout is not included in the responses—crankshaft end play—response A is right.

**14. A**   A drive shaft vibration changes with acceleration and deceleration. Therefore, B is wrong.

Sagged engine mounts cause improper drive shaft angles in a rear-wheel-drive car. Thus, A is right.

**15. C**   Excessive second gear end play, or a weak detent spring on the second gear shift fork, may cause the transmission to jump out of second gear. Thus, Technicians A and B are both right, making C is the correct answer.

**16. A**   Most linkage adjustments are performed with the gear shift lever in neutral. Therefore, B, C, and D are wrong, and A is right.

**17. A**   Excessive output shaft, or input shaft, end play results in lateral shaft movement that would not adversely affect the seal. A worn output shaft bearing would not cause premature extension housing seal failure. Therefore, B, C, and D are wrong.

A scored drive shaft yoke results in excessive extension housing seal wear. Thus, A is right.

**18. D**   The drive shaft may be installed in the original position on the differential flange. The transmission weight should be supported on a transmission jack, and the engine support fixture must be installed before the transmission to engine bolts are loosened. Therefore, A, B, and C are wrong.

A clutch plate alignment tool must be used before transmission installation. Thus, D is right.

**19.  C**    A synchronizer hub that does not slide smoothly over the blocking ring causes the hub to jam, resulting in hard shifting, and the synchronizer hub and sleeve should be marked prior to disassembly. Technicians A and B are both right, making C the correct answer.

**20.  B**    A bent shift rail would not cause the transmission to jump out of gear, or result in gear clash or noise. Hard shifting may be caused by a bent shift rail; thus, B is right.

**21.  C**    A worn pilot bearing contact area on the input shaft, or a worn pilot bearing could result in noise with the clutch pedal depressed. Therefore, A and B are wrong.

The main shaft is not turning with the engine idling and the clutch released with the transmission in neutral, so D is wrong.

Because the input shaft is only turning with the clutch released, a rough input shaft roller bearing or needle bearings would result in a growling noise under this condition. Thus, C is right.

**22.  C**    A worn first-speed gear blocking ring or synchronizer sleeve may cause hard shifting, but they would not result in noise while driving in first gear. Therefore, A and B are wrong. A worn, rough main shaft bearing would cause a growling noise in all gears, so D is wrong.

Chipped and worn first-speed gear teeth would cause a growling noise in first gear. Thus, C is right.

**23.  B**    The blocking ring dog teeth tips should be pointed with smooth surfaces. Therefore, A is wrong. Clearance between the blocking ring and matching gear dog teeth is important for proper shifting, and the synchronizer sleeve must slide freely on the hub. Therefore, C and D are wrong.

Threads in the cone area of the blocking ring must be sharp. Thus, B is right.

**24.  B**    If the clearance between the blocking ring and the fourth-speed gear dog teeth is less than specified, the blocking ring is worn, which results in hard shifting. This problem would not result in noise while driving in fourth gear. Thus, A is wrong and B is right.

**25.  C**    Because the counter gear is turning with the clutch pedal released in neutral and also in any gear, damaged counter gear bearings cause a growling noise under both of these conditions.

Thus, Technicians A and B are both right, making C the correct answer.

**26.  D**    Because the reverse idler gear is only in mesh with other gears in reverse, this gear rotates only in reverse. Thus, D is right.

**27.  C**    In fourth gear, the 1-2 synchronizer is moved ahead so the synchronizer hub is meshed with the dog teeth on the fourth-speed gear on the input shaft. Excessive input shaft end play would cause the transmission to jump out of fourth gear. Thus, C is right.

**28.  A**    Worn clutch facings and pressure plate friction surface may cause clutch slipping, but this problem does not cause a clunking noise during acceleration and deceleration. Thus, B is wrong.

Excessive main shaft end play may cause a clunking noise during acceleration and deceleration. Therefore, A is right.

**29.  B**    A worn speedometer drive and driven gears may cause erratic speedometer operation, but this problem would not result in premature extension housing bushing wear. Thus, A is wrong.

A plugged transmission vent may cause excessive transmission pressure, and fluid leaks, but this problem would not affect extension housing bushing wear. Therefore, C is wrong.

Excessive main shaft end play causes lateral shaft movement, but this does not cause excessive extension housing bushing wear, so D is wrong.

Metal burrs between the extension housing and the transmission case cause misalignment of the extension housing, which forces the extension housing bushing against the drive shaft yoke in one location, resulting in bushing wear. Thus, B is right.

**30.  A**    A worn extension housing bushing may cause premature extension housing seal wear and fluid leaks, so B is wrong.

Excessive main shaft end play has no effect on speedometer operation; thus, C is wrong.

The speedometer drive shown in Figure 3–21 contains a cable to drive the speedometer. Therefore, an electrical problem in the distance sensor would not affect speedometer operation; thus, D is wrong.

A worn adapter bushing and speed distance sensor allows the speedometer cable to jump around causing erratic speedometer operation. Thus, A is right.

**31.  B**   Component 19 in Figure 3–22 is a computer-aided gear select solenoid. Therefore, A, C, and D are wrong, and B is right.

**32.  A**   Because thicker gear oils have higher numbers, B is wrong.

ATF is much thinner compared to 90 W gear oil, and therefore it does reduce friction and improve horsepower and economy. Thus, A is right.

**33.  B**   A worn fourth-speed synchronizer would only affect shifting in fourth gear, so A is wrong.

Excessive main shaft end play would not result in gear clash in all gears. A worn 3 - 4 shift fork may cause shifting problems in third and fourth gear, but this problem would not result in gear clash in all gears. Therefore, C and D are wrong.

A clutch disc sticking on the input shaft would cause the clutch not to release properly, resulting in gear clash in all gears. Thus, B is right.

**34.  C**   An improper shift linkage adjustment may cause hard shifting or sticking in gear. Technicians A and B are both right, making C the correct answer.

**35.  A**   A worn outer drive axle joint may cause a clicking noise while cornering at low speed, but this problem would not cause repeated drive axle seal failure. Thus, B is wrong.

A plugged transaxle vent may cause excessive transaxle pressure and repeated drive axle seal failure. Therefore, A is right.

**36.  C**   A misaligned engine and transaxle cradle may cause drive axle vibrations. Because the lower control arms are connected to this cradle, misalignment of the cradle may cause improper front suspension angles. Technicians A and B are both right, making C the correct answer.

**37.  B**   After the case halves are separated, the differential may be removed first on some transaxles. Therefore, A is wrong.

During service procedures the transaxle case must be supported on a special fixture and shims, so C is wrong.

A special driving tool and a hydraulic press are required to remove the input and output shafts, making D wrong.

The input and output shafts are retained with snap rings under the transaxle cover. Thus, B is right.

**38.  A**   Worn dog teeth on the fourth-speed gear would not cause the transaxle to jump out of third gear. Thus, B is wrong.

A weak detent spring on the 3 - 4 shift rail may cause the transaxle to jump out of third gear, so A is right.

**39.  A**   Worn dog teeth on the third-speed gear or blocking ring may cause hard shifting or jumping out of third gear, but this problem would not cause a growling noise while driving in third gear. Therefore, B and C are wrong.

Worn threads in the third-speed blocking ring may cause hard shifting, but this wear would not cause a growling noise in third gear. Thus, D is wrong.

Worn, chipped teeth on the third speed gear could result in a growling noise while driving in third gear, so A is right.

**40.  D**   Worn dog teeth on the second-speed gear and blocking ring would not result in a growling noise while driving, or accelerating, in second gear. Therefore, A and B are wrong.

This wear problem would not cause hard shifting in second and third gear, so C is wrong.

Worn dog teeth on the second speed gear and blocking ring may cause the transaxle to jump out of second gear. Thus, D is right.

**41.  D**    Synchronizer hubs are not reversible on the shaft, and synchronizer sleeves are not reversible on their hub. Both A and B are wrong, and D is the correct answer.

**42.  A**    Worn reverse idler teeth may cause a growling noise while driving in reverse, but this problem would not cause failure to shift into reverse. Therefore, B is wrong.

A broken reverse shifter fork may cause failure of the transmission to shift into reverse without any noise. Thus, A is right.

**43.  A**    Loctite PST is not used to repair transaxle cases, so B is wrong. An epoxy-based sealer may be used for some transaxle case cracks. Thus, A is right.

**44.  C**    In some transaxles the speedometer drive gear is mounted on the differential case. Thus, C is right.

**45.  D**    The PCM uses the VSS signal to operate the reverse lockout system, making answer A wrong.

The reverse lockout system prevents the driver from shifting into reverse at speeds above 16 mph. Therefore, B is also wrong. Because both technicians are wrong, D is the right answer.

**46.  B**    Damaged ring gear teeth would cause a clicking noise while the vehicle is in motion. This problem would not cause differential chatter. Thus, A is wrong. Improper preload on differential components, such as side bearings, may cause differential chatter, and so B is right.

**47.  D**    In the transaxle shown in Figure 3–37, differential side bearing preload is adjusted with a shim behind one of the side bearing cups. Therefore, A and B are wrong, and the correct answer is D.

**48.  C**    The side gear end play must be measured individually on each side gear with the thrust washers removed. Therefore, A and B are wrong.

Side gears with the specified thrust washer have a slight end play but no preload. Therefore, D is wrong and C is right.

**49.  D**    A side bearing preload adjustment shim is positioned behind one of the side bearing cups, so A is wrong.

The speedometer drive gear will slide over the side bearing so the bearing does not have to be removed first. Therefore, A and B are both wrong; D is the correct answer.

**50.  B**    The selective shims are positioned behind the outer bearing races in the case to obtain the proper preload on the input and output shafts. Therefore, Technician A is wrong and Technician B is right. B is the correct answer.

**51.  C**    Wrong transaxle lubricant may cause burned output shaft bearings. Therefore, A is right. A broken oil feeder behind the front output shaft bearing results in improper output shaft bearing lubrication and burned bearings. Therefore, Technicians A and B are both right, making C the correct answer.

**52.  B**    While measuring the differential side play to determine the required side bearing shim thickness, a new bearing cup is installed in the transaxle case without the shim. Therefore, A is wrong.

The proper shim thickness is equal to the differential end play plus a specified thickness for bearing preload, so C is wrong.

While measuring the differential end play, the transaxle case bolts must be tightened to the specified torque. Therefore, D is wrong.

A medium load should be applied to the differential in the upward direction while measuring the differential end play. Thus, B is the correct answer.

**53.  B**    A worn U-joint may cause a squeaking noise that increases in relation to vehicle acceleration. Therefore, A is wrong.

If the centering ball and socket is worn in a double Cardan U-joint, a heavy vibration may occur during acceleration. Thus, B is right.

**54.  A**    A worn front-wheel bearing usually results in a growling noise while cornering or driving straight ahead. B is wrong.

A clunking noise while decelerating may be caused by a worn inner drive axle joint. Thus, A is right.

**55. C**   A worn drive shaft center bearing or an outer rear axle bearing causes a growling noise that is not influenced by acceleration and deceleration. Thus, Technicians A and B are both right, making C the correct answer.

**56. D**   The drive shaft markings in Figure 3–46 are for drive shaft balancing. Therefore, A, B, and C are wrong. D is right.

**57. D**   While measuring drive shaft runout, the dial indicator must be positioned near the center of the drive shaft, and the drive shaft should be replaced if the runout is excessive. Technicians A and B are both wrong; thus, the correct answer is D.

**58. C**   An out-of-balance drive shaft or wheel may cause a vibration that increases in relation to vehicle speed. Thus, Technicians A and B are both right, so the correct answer is C.

**59. C**   After the wheel bearing adjusting nut is tightened to 20 ft.-lbs. and loosened one-half turn, the next step is to tighten the adjusting nut to 10 to 15 in.-lbs. Thus, C is the correct answer.

**60. B**   Because the side gears are only turning while cornering, they do not cause a whining noise while driving straight ahead. Therefore, A is wrong. However, improper ring gear and pinion gear adjustments may cause this problem. Thus, B is right.

**61. A**   A loose pinion nut allows pinion shaft end play, resulting in a clunking noise on acceleration and deceleration. Thus, A is right.

**62. A**   Excessive side bearing preload does not affect ring gear runout. Therefore, B is wrong.

Side gear end play does not affect ring gear runout, making answer C wrong.

If the ring gear bolts are not properly torqued, the bolts may come loose in service and cause differential damage. Therefore, D is wrong.

Differential case runout may cause excessive ring gear runout. Thus, A is right.

**63. A**   The question asks for the response that is not correct. Hunting-type ring and pinion gear sets do not require timing. Because the statements in B, C, and D are correct, they are not the required answer. However, A is not right, making it the correct answer.

**64. C**   The marking on the pinion must be added to the shim thickness to determine the correct pinion depth shim thickness. Therefore, 0.066 in. (answer C) is right.

**65. A**   The reading on the dial indicator must be subtracted from 0.100 in. to obtain the nominal pinion depth shim thickness. Therefore, the nominal shim thickness is 0.43, and the pinion marking of −4 is subtracted from this figure. Therefore, 0.039 in. (answer A) is correct.

**66. A**   The pinion nut must never be loosened to obtain the specified pinion turning torque, so B is wrong.

The pinion bearings should be lubricated when the pinion turning torque is measured. Thus, A is right.

**67. C**   The left side bearing adjusting nut must be loosened and the right side bearing adjusting nut tightened to increase ring gear backlash. Thus, C is right.

**68. D**   With either excessive ring gear toe contact, or low flank contact, the pinion gear must be moved away from the ring gear. Therefore, both A and B are wrong, and D is the correct answer.

**69. B**   Worn pinion bearings may cause a growling noise while driving straight ahead. Therefore, A is wrong.

Because the side gears and bevel pinions rotate only while turning a corner, rough bearing surfaces on these gears and the differential case may cause a vibration while cornering. Thus, B is right.

**70. D**   This question asks for the answer that is not correct. Ring gear runout, and case side play should be measured before removing the ring gear and case assembly, and the side bearing caps should be marked in relation to the case prior to removal. Therefore, statements A, B, and C are right, but they are not the requested answer.

The side bearings should be lubricated before installation, so statement D is wrong, and this is the correct answer.

**71. B** The side bearings must be in good condition prior to the case runout measurement, so A is wrong.

The ring gear runout should be measured before the case runout. Thus, B is right.

**72. A** This question asks for the statement that is not correct. In a limited slip differential, the steel plates are retained in the case, each clutch set contains a preload spring, and a special lubricant is required. Therefore, statements B, C, and D are correct, but they are not the requested answer.

The friction plates are splined to the clutch hub, so statement A is not correct; thus, it is the right answer.

**73. C** Wrong lubricant, or worn friction and steel plates, in a limited slip differential may cause chattering while cornering. Thus, Technicians A and B are both right, making C the correct answer.

**74. C** The measurement in Figure 3–57 determines the proper shim thickness. Therefore, C is right, and A, B, and D are wrong.

**75. A** Excessive axle shaft runout may cause a vibration at 60 mph (96 kmh), but this problem would not cause a vibration when accelerating at lower speed. Therefore, B is wrong, and A is right.

**76. C** This question asks for the statement that is not correct. An axle bearing retainer should be removed by striking it in several places with a hammer and chisel. The bearing should be pressed off the axle shaft with a hydraulic press, and this press should be used to press the bearing and retainer onto the shaft. Therefore, statements A, B, and D are correct, but they are not the requested answer.

The bearing must be replaced after it is pressed off the axle. Because statement C is wrong, this is the requested answer.

**77. B** Excessive runout on the dial indicator in Figure 3–58 could be caused by a bent axle shaft. Therefore, B is right, and A, C, and D are wrong.

**78. B** Worn U-joints may cause a squeaking or clunking noise, and a vibration while driving straight ahead. Therefore, A is wrong.

Worn outer front drive axle joints on a 4WD vehicle may cause a vibration problem while cornering. Thus, B is right.

**79. C** When a vacuum-operated 4WD does not shift into 4WD, the engine vacuum may be low, or the vacuum motor at the front differential may be the problem. Therefore, A and B are both right, and the correct answer is C.

**80. A** The torsion bars and rear torsion bar support may have to be removed prior to transfer case removal. Thus, A is right.

The transfer case lubricant should be drained before the transfer case is removed. Therefore, B is wrong.

**81. D** In this question, you are asked to select the answer that is not right.

The oil pump may be driven by the output shaft, excessive amounts of silicone sealant may plug the oil pickup screen, and compressed air should be used to blow out all lubricant passages.

Therefore, A, B, and C are right.

Improper transfer case lubricant may cause hard shifting. Therefore, D is wrong, making it the requested answer.

**82. A** The annulus gear is locked to the case, so B is wrong.

In 4WD low, the transfer case input shaft is driving the sun gear, which, in turn, is driving the planetary carrier. A is right.

**83. D** This question asks for the statement that is not correct. Worn U-joints, front axle drive joints, and incorrect drive shaft angles may cause a vibration that is more noticeable when changing throttle position. Therefore, statements A, B, and C are correct, but these statements are not the requested answer.

Worn drive shaft slip joints would not cause this type of vibration. Since statement D is wrong, it is the requested answer.

**84. C** This question asks for the statement that is not correct. A snap ring holds the inner tripod joint on the axle shaft, and a special swaging tool may be necessary to tighten the outer boot clamp. A worn outer CV joint may cause a clicking noise while cornering. Therefore, statements A, B, and D are right, but they are not the requested answer.

All the grease supplied with the joint should be installed in the joint. Therefore, statement C is wrong, and this is the requested answer.

**85. D** Neither the automatic locking hubs nor the caps should be packed in grease. Therefore, Technicians A and B are both wrong, making D the correct answer.

**86. C** A plugged transfer case vent may cause seal leakage, and the remote transfer case vent helps to keep water out of the transfer case. Technicians A and B are both right, making C the correct answer.

**87. B** The shift control motor is operated by the GEM. Therefore, A, C, and D are wrong, and B is right.

**88. D** The vehicle speed must be below 3 mph before the transfer case will shift into 4L, and the transmission must be in neutral. Therefore, both A and B are wrong, and D is the right answer.

**89. C** In the A4WD mode, if the front drive shaft begins to turn faster than the rear drive shaft, the computer increases the clutch coil duty cycle. Therefore, A, B, and D are wrong, and C is right.

# Answers to Post Test

**1. A** A worn pilot bushing does not cause clutch pedal pulsations. Thus, Technician B is wrong. A defective pressure plate may cause uneven pressure plate fingers, or diaphragm, and this results in clutch pedal pulsations. Technician A is right, making A the correct answer.

**2. D** A worn, rough clutch release bearing or pilot bearing do not cause clutch slipping. Excessive input shaft end play does not cause clutch slipping. Therefore, A, B, and C are wrong. Oil-soaked clutch facings may cause clutch slipping. Thus, D is the correct answer.

**3. C** Grinding while changing gears indicates improper clutch release. This problem may be caused by low fluid level or air in the clutch hydraulic system. Therefore, Technicians A and B are both right, and C is the correct answer.

**4. A** Lack of clutch pedal free play may cause continual contact between the release bearing and the pressure plate, and this results in release bearing failure. Thus, A is correct. A worn clutch linkage may cause excessive clutch pedal free play. Therefore, B is wrong. A worn pilot bearing, or clutch housing misalignment, does not affect release bearing wear. Thus, C and D are wrong.

**5. C** Clutch facing wear is measured with a depth gauge or a vernier caliper from the clutch facing surface to the rivet heads. Therefore, A and B are wrong. If the clutch disc is installed backwards, the torsion springs contact the flywheel retaining bolts resulting in excessive noise. C is correct. An alignment tool is used to align the clutch disc hub with the pilot bearing, and so D is wrong.

**6. D** Metal burrs between the clutch housing and the engine block cause clutch housing misalignment and clutch chatter and grabbing during clutch engagement. Therefore, A, B, and C are wrong, and D is the correct answer.

**7.  A**   Excessive crankshaft end play may cause improper clutch release because this condition allows the crankshaft to move forward away from the release bearing. Thus, Technician A is right. Excessive crankshaft end play does not cause clutch slipping. Therefore, Technician B is wrong, making A the correct answer.

**8.  A**   An improper shift linkage adjustment may cause the transmission to jump out of gear. Thus, Technician A is right. The shift linkage is usually adjusted until an alignment pin slides freely into an alignment hole in the shifter assembly. Therefore, Technician B is wrong, and A is the correct answer.

**9.  A**   Improper transmission lubricant, a bent interlock plate, or a broken shifter fork do not cause the transmission to jump out of second gear. Therefore, B, C, and D are wrong. A weak detent spring may cause a transmission to jump out of second gear. A is the correct answer.

**10.  B**   Worn, rusted input shaft splines may cause the clutch plate to stick on the splines resulting in improper clutch release and gear clashing. Worn, rusted input shaft splines do not cause a rattling noise in all forward gears, because the clutch plate is clamped between the pressure plate and the flywheel in all forward gears. Technician A is wrong, and Technician B is right. Thus, B is the correct answer.

**11.  C**   Normal gear tooth wear is a polished finish with very little wear on the face of the gear teeth. Thus, A is wrong. The clearance must be measured between each blocking ring and the dog teeth on the matching gear, making B incorrect. Normal gear tooth wear is a polished finish with very little wear on the face of the gear teeth. Therefore, C is the correct answer. End play measurements are not required on synchronizer hubs. Thus, D is wrong.

**12.  C**   The wrong lubricant of a sticking clutch disc on the input shaft can cause hard shifting in all gears. Therefore, A and B are wrong. Worn blocking ring and third-speed gear dog teeth may cause hard shifting only in third-gear. Thus, C is the correct answer. Misalignment between the clutch housing and the engine block may cause a grabbing clutch, therefore D is wrong.

**13.  C**   An improper shift linkage adjustment may cause hard shifting or sticking in gear. Therefore, Technicians A and B are both right, and C is the correct answer.

**14.  A**   Worn, chipped teeth on the third-speed gear may cause a growling, rattling noise when driving in third gear. Thus, A is correct. Worn dog teeth on the third-speed gear may cause hard shifting or jumping out of gear. Therefore, B is wrong. Worn dog teeth or threads on the third-speed blocking ring may cause hard shifting, and so C and D are wrong.

**15.  D**   If bearing bores in a transaxle case are scored and worn, the case must be replaced. Therefore, Technician A is wrong. Cracks in fluid passages in a transaxle case cannot be repaired. Thus, Technician B is wrong, and D is the correct answer.

**16.  B**   Differential end play is measured with a dial indicator. A is right, but this is not the requested answer. The proper preload shim thickness is equal to the differential end play plus 0.007 in. Statement B is not true, making it the requested answer. The selective shim to obtain the proper side bearing preload is installed behind one of the differential side bearing cups in the transaxle case. C is true, and this is not the requested answer. If the differential turning torque is more than specified, a thinner selective shim must be installed. Thus, D is true, but it is not the requested answer.

**17.  A**   A worn outer CV joint causes a clicking noise when cornering at low speeds. Thus, A is correct. A worn wheel bearing causes a growling noise especially at low speeds. Therefore, B is wrong. A worn inner CV joint may cause a clunking noise during deceleration. Thus, C is wrong. Improper wheel balance causes a vibration. Therefore, D is wrong.

**18.  B**   Damaged differential side gears cause a clicking noise while cornering. A is wrong. Damaged ring gear and pinion teeth may cause a clicking noise while driving straight ahead. Thus, B is correct. Reduced differential side bearing preload causes differential chatter. Therefore, C is wrong. Loose ring gear mounting bolts may cause a knocking noise while driving straight ahead. Thus, D is wrong.

**19.  C**    In a vacuum-shifted 4WD system the transfer case and the front axle are shifted into 4WD by vacuum motors. Statements A and B are true, but these statements are not the requested answer. A vacuum motor is located only on one front axle. Statement C is not true and this is the requested answer. A vacuum leak may affect transfer case and front axle shifting. Therefore, D is true, but this is not the requested answer.

**20.  D**    When shifting an electronically shifted transfer case with an automatic transmission from 4H to 4L, the vehicle speed must be below 3 mph, and the transmission must be in neutral. Therefore, Technicians A and B are both wrong, and D is the correct answer.

# 4 Suspension and Steering

## Pretest

The purpose of this pretest is to determine the amount of review that you may require prior to taking the ASE Suspension and Steering Test. If you answer all the review questions correctly, complete the questions and study the information in this chapter to prepare for the ASE Suspension and Steering Test.

If two or more of your answers to the pretest questions are incorrect, complete a thorough study of the questions and information in this chapter. The pretest answers are located at the end of the pretest, and also in the answer sheets supplied with this book.

1. A customer complains about steering wander while driving straight ahead.
   Technician A says the front wheel bearing adjustments may be loose.
   Technician B says the front wheels may have negative caster settings.
   Who is correct?
   A. A only
   B. B only
   C. Both A and B
   D. Neither A nor B

2. When performing a sector lash adjustment on a manual recirculating ball steering gear, use an inch-pound torque wrench and socket to rotate the worm shaft through a 45-degree arc in each direction with the worm shaft positioned
   A. one turn from the full left of its travel.
   B. in the center of its travel.
   C. one-half turn from the full right of its travel.
   D. one turn to the right from the center of its travel.

3. A power rack and pinion steering gear has excessive steering effort in both directions.
   The cause of this problem could be
   A. the rack piston seal is leaking.
   B. the inner rack seal is leaking.
   C. the outer rack bulkhead seal is leaking.
   D. the rack bearing adjustment is too loose.

4. During a pressure test, a power steering pump has less than the specified pressure.
   Technician A says the steering gear may be the problem.
   Technician B says the pump pressure relief valve may be sticking.
   Who is correct?
   A. A only
   B. B only
   C. Both A and B
   D. Neither A nor B

5. A power rack and pinion steering system has poor steering wheel returnability. The most likely cause of this problem could be
   A. a bent or damaged rack in the steering gear.
   B. excessive negative caster on both front wheels.
   C. worn inner tie rod ends in the steering gear.
   D. reduced front suspension curb riding height.

6. During a road test a Technician discovers a vehicle has bump steer. The vehicle has power rack and pinion steering.
   Technician A says the steering gear mounting bushings may be worn.
   Technician B says the upper strut mounts may be worn.
   Who is correct?
   A. A only
   B. B only
   C. Both A and B
   D. Neither A nor B

7. The front suspension trim height on a computer-controlled air suspension system is less than specified.
   Technician A says the compressor vent valve may be open.
   Technician B says the front height sensors may require adjusting.
   Who is correct?
   A. A only
   B. B only
   C. Both A and B
   D. Neither A nor B

8. While parking a car with four-wheel steering (4WS) the rear wheels
   A. steer about 5 degrees in the same direction as the front wheels.
   B. remain centered and do not steer in either direction.
   C. steer about 5 degrees in the opposite direction to the front wheels.
   D. steer about 1 degree in the same direction as the front wheels.

9. Fluid foaming occurs in a power steering pump reservoir.
   Technician A says the power steering belt may be too loose.
   Technician B says the power steering fluid may be contaminated.
   Who is correct?
   A. A only
   B. B only
   C. Both A and B
   D. Neither A nor B

10. A variable assist power steering system provides
    A. reduced pump pressure at low vehicle speeds.
    B. reduced pump pressure at low engine speeds.
    C. increased pump pressure at low vehicle speeds.
    D. increased pump pressure at high vehicle speeds.

11. A customer complains about front-wheel shimmy. All of the following could be the cause of the problem EXCEPT
    A. improper dynamic wheel balance.
    B. excessive toe-in on the front wheels.
    C. excessive positive caster on the front wheels.
    D. sagged springs in the rear suspension.

12. The steering on a vehicle pulls to the right while driving straight ahead. The cause of this problem could be
    A. more toe-out on the right rear wheel than on the left rear wheel.
    B. more positive camber on the left front wheel than on the right front wheel.
    C. more toe-out on turns on the left front wheel than on the right front wheel.
    D. more positive caster on the left front wheel than on the right front wheel.

13. When performing a shock absorber bounce test, the left side of the front bumper is pushed downward with considerable weight and then released. The chassis completes one free upward bounce before the vertical chassis movement stops. This action indicates
    A. the coil spring is weak.
    B. the stabilizer bar is weak.
    C. the control arm bushings are worn.
    D. the shock absorber is satisfactory.

14. On a vehicle with a MacPherson strut front suspension, the steering pulls to the left while braking.
    Technician A says there may be excessive difference between the left and right SAI readings.
    Technician B says the left rear wheel may have excessive positive camber.
    Who is correct?
    A. A only
    B. B only
    C. Both A and B
    D. Neither A nor B

# Answers to Pretest

**1. C**   Steering wander may be caused by a loose front wheel bearing adjustment or negative caster settings on the front wheels. Therefore, Technicians A and B are both right, and C is the correct answer.

**2. B**   When performing the sector lash adjustment on a recirculating ball steering gear, the worm shaft is positioned in the center of its travel. Therefore, A, C, and D are wrong, and B is the correct answer.

**3. A**   A leaking inner or outer rack seal may increase steering effort in one direction. Therefore, B and C are wrong. A loose rack bearing adjustment reduces steering effort, and so D is wrong. A leaking rack piston seal causes increased steering effort in both directions. Thus, A is the correct answer.

**4. B**   During the power steering pump pressure test, a valve in the tester momentarily closes the hose between the pump and the gear. Therefore, the steering gear does not affect the pressure. Technician A is wrong. A sticking relief valve may reduce the pressure during a power steering pressure test. Thus, Technician B is right, and B is the correct answer.

**5. B**   A bent or damaged rack may cause erratic steering effort. Therefore, A is wrong. Worn inner tie rod ends cause excessive steering wheel free play, and so C is wrong. Reduced front suspension curb riding height causes harsh riding. Thus, D is wrong. Excessive negative caster on the front wheels causes poor steering wheel returnability, making B correct.

**6. A**   Bump steer is caused by unequal tie rod height. Worn upper strut mounts do not cause bump steer. Thus, Technician B is wrong. Worn steering gear mounting bushings may cause unequal tie rod height and bump steer. So, Technician A is the right, making A the correct answer.

**7.  B**    To exhaust air from the air springs, the vent valve and one or more of the air spring sole-noids must be open. An open vent valve alone does not vent air from the springs and cause low trim height. Technician A is wrong. Improper height sensor adjustment may cause improper trim height. Thus, Technician B is right, making B the correct answer.

**8.  C**    When parking a vehicle with 4WS, the rear wheels steer about 5 degrees in the opposite direction to the front wheels. Therefore, C is correct and A, B, and D are wrong.

**9.  B**    Fluid foaming in the power steering reservoir may be caused by contaminated power steer-ing fluid. A loose power steering belt does not cause this problem. Therefore, Technician A is wrong and Technician B is right, making B the correct answer.

**10.  C**    A variable assist power steering system provides increased pump pressure at low vehicle speeds. Therefore, A, B, and D are wrong, and C is correct.

**11.  B**    Improper dynamic wheel balance, or excessive positive caster on the front wheels may cause wheel shimmy. A and C are true, but neither of these is the requested answer. Sagged rear springs increase front wheel positive caster and may cause front wheel shimmy. Therefore, D is true but this is not the requested answer. Excessive front wheel toe-in does not cause front wheel shimmy. Therefore, B is not true, and this is the requested answer.

**12.  D**    Excessive toe-out on the right rear wheel causes steering pull to the left. A is wrong. Excessive positive camber on the left front wheel causes steering pull to the left. B is wrong. More toe-out-on turns on the left front wheel may cause tire squeal when cornering. C is wrong. More positive caster on the left front wheel may cause steering pull to the right. Thus, D is correct.

**13.  D**    If the chassis completes one free upward bounce before chassis movement stops, the shock absorber is satisfactory. Therefore, A, B, and C are wrong, and D is correct.

**14.  A**    Excessive difference between the SAI readings on the front suspension may cause steering pull when braking. Excessive positive camber on the left rear wheel does not cause steering pull when braking. Therefore, Technician A is right, and Technician B is wrong, making A the correct answer.

# Steering Systems Diagnosis and Repair, Steering Columns

## ASE Tasks, Questions, and Related Information

In this chapter, each task in the Suspension and Steering category is provided followed by a question and some information related to the task. If you answer any question incorrectly, study this information very carefully until you understand the correct answer. Question answers and analysis are provided at the end of this chapter and in the answer sheets provided with this book.

**Task 1**  **Diagnose steering column noises and steering effort concerns (including manual and electronic tilt and telescoping mechanisms); determine need-ed repairs**

1. With the steering column mounted in the vehicle and all linkages connected, a steering wheel has 2.35 in. (60 mm) of free play.
Technician A says the flexible coupling may be worn.
Technician B says the steering gear mounting bolts may be loose.

Who is correct?
A. A only
B. B only
C. Both A and B
D. Neither A nor B

**Hint**    *Excessive steering wheel free play may be caused by a worn flexible coupling, loose steering gear mounting bolts, and worn steering linkage components such as inner or outer tie rod ends. A worn flexible coupling may also cause a rattling noise when driving on irregular road surfaces.*

**Task 2**   **Inspect and replace steering column, steering shaft U-joint(s), flexible coupling(s), collapsible columns, steering wheels (includes steering wheels and columns equipped with air bags and/or other steering wheel/column mounted controls, sensors, and components).**

2. A mounting bolt in a collapsible steering column bracket is illustrated in Figure 4–1. Technician A says if the bolt head is touching the bracket, the bracket should be replaced.

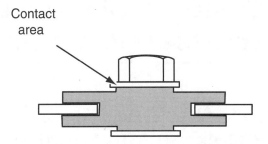

Contact area

**Figure 4–1** Collapsible steering column bracket.

Technician B says if the bolt head is touching the bracket, the shear load is too low.
Who is correct?
A. A only
B. B only
C. Both A and B
D. Neither A nor B

**Hint**    *Before servicing the steering column in an air bag–equipped vehicle, always disconnect the negative battery cable and wait for the time period specified by the vehicle manufacturer. This action prevents accidental air bag deployment. When the air bag module is removed from the steering wheel, always place it facing upward on the work bench.*

**Task 3**   **Disarm, enable, and properly handle air bag system components during vehicle service following manufacturers' procedures.**

3. All of the following statements about air bag system disarming and air bag handling are true EXCEPT
   A. Some air bag systems are disarmed by disconnecting the battery cable and waiting for the specified time period.
   B. Air bag system disarming on some vehicles includes disconnecting the air bag system fuse.
   C. The surface of deployed air bags may be coated with sodium bicarbonate.
   D. After the air bag system has been disarmed and enabled, the air bag system warning light operation indicates proper system operation.

*Hint*      *On many air bag systems, the air bag disarming procedure involved disconnecting the nega-tive battery cable and waiting for the time period specified by the vehicle manufacturer. This time period was usually two to ten minutes depending on the vehicle. The wait period was necessary to allow the backup power supply to power down in the air bag system and prevent accidental air bag deployment when servicing the air bag system. Always wear protective equipment when handling deployed air bags because they may be coated with sodium hydroxide which is a skin and eye irritant.*

*On some later model vehicles, the manufacturer's air bag system disarming procedure requires turning off the ignition switch, removing the ignition key, and disconnecting the air bag system fuse from the fuse block. After these steps are taken, disconnect the two-wire connectors at the base of the steering column (driver's side air bag module), the passenger's side air bag module, the driver's side impact inflator module under the driver's seat, and the passenger's side air bag inflator module under the passenger's seat. After the service work is completed on the air bag system, the air bag fuse is reinstalled and the two-wire connectors are connected to enable the air bag system. Turn on the ignition switch and observe the air bag system warning light for proper operation.*

# Steering Systems Diagnosis and Repair, Steering Units

## ASE Tasks, Questions, and Related Information

**Task 1**    Diagnose steering gear (non-rack and pinion type) noises, binding, vibra-tion, freeplay, steering effort, steering pull (lead), and leakage concerns; determine needed repairs.

   4. Excessive steering effort on a manual steering gear (non-rack and pinion type) may be caused by
      A.  a loose worm bearing preload adjustment.
      B.  an overfilled steering gear.
      C.  less-than-specified positive caster.
      D.  a tight sector lash adjustment.

*Hint*      *Excessive steering effort on a manual steering gear (non-rack and pinion type) may be caused by a tight worm bearing preload or sector lash adjustment. This problem may also be caused by lack of steering gear lubricant or the wrong lubricant. Binding steering linkages such as an idler arm may also cause excessive steering effort.*

**Task 2**    Diagnose rack and pinion steering gear noises, binding, vibration, freeplay, steering effort, steering pull (lead), and leakage concerns; deter-mine needed repairs.

   5. The cause of poor returnability on a manual rack and pinion steering gear could be
      A.  a loose rack bearing adjustment.
      B.  insufficient or improper lubricant in the steering gear.
      C.  loose steering gear mounting bolts.
      D.  excessive positive camber on both front wheels.

*Hint*    *Hard steering on a manual rack and pinion steering gear may be caused by a tight rack bearing adjustment, or lack of steering gear lubrication. Loose steering may be caused by worn inner or outer tie rod ends, loose steering gear mounting bolts, or worn steering gear mounting bushings.*

*Excessive engine vibration may be transmitted to the steering wheel by a rough engine idle problem, worn engine mounts, or improper steering gear mounting.*

## Task 3   Inspect power steering fluid level and condition; determine fluid type and adjust fluid level in accordance with vehicle manufacturers' recommendations.

6. Many vehicle manufacturers recommend checking the power steering fluid level with the fluid temperature at
   A. 60°F (15.5°C).
   B. 100°F (38°C).
   C. 140°F (60°C).
   D. 175°F (80°C).

*Hint*    *The power steering fluid level should be checked with the power steering fluid at the normal operating temperature of 175°F (80°C). Inspect the power steering fluid for foaming, discoloration, and contamination. Foaming of the power steering fluid usually indicates a low fluid level or air in the system. If the fluid is discolored or contaminated, the system should be flushed and filled with new fluid. Discolored power steering fluid may be caused by excessive heat or the wrong type of fluid. If the fluid is contaminated with metal particles, the pump or steering gear are worn.*

## Task 4   Inspect, adjust, align, and replace power steering pump belt(s) and tensioners.

7. When checking the power steering pump belt tension, for every foot of free span, the belt deflection should be
   A. 1/2 in.
   B. 1 in.
   C. 1.5 in.
   D. 1.75 in.

8. Technician A says a belt tension gauge may be used to measure the tension on a ribbed V-belt.
   Technician B says some ribbed V-belts have a spring-loaded tensioner with a belt length scale.
   Who is correct?
   A. A only
   B. B only
   C. Both A and B
   D. Neither A nor B

*Hint*    *Reduced power steering pump belt tension may cause increased or erratic steering effort. Power steering pump belt tension usually is measured with a belt tension gauge. Belt tension may be checked by measuring the belt deflection. The specified deflection is usually 1/2 in. per foot of belt free span (Figure 4–2). If the power steering belt must be tightened, loosen the bracket bolt and pry on the pump ear with a pry bar to tighten the belt. Hold the pump so the belt has the proper tension, and tighten the bracket bolt.*

## Task 5   Diagnose power steering pump noises, vibration, and fluid leakage; determine needed repairs.

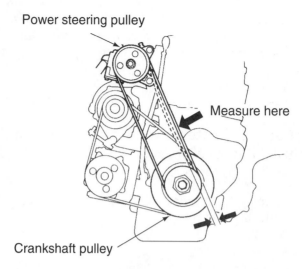

Power steering pulley

Measure here

Crankshaft pulley

**Figure 4–2** Power steering pump belt deflection. *(Courtesy of American Honda Motor Co., Inc.)*

9. Technician A says a squealing noise during acceleration may be caused by a loose power steering belt.

   Technician B says premature power steering pump belt wear may be caused by a misaligned power steering pump. Who is correct?
   A. A only
   B. B only
   C. Both A and B
   D. Neither A nor B

*Hint*     *With the engine running, a growling noise from the power steering pump may be caused by low power steering fluid level and air in the power steering fluid. A growling noise from the power steering pump with the engine running may also be caused by defective pump bearings. A squealing noise during acceleration only may be caused by a loose power steering belt. If the power steering pump has a V-belt, squeaking noise with the engine idling may be caused by dry friction surfaces on the belt. A rattling noise from the power steering pump may be caused by loose, worn power steering pump mountings.*

## Task 6 Remove and replace power steering pump; inspect pump mounting and attaching brackets; remove and replace power steering pump pulley.

10. The tool in Figure 4–3 is used to
    A. remove a pressed-on power steering pump pulley.
    B. remove a bolt-on power steering pump pulley.
    C. install a pressed-on power steering pump pulley.
    D. remove the power steering pump pulley retaining nut.

*Hint*     *After the power steering pump is removed, the pump mounting bolts and bolt openings must be inspected for wear. Worn pump mounting bolts or bolt openings cause power steering pump pulley misalignment, which results in excessive belt wear. Worn pump mountings must be replaced. After the power steering pump is reinstalled, tighten the belt to the specified tension, and be sure the pump pulley is properly aligned. All pump mounting bolts must be tightened to the specified torque. A special puller must be used to remove a pressed-on power steering pump pulley. This type of pulley also requires a special installation tool. If the power steering pump pulley is retained with a nut, the pulley must be held from turning with a special tool while the nut is loosened.*

## Task 7 Inspect and replace power steering pump seals, gaskets, reservoir and valves.

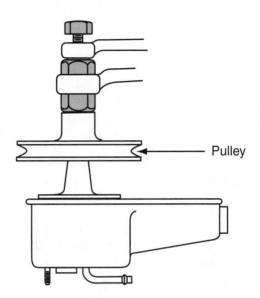

**Figure 4–3** Power steering pump service tool.

11. A power steering pump with an integral reservoir is leaking fluid between the reservoir and the pump housing.

    Technician A says the drive shaft seal may require replacement.

    Technician B says the vent in the pump cap may be restricted.

    Who is correct?

    A. A only

    B. B only

    C. Both A and B

    D. Neither A nor B

*Hint*    *If the power steering pump is leaking around the shaft behind the pulley, the input shaft seal is leaking. When the input shaft seal is leaking, a streak of oil may be sprayed on the underside of the vehicle hood. If the power steering pump is leaking between the reservoir and the pump body, the large reservoir O-ring may be leaking. Fluid leaks may also occur where the high pressure hose is connected to the pump fitting.*

**Task 8** **Perform power steering system pressure and flow tests; determine needed repairs.**

12. During the power steering pump pressure test, the valve shown in Figure 4–4 should be closed for

    A. Ten seconds.

    B. Fifteen seconds.

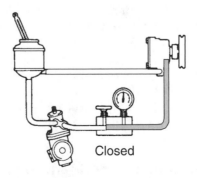

Closed

**Figure 4–4** Power steering pump pressure test. (*Reprinted with permission*)

C. Twenty seconds.

D. Thirty seconds.

**Hint**    *The power steering pump pressure test must be performed with the fluid at normal operating temperature and no air in the system. A pressure gauge and valve assembly is connected in the pressure hose between the power steering pump and gear. With the engine idling, close the valve for a maximum of ten seconds and read the pump pressure. If the power steering pump pressure is less than specified, the pump pressure relief valve may be sticking or the pump may be the problem.*

## Task 9    Inspect and replace power steering hoses, fittings, O-rings, coolers, and filters.

13. During a power steering pump pressure test, the pump pressure is satisfactory, but the steering wheel turning effort is excessive.

    Technician A says the steering gear may be the problem.

    Technician B says the high-pressure hose may be restricted.

    Who is correct?

    A. A only

    B. B only

    C. Both A and B

    D. Neither A nor B

**Hint**    *Power steering hoses should be checked for leaks, dents, restrictions, sharp bends, cracks, and contact with other components. Restricted power steering hoses reduce pressure supplied to the steering gear and increase steering effort.*

## Task 10    Remove and replace steering gear (non-rack and pinion type).

14. If the steering wheel is rotated when the steering gear is disconnected from the steering shaft

    A. the steering gear may be misaligned when it is reinstalled.

    B. the clockspring under the steering wheel may not be centered when the steering gear is reinstalled.

    C. the steering effort may not be equal during right and left turns.

    D. the steering shaft may have a binding condition in the steering column.

**Hint**    *If the vehicle is air bag–equipped, disconnect the negative battery cable and wait for the time period specified by the vehicle manufacturer before proceeding with the steering gear removal procedure.*

*Before removing the steering gear, center the front wheels and remove the key from the ignition switch to lock the steering column. Do not rotate the steering column after the steering gear is removed. This procedure centers the clockspring mounted under the steering wheel. The clockspring maintains electrical contact between the air bag system and the air bag module mounted in the steering wheel. Mark the steering shaft in relation to the stub shaft in the steering gear prior to removing the steering gear from the steering shaft. When the steering gear is reinstalled, be sure the front wheels are centered and the marks on the steering shaft and steering gear stub shaft are aligned. This maintains the clockspring in the centered position. If the clockspring is not installed in the centered position, this component may be broken when the steering wheel is rotated fully in either direction.*

*Always inspect the steering gear mounting bolts for wear or a bent condition. Worn or bent steering gear mounting bolts must be replaced. Replacement steering gear bolts must be the same as the original bolts.*

## Task 11    Remove and replace rack and pinion steering gear; inspect and replace mounting bushings and brackets.

15. When removing the steering gear on an air bag–equipped vehicle, Technician A says rotating the steering wheel with the steering gear disconnected may damage the clockspring electrical connector.

    Technician B says after the negative battery cable is disconnected, immediately begin the steering gear removal procedure.

    Who is correct?

    A. A only

    B. B only

    C. Both A and B

    D. Neither A nor B

**Hint**   *On an air bag–equipped vehicle, the steering wheel should be locked in the centered position before the steering wheel or steering gear is removed. This action maintains the clockspring electrical connector in the centered position. Before starting the steering gear removal procedure on an air bag–equipped vehicle, disconnect the negative battery cable and wait for the time specified by the vehicle manufacturer. Always punchmark the lower universal joint in relation to the steering gear shaft prior to steering gear removal.*

## Task 12   Adjust steering gear (non-rack and pinion type) worm bearing preload and sector lash.

16. In the power-steering gear adjustment in Figure 4–5, Technician A says after the adjuster plug is bottomed it should be tightened to 20 ft.-lbs.

    Technician B says the adjuster plug should be backed off one-quarter turn from the bottomed position.

    Who is correct?

    A. A only

    B. B only

    C. Both A and B

    D. Neither A nor B

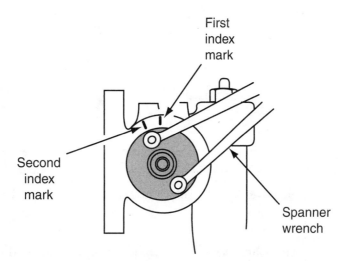

First
index
mark

Second
index
mark

Spanner
wrench

**Figure 4–5** Power recirculating ball steering gear adjustment.

**Hint**   *When adjusting the worm shaft bearing preload on some power-recirculating ball steering gears, the worm shaft bearing adjuster plug must be bottomed and then tightened to 20 ft.-lbs.*

*An index mark should be placed on the housing next to, or on, the adjuster plug holes. Place a second index mark 0.050 in. from the original index mark in a counterclockwise direction. Rotate the adjuster plug until the hole in the adjuster plug is aligned with the second index mark.*

*Prior to the sector shaft lash adjustment, the lash adjuster screw is turned fully counterclockwise. This screw is then rotated one turn clockwise. With the worm shaft in the center position, record the turning torque required to rotate the steering shaft. Turn the sector lash adjustment screw clockwise until the turning torque is 4 to 10 in.-lbs. more than the turning torque recorded with the sector lash screw backed off. Tighten the locknut on the sector shaft lash adjusting screw to the specified torque.*

## Task 13    Inspect and replace steering gear (non-rack and pinion type) seals and gaskets.

17. When servicing the bearing and seal in the adjuster plug of a power-recirculating ball steering gear (Figure 4–6), Technician A says the part number on the needle bearing should face the driving tool when installing this bearing.

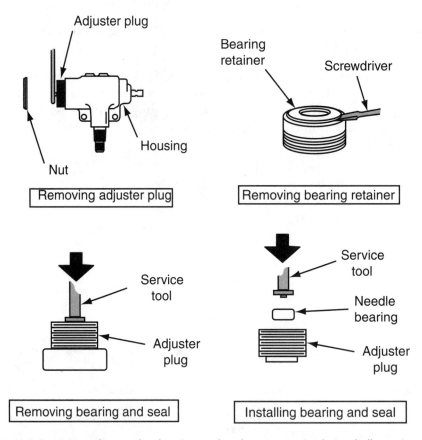

**Figure 4–6** Servicing adjuster plug bearing and seal, power recirculating ball steering gear.

Technician B says the worm shaft bearing preload must be adjusted after the bearing and seal installation in the adjuster plug.

Who is correct?

A.  A only

B.  B only

C.  Both A and B

D.  Neither A nor B

**Hint**    *Possible leak locations on a power-recirculating ball steering gear include the pitman shaft seal, the side cover O-ring seal, the adjuster plug seal, the top cover seal, and the pressure line fitting.*

## Task 14    Adjust rack and pinion steering gear.

18. Poor steering wheel returnability is experienced on a power rack and pinion steering gear.
    Technician A says the steering gear may be misaligned on the chassis.
    Technician B says the rack bearing adjustment may be too tight.
    Who is correct?
    A. A only
    B. B only
    C. Both A and B
    D. Neither A nor B

**Hint**    *When the rack bearing adjustment is performed, a torque wrench and socket is installed on top of the pinion shaft. Turn the torque wrench back and forth to rotate the rack in both directions.*

*Rotate the rack bearing adjuster plug until the specified turning torque is obtained on the torque wrench. Hold the adjuster plug in this position, and tighten the adjuster plug locknut to the proper torque.*

## Task 15    Inspect and replace rack and pinion steering gear bellows/boots.

19. In Figure 4–7, component 13 is
    A. a spacer.
    B. a rack bushing.
    C. a shock dampener.
    D. a rack seal.

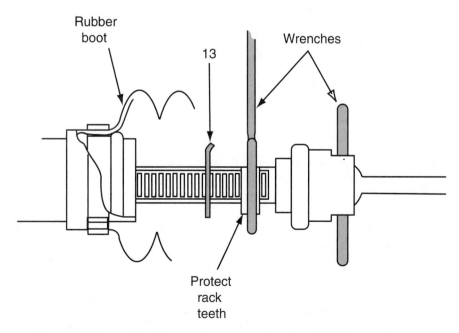

**Figure 4–7** Inner tie rod components, power rack, and pinion steering gear.

*Hint*    *Some inner tie rod ends are threaded onto the rack. On this type of gear, the rack must be held from rotating while the inner tie rods are loosened or tightened. These inner tie rod ends must be tightened to the specified torque. Some inner tie rods are staked in place after they are properly torqued. A claw washer retains the inner tie rods on other steering gears. Some inner tie rod ends are retained on the rack with a steel pin.*

## Task 16  Flush, fill, and bleed power steering systems.

20. While bleeding air from a power steering system, Technician A says if foaming is present in the reservoir after the bleeding process, the bleeding procedure should be repeated.

    Technician B says each time the steering wheel is rotated fully right or left, it should be held in this position for two or three seconds.

    Who is correct?

    A.  A only

    B.  B only

    C.  Both A and B

    D.  Neither A nor B

*Hint*    *A power steering system should be drained through the return hose. If the system has a remote reservoir, disconnect the return hose from this reservoir to the steering gear and plug the reservoir outlet. Place the disconnected return hose in a drain pan. Flush the fluid out of the system by starting the engine and allowing it to idle. Shut the engine off and add more fluid to the reservoir to continue flushing the system.*

*With the reservoir filled to the proper level, operate the engine at 1,000 rpm and rotate the steering wheel fully right and left three or four times to bleed air from the system. Hold the steering wheel in the fully right and left position for two or three seconds. Be sure the reservoir is filled to the proper level. If foaming is still present in the reservoir, repeat the bleeding procedure.*

## Task 17  Diagnose, inspect, repair or replace components of variable-assist steering systems.

21. In the variable-assist power steering system in Figure 4–8, Technician A says the power steering assist is increased at speeds above 50 mph.

    Technician B says the power steering assist is increased when the steering wheel rotation exceeds 15 rpm.

    Who is correct?

    A.  A only

    B.  B only

    C.  Both A and B

    D.  Neither A nor B

*Hint*    *The control module in some variable-assist power steering systems receives input signals from the vehicle speed sensor (VSS) and the steering wheel rotation sensor. In response to these inputs, the control module operates an actuator in the power steering pump to control pump pressure and steering assist.*

*When the vehicle speed is below 10 mph, the control unit positions the actuator to provide full-power steering assist. If the vehicle speed is between 10 mph and 25 mph, the control unit gradually reduces the power steering pump pressure and steering assist. Above 25 mph, the control unit provides a reduced power steering assist.*

*When the steering wheel rotation is above 15 rpm, the control unit provides increased power steering pump pressure and steering assist. If the steering wheel rotation is below 15 rpm, the control unit supplies a reduced power steering assist. Some variable assist power steering systems do not have a steering wheel rotation sensor, and these systems are only vehicle speed sensitive.*

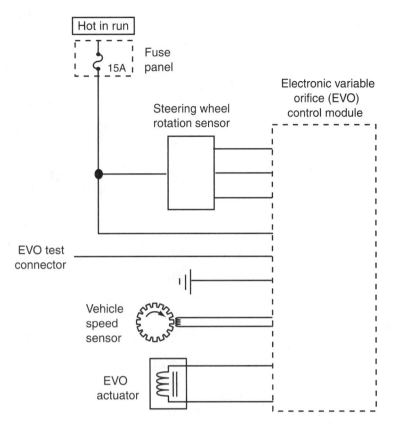

**Figure 4–8** Variable assist power steering system.

# Steering Systems Diagnosis and Repair, Steering Linkage

## ASE Tasks, Questions, and Related Information

**Task 1**  **Inspect and adjust (where applicable) front and rear steering linkage geometry (including parallelism and vehicle ride height).**

22. The measurement in Figure 4–9 is more than specified.
    This problem may result in
    A. front wheel shimmy.
    B. steering pull to the left.
    C. steering pull to the right.
    D. excessive steering effort.

*Hint*   *The steering arms must be parallel to the lower control arms. In a parallelogram steering linkage, the idler arm and pitman arm maintain the proper steering arm position. The steering gear mounting positions the steering arms parallel to the lower control arms in a rack and pinion steering system. If the steering arms are not parallel to the lower control arms, a condition called*

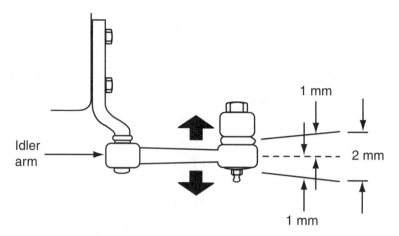

**Figure 4–9** Idler arm movement check.

*bump steer may occur. Bump steer may be defined as the tendency of the steering to veer suddenly in one direction when one or both front wheels strike a bump. A worn idler arm may cause wheel shimmy, bump steer, and steering looseness.*

## Task 2    Inspect and replace pitman arm.

23. Bump steer is experienced during a road test for steering diagnosis.
    Technician A says the steering gear may require adjustment.
    Technician B says the pitman arm may be bent.
    Who is correct?
    A. A only
    B. B only
    C. Both A and B
    D. Neither A nor B

**Hint**    *Bent steering linkage components must be replaced. Never attempt to heat or straighten bent steering components such as pitman arms.*

## Task 3    Inspect and replace center link (relay rod/drag link/intermediate rod).

24. During a front suspension inspection, the Technician discovers a bent relay rod.
    Technician A says this problem changes the front wheel toe setting.
    Technician B says this problem may cause front tire wear on the inside edges.
    Who is correct?
    A. A only
    B. B only
    C. Both A and B
    D. Neither A nor B

**Hint**    *A bent relay rod moves the front wheel toe setting to a toe-out position. This condition may cause feathered wear across the front tire treads.*

## Task 4    Inspect, adjust (where applicable), and replace idler arm(s) and mountings.

25. A vehicle requires excessive steering effort. The power steering belt is tight and the reservoir is filled to the specified level. The cause of this problem could be
    A. worn lower ball joints.
    B. weak front springs.

C. a seized idler arm.

D. a weak stabilizer bar.

**Hint**    *Excessive steering effort may be caused by a slipping power steering belt, low fluid level, air in the power steering system, tight steering gear adjustments, worn steering gear, or a seized steering linkage component such as an idler arm.*

## Task 5   Inspect, replace, and adjust tie rods, tie rod sleeves/adjusters, clamps, and tie rod ends (sockets/bushings).

26. While discussing tie rod sleeve adjustments, Technician A says the tie rod sleeves may be rotated to adjust front wheel toe.

    Technician B says the tie rod sleeves may be rotated to center the steering wheel.

    Who is correct?

    A. A only

    B. B only

    C. Both A and B

    D. Neither A nor B

**Hint**    *A special tie rod sleeve adjusting tool must be used to rotate the tie rod sleeves. Never rotate these sleeves with vise grips or a pipe wrench. Loosen the tie rod sleeve clamp bolts before rotating the sleeves. These sleeves may be rotated to adjust the front wheel toe or center the steering wheel. After the tie rod sleeves are rotated, the sleeve clamps must be positioned with the clamp openings away from the slots in the sleeve. After the clamp bolts are tightened to the specified torque, the outer clamp ends may touch, but some gap must be present between the clamp ends next to the sleeve (Figure 4–10).*

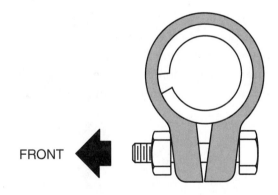

FRONT

**Figure 4–10** Proper clamp position on a tie rod sleeve.

## Task 6   Inspect and replace steering linkage damper(s).

27. A 4WD vehicle has excessive road shock on the steering wheel while driving on irregular road surfaces.

    Technician A says the power steering gear cylinder may be scored.

    Technician B says the steering damper may be worn out.

    Who is correct?

    A. A only

    B. B only

    C. Both A and B

    D. Neither A nor B

**Hint**    *The steering damper usually is connected from the relay rod to the chassis. The steering damper acts like a shock absorber to reduce the transfer of road shock from the steering linkage to the steering wheel.*

# Suspension Systems Diagnosis and Repair, Front Suspensions

## ASE Tasks, Questions, and Related Information

**Task 1**  Diagnose front suspension system noises, body sway/roll, and ride height concerns; determine needed repairs.

28. The front suspension ride height is less than specified on a MacPherson strut suspension system. The cause of this problem could be
   A. worn-out struts.
   B. worn upper strut mounts.
   C. worn steering gear mounting bushings.
   D. worn lower control arm bushings.

**Hint**    *Improper ride height affects many other wheel alignment angles. Therefore, ride height measurement is always one of the first checks when performing a wheel alignment. Reduced ride height may be caused by sagged springs, worn lower control arm bushings, or bent suspension components.*

**Task 2**  Inspect and replace upper and lower control arms, bushings and shafts.

29. Tool A in Figure 4–11 is used to
   A. compress the coil spring.
   B. position the shock absorber rod.
   C. measure ball joint movement.
   D. measure lower control arm bushing wear.

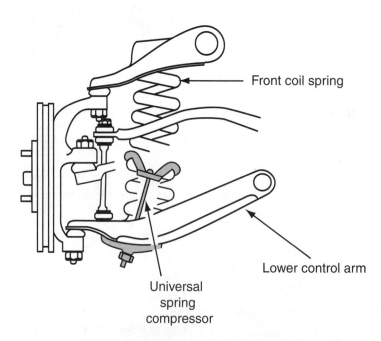

Front coil spring

Lower control arm

Universal
spring
compressor

**Figure 4–11** Front suspension service tool.

*Hint*    *Worn upper or lower control arm bushings on a front suspension may cause improper camber or caster angles. These improper angles may result in steering pull and/or tire tread wear. A rattling or squeaking noise on irregular road surfaces may be caused by worn control arm bushings.*

## Task 3    Inspect and replace rebound and jounce bumpers.

30. Both rebound bumpers are severely damaged on a front suspension system. The most likely cause of this problem is
    A. heavy-duty coil springs have been installed on the front suspension.
    B. worn upper and lower ball joints.
    C. worn front-wheel bearings.
    D. defective front shock absorbers.

*Hint*    *Rebound and jounce bumpers should be inspected for cracking, damage, and looseness. If the rebound bumpers are severely damaged, the suspension curb riding height may be less than specified, or the shock absorbers may be defective.*

## Task 4    Inspect, adjust, and replace strut rods/radius arms (compression/tension), and bushings.

31. The steering on a vehicle pulls to the right while braking, and all the brake components are in satisfactory condition.
    Technician A says the right front strut rod bushing may be worn.
    Technician B says there may be excessive negative camber on the right front wheel.
    Who is correct?
    A. A only
    B. B only
    C. Both A and B
    D. Neither A nor B

*Hint*    *The strut rods are connected from the lower control arm to the chassis. A large rubber bushing is mounted between the front of the strut rod and the chassis. The strut rods prevent fore and aft lower control arm movement. A worn strut rod bushing may allow the outer end of the lower control arm to move rearward while braking. This action provides reduced positive caster on that side of the front suspension. Since the steering pulls to the side with the least positive caster, a worn strut rod bushing may cause steering pull while braking.*

## Task 5    Inspect and replace upper and lower ball joints (with or without wear indicators).

32. When unloading the spring tension and vehicle weight from the ball joints, prior to ball joint wear measurement on a long-and-short arm front suspension system with the coil spring positioned between the lower control arm and the chassis
    A. a jack must be placed under the chassis.
    B. the shock absorber must be disconnected.
    C. the ride height must be within specifications.
    D. a jack must be placed under the lower control arm.

*Hint*    *On a long-and-short arm suspension system, the spring tension and vehicle weight must be unloaded from the ball joints before the ball joint wear is measured. To unload the ball joints on a long-and-short arm suspension system with the coil spring between the lower control arm and the chassis, place a jack under the lower control arm. Operate the jack until the front tire is lifted off the shop floor.*

*When the coil spring is mounted between the upper control arm and the chassis, place a steel spacer between the upper control arm and the chassis. When this spacer is in place, position a jack under the chassis. Operate the jack until the front tire is lifted off the floor.*

*To measure vertical ball joint wear, position a dial indicator stem against the top of the ball joint stud, and use a pry bar to apply a lifting force under the tire. Place a dial indicator stem against the inside of the front wheel rim, and pull the tire and wheel inward and outward to measure horizontal ball joint wear.*

## Task 6    Inspect non-independent front axle assembly for bending, warpage, and misalignment.

33. On a truck with a non-independent front axle, the right front tire tread is worn on the inside edge.

    Technician A says to install a different shim between the front axle and the leaf spring.

    Technician B says to replace the rear spring shackle.

    Who is correct?

    A. A only
    B. B only
    C. Both A and B
    D. Neither A nor B

**Hint**    *A bent front axle is usually indicated by lack of steering control, and abnormal front tire wear. If the front tires are worn on one edge, the king pins and bushings may be worn or the axle may be bent. Use a dial indicator to measure the king pin and bushing wear and perform a wheel alignment to locate the source of the problem.*

## Task 7    Inspect and replace front steering knuckle/spindle assemblies and steering arms.

34. A vehicle has excessive tire squeal while cornering.

    The cause of this problem could be
    A. a bent steering arm.
    B. excessive negative caster.
    C. worn stabilizer bushings.
    D. worn-out front struts.

**Hint**    *A bent steering arm affects the toe-out-on turns angle. Improper toe-out-on turns, or an incorrect toe setting may cause excessive tire squeal while cornering. A bent steering knuckle may change the front wheel camber setting. In many suspension systems, the steering knuckle and arm are combined in one component. Bent steering arms or knuckles must be replaced.*

## Task 8    Inspect and replace front suspension system coil springs and spring insulators (silencers).

35. The coil spring in Figure 4–12 is
    A. a linear-rate spring.
    B. a variable-rate spring.
    C. a heavy-duty spring.
    D. a conventional spring.

36. While using a spring compressor to remove a coil spring from a strut (Figure 4–13), Technician A says the spring should be taped in the areas where the compressor contacts the spring.

    Technician B says all the spring tension must be removed from the upper strut mount before loosening the strut rod nut.

**Figure 4–12** Coil spring. *(Courtesy of Sealed Power Corporation)*

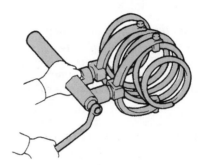

**Figure 4–13** Coil spring compressor. *(Reprinted with permission)*

Who is correct?
A. A only
B. B only
C. Both A and B
D. Neither A nor B

**Hint**    *Linear-rate or conventional coil springs have equal spacing between the coils and one basic shape with consistent wire diameter. Many variable rate springs have a cylindrical shape and unequally spaced coils with consistent wire diameter. Heavy-duty springs are designed to carry 3-percent to 5-percent greater loads compared to conventional springs. The wire diameter may be up to 0.100 in. greater in a heavy-duty spring than in a conventional spring.*

*When removing a coil spring from a strut, the spring should be taped in the areas contacted by the spring compressor. If the coating on a coil spring is chipped, the spring may break. Before loosening the strut rod, the spring compressor must be operated until all spring tension is removed from the upper strut mount. Coil springs must be replaced in pairs.*

## Task 9   Inspect and replace front suspension system leaf spring(s), leaf spring insulators (silencers), shackles, brackets, bushings, and mounts.

37. The steering pulls to the right while driving straight ahead on a truck with a long-and-short arm front suspension and a leaf spring rear suspension. All of the following could be the cause of the problem EXCEPT
    A. a broken center bolt in the left rear spring.
    B. more positive caster on the left front wheel than the right front wheel.
    C. more positive camber on the right front wheel than the left front wheel.
    D. excessive toe-in on the front wheels.

**Hint**    *Worn leaf spring silencers may cause creaking and squawking noises while driving on road irregularities. A broken center bolt in a rear leaf spring may allow the differential housing to move rearward on one side. This problem affects vehicle tracking and causes steering pull.*

## Task 10    Inspect, replace, and adjust front suspension system torsion bars and mounts.

38. On a torsion bar front suspension system, the ride height is below specifications on the right front side of the chassis. The ride height is satisfactory on the left front of the chassis.

   Technician A says the lower control arm bushing may be worn.

   Technician B says the right front torsion bar anchor bolt may need adjusting.

   Who is correct?

   A. A only

   B. B only

   C. Both A and B

   D. Neither A nor B

*Hint*    *Torsion bars replace the coil springs in some suspension systems. Weak torsion bars or worn torsion bar mounting bushings may cause a reduced chassis riding height. On many torsion bar suspension systems, the torsion bar anchor bolt may be adjusted to change the ride height.*

## Task 11    Inspect and replace front stabilizer bar (sway bar) bushings, brackets, and links.

39. A vehicle has excessive body sway while cornering. All of the following could be the cause of the problem EXCEPT

   A. worn strut rod bushing.

   B. a weak stabilizer bar.

   C. worn stabilizer bar bushings.

   D. a broken stabilizer link.

*Hint*    *The stabilizer bar prevents excessive body sway while cornering. The stabilizer bar is mounted to the chassis and the control arms with heavy rubber bushings.*

## Task 12    Inspect and replace front strut cartridge or assembly.

40. In Figure 4–14, the technician is

   A. adjusting front wheel camber to correct tire wear.

   B. loosening the strut rod nut prior to strut removal.

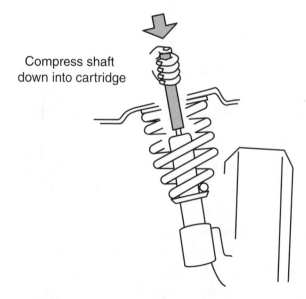

Compress shaft
down into cartridge

**Figure 4–14** Front suspension service procedure.

C. pushing the strut rod downward prior to cartridge removal.

D. aligning the strut rod during strut installation.

*Hint*     *A new cartridge may be installed in some front struts with the strut installed in the vehicle. Other struts must be removed to allow cartridge installation. On some struts, there is no provision for cartridge installation. Prior to strut removal from the vehicle, the upper strut mounting nuts and the strut-to-steering knuckle bolts must be removed. A spring compressor must be used to compress the spring before the spring is removed from the strut.*

## Task 13   Inspect and replace front strut bearing and mount.

41. A customer complains about steering chatter while cornering on a MacPherson strut front suspension system. With the vehicle parked on the shop floor, the Technician can feel a binding and releasing action on the left-front spring as the steering wheel is turned.

    Technician A says the upper strut mount may be the problem.

    Technician B says the lower ball joint may have excessive wear.

    Who is correct?

    A. A only

    B. B only

    C. Both A and B

    D. Neither A nor B

*Hint*     *A worn upper strut mount may cause steering chatter while cornering and poor steering wheel return. A damaged upper strut mount may result in improper camber or caster angles on the front suspension.*

# Suspension Systems Diagnosis and Repair, Rear Suspensions

## ASE Tasks, Questions, and Related Information

## Task 1   Diagnose rear suspension system noises, body sway/roll, and ride height concerns; determine needed repairs.

42. A customer complains about harsh riding and bottoming of the rear suspension.

    Technician A says the rear struts may be the problem.

    Technician B says the rear suspension ride height may be less than specified.

    Who is correct?

    A. A only

    B. B only

    C. Both A and B

    D. Neither A nor B

*Hint*     *A squeaking noise in the rear suspension may be caused by suspension bushings, damaged struts or shock absorbers, or broken springs or spring insulators. Excessive rear suspension lateral movement may be caused by a worn track bar or bushings. Harsh riding may be caused by reduced rear suspension ride height, and damaged struts or shock absorbers. Because rear suspension ride height affects some of the front suspension angles, the ride height must be measured before an alignment procedure.*

## Task 2   Inspect and replace rear suspension system coil springs and spring insulators (silencers).

43. A vehicle has excessive rear suspension oscillations.
    Technician A says the rear struts may be the problem.
    Technician B says the rear coil springs may be weak.
    Who is correct?
    A. A only
    B. B only
    C. Both A and B
    D. Neither A nor B

**Hint**      *Weak coil springs cause harsh riding and reduced curb riding height. Broken springs or spring insulators cause a rattling noise while driving on irregular road surfaces. Worn-out struts or shock absorbers result in chassis oscillations and harsh riding.*

## Task 3   Inspect and replace rear suspension system lateral links/arms (track bars), control (trailing) arms, stabilizer bars (sway bars), bushings, and mounts.

44. While servicing the rear suspension system in Figure 4–15, Technician A says the ball joint nut may be loosened to align the cotter key hole with the nut castellations.

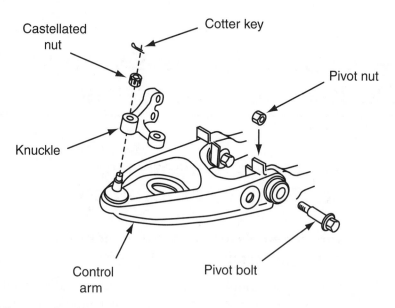

**Figure 4–15** Rear suspension system with lower control arm and ball joint.

Technician B says the lower control arm retaining bolts should be torqued with the vehicle weight on the suspension.
Who is correct?
A. A only
B. B only
C. Both A and B
D. Neither A nor B

*Hint*    *Worn track bar bushings or mounts cause excessive lateral chassis movement. Excessive rear body sway and harsh riding may be caused by a weak stabilizer bar or worn bushings, links, and mounts. Worn lower control arm bushings or ball joint may cause improper rear-wheel toe and camber settings. This condition may result in tire wear or steering pull.*

**Task 4**    **Inspect and replace rear suspension system leaf spring(s), leaf spring insulators (silencers), shackles, brackets, bushings, and mounts.**

45. On a rear suspension with two longitudinally mounted leaf springs, the left rear spring is sagged and the left rear chassis curb riding height is less than specified. This problem could result in
    A. steering pull to the right while driving straight ahead.
    B. excessive left rear tire tread wear.
    C. excessive steering wheel free play.
    D. excessive left front tire tread wear.

*Hint*    *Sagged rear springs increase positive caster on the front suspension. When both rear springs are sagged, positive caster is increased on both front wheels. Under this condition, steering effort, and steering wheel returning force after a turn, are increased. Increased positive caster may cause harsh riding.*

*When one rear spring is sagged, the positive caster is increased on that side of the front suspension.*

*Because the steering tends to pull to the side with the least positive caster, this condition may cause the steering to pull to the opposite side from the sagged spring.*

**Task 5**    **Inspect and replace rear rebound and jounce bumpers.**

*Hint*    *Severely damaged rebound and jounce bumpers indicate reduced rear suspension ride height, or worn out rear shock absorbers or struts. Damaged rebound and jounce bumpers should be replaced, and the cause of this damage corrected.*

**Task 6**    **Inspect and replace rear strut cartridge or assembly, and upper mount assembly.**

46. All of the following may cause a rattling noise in the rear suspension EXCEPT
    A. a broken coil spring.
    B. a broken coil spring insulator.
    C. a bent rear strut.
    D. a worn track bar bushing.

*Hint*    *A similar procedure is used to remove and replace front or rear struts. A spring compressor must be used to remove all the spring tension from the upper strut mount before loosening the strut rod nut. A strut cartridge may be installed in some rear struts after they are removed from the chassis.*

**Task 7**    **Inspect non-independent rear axle assembly for bending, warpage, and misalignment.**

47. A non-independent rear axle is offset as indicated in Figure 4–16. The result of this problem could be
    A. steering wander while driving straight ahead.
    B. steering pull to the right while driving straight ahead.
    C. poor steering wheel returnability.
    D. steering pull to the left during hard acceleration.

*Hint*    *A non-independent rear axle may be checked for bending, warpage, and misalignment by measuring the rear wheel tracking. This operation may be performed with a track bar or a*

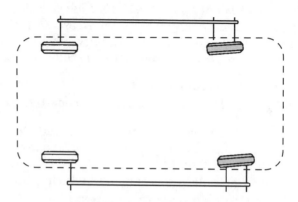

**Figure 4–16** Rear axle offset.

*computer wheel aligner with four-wheel capabilities. A track bar measures the position of the rear wheels in relation to the front wheels. A computer wheel aligner displays the thrust angle, which is the difference between the vehicle thrust line and geometric centerline of the vehicle.*

## Task 8    Inspect and replace rear ball joints and tie rod/toe link assemblies.

48. While discussing the ball joint in Figure 4–17, Technician A says the ball joint is worn and should be replaced.
    Technician B says the ball joint may cause an improper rear wheel camber setting.
    Who is correct?
    A. A only
    B. B only
    C. Both A and B
    D. Neither A nor B

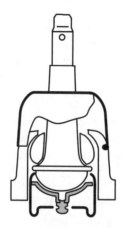

**Figure 4–17** Rear ball joint.

49. The left rear wheel tie rod in Figure 4–18 is longer than specified. This problem could cause
    A. excessive toe-out.
    B. excessive positive camber.
    C. excessive wear on the inside edge of the tire tread.
    D. steering pull to the left.

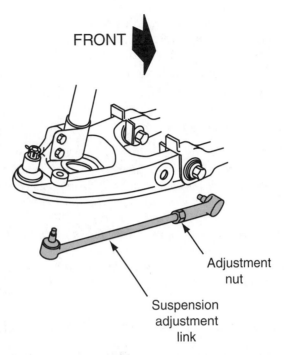

FRONT

Adjustment
nut

Suspension
adjustment
link

**Figure 4–18** Rear suspension tie rod.

*Hint*      *Rear ball joints should be inspected in much the same way as front ball joints. Some rear suspension tie rods are connected from the knuckle to the lower control arm. The length of the tie
rod determines the rear wheel toe setting.*

## Task 9   Inspect and replace rear knuckle/spindle assembly.

50. A front wheel drive car with an independent rear suspension system experiences
rapid tread wear with a feathered appearance on the right-rear tire.
Technician A says the right-rear wheel may require a toe adjustment.
Technician B says the right-rear spindle may be bent.
Who is correct?
A.  A only
B.  B only
C.  Both A and B
D.  Neither A nor B

*Hint*      *A bent steering spindle may change the rear wheel camber or toe setting. Bent spindles must
be replaced.*

# Related Suspension and Steering Service

## ASE Tasks, Questions,
## and Related Information

## Task 1   Inspect and replace shock absorbers, mounts, and bushings.

51. When one side of the front or rear bumper is pushed downward with considerable weight and then released, the bumper makes two free upward bounces before the vertical chassis movement stops. This action indicates
    A. an inoperative shock absorber.
    B. a weak coil spring.
    C. a broken spring insulator.
    D. worn stabilizer bushings.

**Hint**    *When one side of the bumper is pushed downward with considerable weight and then released, a satisfactory shock absorber or strut should complete one free upward bounce before the vertical chassis movement stops. If the bumper completes more than one free upward bounce, the shock absorber is inoperative, or the shock absorber mountings are loose. This is commonly called the shock absorber bounce test.*

*A manual test may be performed on shock absorbers. During this test, the lower end of the shock absorber is disconnected, and hand pressure is used to extend and compress the shock. A satisfactory shock absorber should provide a strong, steady resistance to movement on the compression and rebound strokes. The resistance may be different on the compression and rebound strokes.*

## Task 2    Diagnose and service front and/or rear wheel bearings.

52. After servicing and lubricating the rear-wheel bearings in a front-wheel-drive car, the bearing adjusting nut is tightened to 20 ft.-lbs., and loosened one-half turn. The next step in the bearing adjusting procedure is to
    A. install the nut retainer and cotter key.
    B. tighten the adjusting nut so the cotter key hole lines up properly.
    C. tighten the adjusting nut to 10 to 15 in.-lbs.
    D. check the wheel bearing for free play.

**Hint**    *Loose front-wheel bearings may cause steering wander. When two tapered roller bearings are mounted in the front- or rear-wheel hub, a typical bearing adjustment procedure consists of the following:*
    *1. Tighten the bearing adjustment nut to 17 to 25 ft.-lbs.*
    *2. Back off the adjustment nut one-half turn.*
    *3. Tighten the bearing adjustment nut to 10 to 15 in.-lbs.*
    *4. Install the nut retainer and cotter key.*

## Task 3    Diagnose, inspect, adjust, repair, or replace components (including sensors, switches, and actuators) of electronically controlled suspension systems (including primary and supplemental air suspension and ride control systems).

53. The electronic suspension switch in Figure 4–19 must be in the off position under all of the following conditions EXCEPT
    A. when diagnosing the system with a scan tester.
    B. when jacking up the vehicle to change a tire.
    C. when hoisting the vehicle for undercar service.
    D. when towing the vehicle with a tow truck.

**Hint**    *There are several different types of electronically controlled suspension systems on cars today.*

*Some computer-controlled air suspension systems have an air spring in place of each coil spring.*

*On these systems, the computer operates a compressor that supplies air into each air spring until the specified curb riding height is obtained. Most of these systems have an air suspension switch in the trunk that must be off while jacking, towing, or hoisting the vehicle. In the off*

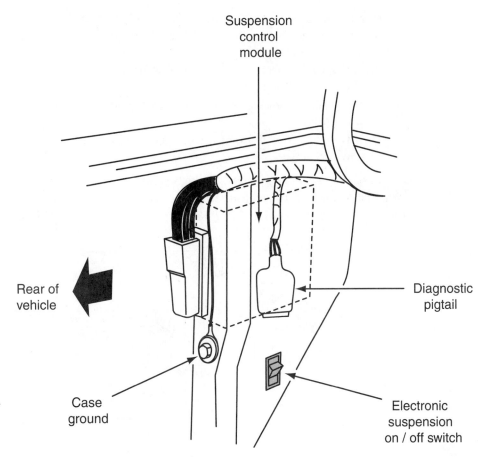

**Figure 4–19** Electronically controlled suspension system components mounted in the trunk.

*position, this switch shuts off the system voltage to the air suspension computer to make the system inoperative.*

## Task 4   Inspect and repair front and/or rear cradle (crossmember/subframe) mountings, bushings, brackets, and bolts.

54. While discussing front cradle alignment, Technician A says the cradle may be measured at various locations to determine if it is bent.

    Technician B says on some cradles an alignment hole in the cradle must be aligned with a hole in the chassis.

    Who is correct?

    A. A only
    B. B only
    C. Both A and B
    D. Neither A nor B

**Hint**    *A bent or improperly positioned front cradle may affect camber, caster, or setback angles on the front suspension. The cradle may be measured at various locations to determine if it is bent. If the cradle is properly positioned on some vehicles, an alignment hole in the cradle must be aligned with a hole in the chassis. A bent or improperly positioned rear subframe may cause improper rear-wheel alignment.*

## Task 5   Diagnose, inspect, adjust, repair or replace components (including sensors, switches, and actuators) of electronically controlled steering systems; initialize system as required.

55. In an electronic power steering system (EPS)
   A. the electric motor that provides steering assist is mounted separately from the steering gear.
   B. the vehicle speed sensor (VSS) voltage signals are sent directly from the VSS to the EPS control unit.
   C. the steering sensor is mounted in the steering gear housing so it senses rack movement.
   D. if the car has a manual transaxle, a differential speed sensor is used in place of the VSS.

**Hint**    *Some cars have an EPS system. The EPS system has a rack and pinion steering gear with electric assist in place of hydraulic assist. The electric motor that provides steering assist is designed into the steering gear so it can help to move the rack (Figure 4–20). A steering sensor mounted*

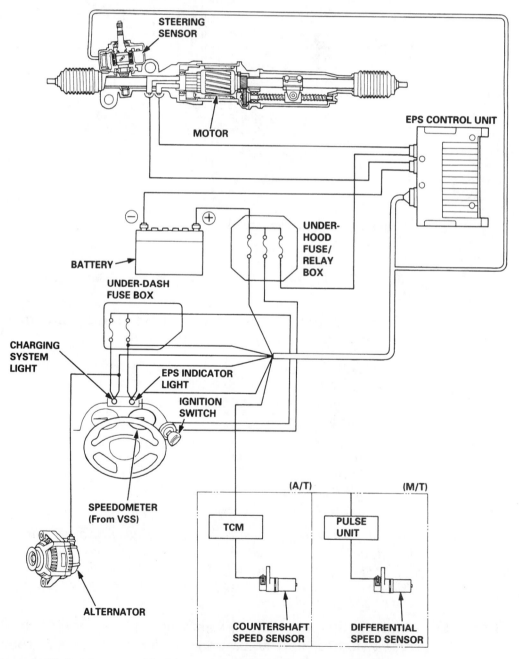

**Figure 4–20** Steering gear with electric assist and related system components. *(Courtesy of American Honda Motor Co., Inc.)*

*in the pinion shaft assembly sends input voltage signals to the EPS control unit in relation to the direction, amount, and torque of the steering wheel rotation.*

*A VSS in the automatic transaxle sends input voltage signals to the transmission control module (PCM/TCM) in relation to vehicle speed. These voltage signals are transmitted from the PCM/TCM to the EPS control unit. If the car has a manual transaxle, a differential speed sensor is used in place of the VSS. The speedometer also sends voltage signals to the EPS control unit in relation to vehicle speed. These voltage signals from the speedometer act as a backup if the VSS or differential speed sensor fails.*

*The EPS control unit is mounted above the steering gear. This control unit receives input signals from the steering sensor, and the VSS or differential speed sensor. When these signals are received, the EPS control unit calculates the proper amount and direction of steering assist. The EPS control unit then commands a power module in the EPS control unit to drive the electric motor in the steering gear and provide the proper direction and amount of steering assist. The EPS also contains self-diagnostic capabilities.*

**Task 6**  **Diagnose, inspect, repair or replace components of power steering idle speed compensation systems.**

56. An engine experiences stalling during deceleration and idle when the steering wheel is turned. The engine idles properly during all other operating conditions, and there are no other drivability complaints. All of the following defects may be the cause of the problem EXCEPT
    A. a defective power steering pressure switch.
    B. an open wire between the power steering pressure switch and the PCM.
    C. a defective PCM.
    D. a defective IAC motor.

**Hint**  *Some power steering systems have a power steering pressure switch (PSPS). When the steering wheel is turned, and the power steering pressure exceeds a predetermined value, the PSPS sends a signal to the powertrain control module (PCM). When this signal is received, the PCM shuts off the AC compressor, and if the engine is idling, the PCM operates the idle air control (IAC) motor to increase the idle speed. This action by the PCM prevents engine stalling during deceleration and idle when the additional load of the power steering pump is placed on the engine.*

# Wheel Alignment Diagnosis, Adjustment, and Repair

## ASE Tasks, Questions, and Related Information

**Task 1**  **Diagnose vehicle wander, drift, pull, hard steering, bump steer (toe curve), memory steer, torque steer, and steering return concerns; determine needed repairs.**

57. The steering on a vehicle pulls to the left while driving straight ahead. The cause of this problem could be
    A. more positive camber on the left front wheel than the right front wheel.
    B. more positive caster on the left front wheel than the right front wheel.
    C. improper steering axis inclination on the left front wheel.
    D. improper toe-out on turns on the right front wheel.

**Hint**    *Steering pull may be caused by improper camber or caster adjustment on one front wheel. Improper rear wheel toe and/or thrust line position also causes steering pull. Steering wander or drift may be caused by reduced positive caster on the front wheels, or loose steering linkages. Excessive steering wheel return force may be caused by excessive positive caster on the front wheels. Torque steer may be caused by unequal length front drive axles on some front wheel drive cars. Bump steer may be caused by unequal steering gear tie rod height. Memory steer may be caused by reduced positive caster adjustment on the front wheels. Memory steer may also be caused by binding in the steering column, gear, or linkages.*

## Task 2  Measure vehicle ride height; determine needed repairs.

58. A customer complains about increased steering effort and rapid steering wheel return after turning a corner.

    Technician A says both front wheels may have a negative camber setting.

    Technician B says the rear suspension curb riding height may be reduced.

    Who is correct?

    A. A only

    B. B only

    C. Both A and B

    D. Neither A nor B

**Hint**    *Reduced curb riding height affects many of the front suspension angles and causes harsh riding. For example, reduced curb riding height on the rear suspension causes excessive positive caster on the front suspension. This problem results in excessive steering effort and rapid steering wheel return.*

## Task 3  Measure front and rear wheel camber; determine needed repairs

59. The rear suspension system in Figure 4–21 requires a camber adjustment.

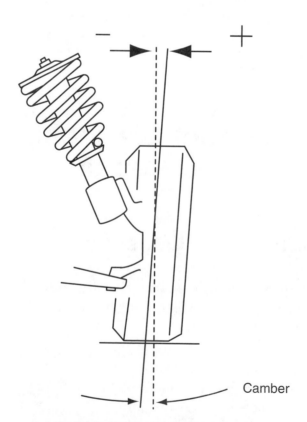

Camber

**Figure 4–21** Rear suspension system.

Technician A says the lower strut-to-knuckle bolt hole in the strut may be elongated with a file to adjust camber.

Technician B says the upper strut mount may be moved in elongated slots to correct the camber setting.

Who is correct?

A. A only

B. B only

C. Both A and B

D. Neither A nor B

**Hint**     *When a vehicle manufacturer does not provide a camber adjustment, replacement components such as upper strut mounts with adjustment capabilities are often available from automotive parts suppliers.*

## Task 4  Adjust front and/or rear wheel camber on suspension systems with a camber adjustment.

60. A long-and-short arm front suspension system has adjustment shims for adjustment of front suspension angles (Figure 4–22). When shims of equal thickness are added on both upper control arm mounting bolts

A. the camber is moved toward a negative position.

B. the caster becomes more positive.

C. the caster is moved toward a negative position.

D. the steering axis inclination angle decreases.

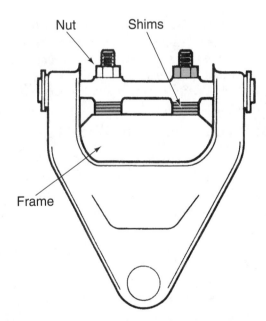

**Figure 4–22** Long-and-short arm front suspension system with adjustment shims. *(Courtesy of Hunter Engineering Company)*

**Hint**     *When adjustment shims are positioned between the upper control arm mounting and the inside of the frame, increasing the shim thickness equally on both mounting bolts moves the camber toward a more negative position. Increasing the shim thickness on the rear bolt and decreasing the shim thickness on the front bolt moves the caster toward a positive position.*

*If the adjustment shims are positioned between the upper control arm and the outside of the frame, increasing the shim thickness equally on both bolts increases the positive camber. When the shim thickness is increased on the rear bolt and decreased on the front bolt, the caster is moved toward a negative position.*

## Task 5   Measure caster; determine needed repairs.

61. A customer complains about harsh riding on the front suspension, excessive steering effort, and rapid steering wheel return.

    Technician A says the front suspension may have excessive positive caster.

    Technician B says the included angle may be more than specified on the front suspension.

    Who is correct?
    A. A only
    B. B only
    C. Both A and B
    D. Neither A nor B

*Hint*    *Excessive positive caster may cause harsh riding, rapid steering wheel return, and excessive steering effort. Negative caster causes reduced directional stability while driving straight ahead.*

## Task 6   Adjust caster on suspension systems with a caster adjustment.

62. All of the following statements about caster adjustment are true EXCEPT
    A. The caster angle is measured with the front wheels straight ahead.
    B. The front wheels are turned 20 degrees outward and then 20 degrees inward to read the caster angle.
    C. The brakes must be applied with a brake pedal jack before reading the caster angle.
    D. The front suspension should be jounced several times before reading the caster angle.

*Hint*    *On front suspension systems without a caster adjustment, some vehicle manufacturers recommend inspecting suspension components for wear or damage to determine the cause of the improper caster angle. Damaged or worn components such as the lower control arm, upper strut mount, or engine cradle may cause an improper caster angle. The damaged or worn component should be replaced. Some parts manufacturers supply suspension components that provide caster adjustment capabilities.*

## Task 7   Measure and adjust front wheel toe.

63. All of the following statements about front wheel toe adjustment are true EXCEPT
    A. The front wheels must be in the straight-ahead position when measuring front wheel toe.
    B. The front wheel toe should be adjusted before the caster angle on the front suspension.
    C. After the toe adjustment, the steering wheel must be centered with the front wheels straight ahead.
    D. While adjusting front wheel toe, a tie rod sleeve rotating tool must be used to turn the tie rod sleeves.

*Hint*    *On parallelogram steering linkages, the tie rod sleeves must be rotated to adjust front wheel toe. On a rack and pinion steering gear, loosen the tie rod end locknut and the outer bellows boot clamp. Then rotate the tie rod to adjust the front wheel toe. After the toe adjustment, the steering wheel must be centered with the front wheels in the straight-ahead position.*

## Task 8   Center steering wheel.

64. On a vehicle with a parallelogram steering linkage mounted behind the front suspension, after the front wheel toe is adjusted, the left spoke on a steering wheel is 2 in. low on the left side when driving the vehicle straight ahead.

    Technician A says the left tie rod should be lengthened and the right tie rod should be shortened.

Technician B says each tie rod sleeve should be rotated 2 turns.

Who is correct?

A. A only

B. B only

C. Both A and B

D. Neither A nor B

**Hint**    *When the steering wheel spoke is low on the left side, rotate the tie rod sleeves to shorten the left tie rod and lengthen the right tie rod. One quarter turn on the tie rod sleeves moves the steering wheel approximately 1 inch.*

## Task 9  Measure toe-out-on-turns (turning radius/angle); determine needed repairs.

65. The turning radius on the right front wheel is not within specifications. The cause of this problem could be
    A. a worn lower right ball joint.
    B. a loose outer right tie rod end.
    C. a worn lower control arm bushing.
    D. a worn right stabilizer bushing.

**Hint**    *On many steering systems, the turning radius on the wheel on the inside of the turn is 2 degrees more than the turning radius on the wheel on the outside of the turn. If the steering radius is not within specifications, the steering arms are bent or steering linkage components such as tie rod ends are worn.*

## Task 10  Measure SAI/KPI (steering axis inclination/king pin inclination); determine needed repairs.

66. Technician A says the SAI line is viewed from the side of the vehicle.
    Technician B says a 6-degree difference between the left and right SAI is acceptable.
    Who is correct?
    A. A only
    B. B only
    C. Both A and B
    D. Neither A nor B

**Hint**    *On MacPherson strut front suspensions, SAI is the angle between the true vertical center line of the tire and a line through the center of the lower ball joint and upper strut mount. On short- and long-arm front suspensions, the SAI line extends through the upper and lower ball joints. On I-beam front suspensions the KPI line extends through the center of the king pins. The SAI/KPI is always viewed from the front of the vehicle. SAI/KPI is not adjustable. If this angle is not within specifications, some front suspension components are bent or worn. For example, improper SAI on a MacPherson strut suspension may be caused by an upper strut tower that is out of position because of collision damage. Improper KPI on an I-beam front suspension may be caused by a bent front axle or worn king pin bushings. The difference between the SAI/KPI on the right and left sides of the front suspension must be within the specified limits. Improper SAI/KPI may cause steering pull during hard braking.*

## Task 11  Measure included angle; determine needed repairs.

67. A front-wheel-drive vehicle has 3-degrees difference in the included angles on the front wheels. This problem could be caused by
    A. a bent engine cradle.
    B. worn steering gear mounting bushings.

C. loose front-wheel bearings.

D. loose inner tie rod ends.

*Hint*    *When the front wheel camber is positive, the camber is added to the steering axis inclination (SAI) to obtain the included angle. If the front wheel camber is negative, the camber is subtracted from the SAI to calculate the included angle. The difference between the included angle on the front wheels should not exceed 1.5 degrees. SAI is not considered adjustable. Improper SAI angles may be caused by a strut tower out of position, a bent lower control arm or center crossmember, or an engine cradle out of position.*

## Task 12    Measure rear wheel toe; determine needed repairs or adjustments.

68. A front-wheel-drive car has excessive toe-out on the left rear wheel. Adjustment shims are positioned between the rear spindles and the spindle mounting surfaces.

    Technician A says a thicker shim should be installed on the front bolts in the left rear spindle.

    Technician B says this problem may cause the steering to pull to the right.

    Who is correct?

    A. A only

    B. B only

    C. Both A and B

    D. Neither A nor B

*Hint*    *On some rear suspension systems, rear-wheel toe is adjusted with shims between the spindle and spindle mounting surface. A thicker shim on the front spindle mounting bolts increases toe-out. On other rear suspension systems, rear-wheel toe is adjusted by shifting the position of the inner end on the lower control arm. The tie rod length may be adjusted to correct the toe on some rear wheels.*

## Task 13    Measure thrust angle; determine needed repairs or adjustments.

69. The thrust angle on a front-wheel-drive vehicle is more than specified, and the thrust line is positioned to the left of the geometric centerline. This problem could be caused by excessive

    A. toe-out on the left rear wheel.

    B. toe-out on the right rear wheel.

    C. positive camber on the left rear wheel.

    D. wear in the left lower ball joint.

*Hint*    *The thrust angle is the angle between the thrust line and the geometric centerline of the vehicle.*

*If the thrust angle is not within specifications, the steering pulls to one side. For example, if the thrust line is positioned to the left of the geometric centerline, the steering pulls to the right.*

*When the thrust angle is not within specifications on a front-wheel-drive car, the rear wheel toe may require adjusting. On a rear-wheel-drive car with a non-independent rear axle assembly, improper rear-wheel toe may be caused by a bent rear axle assembly. On this type of rear axle, the complete rear axle assembly may be turned slightly in either direction, causing the thrust angle to be more than specified. This condition is called rear axle offset.*

## Task 14    Measure front wheelbase setback/offset; determine needed repairs or adjustments.

70. While discussing front-wheel setback, Technician A says front-wheel setback usually is caused by worn suspension components.

    Technician B says a slight front-wheel setback causes steering pull.

    Who is correct?

    A. A only

    B. B only

C. Both A and B

D. Neither A nor B

**Hint**    *Setback occurs when one front wheel is driven rearward in relation to the opposite front wheel. Setback usually is caused by collision damage, and it does not affect steering unless it is extreme.*

## Task 15   Check front and/or rear cradle (crossmember/subframe) alignment; determine needed repairs or adjustments.

71. A front-wheel-drive car has excessive positive camber on the left-front wheel and excessive negative camber on the right-front wheel.

    Technician A says the front cradle may be shifted to the right.

    Technician B says the left-front, lower control arm may be bent.

    Who is correct?

    A. A only

    B. B only

    C. Both A and B

    D. Neither A nor B

**Hint**    *A bent or improperly positioned front cradle may affect camber, caster, or setback angles on the front suspension. The cradle may be measured at various locations to determine if it is bent. If the cradle is properly positioned on some vehicles, an alignment hole in the cradle must be aligned with a hole in the chassis.*

# Wheel and Tire Diagnosis and Service

## ASE Tasks, Questions, and Related Information

### Task 1   Diagnose tire wear patterns; determine needed repairs.

72. The tire tread wear shown in Figure 4–23 could be caused by

    A. excessive positive camber.

    B. excessive positive caster.

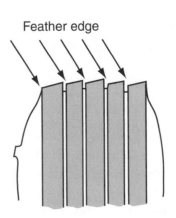

Feather edge

**Figure 4–23** Excessive tire tread wear.

        C. excessive setback

        D. improper toe adjustment.

**Hint**    *Feathered wear across the tire tread usually indicates an improper toe adjustment. Wear on one side of the tire tread usually indicates an improper camber setting. Cupped tread wear may indicate improper wheel balance.*

## Task 2  Inspect tire condition, size, and application (load and speed ratings).

73. The rear tires on a vehicle are replaced with tires that are considerably larger than the tires specified by the vehicle manufacturer. The most likely result of this tire change is
    A. reduced front-wheel positive caster and directional stability.
    B. increased positive camber on the front wheels and increased front-tire tread wear.
    C. increased steering effort.
    D. rapid steering wheel return.

**Hint**    *Replacement tires must have the size and ratings specified by the vehicle manufacturer. Improper tire size may adversely affect wheel alignment, resulting in reduced directional stability and excessive tire tread wear. If the tire ratings are not equal to those specified by the vehicle manufacturer, vehicle safety is compromised, and a tire failure may cause a serious collision.*

## Task 3  Measure and adjust tire air pressure.

74. All of the following statements about tire inflation are true EXCEPT
    A. Overinflation causes excessive wear on the center of the tread.
    B. Underinflation causes excessive wear on both edges of the tread.
    C. Tire pressure should be adjusted when the tires are hot.
    D. Underinflation may cause wheel damage.

**Hint**    *Tire overinflation causes excessive tire tread wear on the center of the tire tread. Underinflation causes excessive wear on the outside edges of the tire tread. Underinflation also reduces fuel economy, increases steering effort, raises tire temperature, and may cause rim damage.*

## Task 4  Diagnose wheel/tire vibration, shimmy, and noise concerns; determine needed repairs.

75. Front-wheel shimmy may be caused by
    A. excessive toe-out.
    B. improper dynamic wheel balance.
    C. excessive front-wheel setback.
    D. excessive positive camber.

**Hint**    *Tire vibration and thumping may be caused by cupped tire treads, excessive tire radial runout, heavy spots in the tire, and improper wheel balance. Front-wheel shimmy may be caused by improper dynamic wheel balance, excessive positive caster, or worn steering linkage components.*

## Task 5  Rotate tires/wheels and torque fasteners according to manufacturers' recommendations.

76. The tire rotation procedure in Figure 4–24 may be used on a vehicle with
    A. radial tires and a compact spare.
    B. radial tires and a conventional spare.
    C. radial or bias ply tires.
    D. bias ply tires with a conventional spare.

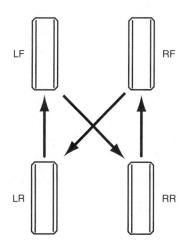

**Figure 4–24** Tire rotation.

**Hint**   *The vehicle manufacturer's tire rotation procedure varies, depending on the type of tires and the type of spare. A different rotation is used for radial or bias ply tires. Because the compact spare cannot be used for extended mileage, this spare is not rotated with the other tires.*

**Task 6**   **Measure wheel, tire, axle flange, and hub runout (radial and lateral); determine needed repairs.**

77.  A vehicle has rear chassis waddle (Figure 4–25).

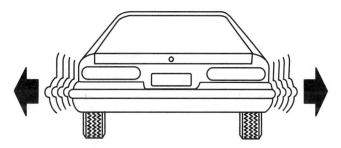

**Figure 4–25** Rear suspension waddle.

Technician A says this problem may be caused by a steel belt in a tire that is not straight.
Technician B says this problem may be caused by a bent rear hub flange.
Who is correct?
A.  A only
B.  B only
C.  Both A and B
D.  Neither A nor B

**Hint**   *Excessive lateral tire runout may cause chassis waddle. This problem may be caused by a steel belt that is not straight in a tire, or a bent wheel or hub.*

**Task 7**   **Diagnose tire pull (lead) problems; determine corrective actions.**

78.  A customer complains about steering pull to the left.
Technician A says the two front tires may have different tread designs.
Technician B says one of the front tires may have a conicity problem.

Who is correct?

A.  A only

B.  B only

C.  Both A and B

D.  Neither A nor B

*Hint*    *Tires of different types, sizes, tread designs, or inflation pressures may cause steering pull. Tire conicity is caused by a tire belt that is wound off center during the manufacturing process. This problem also causes steering pull. Tire conicity is diagnosed with a special tire rotation procedure.*

## Task 8  Dismount and mount tire on wheel.

79.  All of the following statements about tire dismounting and mounting are true EXCEPT

A.  Tire irons should not be used to remove the tire from the rim.

B.  The tire should be partially inflated while dismounting.

C.  The narrow bead ledge should be facing upward on the tire changer.

D.  The bead seating surfaces on the rim must be clean before mounting the tire.

*Hint*    *Always use a tire changer to dismount and mount tires. Do not use hand tools or tire irons for this purpose. Before dismounting the tire, remove the valve core and be sure the tire is completely deflated. Place the tire and wheel on the tire changer with the narrow bead ledge facing upward. Follow the instructions provided by the tire changer manufacturer to force the tire bead away from the rim on both sides of the rim. Push one edge of the top bead into the drop center of the rim. Place the tire changer's bar or lever between the bead and the rim on the opposite side of the rim from where the bead is in the drop center. Operate the tire changer to rotate the bar or lever and move the bead over the top of the rim. Repeat the process to move the lower bead over the top of the rim. The bead contact surfaces on the wheel must be clean, and the tire beads coated with rubber lubricant before mounting the tire.*

## Task 9  Balance wheel and tire assembly.

80.  Dynamic wheel unbalance on a front wheel may cause all of the following problems EXCEPT

A.  excessive wear on the center of the tire tread.

B.  cupped tire tread wear around the tire tread.

C.  front-wheel shimmy while driving at higher speeds.

D.  excessive wear on steering linkage components.

*Hint*    *Static wheel balance refers to the proper balance of a wheel at rest. Dynamic wheel balance refers to the balance of a wheel in motion. Electronic wheel balancers spin the wheel at high speed and indicate the proper position for wheel weight installation to provide proper static and dynamic balance.*

## Task 10  Test and diagnose tire pressure monitoring system; determine needed repairs.

81.  All of these statements regarding tire pressure monitoring systems (TPMS) are true EXCEPT:

A.  The TPMS module is mounted in the passenger compartment.

B.  Each valve stem contains a sensor mounted on the outside of the wheel rim.

C.  Each valve stem sensor acts as a radio transmitter.

D.  Some TPMS indicate the location of the tire with low air pressure.

*Hint*    *A typical TPMS has a pressure sensor attached to each valve stem inside the tire. The valve stem sensors acts as radio antennas and transmit electrical signals to the TPMS module that is located in the passenger compartment. Some TPMS monitor the air pressure in all four tires on the vehicle and the spare tire. Other TPMS do not monitor the spare tire pressure. The valve stem sensors in the four wheels on the vehicle transmit electrical signals every 60 seconds, and the spare tire sensor transmits signals every hour.*

*If the TPMS module receives a signal from a pressure sensor indicating low pressure in a tire, the module provides an audible warning, displays a warning message in the message center, and illuminates the TPMS warning light. The TPMS module may provide a flashing warning light, or continual illumination of this light depending on the system fault. Some TPMS also provide warnings if the tire pressure in any tire is excessive. Other TPMS have the capability to indicate the location of the tire with the low air pressure.*

# Post Test

1. When servicing a steering column in a vehicle equipped with a driver's side air bag
   A. disconnect the negative battery cable and wait the specified time period before working on the steering column.
   B. the air bag inflator module should be placed with the trim cover down on the bench.
   C. hold the trim cover facing your body when carrying the inflator module.
   D. the steering wheel may be removed before the inflator module.

2. When performing a sector lash adjustment on a recirculating ball steering gear, the steering gear is in the center position and the steering shaft is rotated through a 45-degree arc with an inch-pound torque wrench. The sector lash adjustment screw should be rotated in
   A. a clockwise direction until the turning torque is 4 to 10 in.-lbs (0.45 to 1.13 Nm) more than the preliminary turning torque reading.
   B. a counterclockwise direction until the turning torque is 0 in.-lbs.
   C. a clockwise direction until the turning torque is 20 in.-lbs. (2.27 Nm) more than in the preliminary procedure.
   D. a clockwise direction until the turning torque is 5 ft.-lbs. (6.77 Nm) more than in the preliminary procedure.

3. When discussing rack and pinion steering gear mountings, Technician A says worn mounting bushings may cause unequal steering arm heights.
   Technician B says unequal steering linkage heights may cause steering pull when driving straight ahead.
   Who is correct?
   A. A only
   B. B only
   C. Both A and B
   D. Neither A nor B

4. A power steering pump has repeated shaft seal failures. The most likely cause of this problem could be
   A. a worn drive shaft bearing.
   B. discolored dark brown power steering fluid.
   C. A misaligned power-steering pump pulley.
   D. excessive power steering belt tension.

5. While removing and replacing a power rack and pinion steering gear on an air bag–equipped vehicle
   A. align the master splines or flat surfaces on the pinion shaft and lower universal joint.
   B. the steering wheel should be turned fully to the right before gear removal.

C. the clockspring electrical connector may be in any position when installing the gear and steering wheel.

D. disconnect the negative battery cable before removing the gear.

6. In a power rack and pinion steering gear, a rack bearing adjustment that is tighter than specified may cause
   A. excessive wear on the rack and pinion teeth.
   B. damage to the inner rack seal.
   C. wear on the steering gear mounting bushings.
   D. poor steering wheel returnability.

7. When discussing inner tie rod end removal and replacement on a rack and pinion steering gear, Technician A says the inner tie rod should be tightened to the specified torque and then backed off one-quarter turn.

   Technician B says the inner tie rod end should be tightened to the specified torque and then stacked.

   Who is correct?
   A. A only
   B. B only
   C. Both A and B
   D. Neither A nor B

8. While discussing power-steering bleeding and flushing on a system with a remote reservoir, Technician A says that in order to bleed air from the system, the steering wheel should be turned fully in each direction three or four times.

   Technician B says the system should be flushed with the return hose from the reservoir to the pump disconnected.

   Who is correct?
   A. A only
   B. B only
   C. Both A and B
   D. Neither A nor B

9. Bump steer is experienced during a road test for steering diagnosis.
   Technician A says the steering gear may require adjustment.
   Technician B says the pitman arm may be bent.
   Who is correct?
   A. A only
   B. B only
   C. Both A and B
   D. Neither A nor B

10. While discussing steering dampers, Technician A says steering dampers are often used on small compact cars to increase road feel on the steering.
    Technician B says a worn out steering damper may cause steering pull.
    Who is correct?
    A. A only
    B. B only
    C. Both A and B
    D. Neither A nor D

11. The left front tire has excessive wear on the inside edge of the tread.
    Technician A says the inner bushings on the left lower control arm may be worn.
    Technician B says the left outer tie rod end may be loose.
    Who is correct?

   A. A only

   B. B only

   C. Both A and B

   D. Neither A nor B

12. While servicing strut rods mounted in front of the lower control arms and front suspension systems

   A. the strut rod prevents lateral lower control arm movement.

   B. the strut rod prevents fore-and-aft upper control arm movement.

   C. worn strut rod bushings may decrease positive front-wheel caster while braking.

   D. a bent strut rod may decrease positive front-wheel caster.

13. A bent steering arm causes

   A. a change in front-wheel caster setting.

   B. improper toe-out-on turns.

   C. tire squeal while driving straight ahead.

   D. a change in riding height.

14. In a rear suspension system, with longitudinally mounted rear leaf springs, the left rear center bolt is broken and this side of the rear axle is moved rearward.

   Technician A says this defect may cause steering pull to the right.

   Technician B says this defect may cause bump steer on a rough road.

   Who is correct?

   A. A only

   B. B only

   C. Both A and B

   D. Neither A nor B

15. While discussing shock absorber diagnosis and service, Technician A says the relationship between the force required to move the lower end of the shock absorber on the compression and rebound strokes is called shock ratio.

   Technician B says more force is usually required on the rebound or extension stroke compared to the compression stroke.

   Who is correct?

   A. A only

   B. B only

   C. Both A and B

   D. Neither A nor B

16. When diagnosing and servicing front-wheel bearings, loose wheel bearing adjustment may cause

   A. steering wander.

   B. improper caster angle.

   C. improper toe-out-on turns.

   D. steering pull.

17. A left front tire has excessive wear on the outside edge of the tread.

   Technician A says the left front wheel may have excessive positive camber.

   Technician B says the upper-left strut mount may be worn.

   Who is correct?

   A. A only

   B. B only

   C. Both A and B

   D. Neither A nor B

18. When measuring and correcting curb riding height
    A. reduced front curb riding height may reduce directional stability.
    B. reduced front curb riding height may increase front wheel toe-out.
    C. reduced rear curb riding height increases front-wheel positive camber.
    D. excessive rear curb riding height may increase directional stability.

19. The specified front wheel caster is 1.5 degrees. The right front wheel has 4-degrees positive caster and the left front wheel has 1-degree positive caster. These caster settings cause
    A. tire tread wear on the inside edge of the left front tire.
    B. steering pull to the right.
    C. tire tread wear on the outside edge of the right front tire.
    D. steering pull to the left.

20. All of the following statements about turning radius and toe-out-on turns are true EXCEPT
    A. The turning radius on the wheel on the inside of a turn must be 2 degrees more than the turning radius on the outside on the outside wheel.
    B. If the turning radius is not within specifications, one of the steering arms may be bent.
    C. If the turning radius is not within specifications, tire squeal may occur while cornering.
    D. If the turning radius is not within specifications, the steering wheel is not centered while driving straight ahead.

# Answers and Analysis

**1.  C**    A worn flexible coupling or loose steering gear mounting bolts may cause excessive steering wheel free play. Therefore, both A and B are correct, and C is the right answer.

**2.  A**    If the bolt head is touching the steering column bracket, the shear load is too high and the bracket should be replaced. Therefore, B is wrong and A is correct.

**3.  C**    Some air bag systems are disarmed by disconnecting the negative battery cable and waiting for the specified time period. A is true, but this is not the requested answer. Air bag disarming includes disconnecting the air bag system fuse on some vehicles. B is true but this is not the requested answer. Deployed air bags may be coated with sodium hydroxide. Therefore, C is not true, and this is the requested answer. After an air bag system has been disarmed and enabled, the air bag system warning light indicates proper system operation. D is true, but this is not the requested answer.

**4.  D**    A loose worm bearing preload adjustment causes excessive steering wheel free play. An overfilled steering gear does not affect steering effort. When the positive caster is less than specified, steering effort is reduced. Therefore, A, B, and C are wrong. A tight sector lash adjustment increases steering effort. D is the right answer.

**5.  B**    A loose rack bearing adjustment or loose steering gear mounting bolts cause excessive steering wheel free play. Excessive positive camber causes wear on the outside edge of the tire treads. Therefore, A, C, and D are wrong. Insufficient or improper steering gear lubricant may result in poor steering wheel returnability. Thus, B is correct.

**6.  D**    Many vehicle manufacturers recommend checking the power steering fluid level with the fluid temperature at 175°F (80°C). A, B, and C are wrong, and D is correct.

**7.  A**    When checking power steering belt tension, for every foot of free span the belt should have 1/2-in. deflection. Therefore, B, C, and D are wrong, and A is right.

**8.   C**   A belt tension gauge may be used to measure the belt tension on a ribbed V-belt, and some of these belts have a spring-loaded tensioner with a belt length scale. Both A and B are right, and C is the correct answer.

**9.   C**   A squealing noise during acceleration may be caused by a loose power steering belt. Thus, Technician A is right. Premature power steering belt wear may be caused by a misaligned power steering pump, meaning Technician B is also right. Because Technicians A and B are both right, C is the correct answer.

**10.   C**   The tool in Figure 4–3 is used to install a pressed-on power-steering pump pulley. Therefore, A, B, and D are wrong, and C is right.

**11.   D**   When a power-steering pump with an integral reservoir is leaking fluid between the reservoir and the pump housing, the large housing O-ring is leaking. Because Technicians A and B are both wrong, D is the correct answer.

**12.   A**   During the power steering pump pressure test, the pressure gauge valve should be closed for a maximum of ten seconds. Therefore, B, C, and D are wrong, and A is right.

**13.   C**   When the power-steering pump pressure test indicates satisfactory pump pressure, but the steering wheel turning effort is excessive, the steering gear may be the problem, or the high pressure hose may be restricted. Because Technicians A and B are both right, C is the correct answer.

**14.   B**   If the steering wheel is rotated when the steering gear is disconnected from the steering shaft, steering gear alignment, steering effort, or steering column binding are not affected. Therefore, A, C, and D are wrong. However, this action may cause an improperly centered clockspring. Thus, B is right.

**15.   A**   When removing the steering wheel on an air bag–equipped vehicle, the negative battery cable should be disconnected and the technician should wait the time specified by the car manufacturer before working on the vehicle. This time is usually one or two minutes. Therefore, B is wrong. Rotating the steering wheel with the steering gear disconnected may damage the clockspring electrical connector; thus, A is correct.

**16.   A**   When adjusting the worm shaft bearing preload on some steering gears, the adjuster plug should be bottomed and tightened to 20 ft.-lbs. After this procedure, the adjuster plug is backed off 0.050 in. Therefore, Technician B is wrong and Technician A is right, making answer A correct.

**17.   C**   The part number on the needle bearing must face the driving tool, and the worm shaft bearing preload should be adjusted after the bearing and seal are installed. Because Technicians A and B are both right, C is the correct answer.

**18.   B**   Steering wheel returnability is not affected by a misaligned steering gear. Therefore, Technician A is wrong. A tight rack bearing adjustment increases steering effort and causes poor steering wheel returnability, making Technician B correct. Thus, B is the proper answer.

**19.   C**   The component in Figure 4–7 is a shock damper. Therefore, A, B, and D are wrong, and C is correct.

**20.   C**   When bleeding a power-steering system, the steering wheel should be held in the fully right or fully left positions for two to three seconds. If foaming is still present in the reservoir after the bleeding process, the bleeding procedure should be repeated. Because Technicians A and B are both right, C is the correct answer.

**21.   B**   In variable-assist steering systems with a steering wheel rotation sensor, the power steering assist is increased when steering wheel rotation exceeds 15 rpm. Power-steering assist also is increased at low vehicle speeds and decreased at higher speeds. Therefore, Technician A is wrong, and Technician B is correct, making B the appropriate answer.

**22.   A**   Excessive idler arm movement may cause front-wheel shimmy, steering looseness, improper tie rod alignment with the lower control arm, and bump steer. Therefore, B, C, and D are wrong, and A is correct.

**23.**  **B**  Bump steer occurs when the tie rods are not parallel to the lower control arms. This condition may be caused by a bent pitman arm or a loose, improperly adjusted idler arm. Therefore, A is wrong, and B is correct.

**24.**  **A**  A bent relay rod may cause improper front-wheel toe. This condition causes feathered wear across the tire treads. Therefore, B is wrong, and A is right.

**25.**  **C**  Worn ball joints, weak front springs, or a weak stabilizer bar would not cause excessive steering effort. However, this problem may be caused by a seized idler arm. Therefore, A, B, and D are wrong, and C is correct.

**26.**  **C**  The tie rod sleeves must be rotated to adjust front-wheel toe and center the steering wheel. Therefore, both A and B are correct, and C is the right answer.

**27.**  **B**  A scored power-steering gear cylinder increases steering effort, but it does not contribute to road shock on the steering wheel. Therefore, Technician A is wrong. A worn-out steering damper may cause road shock on the steering wheel, making Technician B's assessment correct. Thus, B is the right answer.

**28.**  **D**  Reduced front suspension ride height may be caused by weak or broken springs, or worn control arm bushings. Therefore, A, B, and C are wrong, and D is right.

**29.**  **A**  The tool in Figure 4–11 is used to compress the coil spring on a long-and-short arm suspension system. Therefore, B, C, and D are wrong, and A is correct.

**30.**  **D**  Heavy-duty coil spring, worn ball joints, or worn front-wheel bearings do not cause severely damaged rebound bumpers. Therefore, A, B, and C are wrong. Damaged rebound bumpers may be caused by defective front shock absorbers. Thus, D is correct.

**31.**  **A**  Excessive negative camber on the right front wheel may cause steering pull to the left. Therefore, B is wrong. A worn right front strut rod bushing may result in the lower control arm moving rearward while braking. This action reduces positive caster on the right front wheel. The steering tends to pull to the side with the least positive caster. Therefore, A is right.

**32.**  **D**  When the coil spring is mounted between the lower control arm and the chassis, a jack must be positioned under the lower control arm to unload the ball joints. Therefore, A, B, and C are wrong, and D is right.

**33.**  **D**  Wear on the inside edge of the tire tread is caused by excessive negative camber. The shims between the front axle and the leaf spring are used to adjust caster. Therefore, Technician A is wrong. A worn rear spring shackle does not affect camber, and so Technician B is wrong. Because Technicians A and B are both wrong, D is the correct answer.

**34.**  **A**  Excessive tire squeal while cornering may be caused by improper toe-out-on turns. This problem can be caused by a bent steering arm. Therefore, B, C, and D are wrong, and A is the right answer.

**35.**  **B**  The coil spring in Figure 4–12 is a variable-rate spring. Therefore, A, C, and D are wrong, and B is correct.

**36.**  **C**  While using a spring compressor to remove a coil spring from a strut, the spring should be taped in the spring compressor contact areas, and all the spring tension must be removed from the upper strut mount before loosening the strut rod nut. Therefore, both A and B are correct, and C is the right answer.

**37.**  **D**  In this question, the requested answer does not result in steering pull to the right. A broken center bolt in the left rear spring causes rear axle offset, and this problem may cause steering pull to the right. Because the steering pulls to the side with the least positive caster, excessive positive caster on the left front wheel may cause steering pull to the right. More positive camber on the right front wheel compared to the left front wheel may cause steering pull to the right. Therefore, A, B, and C may cause steering pull to the right, but none of these is the requested answer. Excessive front-wheel toe-in causes feathered tire wear, but this problem does not affect steering pull. Therefore, D is right.

**38.**  **C**  A worn lower control arm bushing or an improperly adjusted front torsion bar anchor bolt may cause reduced ride height. Therefore, both A and B are correct, and C is the right answer.

**39.   A**   In this question, the requested answer does not result in body sway while cornering. A weak stabilizer bar, worn stabilizer bar bushing, or a broken stabilizer link may cause excessive body sway while cornering. Therefore, B, C, and D may cause body sway while cornering, but none of these is the requested answer.

A worn strut rod bushing does not cause this problem. Therefore, A is right.

**40.   C**   In Figure 4–14, the technician is pushing the strut rod downward before removing the strut cartridge. Therefore, A, B, and D are wrong, and C is correct.

**41.   A**   A damaged upper strut mount may result in strut chatter while cornering, but a worn lower ball joint does not cause this problem. Therefore, Technician A is right, and Technician B is wrong, making answer B the appropriate choice.

**42.   C**   Damaged struts or reduced suspension ride height may cause harsh riding. Because Technicians A and B are both right, C is the correct answer.

**43.   A**   Excessive rear suspension oscillations may be caused by damaged struts, but weak coil springs do not cause this problem. Therefore, Technician B is wrong, and Technician A is right, making A the correct answer.

**44.   B**   The ball joint nuts should never be loosened to align the cotter pin hole with the nut castellations. Therefore, Technician A is wrong. On some rear suspension systems, such as the one in Figure 4–15, the lower control arm bolts must be torqued with the vehicle weight on the suspension. Thus, Technician B is right, making answer B correct.

**45.   A**   A sagged left rear leaf spring lowers the left rear ride height and increases the positive caster on the left front wheel. Because the steering pulls to the side with the least positive caster, this problem may cause steering pull to the right. Therefore, B, C, and D are wrong, and A is the right answer.

**46.   C**   In this question, the requested answer does not cause a rattling noise in the rear suspension.

Because a broken spring, spring insulator, or a worn track bar bushing may cause a rattling noise in the rear suspension, these responses are not the requested answer. Therefore, A, B, and D are wrong.

A bent rear strut does not cause a ratting noise in the rear suspension. Therefore, C is correct.

**47.   B**   Rear axle offset as illustrated in Figure 4–16 may cause steering pull to the right. Therefore, A, C, and D are wrong, and B is right.

**48.   C**   Many ball joints have a wear indicator. In these ball joints, the shoulder of the grease fitting must extend a specific distance from the ball joint housing. If this distance is less than specified, the ball joint must be replaced. Because there is no clearance between the grease fitting shoulder and the ball joint housing, Technician A is correct. A worn ball joint may cause improper position of the lower end of the rear knuckle, wheel hub, and wheel. This action may result in improper rear wheel camber. Therefore, Technician B is also correct, making C the right answer.

**49.   A**   If the tie rod in Figure 4–18 is longer than specified, the rear wheel toe-out is excessive. Therefore, A is right, and B, C, and D are wrong.

**50.   C**   An improper rear wheel toe adjustment or a bent rear spindle may cause feathered tread wear on a rear tire. Therefore, Technicians A and B are both right, making C the correct answer.

**51.   A**   When one side of the bumper is pushed downward with considerable force and then released, the bumper should only complete one free upward bounce if the shock absorber or strut is satisfactory. More than one free upward bounce of the bumper indicates damaged shock absorbers or struts. Therefore, B, C, and D are wrong, and A is right.

**52.   C**   After tightening the rear-wheel bearing adjusting nut to 20 ft.-lbs., and backing the nut off one-half turn, this nut should be tightened to 10 to 15 ft.-lbs. Therefore, A, B, and D are wrong, and C is correct.

**53.   A**   In this question, we are asked for the condition when the electronic suspension switch should not be in the off position. The electronic suspension switch should be off when hoisting, towing, or jacking the vehicle. Therefore, B, C, and D are not the requested answers. The electronic suspension switch must be on while diagnosing the suspension system with a scan tool. Therefore, A is the right answer.

**54.  C**   The front cradle must be measured at various locations to determine if it is bent. On some vehicles, an alignment hole in the cradle must be aligned with a matching hole in the chassis. Because Technicians A and B are both right, C is the correct answer.

**55.  D**   In an electronic power-steering gear, the electric motor is integral with the steering gear. Therefore, A is wrong. The VSS sensor voltage signals are sent to the PCM/TCM and then transmitted to the EPS control unit. Thus, B is wrong. The steering sensor is mounted in the pinion shaft assembly, making answer C wrong.

If the car has a manual transaxle, a differential speed sensor is used in place of the VSS sensor. Thus, D is correct.

**56.  D**   A defective power-steering pressure switch, an open wire between the power steering pressure switch and the PCM, or a defective PCM may cause stalling during deceleration or at idle. Therefore, A, B, or C are true, making none of these statements the requested answer. A defective IAC motor also causes improper idle speed during other engine operation conditions, so D is not true. Thus, D is the requested answer.

**57.  A**   The steering pulls to the side with the least positive caster. Therefore, B is wrong. SAI or toe-out on turns do not affect steering pull while driving straight ahead; thus, C and D are wrong. The steering tends to pull to the side with the most positive camber. Therefore, A is the right answer.

**58.  B**   A negative camber setting does not cause rapid steering wheel return and increased steering effort. When the rear suspension ride height is reduced, the positive caster is increased on the front suspension. Because excessive positive caster causes rapid steering wheel return and increased steering effort, B is correct.

**59.  A**   On the rear suspension system in Figure 4–21, the lower strut-to-knuckle bolt hole in the strut must be filed to adjust camber. Therefore, Technician B is wrong, and Technician A is right, making answer A correct.

**60.  A**   When shims of equal thickness are added on both upper control arm bolts in Figure 4–22, the top of the tire is moved inward and the camber is moved toward a negative position. Therefore, B, C, and D are wrong, and A is the right answer.

**61.  A**   Harsh riding, excessive steering effort, and rapid steering wheel return may be caused by excessive positive caster. Therefore, Technician B is wrong and Technician A is right, making A the correct answer.

**62.  A**   This question asks for the statement that is not true. The front suspension should be jounced several times before reading any suspension angle. The brakes should be applied with a brake jack before reading caster. The front wheels are turned outward 20 degrees and then inward 20 degrees to read the caster angle. Therefore, statements B, C, and D are correct, but none of these is the requested answer. The caster angle is not read with the front wheels straight ahead. Therefore, statement A is not true, and this is the requested answer.

**63.  B**   When front wheel toe is measured, the front wheels must be straight ahead. After the toe adjustment is completed, the steering wheel must be centered with the front wheels straight ahead. A tie rod rotating tool must be used to rotate the tie rod sleeves. Therefore, statements A, C, and D are true, and none of these is the requested answer. The front wheel toe should be measured after the caster measurement and adjustment.

Therefore, statement B is not true, and this is the requested answer.

**64.  D**   If the steering wheel spoke is 2 in. low on the left side while driving the vehicle straight ahead, the left tie rod should be shortened, and the right tie rod lengthened. One-quarter turn on the tie rod sleeves moves the steering wheel about one inch. Therefore, both A and B are wrong, and D is the right answer.

**65.  B**   When the turning radius is not within specifications, a steering arm may be bent or a tie rod end may be loose. Therefore, A, C, and D are wrong, and B is the right answer.

**66. D** The SAI line is viewed from the front of the vehicle. Therefore, A is wrong. The maximum variation between left and right SAI angles is usually 1.5 degrees, and so B is wrong. Because A and B are both wrong, D is the right answer.

**67. A** When the included angles on the front wheels have 3-degrees difference, the SAI angle is probably improper on one of the front wheels. Improper SAI may be caused by an improperly positioned strut tower or engine cradle. Therefore, B, C, and D are wrong, and A is correct.

**68. B** Excessive toe-out in the left rear wheel may be corrected by installing a thinner shim on the front spindle bolts. Therefore, Technician A is wrong. Excessive toe-out on the left rear wheel may cause the steering to pull to the right. Thus, Technician B is correct, making B the right answer.

**69. A** If the thrust angle is excessive and the thrust line is positioned to the left of the geometric centerline, the left rear wheel may have excessive toe-out. Therefore, B, C, and D are wrong, and A is right.

**70. D** Front-wheel setback usually is caused by collision damage, and this problem does not cause steering pull unless it is excessive. Therefore, Technicians A and B are both wrong, making D the correct answer.

**71. A** A bent left front lower control arm affects the camber on the left-front wheel, but this defect does not cause improper camber on the right front wheel. Thus, Technician B is wrong. If the cradle is shifted to the right, the left-front camber may become positive, and the right-front camber may become negative. Therefore, Technician A is right, making A the correct answer.

**72. D** The feathered tire wear in Figure 4–23 may be caused by improper toe adjustment.
Therefore, A, B, and C are wrong, and D is right.

**73. A** If the rear tires are considerably larger than the tires specified by the vehicle manufacturer, the front-wheel caster becomes less positive. This reduces directional stability and decreases steering effort and steering wheel returning force. Therefore, A is the correct answer, and B, C, and D are wrong.

**74. C** This question asks for the statement that is not true. Overinflation causes wear on the center of the tire tread, and underinflation causes wear on the edges of the tread. Underinflation may also cause wheel damage. Therefore, statements A, B, and D are true, but none of these is the requested answer.
Tire pressure should be adjusted when the tires are cool. Thus, statement C is not true, making it the requested answer.

**75. B** Front-wheel shimmy may be caused by improper dynamic wheel balance, excessive positive caster, or loose steering linkage components. Therefore, A, C, and D are wrong, and B is correct.

**76. A** The tire rotation procedure in Figure 4–24 may be used on a vehicle with radial tires and a compact spare. Therefore, A is the right answer.

**77. C** Excessive rear chassis waddle may be caused by a steel belt in a tire that is not straight, or a bent rear hub flange. Therefore, Technicians A and B are both right, making C the correct answer.

**78. C** Steering pull may be caused by front tires with different tread designs, or a front tire with a conicity problem. Therefore, Technicians A and B are both right, making C the correct answer.

**79. B** When dismounting and mounting tires, tire irons should not be used, the narrow bead ledge should be facing up on the tire changer, and the bead sealing surfaces must be clean before mounting the tire. Therefore, statements A, C, and D are true, but none of these statements is the requested answer. Before dismounting the tire, it must be completely deflated. Therefore, statement B is not true, and this is the requested answer.

**80. A** This question asks for the problem that is not caused by dynamic wheel unbalance.
Cupped tire treads, wheel shimmy, or excessive steering linkage wear may be caused by dynamic wheel unbalance. Therefore, B, C, and D are not the requested answer.
Excessive wear on the center of the tire tread is not a result of improper dynamic wheel balance.

**81. B**   This question asks for the statement that is not true. The TPMS module is mounted in the passenger compartment, each valve stem sensor acts as a radio transmitter, and some TPMS indicate the location of the tire with low air pressure. Therefore, A, C, and D are true, and none of these is the requested answer. The valve stem sensor is mounted on the wheel rim inside of the tire. Thus statement B is not true making it the requested answer.

# Answers to Post Test

**1. A**   If the inflator module is placed on the bench, the trim cover should be facing upward; thus, B is wrong. The inflator module should be carried with the trim cover facing away from your body. Therefore, C is wrong. The inflator module must be removed before the steering wheel, making D incorrect. On some vehicles, you should disconnect the negative battery cable and wait two minutes before removing the inflator module. Thus, A is correct.

**2. A**   The sector lash adjusting screw should be turned clockwise until the turning torque is 4 to 10 in.-lbs. (0.45 to 1.13 Nm) more than it was in the preliminary procedure. Therefore, A is correct, and B, C, and D are wrong.

**3. D**   Worn steering mounting bushings may cause unequal tie rod heights, but this problem does not affect steering arm height. Unequal steering linkage heights may cause the steering to veer suddenly in one direction, but this problem does not cause steering pull. Therefore, Technicians A and B are both wrong, making D the correct answer.

**4. A**   Discolored dark brown power steering fluid has been overheated, but this should not affect seal life. Thus, B is wrong. A misaligned power steering pump pulley causes excessive belt wear, making C incorrect. Excessive power-steering pump belt tension places extra stress on the drive shaft bearing, but this should not cause repeated seal failure. Thus, D is wrong. A worn drive shaft bearing allows vertical shaft movement and repeated seal failure. Therefore, A is correct.

**5. A**   The lower U-joint should be punchmarked in relation to the pinion shaft. Thus, A is correct. The steering wheel should be centered before removing the gear, making B wrong. The clockspring electrical connector should be centered when installing the gear and steering wheel; thus, C is wrong. Disconnect the battery negative cable and wait two minutes before removing the gear. Thus, D is wrong as well.

**6. D**   A tight rack bearing adjustment does not affect wear on the rack and pinion teeth, inner rack seal, or steering gear mounting bushings. Therefore, A, B, and C are wrong. A tight rack bearing adjustment may cause poor steering wheel returnability. D is correct.

**7. B**   The inner tie rod end should be tightened to the specified torque and then stacked. Therefore, Technician A is wrong and Technician B is right. B is the correct answer.

**8. A**   Air should be bled from the power-steering system by turning the steering wheel fully in each direction three or four times. Thus, Technician A is right. The power-steering system should be flushed with the hose from the gear to the reservoir disconnected. Therefore, Technician B is wrong, making A the correct answer.

**9. B**   Bump steer occurs when the tie rods are not parallel to the lower control arms. Therefore, Technician A is wrong. A bent pitman arm may cause the tie rods not to be parallel to the lower control arms. Therefore, Technician B is right, making B the correct answer.

**10. D**   Steering dampers are used on 4WD vehicles that may be driven on rough roads, so Technician A is wrong. Worn steering dampers cause excessive road shock on the steering wheel, so Technician B is wrong. Because Technicians A and B are both wrong, D is the correct answer.

**11. A**   Worn lower control arm bushings allow this arm to move outward and increase negative camber on the left front wheel, resulting in wear on the inside edge of the tread. Thus, Technician A is correct. A loose tie rod end may cause improper toe, steering looseness, and feathered tire tread wear. Therefore, Technician B is wrong, making A the correct answer.

**12.  C**    The strut rods prevent fore-and-aft lower control arm movement. Thus, A and B are wrong. Worn strut rod bushings may decrease positive front-wheel caster while braking, so C is correct. A bent strut rod pulls the lower control arm forward, and this action increases positive front wheel caster, making answer D wrong as well.

**13.  B**    A bent steering arm does not affect front-wheel caster or riding height. Thus, A and D are wrong. A bent steering arm causes improper toe-out-on turns and tire squeal while cornering. Therefore, B is correct, and C is wrong.

**14.  A**    If the left side of the rear axle is moved rearward, the steering may pull to the right, making Technician A correct. Bump steer is caused by unequal tie rod heights, so Technician B is wrong. Thus, A is the correct answer.

**15.  C**    The relationship between the force required to move the lower end of the shock absorber on the compression and rebound strokes is called shock ratio, and more force is usually required on the rebound or extension stroke. Therefore, Technicians A and B are both right, making C the correct answer.

**16.  A**    Loose front-wheel bearing adjustment may cause steering wander. Thus, A is correct. This problem does not cause improper caster angle, toe-out-on turns, or steering pull. B, C, and D are wrong.

**17.  A**    Wear on the outside edge of the tire tread may be caused by excessive positive camber. Thus, Technician A is correct. A worn upper strut mount moves the front-wheel camber to a more negative position. Technician B is wrong, and A is the correct answer.

**18.  A**    Reduced front curb riding height decreases positive caster and reduces directional stability. Therefore, A is correct. Reduced front curb riding height increases front-wheel toe, so B is wrong. Reduced rear curb riding height does not affect front-wheel camber, making C is wrong. Excessive rear curb riding height decreases front-wheel positive caster and decreases directional stability. Therefore, D is wrong.

**19.  D**    Improper caster setting does not affect tire tread wear. Thus, A and C are wrong. The steering pulls to the side with the least positive caster. Therefore, B is wrong and D is correct.

**20.  D**    The turning radius on the wheel on the inside of a turn is 2 degrees more than the turning radius on the outside wheel. Thus, A is true and not the requested answer. When the turning radius is not within specifications, one of the steering arms may be bent, and this causes tire squeal while cornering. Therefore, B and C are also true, meaning they are not the requested answer either. Improper turning radius does not affect steering wheel centering. Thus, D is not true, making it the requested answer.

# 5 Brakes

## Pretest

The purpose of this pretest is to determine the amount of review that you may require prior to taking the ASE Brakes Test. If you answer all the pretest questions correctly, complete the questions and study the information in this chapter to prepare for the ASE Brakes Test. If two or more of your answers to the pretest questions are incorrect, complete a thorough study of the questions and information in this chapter. The pretest answers are located at the end of the pretest. These answers are also in the answer sheets supplied with this book. Unless stated otherwise, the pretest questions apply to vehicles without antilock brake systems (ABS).

1. During a brake application, a vehicle experiences wheel lockup on the right rear (R/R) wheel. The cause of this problem could be
   A. weak brake shoe return springs on the R/R wheel.
   B. less-than-specified brake pedal free play.
   C. brake linings contaminated with grease on the R/R wheel.
   D. swollen rubber cups in the master cylinder.

2. A vehicle experiences brake drag on all four wheels.
   Technician A says the master cylinder compensating ports may be plugged.
   Technician B says the rubber cups in the master cylinder may be swollen.
   Who is correct?
   A. A only
   B. B only
   C. Both A and B
   D. Neither A nor B

3. A brake hose on the left front (L/F) wheel is soft and spongy in one location, but there are no visible brake fluid leaks. This problem may cause
   A. brake pedal fade during a brake application.
   B. brake drag on the L/F wheel.
   C. premature lockup on the L/F wheel.
   D. steering pull to the left during a brake application.

4. While discussing brake fluids, Technician A says DOT 3 brake fluid is hygroscopic.
   Technician B says DOT 4 and DOT 5 brake fluid may be mixed.
   Who is correct?
   A. A only
   B. B only
   C. Both A and B
   D. Neither A nor B

5. The brake metering valve
   A. reduces fluid pressure to the rear brakes during moderate brake applications.
   B. reduces fluid pressure to the rear brakes during hard brake applications.

    C. delays fluid pressure to the front brakes during light brake applications.

    D. reduces fluid pressure to the front brakes during moderate brake applications.

6.  The red brake warning light is illuminated continually with the ignition switch on.

    Technician A says the brake pedal may be binding.

    Technician B says there may be a fluid leak at the L/R wheel.

    Who is correct?

    A. A only

    B. B only

    C. Both A and B

    D. Neither A nor B

7.  While pressure bleeding the brakes on a vehicle with front disc brakes and rear drum brakes, Technician A says the metering valve should be closed.

    Technician B says the wheels may be bled in any sequence.

    Who is correct?

    A. A only

    B. B only

    C. Both A and B

    D. Neither A nor B

8.  A vehicle with front disc and rear drum brakes experiences excessive pedal fade after several hard brake applications. The cause of this problem could be

    A. the diameter of the brake drums exceeds the maximum specification.

    B. the pistons are seized in the rear wheel cylinders.

    C. there is air trapped in the rear wheel cylinders.

    D. the rear brake shoe return springs are weak.

9.  When the master cylinder is removed the vacuum brake booster contains some brake fluid.

    Technician A says this problem may be caused by a leaking primary piston cup.

    Technician B says this problem may be caused by an inoperative one-way check valve in the booster vacuum hose.

    Who is correct?

    A. A only

    B. B only

    C. Both A and B

    D. Neither A nor B

10.  On a vehicle with drum brakes, grabbing occurs on the R/F wheel during a brake application. The cause of this problem could be

    A. a restricted R/F brake hose.

    B. a leaking brake line near the L/F wheel.

    C. hard spots on the R/F brake drum.

    D. swollen cups in the R/F wheel cylinder.

11.  While discussing brake rotor measurement, Technician A says brake rotor runout should be measured while rotating the wheel with a dial indicator positioned against the rotor friction surface.

    Technician B says rotor parallelism should be measured with a micrometer at six or more locations around the rotor.

    Who is correct?

    A. A only

    B. B only

    C. Both A and B

    D. Neither A nor B

12. On an integral ABS with a high-pressure accumulator, poor stopping ability and a hard brake pedal may be caused by
    A. an open wheel speed sensor winding.
    B. a malfunctioning solenoid valve.
    C. low accumulator pressure.
    D. improper wheel speed sensor adjustment.

13. The amber ABS warning light is illuminated with the engine running on an integral ABS with a high-pressure accumulator. All of the following could be the cause of the problem EXCEPT
    A. a problem in the wheel speed sensor.
    B. a problem in the pump motor.
    C. the parking brake has been partially applied.
    D. a problem in the ABS computer.

14. While obtaining diagnostic trouble codes (DTCs) to diagnose ABSs
    A. DTCs indicate a problem in a specific component.
    B. some ABS systems do not provide flash codes.
    C. on some systems, cycle the ignition switch three times to obtain DTCs.
    D. flash codes must be obtained with the engine running.

# Answers to Pretest

**1. C**   Weak brake shoe return springs may cause a dragging brake, so A is wrong. Brake pedal free play that is less than specified may cause pressure buildup in the brake system, so B is wrong. Contaminated brake linings may cause wheel lockup; thus, C is correct. Swollen master cylinder cups may cause pressure buildup in the brake system, so D is wrong.

**2. C**   Plugged compensating ports in the master cylinder or swollen master cylinder cups may cause pressure buildup in the brake system and brake drag on all four wheels. Therefore, Technicians A and B are both right, making C the correct answer.

**3. A**   A soft spongy spot in a brake hose may cause brake pedal fade during a brake application. Therefore, B, C, and D are wrong, and A is correct.

**4. A**   DOT 3 brake fluid is hygroscopic, meaning it absorbs moisture, so Technician A is right. Different types of brake fluid should never be mixed. Thus, Technician B is wrong, making A the correct answer.

**5. C**   The metering valve delays fluid flow to the front brakes during light brake applications. Therefore, A, B, and D are wrong, and C is correct.

**6. B**   A fluid leak at one wheel may cause one of the master cylinder reservoirs to be low on fluid, and this illuminates the red brake warning light, so Technician B is right. A binding brake pedal may cause pressure buildup in the brake system, but this condition does not illuminate the red brake warning light. Thus, Technician A is wrong, making B the correct answer.

**7. D**   When bleeding the brakes, the metering valve should be held open, and the wheels should be bled in the sequence recommended by the vehicle manufacturer. Therefore, Technicians A and B are both wrong, and D is the correct answer.

**8. A**   Seized caliper pistons cause a hard brake pedal, so B is wrong. Air in the brake system causes a spongy brake pedal, so C is wrong. Weak brake shoe returns springs may cause dragging brakes; thus, D is wrong. Brake pedal fade may be caused by brake drums that exceed the maximum specified diameter, making A correct.

**9. A**   A leaking master cylinder primary piston cup may cause brake fluid accumulation in the vacuum brake booster, so Technician A is right. An inoperative one-way check valve in the brake booster vacuum hose does not cause brake fluid accumulation in the brake booster. Thus, Technician B is wrong, and A is the correct answer.

**10.  C**    A restricted R/F brake hose may cause grabbing on the L/F wheel, so A is wrong. A leaking brake line near the L/F wheel causes loss of fluid from one section of the master cylinder, and this may result in loss of braking on two wheels, low pedal, and poor stopping ability, so B is wrong. Hard spots on the R/F brake drum cause brake grabbing on the R/F wheel, making C correct. Swollen cups in the R/F wheel cylinder may cause brake drag on this wheel, so D is wrong.

**11.  C**    Rotor runout should be measured by rotating the rotor with a dial indicator positioned against the rotor machined surface, so Technician A is right. Rotor parallelism should be measured with a micrometer at six or more locations around the rotor; thus, Technician B is right. Because Technicians A and B are both right, C is the correct answer.

**12.  C**    A malfunctioning solenoid valve, an open wheel speed sensor winding, or an improper wheel sensor adjustment cause improper ABS operation, but these defects do not cause a hard brake pedal. Therefore, A, B, and D are wrong. Low accumulator pressure causes a hard brake pedal, so C is the correct answer.

**13.  C**    A problem in a wheel speed sensor, pump motor, or ABS computer will illuminate the amber ABS warning light. Therefore, A, B, and D are a cause of the problem, but none of these responses are the requested answer. If the parking brake is partially applied, the red brake warning light should be on but the amber ABS warning light is off. Therefore, C is not a cause of the problem, making it the requested answer.

**14.  B**    ABS DTCs indicate a problem in a certain area but not in a specific component, so A is wrong. DTCs on ABS systems are not obtained by cycling the ignition switch three times, so C is wrong. ABS flash codes may be obtained with the ignition switch on; thus, D is wrong. Many ABS systems do not provide flash codes, and a scan tool must be used to diagnose these systems, so B is correct.

# Hydraulic System Diagnosis and Repair, Master Cylinder

## ASE Tasks, Questions, and Related Information

In this chapter, each task in the Brakes category is provided followed by a question and some information related to the task. If you answer any question incorrectly, study this information very carefully until you understand the correct answer. Question answers and analysis are provided at the end of this chapter and in the answer sheets provided with this book.

**Task 1**    **Diagnose poor stopping, dragging, high or low pedal, hard or spongy pedal caused by problems in the master cylinder; determine needed repairs.**

1.  While discussing a vehicle with dragging brakes, Technician A says fluid may be leaking past the master cylinder cups.

Technician B says the stoplight and cruise control switch may require adjusting.

Who is correct?

A.  A only

B.  B only

     C. Both A and B

     D. Neither A nor B

**Hint**     *Dragging brakes may be caused by swollen master cylinder cups, or the compensation port covered in the master cylinder. This problem also may be caused by reduced brake pedal free play, or an improper adjustment of the stoplight and cruise control switch. If brake dragging occurs only on one front caliper, the quick take-up valve may not be allowing fluid movement from the reservoir into the primary piston area. Excessive pedal effort may be caused by swollen master cylinder cups or a corroded master cylinder.*

     *Brake pull may be caused by contaminated brake linings, restricted brake lines or hoses, or seized caliper or wheel cylinder pistons.*

     *Common causes of excessive brake pedal effort are glazed brake linings, seized caliper or wheel cylinder pistons, a restricted vacuum hose connected to the brake booster, or scored brake drums and rotors.*

     *Brake squeal may be caused by an improper type of brake linings; a loose backing plate, brake anchor, or caliper; loose wheel bearings; and weak or broken hold-down springs.*

     *A pulsating brake pedal may be caused by drums that are out-of-round, or rotors with excessive runout.*

## Task 2   Diagnose problems in the step bore master cylinder or internal valves (e.g., volume control devices, quick take-up valve, fast-fill valve, pressure regulating valve); determine needed repairs.

   2. The brake pedal is low and spongy, and all brake adjustments are completed as specified by the vehicle manufacturer. The cause of this problem could be

     A. a binding pedal linkage.

     B. dented brake lines.

     C. a plugged compensating port.

     D. a weak hydraulic brake hose.

**Hint**     *A low, spongy brake pedal may be caused by air in the hydraulic system, or a low brake fluid level in the master cylinder. If the brake drums are machined so they are thinner than specified by the manufacturer, the brake pedal may be low and spongy. Weak hydraulic hoses that expand under pressure also may cause a low, spongy brake pedal.*

     *A hard brake pedal may be caused by contaminated brake fluid, swollen master cylinder cups, dented brake lines, restricted brake hoses, or a plugged quick take-up valve. Glazed brake linings or a defective brake booster also increase pedal effort.*

## Task 3   Measure and adjust master cylinder pushrod length.

   3. A vehicle has no brake pedal free play. This problem may cause

     A. a low brake pedal.

     B. a spongy brake pedal.

     C. pressure buildup and dragging brakes.

     D. fluid leaking past the primary piston cups.

**Hint**     *The specified brake pedal free play is usually 1/16 to 1/4 in. Excessive brake pedal free play causes a low brake pedal. When the brake pedal free play is less than specified, the compensating ports in the master cylinder may be closed by the piston cups when the brake pedal is released. This action results in fluid pressure buildup and dragging brakes.*

## Task 4   Check master cylinder for failures by depressing the brake pedal; determine needed repairs.

   4. With the engine running and the brake pedal applied and held with medium foot pressure, the brake pedal slowly moves downward.

     Technician A says brake fluid may be leaking past the master cylinder piston cups.

Technician B says brake fluid may be leaking from a brake hose at one of the front wheels.

Who is correct?

A.  A only

B.  B only

C.  Both A and B

D.  Neither A nor B

**Hint**    *When the brakes are held in the applied position with the engine running, the pedal should feel firm. On most vehicles, the brake pedal should remain 2 in. from the floor for manual brakes or 1 in. from the floor for power brakes. On a manual brake system, it is not necessary to have the engine running. If the pedal is low, the brakes may require adjusting, or the self-adjusters may not be operating. Excessive brake pedal free play also causes a low pedal. A spongy pedal may be caused by air in the hydraulic system or drums that have been machined until they are too thin. If the brake pedal slowly moves downward, fluid may be leaking past the master cylinder piston cups, or there may be an external leak in a brake hose or line.*

## Task 5    Diagnose the cause of master cylinder external fluid leakage.

5.  There are signs of paint removal on the power brake booster below the master cylinder.

   The cause of this problem could be

   A.  a leaking secondary cup on the primary piston.

   B.  a defective diaphragm in the brake booster.

   C.  a leaking secondary cup on the secondary piston.

   D.  reduced brake free play and fluid pressure buildup.

**Hint**    *If there is an indication of paint removal and fluid leakage on the brake booster below the master cylinder, the secondary cup on the primary piston is leaking. If this leakage is present, there probably will be brake fluid in the brake booster when the master cylinder is removed. Use a suction gun to remove this fluid.*

## Task 6    Remove and replace master cylinder; bench bleed and test operation and install master cylinder; verify master cylinder function.

6.  After the master cylinder is bled and installed, excessive brake pedal effort is required during brake applications.

   Technician A says the vacuum hose connected to the brake booster may be restricted.

   Technician B says the pistons may be seized in the front calipers.

   Who is correct?

   A.  A only

   B.  B only

   C.  Both A and B

   D.  Neither A nor B

7.  The service procedure illustrated in Figure 5–1 is

   A.  checking the master cylinder piston travel.

   B.  measuring the master cylinder displacement volume.

   C.  checking the master cylinder piston return action.

   D.  bleeding air from the master cylinder.

**Hint**    *After the master cylinder is bled and installed, apply and hold the brake pedal. If the vehicle has power brakes, apply the brake pedal with the engine running. The brake pedal should be firm and remain at the proper height. Check the operation of the brakes during a road test. When the brakes are applied, the steering should not pull in either direction, and the vehicle should stop*

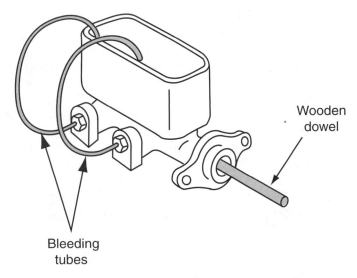

**Figure 5–1** Master cylinder service procedure.

*with a normal amount of pedal effort. During a brake application, there should be no abnormal noises such as brake squeals and rattles. While the brakes are applied, the brake pedal should remain firm without pulsating.*

*To bench bleed the master cylinder, bleeder tubes are connected from the master cylinder outlet ports to the reservoirs. The reservoir should be filled to the proper level with the specified type of brake fluid. A wooden dowel should be used to slowly depress and release the master cylinder pistons. Continue this action until there are no air bubbles coming through the bleeder tubes. Remove the bleeder tubes and plug the master cylinder outlets.*

# Hydraulic System Diagnosis and Repair, Lines and Hoses

## ASE Tasks, Questions, and Related Information

**Task 1** **Diagnose poor stopping, pulling, or dragging caused by problems in the lines and hoses; determine needed repairs.**

8. A vehicle pulls to the left during a brake application. The cause of this problem could be
   A. the right front brake linings are contaminated with grease.
   B. the piston is seized in the right front caliper.
   C. the master cylinder pistons are swollen from contaminated fluid.
   D. the secondary compensating port is plugged in the master cylinder.

**Hint**    *Poor stopping ability may be caused by glazed brake linings, contaminated brake fluid and swollen master cylinder cups, restricted master cylinder tubes, or contaminated master cylinder bores.*

*The most common causes of brake pull are seized caliper or wheel cylinder pistons, contaminated brake linings, scored or burned drums or rotors, and worn suspension components. If one front caliper is seized, the vehicle pulls to the opposite side from the seized caliper. When the brake linings are contaminated on a front wheel, the brake grabs on the wheel with the contaminated linings. This action causes steering pull to the side with the contaminated linings.*

*Brake drag may be caused by plugged master cylinder compensating ports, improper pedal free play, contaminated brake fluid and swollen master cylinder cups, or a binding brake pedal. If the brakes drag on one wheel, the brake line or hose may be restricted, or the piston may be sticking in the caliper or wheel cylinder.*

## Task 2    Inspect brake lines and fittings for leaks, dents, kinks, rust, cracks, or wear, inspect for loose fittings and supports; determine needed repairs.

9. While discussing brake lines, Technician A says a damaged brake line may be repaired with a short piece of line and compression fittings.

   Technician B says the necessary brake line bends should be made with a tubing bending tool.

   Who is correct?

   A. A only

   B. B only

   C. Both A and B

   D. Neither A nor B

**Hint**    *Brake lines should be inspected for leaks, cracks, rust, kinks, flattened areas, and splits.*
*Damaged brake lines should be replaced rather than repaired. A tubing bending tool should be used to complete the necessary brake line bends.*

## Task 3    Inspect flexible brake hoses for leaks, kinks, cracks, bulging, wear, or corrosion; inspect for loose fittings and supports; determine needed repairs.

10. While discussing brake hose replacement, Technician A says the sealing washer on the male end of a brake hose may be reused.

    Technician B says the brake line fitting should be installed and tightened in the female end of the brake hose before the male end is installed.

    Who is correct?

    A. A only

    B. B only

    C. Both A and B

    D. Neither A nor B

**Hint**    *Flexible brake hoses allow for movement between the suspension and the chassis. These brake hoses should be inspected for cracks, leaks, twists, bulges, loose supports, and internal restrictions.*
*Each time a brake hose is removed, the sealing washer on the male end should be replaced. When a brake hose is installed, always install and tighten the male end first.*

## Task 4    Replace brake lines, hoses, fittings, and supports; fabricate brake lines using proper material and flaring procedures (double flare and ISO types).

11. The brake line flare in Figure 5–2 is

    A. a double inverted flare.

    B. an ISO flare.

    C. a single flare.

    D. an SAE flare.

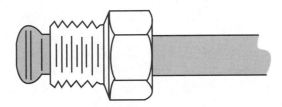

**Figure 5–2** Brake tube flare.

*Hint*     *Brake tubing may have double inverted flares or international standards organization (ISO) flares. Tube and hose mating surfaces should be clean and free from metal burrs.*

12. A brake line is not properly clamped to the frame and this line is now positioned so it contacts the catalytic converter.

   Technician A says this condition may cause the brake pedal height to slowly decrease during a brake application.

   Technician B says this condition may cause a spongy brake pedal.

   Who is correct?

   A. A only

   B. B only

   C. Both A and B

   D. Neither A nor B

*Hint*     *Brake hoses and lines should be inspected for contact with other components. Such contact may wear a hole in the brake line or hose, resulting in a brake fluid leak. Brake lines must be clamped to the chassis at regular intervals to eliminate brake line movement and contact with other components. They must also be kept away from hot components such as catalytic converters. Excessive heat applied to the brake line may cause the brake fluid in the line to boil and change to a vapor. This action will result in a spongy brake pedal.*

# Hydraulic System Diagnosis and Repair, Valves and Switches

## ASE Tasks, Questions, and Related Information

**Task 1**   **Diagnose poor stopping, pulling, or dragging caused by problems in the hydraulic system valve(s); determine needed repairs.**

13. The purpose of the component in Figure 5–3 is to

   A. provide simultaneous front disc and rear drum brake application.

   B. prevent premature front-wheel lockup during hard brake applications.

   C. prevent front brake drag when the brakes are released.

   D. reduce the front brake pad temperature and provide longer pad life.

*Hint*     *The metering valve is used on systems with front disc and rear drum brakes. During a brake application, the brake fluid pressure to the rear brakes has to overcome the force of the brake shoe return springs and force these shoes outward. Therefore, a brief time interval is required to apply the rear brakes, whereas the front caliper pistons move the brake pads quickly against the rotors.*

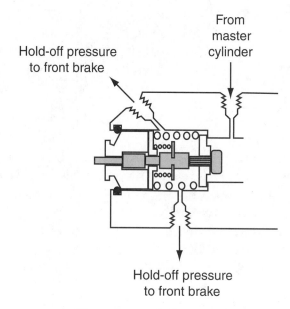

**Figure 5–3** Metering valve.

*The metering valve delays front brake fluid pressure during light brake applications so both front and rear brake applications occur at the same time. This action prevents front-wheel lock-up and skidding during light brake applications on slippery road surfaces. During hard brake applications, the metering valve remains open and does not affect brake operation.*

## Task 2    Inspect, test, and replace metering, proportioning, pressure differential, and combination valves.

14. The pressure gauges connected to the proportioning valve inlet and outlet (Figure 5–4) should indicate

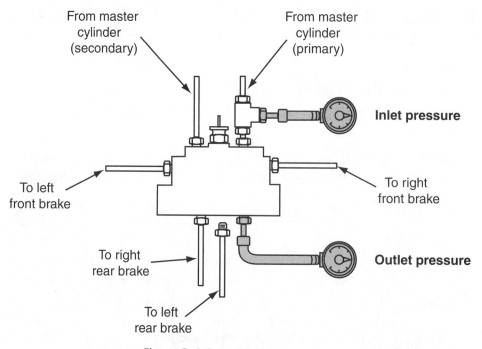

**Figure 5–4** Proportioning valve test.

A. higher pressure at the outlet compared to the inlet during a moderate brake application.

B. the same pressure at the inlet and outlet during a moderate brake application.

C. lower pressure at the outlet compared to the inlet during a moderate brake application.

D. lower pressure at the outlet compared to the inlet during a light brake application.

**Hint** *During a moderate brake application, the proportioning valve reduces pressure to the rear wheel brakes to compensate for the transfer of vehicle weight to the front wheels. This action prevents premature rear wheel lockup of the rear wheels. During a light brake application, the proportioning valve does not affect rear wheel pressure. When the brakes are applied with high pedal pressure, the proportioning valve opens and allows full pressure to the rear wheels. When pressure gauges are connected to each proportioning valve inlet and outlet, the specified pressure difference between the inlet and outlet should be present during a moderate brake application.*

*The metering valve stem should move slightly when the brake pedal is applied and released.*

*If the metering valve is operating normally, a slight bump may be felt on the brake pedal when the pedal has been depressed about one inch.*

## Task 3 Inspect, test, replace, and adjust load or height sensing-type proportioning valve(s).

15. While discussing a load-sensing proportioning valve, Technician A says this valve reduces brake pressure to the rear wheels as the rear suspension load is increased.

Technician B says this valve has a linkage connected from the valve to the rear chassis.

Who is correct?

A. A only

B. B only

C. Both A and B

D. Neither A nor B

**Hint** *The height-sensing proportioning valve is mounted on the chassis, and a linkage is connected from the valve to the rear axle. When the load is light on the rear wheels, the linkage positions the internal valve so brake pressure is reduced to the rear wheels during moderate brake applications.*

*A heavy load on the rear suspension reduces rear chassis height. This action causes the linkage to move the valve in the height-sensing proportioning valve. Under this condition, the proportioning valve does not reduce pressure to the rear brakes during moderate brake applications.*

## Task 4 Inspect, test, and replace brake warning light, switch, sensor and circuit.

16. With the brake warning switch unit positioned as shown in Figure 5–5

A. the brake warning light is illuminated with the ignition switch on.

B. the pressure is higher in the secondary master cylinder section than in the primary section.

C. the circuit is open between the switch terminal and ground.

D. the pressure is equal in the primary and secondary sections of the master cylinder.

**Hint** *When the pressure is equal in the primary and secondary sections of the master cylinder, the warning switch piston remains centered. In this position, the switch piston does not touch the switch pin. If the pressure is unequal in the primary and secondary master cylinder sections, the pressure difference moves the switch piston to one side. In this position, the switch piston pushes the spring-loaded switch pin upward and closes the warning light switch. This action illuminates the brake warning light.*

*The brake warning light circuit may be tested by grounding the warning switch wire with the ignition switch on. Under this condition, the bulb should be illuminated. If the bulb is not illuminated, check the fuse, bulb, and connecting wires.*

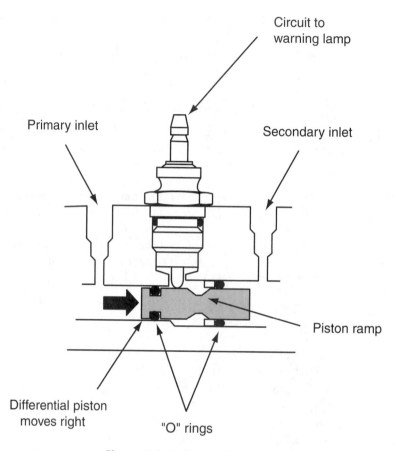

**Figure 5–5** Brake warning switch.

*After brake repairs are completed to restore equal pressure in the primary and secondary sections of the master cylinder, a hard brake pedal application usually centers the warning switch piston. If this action does not center the piston and put the light out, apply light brake pedal pressure and loosen a bleeder screw in the side of the brake system that had high pressure during the failure.*

# Hydraulic System Diagnosis and Repair, Bleeding, Flushing, and Leak Testing

## ASE Tasks, Questions, and Related Information

**Task 1**  **Diagnose poor stopping, pulling, or dragging caused by problems in the brake fluid; determine needed repairs.**

17. A spongy brake pedal occurs after several hard stops in a short time period.
    Technician A says the brake fluid may be contaminated.
    Technician B says the vacuum hose to the brake booster may be restricted.

Who is correct?

A. A only

B. B only

C. Both A and B

D. Neither A nor B

**Hint** *Contaminated brake fluid may cause swollen master cylinder cups that result in brake drag. Improper or contaminated brake fluid may boil at a lower-than-specified temperature. This action may cause vapor in the hydraulic brake system especially after repeated hard braking. Vapor in the brake system causes a spongy brake pedal.*

## Task 2    Bleed and/or flush hydraulic system (manual, pressure, vacuum, or surge method).

18. While discussing pressure brake bleeding, Technician A says the metering valve should be closed.

Technician B says the pressure in the brake bleeder should be 20 psi.

Who is correct?

A. A only

B. B only

C. Both A and B

D. Neither A nor B

19. All of these statements about vacuum brake bleeding are true EXCEPT

A. The vacuum pump handle should be operated two to four times before the bleeder screw is opened.

B. The vacuum pump creates a vacuum in the tester reservoir.

C. The bleeder screw should be opened until there is approximately 1 in. of brake fluid in the reservoir.

D. A one-way check valve is required in the line from the bleeder screw to the reservoir.

20. All of the following statements about manual brake bleeding procedure are true EXCEPT

A. A hose is connected to the bleeder screw, and the opposite end of this hose submerged in a container of brake fluid.

B. Apply the brake pedal with moderate pressure and then open the bleeder screw.

C. When the bleeder screw is opened and the pedal goes to the floor, release the pedal and close the screw.

D. Repeat the bleeding procedure until the fluid escaping from the bleeder hose is free of air bubbles.

**Hint** *During a manual brake bleeding procedure, connect a bleeder hose from a bleeder screw into a container partially filled with brake fluid. Keep the end of this hose submerged below the level of brake fluid in the container. Each wheel caliper or cylinder must be bled in the vehicle manufacturer's specified sequence in any bleeding procedure. Wheel calipers may be tapped with a soft hammer to help remove air bubbles.*

*During a manual brake bleeding procedure, apply the brake pedal with moderate force and open the bleeder screw. When the brake pedal goes down to the floor, close the bleeder screw and release the pedal. Repeat the procedure until there are no air bubbles escaping from the bleeder hose.*

*A pressure bleeder has an adapter connected to the top of the master cylinder reservoir. A hose is connected from this adapter to the pressure bleeder fluid chamber in the top of the pressure bleeder. The pressure bleeder has an air chamber below the fluid chamber, and a diaphragm separates the air and fluid chambers. Shop air is used to pressurize the air chamber to 15 to 20 psi.*

*If the brake system has a metering valve, this valve must be held open with a special tool. The bleeder hose is connected from each bleeder screw into a container partially filled with brake fluid. Open the bleeder screw until a clear stream of brake fluid is discharged.*

*The surge bleeding procedure may be used with manual or pressure bleeding. Bleed the brake system in the conventional manner. Surge bleeding involves pumping the brake pedal quickly several times with the bleeder screw open; then close the bleeder screw. After surge bleeding, use the conventional bleeding procedure to bleed the system one more time.*

*During the vacuum bleeding procedure, the vacuum pump is connected to a sealed container.*

*Another hose is connected from this container to the bleeder screw. A one-way check valve is connected in the hose from the bleeder screw to the container. Operate the vacuum pump handle 10 to 15 times to create vacuum in the container. Open the bleeder screw until about 1 in. of brake fluid is pulled into the container. Repeat the procedure until the fluid coming into the container is free of air bubbles.*

*Flushing a brake system is a continuation of the bleeding procedure. During the flushing procedure, each bleeder screw is opened until all the contaminated fluid is removed. Never reuse fluid from a bleeding or flushing procedure. Discard this fluid according to environmental regulations.*

## Task 3  Pressure test brake hydraulic system.

21. When pressure testing a hydraulic brake system that is equipped with a metering valve and proportioning valves
    A. during a moderate brake application, the pressure at the master cylinder should exceed the pressure at the front-wheel calipers.
    B. below the split point pressure, the pressure at the master cylinder outlet should exceed the pressure at the rear-wheel cylinders.
    C. during a moderate brake application, the pressure should be equal at the front and rear wheels.
    D. above the split point pressure, the pressure at the master cylinder outlet should exceed the pressure at the rear-wheel cylinders.

**Hint**    *A pressure test may be performed with the moderate pressure applied to the brake pedal. If the vehicle has power brakes, the engine should be running. The brake pedal should be firm at the specified pedal height. If the pedal slowly moves toward the floor, there is an internal master cylinder leak or an external leak in the lines, fittings, hoses, wheel cylinders, or calipers. When there are no external leaks, the master cylinder must be leaking internally.*

*A pressure test may be performed to check the brake booster operation. Connect a pressure gauge to one of the master cylinder outlets, and apply the brake pedal with moderate pressure.*

*Maintain this pressure and start the engine. There should be a significant increase in master cylinder pressure if the brake booster operation is satisfactory.*

*A pressure gauge may be connected at each front-wheel caliper. When the brakes are applied with moderate pressure, the gauge readings should be equal. When the pressure is lower at one front wheel, the line to that wheel is probably restricted.*

*The proportioning valves connected to the rear wheels may be tested by connecting pressure gauges on the master cylinder side and wheel cylinder side of each proportioning valve. When the brake pedal is applied with light pressure and the master cylinder pressure is below the specified split point pressure, both gauges should indicate the same pressure. If the brake pedal pressure is increased and the master cylinder pressure exceeds the split point pressure, the master cylinder pressure should exceed the wheel cylinder pressure by the specified amount.*

## Task 4  Select, handle, store, and install proper brake fluids (including silicone fluids).

22. While discussing brake fluids, Technician A says silicone-based brake fluids are hygroscopic.
    Technician B says DOT 3 and DOT 4 brake fluids tend to absorb moisture.

Who is correct?

A. A only

B. B only

C. Both A and B

D. Neither A nor B

**Hint**    *Brake fluid containers should be stored in a clean, dry location. When brake fluid containers are not in use, seal them tightly. Silicone-based fluid is classified as DOT 5 fluid. This type of fluid is nonhydroscopic, which means it does not absorb water. Silicone-based brake fluid has a long-term shelf life and does not damage paint finishes.*

*Commonly used nonpetroleum-based brake fluids are classified as DOT 3 or DOT 4. These brake fluids tend to absorb moisture from the air. Paint surfaces are damaged by DOT 3 or DOT 4 brake fluids. Compared to DOT 3 brake fluid, DOT 4 brake fluid has a higher boiling point and absorbs moisture more slowly.*

# Drum Brake Diagnosis and Repair

## ASE Tasks, Questions, and Related Information

**Task 1**    Diagnose poor stopping, pulling, or dragging caused by drum brake hydraulic problems; determine needed repairs.

23.  A vehicle with drum brakes experiences poor stopping, and the pedal feels springy and spongy during a brake application.

Technician A says the vents in the master cylinder cover may be plugged.

Technician B says there may be air in the hydraulic system.

Who is correct?

A. A only

B. B only

C. Both A and B

D. Neither A nor B

**Hint**    *Poor stopping ability may be caused by hydraulic problems such as contaminated brake fluid, air in the hydraulic system, or plugged master cylinder vents.*

*Brake pull may be a result of hydraulic problems such as a severely restricted brake line or hose, or a wheel cylinder size that is different on opposite sides of the vehicle.*

*Brake drag may be caused by hydraulic problems such as contaminated brake fluid, inferior rubber cups in the master cylinder or wheel cylinders, or plugged compensating ports in the master cylinder. If one wheel is dragging, the wheel cylinder rubber cups may be swollen or the wheel cylinder pistons could be sticking. An obstructed brake line or hose may also be the cause of brake drag.*

**Task 2**    Diagnose poor stopping, noise, pulling, grabbing, dragging, or pedal pulsation caused by drum brake mechanical problems; determine needed repairs.

24.  A vehicle experiences brake squeal during brake applications.

Technician A says the drums may be distorted.

Technician B says the backing plates may be bent.

Who is correct?

A.  A only

B.  B only

C.  Both A and B

D.  Neither A nor B

**Hint**    *Reduced stopping ability may be caused by mechanical brake problems such as improper brake adjustment; incorrect, glazed, or oil-soaked linings; seized wheel cylinder pistons; or bell-mouthed, barrel-shaped, or scored drums. Poor stopping ability combined with brake pedal fade may be caused by drums that have been machined until they are thinner than specified.*

*Brake squeal may be caused by bent backing plates, distorted drums, linings loose at shoe ends, improper lining position on the shoes, weak or broken hold-down springs, or loose wheel bearings. Brake chatter may be caused by improper brake adjustment, loose backing plates, contaminated linings, out-of-round, tapered or barrel-shaped drums, cocked or distorted shoes, or loose wheel bearings.*

*Brake pull may be caused by improper tire inflation, unequal tire tread wear, or different tread designs on opposite sides. Suspension and wheel alignment problems may cause brake pull. This problem also may be caused by different wheel cylinder sizes on opposite sides, seized wheel cylinder pistons, damaged or contaminated brake linings on one side, weak or broken retractor springs, or a scored brake drum on one side.*

*Brake grabbing may be caused by contaminated linings; shoes not centered in the drums; loose, distorted backing plates; or scored, hard-spotted, or out-of-round drums.*

*Dragging brakes may be caused by a binding brake pedal, swollen rubber parts resulting from incorrect or contaminated brake fluid, or plugged master cylinder compensating ports. If only the rear wheels drag, the parking brake cables may be seized. When one wheel drags, that wheel may have a seized caliper or wheel cylinder piston, weak or broken shoe return springs, bent or distorted shoes, loose wheel bearings, improper shoe adjustment, or a restricted brake hose or line.*

*Pedal pulsations may be caused by out-of-round brake drums, or rotors with improper parallelism.*

## Task 3    Remove, clean, inspect, and measure brake drums; follow manufacturers' recommendations in determining need to machine or replace.

25.  All of the following statements are true about brake drum inside diameter measurement with a brake drum micrometer EXCEPT

A.  The drum should be cleaned before measuring the diameter.

B.  If the drum diameter is more than specified, replace the drum.

C.  The diameter should be measured at two locations around the drum.

D.  The drum diameter variation should not exceed 0.0035 in.

**Hint**    *Brake drums should be inspected for cracks, heat checks, out-of-round, bell-mouthed shapes, scoring, and hard spots. The inside drum diameter should be measured with a drum micrometer. If the drum diameter exceeds the maximum limit specified by the manufacturer, replace the drum. The drum diameter should be measured every 45 degrees around the drum. The maximum allowable out-of-round specified by some manufacturers is 0.0035 in. If the out-of-round exceeds specifications, the drum may be machined if this machining does not cause the drum to exceed the maximum allowable diameter.*

## Task 4    Machine drums according to manufacturers' procedures and specifications.

26.  While machining a brake drum

A.  tool chatter marks may be caused by excessive damping belt tension.

B.  the tool bit depth for a rough cut should be 0.005 to 0.010 in.

C. the tool bit depth for a rough cut should be 0.010 to 0.020 in.

D. the tool bit depth of a finish cut should be 0.008 to 0.010 in.

**Hint**   *When machining a brake drum, the drum must be installed securely and centered on the lathe. A dampening belt must be installed tightly around the outside of the drum to prevent the cutting tool from chattering on the drum. Many manufacturers recommend a rough cut tool depth of 0.005 to 0.010 in., and a finish cut tool depth of 0.005 in. After the machining procedure, the drum should be sanded to remove minor irregularities. Many manufacturers recommend drum machining until the drum diameter is within 0.030 in. of the maximum allowable diameter.*

*This 0.030 in. is necessary to allow for drum wear.*

**Task 5**   **Using proper safety procedures, remove, clean, and inspect brake shoes/linings, springs, pins, self-adjusters, levers, clips, brake backing (support) plates and other related brake hardware; determine needed repairs.**

27. While discussing drum brake hardware, Technician A says if item 17 in Figure 5–6 is omitted during brake hardware assembly, the brake will grab on that wheel.

    Technician B says if item 8 in Figure 5–6 is omitted during brake hardware assembly, there is no self-adjusting action.

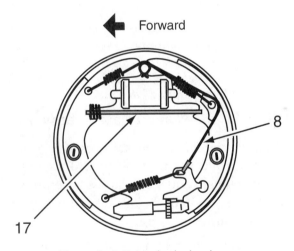

**Figure 5–6** Drum brake hardware.

Who is correct?

A. A only

B. B only

C. Both A and B

D. Neither A nor B

28. While discussing brake backing plates, Technician A says a bent backing plate may cause brake grabbing.

    Technician B says a loose anchor bolt may cause brake chatter.

    Who is correct?

    A. A only

    B. B only

    C. Both A and B

    D. Neither A nor B

*Hint*    *All brake return springs should be inspected for distortion and stretching. Brake shoes should be cleaned with a shop towel and inspected for broken welds, cracks, wear, and distortion. If the wear pattern on the brake shoes is uneven, the shoes are distorted. Check all clips and levers for wear and bending. Inspect the brake linings for contamination with oil, grease, or brake fluid. Clean and lubricate adjusting and self-adjusting mechanisms.*

*The backing plate should be cleaned with an approved brake cleaner that does allow the release of asbestos dust into the shop air. Inspect the backing plates for distortion, cracks, rust damage, and wear in the shoe contact areas. A distorted backing plate may cause brake grabbing.*

*Check the anchor bolt for looseness which could result in brake chatter.*

## Task 6    Lubricate brake shoe support pads on backing (support) plate, self-adjuster mechanisms, and other brake hardware.

29. While discussing brake hardware service, Technician A says dry shoe ledges on the backing plates may cause a squeaking noise during brake applications.

    Technician B says slightly scored shoe ledges on the backing plates may be resurfaced and lubricated with high-temperature grease.

    Who is correct?

    A. A only

    B. B only

    C. Both A and B

    D. Neither A nor B

*Hint*    *The backing plates should be cleaned with an approved brake cleaning method. Because a warped backing plate may cause uneven shoe wear and brake grabbing, inspect the backing plates for this condition. If the shoe ledges are lightly scored, they may be resurfaced. The backing plate should be replaced if the shoe ledges are severely scored or worn. Lubricate the shoe ledges with high-temperature lubricant.*

## Task 7    Install brake shoes and related hardware.

30. While assembling the brake shoes and related hardware

    A. the secondary shoe faces toward the front of the vehicle.

    B. the primary and secondary shoe return springs are interchangeable.

    C. the adjuster must be installed in the proper direction.

    D. the adjuster cable usually is mounted on the primary shoe.

*Hint*    *The secondary shoe must face toward the rear of the vehicle, and the primary shoe faces the front of the vehicle. The secondary shoe usually has a longer lining and the linings on the primary and secondary shoes usually are made of different materials. The primary and secondary shoes have different return springs. Brake adjusters must be installed in the proper direction so the star wheel is behind the backing plate adjustment opening. The adjuster cable and related mechanism usually is installed on the secondary shoe.*

## Task 8    Pre-adjust brake shoes and parking brake before installing brake drums or drum/hub assemblies and wheel bearings.

31. Technician A says that in Figure 5–7 the brake shoes are being adjusted to match the drum size.

    Technician B says that in Figure 5–7 the drum and brake shoes are being measured for wear.

    Who is correct?

    A. A only

    B. B only

    C. Both A and B

    D. Neither A nor B

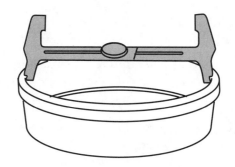

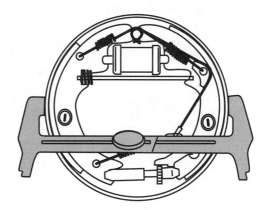

**Figure 5–7** Brake shoe service.

32. While adjusting the parking brake in Figure 5–8, Technician A says the parking brake should be applied until a pin in hole A contacts the parking brake outer flange.

Technician B says the parking brake shown in Figure 5–8 should be released during the adjustment.

Who is correct?

A.  A only

B.  B only

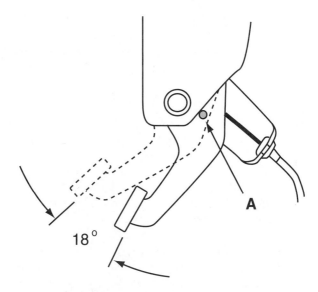

**Figure 5–8** Parking brake adjustment.

C.  Both A and B

D.  Neither A nor B

*Hint*    *A special tool is used to measure the brake drum inside diameter. The other side of the tool is then installed on the outside of the brake shoes, and the shoe adjuster is turned until the shoes contact the tool. This procedure adjusts the shoes properly in relation to the drum size before installing the drum.*

*The brake shoes must be properly adjusted before the parking brake adjustment. Some manufacturers recommend adjusting the parking brake with the brake released. Tighten the parking brake cable adjusting nut until there is a light drag while rotating the rear wheels. Loosen the parking brake cable adjusting nut until the rear wheels rotate freely, and then loosen the parking brake cable adjusting nut two turns.*

## Task 9    Reinstall wheel, torque lug nuts, and make final checks and adjustments.

33.  While discussing the brake adjustment in Figure 5–9, Technician A says the thin tool is used to position the star wheel so it is accessible with the adjusting tool.

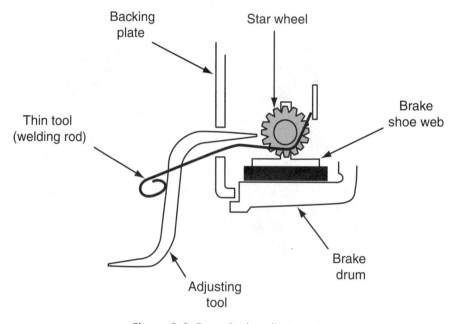

**Figure 5–9** Drum brake adjustment.

Technician B says the star wheel should be rotated until there is wheel drag and then back off the star wheel until the wheel rotates freely.

Who is correct?

A.  A only

B.  B only

C.  Both A and B

D.  Neither A nor B

*Hint*    *The lug nuts must be tightened in the proper sequence with a crisscross pattern. Tighten the lug nuts to one-half the specified torque, and then tighten them to the specified torque. An improper lug nut tightening sequence, or excessive torque, may cause warped brake drums and grabbing brakes.*

*If a brake adjustment is necessary, a thin tool may be inserted through the backing plate opening to release the self-adjusting mechanism. After this tool is in place, a brake adjusting tool may be used to rotate the star wheel until a slight wheel drag is felt while rotating the wheel. Rotate the star wheel in the opposite direction until the wheel rotates freely.*

# Disc Brake Diagnosis and Repair

## ASE Tasks, Questions, and Related Information

**Task 1**  Diagnose poor stopping, pulling, or dragging caused by disc brake hydraulic problems; determine needed repairs.

34. All of the following problems may cause the car to pull to one side while braking EXCEPT
   A. incorrect or loose brake pads.
   B. a loose caliper mounting bracket.
   C. seized master cylinder pistons.
   D. a sticking caliper piston.

**Hint**  *Poor stopping may be caused by these hydraulic problems; restricted brake lines or hoses, sticking master cylinder pistons; or sticking caliper pistons.*

*Pulling to one side while braking may be caused by a sticking caliper piston or a restricted line or hose.*

*Dragging brakes may be caused by soft or swollen rubber components due to incorrect or contaminated brake fluid. This problem also may be caused by plugged master cylinder compensating ports or a restricted hose or line.*

**Task 2**  Diagnose poor stopping, noise, pulling, grabbing, dragging, pedal pulsation or pedal travel caused by disc brake mechanical problems; determine needed repairs.

35. Excessive brake pedal pulsations are experienced while braking.
   Technician A says the outer drive axle joint may be worn.
   Technician B says the front wheel bearing may be worn and loose.
   Who is correct?
   A. A only
   B. B only
   C. Both A and B
   D. Neither A nor B

**Hint**  *Poor stopping may be caused by contaminated brake linings, a malfunctioning power brake booster or low vacuum supply, or a malfunctioning power brake vacuum check valve.*

*Pulling to one side while braking may be caused by the following mechanical problems: incorrect or loose pads, contaminated brake linings, loose caliper mounting, or damaged suspension parts.*

*Dragging brakes may be caused by a binding brake pedal, loose or worn wheel bearings, a sticking caliper piston, or a restricted line or hose.*

*Braking noise may be caused by bent, damaged, or loose pads, worn-out linings, foreign material embedded in the lining, or loose caliper mounting.*

*Grabbing brakes may be caused by contaminated linings, incorrect or loose pads, or loose caliper mounting.*

*Pedal pulsations may be caused by excessive rotor runout, and worn or loose wheel bearings.*

**Task 3**  Retract integral parking brake caliper piston(s) according to manufacturers' recommendations.

36. When discussing parking brake assemblies on rear disc brakes, Technician A says the parking brake lever on the caliper rotates a spindle inside the caliper.

   Technician B says the caliper pistons may be in any position before installing the caliper over the brake pads.

   Who is correct?

   A. A only

   B. B only

   C. Both A and B

   D. Neither A nor B

**Hint**    *On rear disc brakes, the parking brake cable is connected to levers on each rear brake caliper (Figure 5–10). Each parking brake lever is connected to a spindle in the caliper. When the parking brake is applied, the lever rotates the spindle. This action forces a connecting rod against an internal adjusting screw that is threaded into a sleeve nut in the caliper piston. The rotating action of the parking brake lever and spindle forces the connecting rod against the internal adjusting screw and moves the piston outward against the inboard brake pad to apply the parking brakes. When the inboard brake pad lining is forced against the rotor, the caliper housing slides inward to force the outboard pad against the rotor.*

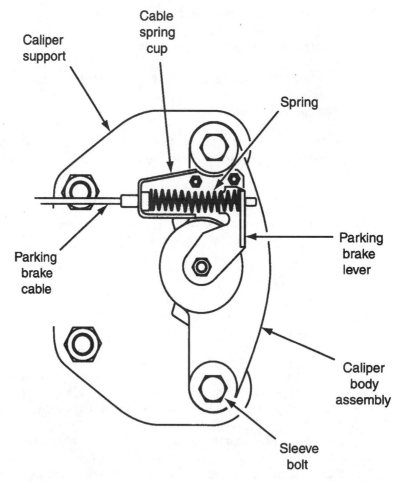

**Figure 5–10** Rear disc brake caliper with parking brake mechanism.

*As the brake linings wear, the caliper piston moves outward to maintain proper lining-to-rotor clearance. The parking brake adjusts automatically through an internal sleeve nut that rotates and moves with the piston. However, a parking brake cable adjustment is possible.*

*After the brake pads and/or caliper have been replaced, the caliper piston must be rotated with a suitable spanner tool until the piston is bottomed. The caliper piston slots must be positioned properly before installing the caliper over the brake pads (Figure 5–11).*

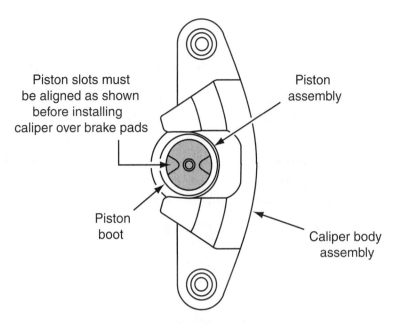

**Figure 5–11** Proper piston position in rear disc brake caliper.

## Task 4  Remove caliper assembly from mountings; inspect for leaks and damage to caliper housing.

37. After a brake application, the caliper piston is returned by
    A. the twisting action of the seal.
    B. a return spring.
    C. brake fluid pressure.
    D. atmospheric pressure.

**Hint**    *Inspect the caliper for evidence of brake fluid in and around the boot area. If brake fluid is evident, the caliper seal must be replaced. Replace the boot if it is torn, damaged, or cracked. If the caliper ears are worn, broken, or elongated, replace the caliper.*

## Task 5  Clean, inspect, and measure caliper mountings and slides/pins for wear and damage.

38. On a single piston floating caliper, the inside brake pad lining is worn out, but there is very little wear on the outside pad lining. The cause of this problem could be
    A. worn caliper pins and bushings.
    B. a leaking caliper piston seal.
    C. a leaking brake hose.
    D. excessive rotor lateral runout.

**Hint**    *Inspect the surfaces of the abutments on the caliper and anchor plate. Smooth these surfaces with emery cloth if they are rusty, burred, or corroded. Coat these surfaces with anti-seize lubricant before installing the caliper. Check for wear on caliper pins, slides, springs, and bushings. Worn components must be replaced.*

## Task 6  Remove, clean, and inspect pads and retaining hardware; determine needed repairs, adjustments, and replacements.

39. On a vehicle with front disc and rear drum brakes, a scraping noise is present in one front wheel while driving the vehicle. The cause of this problem could be
    A. worn caliper pins and bushings.
    B. the pad wear sensor contacting the rotor.

C. loose caliper mounting bolts.

D. loose pad mounting in the caliper.

*Hint*    *Replace the brake pads if the linings are worn beyond the specified limit. Replacement also is necessary if the wear indicators are touching the rotor. The most common type of wear indicator is a metal tab attached to the edge of the pad. When the brake lining is worn to the specified limit, this metal tab contacts the rotor and produces a scraping noise while driving the vehicle.*

## Task 7    Clean caliper assembly; inspect external parts for wear, rust, scoring, and damage; replace any damaged or worn parts; determine the need to repair or replace caliper assembly.

40. After honing a brake caliper, the maximum increase in caliper bore diameter is
    A. 0.001 in.
    B. 0.002 in.
    C. 0.005 in.
    D. 0.008 in.

41. While reassembling a brake caliper
    A. the boot should be installed and seated followed by the seal.
    B. coat the piston seal and boot with clean brake fluid.
    C. leave the piston dry and install it through the boot and seal until it bottoms.
    D. plug the bleeder screw hole and the high-pressure inlet while installing the piston.

*Hint*    *Caliper pistons may be removed by pumping the brake pedal slowly with the caliper connected to the brake hose. Inspect steel pistons for worn surfaces, rust, pitting, scoring, or corrosion.*

*Check phenolic pistons for swelling, cracks, chips, and gouges. Remove minor scratches from the caliper bore with crocus cloth. The caliper bore may be honed with the proper brake hone lubricated with brake fluid. Honing must not increase the caliper bore more than 0.001 in. Clean all parts with denatured alcohol or clean brake fluid. Replace the caliper boot and seal.*

*Be sure the caliper bore and seal groove are perfectly clean. The seal should be lubricated with clean brake fluid and seated in the groove. Lubricate the boot with clean brake fluid and install it in the proper bore groove. Coat the piston with clean brake fluid or the specified lubricant and install it through the boot and seal. Do not plug the high-pressure inlet and bleeder screw hole while installing the piston. Install the piston until it bottoms and then plug the bleeder screw hole and high-pressure inlet to keep out foreign material.*

## Task 8    Clean, inspect, and measure rotor with a dial indicator and a micrometer; follow manufacturers' recommendations in determining the need to index, machine or replace the rotor.

42. In the brake rotor measurement in Figure 5–12, Technician A says the brake rotor is being measured for runout.

    Technician B says this measurement should be made at three locations around the rotor.

    Who is correct?

    A. A only

    B. B only

    C. Both A and B

    D. Neither A nor B

*Hint*    *With the rotor mounted on the vehicle, position a dial indicator against the rotor friction surface and rotate the rotor for one revolution (Figure 5–13). If the runout exceeds the manufacturer's specifications, replace the rotor. Excessive runout causes too much piston and pad-to-rotor clearance that results in excessive pedal travel and brake chatter.*

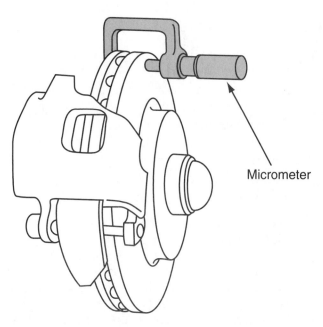

**Figure 5–12** Brake rotor measurement with a micrometer.

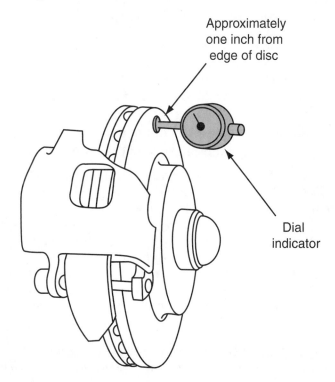

**Figure 5–13** Brake rotor measurement with a dial indicator.

*Use a micrometer to measure variations in rotor thickness or parallelism at six to twelve loca-tions around the rotor. These readings should be taken near the center of the friction surface.*

*Replace the rotor if the thickness variations exceed the manufacturer's specifications. Excessive thickness variations may cause pedal pulsations and brake grabbing. Most rotors have a minimum or discard thickness cast into them. After machining a rotor, the thickness must exceed the minimum thickness. Some manufacturers specify that the rotor thickness after machining must be at least 0.030 in. more than the minimum or discard thickness.*

## Task 9    Remove and replace rotor.

43. When removing and replacing a brake rotor, Technician A says the caliper may be suspended on the end of the brake hose.

    Technician B says the wheel bearings must be repacked and adjusted during the rotor service or replacement.

    Who is correct?

    A. A only

    B. B only

    C. Both A and B

    D. Neither A nor B

**Hint**    *Prior to removing the rotor, the caliper must be removed. After the caliper is removed, suspend the caliper from a chassis member with a piece of Technician's wire. Do not allow the caliper to hang on the end of the brake hose. The hub must be thoroughly cleaned before machining the rotor. After the rotor is machined, the hub must be thoroughly cleaned again to remove all metal filings. Clean and inspect the wheel bearings, and repack the wheel bearings with the proper wheel bearing grease. After the hub and rotor are installed, the wheel bearings must be adjusted following the vehicle manufacturer's recommended procedure. Install the caliper and tighten all fasteners to the specified torque.*

## Task 10    Machine rotor, using on-car or off-car method, according to manufacturers' procedures and specifications.

44. All of these statements about rotor machining are true EXCEPT

    A. A vibration damper must be placed around the outside diameter of the rotor to prevent chatter marks.

    B. With fixed caliper rotors, unequal amounts of metal may be machined from each side of the rotor.

    C. Machine both sides of the rotor before removing it from the rotor lathe.

    D. Use a sanding pad to sand the rotor surfaces after machining is completed.

**Hint**    *Some rotors may be removed separately from the wheel hub. Most rotors and wheel hubs are removed as an assembly. The brake caliper must be removed prior to hub and rotor removal. The dust cap, cotter key, and wheel bearing retaining nut must be removed prior to hub and rotor removal.*

   *The grease must be removed from the hub before mounting the hub and rotor on the lathe.*

   *Place a vibration damper around the outside diameter of the rotor, and install the hub and rotor securely on the lathe so there is no movement between the hub and lathe. Machine both sides of the rotor. On fixed caliper brake rotors, equal amounts of metal must be machined from each side of the rotor. Use a sanding pad to sand the rotor after machining. After sanding, clean the rotor surface with a shop towel saturated with denatured alcohol to clean the rotor surfaces. Be sure all metal cuttings are removed from the hub.*

## Task 11    Install pads, calipers, and related attaching hardware; lubricate components following manufacturers' procedures and specifications; bleed system.

45. While discussing brake pad installation, Technician A says the inboard and outboard brake pads are interchangeable in many calipers.

    Technician B says brake pad rattle may occur if there is any clearance between the pad retainer and the caliper retainer ledge.

    Who is correct?

    A. A only

    B. B only

    C. Both A and B

    D. Neither A nor B

*Hint*    *The manufacturer's recommended assembly procedure must be followed for each type of brake caliper. Install the inboard pad in the caliper. Install anti-rattle springs if so equipped. Position the outboard shoe in the caliper and be sure there is no clearance between the pad retainer flange and the caliper-machined retainer ledge (Figure 5–14). If there is clearance at this location, remove the pad and bend the pad retainer flanges as necessary. Clearance between the pad retainer flanges and the caliper retainer ledges may cause pad rattle. Be sure the pads are properly positioned and slide the caliper over the rotor. Install and tighten the caliper mounting bolts.*

*Next, install a new sealing washer on the brake hose and tighten the brake hose into the caliper. Afterward, install and tighten the brake line fitting in the brake hose.*

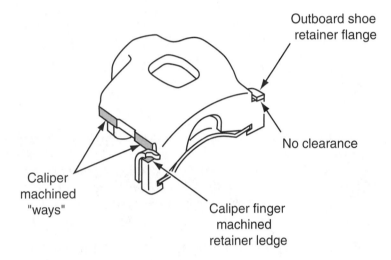

**Figure 5–14** Brake caliper assembly.

## Task 12    Adjust calipers with integrated parking brakes according to manufacturers' recommendations.

46. While discussing the rear brake caliper and parking brake mechanism in Figure 5–15
    A. during a parking brake application, the caliper is moved toward the rotor surface.
    B. the parking brake should be depressed 1.5 in. before the cable adjustment check.

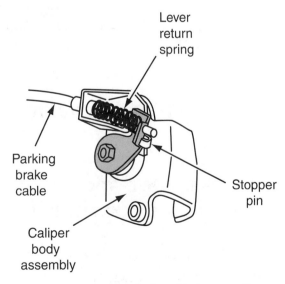

**Figure 5–15** Parking brake cable in rear caliper.

   C. the parking brake cable should be adjusted so the stopper pin is just touching the stop.

   D. the parking brake cable should be adjusted so there is 0.050 in. between the stopper pin and the stop.

**Hint**    *Before the cable adjustment, the cables and the parking brake levers in the rear calipers must be working freely. With the parking brake released, the cable adjuster should be rotated until the stopper pins just contact the stops in the rear calipers.*

## Task 13    Fill master cylinder to proper level with recommended fluid; inspect caliper for leaks.

47. In most cast-iron master cylinders, the distance from the brake fluid level to the top of the reservoir casting should be

   A. 0 in.

   B. 1/4 in.

   C. 3/4 in.

   D. 1 in.

**Hint**    *After the brakes are bled, the master cylinder reservoir should be filled with the manufacturer's specified brake fluid to the proper level. Many cast-iron master cylinder reservoirs are filled to within 1/4 in. of the reservoir top casting. Aluminum master cylinders with transparent plastic reservoirs are filled to the full mark on the reservoir.*

   *After several firm brake applications, check the calipers for leaks in the piston and hose attachment areas.*

## Task 14    Reinstall wheel, torque lug nuts, and make final checks and adjustments.

48. While discussing wheel and lug nut installation, Technician A says excessive lug nut torque may cause rotor runout.

   Technician B says an impact wrench may be used to tighten lug nuts.

   Who is correct?

   A. A only

   B. B only

   C. Both A and B

   D. Neither A nor B

**Hint**    *Excessive lug nut torque may distort the brake rotor and cause excessive rotor runout. Tighten the lug nuts to the specified torque following the manufacturer's recommended lug nut tightening sequence. Do not use an impact wrench to tighten the lug nuts.*

# Power Assist Units Diagnosis and Repair

## ASE Tasks, Questions, and Related Information

## Task 1    Test pedal free travel with and without engine running to check power booster operation.

49. With the engine stopped, the technician pumps the brake pedal several times and then holds the brake pedal on and starts the engine. The pedal moves slightly downward when the engine is started. This action indicates
    A. a restriction in the power booster vacuum hose.
    B. a defective check valve in the power booster vacuum hose.
    C. a normal vacuum supply to the power booster.
    D. a low-intake manifold vacuum.

**Hint**   *With the engine stopped, pump the brake pedal several times to release the vacuum stored in the brake booster. Push the brake pedal lightly downward until resistance is felt. The free play is the pedal movement from this point to the released position. Adjust the master cylinder pushrod until the pedal free play is within specifications. With the engine stopped, pump the brake pedal several times, and hold the pedal in the applied position. When the engine is started, the pedal should move slightly downward if the vacuum supply to the brake booster is normal. If the pedal does not move slightly downward, check the vacuum hose and the one-way check valve to the brake booster.*

## Task 2  Check vacuum supply (manifold or auxiliary pump) to vacuum-type power booster.

50. While checking the vacuum supply to the power brake booster, Technician A says the vacuum gauge should be connected between the one-way check valve and the brake booster.
    Technician B says that with the engine idling, the vacuum supplied to the brake booster should be 8 to 10 in. Hg.
    Who is correct?
    A. A only
    B. B only
    C. Both A and B
    D. Neither A nor B

**Hint**   *Connect a vacuum gauge with a T connection in the hose between the one-way check valve and the brake booster. With the engine idling, the vacuum should be 17 to 21 in. Hg. If the vacuum is low, connect the vacuum gauge directly to the intake manifold. When the vacuum in the intake is within specifications, check the one-way check valve and the hoses from the intake manifold to the brake booster. If the intake manifold is low, check engine compression and the intake manifold vacuum leaks. When the vacuum supply to the brake booster is low on a vehicle with a vacuum pump, check the pump and hoses connected to the brake booster.*

## Task 3  Inspect the vacuum-type power booster unit for vacuum leaks and proper operation; inspect the check valve for proper operation; repair, adjust, or replace parts as necessary.

51. There is evidence of engine oil in the vacuum brake booster. The cause of this problem could be
    A. a damaged vacuum hose to the brake booster.
    B. an inoperative one-way check valve in the booster vacuum hose.
    C. an inoperative PCV valve with excessive restriction.
    D. a partially restricted air cleaner element.

**Hint**   *To check the one-way check valve in the brake booster vacuum hose, operate the engine at 2,000 rpm and then allow the engine to idle. Shut the engine off and wait 90 seconds. Pump the brake pedal five or six times. The first two pedal applications should be power-assisted. If the first two brake applications are not power-assisted, the one-way check valve is defective in the brake booster vacuum hose. An inoperative one-way check valve in the brake booster vacuum hose may cause an accumulation of engine oil in the brake booster.*

*To check the vacuum brake booster for air tightness, run the engine for two minutes and then shut it off. Pump the brake pedal several times with normal braking pressure. If the brake booster is operating normally, the pedal should go down normally on the first brake application.*

*The pedal should then gradually become higher with each brake application. With the engine running, apply the brakes and then shut off the engine while maintaining the pedal pressure for 30 seconds. When the pedal height does not change, the brake booster is not leaking. If the pedal height gradually rises, the brake booster is leaking.*

## Task 4    Inspect and test hydro-boost system and accumulator for leaks and proper operation; repair or replace parts as necessary; refill system.

52. The power steering operates normally on a hydro-boost–equipped vehicle. During a hydroboost brake test, the brake pedal is pumped several times with the engine not running. When medium pressure is applied to the brake pedal and the engine is started, the brake pedal height remains unchanged.

    Technician A says the power steering pump pressure may be low.

    Technician B says the accumulator may be inoperative in the hydro-boost unit.

    Who is correct?

    A. A only

    B. B only

    C. Both A and B

    D. Neither A nor B

**Hint**    *Pump the brake pedal several times with the engine not running to begin the hydro-boost test.*

*Apply medium foot pressure to the brake pedal and start the engine. If the hydro-boost unit is operating normally, the brake pedal should move downward followed by a slight push back on the pedal. When the pedal does not have this action, the hydro-boost unit may be inoperative.*

*Because the hydro-boost unit depends on power-steering pump pressure, always be sure the power-steering system is operating normally before testing or servicing the hydro-boost unit.*

*To test the accumulator, operate the engine and turn the steering wheel until the wheels lightly touch their stops. Hold the steering wheel in this position for five seconds, and then release the steering wheel. Shut off the engine and pump the brake pedal several times. Several power-assisted brake applications should be available if the accumulator is properly charged. Start the engine and hold the steering wheel with the wheels against the stops to recharge the accumulator. A slight hissing sound should be heard while the accumulator is being charged. Shut the engine off and wait for one hour. If there are no leaks in the accumulator or hydro-boost system, several power-assisted brake applications should be available without starting the engine.*

# Miscellaneous Systems (Pedal linkage, Wheel Bearings, Parking Brakes, Electrical, etc.) Diagnosis and Repair

## ASE Tasks, Questions, and Related Information

**Task 1**    Diagnose wheel bearing noises, wheel shimmy, and vibration problems; determine needed repairs.

53. A rear-wheel-drive car experiences a growling noise only on deceleration. The cause of this problem could be
    A. a damaged differential ring gear and pinion gear.
    B. a damaged rear-wheel bearing.
    C. a damaged front-wheel bearing.
    D. a damaged differential side bearing.

**Hint**   *A damaged wheel bearing causes a continual growling noise that is most noticeable at low speeds. Wheel bearing noise is not influenced by acceleration and deceleration. A growling noise caused by a damaged front-wheel bearing may be more noticeable while cornering. Wheel shimmy may be caused by loose front-wheel bearings, improper front-wheel alignment, or improperly balanced front wheels.*

## Task 2   Remove, clean, inspect, repack wheel bearings, or replace wheel bearings and races; replace seals; replace hub and bearing assemblies, adjust wheel/hub bearings according to manufacturers' specifications.

54. A front-wheel-drive vehicle has two tapered roller bearings in the rear-wheel hubs. While adjusting the rear-wheel bearings, the bearing adjusting nut is tightened to 25 ft.-lbs. The bearing nut should then be
    A. loosened one turn and tightened to 50 in.-lbs.
    B. loosened one and one-half turns and tightened to 75 in.-lbs.
    C. loosened one-half turn and tightened to 10 to 15 in.-lbs.
    D. loosened one-half turn and tightened to 40 in.-lbs.

**Hint**   *Wheel bearings should be cleaned in an approved cleaning solvent. After the bearings are cleaned, they should be repacked with the manufacturer's recommended wheel bearing grease.*
   *The bearings may be repacked by hand or with a bearing repacking tool. After the bearings are installed in the hub, use the proper seal driver to install a new hub seal. Tighten the wheel bearing adjusting nut to 25 ft.-lbs. while rotating the wheel. Loosen the adjusting nut one-half turn and then tighten the nut to 10 to 15 in.-lbs. Install a new cotter key, and then the dust cap.*

## Task 3   Check parking brake system; inspect cables and parts for wear, rust, and corrosion; clean or replace parts as necessary; lubricate assembly.

55. While discussing parking brake adjustment, Technician A says the brake shoes should be properly modified before the parking brake adjustment.
    Technician B says after the cable adjusting nut is tightened so the rear wheels have a slight drag, loosen this nut four turns.
    Who is correct?
    A. A only
    B. B only
    C. Both A and B
    D. Neither A nor B

**Hint**   *The brake shoes should be adjusted properly, and the parking brake cables must be operating freely before a parking brake adjustment. Tighten the parking brake cable adjusting nut until there is a slight drag while rotating the rear wheels. Loosen the cable adjusting nut until the rear wheels rotate freely, and then loosen this nut two more turns. Apply the parking brake several times and be sure the rear wheels are firmly held. After each application, the rear wheels must rotate freely.*

## Task 4   Adjust parking brake assembly; check operation.

56. In the measurement shown in Figure 5–16
    A. the drive shaft center support bearing wear is measured.
    B. the propeller shaft parking brake is adjusted.

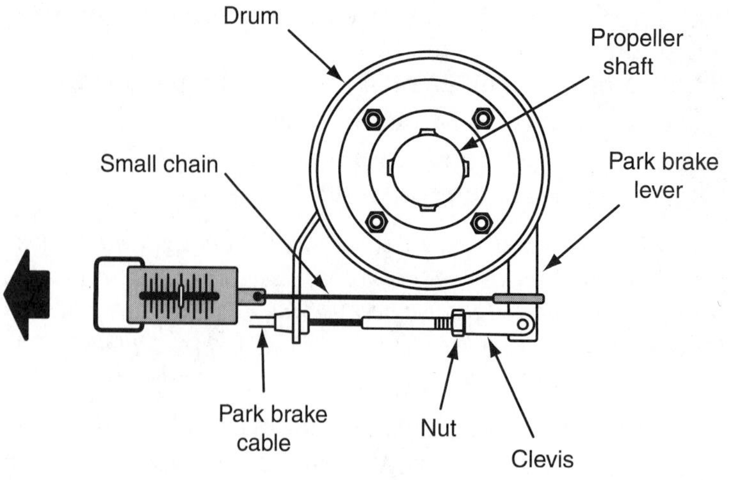

Figure 5–16 Adjustment with pull-scale on propeller shaft mechanism.

C. the brake shoes must be adjusted before this measurement.

D. the parking brake is released during this measurement.

**Hint**    *During the propeller parking brake adjustment, the clevis pin is removed from the parking brake lever at the brake assembly. The parking brake is applied until the specified round gauge fits in an opening in the parking brake mechanism and the gauge is contacting the parking brake outer flange. A pull-scale-type gauge is connected to the parking brake lever, and the parking brake tightening device is rotated until a specific reading is obtained on the gauge. The cable adjuster is then rotated until the clevis pin slides freely through the clevis and parking brake lever opening.*

## Task 5    Test parking brake indicator light, switch and wiring.

57. The brake warning light is illuminated continually with the ignition switch on in Figure 5–17. All of the following may be the cause of the problem EXCEPT
    A. an open circuit in the wire to the parking brake switch.
    B. the wire to the parking brake switch touching the vehicle ground.
    C. a continually closed parking brake switch.
    D. low fluid level in the secondary section of the master cylinder reservoir.

**Hint**    *When the ignition switch is turned on, voltage is supplied through a fuse to one terminal on the brake warning light. When the parking brake is applied, the parking brake switch connects the other side of the brake warning bulb to ground. This action illuminates the brake warning light. The brake warning switch also may ground the brake warning light if there is unequal pressure between the primary and secondary sections of the master cylinder.*

*If the brake warning light is illuminated continually with the ignition switch on, check the master cylinder brake fluid level. If the brake fluid level is satisfactory, disconnect the wire from the brake warning light switch and the parking brake switch. If the brake warning light goes out when one of these wires is disconnected, that switch is faulty. When the brake warning light is illuminated with wires disconnected from these switches, check for a grounded condition on the wires from the switches to the bulb.*

## Task 6    Test, adjust, repair or replace brake stop light switch, lamps, and related circuits.

58. When the brake pedal is depressed, the rear stoplights are not illuminated, but the high-mounted stoplights operate normally (Figure 5–18 and Figure 5–19). The turn signals and hazard warning circuits operate normally.

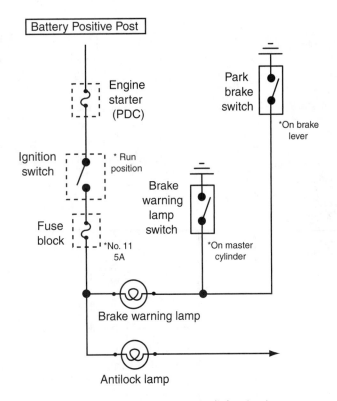

**Figure 5–17** Brake warning light circuit.

Technician A says the cause of this problem may be a blown stop-hazard fuse. Technician B says there may be an open circuit between junction S206 and the turn signal switch.

Who is correct?

A. A only

B. B only

C. Both A and B

D. Neither A nor B

**Hint**    *The stoplight switch may be adjusted on many vehicles. In most applications, this switch should be adjusted so the stoplights are illuminated when the brake pedal is depressed 1/4 in.*

*An improper stoplight switch adjustment or a binding brake pedal may cause the stoplights to be illuminated continually with the ignition switch on. Stoplight circuits vary, depending on the vehicle make and model year, so the technician must use a wiring diagram to diagnose each circuit.*

*When diagnosing a stoplight circuit, the stoplight fuse should be tested first. With the wires disconnected from the stoplight switch, connect a pair of ohmmeter leads to the switch terminals.*

*With the brake pedal released, the ohmmeter should provide an infinity reading. When the brake pedal is depressed, the ohmmeter should indicate 0.5 ohm or less.*

**Task 7**    **Inspect and test brake pedal linkage for binding, looseness, and adjustment; determine needed repairs.**

59. The brake pedal free play is 1.25 in. This condition causes

A. brake dragging.

B. a spongy brake pedal.

C. a pulsating brake pedal.

D. a low brake pedal.

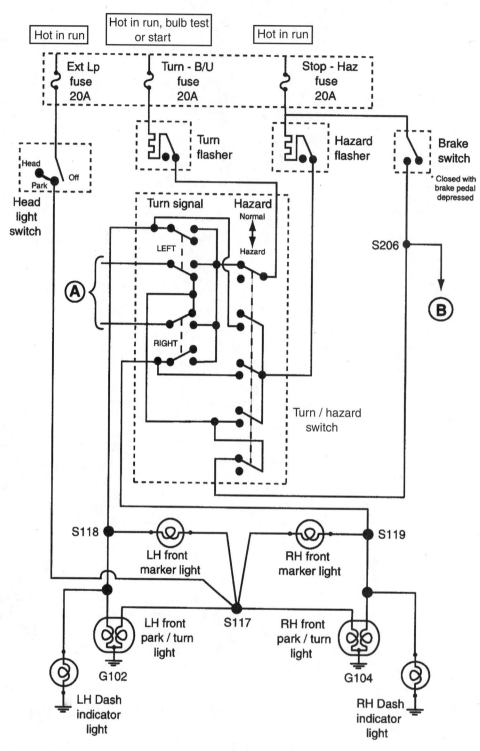

**Figure 5–18** Stop and turn signal wiring diagram.

*Hint*    *Inspect the brake pedal linkage for wear, looseness, or binding. Be sure all pins and fasteners in the linkage are properly retained. The specified brake pedal free play is usually between 1/16 to 1/4 in. Excessive brake pedal free play causes a low brake pedal. When the brake pedal free play is less than specified, the compensating ports in the master cylinder may be closed by the piston cups when the brake pedal is released. This action results in fluid pressure buildup and dragging brakes.*

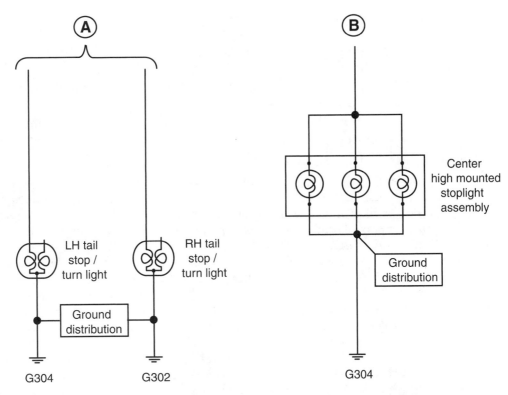

**Figure 5–19** Stop and turn signal wiring diagram (continued).

# Electronic Brake Control Systems; Antilock Brake Systems (ABS) and Traction Control Systems (TCS) Diagnosis and Repair

## ASE Tasks, Questions, and Related Information

**Task 1** Follow manufacturers' service and safety precautions when inspecting, testing and servicing ABS/TCS hydraulic, electrical, and mechanical components.

60. While discussing ABS/TCS service, Technician A says the high-pressure accumulator must be discharged before a brake line is disconnected.

Technician B says some manufacturers recommend relieving the accumulator gas pressure before accumulator disposal.

Who is correct?

A. A only

B. B only

C. Both A and B

D. Neither A nor B

61. All of the following statements about a TCS system with hydraulic functions are true EXCEPT
    A. During TCS operation, the TCS system applies the brakes on the fastest spinning wheel.
    B. During TCS operation, the EBCM starts the pump motor in the ABS/TCS solenoid assembly.
    C. During TCS operation, vehicle stability and control are improved.
    D. During TCS operation, the EBCM does not isolate the ABS system.

**Hint**    *Before a brake line is removed, the brake pedal should be applied and released 25 to 30 times to relieve the brake fluid pressure in the high-pressure accumulator. When the amber ABS warning light is illuminated, the antilock function is cancelled because the ABS computer has sensed an electrical problem in the system. Under this condition, normal power-assisted braking is available, and the car may be driven to an automotive repair facility as soon as possible so the necessary repairs can be completed. When the red brake warning light is illuminated there may be a serious braking problem, such as low brake fluid level. Under this condition, the vehicle should not be driven.*

*On ABS/TCS systems, the ABS computer operates the ABS and TCS systems. On some TCS systems, if the wheel speed sensors indicate that one drive wheel is rotating faster than the opposite drive wheel, the ABS/TCS computer transmits wheel spin data through the data links to the powertrain control module (PCM). When this data is received, the PCM reduces the spark advance and fuel delivered by the injectors. The PCM also upshifts the transmission on some vehicles depending on the gear that the transmission is operating in. This action by the PCM reduces engine power and prevents drive wheel spin. On other vehicles, when the EBCM receives drive wheel spin data from the wheel speed sensors, the EBCM operates two TCS solenoids in the ABS/TCS solenoid assembly. The EBCM can operate the appropriate TCS solenoid to apply the brakes on the drive wheel that is spinning faster. The EBCM also starts the pump motor in the ABS/TCS solenoid assembly, and this motor supplies fluid pressure to the TCS system. This action prevents drive wheel spin to improve traction and vehicle stability. When the EBCM operates a TCS solenoid to provide traction control, the EBCM also operates an ABS solenoid to isolate the ABS system.*

## Task 2   Diagnose poor stopping, wheel lockup, pedal feel and travel, pedal pulsation, and noise concerns associated with the ABS/TCS; determine needed repairs.

62. While road testing a vehicle with ABS/TCS, a clicking noise is heard for a short time when the engine is started and the vehicle starts off at low speed. During a normal stop when the ABS function is not operating, pedal pulsations are experienced.

    Technician A says the clicking action may be caused by defective solenoids in the ABS hydraulic control unit.

    Technician B says the pedal pulsations are typical on an ABS during a normal stop when the ABS is inoperative.

    Who is correct?

    A. A only
    B. B only
    C. Both A and B
    D. Neither A nor B

**Hint**    *In many ABSs, the computer enters a prove-out mode each time the ignition switch is turned on and the engine is started. During this mode, the computer scans the ABS electrical system for problems. On some ABSs, the computer prove-out check includes momentarily energizing all the solenoids in the system. This computer action may cause an audible clicking action when the vehicle is driven at low speeds. If the brakes are applied during the prove-out mode, momentary pedal pulsations may be felt.*

*On many ABSs, pedal pulsations are felt while the brakes are operating in the ABS mode. On some ABSs, an increase in brake pedal height may be experienced during the ABS mode. If pedal pulsations are experienced during the normal braking mode, a brake drum is out-of-round, or a rotor has excessive runout.*

*Poor stopping may be caused by contaminated or glazed brake linings, scored drums or rotors, or malfunctioning brake booster operation. In some ABSs, the accumulator pressure supplies power assist. In these systems, low accumulator pressure causes poor stopping ability.*

*Brake grabbing may be caused by contaminated linings, bent backing plates, or distorted drums or rotors.*

## Task 3   Observe ABS/TCS warning light(s) at startup and during road test; determine if further diagnosis is needed.

63. All of the following conditions may cause illumination of the red brake warning light EXCEPT
    A. parking brake engagement.
    B. low fluid level in the master cylinder.
    C. an open wheel speed sensor winding.
    D. an accumulator pressure below 1,500 psi.

64. While discussing ABS/TCS system diagnosis, Technician A says the TCS on light is illuminated if the EBCM is providing a TCS function.
    Technician B says the TCS off light is illuminated if the EBCM detects an electrical defect in the ABS/TCS system.
    Who is correct?
    A. A only
    B. B only
    C. Both A and B
    D. Neither A nor B

**Hint**   *The red brake warning light is normally illuminated while cranking the engine and for a few seconds after the engine is started. This light may be illuminated with the engine running if the parking brake is applied, the master cylinder fluid level is low, or the accumulator pressure drops below 1,500 psi.*

*The amber ABS brake warning light is illuminated while cranking the engine and for a few seconds after the engine is started. When the amber ABS light is illuminated with the engine running, the computer has sensed an electrical problem in the ABS system.*

*Many TCS systems have a low traction or TCS on light and a TCS off light in the instrument panel. The electronic brake control module (EBCM) controls both the ABS and TCS systems. At the beginning of each ignition cycle on some TCS systems, the EBCM illuminates the TCS on and TCS off lights for three seconds to prove these indicator lights are functioning. The EBCM illuminates the TCS on light when the EBCM detects drive wheel slip and operates the TCS system to prevent this action. The EBCM illuminates the TCS off light if the TCS off button on the dash is depressed. The EBCM also illuminates the TCS off light if the parking brake is on or when the EBCM detects an electrical defect in the ABS/TCS system.*

## Task 4   Diagnose ABS/TCS electronic control(s), components, and circuits using on-board diagnosis and/or recommended test equipment; determine needed repairs.

65. While discussing ABS diagnosis, Technician A says a DTC indicates a problem in a specific component.
    Technician B says data links are connected to the on-board computers and to the DLC.
    Who is correct?
    A. A only
    B. B only

C. Both A and B
D. Neither A nor B

**Hint**    *Cars and light-duty trucks manufactured since 1996 are equipped with on-board diagnostic II (OBD II) systems. These systems have a standard 16-terminal data link connector (DLC) mounted under the dash near the steering column (Figure 5–20). Data links are connected between the various computers on the vehicle, and these data links are also connected to the DLC. A scan tool may be connected to the DLC to obtain diagnostic trouble codes (DTCs) and other data from the on-board computers, including the ABS computer. A DTC represents an electrical problem in a certain area. For example, a DTC representing the left front (L/F) wheel speed sensor indicates a problem in this sensor or the connecting wires. In most cases, voltmeter or ohmmeter tests must be performed to locate the exact cause of the problem. On some vehicles manufactured before 1996, the ABS warning light would flash DTCs when the specified terminals in the DLC were connected with a jumper wire.*

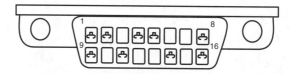

| Pin | Function |
|-----|----------|
| 1 | Serial data line - UART |
| 2 | Serial data line - Class 2 |
| 3 | Not used |
| 4 | Ground |
| 5 | Ground |
| 6-7 | Not used |
| 8 | Keyless entry enable |
| 9 | Serial data line - UART |
| 10 | Not used |
| 11 | Magnetic variable steering |
| 12-13 | Not used |
| 14 | Serial data line - E&C |
| 15 | Not used |
| 16 | Fused battery output |

**Figure 5–20**  Data link connector (DLC).

## Task 5  Bleed and/or flush the ABS/TCS hydraulic system following manufacturers' procedures.

66. All of the following statements about bleeding an ABS with a high-pressure accumulator are true EXCEPT
    A. The front and rear brakes may be bled with a pressure bleeder.
    B. Be sure the ignition switch is on when using a pressure bleeder.
    C. Be sure the brake pedal is released when using a pressure bleeder.
    D. The rear brakes may be bled with a fully charged accumulator.

**Hint**    *On an ABS with a high-pressure accumulator, the front or rear brakes may be bled with a pressure bleeder. If a pressure bleeder is attached to the master cylinder for bleeding purposes, maintain the bleeder pressure at 35 psi (240 kPa). Be sure the ignition switch is off and the brake pedal is released. Connect a hose from the bleeder screw into a plastic container, and open the bleeder screw for ten seconds. Tighten the bleeder screw and repeat the procedure at each wheel.*

Be sure the bleeder pressure is maintained. Adjust the master cylinder fluid level to the maximum fill line when the bleeding procedure is completed.

Because accumulator pressure is supplied to the rear wheels, the rear brakes may be bled with
a fully charged accumulator.

On an integral ABS with a high-pressure accumulator, the brake fluid level in the master
cylinder should be checked with a fully charged accumulator. To fully charge the accumulator,
turn on the ignition switch and pump the brake pedal several times until the ABS pump motor
starts. When the pump motor stops, the accumulator is fully charged. Under this condition, the
brake fluid should be at the specified level in the master cylinder reservoir. If the accumulator is
discharged, it is normal for the brake fluid to be above the specified level in the master cylinder.

## Task 6   Remove and install ABS/TCS components following manufacturers' pro-
cedures and specifications; observe proper placement of components and
routing of wiring harness.

67. When servicing the integral ABS unit with a high-pressure accumulator in Figure 5–21,
Technician A says the accumulator pressure should be relieved before removing the
solenoid valve body.

Technician B says the accumulator pressure should be relieved before removing the
reservoir.

Who is correct?

A. A only

B. B only

C. Both A and B

D. Neither A nor B

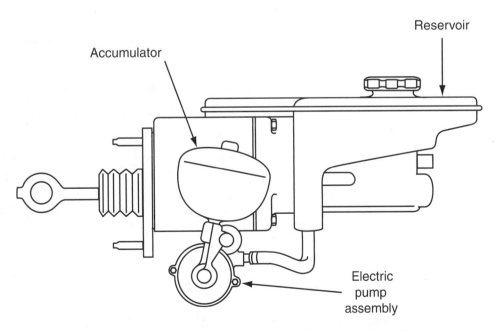

**Figure 5–21** Integral ABS assembly with high-pressure accumulator.

*Hint*        *Always be sure the accumulator pressure is relieved before removing the ABS assembly or any
component in this assembly. On many ABSs, the high-pressure accumulator is depressurized by
pumping the brake pedal 25 times with the ignition switch off. On other systems, the accumulator pressure is relieved by connecting a bleeding tool to the bleeder screw on the hydraulic control unit and loosening the bleeder screw with the ignition switch off. When an accumulator is
replaced, the gas pressure should be relieved in the old accumulator before discarding it. After the
accumulator is removed, a pressure relief screw in the accumulator is loosened the specified number*

*of turns until the gas pressure is relieved. If the accumulator is removed, always install a new O-ring under the accumulator. Coat the new O-ring with clean brake fluid before installation. Turn off the ignition switch and disconnect the negative battery cable before disconnecting any wiring connectors in the ABS. If the vehicle is equipped with an air bag or bags, wait the length of time specified by the manufacturer after the negative battery cable is disconnected before proceeding with the service or diagnosis procedure.*

**Task 7**  **Test, diagnose and service ABS/TCS speed sensors (digital or analog), toothed ring (tone wheel), magnetic encoder, and circuits following manufacturers' recommended procedures (includes output signal, resistance, shorts to voltage/ground, and frequency data).**

68. A diagnostic trouble code (DTC) representing the L/R wheel speed sensor is obtained in the ABS in Figure 5–22. This DTC may be caused by
    A. an open circuit at terminal 6 on the ABS computer connector.
    B. a grounded circuit on the #885 red wire connected to terminal 10 on the ABS computer.
    C. a larger tire than specified by the manufacturer on the L/R wheel.
    D. an open circuit in the electronic brake control relay winding.

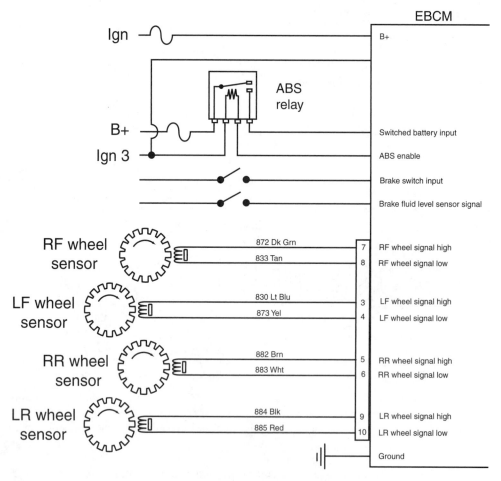

**Figure 5–22** ABS wiring diagram.

**Hint**    *Some wheel speed sensors have a paper shim installed on the inner end of the sensor. If this type of sensor is removed, a new shim should be installed on the sensor. The sensor should be installed until the paper shim lightly contacts the toothed ring. Hold the sensor in this position, and tighten the sensor retaining bolt. On other sensors, the clearance between the sensor tip and* *the toothed ring is measured with a feeler gauge. Rotate the wheel one revolution and measure the sensor clearance to be sure this clearance is uniform. If necessary, loosen the sensor retaining bolt and move the sensor to adjust the clearance.*

*If a DTC is obtained representing a wheel speed sensor, the sensor and connecting wires may be tested with an ohmmeter. Disconnect the ABS computer wires and connect the ohmmeter leads to the appropriate sensor wires to test the sensor for an open circuit. When an infinity reading is obtained, the sensor or connecting wires are open. Disconnect the wiring connector at the sensor, and connect the ohmmeter leads to the sensor terminals to determine if the open circuit is in the sensor winding or connecting wires. When the ohmmeter leads are connected from one of the wheel speed sensor terminals at the ABS computer connector to ground, an infinity reading should be obtained. If the ohmmeter provides a low reading, the sensor winding or connecting wires are shorted to ground.*

## Task 8   Diagnose ABS/TCS braking concerns caused by vehicle modifications (wheel/tire size, curb height, final drive ratio, etc.) and other vehicle mechanical and electrical/electronic modifications (communication, security, and radio, etc.).

69. An ABS-equipped rear-wheel-drive vehicle experiences lockup on both rear wheels during the antilock brake mode.

    Technician A says larger-than-specified tires may be installed on the rear wheels.

    Technician B says the rear tires may not be the same size on each rear wheel.

    Who is correct?

    A. A only

    B. B only

    C. Both A and B

    D. Neither A nor B

**Hint**    *On an ABS-equipped vehicle the original tire size and differential ratio must be maintained. When a different tire size is installed, the wheel speed is changed in relation to vehicle speed. Under this condition, different wheel speed sensor signals are sent to the ABS computer. This action may reduce ABS effectiveness or cause wheel lockup in the antilock brake mode.*

## Task 9   Repair wiring harness and connectors following manufacturers' procedures.

70. While discussing wiring repairs, Technician A says a broken pigtail lead on a wheel speed sensor may be spliced.

    Technician B says a wiring harness designed to reduce electrical noise may have a drain wire.

    Who is correct?

    A. A only

    B. B only

    C. Both A and B

    D. Neither A nor B

**Hint**    *If pigtail lead wires are damaged on ABS components such as wheel speed sensors, the wheel speed sensor and lead wire assembly must be replaced. Do not attempt to repair these wires.*

*Splice sleeves may be used to join two copper wires during a wiring harness repair. Use the proper size opening in a wire stripper to strip 5/16 in. of insulation from the end of each wire to be joined. Place the insulating sleeve over one of the wires to be joined. Select the proper size of the splice sleeve, and insert one of the stripped wire ends into the splice sleeve until it bottoms on the stop in the center of the sleeve. Insert the other wire end into the splice sleeve until it bottoms. Be sure to maintain the wires in the bottomed position. Place the proper opening in the wire strippers over the center of area 1 on the splice sleeve, and squeeze the wire stripper handles until these handles open when released (Figure 5–23). Repeat this procedure with the wire strippers centered on area 2 on the splice sleeve. Slide the insulating sleeve over the splice sleeve, and heat the insulating sleeve with a heat torch starting at the crimped area and moving to the open end of the insulating sleeve. This sleeve will shrink tightly over the splice sleeve.*

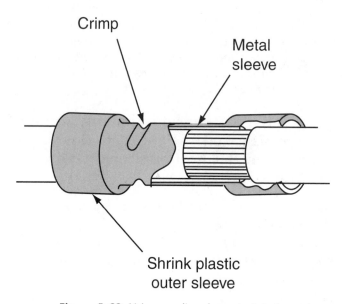

**Figure 5–23** Using a splice sleeve to join two wires.

*Some wires contain two conductors with a small diameter drain wire. This type of wire is used to protect the wiring from electromagnetic interference (EMI), which may cause electrical noise.*

*When splicing this type of wire, cut the outer jacket from the wire. Unwrap the Mylar tape from the wire, but do not cut this tape. Splice the two conductors in the wire with splice sleeves as explained previously. Cut the wires so the splice sleeves centers are 2.5 in. apart. Rewrap the Mylar tape over the conductors, and use a splice sleeve to splice the drain wire (Figure 5–24).*

*Apply Mylar tape over the spliced area with the aluminum side of the tape facing inward to ensure good electrical contact with the drain wire.*

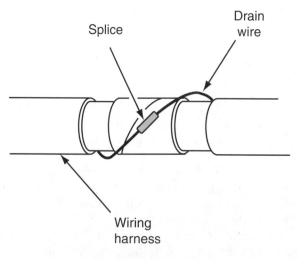

**Figure 5–24** Splicing a drain wire.

# Post Test

1. A vehicle requires excessive brake pedal effort.
   Technician A says the master cylinder cups may be swollen.
   Technician B says the master cylinder compensating port may be plugged.
   Who is correct?
   A. A only
   B. B only
   C. Both A and B
   D. Neither A nor B

2. A vehicle has a low, spongy brake pedal. The cause of this problem could be
   A. restricted brake hoses.
   B. glazed brake linings.
   C. air in the wheel cylinders.
   D. a defective brake booster.

3. During a brake application, the brake pedal feels harder than normal. The most likely cause of this problem is
   A. a binding pedal linkage.
   B. glazed brake linings.
   C. a plugged compensating port.
   D. a weak hydraulic brake hose.

4. When bleeding a master cylinder before installation, Technician A says plastic bleeder tubes should be installed from the master cylinder outlets to a clean container.
   Technician B says the master cylinder pistons should be depressed and released slowly until there are no bubbles through the bleeder tubes.
   Who is correct?
   A. A only
   B. B only
   C. Both A and B
   D. Neither A nor B

5. While discussing brake fluids, Technician A says hygroscopic brake fluids absorb moisture from the air.
   Technician B says hydroscopic brake fluids have the ability to keep the hydraulic brake system clean, and that nonhydroscopic fluids do not have this capability.
   Who is correct?
   A. A only
   B. B only
   C. Both A and B
   D. Neither A nor B

6. When diagnosing and servicing metering valves, it's important to know that
   A. a metering valve is used with four-wheel drum brakes.
   B. the metering valve delays fluid pressure to the front brakes.
   C. the metering valve prevents rear-wheel lockup during light brake applications.
   D. the metering valve remains open with the brake pedal released.

7. While discussing the height-sensing proportioning valve, Technician A says this valve is mounted on the rear suspension and is linked to the rear axle.

Technician B says if the rear suspension load is heavy, this valve reduces brake fluid pressure to the rear wheels during moderate brake applications. Who is correct?

A.  A only

B.  B only

C.  Both A and B

D.  Neither A nor B

8.  A vehicle has front disc and rear drum brakes and a diagonally connected hydraulic brake system. When performing a manual brake bleeding procedure on this vehicle, start with the

A.  left rear wheel.

B.  left front wheel.

C.  right front wheel.

D.  right rear wheel.

9.  A vehicle with four-wheel drum brakes pulls to the right during a brake application.

Technician A says the front wheel cylinders may be different sizes.

Technician B says the left-front flexible brake hose may be restricted.

Who is correct?

A.  A only

B.  B only

C.  Both A and B

D.  Neither A nor B

10.  Rear brake drag occurs on a vehicle with front disc brakes and rear drum brakes. The most likely cause of this problem is

A.  a binding brake pedal.

B.  seized parking brake cables.

C.  a restricted left-rear brake line.

D.  contaminated brake fluid.

11.  All of the following statements about brake drum inside diameter measurement are true EXCEPT

A.  The drum should be cleaned before measuring the diameter.

B.  If the drum diameter is more than specified, replace the drum.

C.  When the drum out-of-round is more than specified, brake grabbing may occur.

D.  If the drum out-of-round is more than specified, pedal pulsations may occur.

12.  When machining brake drums, Technician A says the dampening belt must be placed tightly around the drum to prevent cutting tool chatter.

Technician B says if the drum is machined so the drum diameter is equal to the maximum allowable diameter, the drum is satisfactory.

Who is correct?

A.  A only

B.  B only

C.  Both A and B

D.  Neither A nor B

13.  A vehicle with front disc and rear drum brakes has a low, firm pedal during a brake application.

Technician A says the self adjusting mechanisms in the rear wheels may be bent or worn.

Technician B says the shoe adjusters in the rear wheels may be seized.

Who is correct?
A. A only
B. B only
C. Both A and B
D. Neither A nor B

14. A vehicle with four-wheel disc brakes has poor stopping ability with no pull to either side. The most likely cause of this problem is that
A. one front brake caliper piston is seized.
B. one front brake flexible hose is restricted.
C. both front brake caliper pistons are seized.
D. the brake pedal is binding.

15. A vehicle with front disc and rear drum brakes has a scraping noise coming from one front wheel during a brake application, but the noise disappears when the brakes are released.
Technician A says the front caliper mounting bolts may be loose.
Technician B says the front brake pad linings may be worn.
Who is correct?
A. A only
B. B only
C. Both A and B
D. Neither A nor B

16. When using a micrometer to measure a brake rotor for thickness variation or parallelism, Technician A says to measure the rotor at a minimum of three locations around the rotor.
Technician B says excessive thickness variation on a rotor may cause brake pull.
Who is correct?
A. A only
B. B only
C. Both A and B
D. Neither A nor B

17. When installing brake pads and calipers
A. the caliper must be installed over the rotor before the brake pads are installed.
B. the specified clearance must be available between the pad retainer flange and the caliper ledge.
C. a C-clamp may be used to move the piston toward the bottom of the caliper bore.
D. in many calipers the inboard and outboard brake pads are interchangeable.

18. A technician has replaced the brake calipers and wheel cylinders, installed new linings, and bled the brakes. During a firm brake application with the engine running, the brake pedal goes slowly downward. The most likely cause of this problem is
A. air in the hydraulic brake system.
B. the metering valve is stuck open.
C. a leak in one of the brake lines.
E. the proportioning valve is stuck open.

19. When checking a vacuum brake booster, a technician shuts the engine off and waits 90 seconds. The technician applies and releases the brake pedal six times and there is no power assist on any of the brake applications.
Technician A says the one-way check valve between the brake booster and the intake manifold may be defective.
Technician B says the intake manifold or throttle body gaskets may be leaking.

Who is correct?
A. A only
B. B only
C. Both A and B
D. Neither A nor B

20. The ABS warning light is on with the engine running. The most likely cause of this problem is
A. a fluid leak at one of the wheel calipers.
B. a restricted brake line or hose.
C. excessive brake pedal free-play.
D. an open circuit in a wheel speed sensor winding.

# Answers and Analysis

**1.  B**   Brake fluid leaking past the master cylinder cups may cause the brake pedal to slowly move downward during a brake application; thus, A is wrong. An improperly adjusted stoplight and cruise control switch may keep the brake pedal partially depressed, and this action results in brake drag. Therefore, B is correct.

**2.  D**   A binding pedal linkage may cause improper pedal return and brake drag. Dented brake lines may cause pull to one side and brake drag. A plugged master cylinder compensating port causes dragging brakes. Therefore, A, B, and C are wrong, and D is right.

**3.  C**   Insufficient brake pedal free play causes pressure buildup and brake drag. Therefore, A, B, and D are wrong, and C is correct.

**4.  C**   When the brake pedal moves downward slowly during a brake application, the brake fluid may be leaking past the master cylinder cups, or there may be a leak in a brake hose or line. Therefore, Technicians A and B are both right, making C the correct response.

**5.  A**   A leaking secondary cup on the primary master cylinder piston may allow fluid to leak and run down the front of the vacuum brake booster. Therefore, B, C, and D are wrong, and A is right.

**6.  C**   A restricted brake booster vacuum hose causes excessive brake pedal effort. Technician A is right. Seized front caliper pistons also result in excessive brake pedal effort. Therefore, both A and B are right, and C is the correct answer.

**7.  D**   The service procedure in Figure 5–1 is bleeding air from the master cylinder; thus, A, B, and C are wrong, and D is correct.

**8.  B**   When both front brake linings are contaminated with grease, both front wheels may grab, and pedal effort may be increased. Therefore, A is wrong. If the master cylinder piston cups are swollen, the compensating ports may be covered, resulting in brake drag. A plugged secondary compensating port causes brake drag on the brakes supplied from the secondary section of the master cylinder. Therefore, C and D are wrong.

If the right front caliper piston is seized, there is very little or no braking action on this wheel. With normal braking action on the left front wheel, the vehicle may pull to the left during a brake application. Therefore, B is right.

**9.  B**   Damaged brake lines should be replaced, and so A is wrong. A tubing bender should be used to make the necessary brake tubing bends without kinking the brake tubing. Therefore, B is correct.

**10.  D**   The male end of the brake hose should be installed and tightened before the female end. The brake hose sealing washer should be replaced when installing the brake hose. Therefore, Technicians A and B are both wrong, making D the correct response.

**11.  B**   The brake line flare in Figure 5–2 is an ISO flare. Therefore, A, C, and D are wrong, and B is right.

**12.  B**  Excessive heat on a brake line does not cause the pedal height to slowly decrease during a brake application. Thus, Technician A is wrong. Excessive heat applied to a brake line may cause the brake fluid to boil in the line. When the brakes are applied, the brake fluid vapor is compressed, resulting in a spongy pedal. Therefore, Technician B is right, making answer B correct.

**13.  A**  The metering valve in Figure 5-3 delays brake fluid movement to the front wheels to provide simultaneous application of the front and rear brakes. Therefore, B, C, and D are wrong, and A is correct.

**14.  C**  Because the proportioning valve reduces pressure to the rear brakes during moderate brake applications, the pressure gauges connected to this valve should indicate lower pressure at the valve outlet compared to the inlet. Therefore, A, B, and D are wrong, and C is right.

**15.  D**  The load-sensing proportioning valve allows more pressure to the rear brakes as the rear suspension load is increased. This valve is mounted on the chassis and has a linkage to the rear suspension. Therefore, Technicians A and B are both wrong, making D the correct response.

**16.  A**  With the brake warning switch positioned, as shown in Figure 5–5, the pressure is higher in the primary master cylinder section than in the secondary section. Therefore, B is wrong.

The circuit is completed from the switch terminal to ground, and pressure is higher in the primary master cylinder section. Therefore, C and D are wrong.

The brake warning light is illuminated with the ignition switch on or the engine running; thus, A is correct.

**17.  A**  If the brake pedal becomes spongy after several hard brake applications in a short time period, the brake fluid may be contaminated. Thus, Technician A is right. A restricted vacuum hose connected to the brake booster causes excessive brake pedal effort during a brake application. Therefore, Technician B is wrong, and A is the correct answer.

**18.  B**  When using a pressure bleeder, the metering valve must be held open with a special tool; thus, A is wrong. The pressure in the pressure bleeder should be maintained at 20 psi, so B is correct.

**19.  A**  This question asks for the statement that is not true. The vacuum pump creates a vacuum in the tester reservoir, and the bleeder screw should be opened until there is about 1 in. of brake fluid in the reservoir. A one-way check valve is required in the hose from the bleeder screw to the reservoir. Therefore, B, C, and D are right, so none of these is the requested answer.

The vacuum pump handle should be operated at least ten times before the bleeder screw is opened. Therefore, statement A is wrong, making it the requested answer.

**20.  C**  This question asks for the statement that is not true. A hose is connected from the bleeder screw while the other end is submerged in a container of brake fluid. The brake pedal should be applied with moderate pressure and then the bleeder screw is opened. The bleeding procedure should be repeated until the fluid escaping from the bleeder hose is free of bubbles.

Therefore, A, B, and D are correct, so none of these is the requested answer.

When the bleeder screw is opened and the pedal goes to the floor, the bleeder screw must be closed before releasing the brake pedal. Therefore, statement C is wrong, making it the requested answer.

**21.  D**  During a moderate brake application, the master cylinder pressure should be equal to the front-wheel pressure. Therefore, A is wrong. Below the split point pressure of the proportioning valve, the master cylinder pressure and the rear-wheel pressure should be equal; thus, B is wrong. During a moderate brake application, the master cylinder pressure should be higher than the pressure at the rear wheels, so C is wrong.

Above the split point pressure of the proportioning valve, the master cylinder pressure should be higher than the rear-wheel pressure. Therefore, D is correct.

**22.  B**  Silicone brake fluids are nonhydroscopic, which means they do not absorb moisture.

Because DOT 3 and DOT 4 brake fluids are both hydroscopic, they tend to absorb moisture from the atmosphere. Thus, B is the proper response.

**23.  C**    If the master cylinder vents are plugged, air cannot enter between the cover and the cover gasket to allow this gasket to move downward and replace the brake fluid forced from the master cylinder to the wheel cylinders or calipers. This may cause a spongy brake pedal. Air in the hydraulic brake system also may cause this problem. Therefore, Technicians A and B are both correct, making C the correct choice.

**24.  C**    Brake squeal may be caused by distorted brake drums or bent backing plates. Thus, Technicians A and B are both correct, making C the right answer.

**25.  C**    This question asks for the statement that is not true. The drum should be cleaned before diameter measurement, and the drum should be replaced if the diameter exceeds specifications.

Maximum drum diameter variation should not exceed 0.0035 in. Therefore, statements A, B, and D are right, but these are not the requested answer.

The drum diameter should be measured at eight locations around the drum. Therefore, statement C is not correct, making it the requested answer.

**26.  B**    If tool chatter marks occur on the brake drum friction surface, the damping belt may be too loose. The tool bit depth for a rough cut should be 0.005 to 0.010 in., and the tool bit depth for a finish cut should be 0.003 to 0.005 in. Therefore, A, C, and D are wrong, and B is right.

**27.  B**    If item 17 in Figure 5–6 is omitted, there is no parking brake action. Therefore, Technician A is wrong. When item 8 is omitted, there is no self-adjusting action. Thus, Technician B is correct, making B the right answer.

**28.  C**    A bent backing plate may cause brake grabbing, and a loose anchor bolt may cause brake chatter. Therefore, Technicians A and B are both correct, making C the right answer.

**29.  C**    Dry brake shoe ledges on the backing plate may cause a squeaking noise during brake applications.

Slightly scored brake shoe ledges on the backing plate may be resurfaced and lubricated with high-temperature grease. Therefore, Technicians A and B are both correct, making C the right answer.

**30.  C**    The secondary shoe faces toward the rear of the vehicle, and the primary and secondary shoe return springs are not interchangeable. The adjuster cable is usually mounted on the secondary shoe. Therefore, A, B, and D are wrong.

The adjuster must be installed in the proper direction so the star wheel is accessible through the backing plate opening; thus, C is the correct answer.

**31.  A**    In Figure 5–7, the brake shoes are being adjusted to match the drum size. Therefore, B is wrong, and A is correct.

**32.  A**    When adjusting the parking brake shown in Figure 5–8, it should be applied until the pin in hole A contacts the parking brake outer flange. Therefore, B is wrong and A is correct.

**33.  B**    In Figure 5–9, the thin tool is used to release the self-adjusting mechanism, so Technician A is wrong.

The star wheel should be rotated until the wheel drags, and then the star wheel is rotated in the opposite direction until the wheel rotates freely. Thus, Technician B is right.

**34.  C**    In this question, we are asked for the statement that is not true. Incorrect or loose brake pads, a loose caliper mounting, or a sticking caliper piston may cause pull to one side while braking. Therefore, statements A, B, and D are correct, so none of these is the requested answer.

Seized master cylinder pistons do not contribute to pull while braking. Therefore, statement C is not true, making it the requested answer.

**35.  B**    A worn outer drive axle joint does not cause brake pedal pulsations; thus, A is wrong. Brake pedal pulsations may be caused by worn or loose wheel bearings, making B correct.

**36.  A**    The caliper pistons must be in the specified position before installing the caliper. Therefore, B is wrong. The parking brake lever rotates a spindle inside the caliper, so A is the right answer.

**37.**  **A**  After a brake application, the caliper piston is returned by the twisting action of the seal. Therefore, B, C, and D are wrong, and A is right.

**38.**  **A**  Excessive wear on the inside brake pad lining compared to the outside lining may be caused by worn caliper pins and bushings, which cause a sticking caliper. Therefore, B, C, and D are wrong, and A is correct.

**39.**  **B**  A scraping noise while braking may be caused by a pad wear sensor contacting the rotor. Therefore, A, C, and D are wrong, and B is right.

**40.**  **A**  When honing a brake caliper, the maximum increase in caliper bore is 0.001 in., so B, C, and D are wrong, and A is correct.

**41.**  **B**  When reassembling a brake caliper, the seal should be installed followed by the boot. Therefore, A is wrong. The piston should be lubricated with clean brake fluid before it is installed, and the bleeder screw hole or high-pressure inlet should be open when installing the piston. Therefore, C and D are wrong.

The piston seal and boot should be lubricated with clean brake fluid prior to installation; thus, B is the right answer.

**42.**  **D**  In Figure 5–12, the rotor is measured for parallelism or thickness variation; this measurement should be made at six to twelve locations around the rotor. Therefore, both A and B are wrong, and D is the correct response.

**43.**  **B**  A piece of Technician's wire should be used to suspend the caliper from a chassis component. The caliper must not be suspended on the brake hose, So Technician A is wrong. The wheel bearings must be repacked and adjusted during rotor service and replacement, making Technician B correct.

**44.**  **B**  In this question, we are asked for the statement that is not true. During the machining process, a vibration damper must be positioned around the outside diameter of the rotor.

Both sides of the rotor should be machined before removing it from the lathe, and a sanding pad is used to sand the rotor surface after machining. Therefore, A, C, and D are right, so these are not the requested answer.

Equal amounts of metal must be removed from each side of the rotor on fixed caliper rotors. Therefore, B is not true, making this the requested answer.

**45.**  **B**  The inboard and outboard brake pads are not interchangeable in many calipers, so Technician A is wrong. If there is clearance between the pad retainer and the caliper retainer ledge, brake pad rattle may occur. Thus, Technician B is right, making answer B the proper choice.

**46.**  **C**  In the rear caliper in Figure 5–15, the parking brake mechanism moves the caliper piston toward the rotor surface. In addition, the parking brake should be released during an adjustment.

Therefore, A and B are wrong. The parking brake cable should be adjusted so the stopper pin just contacts the stop. Therefore, D is wrong and C is right.

**47.**  **B**  On many cast-iron master cylinders, the brake fluid level should be 1/4 in. below the top of the casting surface. Therefore, A, C, and D are wrong, and B is correct.

**48.**  **A**  An impact wrench should not be used to tighten wheel lug nuts because the excessive torque from this procedure may distort drums and cause excessive rotor runout. Therefore, Technician B is wrong and Technician A is correct, making A the right answer.

**49.**  **C**  The brake pedal action in this question indicates a normal vacuum supply to the power brake booster; thus, A, B, and D are wrong, and C is right.

**50.**  **A**  The vacuum gauge should be installed between the one-way check valve and the brake booster to check the vacuum supply, and with the engine idling the vacuum should be 16 to 18 in. Hg. Therefore, Technician B is wrong and Technician A is correct, making A the right answer.

**51.**  **B**  If there is evidence of oil in the vacuum brake booster, the one-way check valve in the booster vacuum hose is defective. Therefore, A, C, and D are wrong, and B is right.

**52.   B**    If the brake pedal height remains unchanged, as described in the question, there is no accumulator pressure to supply power assist. This may be caused by low power-steering pump pressure, but this problem also reduces steering assist. Because the question informs us that power-steering operation is normal, low power-steering pump pressure is not the cause of the complaint. Therefore, Technician A is wrong.

An inoperative accumulator may cause reduced power brake assist, so Technician B is correct, making B the right answer.

**53.   A**    The noise from a damaged front- or rear-wheel bearing would not be affected by acceleration or deceleration. Therefore, B and C are wrong. Acceleration and deceleration has very little effect on damaged differential side bearing noise, so D is wrong.

Noise from a damaged differential ring and pinion gear usually is affected by acceleration and deceleration, so A is right.

**54.   C**    After tightening the wheel bearing nut to 25 ft.-lbs., the wheel bearing adjusting nut should be loosened one-half turn and tightened to 10 to 15 in.-lbs. Therefore, A, B, and D are wrong, and C is right.

**55.   A**    On many vehicles, the parking brake cable should be adjusted so there is a slight rear wheel drag, and then loosened two turns. Thus, B is wrong. The brake shoes should be properly adjusted before a parking brake adjustment, making Technician A correct. Therefore, A is the right answer.

**56.   B**    In Figure 5–16, the propeller parking brake adjustment is performed. Brake shoe adjustment does not affect this type of parking brake. During the parking brake adjustment, the specified pin is installed in a hole in the parking brake pedal mechanism. The parking brake is applied until this pin contacts the outer flange on the pedal mechanism. Therefore, A, C, and D are wrong, and B is right.

**57.   A**    In this question, we are asked for the answer that is not the cause of the problem. A continually illuminated brake warning light may be caused by a grounded wire to the parking brake switch, a continually closed parking brake switch, or low fluid level in one section of the master cylinder reservoir. Therefore, B, C, and D may be the cause of the problem, making none of these the requested answer.

If an open circuit occurs in the wire to the parking brake switch, the brake warning light is not illuminated when the parking brake is applied. Therefore, A is not the cause of a continually illuminated brake warning light, so this is the correct answer.

**58.   B**    If the stop hazard fuse is blown, the stoplights and hazard warning lights are inoperative. Therefore, Technician A is wrong. An open circuit between junction S206 and the turn signal switch causes inoperative stoplights. Because the high-mounted stoplights are connected to junction S206, they work normally. Therefore, Technician B is correct, making B the right answer.

**59.   D**    A brake pedal free play of 1.25 in. is excessive, and this results in a low brake pedal. Thus, A, B, and C are wrong, and D is the correct answer.

**60.   C**    The high-pressure accumulator must be discharged before a brake line is disconnected, and some manufacturers recommend relieving accumulator gas pressure before accumulator disposal. Therefore, Technicians A and B are both correct, making C the right answer.

**61.   D**    In a TCS system with hydraulic functions the TCS applies the brakes on the fastest spinning drive wheel. A is true, so this is not the requested answer. During TCS operation, the EBCM starts the pump motor, and TCS operation improves vehicle stability and control. Therefore, B and C are true, so they also are not the requested answer. During TCS operation, the EBCM isolates the ABS system. D is not true, so this is the requested answer.

**62.   D**    The clicking action during initial driving is a result of the ABS computer prove-out mode in which the computer momentarily energizes the solenoids in the ABS. Therefore, A is wrong.

On many ABSs, pedal pulsations are normal during the ABS function. However, pedal pulsations during a normal stop when the ABS function is not operating may be caused by out-of-round drums or rotors with excessive runout. Therefore, B also is wrong, and the correct answer is D.

**63.  C**    In this question, we are asked for the answer that is not the cause of red brake warning light illumination. A parking brake application, low fluid level in the master cylinder, or accumulator pressure below 1,500 psi, may cause red brake warning light illumination.

Therefore, A, B, and D are not the required answer.

An open wheel speed sensor winding does not cause illumination of the red brake warning light, but this problem results in illumination of the amber ABS light. Therefore, C is correct.

**64.  C**    The TCS on light is illuminated if the EBCM is providing a TCS function. Technician A is right. The TCS off light is illuminated if the EBCM detects an electrical defect in the ABS/TCS system. Thus, Technician B is right. Because Technicians A and B are both right, C is the correct answer.

**65.  B**    A DTC indicates a problem in a specific area. Thus, Technician A is wrong. Data links are connected to the on-board computers and to the DLC, making Technician B is right. Therefore, answer B is the correct choice.

**66.  B**    In this question, we are asked for the statement that is not true about bleeding an ABS with a high-pressure accumulator. On these systems, the front and rear brakes may be bled with a pressure bleeder, the brake pedal should be released when using a pressure bleeder, and the rear brakes may be bled with a fully charged accumulator. Therefore, statements A, C, and D are true and none of these is the requested response.

The ignition switch should be off when using a pressure bleeder on these systems.

Therefore, statement B is not true, and this is the requested answer.

**67.  C**    In an integral ABS with a high-pressure accumulator, the accumulator pressure should be relieved before removing the solenoid valve body or the reservoir. Therefore, Technicians A and B are both correct, making C the right answer.

**68.  B**    Since terminal 6 on the ABS computer connector is connected to the R/R wheel speed sensor, an open circuit at this terminal does not cause a code representing the L/R wheel speed sensor. Therefore, A is wrong.

A larger tire than specified on the L/R wheel may affect the ABS operation, but it does not result in this type of DTC, so C is wrong. An open circuit at the electronic brake control relay winding causes an inoperative ABS, but this problem does not result in a DTC representing the L/R wheel speed sensor. Therefore, C and D are wrong.

Because terminal 10 on the ABS computer is connected to the L/R wheel speed sensor, a grounded circuit at this location causes a DTC representing the L/R wheel. Therefore, B is correct.

**69.  A**    Rear tires of unequal size may result in wheel lockup on one rear wheel during the ABS mode, so Technician B is wrong. Larger-than-specified tires on both rear wheels may cause lockup on both rear wheels during the ABS mode. Thus, Technician A is right, making answer A the correct choice.

**70.  B**    A broken pigtail lead wire on a wheel speed sensor should not be repaired. Thus, Technician A is wrong. A wiring harness designed to reduce electrical noise may have a drain wire, making Technician B correct. Therefore, B is the right answer.

# Answers to Post Test

**1.  A**    Swollen master cylinder cups may cause excessive brake pedal effort. Technician A is right. A plugged compensating port may cause dragging brakes. Therefore, Technician B is wrong, and A is the correct answer.

**2.  C**    Restricted brake hoses cause increased braking effort. Thus, A is wrong. Glazed brake linings cause increased braking effort. Thus, B is wrong. Air in the wheel cylinders causes a low, spongy brake pedal, so C is correct. A defective brake booster causes increased braking effort. Thus, D is wrong.

3.  B    A binding brake pedal linkage causes improper pedal return and dragging brakes. Thus, A is wrong. Glazed brake linings cause increased brake pedal effort and a hard pedal, so B is correct. A plugged compensating port causes brake drag, so C is wrong. A weak hydraulic hose may cause a low, spongy pedal, making answer D wrong.

4.  B    Plastic bleeder tubes should be installed from the master cylinder outlets to the master cylinder reservoir. Thus, Technician A is wrong. The master cylinder pistons should be depressed and released slowly until there are no bubbles in the bleeder tubes. Therefore, Technician B is correct, making B the right answer.

5.  A    Hygroscopic brake fluids absorb moisture from the air, so Technician A is correct. There is no difference in the cleaning capabilities of hydroscopic and non-hydroscopic brake fluids. Thus, Technician B is wrong, making A the correct answer.

6.  B    A metering valve is used with front disc and rear drum brakes, so A is wrong. The metering valve delays fluid pressure to the front brakes to provide simultaneous front and rear brake application. Thus, B is correct. The metering valve does not prevent rear-wheel lockup during light brake applications, and the metering valve is closed with the brake pedal released. Therefore, C and D are wrong.

7.  D    The height-sensing proportioning valve is mounted on the chassis and linked to the rear axle. Thus, Technician A is wrong. If the rear suspension is light, the proportioning valve reduces fluid pressure to the rear brakes during moderate brake applications, so Technician B is wrong. Because Technicians A and B are both wrong, D is the correct answer.

8.  D    When performing a manual bleeding procedure on a vehicle with a diagonally connected brake system, start with the wheel farthest from the master cylinder, which is the right rear wheel. Thus, D is the correct answer.

9.  C    If the front-wheel cylinders have different sizes, brake pull may result. Thus, Technician A is right. If the left-front flexible brake hose is restricted, pull to the right may occur during a brake application, so Technician B is right, too. Because Technicians A and B are both right, C is the correct answer.

10.  B    A binding brake pedal may cause brake drag on all the wheels, so A is wrong. Seized parking brake cables may cause rear brake drag, so B is correct. A restricted left rear brake line does not cause rear brake drag. Thus, C is wrong. Contaminated brake fluid may cause brake drag on all four wheels, making answer D wrong.

11.  C    The brake drum should be cleaned before measuring the diameter. A is true, but this is not the requested answer. If the drum diameter is more than specified, the drum should be replaced. Therefore, B is true, but this is not the requested answer. When the drum out-of-round is more than specified, pedal pulsations may occur, but this condition does not cause brake grabbing. Therefore, D is true, and C is not true, making C the requested answer.

12.  A    The dampening belt must be placed tightly around the drum to prevent cutting tool chatter. Thus, Technician A is correct. If the drum is machined to a drum diameter equal to the maximum allowable diameter, the drum should be replaced. Therefore, Technician B is wrong, making A the correct answer.

13.  C    A low, firm brake pedal may be caused by worn or bent self-adjusting mechanisms in the rear brakes. Thus, Technician A is right. A low, firm brake pedal may also be caused by seized shoe adjusters in the rear brakes. Thus, Technician B is also right. Because Technicians A and B are both right, C is the correct answer.

14.  C    A seized front brake caliper piston or a restricted flexible brake hose may cause pull to one side when braking. Therefore, A and B are wrong. Both front brake caliper pistons being seized cause poor stopping ability without any pull. Thus, C is correct. A binding brake pedal may cause brake drag, so D is wrong.

15.  B    Loose caliper mounting bolts may cause a clunking noise when the brakes are applied, so Technician A is wrong. Completely worn front brake pad linings may allow the metal backing

in the brake pad to contact the rotor, resulting in a scraping noise. Thus, Technician B is right, making B the correct answer.

**16.  D**    When measuring rotor thickness variation, the rotor should be measured at six locations. Thus, Technician A is wrong. Excessive thickness variation may cause pedal pulsations, but this problem does not cause brake pull. Therefore, Technician B is wrong. Because Technicians A and B are both wrong, D is the correct answer.

**17.  C**    The brake pads must be installed in the caliper before the caliper is installed over the rotor. Thus, A is wrong. There should be no clearance between the pad retainer flange and the caliper ledge, so B is wrong. A C-clamp may be used to move the caliper piston toward the bottom of the bore. Thus, C is correct. In many calipers, the in-board and out-board pads are not interchangeable, so D is wrong.

**18.  C**    Air in the hydraulic brake system, a stuck metering valve, or a stuck proportioning valve do not cause the brake pedal to move slowly downward during a firm brake application. Therefore, A, B, and D are wrong. A leak in one of the brake lines causes the brake pedal to move slowly downward during a firm brake application. Thus, C is the correct answer.

**19.  A**    The one-way check valve should maintain vacuum in the booster to provide power assist for one or two brake applications. Because there was no power assist on any of the six brake applications, the one-way check valve may be defective. Thus, Technician A is right. Leaking intake manifold or throttle body gaskets do not cause this problem. Therefore, Technician B is wrong, and A is the correct answer.

**20.  D**    A fluid leak at a wheel caliper, a restricted brake line or hose, or excessive brake pedal free play do not cause the ABS warning light to be illuminated. Thus, A, B, and C are wrong. An open circuit in a wheel speed sensor causes the ABS computer to illuminate the ABS warning light, so D is the correct answer.

# 6 Electrical/Electronic Systems

## Pretest

The purpose of this pretest is to determine the amount of review that you may require prior to taking the ASE Electrical/Electronic Systems Test. If you answer all the pretest questions correctly, complete the questions and study the information in this chapter to prepare for the ASE Electrical/Electronic Systems Test. If two or more of your answers to the pretest questions are incorrect, complete a thorough study of the questions and information in this chapter.

The pretest answers are located at the end of the pretest, and also in the answer sheets supplied with this book.

1. In a series electrical circuit
   A. the same amount of current flows through each resistance.
   B. each resistance is a separate path for current flow.
   C. the same voltage is dropped across each resistance.
   D. a resistance increase results in more current flow.

2. A 12 V light circuit has a short to ground (Figure 6–1).
   Technician A says the current flow through the lamp is higher than normal.
   Technician B says the light cannot be turned off with the switch.
   Who is correct?
   A. A only
   B. B only
   C. Both A and B
   D. Neither A nor B

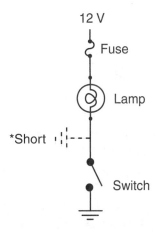

**Figure 6–1** Short to ground in a 12 V light circuit.

3. A shorted condition in an electromagnet causes
    A. an increase in coil resistance.
    B. an increase in the effective number of coil turns.
    C. an increase in coil current flow.
    D. a weaker magnetic field surrounding the coil.

4. The tester in Figure 6–2 is connected to test
    A. starting motor current draw.
    B. battery capacity.
    C. positive cable voltage drop.
    D. negative cable voltage drop.

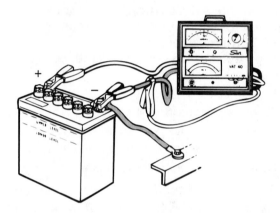

**Figure 6–2** Volt-ampere tester. *(Reprinted with permission)*

5. While discussing a battery state-of-charge test on a battery with a built-in hydrometer, Technician A says the battery may be charged quickly if the hydrometer is yellow or clear.

    Technician B says the battery may be tested if the hydrometer is green.

    Who is correct?
    A. A only
    B. B only
    C. Both A and B
    D. Neither A nor B

6. The voltmeter in Figure 6–3 is connected to test
    A. battery voltage while cranking.
    B. voltage drop on the positive cable.
    C. voltage drop across the starting motor.
    D. voltage drop across the solenoid contacts.

7. When the ohmmeter is connected as illustrated in Figure 6–4, a low reading is obtained.

    Technician A says the field winding may have an open circuit.

    Technician B says the turns may be shorted together in the field coil.

    Who is correct?
    A. A only
    B. B only
    C. Both A and B
    D. Neither A nor B

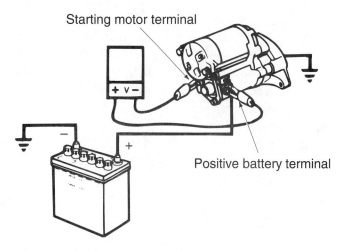

**Figure 6–3** Starting circuit test. *(Reprinted with permission)*

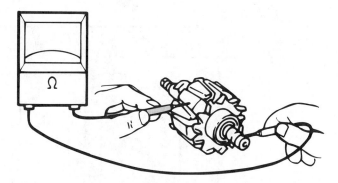

**Figure 6–4** Field winding test. *(Reprinted with permission)*

8. When an ohmmeter is connected to the stator leads as shown in Figure 6–5, an infinity reading is obtained. This reading indicates the stator windings
    A. have an open circuit.
    B. are shorted together.
    C. are grounded to the stator frame.
    D. have a high-resistance problem.

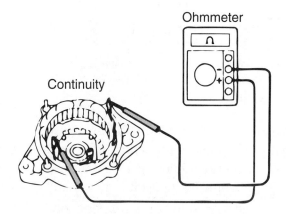

**Figure 6–5** Alternator stator test. *(Reprinted with permission)*

9. A vehicle has four high-beam headlights, and two low-beam headlights. The headlights cycle off and on at regular intervals while driving with the lights on high beam. The problem does not occur on low beam.

   Technician A says the dimmer switch may be the problem.

   Technician B says the headlight switch contacts may be the problem.

   Who is correct?

   A. A only

   B. B only

   C. Both A and B

   D. Neither A nor B

10. A dome light is illuminated continually with the doors closed (Figure 6–6). The cause of this problem could be

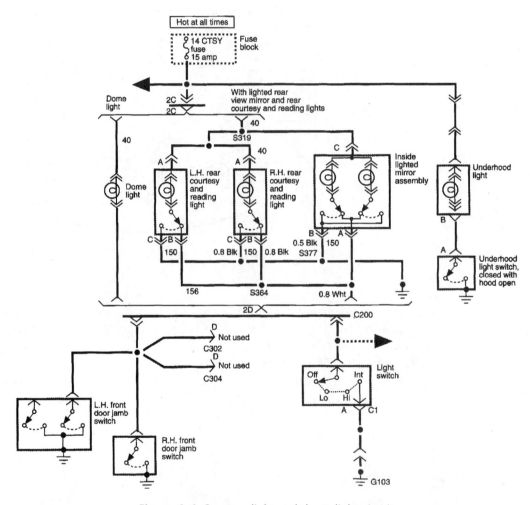

**Figure 6–6** Courtesy light and dome light circuit.

   A. the right-hand door jamb switch has a grounded condition.

   B. the light switch internal (int) contact is grounded.

   C. the left-hand door jamb switch has an open circuit.

   D. the dome light circuit is grounded between the bulb and the fuse.

11. The left-side headlight is dim only on high beam (Figure 6–7). The other headlights operate normally.

Technician A says there may be high resistance in the left-side headlight ground. Technician B says there may be high resistance in the dimmer switch high-beam contacts.

Who is correct?

A. A only

B. B only

C. Both A and B

D. Neither A nor B

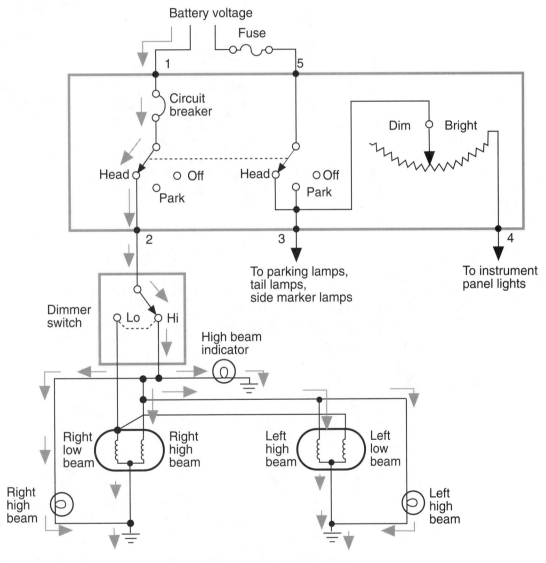

**Figure 6–7** Headlight circuit.

12. All the gauges read higher than normal in a vehicle with an instrument voltage limiter and thermal-electric gauges. The cause of this problem could be

A. an open circuit at the instrument voltage limiter contacts.

B. high resistance in the instrument panel ground.

C. a grounded fuel gauge sending unit wire.

D. a blown gauge fuse in the fuse panel.

13. A horn blows continually (Figure 6–8). All of the following could be the cause of the problem EXCEPT
   A. a short to ground between the horn relay and the horn switch.
   B. the horn relay contacts stuck closed.
   C. a short to ground at the horn slip ring.
   D. a short to ground at the right-hand horn.

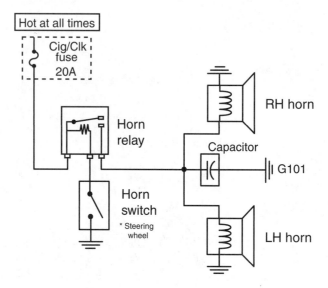

**Figure 6–8** Horn circuit.

14. An air bag warning light in the instrument panel is illuminated with the engine running.
   Technician A says the air bag system may be inoperative and the customer should be advised not to drive the vehicle.
   Technician B says the air bag system should be checked for diagnostic trouble codes.
   Who is correct?
   A. A only
   B. B only
   C. Both A and B
   D. Neither A nor B

# Answers to Pretest

**1.  A**   In a series circuit, the same amount of current flows through all the resistances. Thus, A is right. Each resistance is not a separate path for current flow, so B is wrong. Part of the voltage is dropped across each resistance, so C is wrong. A resistance increase results in a current decrease, making answer D wrong.

**2.  B**   Because the resistance of the bulb is still in the circuit, the current is unchanged, and Technician A is wrong. Since the short to ground is before the switch, the switch cannot turn the light off. Thus, Technician B is right, and B is the correct answer.

**3.  C**   A shorted condition in an electromagnet causes a decrease in coil resistance and the number of turns, an increase in current flow, and the same magnetic strength. Therefore, A, B, and D are wrong, and C is the correct answer.

**4.  B**   The tester is connected to test battery capacity. Therefore, A, C, and D are wrong, and B is the right answer.

**5. B**   If the built-in hydrometer is yellow or clear, the battery should not be fast charged. Thus, Technician A is wrong. The battery may be tested if the hydrometer is green. Therefore, Technician B is right, and B is the correct answer.

**6. D**   The voltmeter is connected to a test voltage drop across the solenoid contacts. D is the correct answer, and A, B, and C are wrong.

**7. D**   With the ohmmeter connected as shown in Figure 6–4, a low reading indicates a grounded field coil. Therefore, Technicians A and B are both wrong, and D is the right answer.

**8. A**   When the ohmmeter is connected (as illustrated in Figure 6–5), an infinite reading indicates an open stator winding. Therefore, A is the correct answer and B, C, and D are wrong.

**9. D**   If the headlights cycle off and on at regular intervals, the circuit breaker is opening. This may be caused by high current flow in the circuit or a defective breaker. Thus, Technicians A and B are both wrong, and D is the correct answer.

**10. A**   If the dome light is illuminated continually with the doors closed, one of the door jamb switches or a wire to these switches has a grounded condition. Therefore, A is right, and B, C, and D are wrong.

**11. D**   If one headlight is dim only on high beam, there must be high resistance between the dimmer switch and the left-side high beam headlight. High resistance in the left-side headlight ground causes both the low and high beams to be dim in that light. High resistance in the dimmer switch causes both high beam headlights to be dim. Thus, Technicians A and B are both wrong, and D is the correct answer.

**12. B**   The voltage limiter heating coil is grounded to the limiter case which in turn is grounded on the instrument panel. High resistance in the panel ground reduces heating coil current and allows the limiter contacts to remain closed longer. This action supplies higher voltage to the gauges and causes the high readings. Thus, B is the correct answer, and A, C, and D are wrong.

**13. D**   If the horn blows continually, there may be a short to ground between the relay and the switch, the relay contacts may be stuck closed, or there may be a short to ground at the horn slip ring. Therefore, A, B, and C are true, so none of these is the requested answer. A short to ground at the right-hand horn only makes this horn inoperative. Thus, D is not true, making it the requested answer.

**14. C**   If the air bag system warning light is on with the engine running, there is an electrical defect in the air bag system, and the customer should not drive the vehicle because the air bag(s) may fail to deploy in a collision. A diagnostic trouble code(s) stored in the air bag computer should indicate the area in which the electrical defect is located. So, Technicians A and B are both right, making C the correct answer.

# General Electrical/Electronic System Diagnosis

## ASE Tasks, Questions, and Related Information

In this chapter, each task in the Electrical/Electronics category is provided, followed by a question and some information related to the task. If you answer any question incorrectly, study this information very carefully until you understand the correct answer. Question answers and analysis are provided at the end of this chapter and in the answer sheets provided with this book.

## Task 1    Check electrical circuits with a test light; determine needed repairs.

1.  When an open circuit occurs at the connector in Figure 6–9, and 12 V is supplied from the battery to the circuit
    A.  the test light is illuminated when connected as shown in Figure 6–9.
    B.  the test light is illuminated when connected to the motor side of the open circuit.
    C.  the current continues to flow from the battery through the motor.
    D.  when a voltmeter is connected across the motor, the voltage drop across the motor will be 11 V.

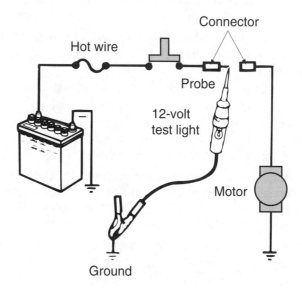

**Figure 6–9** Circuit continuity diagnosis with a test light. *(Reprinted with permission)*

**Hint**    *Continuity in an electric circuit may be tested with a 12 V test lamp. Connect the test light lead to ground. With voltage supplied to the circuit, begin at the battery and connect the test light to various terminals in the circuit. When the test light is not illuminated, the open circuit is between the terminal where the test light is connected and the last terminal where the test light was illuminated.*

## Task 2    Check voltages and voltage drops in electrical/electronic circuits; interpret readings and determine needed repairs.

2.  The battery in Figure 6–10 is fully charged and the switch is closed. The voltage drop across the light indicated on the voltmeter is 9 V.
    Technician A says there may be a high resistance problem in the light.
    Technician B says the circuit may be grounded between the switch and the light.
    Who is correct?
    A.  A only
    B.  B only
    C.  Both A and B
    D.  Neither A nor B

**Hint**    *A voltmeter may be connected across a component in a circuit to measure the voltage drop across the component. Current must be flowing through the circuit during the voltage drop test.*
    *The amount of voltage drop depends on the resistance in the component, and the amount of current flow.*

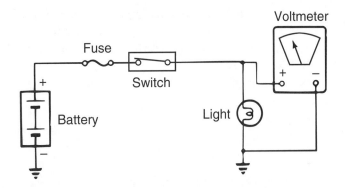

**Figure 6–10** Measuring circuit voltages and voltage drops with a voltmeter. *(Reprinted with permission)*

## Task 3   Check current flow in electrical/electronic circuits and components; interpret readings and determine needed repairs.

3. As indicated on the ammeter in Figure 6–11 the current flow through the light bulb is higher than specified. The cause of the high current flow could be
   A. the fuse has an open circuit.
   B. the battery voltage is low.
   C. the light bulb filament is shorted.
   D. the light bulb filament has high resistance.

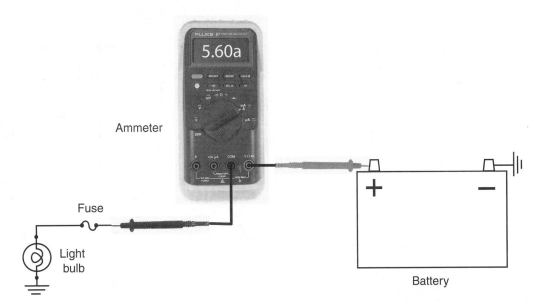

**Figure 6–11** Ammeter connected to measure current flow in a circuit.

*Hint*  *An ammeter has low internal resistance, and this meter must be connected in series in a circuit.*

*Some ammeters have an inductive clamp that fits over a wire in the circuit. These ammeters measure the current flow from the strength of the magnetic field surrounding the wire. High current flow is caused by high voltage or low resistance. Conversely, low current flow results from high resistance or low voltage.*

## Task 4   Check continuity and resistances in electrical/electronic circuits and components; interpret readings and determine needed repairs.

4. While discussing resistance measurement with an ohmmeter, Technician A says an ohmmeter may be connected to a circuit in which current is flowing.

   Technician B says when testing a spark plug wire with 20,000 Ω resistance, use the X100 meter scale.

   Who is correct?

   A. A only

   B. B only

   C. Both A and B

   D. Neither A nor D

**Hint**    *An ohmmeter has an internal power source. Meter damage may result if this meter is connected to a live circuit. The proper scale on the meter must be selected for the component being tested. For example, when testing a component with 10,000 Ω, select the X1000 scale on the meter.*

## Task 5    Check electronic circuit waveforms; interpret readings and determine needed repairs.

5. Technician A says the waveform in Figure 6–12 indicates a defective throttle position sensor (TPS).

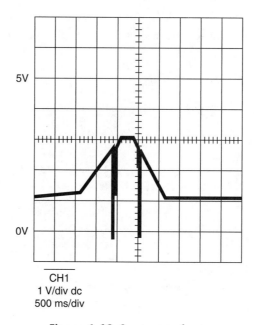

**Figure 6–12** Scope waveform.

   Technician B says the defect shown in Figure 6–12 may cause a hesitation during low speed acceleration.

   Who is correct?

   A. A only

   B. B only

   C. Both A and B

   D. Neither A nor B

**Hint**    *A lab scope reacts many times faster compared to an ordinary voltmeter. Therefore, using a lab scope to observe the voltage waveforms from various electronic components will display component defects that may not be detected with a voltmeter. A graphing voltmeter also* *displays voltage waveforms, but this meter does not react as fast as a lab scope. A lab scope or graphing voltmeter may also be used to display graphs of the current flow in a circuit or component.*

## Task 6   Use scan-tool data to diagnose electronic systems; interpret readings and determine needed repairs.

6. A customer complains about rough idle operation. The engine compression is satisfactory, and the scope test indicates satisfactory ignition operation. An injector balance test indicates normal injector operation.

   Technician A says to use a scan tool to test the grams of air flow reading on the mass air flow (MAF) sensor.

   Technician B says to use the scan tool to determine the command from the powertrain control module (PCM) to the torque converter clutch.

   Who is correct?

   A. A only
   B. B only
   C. Both A and B
   D. Neither A nor B

*Hint*    *A scan tool displays data from various sensors and also indicates output commands sent from on-board computers to various outputs. In addition, the scan tool displays data that is transmitted via the data links that interconnect various on-board computers. All of this data is very helpful to the technician when diagnosing electronic system problems. For example, the scan tool may display the grams per second of air flow through a mass air flow (MAF) sensor. If the grams per second reading is erratic, the sensor hot wire may be coated with pollutants, or the sensor may be defective. This defect causes rough engine idle and surging at low speeds. The scan tool also displays diagnostic trouble codes (DTCs) that represent electrical/electronic defects in certain areas of electronic systems. For example, the scan tool may display a DTC representing the engine coolant temperature (ECT) sensor. This DTC may be caused by a defective ECT sensor, a defect in the ECT sensor wiring, or the powertrain control module (PCM) may be unable to receive the ECT signal.*

## Task 7   Check electrical/electronic circuits with jumper wires; determine needed repairs.

7. The light bulb is inoperative in Figure 6–13. A jumper wire is connected from the battery's positive terminal to the light bulb with the switch on, and the bulb is not illuminated. With the switch on, the bulb is illuminated when a jumper wire is

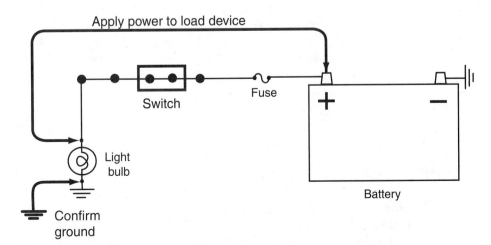

**Figure 6–13** Jumper wire diagnosis of circuit components.

connected from the ground side of the light bulb to ground. The cause of the inoperative light bulb could be

A.  an open circuit in the ignition switch.

B.  an open circuit in the light bulb ground.

C.  an open circuit in the battery ground cable.

D.  a burned-out fuse.

**Hint**    *A jumper wire may be used to bypass a part of a circuit to locate a problem. When a component is bypassed with a jumper wire, and the circuit operation is restored to normal, the bypassed component is the problem.*

**Task 8**   **Find shorts, grounds, opens, and resistance problems in electrical/electronic circuits; determine needed repairs.**

8.  The light bulb in Figure 6–14 is inoperative. A 12 V test light is installed in place of the fuse. When the switch is turned on, the test light is on, but the light bulb remains off. When the connector near the light bulb is disconnected, the 12 V test light remains illuminated.

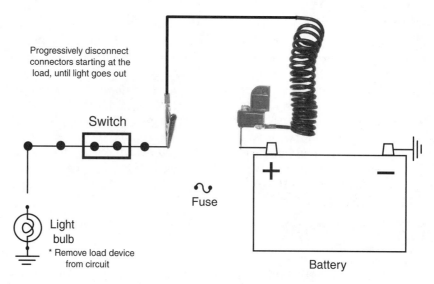

**Figure 6–14** Electrical circuit diagnosis.

Technician A says the circuit may be shorted to ground between the fuse and the disconnected connector.

Technician B says the circuit may be open between the disconnected connector and the light bulb.

Who is correct?

A.  A only

B.  B only

C.  Both A and B

D.  Neither A nor B

**Hint**    *A high-resistance problem may be diagnosed by measuring the voltage drop across various system components. High resistance in a component causes a higher-than-specified voltage drop. A short to ground may be diagnosed by connecting a 12 V test light in place of the circuit fuse.*

*With the circuit switch on, disconnect connectors beginning at the load. When the 12 V test light remains on, the short to ground is between the test light and the disconnected connector.*

*If the test light goes out, the short to ground is between the disconnected connector and the load.*

## Task 9   Measure and diagnose the cause(s) of abnormal key-off battery drain (parasitic draw); determine needed repairs.

9. While performing the battery drain test in Figure 6–15
   A. the switch tester switch should be closed while starting or running the engine.
   B. a battery drain of 125 milliamperes is considered normal and will not discharge the battery.
   C. the actual battery drain is recorded immediately when the switch is opened.
   D. the driver's door should be open while measuring the battery drain.

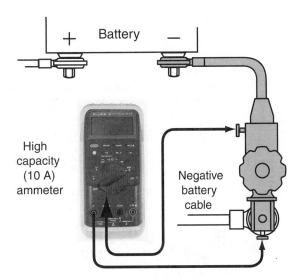

**Figure 6–15** Battery drain test.

*Hint*     *Many car manufacturers recommend measuring battery drain with a tester switch connected in series at the negative battery terminal. The drain test procedure must be followed in the vehicle manufacturer's service manual. A multimeter with a milliampere scale is connected parallel to the tester switch. When the tester switch is open, any current drain from the battery must flow through the tester switch. Some computers require several minutes after the ignition switch is turned off before they enter the sleep mode with a reduced current drain. Therefore, after the ignition switch is turned off and the tester switch is opened, wait for the specified time before recording the milliampere reading. Some vehicle manufacturers specify a maximum battery drain of 50 milliamperes. Other vehicle manufacturers specify the battery drain is calculated by dividing the battery reserve capacity rating by 4.*

## Task 10   Inspect, test, and replace fusible links, circuit breakers, fuses, diodes, and current limiting devices.

10. A circuit breaker is removed from a power seat circuit, and an ohmmeter is connected to the circuit breaker terminals.
    Technician A says the ohmmeter should provide an infinity reading if the circuit breaker is satisfactory.
    Technician B says the ohmmeter current may cause the circuit breaker to open.
    Who is correct?
    A. A only
    B. B only

C. Both A and B

D. Neither A nor B

**Hint**    *When an ohmmeter is connected to a circuit breaker, fuse, or fuse link, the meter should read 0 ohm if the component is satisfactory. An open circuit breaker, fuse, or fuse link, causes an infinity ohmmeter reading. The current flow from an ohmmeter does not cause an automotive circuit breaker to open.*

## Task 11   Read and interpret electrical schematic diagrams and symbols.

11. The component D1 in Figure 6–16 is a

A. transistor.

B. field effect transistor.

C. Zener diode.

D. light-emitting diode.

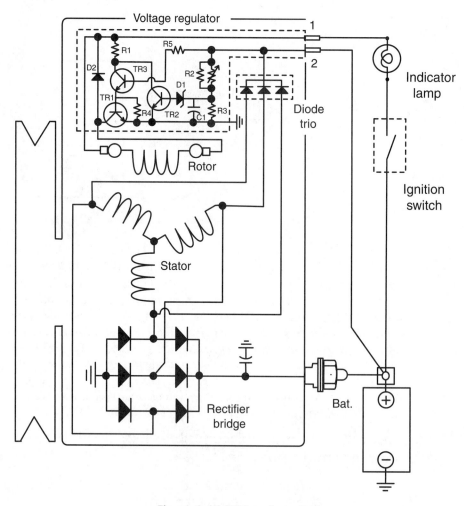

**Figure 6–16**  Wiring diagram.

**Hint**    *Electrical and electronic components are represented by symbols on wiring diagrams. Technicians must understand these symbols to properly understand wiring diagrams and electrical/electronic system operation.*

# Battery Diagnosis and Service

## ASE Tasks, Questions, and Related Information

**Task 1**  Perform battery state-of-charge test; determine needed service.

12. When performing a battery hydrometer test
    A. if the battery temperature is 0°F, 0.050 should be subtracted from the hydrometer reading.
    B. if the battery temperature is 120°F, 0.020 should be subtracted from the hydrometer reading.
    C. the maximum variation in cell hydrometer readings is 0.050 specific gravity points.
    D. the battery is fully charged if all the cell hydrometer readings exceed 1.225.

**Hint**  *When performing a battery state of charge test with a hydrometer, 0.004 specific gravity points should be subtracted from the hydrometer reading for every 10°F of electrolyte temperature below 80°F. During this test, 0.004 specific gravity points must be added to the hydrometer reading for every ten degrees of electrolyte temperature above 80°F. The maximum variation is 0.050 in cell specific gravity readings. When all the cell readings exceed 1.265, the battery is fully charged.*

**Task 2**  Perform battery capacity (load, high-rate discharge) test; determine needed service.

13. While discussing a battery capacity test with the battery temperature at 70°F, Technician A says the battery discharge rate is calculated by multiplying two times the battery reserve capacity rating.
    Technician B says the battery is satisfactory if the voltage remains above 9.6 V.
    Who is correct?
    A. A only
    B. B only
    C. Both A and B
    D. Neither A nor B

**Hint**  *The battery discharge rate for a capacity test is usually one-half of the cold cranking rating. The battery is discharged at the proper rate for 15 seconds, and the battery voltage must remain above 9.6 V with the battery temperature at 70°F or above.*

**Task 3**  Maintain or restore electronic memory functions.

14. The battery voltage is disconnected from the electrical system in a vehicle with several on-board computers. This procedure may cause
    A. damage to all the computers.
    B. failure of the engine to start.
    C. erasure of the computer adaptive memories.
    D. voltage surges in the electrical system.

**Hint**  *If battery voltage is disconnected from a computer, the adaptive memory in the computer is erased. In the case of a powertrain control module (PCM), this action may cause erratic engine operation or erratic transmission shifting when the engine is restarted. After the vehicle is driven* *for five minutes the computer relearns the system, and normal operation is restored. If the vehicle is equipped with memory seats or mirrors, disconnecting battery voltage also erases the memory in the computer that operates these systems. Disconnecting the battery voltage from the electrical system also erases the preprogrammed station memory in the stereo system.*

*A 12 V power supply from a dry-cell battery may be connected to the cigarette lighter to maintain voltage to the electrical system when the battery is disconnected.*

## Task 4    Inspect, clean, fill, or replace battery.

15. A maintenance-free battery is low on electrolyte, and the built-in hydrometer indicates light yellow.

    Technician A says this problem may be caused by an inoperative voltage regulator.

    Technician B says this problem may be caused by a loose alternator belt.

    Who is correct?

    A. A only

    B. B only

    C. Both A and B

    D. Neither A nor B

**Hint**    *A battery may be cleaned with a baking soda and water solution. If the built-in hydrometer indicates light yellow or clear, the electrolyte level is low, and the battery should be replaced. The low electrolyte level may be caused by a high-voltage regulator setting that causes overcharging.*

*When disconnecting battery cables, always disconnect the negative battery cable first.*

## Task 5    Perform slow/fast battery charge in accordance with manufacturer's recommendations.

16. While charging batteries

    A. battery charge time is the same for batteries with different capacities.

    B. the battery temperature should not exceed 125°F while charging.

    C. a high charging rate may be used to charge a battery at –20°F battery temperature.

    D. the battery may be fully charged at a high rate on a fast charger.

**Hint**    *If the battery is charged in the vehicle, the battery cables should be disconnected during the charging procedure. The charging time depends on the battery state of charge and the battery capacity. If the battery temperature exceeds 125°F while charging, the battery may be damaged.*

*When fast charging a battery, reduce the charging rate when the specific gravity reaches 1.225 to avoid excessive battery gassing. The battery is fully charged when the specific gravity increases to 1.265. Do not attempt to fast charge a cold battery.*

## Task 6    Inspect, clean, and repair or replace battery cables, connectors, clamps and holddowns.

17. All of the following statements about servicing the battery and electrical system in an air bag–equipped vehicle are true EXCEPT

    A. Disconnect the negative battery cable first.

    B. Reconnect the positive battery cable first.

    C. Begin servicing the electrical system immediately after the battery cables are disconnected.

    D. Connect a 12 V source to the cigarette lighter socket before disconnecting the battery cables.

**Hint**    *Prior to disconnecting the battery cables, a 12 V source should be plugged into the cigarette lighter socket. This action prevents computer memories from being erased, and the deprogramming of stereo equipment, when the battery cables are disconnected. When disconnecting the battery cables, always disconnect the negative battery cable first. Under this condition, if a* *wrench slips and connects the positive cable to ground, no current will flow through the wrench because the negative cable is disconnected. When reconnecting the battery cables, always connect the positive cable first. After the battery cables are disconnected, wait for the time specified by the vehicle manufacturer before servicing the electrical system. This time is usually two to ten minutes. Waiting for this time period allows the backup power supply in the air bag system to power down, and this action prevents accidental air bag deployment. Some vehicle manufacturers now recommend removing the air bag system fuse and the air bag system major components rather than disconnecting the negative battery cable prior to air bag system service. Follow the vehicle manufacturer's service information regarding the air bag system components to be disconnected.*

## Task 7   Jump start a vehicle with jumper cables and a booster battery or auxiliary power supply.

18. While jump-starting a vehicle with a booster battery, Technician A says the accessories should be on in the boost vehicle while starting the vehicle being boosted.

    Technician B says the negative booster cable should be connected to an engine ground on the vehicle being boosted.

    Who is correct?
    A. A only
    B. B only
    C. Both A and B
    D. Neither A nor B

**Hint**   *The accessories must be off in both vehicles during the boost procedure. The negative booster cable must be connected to an engine ground in the vehicle being boosted. Always connect the positive booster cable followed by the negative booster cable, and complete the negative cable connection last on the vehicle being boosted. When disconnecting the booster cables, remove the negative booster cable first on the vehicle being boosted.*

# Starting System Diagnosis and Repair

## ASE Tasks, Questions, and Related Information

## Task 1   Perform starter current draw test; determine needed repairs.

19. During a starter current draw test, the current draw is more than specified, and the cranking speed and battery voltage are less than specified. The cause of this problem may be
    A. worn bushings in the starting motor.
    B. high resistance in the field windings.
    C. high resistance in the battery positive cable.
    D. a burned solenoid disc and terminals.

**Hint**   *High starter current draw, low cranking speed, and low cranking voltage usually indicate a damaged starter. This condition also may be caused by internal engine problems such as partially seized bearings.*

   *Low current draw, low cranking speed, and high cranking voltage usually indicate excessive resistance in the starting circuit.*

# Task 2  Perform starter circuit voltage drop tests; determine needed repairs.

20. In Figure 6–17, the voltmeter is connected to test the voltage drop across
    A. the positive battery cable.
    B. the starter solenoid windings.
    C. the starter ground circuit.
    D. the starter solenoid disc and terminals.

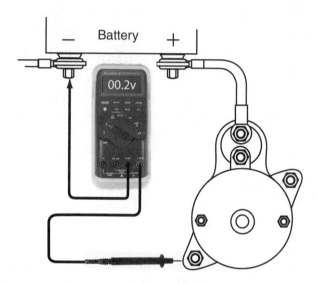

**Figure 6–17** Starter circuit voltage drop test.

**Hint**    *Measure the voltage drop across each component in the starter circuit to check the resistance in that part of the circuit. The ignition and fuel systems must be disabled while making these tests. Read the voltage drop across each component while the starting motor is operating. For example, connect the voltmeter leads to the positive battery terminal and the positive cable on the starter solenoid, and crank the engine to measure the voltage drop across the positive battery cable.*

# Task 3  Inspect, test, and repair or replace switches, connectors, and wires of starter control circuits.

21. In the starter circuit in Figure 6–18, the battery is fully charged and the starter relay and solenoid are completely inoperative when the ignition switch is in the start position. The cause of this problem could be

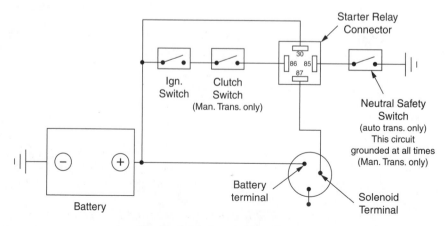

**Figure 6–18** Starter relay circuit.

    A. a grounded circuit at terminal 85 on the starter relay.

    B. an open circuit at terminal 86 on the starter relay.

    C. a continually closed neutral safety switch.

    D. a slight resistance at terminal 87 on the starter relay.

**Hint**    *Relays and switches in the starting motor circuit may be tested with an ohmmeter. When an ohmmeter is connected across the relay or switch contacts, the meter should provide an infinity reading if the contacts are open. If the relay or switch contacts are closed, the ohmmeter reading should be at or near zero. When the ohmmeter leads are connected across the terminals connected to the relay winding, the meter should indicate the specified resistance. A resistance below the specified value indicates a shorted winding, whereas an infinity reading proves the winding is open.*

## Task 4    Inspect, test, and replace starter relays and solenoids.

22. While discussing the solenoid winding test in Figure 6–19, Technician A says the ohmmeter is connected to test the solenoid pull-in winding.

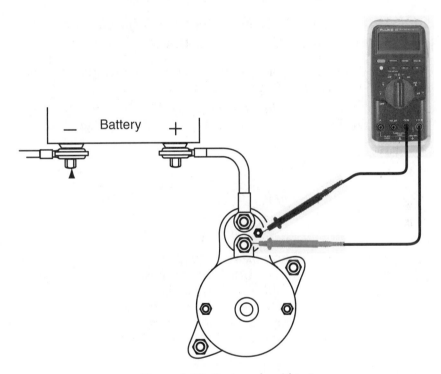

**Figure 6–19** Starter solenoid test.

Technician B says that if the winding is satisfactory, the ohmmeter should provide an infinity reading.

Who is correct?

    A. A only

    B. B only

    C. Both A and B

    D. Neither A nor B

**Hint**    *The ohmmeter leads must be connected across the solenoid terminal and the field coil terminal to test the pull-in winding. Connect the ohmmeter leads from the solenoid terminal to ground to test the hold-in winding.*

## Task 5    Remove and replace starter.

23. A starting motor does not disengage properly and continues to run with the engine for a few seconds after the engine starts.

    Technician A says the starting motor solenoid may be the problem.

    Technician B says the starting motor may be misaligned with the flywheel housing.

    Who is correct?

    A. A only

    B. B only

    C. Both A and B

    D. Neither A nor B

**Hint**    *If shims are located between the starting motor mounting flange and the flywheel housing, these shims must be reinstalled in their original position when the starting motor is installed.*

*The mounting surfaces on the starting motor and the flywheel housing must be clean before installing the starter. Dirt or paint on these surfaces may cause starting motor misalignment. If a ground cable is connected to one of the starting motor mounting bolts, be sure this cable is reconnected when the starting motor is installed.*

**Task 6**    **Differentiate between electrical and engine mechanical problems that cause a slow crank or no crank condition.**

24. When an engine is at normal operating temperature, it kicks back against the starting motor and intermittently cranks slowly.

    Technician A says the crankshaft bearings may be seizing.

    Technician B says the base ignition timing may be excessively advanced.

    Who is correct?

    A. A only

    B. B only

    C. Both A and B

    D. Neither A nor B

**Hint**    *Worn engine components such as crankshaft bearings that are beginning to seize up may cause slow cranking of the engine and high starting motor current draw. If this problem is suspected, turn the ignition switch off and attempt to turn the flywheel with a pry bar installed in the ring gear teeth. If the engine is harder than normal to rotate, a mechanical problem exists in the engine. Excessively advanced ignition timing may cause the engine to kick back against the starter while cranking, resulting in slow cranking and hard starting. This problem usually occurs when the engine is at normal operating temperature.*

# Charging System Diagnosis and Repair

## ASE Tasks, Questions, and Related Information

**Task 1**    **Diagnose charging system problems that cause an undercharge, a no-charge, or an overcharge condition.**

25. When discussing an alternator with zero output, Technician A says the alternator field circuit may have an open circuit.

Technician B says the fuse link may be open in the alternator to battery wire.

Who is correct?

A. A only

B. B only

C. Both A and B

D. Neither A nor B

*Hint*    *A low-charging voltage caused by a malfunctioning voltage regulator or alternator results in a reduced charging rate and an undercharged battery. This problem also may be caused by a loose alternator belt or excessive resistance in the wire from the alternator battery terminal to the positive battery terminal. An overcharged battery usually is caused by a malfunctioning voltage regulator that allows high charging circuit voltage. A no-charge condition may be caused by an open alternator field circuit, or an open fuse link in the wire from the alternator battery terminal to the positive battery terminal.*

## Task 2   Inspect, adjust, and replace generator (alternator) drive belts, pulleys, and tensioners.

26. An alternator with a 90 ampere rating produces 45 amperes during an output test. The alternator is driven with a V-belt and the belt has the specified tension.
    Technician A says the V-belt may be worn and bottomed in the pulley.
    Technician B says the alternator pulley may be misaligned with the crankshaft pulley.
    Who is correct?
    A. A only
    B. B only
    C. Both A and B
    D. Neither A nor B

*Hint*    *An undercharged battery may be caused by a slipping alternator belt. A slipping belt could be the result of insufficient belt tension, or a worn, glazed, or oil-soaked belt. Belt tension may be tested with a belt tension gauge or by measuring the belt deflection in the center of the belt span. A belt should have 1/2 in. of deflection for every foot of free span. Many ribbed V-belts have an automatic spring-loaded tensioner with a belt wear scale.*

## Task 3   Perform charging system voltage output test; determine needed repairs.

27. While discussing a charging system output test, Technician A says the vehicle accessories should be on during the test.
    Technician B says the charging system voltage should be limited to 17 V.
    Who is correct?
    A. A only
    B. B only
    C. Both A and B
    D. Neither A nor B

*Hint*    *The alternator belt tension and condition should be checked before an output test is performed.*
    *Turn off the vehicle accessories during the test. If the alternator is full-fielded during the output test, a carbon pile load in the volt-ampere tester must be used to maintain the voltage below 15 V. The alternator output may be tested by lowering the voltage to the voltage specified by the vehicle manufacturer.*

## Task 4   Perform charging system current output test; determine needed repairs.

28. During an output test using the full-field method, a 100-ampere alternator with an integral electronic regulator produces 30 amperes. The cause of the low alternator output could be
   A. a shorted diode in the alternator.
   B. a broken brush lead wire in the alternator.
   C. an open circuit in the voltage regulator.
   D. an inoperative alternator capacitor.

*Hint*    *When the alternator output is zero during an output test, the field circuit is probably open. This problem may be caused by worn brushes or an open field winding in the rotor. If the alternator output is less than specified, there is probably a problem in the diodes or stator. A high resistance in the field winding also reduces output. When the alternator is full-fielded to test output, the voltage regulator is bypassed and does not affect output.*

## Task 5    Inspect and test generator (alternator) control circuit; determine needed repairs.

29. The charging system voltage on a vehicle is 16.2 V. This condition may cause all the following problems EXCEPT
   A. an overcharged battery.
   B. burned-out electrical components.
   C. electrolyte gassing in the battery.
   D. reduced headlight brilliance.

*Hint*    *When the alternator voltage is erratic or too low, the alternator may be full-fielded to determine the cause of the problem. When the alternator is full-fielded and the alternator current and voltage output are normal, the voltage regulator is probably the problem. If the charging system voltage is higher than specified, the voltage regulator probably is the problem. On a charging system with an external regulator, this problem may be caused by excessive resistance in the field circuit between the ignition switch and the regulator.*

## Task 6    Perform charging circuit voltage drop tests; determine needed repairs.

30. With the ammeter and voltmeter connected to the charging system, as indicated in Figure 6–20, the voltmeter indicates 2 V and the ammeter reads 10 amperes.
   Technician A says this condition may cause an undercharged battery.
   Technician B says this condition may result in head lamp flare-up during acceleration.
   Who is correct?
   A. A only
   B. B only
   C. Both A and B
   D. Neither A nor B

*Hint*    *A voltmeter may be connected from the alternator battery wire to the positive battery terminal to measure voltage drop in the charging circuit. Many car manufacturers recommend a 10-ampere charging rate while measuring this voltage drop. When the voltage drop is more than specified, the circuit resistance is excessive. High charging circuit resistance between the alternator battery terminal and the positive battery terminal may cause an undercharged battery.*

## Task 7    Inspect, repair, or replace connectors and wires of charging circuits.

31. A replacement fusible link must be
   A. at least 15 in. (38 cm) long.
   B. the same gauge size as the wire it is protecting.
   C. two gauge sizes smaller than the wire it is protecting.
   D. four gauge sizes smaller than the wire it is protecting.

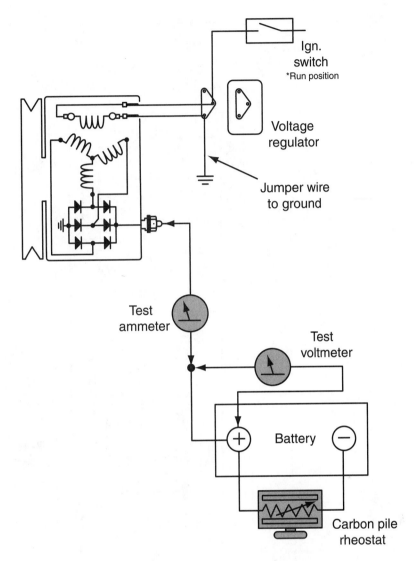

**Figure 6–20** Charging system voltage test.

32. The charging system in Figure 6–21 has a glowing charge indicator bulb with the engine running. An alternator output test indicates satisfactory output.

    Technician A says there may be excessive resistance in the wire from the alternator battery terminal to the positive battery terminal.

    Technician B says there may be excessive resistance in the wire from the alternator L terminal to the charge indicator bulb.

    Who is correct?

    A. A only

    B. B only

    C. Both A and B

    D. Neither A nor B

*Hint*    *Always disconnect the negative battery cable before servicing charging circuit wires or components.*

*After disconnecting the negative battery cable on an air bag–equipped vehicle, wait for the time period specified by the vehicle manufacturer before servicing charging circuit wires or components.*

*When joining wires with a solder-type clip, there must be solder fillets between the wires and the clip, and solder should be visible in the clip window (Figure 6–22). When replacing a fusible*

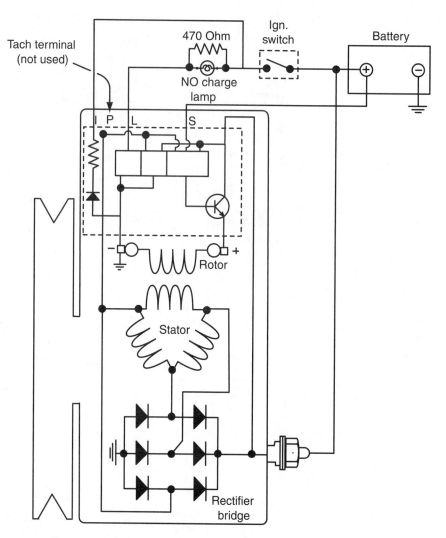

**Figure 6–21** Charging system with integral electronic regulator.

**Figure 6–22** Properly soldered electrical connection.

*link, the length of the link must not exceed 9 in. (225 mm), and the link should be four gauge sizes smaller than the wire it is protecting (Figure 6–23).*

## Task 8   Remove, inspect, and replace generator (alternator).

33. When discussing alternator removal and replacement, Technician A says excessive alternator belt tension may cause premature alternator bearing failure.

    Technician B says worn alternator mounting bolt openings may cause rapid alternator belt wear.

    Who is correct?

    A. A only

    B. B only

    C. Both A and B

    D. Neither A nor B

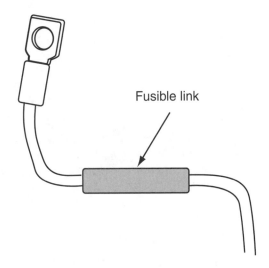

**Figure 6-23**  Fusible link.

*Hint*     *Always disconnect the negative battery cable before removing an alternator. After disconnecting the negative battery cable on an air bag–equipped vehicle, wait for the time period
specified by the vehicle manufacturer before servicing the alternator. When the alternator is
removed, inspect all the alternator mounting bolts, bolt openings, and brackets for wear.
Worn mounting bolts, bolt openings, or brackets may cause alternator belt misalignment and
rapid belt wear.*

*Replace all worn mounting bolts or alternator housings with worn bolt openings. After the
alternator is reinstalled, the alternator drive belt must have the specified tension, and all mounting bolts must be tightened to the specified torque.*

*Connect the ohmmeter leads across each diode, and then reverse the leads to test the
diodes.*

*A satisfactory diode provides one low and one high ohmmeter reading. A shorted diode is indi-
cated by two low meter readings, and an open diode provides two infinity readings.*

# Lighting Systems Diagnosis and Repair, Headlights, Parking Lights, Taillights, Dash Lights, and Courtesy Lights

## ASE Tasks, Questions, and Related Information

**Task 1**  **Diagnose the cause of brighter than normal, intermittent, dim, continuous or no operation of headlights.**

34. The headlights on a vehicle go out intermittently and come back on after a few
    minutes.
    Technician A says this problem may be caused by an intermittent short to ground.
    Technician B says this problem may be caused by a high charging system voltage.

Who is correct?

A.  A only

B.  B only

C.  Both A and B

D.  Neither A nor B

**Hint**    *If the headlights are inoperative, check the circuit breakers or fuses. Many headlight circuits have a circuit breaker in the headlight switch. If the headlights are dim, check for resistance in the headlight circuit or for low charging system voltage. When the headlight operation is intermittent, check for an intermittent open circuit in the headlight wiring, dimmer switch, or headlight switch. Intermittent headlight operation also may be caused by a shorted condition or a short to ground. Either of these conditions cause excessive current flow in the circuit that may cause the circuit breaker to open. This action turns off the headlights. When the circuit breaker cools, the headlights come back on.*

## Task 2    Inspect, test, and repair daytime running light systems.

35.  Technician A says the daytime running light system turns on the high beam headlights when driving during the day.

Technician B says the daytime running lights may be operated by a module that supplies a pulsating voltage to the headlights.

Who is correct?

A.  A only

B.  B only

C.  Both A and B

D.  Neither A nor B

**Hint**    *Some vehicles are equipped with daytime running lights. In these systems, the low-beam headlights are on when the ignition switch is activated. Some daytime running light circuits supply voltage to the low-beam headlights through a resistor so these lights are illuminated at partial brilliance. Other daytime running lights are controlled by a module that supplies a pulse width modulated (PWM) voltage to the low-beam headlights to supply partial brilliance. Some daytime running lights have an override feature that allows technicians to turn off these lights when working on the vehicle in the shop. Some daytime running lights may be shut off by applying the parking brake, while other daytime running lights are turned off when the transmission selector is placed in PARK. The daytime running lights are turned on when the gear selector is placed in DRIVE.*

## Task 3    Inspect, replace, and aim headlights/bulbs, and auxiliary lights (fog lights/driving lights).

36.  All of the following statements about halogen headlight bulb replacement are true EXCEPT

A.  Handle the bulb only by its base.

B.  Do not drop or scratch the bulb.

C.  Change the bulb with the headlights on.

D.  Keep moisture away from the bulb.

**Hint**    *Always turn off the headlights and allow the bulb to cool before changing a halogen bulb.*

*Keep moisture away from the bulb and handle the bulb by its base. Do not scratch or drop the bulb. Many manufacturers recommend using a headlight screen to align the headlights. The vehicle is parked 25 ft. from the screen on a level floor (Figure 6–24). Many vehicle manufacturers recommend aligning the low beams, and the high beams are considered nonadjustable. On some older vehicles, the manufacturers recommend aligning the high beams.*

*On some older vehicles, the manufacturers recommend measuring headlight alignment with headlight aimers (Figure 6–25). These aimers are held on the headlights with a suction cup. The*

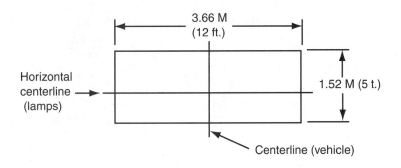

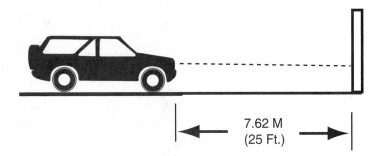

**Figure 6–24** Headlight alignment screen.

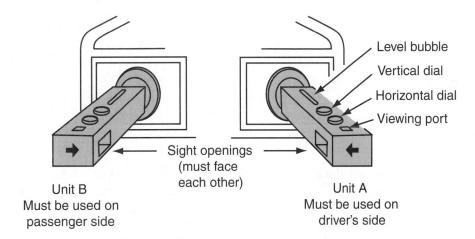

**Figure 6–25** Headlight aimers.

*headlights are adjusted until the level bubble in the aimers is in the specified position. Some vehicle manufacturers recommend measuring headlight alignment with a photoelectric aimer that measures headlight beam intensity. Headlight adjustments are provided on each headlight unit (Figure 6–26 and Figure 6–27).*

## Task 4   Inspect, test, and repair or replace headlight and dimmer switches, relays, control units, sensors, sockets, connectors, and wires of headlight circuits.

37.  An open circuit in fuse number 12 in Figure 6–28 could result in
   A.  inoperative taillights.
   B.  inoperative stoplights.
   C.  inoperative instrument panel lights.
   D.  inoperative low-beam headlights.

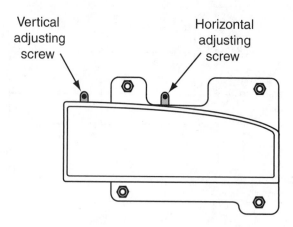

**Figure 6–26** Composite headlight adjusting screws.

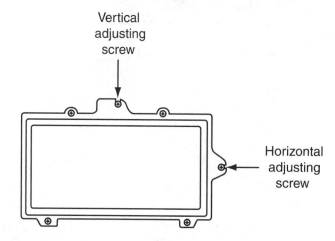

**Figure 6–27** Sealed beam headlight adjustment screws.

**Hint**     *Many headlight switches contain a circuit breaker that is connected in the headlight circuit. Other light circuits such as the taillights, stoplights, or instrument panel lights have separate fuses. The dimmer switch usually is part of the multifunction switch in the steering column. The dimmer switch is connected in series between the headlight switch and the headlights. When the headlights are turned on, the dimmer switch directs the current flow to the low beam or high beam headlights.*

## Task 5    Diagnose the cause of brighter than normal, intermittent, dim, continuous or no operation of parking lights, taillights, and/or auxiliary lights (fog lights/driving lights).

38.  The right front parking light is dim, but all the other lights have normal brilliance.
Technician A says there may be a high resistance problem at the light switch.
Technician B says the right front parking light ground may have high resistance.
Who is correct?
A.  A only
B.  B only
C.  Both A and B
D.  Neither A nor B

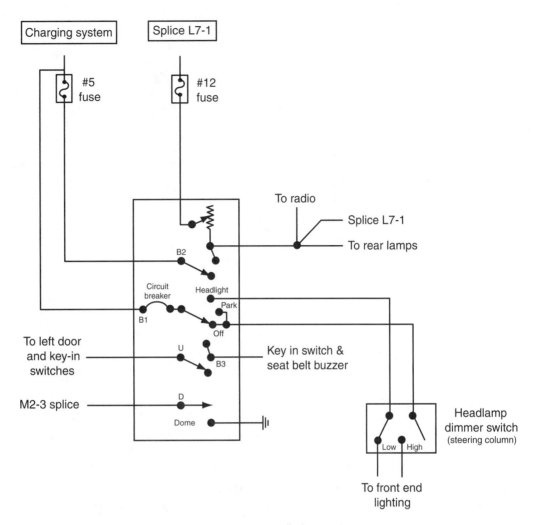

**Figure 6–28** Headlight circuit.

*Hint*      *Intermittent parking lights or taillights are usually caused by an intermittent open circuit in the affected light(s). When these lights are dim, a high-resistance problem is usually present in the circuit. The high-resistance problem may be on the input or ground side of the affected light(s). Inoperative parking lights and/or taillights are usually caused by a blown fuse. The most likely cause of a blown fuse is a short to ground.*

**Task 6**   **Inspect, test, and repair or replace switches, relays, bulbs, sockets, connectors, wires, and controllers of parking light, taillight circuits, and auxiliary light circuits (fog lights/driving lights).**

39. The rear light ground connection on the left side of Figure 6–29 has an open circuit. The ground connection on the right side of Figure 6–29 is satisfactory. This problem could result in
    A. inoperative left rear tail, stop, and side marker lights.
    B. no change in the rear light operation.
    C. inoperative left rear tail and stoplights.
    D. inoperative backup lights.

*Hint*      *Many rear light bulbs are a combination bulb containing stop and taillight filaments. Backup light bulbs and side marker bulbs are single filament. When the headlight switch is turned to the park or headlight position, voltage is supplied from this switch to the taillight bulbs and side marker bulbs. Most rear lights share a common ground connection. When the brakes are applied,*

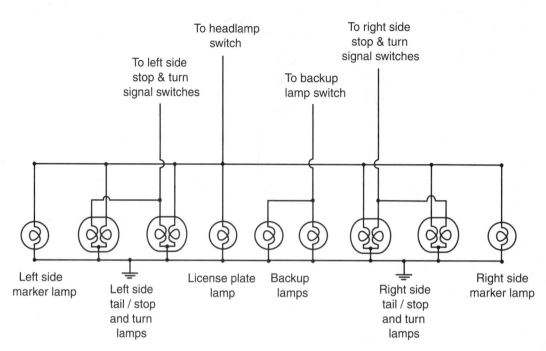

**Figure 6-29** Rear lighting circuit.

*voltage is supplied from the brake light switch to the stoplight bulbs. If the gear selector is placed in reverse, voltage is supplied from the backup light switch to the backup light bulbs.*

*If any bulb is dim, there is a resistance problem in the voltage supply wire or ground wire connected to the bulb. When the engine is accelerated and all the lights are brighter than normal, the charging system voltage is higher than specified.*

## Task 7    Diagnose the cause of intermittent, dim, no lights, continuing operation or no brightness control of instrument lighting circuits.

40. All the instrument panel lights intermittently go dim. The rheostat in the instrument panel light circuit is tested and proven to be in satisfactory condition.

    Technician A says to check the instrument panel ground circuit.

    Technician B says to check each instrument panel bulb for a loose connection.

    Who is correct?

    A. A only

    B. B only

    C. Both A and B

    D. Neither A nor B

**Hint**    *In an instrument panel light circuit, each bulb is connected parallel to the battery. A printed circuit board is located in a conventional instrument panel, and all the light bulbs and components are connected to the printed circuit board. A ground wire is usually connected from the printed circuit board ground connection to the chassis sheet metal. High resistance in the printed circuit ground causes all the instrument panel bulbs to be dim.*

## Task 8    Inspect, test, and repair or replace switches, relays, bulbs, sockets, connectors, wires, controllers, and printed circuit boards of instrument lighting circuits.

41. While discussing the instrument panel lights in Figure 6–30, Technician A says an open circuit in the rheostat may cause all the bulbs to be inoperative.

    Technician B says an open circuit in one of the bulbs may cause all the bulbs to be inoperative.

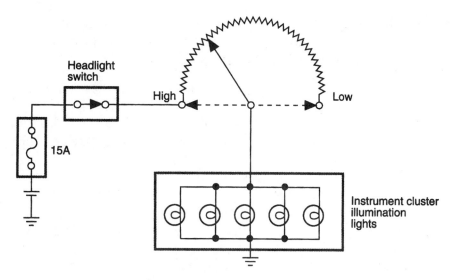

**Figure 6-30** Instrument panel light circuit.

Who is correct?

A. A only

B. B only

C. Both A and B

D. Neither A nor B

**Hint**    *A rheostat is connected in series with the instrument panel bulbs. This rheostat is operated by the headlight switch knob or by a separate control knob. When the rheostat control knob is rotated, the voltage to the instrument panel bulbs is reduced. This action lowers the current flow and reduces the brilliance of the bulbs. The instrument panel bulbs are connected parallel to the battery.*

*If one bulbs burns out, the other bulbs remain illuminated.*

## Task 9   Diagnose the cause of intermittent, dim, continuous, or no operation of courtesy lights (dome, map, vanity, cargo, trunk, and hood light).

42. The courtesy lights turn on intermittently when driving a vehicle on the road.

    Technician A says to check the door jamb switch on the driver's door.

    Technician B says to check the courtesy light fuse.

    Who is correct?

    A. A only

    B. B only

    C. Both A and B

    D. Neither A nor B

**Hint**    *Many courtesy light circuits are operated by door jamb switches. In many of these circuits, the door jamb switches provide a ground for the courtesy lights when either front door is opened.*

*Some door jamb switches are located on the input side of the courtesy lights, and these switches supply voltage to the courtesy lights when either front door is opened. In this type of circuit, the ground side of the courtesy light bulbs is connected directly to ground. Some cars have courtesy light switches on all four doors. In most vehicles, a manual switch in the instrument panel allows the driver to operate the courtesy lights with the doors closed. Because the driver's door is usually opened more frequently than the other doors, the door jamb switch in this door often wears out first.*

## Task 10    Inspect, test, and repair or replace switches, relays, bulbs, sockets, connectors, wires, and controllers of courtesy light (dome, map, vanity, cargo, trunk, and hood light) circuits.

43. Circuit 156 is shorted to ground at terminal S363 in Figure 6–31. This problem may cause
    A. continual operation of the courtesy lights.
    B. no operation of the courtesy lights and lighted mirror.
    C. continual operation of the underhood light.
    D. a burned-out courtesy light fuse.

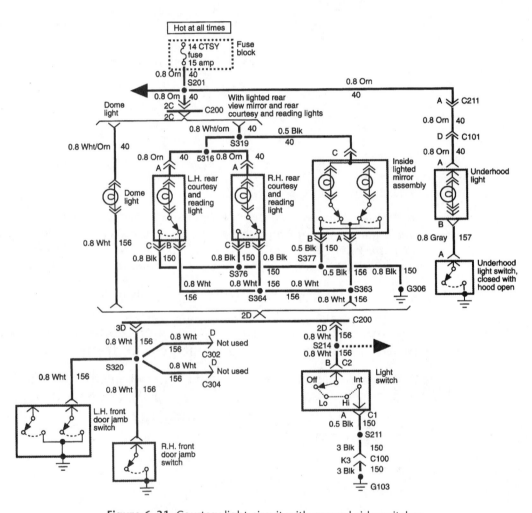

**Figure 6–31** Courtesy light circuit with ground-side switches.

**Hint**    *Some courtesy light circuits have ground-side switches. In these circuits, voltage is supplied from the positive battery terminal through a fuse to the courtesy light bulbs. When a door is opened, one of the door jamb switches closes. This switch provides a ground for the courtesy light bulbs.*

*In other courtesy light circuits, the switches are connected on the insulated side of the circuit between the battery positive terminal and the courtesy light bulbs. A ground wire is connected from each bulb to ground (Figure 6–32).*

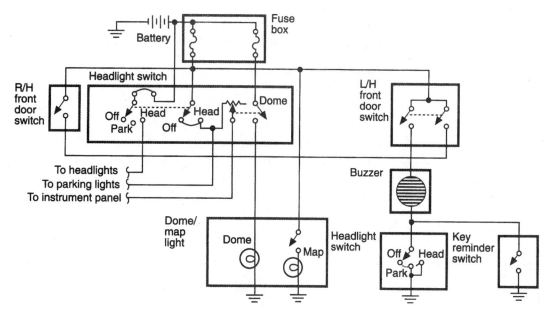

**Figure 6–32** Courtesy light circuit with insulated-side switches.

# Lighting Systems Diagnosis and Repair, Stoplights, Turn Signals, Hazard Lights, and Backup Lights

## ASE Tasks, Questions, and Related Information

**Task 1**　**Diagnose the cause of intermittent, dim, continuous or no operation of stoplights (brake lights).**

44. A vehicle experiences repeated blowing of the stoplight fuse.

    Technician A says there may be a short to ground on the ground side of one stoplight bulb.

    Technician B says there may be an intermittent open circuit in the stoplight circuit.

    Who is correct?

    A. A only

    B. B only

    C. Both A and B

    D. Neither A nor B

***Hint***　*If both stoplights are inoperative, check the stoplight fuse. If this fuse blows repeatedly, test the stoplight circuit for a short to ground. When only one stoplight is inoperative, test the bulb in the inoperative light. If one stoplight is dim, be sure the proper bulb is installed in the light, and test for high resistance on the ground side and insulated side of the bulb. Intermittent stoplight operation is usually caused by an intermittent open in the stoplight circuit.*

## Task 2    Inspect, test, adjust, and repair or replace switch, bulbs, sockets, connectors, wires, and controllers of stoplight (brake light) circuits.

45. The cigar lighter fuse is blown in the stoplight circuit in Figure 6–33. The result of this problem could be
    A. the courtesy and dome lights come on dimly when the cigar lighter is pushed in.
    B. the stop and dome lights are completely inoperative.
    C. the parking lights, taillights, and instrument panel lights are inoperative.
    D. a battery drain occurs with all the light switches in the off position.

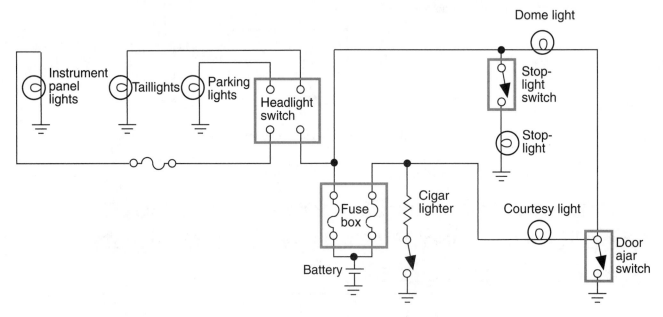

**Figure 6–33** Stoplight and cigar lighter circuits.

**Hint**    *If the cigar lighter is pushed in and the cigar lighter fuse is blown, current flows through the dome light, courtesy light, and cigar lighter to ground (Figure 6–34). Because these lights are now in series with the cigar lighter, the lights glow dimly.*

*In many stoplight circuits, voltage is supplied to the brake light switch from the battery positive terminal. When the brakes are applied, brake pedal movement closes the stoplight switch.*

*This action supplies voltage to the stoplights and the collision avoidance light (Figure 6–35). In many stoplight systems, the stoplight filaments are mounted in the same bulb as the taillight filaments.*

## Task 3    Diagnose the cause of no turn signal and/or hazard lights, or lights with no flash on one or both sides.

46. In the signal light circuit in Figure 6–36, the left rear signal light bulb is dim compared to the right rear signal light bulb.
    Technician A says there may be a high-resistance problem in wire D2 18RD.
    Technician B says there may be a high-resistance problem in wire DB 180G RD.
    Who is correct?
    A. A only
    B. B only
    C. Both A and B
    D. Neither A nor B

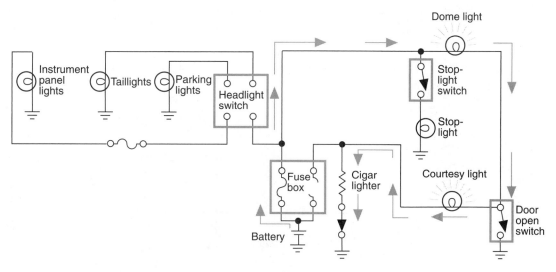

**Figure 6–34** Stoplight and cigar lighter circuits with blown fuse.

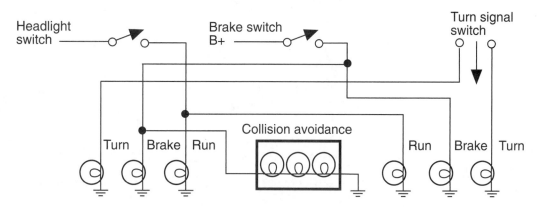

**Figure 6–35** Stoplight circuit with collision avoidance light.

*Hint*    *If the signal lights do not flash on either side, test the signal light fuse and the flasher. When the signal lights only flash on one side, test the front and rear bulbs on that side and the connecting wires. If one signal light is dim, test for high resistance on the ground side and input side of the dim bulb.*

**Task 4** **Inspect, test, and repair or replace switches, flasher units, bulbs, sockets, connectors, wires, and controllers of turn signal and hazard light circuits.**

47. In the signal light circuit in Figure 6–36, the right rear signal light is dim, and all the other lights work normally. The cause of this problem may be
    A. high resistance in the DB 180G RD wire from the signal light switch to the rear lamp wiring.
    B. a short to ground in the DB 180G RD wire from the signal light switch to the rear lamp wiring.
    C. high resistance in the D7 18BR RD wire from the signal light switch to the rear lamp wiring.
    D. high resistance in the D2 18 RD wire from the signal light flasher to the switch.

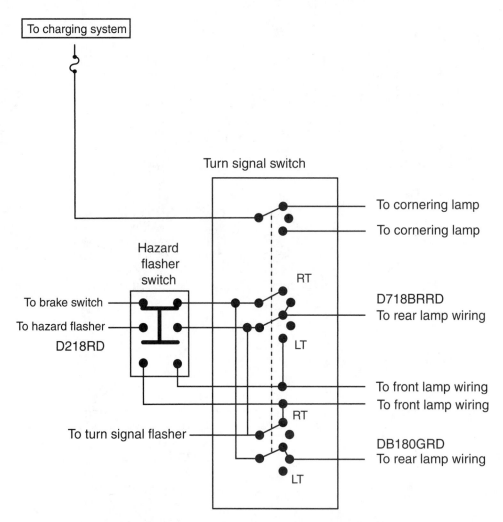

**Figure 6–36** Signal light circuit in the right turn position.

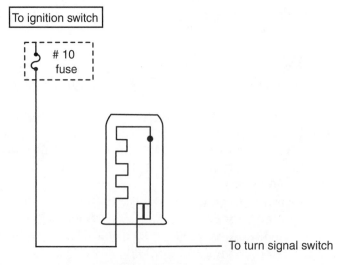

**Figure 6–37** Signal light flasher.

**Hint**    *The signal light flasher contains a bimetallic heating strip surrounded by a heating coil. One flasher contact is mounted on the bimetallic strip, and the other contact is stationary. When the ignition switch is turned on, voltage is supplied through the flasher contacts to the signal light switch (Figure 6–37). This switch directs the voltage to the left or right signal light bulbs depending on the signal light lever position selected by the driver. When current starts flowing through the flasher, the heat from the heating coil bends the bimetallic strip and opens the flasher contacts.*

*The bimetallic strip cools, allowing the contacts to close, and this action repeats to provide a flashing action. If the brake pedal is applied during a right turn, the left brake light is illuminated.*

*When the hazard switch is pressed, voltage is supplied from the hazard flasher through the hazard switch and signal light switch to the front and rear signal lights. The hazard flasher has the same internal design as the conventional flasher.*

## Task 5    Diagnose the cause of intermittent, dim, improper, continuous or no operation of backup lights.

48. The backup lights are illuminated continually.

    Technician A says the backup light switch may be the problem.

    Technician B says the wire from the backup light switch to the backup lights may be shorted to 12 V.

    Who is correct?

    A. A only

    B. B only

    C. Both A and B

    D. Neither A nor B

**Hint**    *If the backup lights are inoperative, test the backup light fuse and the switch. The backup light switch is often mounted on the steering column and activated by the gearshift linkage. Some backup light switches are mounted in the transmission. If one backup light is inoperative, test the bulb in the inoperative light. When one backup light is dim, test for high resistance on the ground side and input side of the dim bulb.*

## Task 6    Inspect, test, and repair or replace switch, bulbs, sockets, connectors, wires, and controllers of backup light circuits.

49. The right-hand backup light circuit is grounded on the switch side of the bulb (Figure 6–38).

    Technician A says this condition may blow the backup light fuse.

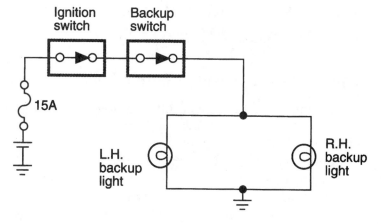

**Figure 6–38** Backup light circuit.

Technician B says the left-hand backup light may work normally while the right-hand backup light is inoperative.

Who is correct?

A. A only

B. B only

C. Both A and B

D. Neither A nor B

**Hint**    *When the ignition switch is on and the backup light switch is closed, voltage is supplied through these switches to the backup lights. The backup light switch is operated by the gear selector linkage. This switch may be mounted on the steering column or transmission.*

## Task 7    Inspect, test, and repair or replace trailer wiring harness and connector.

50. A trailer is being towed regularly, but the trailer battery requires frequent charging with a battery charger. The trailer lights operate normally. The most likely cause of this problem is

    A. high resistance in the ground wire through the 7-wire connector.

    B. a grounded connection in the trailer taillights.

    C. an open circuit in the trailer backup light circuit.

    D. a blown fuse in the battery wire from the positive battery terminal on the tow vehicle to the 7-wire connector.

**Hint**    *Many trailers have a 7-wire connector between the trailer and the vehicle. A fused wire from the positive battery terminal on the tow vehicle is connected through the 7-wire connector to charge the battery in the trailer when towing the trailer. A ground wire from the trailer chassis to the tow vehicle chassis is also connected through the 7-wire connector. This ground wire eliminates the possibility of having to provide a ground connection through the hitch. The other wires in the 7-wire connector supply voltage to the stoplights, taillights, signal lights, and backup lights. Two separate terminals in the 7-wire connector are required to supply voltage to the right and left trailer signal lights.*

# Gauges, Warning Devices, and Driver Information Systems Diagnosis and Repair

## ASE Tasks, Questions, and Related Information

## Task 1    Diagnose the cause of intermittent, high, low, or no gauge readings.

(Note: Diagnosing causes of abnormal charging system gauge readings are limited to dash units and their electrical connections; other causes of abnormal charging system gauge readings are covered in category D.)

51. An instrument panel is equipped with an instrument voltage limiter and thermal electric gauges. All the gauges intermittently and simultaneously provide high readings.

    Technician A says to test the resistance in the instrument panel ground.

    Technician B says to check for a short to ground between the fuel gauge and the fuel gauge sending unit.

Who is correct?

A. A only

B. B only

C. Both A and B

D. Neither A nor B

**Hint**   *If voltage is supplied through an instrument voltage limiter to the gauges, high or low gauge readings may be caused by an inoperative instrument voltage limiter. Because the instrument voltage limiter must be properly grounded, an intermittent resistance problem in the instrument panel ground causes erratic gauge operation. A high gauge reading on one gauge may be caused by a short to ground between the gauge and the sending unit on a thermal electric gauge.*

## Task 2   Inspect, test, and repair or replace gauges, gauge sending units, connectors, wires, controllers, and printed circuit boards of gauge circuits.

52. The fuel gauge in Figure 6–39 reads lower than the actual level of fuel in the tank. All the other gauges operate normally. The cause of this problem may be

A. high resistance in the sending unit ground wire.

B. high resistance between the instrument voltage limiter and the gauge.

C. a short to ground between the gauge and the sending unit.

D. an open circuit in the wire from the gauge to the sending unit.

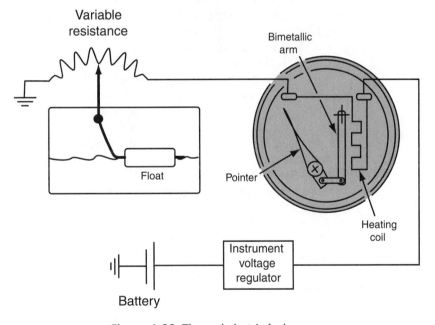

**Figure 6–39** Thermal-electric fuel gauge.

53. All the gauges are erratic in an instrument panel with thermal-electric gauges and an instrument voltage limiter.

Technician A says the alternator may be the problem.

Technician B says the instrument voltage limiter may be the problem.

Who is correct?

A. A only

B. B only

C. Both A and B

D. Neither A nor B

54. On a two-coil temperature gauge, the gauge pointer remains in the hot position regardless of the engine temperature (Figure 6–40). The cause of this problem could be
    A. the wire may be open from the hot coil to the sending unit.
    B. the wire may be open from the cold coil to ground.
    C. the sending unit may have excessive internal resistance.
    D. an excessively high resistance between the sending unit and ground.

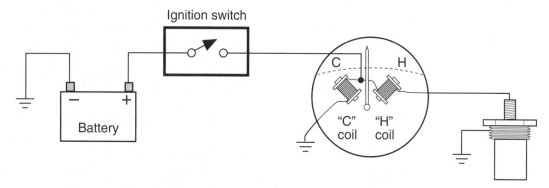

**Figure 6–40** Two-coil temperature gauge.

**Hint**    *Many vehicles are equipped with thermal-electric gauges. These gauges contain a bimetallic strip surrounded by a heating coil. The pivoted gauge pointer is connected to the bimetallic strip.*

*The sending unit contains a variable resistor. In a fuel gauge, this variable resistor is connected to a float in the fuel tank. If the tank is filled with fuel, the sending unit resistance decreases, and current flow through the bimetallic strip increases. This increased current flow heats the bimetallic strip and pushes the pointer toward the full position.*

*The voltage limiter supplies about 5 V to the gauges regardless of the charging system voltage.*

*If the voltage limiter voltage is higher than specified, all the gauges will have high readings. An inoperative voltage limiter may also cause low or erratic readings on all the gauges. The voltage limiter requires a ground connection through the instrument panel. High resistance in the instrument panel ground reduces heating coil current in the voltage limiter. This action allows the limiter contacts to remain closed longer. Under this condition, voltage output from the limiter increases and gauge readings are higher.*

*Some gauges contain two coils, and the pointer is mounted on a magnet under these coils. In a temperature gauge, the sending unit is connected to the hot coil and the cold coil is grounded. If the coolant is cold, the sending unit has high resistance. Under this condition, current flows through the lower resistance of the cold coil. Coil magnetism around the cold coil attracts the magnet and pointer to the cold position. As the coolant temperature increases the sending unit resistance decreases. When the engine is at normal operating temperature, the current flows through the lower resistance of the hot coil and sending unit. This action attracts the magnet and pointer to the hot position.*

**Task 3**    **Diagnose the cause(s) of intermittent, high, low, or no readings on electronic instrument clusters.**

55. An electronic instrument cluster is completely inoperative.
    Technician A says to replace the instrument cluster.
    Technician B says to test the instrument panel fuse.
    Who is correct?
    A. A only
    B. B only
    C. Both A and B
    D. Neither A nor B

*Hint*     *If one segment on an electronic instrument cluster is inoperative, the cluster must be repaired or replaced. Electronic instrument cluster repairs are done by electronic specialty shops. In most cases, the electronic instrument cluster is sent through the OEM dealer to the specialty repair facility.*

*If none of the segments are illuminated in an electronic instrument cluster, test the fuse and voltage supply circuit.*

## Task 4   Inspect, test, repair or replace sensors, sending units, connectors, wires, and controllers of electronic instrument circuits.

56. When the ignition switch is turned on, most of the electronic instrument displays are brightly illuminated, but a few of the displays are not illuminated.

    Technician A says the inputs for the non-illuminated displays may be the problem.

    Technician B says the electronic instrument display may be the problem.

    Who is correct?

    A. A only

    B. B only

    C. Both A and B

    D. Neither A nor B

*Hint*     *Many electronic instrument displays provide an initial illumination of all segments when the ignition switch is turned on. This illumination proves the operation of the display segments.*

*During this initial display, all the segments in the electronic instrument displays should be brightly illuminated for a few seconds. If some of the segments are not illuminated, replace the electronic instrument cluster. When none of the segments are illuminated, check the fuses and voltage supply to the display.*

*Many electronic instrument displays have self-diagnostic capabilities. In some electronic instrument displays, a specific gauge illumination or digital display indicates certain problems in the display. Other electronic instrument displays may be diagnosed with a scan tool.*

## Task 5   Diagnose the cause of constant, intermittent, or no operation of warning light, indicator lights, and other driver information systems.

57. The oil pressure warning light is illuminated with the engine running.

    Technician A says the first step should be to test the engine oil pressure with a test gauge.

    Technician B says the first step should be to remove the oil pan, and then measure the crankshaft bearing clearance.

    Who is correct?

    A. A only

    B. B only

    C. Both A and B

    D. Neither A nor B

*Hint*     *When an oil pressure warning light is located in the instrument panel, the oil pressure sending unit contains a diaphragm and a set of contacts. When the oil pressure supplied to the sending unit is less than 3 to 5 psi, the contacts remain closed. These closed contacts provide a ground connection for the oil pressure warning light, and under this condition the light is illuminated.*

*If the engine is running and the oil pressure exceeds 3 to 5 psi, the oil pressure supplied to the sending unit moves the diaphragm and opens the contacts. When this action occurs, the oil pressure warning light is turned off. If the oil pressure warning light is illuminated with the engine running, the oil level and/or pressure may be low, the oil pressure sending unit may be the problem, or the wire may be grounded between the oil pressure warning light and the sending unit.*

*When the oil pressure warning light is illuminated with the engine running, the engine should be shut off immediately, and the oil level and condition should be checked. If the oil level and condition are satisfactory, the next diagnostic step is to remove the oil pressure sending unit and install a test gauge in the sending unit opening to test the oil pressure. When the oil pressure is within specifications, check the sending unit and the wire from the sending unit to the oil pressure warning light.*

## Task 6    Inspect, test, and repair or replace bulbs, sockets, connectors, wires, electronic components, and controllers of warning light and driver information system circuits.

58. The door ajar light in the message center is illuminated continually with the ignition switch on (Figure 6–41). The cause of this problem may be
    A. an open circuit in wire between BCM terminal 3 and the door ajar switches.
    B. a short to ground on the BCM side of the RF door ajar switch.
    C. an inoperative LR door ajar switch that never moves to the closed position.
    D. an open circuit from all the door ajar switches to the chassis ground.

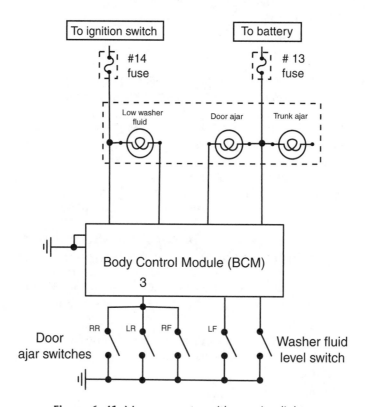

**Figure 6–41** Message center with warning lights.

**Hint**    *Some warning lights are operated by the body control module (BCM). The door ajar switches and the low washer fluid switch send an input signal to the BCM if a door is ajar or the washer reservoir is low. When one of these signals is received, the BCM grounds the appropriate bulb.*

## Task 7    Diagnose the cause of constant, intermittent, or no operation of audible warning devices.

59. The seat belt buzzer and the seat belt light operate continually with the ignition switch on and the driver's seat belt buckled in the seat belt and key buzzer system (Figure 6–42). The cause of this problem may be
    A. the timer contacts are stuck in the closed position.
    B. the circuit is shorted to ground at terminal 3 on the buzzer.

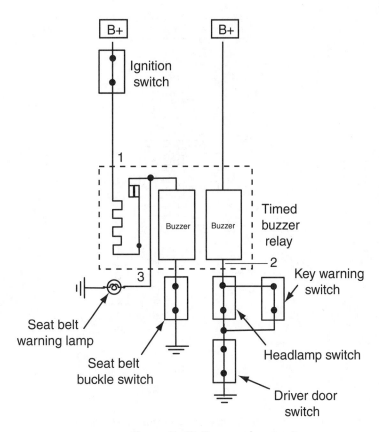

**Figure 6–42** Buzzer relay circuit.

C. the timer contacts and the seat belt switch are stuck closed.

D. the circuit is open at terminal 2 on the buzzer relay.

***Hint***     *Some buzzer relays contain the seat belt and the key buzzer. If the key is left in the ignition switch or the headlights are on, and the driver's door is opened, the circuit is completed from buzzer terminal 1 through the key warning switch, or headlight switch, and the driver door switch to ground. Under this condition, current flows through the circuit and the buzzer is activated.*

    *When the ignition switch is turned on, the current flows through the timer, seat belt buzzer, and seat belt buckle switch to ground. Under this condition, the buzzer is activated. Current also flows from the timer through the seat belt warning light to ground. When the driver's seat belt is buckled, the buzzer circuit is open and the buzzer is deactivated. The heater opens the timer contacts after eight seconds and the light goes out.*

**Task 8**   **Inspect, test, and repair or replace switches, relays, sensors, timers, electronic components, controllers, printed circuits, connectors, and wires of audible warning device circuits.**

60. The seat belt warning light and the seat belt buzzer in Figure 6–43 are inoperative.

    Technician A says the dedicated 10A fuse may be blown.

    Technician B says ground circuit number 7 may have an open circuit.

    Who is correct?

    A.  A only

    B.  B only

    C.  Both A and B

    D.  Neither A nor B

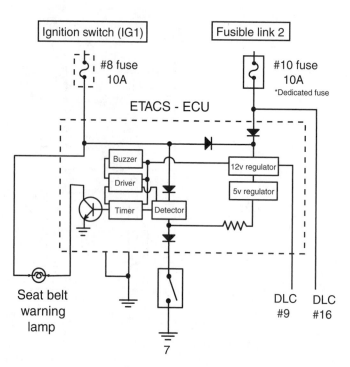

**Figure 6–43** Seat belt warning light and buzzer circuit.

**Hint**    *Many automotive electrical and electronic systems contain timers and controllers. For example, most electric-drive cooling fan circuits contain a timer to limit the fan run time with the engine shut off. Interior light circuits contain a timer to limit the on time of the interior lights if it is dark, the engine is shut off, and the doors are locked. Seat belt circuits contain a timer that limits the buzzer operation time if the driver's seat belt is not buckled. In most of these circuits, the seat belt warning light is still illuminated after the buzzer shuts off with the driver's seat belt unbuckled.*

# Horn and Wiper/Washer Diagnosis and Repair

## ASE Tasks, Questions, and Related Information

**Task 1**  **Diagnose the cause of constant, intermittent, or no operation of horn(s).**

61. In the horn circuit illustrated in Figure 6–44, the horn blows continuously.

    Technician A says the wire connected from the horn relay to the horn brush/slip ring may be shorted to ground.

    Technician B says the wire connected from the horn relay to the LH horn may be shorted to ground.

    Who is correct?

    A.  A only

    B.  B only

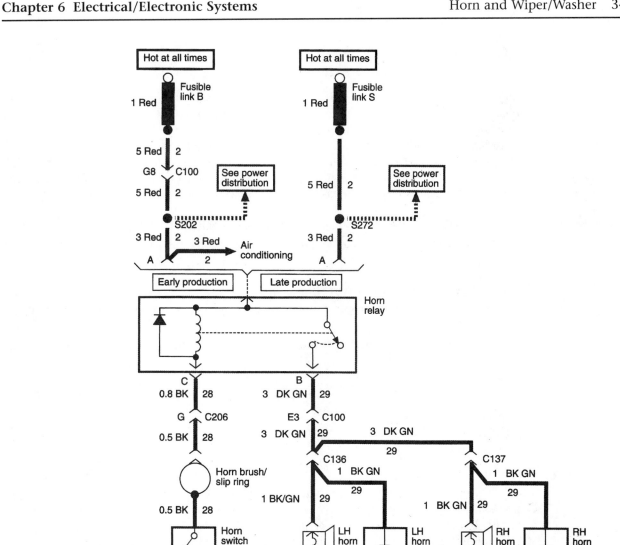

**Figure 6–44** Horn circuit.

C. Both A and B

D. Neither A nor B

**Hint**    *If the horns are inoperative, test the voltage input to the horn relay contacts and the horn relay winding. Both of these voltage inputs should be 12 V. If either of these inputs is 0 V, test the fusible link. An open circuit in the horn relay winding or in the horn relay contacts will also cause inoperative horns. The most likely cause of the horns blowing continuously is a short to ground in the wire from the horn relay winding to the horn brush/slip ring in the steering column.*

**Task 2**  **Inspect, test, and repair or replace horn(s), horn relay, horn button (switch), connectors, wires, and controllers of horn circuits.**

62. The horn circuit is inoperative (see Figure 6–44). All of the following may be the cause of the problem EXCEPT

A. an open ground circuit on the horn relay.

B. an open circuit in the horn relay winding.

C. an open circuit at the horn brush/slip ring.

D. an open fuse link in the relay power wire.

*Hint*    *Many horn circuits contain a relay. Voltage is supplied from the positive battery terminal through a fuse link to the relay winding and contacts. When the horn switch is closed on top of the steering column, the relay winding is grounded through the switch. This action closes the relay contacts, and voltage is supplied through these contacts to the horns.*

*When the horn switch is closed in some circuits, voltage is supplied through the horn switch to the horns. A relay is not used in these circuits. Many vehicles have a low pitch and a high pitch horn. Some horns have a pitch adjustment screw.*

## Task 3    Diagnose the cause of wiper problems including constant operation, intermittent operation, poor speed control, no parking, or no operation of wiper.

63. The wiper circuit has an open shunt coil.

    Technician A says this problem may cause the wiper motor to operate only at low speed.

    Technician B says this problem may cause the wiper motor to park with the wipers up on the windshield.

    Who is correct?

    A. A only

    B. B only

    C. Both A and B

    D. Neither A nor B

*Hint*    *Some wiper motors contain a series field coil, a shunt field coil, and a relay. When the wiper switch is turned on, the relay winding is grounded through one set of switch contacts. This action closes the relay contacts, and current is supplied through these contacts to the series field coil and armature. Under this condition, the wiper motor starts turning. If the wiper switch is in the high speed condition, the shunt coil is not grounded and the motor turns at high speed.*

*When the wiper switch is in the low-speed position, the shunt coil is grounded through the second set of wiper switch contacts. Under this condition, current flows through the shunt coil and the wiper switch to ground. Current flow through the shunt coil creates a strong magnetic field that induces more opposing voltage in the armature windings. This opposing voltage in the armature windings reduces current flow through the series coil and armature windings to slow the armature.*

*If the wiper motor fails to park, or parks in the wrong position, the parking switch or cam probably is the problem.*

*Some wiper motors have permanent magnets in place of the field coils. These motors have a low-speed and a high-speed brush. In some of these motors, the low-speed brush is directly opposite the common brush, and the high-speed brush is positioned in between these two brushes.*

## Task 4    Inspect, test, and replace intermittent (pulsing) wiper controls.

64. An intermittent wiper system does not operate on low speed or in the intermittent mode (Figure 6–45). The wipers operate normally at high speed.

    Technician A says there may be an open circuit between the low-speed relay winding and the wiper module.

    Technician B says there may be an open circuit at the ground connection on the wiper module.

    Who is correct?

    A. A only

    B. B only

    C. Both A and B

    D. Neither A nor B

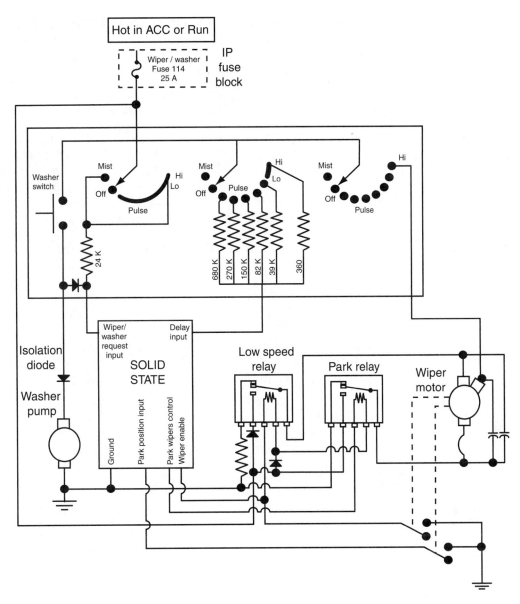

**Figure 6–45** Intermittent wiper system.

*Hint*   *Many vehicles are equipped with intermittent or interval wiper motor circuits. In most of these circuits, a driver rotates a control knob to adjust the wiper delay interval. A variable resistor in the intermittent wiper control provides a voltage input to the intermittent wiper module. This module operates the wiper motor to provide the proper delay interval. In some intermittent wiper controls, various resistors are connected in the circuit as the switch is rotated.*

*When the wiper switch is placed in the high-speed position, voltage is supplied through the high-speed switch contact to the high-speed brush in the wiper motor.*

*If the wiper switch is placed in the low-speed position, a voltage signal is sent from the wiper switch to the intermittent wiper module. When this signal is received, the module grounds the low-speed relay winding. This action closes the relay contacts and voltage is supplied through these contacts to the low-speed brush in the wiper motor.*

*If the wiper switch is placed in one of the intermittent positions, a unique voltage signal is sent to the intermittent wiper module. When this signal is received, the module opens and closes the ground circuit on the low-speed relay winding to provide the proper delay interval.*

**Task 5**  Inspect, test, and replace wiper motor, resistors, switches, relays, controllers, connections, and wires of wiper circuits.

65. The intermittent wiper motor circuit in Figure 6–45 has an open circuit in the 270 K resistor. This defect causes
   A. no high-speed wiper operation.
   B. no low-speed wiper operation.
   C. the wiper blades to park part way up on the windshield.
   D. failure of the wiper motor to operate in one of the intermittent positions.

**Hint**    *When diagnosing wipe motor circuits, the first step is to determine the exact problem with the wiper operation. The next step in the diagnostic procedure is to inspect the wiper system mechanical components and electrical circuits. Be sure the wiper motor fuse is not blown, and inspect all the wiper motor wiring connections, including the ground terminals, for looseness and corrosion. If no problems are located in the basic inspection, voltmeter, ohmmeter, and/or ammeter tests must be performed to locate the root cause of the problem.*

## Task 6    Diagnose the cause of constant, intermittent, or no operation of window washer.

66. When the windshield wipers are turned on in any mode, the windshield washer pump in Figure 6–45 operates continuously. If the wipers are turned off, the washer pump also stops operating.
   Technician A says the wire from the wiper/washer switch to the washer motor may be shorted to 12 V.
   Technician B says the diode between terminals H and E in the wiper/washer switch may be shorted.
   Who is correct?
   A. A only
   B. B only
   C. Both A and B
   D. Neither A nor B

**Hint**    *If the wipers and washer pump are inoperative, test the fuse that supplies voltage to the wiper/washer switch assembly. When the washers operate intermittently, test for an intermittent open circuit in the washer switch and the wire from the switch to the washer pump motor. Be sure the washer pump has a resistance-free ground connection. If the washer pump operates continually when the wipers are turned on in any mode, the diode connected between terminals H and E in the wiper switch may be shorted.*

## Task 7    Inspect, test, and repair or replace washer motor, pump assembly, relays, switches, connectors, and wires of washer circuits.

67. There is no operation from the windshield washer system (see Figure 6–45). The wiper motor operation is normal.
   Technician A says the wiper/washer fuse may have an open circuit.
   Technician B says the isolation diode may have an open circuit.
   Who is correct?
   A. A only
   B. B only
   C. Both A and B
   D. Neither A nor B

**Hint**    *Many windshield washer systems have an electric pump mounted in the bottom of the washer fluid reservoir. When the washer button is pressed, voltage is supplied through the switch to the washer motor. This motor operates a pump that forces washer fluid through the hoses to the nozzles in front of the windshield.*

# Accessories Diagnosis and Repair, Body

## ASE Tasks, Questions, and Related Information

**Task 1** Diagnose the cause of slow, intermittent, or no operation of power windows.

68. A power window operates normally from the master switch, but the window does not work using the window switch (Figure 6–46). The cause of this problem may be
   A. an open circuit between the ignition switch and the window switch.
   B. an open circuit in the window switch's movable contacts.
   C. an open circuit in the master switch's ground wire.
   D. a short to ground at the circuit breaker in the motor.

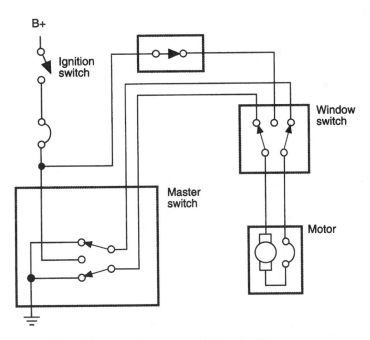

**Figure 6–46** Power window circuit.

**Hint**   *Power window circuits usually contain a master switch, individual window switches, and a window motor in each door. Some power window circuits have a window lockout switch to prevent operation of the window switches. When the master switch is placed in the down position, voltage is supplied from the center contact in this switch through the movable switch contact to the brush on the lower side of the commutator. The other brush is grounded through the master switch. Under this condition, the motor moves the window to the down position.*

*When the up position is selected in the master switch, current flow through the motor is reversed. Voltage is supplied from the ignition switch circuit breaker and lockout switch to the window switch. Pressing the window switch has the same effect as pressing the master switch.*

**Task 2** Inspect, test, and repair or replace regulators (linkages), switches, controllers, relays, motors, connectors, and wires of power window circuits.

69. There is no operation from the right rear power quarter window (Figure 6–47). The left rear power quarter window operates normally.

Technician A says there maybe an open circuit between the right rear power window switch and ground.

Technician B says the upper contacts in the right rear power window switch may be open.

Who is correct?

A. A only

B. B only

C. Both A and B

D. Neither A nor B.

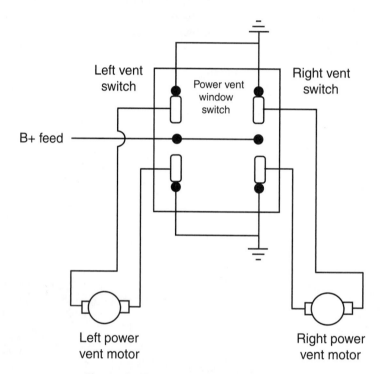

**Figure 6–47** Rear power window circuit.

*Hint*        *Some vans are equipped with power windows in the rear quarter panels. In these circuits, the left and right window switches share a common power supply and ground wire. When one of the switches is placed in the down position, current is supplied through the window motor in the proper direction to lower the window. Window motor current is reversed to raise the window if the switch is placed in the up position.*

**Task 3**  **Diagnose the cause of slow, intermittent, or no operation of power seat and driver memory controls.**

70. The power seat in Figure 6–48 is completely inoperative.

Technician A says the body ground may be open behind the left cowl trim panel.

Technician B says fuse number 7 may have an open circuit.

Who is correct?

A. A only

B. B only

C. Both A and B

D. Neither A nor B

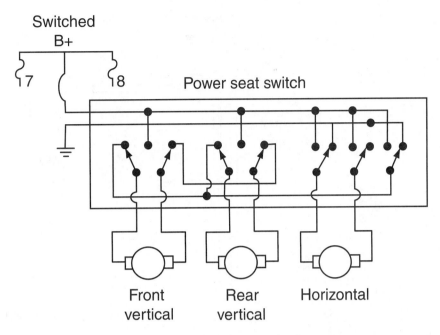

**Figure 6–48** Power seat circuit.

**Hint**    *If the power seat is inoperative in any position, test the fuse and wire that supplies voltage to the power seat switch. Be sure the ground connection from the switch to ground is satisfactory.*
*When the power seat only operates in some positions, test the wire from the switch to the motor and the inoperative motor. Also be sure there is nothing jammed between the inoperative seat motor and its track.*

**Task 4    Inspect, test, adjust, and repair or replace power seat gear box, cables, switches, controllers, sensors, relays, solenoids, motors, connectors, and wires of power seat circuits and driver memory controls.**

71. A six-way power seat moves vertically at the front and rear, but there is no horizontal seat movement (see Figure 6–48). All of the following may be the cause of the problem EXCEPT
   A. a newspaper jammed in the seat track mechanism.
   B. an open circuit between the switch and the horizontal motor.
   C. an open circuit in the circuit from the switch assembly to ground.
   D. burned contacts in the horizontal seat switch.

**Hint**    *A six-way power seat moves vertically at the front and rear, and horizontally forward and rearward. This type of seat has two vertical motors and a horizontal motor. These motors are connected through gear boxes and cables to the seat track mechanisms (Figure 6–49).*
*The front and rear switches have upward and downward positions, and the center switch has forward and rearward positions. When any of the switches are pressed, voltage is supplied to the appropriate motor in the proper direction. The motor moves the seat in the desired direction.*

**Task 5    Diagnose the cause of poor, intermittent, or no operation of rear window defogger.**

72. The rear defogger shown in Figure 6–50 is turned on continuously.
   Technician A says the rear defogger relay contacts may be stuck closed.
   Technician B says there may be an open circuit at ground connection G200.

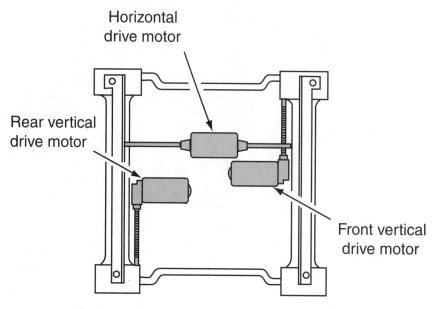

**Figure 6–49** Power seat motors and track mechanisms.

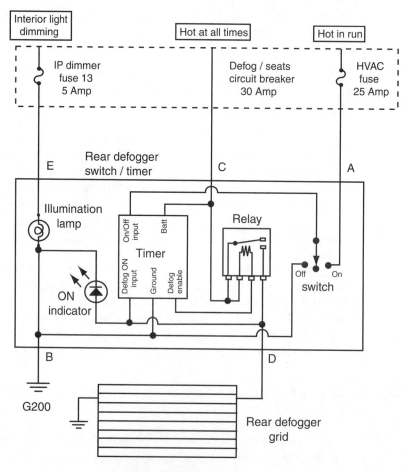

**Figure 6–50** Rear defogger circuit.

Who is correct?
A. A only
B. B only
C. Both A and B
D. Neither A nor B

**Hint**    *If the rear defogger circuit is inoperative, test the fuse and circuit breaker. Be sure the ground connections on the defogger and the switch timer are satisfactory. When the fuse and circuit breaker are satisfactory and 12 V are supplied to the switch timer at terminals A and C, test for voltage at switch timer terminal D with the switch in the on position. If there is no voltage available at terminal D, replace the switch timer assembly. When the defogger is on continuously, the relay, solid state timer, or switch may be the problem in the switch timer.*

## Task 6    Inspect, test, and repair or replace switches, relays, timers, controllers, window grid, connectors, and wires of rear window defogger circuits.

73. When the rear defogger switch is turned on, the rear defogger light is illuminated, but there is no defogger grid operation (see Figure 6–50). The cause of this problem could be
A. an open defogger relay winding.
B. an open circuit at the defogger relay contacts.
C. an open circuit between the switch/timer and the grid.
D. an inoperative defogger on/off switch.

**Hint**    *When the rear defogger switch is pressed, a signal is sent to the solid-state timer. When this signal is received, the timer grounds the relay winding. Under this condition, the relay contacts supply voltage to the defogger grid. When the relay is closed, current flows through the LED indicator to ground. After ten minutes, the timer opens the relay to shut off the grid current.*

*The grid tracks may be tested with a 12 V test light. A special compound is available to repair open circuits in the grid tracks.*

*Some cars have an electric fan motor to circulate air past the rear window for defogging action.*

## Task 7    Diagnose the cause of poor, intermittent, or no operation of electric door lock and hatch/trunk lock.

74. The power door locks in Figure 6–51 are completely inoperative, and the PWR ACC fuse 7 is satisfactory.
Technician A says there may be an open circuit between the PWR ACC fuse and junction S500.
Technician B says there may be an open circuit at ground connection G200.
Who is correct?
A. A only
B. B only
C. Both A and B
D. Neither A nor B

**Hint**    *If the power door locks are completely inoperative, test the fuse that supplies voltage to this circuit.*

*Be sure the ground connection on the power door lock switches is satisfactory. If the power door locks operate only when some of the switches are pressed, check the individual switches, wires, and power door lock motors.*

## Task 8    Inspect, test, and repair or replace switches, relays, controllers, actuators/solenoids, connectors, and wires of electric door lock/hatch/trunk circuits.

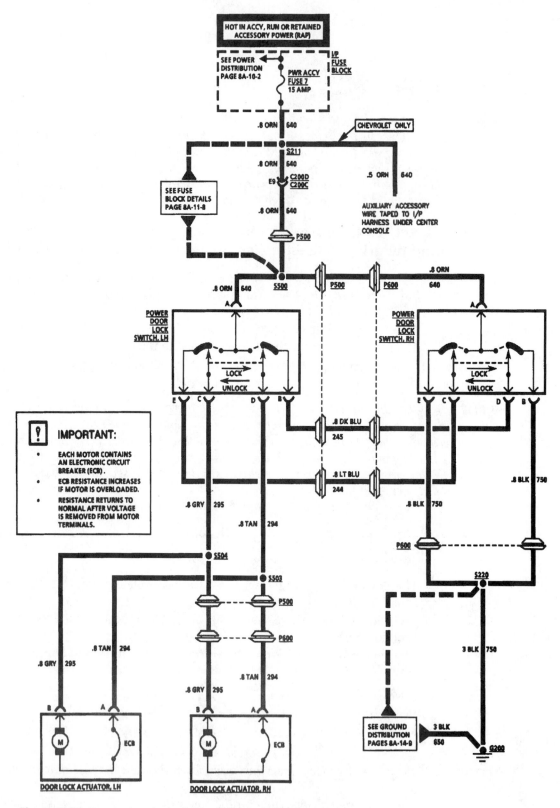

**Figure 6–51** Power door lock circuit. *(Courtesy of Pontiac Motor Division, General Motors Corporation)*

75. When the left-hand door lock is pressed, there is no operation from the power door locks (see Figure 6–51). The power door locks operate normally when the right-hand door lock is pressed.

Technician A says there may be an open circuit between junction S500 and the left-hand door lock switch.

Technician B says there may be an open circuit between terminal C and junction S504 on the left-hand door lock.

Who is correct?
A. A only
B. B only
C. Both A and B
D. Neither A nor B

**Hint**    *Most electric door lock circuits have small electric motors to operate the door locks. When either door lock switch is pushed to the lock position, voltage is supplied to all the door lock motors in the proper direction to provide lock action. If either door lock switch is pushed to the unlock position, voltage is supplied to all the door lock motors in the opposite direction to provide unlock action.*

## Task 9    Diagnose the cause of poor, intermittent, or no operation of keyless and remote lock/unlock devices.

76. A remote keyless entry system does not operate unless the person holding the remote control is right beside the vehicle.

Technician A says to replace the body control module (BCM).

Technician B says to replace the remote control door lock receiver.

Who is correct?
A. A only
B. B only
C. Both A and B
D. Neither A nor B

**Hint**    *In a remote keyless entry system, the remote control is a miniature transmitter that sends unique lock and unlock radio frequency signals to the remote control door lock receiver on the vehicle. The remote control door lock receiver is connected via a serial data wire to the BCM.*

*When the remote control transmits an unlock signal to the remote control door lock receiver, this receiver sends an unlock signal to the BCM. When this signal is received, the BCM operates the door locks to unlock the doors. Many remote controls will lock or unlock the doors when the person holding the remote control is within 100 ft. (30 m) or less of the vehicle. Many remote controls have a replaceable 3V battery. When the remote keyless entry system fails to operate, one of the first steps is to test the battery in the remote control.*

## Task 10    Inspect, test, and repair or replace components, connectors, controllers, and wires of keyless and remote lock/unlock device circuits; reprogram system.

77. There is no operation from the remote keyless entry system (Figure 6–52).

Technician A says the battery may be discharged in the remote transmitter.

Technician B says fuse number 4 may be blown in the fuse block.

Who is correct?
A. A only
B. B only
C. Both A and B
D. Neither A nor B

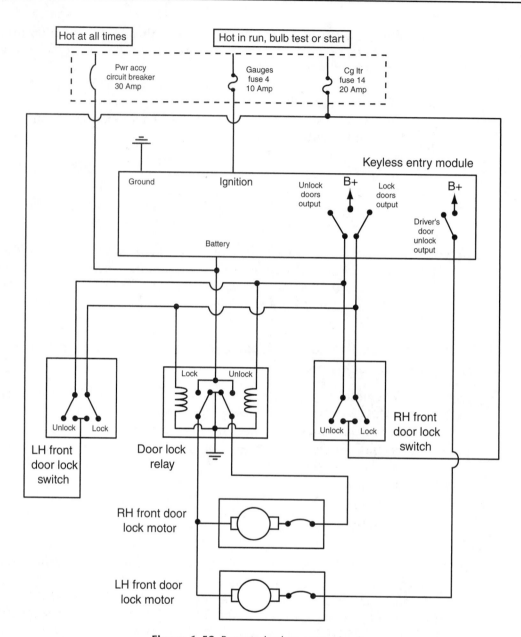

**Figure 6–52** Remote keyless entry circuit.

*Hint*    *The remote keyless entry module is connected to the power door lock circuit. A small remote transmitter sends lock and unlock signals to this module when the appropriate buttons are pressed on the remote control. When the handheld remote transmitter is a short distance from the truck, the module responds to the transmitter signals. When the unlock button is pressed on the remote transmitter, the module supplies voltage to the unlock relay winding to close these relay contacts and move the door lock motors to the unlock position.*

## Task 11  Diagnose the cause of slow, intermittent, or no operation of electrical sunroof and convertible/retractable top.

78. The sunroof opens normally, but does not close when the close switch is pressed (Figure 6–53). All of the following may be the cause of the problem EXCEPT
    A. an open circuit at the close relay contacts.
    B. an open close relay winding.
    C. an open circuit at the close switch.
    D. an open circuit at terminal C on the sunroof switch.

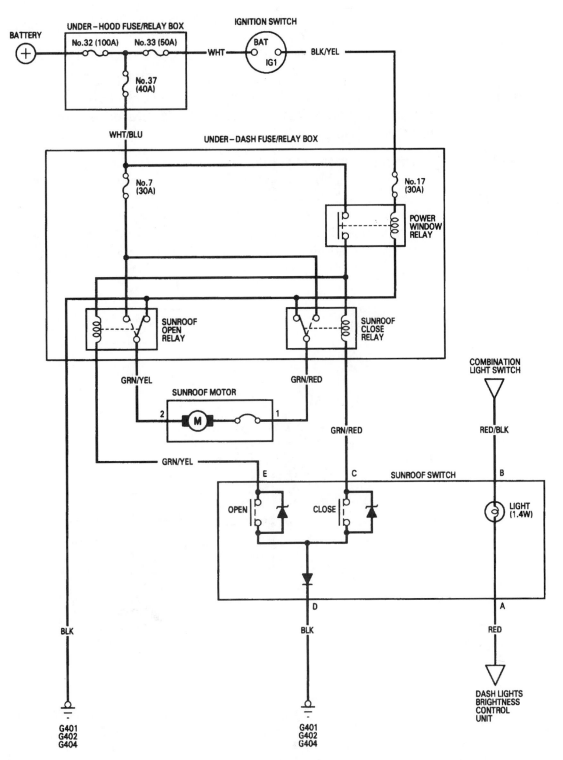

**Figure 6–53** Sunroof circuit. *(Courtesy of American Honda Motor Co., Inc.)*

*Hint*    *When the open sunroof switch is pressed, the open relay winding is grounded through the switch contacts. Under this condition, the relay contacts close and supply voltage to the sunroof motor brush. The other motor brush is connected through the close relay contacts to ground.*

*Current now flows through the motor and the motor opens the sunroof.*

*If the close button is pressed, the close relay winding is grounded through the close switch contacts.*

*Under this condition, the close relay contacts supply voltage to the sunroof motor in the opposite direction to close the sunroof.*

## Task 12   Inspect, test, and repair or replace motors, switches, controllers, relays, connectors, and wires of electrically operated sunroof and convertible/retractable top circuits.

79. The convertible top goes down normally, but there is no motor operation when the up button is pressed (Figure 6–54). The cause of this problem may be
   A. an open circuit breaker in the convertible top motor.
   B. an open ground wire on the convertible top switch.
   C. a jammed linkage mechanism on the convertible top.
   D. an open circuit at terminal C on the convertible top switch.

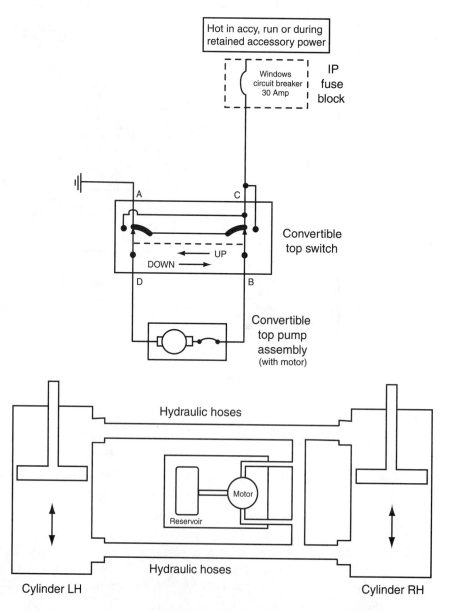

**Figure 6–54** Convertible top electric circuit and hydraulic system.

*Hint*     *The convertible top system contains a dual switch, pump motor, hydraulic cylinders, and linkages from these cylinders to the convertible top. When the down button is pressed, voltage is supplied through these switch contacts to a motor brush. The opposite motor brush is grounded through the up contacts. Under this condition, current flows through the motor, and the motor drives the pump. With this motor rotation, the pump supplies hydraulic pressure to the proper side of the cylinder pistons to move the top downward.*

*If the up button is pressed, motor and pump rotation are reversed and the pump supplies hydraulic pressure to the upward side of the cylinder pistons.*

**Task 13**   **Diagnose the cause of poor, intermittent, or no operation of electrically operated/heated mirror.**

80. There is no operation from the power mirror system (Figure 6–55).

    Technician A says there may be an open circuit in the mirror select switch.

    Technician B says there may be an open circuit between the mirror switch assembly and ground.

    Who is correct?

    A. A only

    B. B only

    C. Both A and B

    D. Neither A nor B

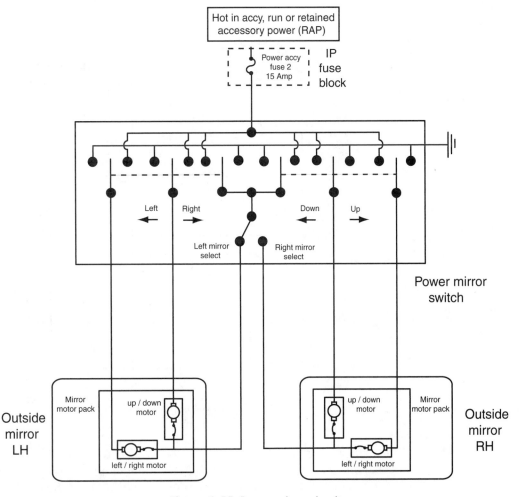

**Figure 6–55** Power mirror circuit.

***Hint***    *Voltage is supplied through a fuse to the power mirror switch assembly. When the mirror select switch is in the left position, it supplies voltage to the left mirror motor. When the left/right switch is pressed to the left position, a ground connection is completed from the right/left motor through the switch to ground. Under this condition, the motor moves the mirror to the left.*

*If the left/right switch is pressed to the right position, voltage is supplied through this switch to the right/left motor. The mirror select switch now provides a ground for the left-side mirror motor. This action reverses the current flow through the left-side mirror motor, and the mirror moves to the right.*

*Up/down mirror operation and operation of the right-side mirror is basically the same. The mirror select switch and the power mirror switches direct current flow through the mirror motors to supply the desired mirror movement.*

## Task 14    Inspect, test, and repair or replace motors, heated mirror grids, switches, controllers, relays, connectors, and wires of electrically operated/heated mirror circuits.

81. There is no operation from the heated mirror element, but the rear defogger operates normally (Figure 6–56). All of the following may be the cause of the problem EXCEPT

    A. an open circuit in the number 4 circuit breaker in the fuse block.

    B. a blown fuse number 1 in the relay center.

    C. an open circuit between the heated mirror element and ground.

    D. an open circuit between the timer relay and fuse number 1.

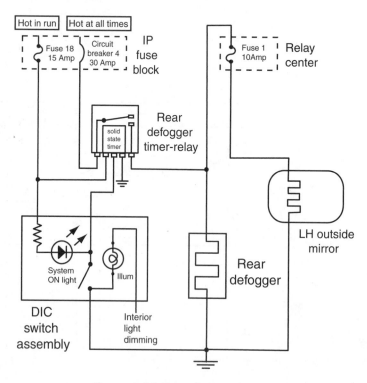

**Figure 6–56** Heated mirror circuit.

***Hint***    *Some vehicles have a heated driver's side mirror. When the rear defogger button is pressed, the timer relay supplies voltage to the rear defogger grid and also to the heated mirror element. After ten minutes, the timer relay shuts off the voltage supply to the defogger grid and the heated mirror element.*

# Accessories Diagnosis and Repair, Miscellaneous

## ASE Tasks, Questions, and Related Information

**Task 1**   Diagnose the cause of poor sound quality, noisy, erratic, intermittent, or no operation of the audio system, remove and reinstall audio system component (unit).

82. A radio has a whining noise that increases with engine speed. When the alternator field wire is disconnected, the noise stops. All of the following may be the cause of the problem EXCEPT
    A. an inoperative stator.
    B. an inoperative diode.
    C. an inoperative capacitor.
    D. an open field winding.

83. While discussing a radio static problem, Technician A says there may be a poor metal-to-metal connection between the hood and other body components.
    Technician B says the suppression coil may be inoperative on the instrument voltage limiter.
    Who is correct?
    A. A only
    B. B only
    C. Both A and B
    D. Neither A nor B

*Hint*   *Radio static may be caused by a malfunctioning alternator or spark plug wires. A damaged ground connection between the engine and the chassis or between chassis components may result in radio static. Defective radio suppression devices, such as a suppression coil on an instrument voltage limiter, may cause static on the radio. A damaged antenna with poor ground shielding may result in radio static. An open circuit in the antenna or lead-in wire may cause weak radio operation.*

*Intermittent radio operation may be caused by an intermittent open circuit in the antenna or lead-in wire. An intermittent open circuit in the voltage supply wire to the radio may cause intermittent radio operation. No radio operation may be caused by a blown fuse, or an open circuit in the voltage supply wire to the radio.*

*A radio noise locator may be made from an antenna lead-in wire. Cut the connector off the antenna end of the lead-in wire and remove a 2-in. length of the coax shield to expose the center conductor. Plug the other end of the lead-in wire into the radio, and place the exposed center conductor near any suspected noise sources. When noise is heard with the end of the lead-in wire near a component, that component is the source of the noise.*

*Some audio systems have self-diagnostic capabilities. When two of the control buttons, such as the number 3 and seek down buttons, are pressed simultaneously for three seconds, the audio system is placed in the diagnostic mode. In this mode, diagnostic trouble codes (DTCs) are shown in the audio system display.*

**Task 2**   Inspect, test, and repair or replace speakers, amplifiers, remote controls, antennas, leads, grounds, connectors, and wires of sound system circuits.

84. All of the following statements about radio antenna diagnosis with an ohmmeter are true EXCEPT
    A. Continuity should be present between the end of the antenna mast and the center pin on the lead-in wire.
    B. Continuity should be present between the ground shell of the lead-in wire and the antenna mounting hardware.
    C. No continuity should be present between the center pin on the lead-in wire and the ground shell.
    D. Continuity should be present between the end of the antenna mast and the antenna-mounting hardware.

**Hint**    *An antenna may be tested with an ohmmeter. Continuity should be present between the end of the antenna mast and the center pin on the lead-in wire. Continuity also should be present between the ground shell on the lead-in wire and the antenna-mounting hardware. No continuity should exist between the center pin on the lead-in wire and the ground shell.*

## Task 3  Inspect, test, and repair or replace switches, relays, motor, connectors, and wires of power antenna circuits.

85. The power antenna goes up when the radio is turned on, but the antenna does not go back down (Figure 6–57). The cause of this problem could be an open circuit in
    A. the radio fuse.
    B. the antenna fuse.
    C. the down limit switch.
    D. the antenna motor.

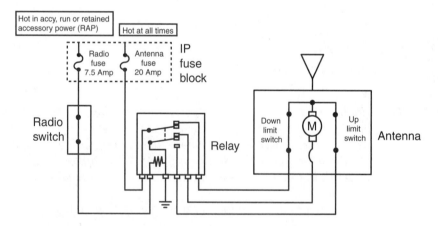

**Figure 6–57** Power antenna circuit.

**Hint**    *When the radio is turned on, voltage is supplied to the relay winding. This action moves the relay points to the up position, and current flows through the motor to move the antenna upward. When the antenna is fully extended, the up limit switch opens and stops current flow through the motor.*
    *If the radio is turned off, current flow through the relay coil stops. Under this condition the relay contacts move to the down position. This action reverses current flow through the motor and moves the antenna downward. When the antenna is fully retracted, the down limit switch opens and stops the antenna movement.*

## Task 4  Inspect, test, and replace noise suppression components.

86. The ohmmeter leads are connected from a radio suppression capacitor lead to the capacitor case. The ohmmeter reading is 65 ohms. This reading indicates the capacitor
   A. is satisfactory.
   B. has a high-resistance problem.
   C. has insulation leakage.
   D. has reduced plate capacity.

*Hint*     *An electric or electronic component with a varying magnetic field may cause radio static. A radio choke coil is connected to some components to reduce radio static (Figure 6–58). In some circuits, a radio suppression capacitor may be connected from the circuit to ground to reduce radio static. An ohmmeter may be connected from the capacitor lead to the case to check the capacitor for insulation leakage between the capacitor plates. When the ohmmeter is placed on the X1000 scale, the meter should provide an infinity reading. A capacitor tester may be used to test the capacitor for leakage, capacity, and resistance.*

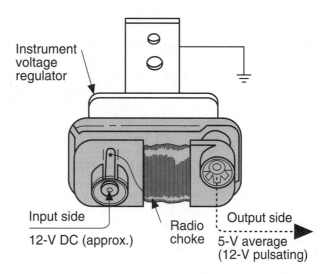

**Figure 6–58** Radio suppression choke.

## Task 5  Inspect, test, and repair or replace case, fuse, connectors, relays, and wires of cigar lighter/power outlet circuits.

87. The cigar lighter and the dome light are both inoperative (Figure 6–59).
   Technician A says there may be an open circuit at terminal B on the cigar lighter.
   Technician B says there may be an open circuit at ground connection G202.
   Who is correct?
   A. A only
   B. B only
   C. Both A and B
   D. Neither A nor B

*Hint*     *Voltage is supplied from the battery positive terminal through a fuse to one terminal on the cigar lighter. The other terminal on the cigar lighter is connected to ground. When the cigar lighter element is pushed inward, the circuit is completed through the lighter to ground. Current flow through the lighter heats the lighter element. When the element is hot, the lighter element moves outward and opens the circuit. On some vehicles, the lighter fuse also supplies voltage to the dome light and these components share a common ground.*

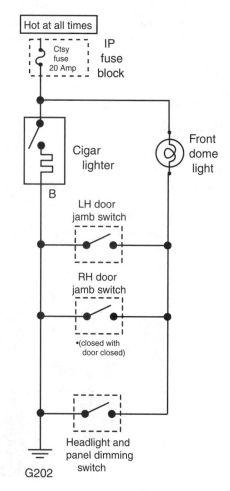

**Figure 6–59** Cigar lighter and dome light circuit.

## Task 6  Inspect, test, and repair or replace clock, connectors, and wires of clock circuits.

88. The clock is inoperative (Figure 6–60). All of the following may be the cause of the problem EXCEPT
    A. a grounded circuit on the SB wire at connector 51M.
    B. an open circuit at body ground B.
    C. an open circuit at terminal B in the clock connector.
    D. a grounded circuit on wire B/R at the clock connector.

*Hint*    *Voltage is supplied from the positive battery terminal through a 10A fuse to the clock. The other clock terminal is grounded. The illumination control contains a variable resistor to control illumination brilliance in the clock display.*

## Task 7  Diagnose the cause of unregulated, intermittent, or no operation of cruise control.

89. A cruise control with a vacuum servo and PCM operated solenoids provides erratic vehicle speed when it is engaged.
    Technician A says to check the vacuum hose from the intake manifold to the servo for leaks.

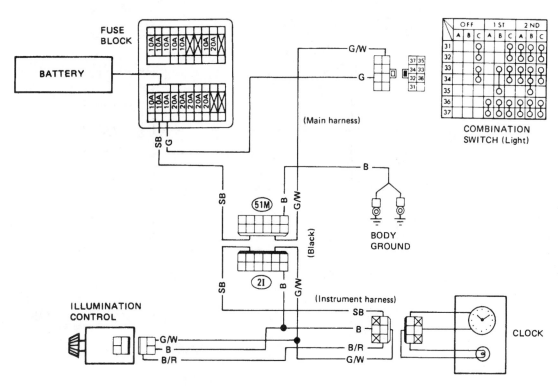

**Figure 6–60** Clock circuit. *(Used with permission from Nissan North America, Inc.)*

Technician B says to check the servo diaphragm for leaks.

Who is correct?

A. A only

B. B only

C. Both A and B

D. Neither A nor B

**Hint**     *If a vacuum-operated cruise control with a vacuum servo and PCM-operated solenoids does not operate or is erratic, one of the first diagnostic steps is to be sure the vacuum hose is connected to the servo. This vacuum hose should also be checked for leaks, kinks, and oil contamination.*

*The servo diaphragm should be checked to be sure it is not leaking. On many vehicles with this type of cruise control, a scan tool may be connected to the data link connector (DLC) to diagnose the cruise control solenoids, switches, and PCM inputs related to cruise control operation.*

**Task 8** **Inspect, test, adjust, and repair or replace regulator, servo, hoses, switches, relays, electronic controller, speed sensors, connectors, and wires of cruise control circuits.**

90. The cruise control is inoperative (Figure 6–61).

    Technician A says the vehicle speed sensor may be the problem.

    Technician B says the 20 amp gauge fuse may be the problem.

    Who is correct?

    A. A only

    B. B only

    C. Both A and B

    D. Neither A nor B

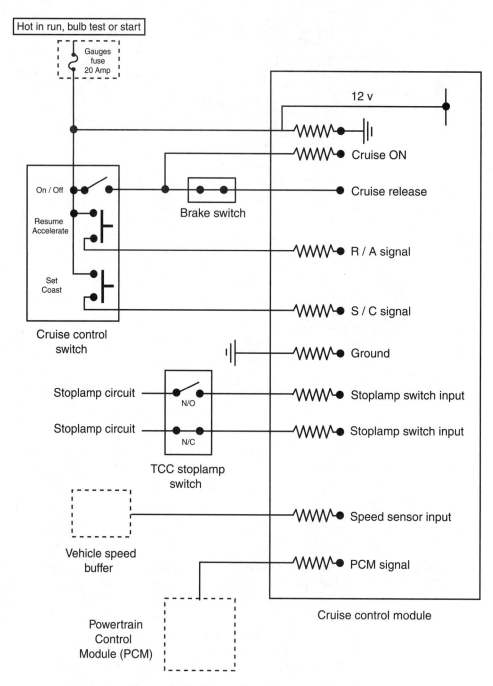

**Figure 6–61** Electronic cruise control circuit.

*Hint*    *Many vehicles have an electronic cruise control. In some of these systems, the control module and the stepper motor are combined in one unit. A cable is connected from the stepper motor to the throttle linkage. The control unit receives inputs from the cruise control switch, brake switch, and vehicle speed sensor (VSS). The control module sends output commands to the stepper motor to provide the desired throttle opening. An inoperative VSS may cause erratic or no cruise control operation. A cruise control cable adjustment is required on these systems. Remove the cruise control cable from the throttle linkage. With the throttle closed and the cable pulled all the way outward, install the cable on the throttle linkage. Turn the adjuster screw on the cruise control cable to obtain 0.197-in. lash in the cable.*

*Some cruise control systems have the control module mounted in the powertrain control module (PCM). The control module is connected to an external servo. This servo contains a vacuum diaphragm that is connected by a cable to the throttle linkage. The servo also contains a vent solenoid and a vacuum solenoid (Figure 6–62). The control module receives the same inputs described in the previous paragraph. In response to these inputs, the control module operates the vent and vacuum solenoids to supply the proper vacuum to the servo diaphragm. Because the servo diaphragm is connected to the throttle, the vacuum supplied to this diaphragm provides the desired throttle opening. In these systems, a leak in the servo diaphragm may cause erratic cruise control operation, or a gradual reduction in the cruise set speed.*

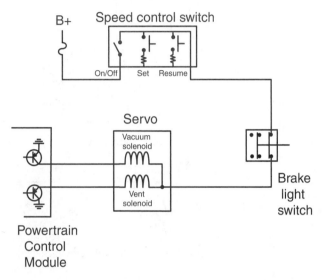

**Figure 6–62** Electronic cruise control with external servo.

## Task 9  Diagnose the cause of false, intermittent, or no operation of anti-theft system.

91. There is no action from the starting motor in a vehicle with a Pass-Key II vehicle theft deterrent (VTD) system (Figure 6–63). The battery is fully charged and the battery cables and connections are satisfactory.

    Technician A says the resistor sensing contacts in the ignition switch may be damaged.

    Technician B says there may be an open circuit in the wire from the BCM to the starter relay winding.

    Who is correct?

    A. A only

    B. B only

    C. Both A and B

    D. Neither A nor B

**Hint**    *Some vehicles have a Pass-Key II VTD system. This system has a resistor pellet built into the ignition key. When the proper key is inserted into the ignition switch, the body control module (BCM) receives a specific resistor pellet code. When this code is received, the BCM sends a 40–60Hz pulse width modulated (PWM) voltage signal to the PCM, and provides a ground for the starter relay winding, which allows starter operation. When the PCM receives the proper signal from the BCM, the PCM begins operating the injectors. When the proper specific resistor pellet code is not received from the ignition key, the BCM disables the starter relay circuit, and the PCM does not allow injector operation. A scan tool connected to the DLC displays Pass-Key II status under "VTD Fuel Disabled." If the Pass-Key II system is operating properly, the scan tool displays "Active." When the Pass-Key II system disables the starter relay and the injectors, the scan tool displays "Inactive."*

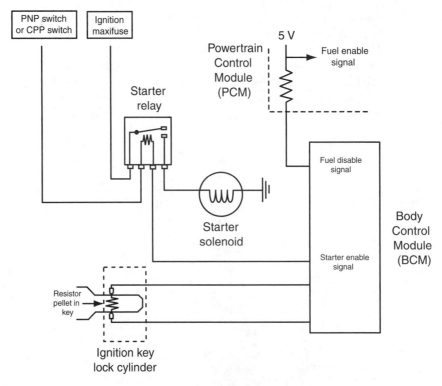

**Figure 6–63** Pass-Key II vehicle theft deterrent (VTD) system.

## Task 10    Inspect, test, and repair or replace components, controllers, switches, relays, connectors, sensors, and wires of anti-theft system circuits.

92. All of the following statements about an anti-theft system are true EXCEPT (Figure 6–64)
    A. The system is triggered by unauthorized operation of the doors, hood, trunk lock, or ignition.
    B. When triggered, the system sounds the horn for three minutes and flashes the park lights for eighteen minutes.
    C. When triggered, the system prevents engine starting by disabling the injectors.
    D. When triggered, the system may be disarmed by disconnecting the battery.

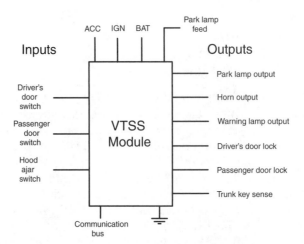

**Figure 6–64** Vehicle anti-theft system.

*Hint*    *The driver arms the anti-theft system by turning off the ignition switch, opening a door, locking the doors using the power door lock switch, and closing the door. These actions do not have to be sequential. After the arming actions are completed, the security alarm light in the instrument panel flashes for fifteen seconds. When this light stops flashing, the system is armed.*

*The system may be triggered by unauthorized operation of doors, hood, trunk lock, or ignition. When the system is triggered, the horn sounds and the park lights flash for three minutes. After this time period if the system is not disarmed, the horn stops blowing, but the park lights continue flashing for an additional fifteen minutes. The system is disarmed by unlocking either front door with the key. Disconnecting the battery does not disarm the system. Once the battery is reconnected, the alarm action continues.*

## Task 11    Diagnose the cause(s) of the supplemental restraint/air bag warning light staying on or flashing.

93. An air bag warning light is illuminated intermittently with the engine running.
    Technician A says the air bag system has an electrical problem.
    Technician B says this problem may cause the air bag to inflate accidentally.
    Who is correct?
    A. A only
    B. B only
    C. Both A and B
    D. Neither A nor B

*Hint*    *In many air bag systems, the warning light is illuminated for five or six seconds after the engine starts. After this time, the air bag warning light should remain off while the engine is running.*

*If the air bag warning light is illuminated with the engine running, a fault is present in the air bag system. In some air bag systems, the air bag warning light begins flashing a diagnostic trouble code (DTC) if an electrical problem occurs in the system. In other systems, the air bag warning light is illuminated continually if an electrical problem occurs within the system. When a scan tester is connected to the air bag diagnostic connector, or data link connector (DLC), the DTCs stored in the air bag module memory are indicated on the scan tester.*

## Task 12    Disarm and enable the air bag system for vehicle service following manufacturers' recommended procedures.

94. While discussing air bag system diagnosis and service, Technician A says if the air bag system is not disarmed, an accidental air bag deployment may occur.
    Technician B says using air bag circuit diagnostic equipment that is not recommended by the vehicle manufacturer may cause an accidental air bag deployment.
    Who is correct?
    A. A only
    B. B only
    C. Both A and B
    D. Neither A nor B

*Hint*    *On many vehicles, the air bag system is disarmed by turning off the ignition switch and disconnecting the negative battery cable. After this cable is disconnected, wait for the time period specified by the vehicle manufacturer before working on the vehicle. This time period is usually two to ten minutes. On some later model vehicles with driver's side, passenger's side, and side air bags, the air bag system is disarmed by performing the following procedure:*
*1. Place the front wheels in the straight-ahead position.*
*2. Turn off the ignition switch and remove the key from the switch.*
*3. Remove the air bag system fuse from the fuse block.*
*4. Disconnect the two-wire connectors from the driver's side air bag connector at the steering column base, passenger's side air bag inflator module, driver's side impact inflator*

*module under the driver's seat, and the passenger's side air bag inflator module under the passenger's seat.*

*After the service is completed on the vehicle, connect all the disconnected connectors and install the air bag system fuse. Turn on the ignition switch and observe the air bag system warning light for proper operation.*

## Task 13  Inspect, test, repair or replace the air bag(s), controller, sensors, connectors, and wires of the air bag system circuit(s).

95. All of the following statements about air bag system service are true EXCEPT (Figure 6–65)

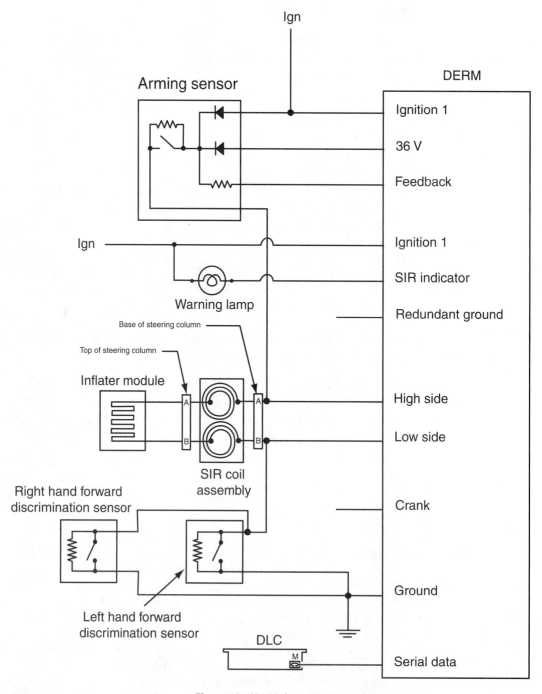

**Figure 6–65** Air bag circuit.

A. The negative battery cable should be disconnected and the manufacturer's recommended waiting period completed.

B. Safety glasses and gloves should be worn when handling deployed air bags.

C. A 12V test light may be used to test continuity between the inflator module and the sensors.

D. Sensor operation may be affected if the sensor brackets are bent or twisted.

**Hint**      *When servicing an air bag system, always disconnect the negative battery cable and wait for the time period specified by the vehicle manufacturer. This time is usually one to two minutes. Never use a 12 V test light to diagnose an air bag system. Diagnose these systems with an ohmmeter, voltmeter, or the manufacturer's recommended equipment. Since deployed air bags may contain small quantities of sodium hydroxide, wear safety glasses and gloves when handling these components.*

*Sensors must always be mounted in their original direction. Most sensors have a directional arrow that must face toward the front of the vehicle. Always store inflator modules face upward on the bench, and carry these components with the trim cover facing away from your body.*

# Post Test

1. An ohmmeter is connected to the relay winding terminals in a horn relay. The ohmmeter indicates the relay winding has 200 ohms resistance and the specified resistance is 250 ohms.

    Technician A says some of the turns of wire in the winding are shorted together.

    Technician B says the current flow through the relay winding will be lower than normal.

    Who is correct?

    A. A only

    B. B only

    C. Both A and B

    D. Neither A nor B

2. When testing battery drain with an ammeter and a tester switch

    A. read the ammeter immediately when the ignition switch is turned off and the tester switch is opened.

    B. the tester switch is connected parallel to the battery negative cable.

    C. The ammeter is connected parallel to the tester switch contacts.

    D. The drain is excessive if the ammeter reading exceeds 15 milliamperes with the tester switch open.

3. When performing a battery capacity (load) test, Technician A says a conclusive load test may be performed if the battery specific gravity is 1150.

    Technician B says after the fifteen-second load test, read the voltage before turning off the load.

    Who is correct?

    A. A only

    B. B only

    C. Both A and B

    D. Neither A nor B

4. A starting motor on a vehicle has low cranking speed. The starter draw is lower than specified, and the cranking voltage is high.

    Technician A says there may be excessive resistance in the starting circuit cables.

Technician B says there may be a short circuit in the starting motor armature.

Who is correct?

A. A only

B. B only

C. Both A and B

D. Neither A nor B

5. The battery is undercharged in a vehicle that is driven regularly. The charge indicator light is off with the engine running. The most likely cause of this problem is

A. a defective voltage regulator allowing high charging circuit voltage.

B. an open field circuit in the alternator.

C. a shorted diode in the alternator.

D. worn out brushes in the alternator.

6. A charging circuit in a vehicle experiences repeated voltage regulator failures. The most likely cause of this problem is

A. an open alternator diode.

B. a shorted alternator stator.

C. high resistance in the alternator battery wire.

D. a shorted alternator field coil.

7. The headlights on a vehicle have normal brilliance with the engine idling. When the engine is accelerated, the headlights become considerably brighter. The most likely cause of this problem is

A. high charging circuit voltage

B. high resistance in the headlight circuit.

C. an open stator winding in the alternator.

D. high resistance in the alternator battery wire.

8. When aligning headlights with a headlight screen, Technician A says to position the vehicle 35 ft. from the headlight screen on the wall.

Technician B says some vehicle manufacturers recommend aligning the headlights on low beam.

Who is correct?

A. A only

B. B only

C. Both A and B

D. Neither A nor B

9. A vehicle experiences repeated blowing of the taillight fuse. This problem could be caused by

A. a grounded stop light wire at the back of the vehicle.

B. an intermittent open circuit in a tail light wire.

C. an intermittent ground in a taillight wire.

D. a loose taillight wiring connector.

10. During a right turn, the right signal lights are flashing and the brakes are applied during the turn.

Technician A says that when the brakes are applied, the left-rear stop and signal light should not be illuminated.

Technician B says the right-rear stop and signal light should be illuminated continually.

Who is correct?

A. A only

B. B only

C. Both A and B

D. Neither A nor B

11. The stop lights on a vehicle are completely inoperative but the signal lights operate normally. All of the following defects could be the cause of the problem EXCEPT.

A. an open circuit at one of the combination stop and signal light bulbs.

B. a blown stoplight fuse.

C. an open circuit between the stoplight fuse and the stoplight switch.

D. an open circuit in the stoplight switch.

12. On a vehicle with an instrument voltage limiter and thermal electric gauges, the fuel gauge reads full regardless of the amount of fuel in the tank. All other gauges provide normal readings. The cause of this problem may be

A. high resistance in the fuel gauge sending unit ground.

B. high resistance between the instrument voltage limiter and the instrument panel.

C. an open circuit in the wire from the fuel gauge to the sending unit.

D. a short to ground in the wire from the fuel gauge to the sending unit.

13. In an instrument panel with thermal electric gauges and an instrument voltage limiter, all the gauges are erratic.

Technician A says the charging circuit may be defective.

Technician B says the instrument voltage limiter may be defective.

Who is correct?

A. A only

B. B only

C. Both A and B

D. Neither A nor B

14. When diagnosing two-speed wiper motors that contain a shunt coil

A. if the shunt coil is energized, the wiper motor operates at high speed.

B. if the wiper blades park in the wrong position, the park switch may be defective.

C. the shunt coil is energized continually in the low-speed and high-speed modes.

D. the wiper switch grounds the series field coils to turn on the wipers.

15. When diagnosing intermittent wiper motor circuits, Technician A says that in many intermittent wiper motor circuits the module opens and closes the low-speed relay winding to provide intermittent wiper operation.

Technician B says that in many intermittent wiper motor circuits various resistors in the wiper switch provide unique input voltage signals to the module.

Who is correct?

A. A only

B. B only

C. Both A and B

D. Neither A nor B

16. When diagnosing a six-way power seat circuit

A. the seat should move vertically at the front and rear.

B. the seat has two horizontal motors.

C. the seat has one vertical motor.

D. the seat has a fusible link in the electrical circuit.

17. All of the following statements about rear defogger circuits are true EXCEPT

A. Many rear defogger circuits are controlled by a timer.

B. When the defogger is on, 9.5V are supplied to the grid.

C. Each track in the grid is parallel to the other tracks.

D. In many rear defogger circuits, the timer energizes the rear defogger relay.

18. While discussing remote keyless entry systems, Technician A says most remote controls are a remote transmitter.
    Technician B says most remote controls contain a replaceable battery.
    Who is correct?
    A. A only
    B. B only
    C. Both A and B
    D. Neither A nor B

19. A radio static problem may be caused by
    A. an open circuit in the alternator field coil.
    B. high constant charging voltage.
    C. an open circuit in the distributor pickup coil.
    D. defective spark plug wires.

20. An air bag system warning light is illuminated continually when the engine is running. Under this condition
    A. there may be a defect in one of the air bag system input sensors.
    B. the air bag will deploy normally if the vehicle is in a collision.
    C. the flashes of the malfunction indicator light (MIL) denote the location of the defect.
    D. there may be a defect in the transmission computer.

# Answers and Analysis

**1.  A**   The test light is illuminated when connected to the battery side of the open circuit, but it is not on if it is connected to the motor side of the open circuit. With an open circuit there is no current flow through the circuit, and the voltage drop across the motor is 0V. Thus, A is right.

**2.  D**   The light bulb is the only resistance in the circuit and nearly all the battery voltage is dropped across this bulb. If the voltage drop across the bulb is 9V, there is high resistance in the switch or wires. This resistance causes some voltage drop across the switch or wires and reduces the voltage drop across the bulb. Therefore, both A and B are wrong and D is the correct answer.

**3.  C**   A shorted condition causes lower resistance and increased current flow. Therefore, A, B, and D are wrong, and C is the right answer.

**4.  D**   An ohmmeter should never be connected to a circuit in which current is flowing. Therefore, Technician A is wrong. When testing a spark plug wire with 20,000 Ω, the X1000 meter scale should be used, so Technician B is also wrong, making D the correct response.

**5.  C**   A defective TPS is illustrated in the waveform in Figure 6–12. Thus, Technician A is right. The defect in Figure 6–12 may cause a hesitation during low speed acceleration, so Technician B is right. Because both Technicians A and B are right, C is the correct answer.

**6.  A**   The rough idle problem may be caused by an erratic grams per second air flow reading through the MAF sensor caused by a contaminated or defective sensor, so Technician A is right. The PCM has to receive a signal from the vehicle speed sensor (VSS) indicating a specific vehicle speed before it can command the torque converter clutch on. Thus, Technician B is wrong, making A the correct answer.

**7.  B**   If the bulb is illuminated when a jumper wire is connected from the ground side of the light bulb to ground, there is an open circuit in the light bulb ground. Therefore, A, C, and D are wrong, and B is correct.

**8.** **A**   When the 12V test light remains illuminated with the connector disconnected at the light bulb, the circuit must be shorted to ground between the fuse and the disconnected connector. Therefore, B is wrong and A is right.

**9.** **A**   Normal battery drain on some vehicles is 50 milliamperes. Therefore, B is wrong. Because some computers require a time period to enter the sleep mode, the drain should not be recorded immediately when the switch is opened. Thus, C is wrong. The doors should be closed and all accessories turned off during the drain test, so D is wrong. The tester switch must be closed when starting or running the engine to carry the starter or alternator current. Therefore, A is the correct response.

**10.** **D**   If the circuit breaker is satisfactory, the ohmmeter reading should be 0 ohm. Therefore, Technician A is wrong. The ohmmeter current is not high enough to open the circuit breaker, so Technician B also is wrong, making D the right answer.

**11.** **C**   Component D1 in Figure 6–15 is a Zener diode. Therefore, A, B, and D are wrong, and C is correct.

**12.** **C**   At 0°F battery temperature, 0.032 point should be subtracted from the hydrometer reading, so A is wrong. If the battery temperature is 120°F, 0.016 point should be added to the hydrometer reading, so B also is wrong. The battery is fully charged when the hydrometer reading is 1.265; thus, D is wrong. The maximum variation in battery cells is 0.050 specific gravity points. The correct answer is C.

**13.** **B**   The battery discharge rate is one-half the cold cranking rating, so A is wrong. The battery is satisfactory if the voltage remains above 9.6V at 70°F. Therefore, B is correct.

**14.** **C**   Disconnecting the battery from the electrical system erases the computer adaptive memories. Therefore, the correct answer is C.

**15.** **A**   If a built-in hydrometer indicates yellow, the battery electrolyte level is low. This problem may be caused by high charging voltage resulting from a defective voltage regulator. A loose alternator belt causes an undercharged battery. Therefore, B is wrong and A is correct.

**16.** **B**   The battery charging time varies for batteries with different capacities, so A is wrong. A high charging rate should not be used on a cold battery. The charging rate should be reduced when the battery reaches 1.225 specific gravity. Therefore, C and D are wrong. During the charging process, the battery temperature should not exceed 125°F. Therefore, B is the correct answer.

**17.** **C**   When servicing the battery and electrical system in an air bag–equipped vehicle, the negative battery cable should be disconnected first, the positive battery cable should be reconnected first, and a 12V source should be connected to the cigarette lighter socket before disconnecting the battery. Therefore, answers A, B, and D are true. After the battery cables are disconnected, the technician must wait the specified time period before servicing the electrical system. Therefore, C is not true, making it the correct answer.

**18.** **B**   When using a booster battery, the accessories should be off on both vehicles. Therefore, Technician A is wrong. The negative booster cable should be connected to an engine ground on the vehicle being boosted. Thus, Technician B is right, making answer B the correct response.

**19.** **A**   If the field windings or positive battery cable have high resistance, starter current draw and cranking speed are reduced. Therefore, B and C are wrong. A burned solenoid disc and terminals may cause a clicking action from the starter solenoid, so D is wrong. Worn starter bushings may cause high current draw and reduced cranking speed. Thus, A is the correct answer.

**20.** **C**   In Figure 6–17, the voltmeter is connected to test the voltage drop in the starter ground circuit. Therefore, A, B, and D are wrong, and C is correct.

**21.** **B**   A grounded circuit at relay terminal 85, or a continually closed neutral safety switch, can cause the starting motor to operate in any gear selector position. Therefore, A and C are wrong.

A slight resistance at relay terminal 86 may reduce the voltage at the solenoid windings, resulting in a clicking action from the solenoid. Therefore, D is wrong.

An open circuit at terminal 86 on the starter relay stops the current flow through the relay winding. This problem causes a completely inoperative starter relay and solenoid. Therefore, B is correct.

**22.  A**    In Figure 6–19 the ohmmeter is connected to test the pull-in winding. A satisfactory winding provides a low ohmmeter reading. Therefore, B is wrong and A is correct.

**23.  C**    A weak or damaged solenoid plunger return spring may cause the starter to run with the engine for a few seconds, or longer when the engine starts. A misaligned starting motor may also cause this problem. Therefore, Technicians A and B are both correct, and C is the correct answer.

**24.  B**    Crankshaft bearings that are beginning to seize may cause slow engine cranking, but this problem should not cause the engine to kick back against the starter. Therefore, Technician A is wrong. Excessively advanced ignition timing may cause intermittent slow cranking, and kicking back against the starting motor. Thus, Technician B is right, making B the correct answer.

**25.  C**    An open fuse link between the alternator battery wire causes zero alternator output.

Therefore, B is correct. If the field circuit is open, there is no magnetic field around the rotor. Therefore, alternator output is zero, and both Technicians A and B are correct. Therefore, C is the correct answer.

**26.  A**    If the alternator belt is bottomed in the pulley, the belt may slip and reduce the alternator output. A misaligned belt may cause rapid belt wear, but the belt should not slip. Therefore, Technician B is wrong and Technician A is correct, making A the right answer.

**27.  D**    When performing an alternator output test, the vehicle accessories should be off and the voltage should be limited to 15V. Therefore, both Technicians A and B are wrong, and D is the right answer.

**28.  A**    A broken brush may cause zero field current and alternator output, so B is wrong. When the alternator is full-fielded, the voltage regulator is bypassed. Therefore, an inoperative regulator does not reduce output and C is wrong. Because an inoperative capacitor usually does not affect output, D is wrong. A shorted diode reduces alternator output, thus A is correct.

**29.  D**    This question asks for the statement that is not a result of the problem. A high charging voltage may cause an overcharged battery, burned out electrical components, or electrolyte gassing from the battery. Therefore, A, B, and C are a result of the problem and none of these responses is right.

A high charging voltage does not reduce headlight brilliance, so D is not a result of the problem, and this is the requested answer.

**30.  A**    In Figure 6–20, the voltmeter is connected to measure the voltage drop from the alternator battery wire to the positive battery cable. A 2V drop in this circuit indicates excessive resistance.

This resistance reduces charging current, and causes an undercharged battery. When this problem is present, the headlights may be dimmer than normal. Therefore, Technician B is wrong and Technician A is correct, making answer A the right choice.

**31.  D**    A replacement fuse link must be four gauge sizes smaller than the wire it is protecting.

Therefore, A, B, and C are wrong, and D is correct.

**32.  C**    When the engine is running, the charge indicator light remains off because there is equal voltage supplied to each side of the bulb. A glowing charge indicator bulb may be caused by unequal voltage on each side of the bulb.

High resistance in the alternator-to-battery wire reduces the voltage on the ignition switch side of the charge indicator bulb, while full voltage is supplied to the L terminal side of the bulb. High resistance in the wire from the L terminal to the bulb reduces the voltage on the L terminal side of the bulb. Therefore, both Technicians A and B are correct, making C the right answer.

**33.  C**   Excessive alternator belt tension may cause premature alternator bearing failure, and worn alternator mounting bolt openings may cause alternator misalignment and rapid belt wear. Therefore, Technicians A and B are both right, making C the correct answer.

**34.  A**   A short to ground may cause excessive current flow in the light circuit. This excessive current flow may cause the circuit breaker to open the circuit and shut off the lights. If the charging system voltage is high enough to open the headlight circuit breaker, many other problems would occur such as damaged electronic components and an overcharged battery. Therefore, Technician B is wrong and Technician A is correct, making A the right answer.

**35.  B**   The daytime running light system supplies voltage to the low-beam headlights, so Technician A is wrong. The daytime running lights may be operated by a module that supplies a pulsating voltage to the headlights. Thus, Technician B is right, making B the correct answer.

**36.  C**   This question asks for the statement that is not true. When replacing halogen bulbs, handle the bulb by the base, do not scratch or drop the bulb, and avoid moisture contact with the bulb. Therefore, statements A, B, and D are true, so they are not the requested answer. The bulb should be changed with the headlights off, so statement C is not true, making it the requested answer.

**37.  C**   Because fuse number 12 is connected only to the instrument panel lights, a blown fuse causes these lights to be inoperative. Therefore, A, B, and D are wrong, and C is right.

**38.  B**   A high-resistance problem at the light switch causes all the parking lights to be dim. Therefore, Technician A is wrong. A high resistance in the right-front parking light ground causes only the right-front parking light to be dim. Thus, Technician B is correct, making B the right answer.

**39.  B**   Because the right-side and left-side ground connections are connected together, an open circuit at one ground connection does not affect rear light operation. Therefore, A, C, and D are wrong, and B is right.

**40.  A**   A loose connection at any instrument panel bulb only causes that bulb to go dim intermittently. Therefore, Technician B is wrong. An intermittent high resistance in the instrument panel ground causes all the instrument panel bulbs to go dim intermittently, so Technician A is correct, making A the right answer.

**41.  A**   An open circuit in the rheostat causes all the instrument panel bulbs to be inoperative. Because these bulbs are connected in parallel, an open circuit in one bulb does not affect the other bulbs. Therefore, Technician B is wrong, making A the correct response.

**42.  A**   If the courtesy light fuse was blown, these lights would be inoperative. A malfunctioning door jamb switch on the driver's side may intermittently ground the courtesy light circuit with the door closed, and this action turns on the courtesy lights. Thus, Technician A is correct, making A the right answer.

**43.  A**   A short to ground at terminal S 363 does not affect the underhood light, so C is wrong. Because current still flows through the courtesy lights and the short to ground, this problem does not increase current flow and blow a fuse. Therefore, D is wrong. This short to ground causes continual operation of the courtesy lights; thus, B is wrong and A is correct.

**44.  D**   A short to ground on the ground side of one stoplight bulb has no effect, because that side of the bulb is normally connected to ground. An intermittent open circuit in the stoplight circuit causes the stoplights to be intermittently inoperative. Therefore, Technicians A and B are both wrong, making D the correct answer.

**45.  A**   If the cigar lighter fuse is blown, the courtesy and dome lights would glow dimly when the cigar lighter is pushed in because the cigar lighter is now in series with these lights. Therefore, B, C, and D are wrong, and A is right.

**46.  B**   Wire D2 18RD is connected to the hazard light flasher. Therefore, high resistance in this wire would affect all the signal lights. Therefore, Technician A is wrong. Because wire DB 180G RD is connected to the left rear signal light, high resistance in this wire causes this signal light bulb to be dim compared to the right signal light. Thus, Technician B is correct, making B the right answer.

**47.  C**    The D7 18BR RD wire is connected from the signal light switch to the right rear signal light. When this is the only dim light, the problem is in this wire. Therefore, A, B, and D are wrong, and C is right.

**48.  C**    An inoperative backup light switch may cause the backup lights to be illuminated continually. If the wire from the backup light switch to the backup lights is shorted to 12 V, the backup lights are on continually. Therefore, Technicians A and B are both right, and C is the correct answer.

**49.  A**    If the right-hand backup light circuit is grounded on the switch side of the bulb, excessive current blows the backup light fuse. This action makes both backup lights inoperative.

Therefore, Technician B is wrong and Technician A is right, making A the correct answer.

**50.  D**    A high resistance in the ground wire through the 7-wire connector may cause intermittent operation of all the trailer lights, so A is wrong. High resistance in the trailer taillight circuit may cause the taillights to be dim. An open circuit in the trailer backup light circuit causes the backup lights to be inoperative. A blown fuse in the wire from the positive battery terminal to the 7-wire connector in the tow vehicle prevents the trailer battery from being charged as the trailer is towed, and this results in a discharged trailer battery. Thus, D is the correct answer.

**51.  A**    A short to ground between the fuel gauge and the fuel gauge sending unit only affects the fuel gauge. Therefore, Technician B is wrong. In an intermittent resistance problem in the instrument panel, the ground circuit causes the instrument voltage limiter to supply higher voltage to the gauges, which causes high gauge readings. Thus, Technician A is correct, making A the right answer.

**52.  A**    High resistance between the voltage limiter and the gauge would probably affect the other gauges, so B is wrong. A short to ground between the gauge and the sending unit causes a high gauge reading, so C is wrong. An open circuit between the gauge and the sending unit causes an empty gauge reading; thus, D is also wrong.

High resistance in the sending unit ground wire reduces gauge current flow and provides a lower gauge reading, making A the right answer.

**53.  B**    Because the voltage limiter produces a constant 5 V regardless of the charging system voltage, an inoperative alternator does not affect the gauge reading. Therefore, Technician A is wrong. An inoperative instrument voltage limiter may cause erratic gauge operation, so Technician B is right. Thus, B is the correct answer.

**54.  B**    An open wire from the hot coil to the sending unit causes the current to flow through the cold coil. This action results in a low gauge reading, so A is wrong. Excessive resistance in the sending unit or sending unit ground reduces the hot coil current, and increases the cold coil current. Under this condition, the gauge reading is lower. Therefore, C and D are wrong.

An open wire from the cold coil to ground causes all the gauge current to flow through the hot coil. This action causes a continual hot gauge reading, so B is right.

**55.  B**    When an electronic instrument cluster is completely inoperative, the instrument panel fuse should be tested before replacing the cluster. Therefore, Technician A is wrong and Technician B is correct. So the right answer is B.

**56.  B**    If some of the electronic instrument panel display segments are not illuminated when the ignition switch is turned on, the display is inoperative and must be replaced. Therefore, Technician A is wrong and Technician B is correct. Thus, B is the right answer.

**57.  A**    If the oil pressure warning light is illuminated with the engine running, the first step is to test the engine oil pressure with a gauge to prove if the engine oil pressure is satisfactory. Therefore, Technician B is wrong and Technician A is correct. So the right answer is A.

**58.  B**    An open circuit in the wire between the BCM terminal 3 and the door ajar switches causes an inoperative door ajar light. An open circuit from the door ajar switches to ground also causes an inoperative door ajar light. A continually open LR door ajar switch causes no door ajar light operation when the LR door is opened. Therefore, A, C, and D are wrong.

A short to ground on the BCM side of the RF door ajar switch causes continual door ajar light operation. Thus, B is the right answer.

**59.  C**   If the timer contacts are stuck closed, the seat belt buzzer would go off once the driver's seat belt is buckled. Therefore, A is wrong. If the circuit is shorted to ground at buzzer terminal 3, the seat belt light is inoperative, and the higher current flow may damage the buzzer contacts.

Therefore, B is wrong. An open circuit at terminal 2 causes zero current flow in the timer heater circuit and continually closed contacts on the timer. However, the seat belt buzzer still goes off when the driver's seat belt is buckled, so D is wrong.

If the timer contacts and seat belt switch are stuck closed, the buzzer and seat belt warning light operate continually. Thus, C is right.

**60.  A**   When ground circuit number 7 has an open circuit, the electronic control unit will not know when the seat belt is buckled, and the seat belt warning light does not go off. Thus, Technician B is wrong. If the dedicated 10 A fuse is blown, the seat belt warning light and buzzer are inoperative. Therefore, Technician A is correct. So A is the right answer.

**61.  A**   If the wire connected from the horn relay to the LH horn is shorted to ground, the LH horn would be inoperative, and fusible link S would likely be burned out. Thus, Technician B is wrong. When the wire from the horn relay to the horn brush/slip ring is shorted to ground, the horn blows continually with the ignition switch on. Therefore, A is correct, making A the right answer.

**62.  A**   In this question, we are asked for the statement that is not the cause of the problem. An open circuit in the horn relay winding, or in the brush/slip ring would cause no horn operation.

An open fuse link in the relay power wire also causes this problem. Therefore, B, C, and D may cause the problem, and these are not the requested answer.

Because the relay does not require a ground, an open circuit in the relay ground has no effect and is not a cause of the problem. Thus, A is right.

**63.  D**   When current flows through the shunt coil, the shunt coil magnetic field produces more opposing voltage in the armature windings. This action reduces armature and wiper speed.

An open shunt coil causes high armature speed, and this problem does not affect wiper parking. Therefore, Technicians A and B are both wrong, making D the correct response.

**64.  C**   An open circuit between the low-speed relay winding and the module, or an open circuit in the module ground would cause no low speed or intermittent wiper operation.

Therefore, Technicians A and B are both correct, and C is the right answer.

**65.  D**   If the 270 K resistor is open in the wiper circuit, one of the intermittent wiper positions is inoperative. Therefore, D is correct. The wiper motor operates normally on low speed, high speed, and while parking. Thus, A, B, and C are wrong.

**66.  A**   If the diode between terminals H and E in the wiper/washer switch is shorted, current has to flow through a 24,000-ohm resistor and then through the washer motor. This resistance reduces the current flow to such a low state that the washer motor does not operate, so Technician B is wrong. If the wire from the wiper/washer switch is shorted to 12 V, the washer pump operates continuously. Thus, Technician A is right, making A the right answer.

**67.  B**   An open circuit in the wiper/washer fuse causes inoperative wipers and washers, so Technician A is wrong. Because the isolation diode is only connected in the washer circuit, an open diode causes no operation of the washer. Therefore, Technician B is correct, making B the right answer.

**68.  A**   An open circuit in the window switch movable contacts causes no window operation from the master switch, so B is wrong. An open circuit in the master switch ground wire also causes no window operation from the master switch, thus C is wrong. A short to ground at the motor circuit breaker causes high current flow in the down position and a blown fuse. Therefore, D is wrong.

If there is an open circuit between the ignition switch and the window switch, there is no voltage supplied to the window switch and the window does not work from this switch. Thus, A is the correct response.

**69. B**     Because the left and right rear window switches both use the same ground, an open circuit in this ground circuit causes no operation of either windows. Therefore, Technician A is wrong.

An open circuit at the upper contacts on the right rear power window switch causes no operation of this window, so Technician B is correct. Thus, B is the right answer.

**70. A**     Fuse number 7 does not supply voltage to the power seat circuit, so Technician B is wrong. An open circuit at the body ground behind the left cowl trim panel causes the power seat to be completely inoperative. Thus, Technician A is correct, making A the right answer.

**71. C**     This question asks for the statement that is not the cause of the problem. A newspaper jammed in the horizontal seat track may cause no horizontal seat operation. An open circuit between the horizontal switch and motor may be the cause of no horizontal seat operation. Burned horizontal switch contacts also may cause this problem. Because A, B, and D may cause the problem, they are not the requested answer.

An open circuit from the switch assembly to ground causes no operation of the seat in any direction. Therefore, this is not the cause of the problem, and so C is the requested answer.

**72. A**     An open circuit at ground connection G200 causes the rear defogger to be inoperative because this connection provides a ground for the solid state timer. Thus, Technician B is wrong. If the relay contacts are stuck closed, the rear defogger operates continually. Therefore, Technician A is correct, making A the right answer.

**73. C**     The defogger relay contacts must close to illuminate the indicator light. An open defogger relay winding, an open circuit at the relay contacts, or an inoperative on/off switch would not allow the relay contacts to close. Therefore, A, B, and D are wrong.

An open circuit between the switch/timer and the grid causes the indicator to be on with no grid operation. Thus, the correct answer is C.

**74. C**     An open circuit between the PWR ACC fuse and junction S500 causes 0 V at the power door lock switches, which results in no action from the door locks. An open circuit at ground connection G200 also causes inoperative door locks, because this connection completes the power lock circuit to ground in the up or down mode. Therefore, Technicians A and B are both right, and C is the correct answer.

**75. A**     An open circuit between terminal C and junction S504 on the left-hand door lock switch prevents operation of the door locks from either switch. Therefore, Technician B is wrong.

An open circuit between junction S500 and the left-hand door lock switch prevents operation of the door locks from the left-hand switch. Therefore, Technician A is the right answer, making A the correct choice.

**76. D**     If a remote keyless entry system does not operate unless the person holding the remote control is right beside the vehicle, the battery in the remote control is likely inoperative. One method would be to test the battery in the remote control. Therefore, Technicians A and B are both wrong, and D is the correct answer.

**77. C**     Discharged remote control batteries, or a blown number 4 fuse prevent the operation of the remote keyless entry system. Therefore, Technicians A and B are both correct, and C is the right answer.

**78. A**     This question asks for the statement that is not the cause of the problem. An open circuit in the close relay winding at the close switch, or at terminal C in the sunroof switch, prevents closure of the sunroof. Therefore, B, C, and D may be causes of the problem and these are not the requested answer.

An open circuit at the close relay contacts prevents sunroof opening, so A is not a cause of the problem and this is the requested answer.

**79. C**     An open circuit breaker in the motor, an open switch ground wire, or an open circuit at switch terminal C will prevent the downward operation of the convertible top. Therefore, A, B, and D are wrong.

A jammed linkage mechanism on the convertible top may allow downward top operation but no upward movement. Therefore, C is correct.

**80.** **C**   An open circuit in the mirror select switch, or an open circuit between the mirror switch and ground may cause no mirror operation. Therefore, Technicians A and B are both correct, and C is the right answer.

**81.** **A**   This question asks for the statement that is not the cause of the problem. A blown fuse number 1 in the relay center, an open circuit between the heated mirror element and ground, or an open circuit between the timer relay and fuse number 1 may prevent operation of the heated mirror element and allow rear defogger operation. Because statements B, C, and D may cause the problem, these are not the requested answer.

An open circuit in the number 4 circuit breaker in the fuse block prevents operation of the rear defogger and the heated mirror element. Therefore, A is not a cause of the problem and this is the requested answer.

**82.** **D**   This question asks for the statement that is not the cause of the problem. An inoperative stator, diode, or capacitor may cause radio noise from the alternator. Therefore, A, B, and C may cause the problem, and these are not the requested answer.

An open field winding reduces the alternator output to zero, so this problem would not cause radio static from the alternator. Therefore, D is the requested answer.

**83.** **C**   A poor metal-to-metal connection between the hood and other body components may cause radio static. An inoperative suppression coil on the instrument voltage limiter also results in this problem. Therefore, Technicians A and B are both correct, and C is right.

**84.** **D**   This question asks for the statement that is not true. There should be continuity between the end of the antenna mast and the center pin in the lead-in wire. Continuity also should be present between the ground shell of the lead-in wire and the antenna hardware. No continuity should be present between the center pin in the lead-in wire and the ground shell. Therefore, statements A, B, and C are correct, so they are not the requested answer.

Continuity should not be present between the end of the antenna mast and the mounting hardware. Therefore, statement D is not true, making it the requested answer.

**85.** **C**   An open circuit in the radio fuse, antenna fuse, or antenna motor would cause no operation from the antenna. Therefore, A, B, and D are wrong. Because the down limit switch only has current flow through it during antenna down operation, an open down limit switch allows upward antenna movement but no downward action. Thus, C is correct.

**86.** **C**   When the ohmmeter leads are connected from the capacitor lead to the case, and a low reading is obtained, the capacitor has insulation leakage between the capacitor plates.

Therefore, A, B, and D are wrong, and C is the right answer.

**87.** **B**   An open circuit at terminal B on the cigar lighter causes an inoperative cigar lighter, and normal dome light operation. Therefore, Technician A is wrong. Because the cigar lighter and the dome light share a common ground connection at terminal G202, an open circuit at this location causes both of these components to be inoperative. Thus, Technician B is correct, making B the right answer.

**88.** **D**   This question asks for the statement that is not the cause of the problem. A grounded circuit on the SB wire causes excessive current flow, which results in a blown clock fuse and inoperative clock. Because body ground B is the clock ground, an open circuit at this location causes an inoperative clock. The clock also is inoperative if there is an open circuit at terminal B in the clock connector. Therefore, A, B, and C may be the cause of the problem, so these are not the requested answers.

Because wire B/R at the clock connector is only in the clock illumination circuit, a grounded circuit at this location does not affect clock operation. Therefore, D is the requested answer.

**89.** **C**   Erratic cruise control operation may be caused by a leak in the vacuum hose connected to the servo or in the servo diaphragm. Therefore, Technicians A and B are both right, and C is the correct answer.

**90.** **C**   Because the 20 A gauges fuse supplies voltage to the cruise control switch, a blown fuse results in an inoperative cruise control. Therefore, Technician B is correct. A malfunctioning vehicle speed sensor also causes an inoperative cruise control. Thus, Technicians A and B are both correct, and C is the right answer.

**91. C**    If the resistor sensing contacts are damaged in the ignition switch, the BCM does not receive the proper signal from the ignition switch, and the BCM does not provide a ground for the starter relay winding, resulting in no action from the starting motor. An open circuit in the starter relay winding also results in no action from the starting motor. Therefore, Technicians A and B are both right, and C is the correct answer.

**92. D**    This question asks for the statement that is not true. The anti-theft system is triggered by unauthorized operation of the doors, hood, trunk lock, or ignition. When triggered, the system sounds the horn for three minutes, flashes the lights for eighteen minutes, and disables the injectors. Therefore, statements A, B, and C are correct, and these are not the requested answer.

When triggered, the system can be disarmed by operating one of the front door locks with the key. Disconnecting the battery does not disarm the system. Therefore, D is not true, so this is the requested answer.

**93. A**    If the air bag warning light is on with the engine running, an electrical problem is present in the system. Because the sensors have to close to deploy the air bag, it is very unlikely that an electrical problem in the system will deploy the air bag. Therefore, Technician B is wrong and Technician A is right, making A the correct answer.

**94. C**    Failure to disarm an air bag system prior to servicing the vehicle, or using test equipment that is not recommended by the vehicle manufacturer, may cause an accidental air bag deployment. Therefore, Technicians A and B are both right, and C is the correct answer.

**95. C**    This question asks for the statement that is not true. Before servicing an air bag system, the negative battery cable should be disconnected and the technician should wait for the time period specified by the vehicle manufacturer. Safety glasses and gloves should be worn when handling deployed air bags. Sensor operation may be affected if the brackets are bent or twisted. Therefore, statements A, B, and D are true, so these are not the requested answer.

A 12 V test light must not be used to diagnose an air bag system. Therefore, statement C is not true, making it the requested answer.

# Answers to Post Test

**1. A**    Shorted turns on the relay winding cause the winding to have less than the specified resistance. Thus, Technician A is right. The current flow through a shorted winding is higher than normal. Therefore, Technician B is wrong, and A is the correct answer.

**2. C**    After the switch is opened, the technician must wait for the on-board computers to enter the sleep mode, and during this time the drain decreases. Thus, A is wrong. The tester switch is in series with the battery negative cable, so B is wrong. The ammeter is parallel to the tester switch contacts, so C is correct. The drain should not exceed the manufacturer's specifications or 50 milliamperes, so D is also wrong.

**3. B**    The battery specific gravity must be at least 1190 before performing a load test. Therefore, Technician A is wrong. At the end of the fifteen-second load test, the voltage must be read before turning off the load. Thus, Technician B is right, and B is the correct answer.

**4. A**    High resistance in the starting circuit cables causes low starter draw and high cranking voltage. Thus, Technician A is right. A short in the starting motor armature causes high starter draw. Therefore, Technician B is wrong, and A is the correct answer.

**5. C**    A defective regulator causing high charging circuit voltage results in an overcharged battery so A is wrong. An open field circuit or worn out brushes cause zero alternator output and the charge indicator light is illuminated with the engine running. Thus, B and D are wrong. A shorted alternator diode reduces alternator output and causes an undercharged battery, so C is the correct answer.

**6.  D**  An open diode reduces alternator output, but this problem does not affect the regulator. Thus, A is wrong. A shorted stator reduces alternator output, but this problem does not affect the regulator, so B is wrong. High resistance in the alternator battery wire reduces the amperes flowing from the alternator to the battery, but this does not affect the regulator. A shorted field winding causes high current flow in the field coil, and this field current also flows through the regulator. So, D is the correct answer.

**7.  A**  A high charging system voltage causes the lights to become brighter when the engine is accelerated. Thus, A is correct. High resistance in the headlight circuit causes reduced headlight brilliance, so B is also wrong. An open stator reduces alternator output, but this problem does not cause the headlights to become brighter as the engine is accelerated, so C is wrong. High resistance in the alternator battery wire reduces the current flow from the alternator through the battery, but this defect does not cause the headlights to become brighter with engine speed. Thus, D is wrong.

**8.  D**  The vehicle should be positioned 25 ft. from the aligning screen on the wall. Thus, Technician A is wrong. The headlights should be aligned on high beam, so Technician B is wrong. Because Technicians A and B are both wrong, D is the correct answer.

**9.  C**  A grounded stop light wire does not affect the taillight circuit, so A is wrong. An intermittent open circuit in a taillight wire causes the taillight to operate intermittently, so B is wrong. An intermittent ground in a taillight wire bypasses the bulb resistance and causes repeated blowing of the taillight fuse. Thus, C is correct. A loose tail light wiring connector may cause intermittent tail light operation, so D is also wrong.

**10.  D**  When the brakes are applied, the left-rear stop and signal light should be illuminated continually. Thus, Technician A is wrong. When the brakes are applied during a right turn, the right-rear stop and signal light should continue flashing normally. Therefore, Technician B is wrong. Because Technicians A and B are both wrong, D is the correct answer.

**11.  A**  An open circuit at one of the combination stop and signal light bulbs prevents operation of the stop and signal light on that side of the vehicle. Therefore, A is not a cause of the problem, making it the requested answer. A blown stop light fuse, an open circuit in the stop light switch, or an open circuit between the stop light fuse and the switch causes the stop lights to be inoperative and the signal lights to work normally. Therefore, B, C, and D could be the cause of the problem, and none of these is the requested answer.

**12.  D**  High resistance in the sending unit ground may cause intermittent or low gauge readings, so A is wrong. High resistance between the instrument voltage limiter and the instrument panel may cause all gauges to read high, so B is wrong. An open circuit in the wire from the fuel gauge to the sending unit causes a very low gauge reading. Thus, C is wrong.

**13.  B**  The instrument voltage limiter supplies 5 V to the gauges regardless of the charging circuit voltage. Thus, Technician A is wrong. A defective instrument voltage limiter may cause all the gauges to be erratic. Therefore, Technician B is right, and B is the correct answer.

**14.  B**  If the shunt coil is energized, the wiper motor operates at low speed, so A is wrong. If the wiper blades park in the wrong position, the park switch may be defective. Thus, B is correct. The shunt coil is energized only in the low-speed mode. Therefore, C is wrong. The wiper switch grounds the relay winding to turn on the wipers. Thus, D is wrong.

**15.  C**  In many intermittent wiper motor circuits, the module opens and closes the low-speed relay winding to provide intermittent operation. Thus, Technician A is right. In these wiper motor circuits, various resistors in the wiper switch provide unique input voltage signals to the module, so Technician B is also right. Because Technicians A and B are both right, C is the correct answer.

**16.  A**  A six-way power seat should move vertically at the front and rear, so A is correct. This type of seat has one horizontal motor and two vertical motors, so B and C are wrong. This type of power seat circuit usually has a circuit breaker in the electrical circuit. Thus, D is wrong.

**17.  B**  Many rear defogger circuits are controlled by a timer. Thus, A is true and is not the requested answer. When the defogger is on, 12 V are supplied to the grid. Thus, B is not true, making it the requested answer. Each grid track is parallel to the other tracks. Therefore, C is true and is not the requested answer. The timer energizes the defogger relay, so D is true, and is not the requested answer.

**18. C**    Most remote controls are a remote transmitter, so Technician A is right. Most remote controls contain a replaceable battery, so Technician B is also right. Because Technicians A and B are both right, C is the correct answer.

**19. D**    An open alternator field circuit causes zero alternator output, but this problem does not cause radio static. Thus, A is wrong. High charging voltage causes an overcharged battery, but this problem does not cause radio static. An open distributor pickup coil causes a no-start condition. so C is wrong. Defective spark plug wires with electrical leakage problems may cause radio static. Thus, D is correct.

**20. A**    A defect in an air bag input sensor causes the air bag warning light to be illuminated with the engine running. Thus, A is correct. If the warning light is illuminated, the air bag(s) may not deploy properly in a collision, so B is wrong. There is no connection between the flashes of the MIL and the air bag warning light because the air bag system has a separate computer, so C is wrong. Because the air bag system has a separate computer, a defective transmission computer does not cause illumination of the air bag system warning light. Thus, D is also wrong.

# 7 Heating and Air Conditioning

## Pretest

The purpose of this pretest is to determine the amount of review that you may require prior to taking the ASE Heating and Air-Conditioning Systems Test. If you answer all the pretest questions correctly, complete the questions and study the information in this chapter to prepare for the ASE Heating and Air-Conditioning Systems Test. If two or more of your answers to the pretest questions are incorrect, complete a thorough study of the questions and information in this chapter. The pretest answers are located at the end of the pretest, and are also included in the answer sheets supplied with this book.

1. Technician A says the line from the condenser to the evaporator should feel warm in the A/C mode.

   Technician B says the accumulator and the compressor suction line should feel cool in the A/C mode.

   Who is correct?
   A. A only
   B. B only
   C. Both A and B
   D. Neither A nor B

2. The sight glass appears clear when an R-12 refrigerant system is operating in the A/C mode. This condition may indicate
   A. air and moisture in the system.
   B. a refrigerant overcharge.
   C. low refrigerant charge.
   D. excessive oil in the system.

3. A refrigerant system has low high-side and high low-side pressure. This could indicate
   A. a malfunctioning A/C compressor.
   B. a low refrigerant charge.
   C. a restricted receiver/dryer.
   D. a restricted TXV valve.

4. Technician A says a flame-type leak detector may be used on an R-134a refrigerant system.

   Technician B says an R-12 refrigerant system may be charged with R-12 containing a red leak detecting dye.

   Who is correct?
   A. A only
   B. B only
   C. Both A and B
   D. Neither A nor B

5. After a refrigerant system is evacuated, the low-side gauge indicates 29 in. Hg. The vacuum gauge reading rises 1 in. Hg in ten minutes.
   Technician A says the refrigerant system has a leak that must be repaired.
   Technician B says to install a partial charge and leak test the system.
   Who is correct?
   A. A only
   B. B only
   C. Both A and B
   D. Neither A nor B

6. During a low-side refrigerant charging procedure on an R-12 system, the low-side manifold gauge set valve should be adjusted to maintain the system pressure at
   A. 20 psi (138 kPa).
   B. 40 psi (275 kPa).
   C. 55 psi (379 kPa).
   D. 75 psi (517 kPa).

7. Technician A says the lubrication in an R-12 system is a polyalkalene glycol (PAG) oil.
   Technician B says a polyalkalene glycol (PAG) oil is hygroscopic.
   Who is correct?
   A. A only
   B. B only
   C. Both A and B
   D. Neither A nor B

8. The fuse in a compressor clutch circuit blows repeatedly. The cause of this problem could be
   A. a loose ground connection on the clutch coil.
   B. a shorted winding in the clutch coil.
   C. an intermittent open circuit in the clutch coil.
   D. a grounded circuit in the clutch control relay winding.

9. The clearance is less than specified between the compressor clutch plate and the pulley friction surface.
   Technician A says another shim is required behind the armature.
   Technician B says another shim is required behind the armature retaining nut.
   Who is correct?
   A. A only
   B. B only
   C. Both A and B
   D. Neither A nor B

10. In the A/C mode, frost is forming on the tube between the condenser and the receiver/dryer. The cause of this problem could be
    A. restricted condenser air passages.
    B. a restricted receiver/dryer.
    C. a restricted fixed orifice tube.
    D. a restriction in the condenser-to-receiver/dryer tube.

11. A refrigerant system has high pressures indicated on the low-side and high-side gauges, and frost is forming on the TXV valve. A shop towel soaked in hot water is placed on the TXV and the system pressures drop but still remain above normal. The most likely cause of this problem is
    A. moisture in the refrigerant system.
    B. a refrigerant overcharge.

   C. a restricted TXV valve.

   D. a restricted evaporator.

12. When retrofitting an R-12 system to an R-134a system
   A. barrier hoses are required.
   B. R-134a service ports must be installed.
   C. mineral oil may be used with R-134a.
   D. the clutch cycling switch must be changed.

13. When using refrigerant recovery/recycling equipment
   A. R-12 and R-134a refrigerants may be mixed in a refrigerant system.
   B. mineral oil and PAG oil may be mixed in a refrigerant system.
   C. the refrigerant container specified by the manufacturer must be used.
   D. the equipment must have an SAE J1930 approval.

14. All of the following statements about R-12 and R-134a refrigerant are true EXCEPT
   A. R-12 is stored in white containers.
   B. R-134a is stored in blue containers.
   C. R-134a and R-12 may be released to the atmosphere.
   D. R-12 is harmful to the earth's ozone layer.

# Answers to Pretest

**1. C**  The line from the condenser to the evaporator should feel warm in the A/C mode. Thus, Technician A is right. The accumulator and the compressor suction line should feel cool in the A/C mode, so Technician B is also right. Because Technicians A and B are both right, C is the correct answer.

**2. B**  Air and moisture in the system cause bubbles in the sight glass, so A is wrong. A refrigerant overcharge provides a clear sight glass, so B is correct. A low refrigerant charge causes bubbles in the sight glass, so C is wrong. Excessive oil in the system causes oil streaks on the sight glass. Thus, D is wrong.

**3. A**  If a refrigeration system has high low-side and low high-side pressures, a defective compressor is indicated. Therefore, B, C, and D are wrong, and A is correct.

**4. B**  A flame leak detector should not be used with R-134a refrigerant. Thus, Technician A is wrong. An R-12 system may be charged with R-12 containing a red dye. Therefore, Technician B is right, making B the correct answer.

**5. D**  Under the conditions described in the question, the low-side gauge should not rise faster than 1 in. Hg. in five minutes. Therefore, a 1-in. Hg. rise in ten minutes indicates that the refrigeration system is not leaking. Thus, Technicians A and B are both wrong, making D the correct answer.

**6. B**  During a low-side charging procedure, the low-side gauge valve should be adjusted to maintain a system pressure of 40 psi (275 kPa). Therefore, A, C, and D are wrong, and B is the correct answer.

**7. B**  A polyalkalene glycol (PAG) oil should not be used in an R-12 system, so Technician A is wrong. A PAG oil is hygroscopic, so Technician B is right, making B the correct answer.

**8. B**  A loose ground connection on the compressor clutch coil decreases current in the clutch circuit, so A is wrong. A shorted compressor clutch winding increases current flow and may cause repeated blowing of the fuse in the circuit, making B correct. An intermittent open in the clutch coil may cause intermittent operation of the compressor, so C is wrong. A grounded circuit in the clutch control relay winding does not cause high current flow through the clutch coil. Thus, D is wrong.

**9.  A**    If the clearance is less than specified between the compressor clutch plate and the pulley friction surface, another shim is required behind the armature. Thus, Technician A is right and Technician B is wrong, making A the correct answer.

**10.  D**   If frost is forming on the tube between the condenser and the receiver/dryer, there is a restriction in this tube. Therefore, D is correct, and A, B, and C are wrong.

**11.  A**   When a shop towel soaked in hot water and placed on the thermostatic expansion valve (TXV) corrects the defective conditions described in the question, there is moisture in the refrigerant system. Thus, A is correct, and B, C, and D are wrong.

**12.  B**   When retrofitting A/C systems, barrier hoses are not required. R-134a refrigerant requires a PAG or ester oil, and the clutch cycling switch does not have to be changed. Therefore, A, C, and D are wrong. The service ports must be changed because R134a service ports are a different size compared to R-12 service ports. Thus, B is correct

**13.  C**   When using refrigerant recovery/recycling equipment, R-12 and R-134a refrigerants must not be mixed. Thus, A is wrong. Mineral oil and PAG oil must not be mixed, so B is wrong. The refrigerant container specified by the manufacturer must be used, so C is correct. The equipment must have a SAE J1990 approval, making answer D wrong.

**14.  C**   R-12 is stored in white containers, so A is true, but this is not the requested answer. R-134a is stored in blue containers, so B is true, but this again is not the requested answer. Neither R-134a nor R-12 should be released to the atmosphere. Thus, C is not true, and is not the requested answer. R-12 is harmful to the earth's ozone layer. Therefore, D is true, and is not the requested answer.

# A/C System Service, Diagnosis, and Repair

## ASE Tasks, Questions, and Related Information

In this chapter, each task in the Heating and Air-Conditioning Systems category is provided followed by a question and some information related to the task. If you answer any question incorrectly, study this information very carefully until you understand the correct answer. Question answers and analysis are provided at the end of this chapter and in the answer sheets provided with this book.

**Task 1**  Identify the system type and conduct performance test on the A/C system; determine needed repairs.

1.  If an A/C system is operating properly at the end of an A/C performance test, the temperature at the dash outlets should be
    A.  20° to 25°F.
    B.  35° to 40°F.
    C.  40° to 50°F.
    D.  60° to 65°F.

*Hint*   *The Technician must know whether the A/C system is an R-12 or R-134a system. Most of the major components in an R-134a refrigerant system have light blue labels indicating these components are designed for operation in an R-134a system. Vehicles with R-134a A/C systems have a light blue underhood A/C label. R-12 refrigerant systems have Schrader-type service valves, but R-134a systems have metric-thread, quick-disconnect service valves. When retrofitting an A/C system with an acceptable alternative refrigerant, the Technician must install a label over the existing label. The replacement label contains the following information: 1) The name and address of the company and technician performing the retrofit, 2) the retrofit date, 3) the refrigerant trade name, amount of charge, and numerical designation where applicable, 4) the lubricant type, amount, and manufacturer, and 5) the words "ozone depleter" listed on the label if the refrigerant depletes the ozone layer The type of compressor should be identified as a variable displacement or constant displacement.*

*When the A/C system is turned on, most variable-displacement compressor clutches are engaged continually, whereas constant-volume compressor clutches cycle on and off.*

*A/C systems may be classified according to the type of expansion device in the evaporator inlet line. Most A/C systems have a thermostatic expansion valve (TXV) (Figure 7–1), or a fixed orifice tube (FOT) (Figure 7–2), in the evaporator inlet line to control refrigerant flow into the evaporator.*

*A/C control systems may be identified as mechanical, semiautomatic, and automatic. Mechanical A/C systems have a slide-type lever or rotary switch to manually control the in-car temperature. In semiautomatic A/C systems, some of the control features, such as in-car temperature, are handled*

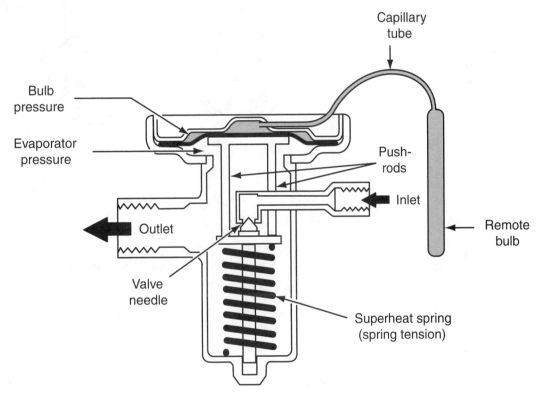

**Figure 7–1** Thermostatic expansion valve (TXV).

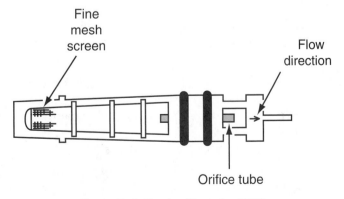

**Figure 7–2** Fixed orifice tube (FOT).

*automatically. An automatic air-conditioning system usually has the temperature reading displayed on a digital reading, and the system automatically provides the requested temperature. The blower speed may be automatically controlled in an automatic A/C system.*

## Task 2    Diagnose A/C system problems indicated by system pressures and/or temperature readings; determine needed repairs.

2. The gauge pressures in Figure 7–3 occur on a TXV valve, cycling clutch A/C system with an ambient temperature of 80°F. The cause of these readings could be
   A. air and moisture in the refrigerant system.
   B. an inoperative A/C compressor.
   C. a low refrigerant charge.
   D. the TXV valve is restricted.

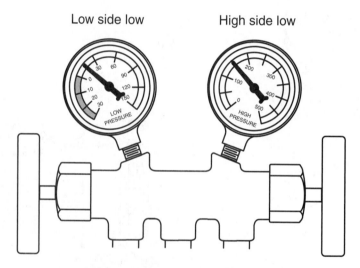

**Figure 7–3** Manifold gauge set readings.

3. In a TXV valve, cycling clutch A/C system with an ambient temperature of 80°F, the low-side pressure is 55 psi (345 kPa) and the high-side pressure is 260 psi (1,793 kPa). There is no indication of frosting on any of the refrigerant system components, and the discharge air is slightly cool.
   Technician A says the cause of this problem may be air and moisture in the refrigerant system.
   Technician B says the cause of this problem may be a restricted TXV valve.
   Who is correct?
   A. A only
   B. B only
   C. Both A and B
   D. Neither A nor B

4. When diagnosing an R-134a refrigeration system, the low-side pressure is 37 psi (42 kPa) and the high-side pressure is 263 psi (1,813 kPa) with the engine at normal operating temperature and an atmospheric temperature of 76°F (24°C). The most likely cause of these readings is
   A. clogged condenser air passages.
   B. a slipping A/C compressor belt.
   C. a restricted evaporator core.
   D. an inoperative compressor valve.

***Hint***      *When diagnosing an A/C refrigerant system from the low-side and high-side refrigerant system pressures, pressure specifications must be available in relation to ambient temperature. Some vehicle or A/C equipment manufacturers provide pressure specifications in relation to condenser and evaporator temperature. Diagnosis of refrigerant systems from the system pressures may be summarized as follows:*

- *Excessive low-side and high-side pressures—refrigerant overcharge, condenser restricted, air and/or moisture in the refrigerant system, TXV valve stuck open, engine overheating, or malfunctioning cooling fan.*
- *Low-side and high-side pressures lower than specified—low refrigerant charge, TXV valve stuck closed, restricted line from the condenser to the evaporator.*

*If the TXV valve is stuck closed or a restriction occurs in the line between the condenser and the evaporator, frosting occurs at the restriction.*

*A damaged compressor may cause a low-side gauge reading that is higher than specified and a high-side gauge reading that is lower than specified. When the low-side pressure is lower than normal and the high-side pressure is normal (Figure 7–4), the TXV valve may be sticking closed or restricted. If the low-side pressure is lower than normal and the high-side pressure is extremely high (Figure 7–5), a restriction is likely present in the high side of the refrigeration system. When*

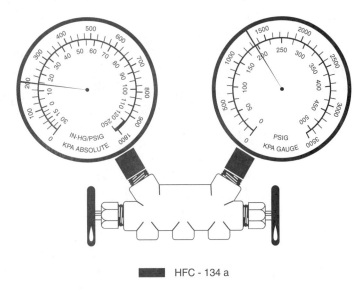

HFC - 134 a

**Figure 7–4** Low-side pressure low, high-side pressure normal

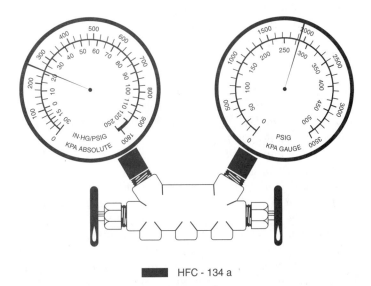

HFC - 134 a

**Figure 7–5** Low-side pressure low, high-side pressure extremely high

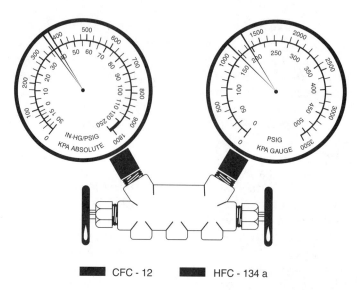

CFC - 12     HFC - 134 a

**Figure 7–6** Low-side pressure high, high-side pressure normal

*the low-side pressure is higher than normal, and the high-side pressure is normal (Figure 7–6), the TXV is sticking open. This condition may be caused by poor contact between the TXV remote bulb and the evaporator outlet.*

## Task 3  Diagnose A/C system problems indicated by sight, sound, smell, and touch procedures; determine needed repairs.

5. An air-conditioning (A/C) compressor has a growling noise only when the compressor clutch is engaged. The cause of this noise could be
   A. a damaged internal compressor bearing.
   B. a damaged pulley bearing.
   C. a low refrigerant charge.
   D. excessive refrigerant system pressure.

6. When a fixed orifice tube (FOT) cycling clutch A/C system is operating at 82°F ambient temperature, the compressor clutch cycles seven times per minute and the evaporator outlet line is warm. There is no frost on any of the A/C system components. The cause of this problem could be
   A. a low refrigerant charge.
   B. a flooded evaporator.
   C. a restricted TXV valve.
   D. a restricted receiver/dryer.

**Hint**     *A squealing noise may be caused by a loose, dry, or worn A/C compressor drive belt. This noise may be worse during fast engine acceleration. Worn or dry blower motor bushings may cause a squealing noise when the blower motor is running; this noise may occur when the engine is first started after sitting overnight.*

*A rattling noise from the compressor may be caused by a loose or worn clutch hub, or loose compressor mounting bolts.*

*A thumping, banging noise from the compressor may be caused by liquid refrigerant entering the compressor, refrigerant system blockage, or incorrect system pressures. Internal compressor damage may cause heavy knocking noises from the compressor. A growling noise with the compressor clutch engaged or disengaged may be caused by a worn compressor pulley bearing. If the growling noise is evident only when the clutch is engaged, the compressor internal bearings may be worn.*

*The refrigerant system components should be visually inspected for frosting. Frost on the receiver/dryer usually indicates an internal restriction in this component. Because the receiver/dryer is connected between the condenser and the evaporator, it should normally feel warm. Frosting of the TXV valve indicates this valve is restricted or sticking closed. Frost formation on the evaporator outlet usually indicates a flooded evaporator caused by excessive refrigerant charge or the TXV valve stuck open. These problems also may cause frost formation on the compressor suction hose. On a refrigerant system with a pilot-operated absolute (POA) valve, frosting of the compressor suction hose is normal.*

*If the refrigerant system has an accumulator, it should feel cold because it is connected between the evaporator and the compressor. Both the evaporator inlet and outlet should feel cold when the refrigerant system is operating normally. If the evaporator outlet is warm, the refrigerant charge may be low.*

*High-side refrigerant systems should normally feel hot or warm, and low-side components should be cold or cool. Because high-side components may be very hot, use caution when touching these components. If the line from the condenser to the TXV valve or FOT is cold, there may be a restriction in the high side.*

*A strong odor similar to a rotten-egg smell in the passenger compartment usually is caused by a plugged drain on the evaporator case. When this drain is plugged, the water collects and stagnates in the evaporator case, resulting in a strong odor.*

## Task 4   Leak test A/C system; determine needed repairs.

7. An oily residue is present on the fittings of the hose connected from the compressor to the condenser.

   Technician A says this residue may be caused by excessive oil in the refrigerant system.

   Technician B says this residue may be caused by a leak at the hose fittings.

   Who is correct?

   A. A only
   B. B only
   C. Both A and B
   D. Neither A nor B

**Hint**   *Refrigerant systems may be leak tested with dye, a flame-type leak detector, or an electronic leak detector. An R-12 system may be charged with R-12 that contains dye. This dye mixes with the system lubricant. After charging, the A/C system must be operated for at least fifteen minutes. If the system has a leak, the dye appears on lines, fittings, or components that are leaking. Longer A/C system operation may be necessary to locate small leaks.*

*A flame-type leak detector may be used on R-12 systems. If the system has a leak, the chlorine in the R-12 causes the halide torch flame to turn green, bright blue, or purple, depending on the size of the leak. Because the burning of R-12 creates a phosgene gas, this type of leak testing must be done in a well-ventilated area. There is no chlorine in R-134a refrigerant, so this method of leak detection does not work on R-134a systems.*

*Electronic leak detectors provide an audible beeping noise when the leak detector probe is placed near the leak source. Because R-12 and R-134a refrigerant contain different chemicals, each of these systems requires a different electronic leak detector, or one that can be switched to test each of these systems.*

## Task 5   Identify A/C system refrigerant and existing change amount.

8. While discussing refrigerant identification, Technician A says that recycled R-12 is sold in white containers marked with a DOT code.

   Technician B says that R-134a refrigerant is sold in blue containers.

   Who is correct?

   A. A only
   B. B only
   C. Both A and B
   D. Neither A nor B

   *Neither R-12 nor R-134a refrigerant may be vented to the atmosphere. Refrigerant recovery and recycling equipment is available to recover, recycle, and recharge R-12 or R-134a systems (Figure 7–7). New or recycled R-12 is sold in white containers, and the recycled*

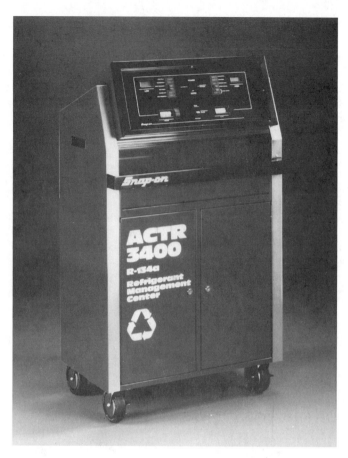

**Figure 7–7** R-134a refrigerant recovery, recycle, and recharging equipment. *(Courtesy of Snap-on Tools Company)*

*R-12 containers have a Department of Transportation (DOT) code. The sale of both refrigerants is restricted to certified A/C service facilities.*

*Recover-only machines are available to recover refrigerant. This type of machine may be used by automotive salvage operations who must recover refrigerant from A/C systems before dismantling a vehicle, but the refrigerant will not be reinstalled. Refrigerant identifier machines are now available to identify the type of refrigerant in A/C systems. These machines are necessary because of the various legal and illegal refrigerant blends. In some cases, flammable refrigerants have been installed in A/C systems, which could endanger Technicians who attempt to service the system. These flammable refrigerants or blends are prohibited in many states.*

## Task 6  Evacuate A/C system.

9. Evacuating a refrigerant system removes
    A. moisture from the system.
    B. rust particles from the system.
    C. aluminum particles from the system.
    D. desiccant particles from the system.

**Hint**    *Prior to the refrigerant system evacuation procedure the manifold gauge set must be connected to the service fittings. A vacuum pump is connected to the center hose on the manifold gauge set, and this pump usually is operated for thirty minutes with the service valves open and the low-side gauge valve open. After five minutes of vacuum pump operation, the low-side gauge should indicate 20 in. Hg (67.6 kPa), and the high-side gauge should read below zero unless it is restricted by a stop pin. If the high-side gauge does not drop below zero, refrigerant system blockage is indicated. When system blockage is indicated, stop the evacuation procedure and repair the blockage.*

*After fifteen minutes of vacuum pump operation, the low-side gauge should indicate 24 to 26 in. Hg (81 to 88 kPa) if there are no leaks in the refrigerant system. When the low-side gauge is less than this value, close the low-side valve and observe the gauge. If the low-side gauge needle rises slowly, a refrigerant system leak is indicated. When this condition is present, stop the evacuation, and install a partial refrigerant charge to locate the leak.*

*When there are no refrigerant system leaks, continue the evacuation for thirty minutes. Close the low-side gauge set valve, shut off the vacuum pump, and disconnect the center gauge set hose from the pump. Replace the protective caps on the pump inlet and outlet fittings.*

*After the evacuation procedure, the low-side gauge should not rise faster than 1 in. Hg. in five minutes. If the low-side gauge rises faster than 1 in. Hg in five minutes, the refrigerant system is leaking. Install a partial refrigerant charge and leak test the system.*

## Task 7  Inspect A/C system components for contamination.

10. Technician A says that some manufacturers recommend the installation of an in-line filter between the condenser and the evaporator as an alterative to refrigerant system flushing.

    Technician B says an in-line filter containing a fixed orifice tube may be installed and the original orifice tube left in the system.

    Who is correct?

    A. A only

    B. B only

    C. Both A and B

    D. Neither A nor B

**Hint**   *Refrigerant systems may be flushed to remove debris such as aluminum particles from a failed compressor. Rather than system flushing, some vehicle manufacturers recommend the installation of an in-line filter between the condenser and the evaporator to remove debris. These in-line filters are available with or without an internal fixed orifice tube (FOT). If an in-line filter containing an FOT is installed, the original FOT must be removed from the system.*

*The refrigerant system must be discharged prior to flushing. The complete system or individual components may be flushed with dry nitrogen. Before flushing individual components, disconnect the refrigerant lines from the component and cap the ends of the lines. When flushing the complete system, remove the receiver/dryer and bypass it with an appropriate hose. After system flushing, the receiver/dryer must be replaced.*

## Task 8  Charge A/C system with refrigerant (liquid or vapor).

11. Technician A says a high-side charging procedure should be completed with the engine running.

    Technician B says if liquid refrigerant enters the compressor, damage to the compressor may result.

    Who is correct?

    A. A only

    B. B only

    C. Both A and B

    D. Neither A nor B

**Hint**   *Before charging a refrigerant system, the recovery and evacuation procedures must be completed.*

*Always follow the charging procedure recommended by the vehicle manufacturer. High-side (liquid) or low-side (vapor) charging procedures may be recommended. The manifold gauge set hoses or hoses from the refrigerant recovery, recycling, and charging equipment must be connected to service fittings. Be sure both low-side and high-side gauge valves are closed. Connect the center hose on the manifold gauge set to the proper refrigerant container. Open the valve on the refrigerant container to charge the center hose with refrigerant. Open the high-side gauge valve, and observe the low-side gauge. Then close the high-side gauge valve. If the low-side gauge does not move from a vacuum to a pressure, the refrigerant system is restricted. Repair the restriction problem before proceeding with the charging procedure.*

*When the system blockage is satisfactory, open the high-side gauge valve to proceed with the high-side charging procedure. Charging is completed when the specified weight of refrigerant has entered the system. Close the high-side gauge valve and the refrigerant container valve. Remove the compressor belt and rotate the compressor by hand for several revolutions to be sure there is no liquid refrigerant in the system. Install and tighten the compressor belt. Start the engine and hold the engine speed at fast idle. Adjust the A/C controls to maximum cooling, and complete an A/C performance test.*

*When low-side charging is recommended, the system is charged with the engine running at the specified rpm for refrigerant system charging. Set the A/C controls to maximum cooling and high blower speed. Open the low-side gauge valve to allow refrigerant to enter the system. Adjust the low-side gauge valve so the low-side pressure does not exceed 40 psi (275 kPa). Continue charging the system until the specified refrigerant weight has entered the system. Close the low-side gauge valve, and complete an A/C system performance test.*

*After either charging operation is completed, back seat the service valves and disconnect the manifold gauge set hoses. Replace all the protective caps and covers.*

## Task 9  Identify A/C system lubricant type and capacity.

12. The oil required in an R-134a refrigerant system is
    A. a polyalkylene glycol (PAG) oil.
    B. a synthetic engine oil.
    C. a synthetic mineral oil.
    D. a 10W-30 engine oil.

**Hint**     *An A/C refrigerant system must contain the specified amount of refrigerant oil to lubricate the compressor components and prevent premature compressor wear. An excessive amount of oil in a refrigerant system reduces the cooling efficiency of the system.*

*An R-12 A/C system requires a mineral oil with a YN-9 designation, whereas an R-134a system with a reciprocating compressor must have a synthetic polyalkylene glycol (PAG) oil designated as YN-12. A different type of PAG oil is used with a rotary A/C compressor. Polyalkylene Glycol Synthetic refrigerant oil is specified for some 134a refrigerant systems. Polyol ester (ester) oil is also recommended in other R-134a refrigerant systems. Always use the refrigerant oil recommended by the original equipment manufacturer (OEM) or compressor manufacturer. If the oils used in R-12 and R-134a systems are interchanged, compressor damage will result. Both types of refrigerant oils are hydroscopic, which means these oils absorb moisture very easily. The PAG oil used in R-134a systems is more hydroscopic than the mineral oil in R-12 systems. Refrigerant oil containers must be kept tightly capped at all times.*

*A dipstick is required to measure the oil level in some compressors. The refrigerant system must be discharged or the compressor isolated before the compressor oil level is measured. Other compressors must be removed and the oil drained before adding new oil. Always follow the vehicle manufacturer's recommended oil level checking procedure and add the specified amount of oil. When individual refrigerant system components are replaced, install the specified amount of the proper refrigerant oil in the component. Refrigerant oil may be added to the system from pressurized cans. The specified amount of refrigerant oil may also be added to the system between the discharging and evacuation procedures.*

## Task 10  Inspect and replace passenger compartment (cabin air, pollen) filter.

13. A customer complains about a low volume of air flow in both the heater and A/C modes. The blower operates at all speeds. The most likely cause of this problem is
    A. restricted refrigerant passages in the evaporator core.
    B. restricted air passages in the heater core
    C. restricted A/C ducts
    D. restricted passenger compartment air filter.

*Hint*    *Most newer vehicles have a passenger compartment air filter that filters the air entering the passenger compartment in the heat or A/C modes. This filter must be replaced at intervals specified by the vehicle manufacturer.*

**Task 11**  **Disarm and enable the air bag system for vehicle service following manufacturers' recommended procedures.**

14. When performing electronic system service on an air bag–equipped vehicle, Technician A says the wait period after the battery negative cable is disconnected allows the air bag control module to enter the backup mode.

Technician B says if the air bag system is not properly disarmed before electronic service is performed on the system, accidental air bag deployment may occur.

Who is correct?

A. A only

B. B only

C. Both A and B

D. Neither A nor B

*Hint*    *On some older vehicles, the air bag system is disarmed by disconnecting the battery negative terminal and waiting for the time specified by the vehicle manufacturer. This wait time is usually between two and ten minutes. The wait time allows the backup power supply to power down in the air bag control module. If the air bag system is not disarmed properly before servicing the electrical/electronic system, accidental air bag deployment may occur. Because disconnecting battery voltage from the electrical/electronic system erases computer memories (causing service complications) on newer vehicles, the manufacturer usually recommends disarming the air bag system by turning the ignition switch off, and removing the ignition key. The next step in the air bag disarming process is to remove the air bag system fuse and then disconnect specific connectors in the system. Typical air bag system connectors that must be disconnected in order to service the air bag system are the following:*

- *The driver's side air bag two-wire connector at the base of the steering column.*
- *The passenger's side air bag two-wire connector behind the passenger's side air bag inflator module.*
- *The driver's side impact inflator module two-wire connector under the driver's seat.*
- *The passenger's side impact inflator module two-wire connector under the passenger's seat.*

*On other newer vehicles, the manufacturer divides the electrical/electronic system on the vehicle into zones, and recommends disconnecting the air bag system connectors in specific zone(s) to disarm the air bag system depending on the service being performed on the vehicle.*

# Refrigeration System Component Diagnosis and Repair, Compressor and Clutch

## ASE Tasks, Questions, and Related Information

**Task 1**  **Diagnose A/C system problems that cause the protection devices (pressure, thermal, and electronic controls) to interrupt system operation; determine needed repairs.**

15. The purpose of the refrigerant system component 5 in Figure 7–8 is to

A. protect the system components from excessive pressure.

B. shut off the compressor if the refrigerant charge is low.

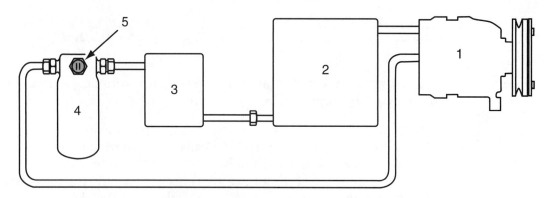

**Figure 7–8** Refrigerant system components.

C. cycle the compressor on and off in relation to system pressure.

D. shut off the compressor if the refrigerant temperature is excessive.

**Hint**    *Some refrigerant systems use a pressure cycling switch to cycle the compressor on and off in relation to the low-side pressure. In cycling clutch orifice tube (CCOT) systems, the pressure cycling switch usually is mounted in the accumulator between the evaporator and the compressor.*

*This switch closes and turns on the compressor when the refrigerant system pressure is at or above 46 psi (315 kPa). The A/C switch in the instrument panel must be turned on to supply voltage to the pressure cycling switch. The pressure cycling switch opens and turns off the compressor when the system pressure decreases to 25 psi (175 kPa). This cycling action maintains the evaporator temperature at 33°F (1°C).*

*Some refrigerant systems have a clutch cycling switch that cycles the compressor on and off in relation to evaporator outlet temperature (Figure 7–9). This control device may be called a thermostatic switch. A capillary tube is connected from the clutch cycling switch to the evaporator outlet pipe.*

*Many refrigerant systems have a low pressure cut-off switch that opens the circuit to the compressor clutch if the system pressure drops below a preset value because the refrigerant charge*

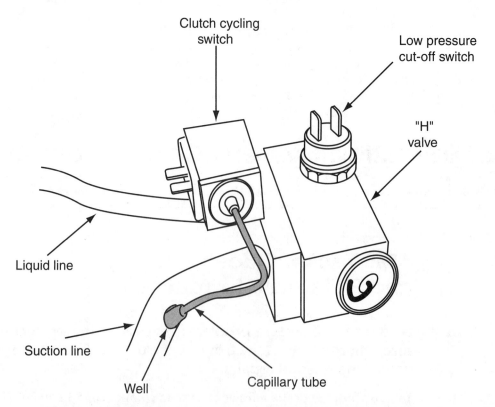

**Figure 7–9** Thermostat clutch cycling switch and low pressure cut-off switch.

*is low. If a large refrigerant leak occurs, the oil may be lost with the refrigerant. When the compressor is allowed to run under this condition, compressor damage may occur.*

*Some refrigerant systems have a high-pressure relief valve that is sometimes mounted in the receiver/dryer. This valve opens and relieves the system pressure if the pressure exceeds 450 to 550 psi (3,100 to 3,792 kPa). When the system pressure decreases below 450 psi (3,100 kPa), the high-pressure relief valve closes. Extremely high refrigerant system pressures may be caused by restricted airflow through the condenser or a refrigerant overcharge.*

*When retrofitting an A/C system with an acceptable alternative refrigerant, if the system has a high-pressure relief valve, a high-pressure shut-off switch must be installed that shuts off the voltage supply to the compressor clutch before refrigerant pressure opens the high-pressure relief valve. This action prevents refrigerant from being released through the high-pressure relief valve to the atmosphere. It is illegal to vent any refrigerant to the atmosphere.*

*Some older model refrigerant systems contained a superheat switch mounted in the compressor and a thermal fuse. If a refrigerant leak occurs, high system temperature closes the superheat switch contacts. Under this condition, excessive current flow blows the thermal fuse and opens the compressor clutch circuit. Because the refrigerant system oil may have been lost with the refrigerant, the blown thermal fuse protects the compressor from running without lubrication.*

*In many computer-controlled A/C systems, the powertrain control module (PCM) operates a relay that supplies voltage to the compressor clutch. All the input sensor signals are sent to the PCM. In some systems, these inputs include a refrigerant pressure signal. If the input signals indicate an abnormal condition, the PCM does not energize the compressor clutch.*

**Task 2   Inspect, test, and replace A/C system pressure and thermal protection devices.**

16. In the A/C system in Figure 7–10, the ignition switch is on, and the A/C switch is in the AUTO position. The ambient temperature is 75°F (24°C), and the compressor clutch is inoperative. There is 12 V at terminals B, C, and S on the thermal fuse. The cause of the inoperative compressor clutch could be

   A. an open thermal fuse.

   B. an open compressor clutch coil.

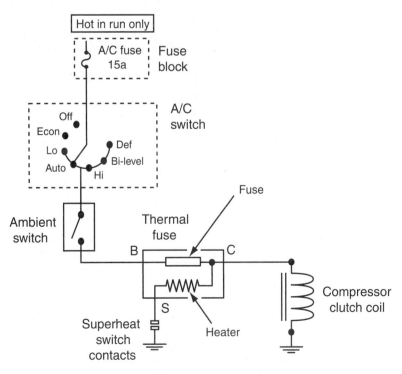

**Figure 7–10** Compressor clutch circuit with thermal fuse and superheat switch.

C. a damaged superheat switch.

D. a damaged thermal fuse heater.

**Hint**    *The thermal fuse and superheat switch may be tested with an ohmmeter or voltmeter. With the ignition switch and A/C switch on, and the ambient switch closed, there should be 12V at terminals B, C, and S on the thermal fuse. If there is 12 V at terminal B and 0V at terminals C and S, the thermal fuse is blown. When there is 12 V at terminals B and C, but a lower voltage at terminal S, the superheat switch contacts are closed, or the wire from the thermal fuse to the superheat switch is shorted to ground. If there is 12 V at terminals B, C, and S on the thermal fuse, but 0V at the compressor clutch, the wire from the thermal fuse to the compressor clutch is open.*

*When the ohmmeter leads are connected to the compressor clutch terminals, an infinity reading indicates an open clutch coil, whereas an ohmmeter reading below the specified value indicates a shorted clutch coil.*

*If there is evidence of oil around the high-pressure relief valve and the system is low on refrigerant, check the system pressures and inspect the condenser for restricted air passages. When the condenser air passages are not restricted and the system pressures are normal, the high-pressure relief valve may be the problem.*

*In some A/C systems a thermal switch is connected in series with the compressor clutch. This switch usually is mounted in the compressor. Many thermal switches open at 257°F (125°C), and close at 230°F (110°C).*

## Task 3  Inspect, adjust, and replace A/C compressor drive belts, pulleys and tensioners.

17. An intermittent squealing noise is heard upon acceleration with the A/C control switch in the on or off position. The most likely cause of this problem is
    A. a loose power steering belt.
    B. a loose A/C compressor belt.
    C. a loose air pump belt.
    D. a worn A/C compressor pulley bearing.

**Hint**    *Because the friction surfaces are on the sides of a V-belt, this type of belt must not be bottomed in the A/C compressor pulley. If the compressor belt is loose or bottomed in the pulley, the belt may slip. Erratic compressor operation and inadequate passenger compartment cooling may be caused by a loose compressor belt. A slipping compressor belt may cause belt squealing especially on acceleration with the A/C on and the compressor clutch engaged. The compressor belt should be inspected for cracked, oil-soaked, glazed, and torn or split conditions.*

*The belt tension should be measured with a belt tension gauge positioned at the center of the longest belt span. V-belt tension may be adjusted by loosening the adjustment bolt and moving the compressor until the proper tension is obtained. Tighten the adjustment bolt to the specified torque. Most ribbed V-belts have a spring-loaded tensioner pulley.*

## Task 4  Inspect, test, service, and replace A/C compressor clutch components or assembly.

18. The measurement in Figure 7–11 is more than specified.

    Technician A says this condition may cause an intermittent scrapping noise with the engine running and the compressor clutch energized.

    Technician B says that to correct this condition another shim should be added behind the pulley armature plate.

    Who is correct?
    A. A only
    B. B only
    C. Both A and B
    D. Neither A nor B

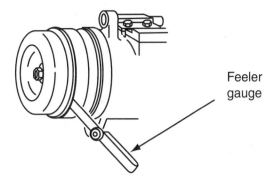

**Figure 7–11** A/C compressor clutch measurement.

**Hint**    *A special tool or two box-end wrenches may be used to hold the pulley while the compressor shaft nut is removed. After this nut is removed, the armature plate and shims may be removed. Remove the pulley snap ring, pulley, and field coil core (Figure 7–12).*

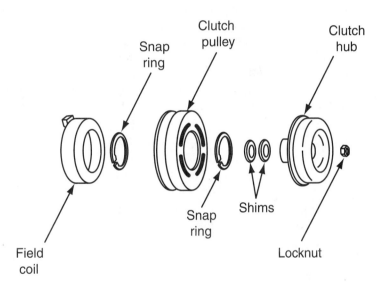

**Figure 7–12** Compressor clutch components.

*The pulley and armature plate frictional surfaces should be inspected for wear and oil contamination. Replace contaminated or worn components. Check the hub bearing for roughness, grease leakage, and looseness. Replace the bearing if any of these conditions are present.*

*When the field coil is installed, the nipple in the back of the coil must be aligned with the locating indentation in the front compressor cover. The pulley snap ring must be installed with the bevelled side facing outward.*

*After the shims and armature are installed, the clearance between the armature and pulley frictional surface must be measured with a feeler gauge. If the clearance is excessive, remove a shim from behind the armature. Recheck the armature to pulley clearance after the shaft nut is tightened to the specified torque.*

*Some vehicle manufacturers recommend checking the compressor clutch circuit with a voltmeter and an ohmmeter. With the compressor clutch engaged, the voltage supplied to the clutch coil must be within 2 V of the battery voltage. If the voltage supplied to the coil is zero, check the clutch fuse link or fuse. If the voltage supplied to the clutch coil is less than specified, check the resistance in the circuit from the battery to the coil.*

*When the clutch is engaged, an ammeter connected in series with the coil should indicate 2.0 to 4.15 amperes. A low ammeter reading indicates excessive resistance in the coil or the ground circuit. If the ammeter reading is higher than specified, the coil is shorted.*

## Task 5    Identify required lubricant type; inspect and correct level in A/C compressor.

19. When replacing an A/C compressor the refrigerant system is discharged and 2 oz. of oil are recovered from the system. When the old compressor is removed, 2 oz. of oil are drained from the compressor. The amount of oil drained from the new compressor is 6 oz. When the new compressor is installed, the amount of oil added to the compressor should be
    A. 1 oz.
    B. 2 oz.
    C. 4 oz.
    D. 6 oz.

**Hint**    *An R-12 A/C system requires a mineral oil with a YN-9 designation, whereas an R-134a system with a reciprocating compressor must have a synthetic polyalkylene glycol (PAG) oil designated as YN-12. A different type of PAG oil is used with a rotary A/C compressor. If the oils used in R-12 and R-134a systems are interchanged, compressor damage will result. Both types of refrigerant oils are hydroscopic, which means these oils absorb moisture very easily. The PAG oil used in R-134a systems is more hydroscopic than the mineral oil in R-12 systems. Refrigerant oil containers must be kept tightly capped at all times.*

*The oil level in some A/C compressors may be checked with a dipstick. When the refrigerant is recovered from a system, the amount of oil recovered should be measured and an equal amount of new oil added to the system. Vehicle manufacturers usually specify the total amount of refrigerant oil required in the system, and the amount of oil required in each component. When any refrigerant system component is replaced, the required amount of refrigerant oil is the total system oil capacity minus the oil capacity of the components that have not been replaced plus the amount of oil recovered during the discharge procedure.*

## Task 6    Inspect, test, service or replace A/C compressor.

20. While diagnosing A/C compressor problems, Technician A says that if both low-side and high-side pressures are more than specified, the compressor may be defective.

    Technician B says if the high-side pressure is low and the low-side pressure is high the compressor may be defective.

    Who is correct?
    A. A only
    B. B only
    C. Both A and B
    D. Neither A nor B

**Hint**    *A damaged compressor may be indicated by excessive noise, seizure, refrigerant leaks, and abnormal A/C system pressure such as high low-side, and low high-side pressure. Always diagnose the system to determine the exact cause of the problem before replacing the compressor.*

*Excessive compressor noise may be caused by loose compressor mountings or liquid refrigerant entering the compressor. Be sure these conditions are not causing a noise problem before replacing the compressor to correct excessive noise. If the compressor is seized, always be sure the refrigerant system contains the proper type and amount of refrigerant oil. When refrigerant pressures indicate a malfunctioning compressor, always be sure some other refrigerant system component is not causing the pressure problem. For example, high low-side pressure and low high-side pressure may indicate a damaged compressor, but a restriction in the low-side may also cause this problem.*

## Task 7    Inspect, repair or replace A/C compressor mountings fasteners.

21. Technician A says component 6 in Figure 7–13 is a seal protector that is placed over the compressor shaft.

    Technician B says the seal seat O-ring must be installed before the compressor shaft seal.

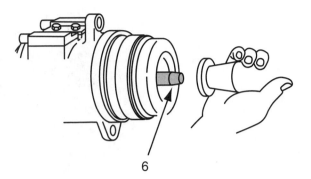

**Figure 7–13** Compressor service tool.

Who is correct?
A.  A only
B.  B only
C.  Both A and B
D.  Neither A nor B

*Hint*     *Inspect the compressor for oil deposits in the pulley area, around the line connections, and at the pressure relief valve. Oil deposits in the pulley area indicate a refrigerant system leak at the shaft seal.*

*Check the compressor belt for proper tension, wear, cracks, or an oil-soaked condition. Inspect all the compressor mounts and mounting bolts for wear and proper torque. A rattling noise may be caused by loose compressor mounts. This noise will likely be worse when the compressor clutch is engaged.*

*A growling noise that occurs only when the compressor is operating is likely caused by a worn bearing in the compressor. A damaged pulley bearing also causes a growling noise with the compressor clutch disengaged.*

*A malfunctioning compressor may be indicated by low high-side pressure, and high low-side pressure on the manifold gauge set.*

*Discharge the refrigerant system or isolate the compressor before removing the compressor. If the refrigerant system has stem-type service valves, these valves may be front-seated to isolate the compressor from the refrigerant system. Always follow the recommended isolation procedure in the vehicle manufacturer's service manual. When any refrigerant system component is disconnected, cap all lines and fittings to prevent moisture entry.*

# Refrigeration System Component Diagnosis and Repair, Evaporator, Condenser, and Related Components

## ASE Tasks, Questions, and Related Information

**Task 1**   Inspect, repair, or replace A/C system mufflers, hoses, lines, filters, fittings, and seals.

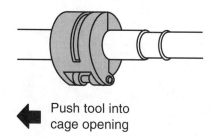

Push tool into
cage opening

**Figure 7–14** Refrigerant line service tool.

22. When servicing refrigerant lines, the tool in Figure 7–14 is used to
    A. disconnect spring lock couplings.
    B. connect spring lock couplings.
    C. connect and disconnect spring lock couplings.
    D. crimp male bare-type fittings.

**Hint**    *A damaged refrigerant line may be replaced with a complete new line. Male and female barbed fittings are available to repair damaged refrigerant lines or hoses (Figure 7–15). Some refrigerant line connections are sealed with O-rings and retained with spring lock couplings (Figure 7–16). A special tool is required to release these spring lock couplings. Other refrigerant line fittings have a ferrule and an O-ring (Figure 7–17).*

*Some refrigeration systems have a filter in the line between the condenser and the evaporator. Some of these filters contain an orifice tube. This type of filter must be installed in the proper direction (Figure 7–18).*

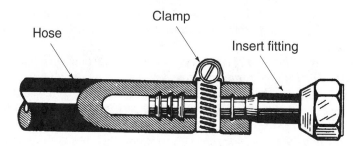

**Figure 7–15** Female barbed fitting for refrigerant hose repair.

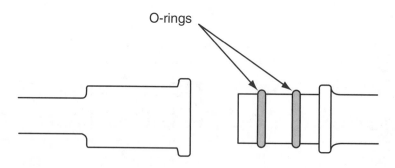

A    Clean fittings
B    Install new O-ring seals
C    Lubricate mating surfaces with refrigerant oil
D    Assemble fittings by pushing with a slight
     twisting motion

**Figure 7–16** Refrigerant line spring lock coupling.

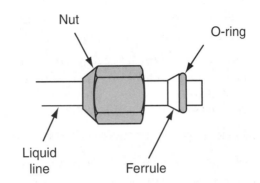

Figure 7–17  Refrigerant line with ferrule and O-ring.

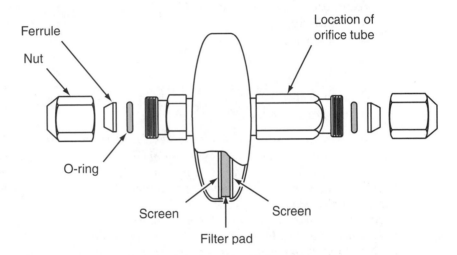

Figure 7–18  Refrigerant filter with orifice tube.

## Task 2    Inspect A/C condenser for misdirected or restricted airflow restrictions.

23. The air passages through an A/C condenser are severely restricted.
    Technician A says this may cause refrigerant discharge from the high-pressure relief valve.
    Technician B says this may cause excessive high-side pressure and low-side pressure.
    Who is correct?
    A. A only
    B. B only
    C. Both A and B
    D. Neither A nor B

*Hint*    *Debris in the condenser air passages causes excessive high-side and low-side pressures and reduced cooling from the A/C system. This problem may also cause the high-pressure relief valve to discharge refrigerant. The condenser air passages may be blown out with compressed air or washed with a water hose. A plastic rod and a soft-bristled brush may be used to remove debris that is stuck tightly in the condenser air passages. Bent condenser fins must be straightened with a plastic rod or needle-nosed pliers.*

## Task 3    Inspect, test, and replace A/C system condenser, mountings, and air seals.

24. Frost is forming on one of the condenser tubes near the bottom of the condenser.
    This problem could be caused by
    A. restricted airflow passages in the condenser.
    B. a refrigerant leak in the condenser.

C. a restricted refrigerant passage in the condenser.

D. a restricted fixed orifice tube.

**Hint**    *The condenser should be inspected for any sign of oil deposits on the inlet or outlet fittings or on the condenser tubing. Oil deposits indicate a refrigerant leak. Frosting on any of the condenser tubing indicates a refrigerant passage restriction. This condition results in excessive high-side and low-side pressures, and inadequate cooling.*

## Task 4    Inspect and replace receiver/dryer or accumulator/dryer.

25. A receiver/dryer is located between the condenser and the evaporator.

    Technician A says the receiver/dryer should be changed if the outlet is colder than the inlet.

    Technician B says the receiver/dryer should be changed if the refrigerant in the sight glass appears red.

    Who is correct?

    A. A only

    B. B only

    C. Both A and B

    D. Neither A nor B

**Hint**    *If the inlet and outlet pipes have a significant temperature difference, the receiver/dryer is restricted and must be replaced. Frost forming on the receiver/dryer indicates an internal restriction, which requires receiver/dryer replacement. Receiver/dryer replacement also is necessary if there is moisture in the refrigeration system indicated by bubbles and foam in the sight glass or rust contamination in the system.*

*Blue or gray particles in the sight glass indicate that the desiccant in the receiver/dryer is disintegrating and circulating through the refrigeration system. This condition also requires receiver/dryer replacement. If the refrigerant in the sight glass is red or yellow, leak-detecting dye has been added to the refrigeration system. This condition does not require corrective action. The refrigeration system must be discharged before receiver/dryer removal.*

## Task 5    Inspect, test, and replace expansion valve(s).

26. An A/C system blows cool air when the vehicle is started with an ambient temperature of 80°F. After the vehicle is driven for about ten miles, the system stops blowing cool air, and the TXV is frosted. When the A/C system is shut off for five minutes and turned on again, it blows cool air for another ten miles. The most likely cause of this problem is

    A. an inoperative TXV capillary tube.

    B. a refrigerant overcharge.

    C. a restricted condenser refrigerant passage.

    D. moisture in the refrigeration system.

**Hint**    *Moisture in the refrigeration system may freeze in the TXV, resulting in intermittent A/C system operation. If the TXV is stuck closed or the inlet screen is contaminated with debris, frosting may occur on the TXV. This condition causes the air discharged from the evaporator to be warm or slightly cool, and the low-side pressure to be low.*

*The TXV may stick in the open position, or an inoperative capillary tube may cause this valve to remain open. Under this condition, the evaporator floods, causing inadequate evaporator cooling, frost on the evaporator suction line, and high low-side pressure.*

*If there is very little cool air from the evaporator and the low-side pressure is low with TXV frosting and normal high-side pressure, place a shop towel soaked in hot water around the TXV valve. When this action causes the low-side pressure to increase to normal, there is moisture in the refrigeration system and ice formation in the TXV.*

*When the hot shop towel around the TXV does not change the low-side pressure, remove the remote bulb and warm it in the hand. If the low-side pressure rises, the remote bulb is improperly positioned. When the low-side pressure does not increase, replace the TXV.*

*If the low-side gauge reading is higher than normal with normal high-side pressure and frosting of the evaporator suction tube, remove the remote bulb and place it in a pan of ice water. When this action causes the low-side pressure to decrease to normal, the remote bulb may be improperly positioned or insulated. Reposition and reinsulate the remote bulb, and check the low-side pressure. If the low-side pressure is still higher than normal, replace the TXV.*

## Task 6    Inspect and replace the orifice tube(s).

27. All of the following statements are true about using the tool shown in Figure 7–19 to remove a complete orifice tube assembly EXCEPT
    A. Pour a small amount of refrigerant oil on top of the orifice tube to lubricate the O-rings.
    B. Rotate the T-handle to engage the notch in the tool in the orifice tube tangs.
    C. Rotate the T-handle to remove the orifice tube from the evaporator inlet pipe.
    D. Hold the T-handle and rotate the outer sleeve on the tool to remove the orifice tube.

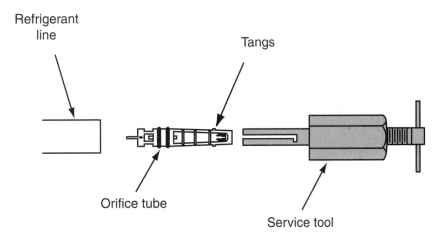

**Figure 7–19** Refrigeration system service tool.

**Hint**    *A restricted orifice tube may cause lower-than-specified low-side pressure, frosting of the orifice tube, and inadequate cooling from the evaporator. When these conditions are present, place a shop towel soaked in hot water around the orifice tube. If the low-side pressure increases, there is moisture freezing in the orifice tube. Under this condition, the refrigeration system must be discharged, evacuated, and recharged.*

*When the orifice tube is warmed with a shop towel and the low-side pressure does not increase, inspect and clean or replace the orifice tube.*

## Task 7    Inspect, test, clean, or replace evaporator(s).

28. Technician A says restricted refrigerant passages in the evaporator may cause frosting of the evaporator outlet pipe.
    Technician B says restricted refrigerant passages in the evaporator may cause much higher-than-specified low-side pressures.
    Who is correct?
    A. A only
    B. B only
    C. Both A and B
    D. Neither A nor B

**Hint**    *Evaporator replacement is necessary if it is leaking or restricted. Evaporator leaks may be detected with an electronic tester with an audible beep. Remove the resistor assembly in the blower circuit from the A/C heater case, and insert the leak tester probe through this opening* *in the case to check for evaporator leaks. Refrigerant leaks in the evaporator also are indicated by evidence of oil in the area of the leak. Evaporator restriction is indicated by a low-side pressure that is considerably lower than specified, and inadequate cooling with a normal TXV or orifice tube.*

*Always discharge the refrigeration system and disconnect the negative battery cable before removing the evaporator. Since the evaporator core usually is mounted in the A/C-heater case, this assembly must be removed from the vehicle to remove the evaporator.*

## Task 8   Inspect, clean and repair evaporator housing, and water drain.

29. The inside of the windshield has an oily film, A/C cooling is inadequate.

   Technician A says this oil film may be caused by a plugged A/C heater case drain.

   Technician B says this oil film may be caused by a leak in the evaporator core.

   Who is correct?

   A. A only

   B. B only

   C. Both A and B

   D. Neither A nor B

**Hint**   *If the water drain is plugged on the A/C-heater case, water collects in the bottom of this case. After a period of time this water becomes very stale and produces a pungent odor in the passenger compartment. In some cases, it may be possible to push a plastic rod up through the A/C heater case drain pipe from the lower end of the pipe and clean out the obstruction. In other cases, it may be necessary to remove the A/C-heater case and clean the water drain opening and pipe.*

*The refrigeration system must be discharged and the coolant drained before the A/C-heater case is removed. If the evaporator core has a leak, an oily film may appear on the inside of the windshield and the discharge air temperature from the evaporator becomes warmer than specified. With the A/C system operating, this leak may be located with an electronic tester.*

## Task 9   Inspect, test, and replace evaporator pressure/temperature control systems and devices.

30. While discussing refrigerant systems with a suction throttling valve (STV), pilot operated absolute (POA) valve, or evaporator pressure regulator (EPR) valve, Technician A says that on some of these refrigerant systems the compressor runs continually in the A/C mode.

   Technician B says an excessive pressure drop across the EPR valve indicates this valve is sticking open.

   Who is correct?

   A. A only

   B. B only

   C. Both A and B

   D. Neither A nor B

**Hint**   *Some refrigeration systems have an evaporator pressure control device connected at the evaporator outlet. This type of pressure control valve may be called a suction throttling valve (STV) or a pilot operated absolute (POA) valve. Some General Motors vehicles have a valve in receiver (VIR) assembly that contains the TXV valve POA valve and receiver/dryer (Figure 7–20). The refrigerant flows from the condenser into the receiver/dryer in the VIR. After leaving the receiver/dryer, the refrigerant flows through the TXV to the evaporator. When the refrigerant returns from the evaporator, it flows through the POA valve to the compressor (Figure 7–21). The POA valve contains a pulsating piston that controls evaporator outlet pressure. In some A/C systems with a POA or STV valve, the compressor runs continually when the driver selects the A/C mode and the ambient temperature is above a specific value.*

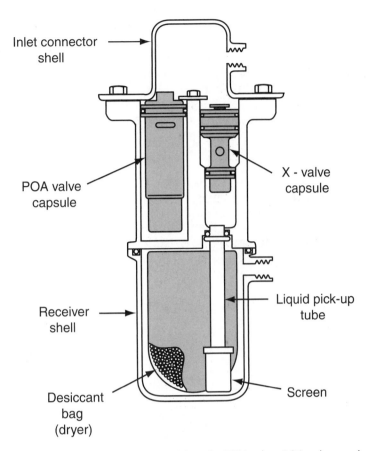

**Figure 7–20** Valves in receiver assembly containing the TXV valve, POA valve, and receiver/dryer.

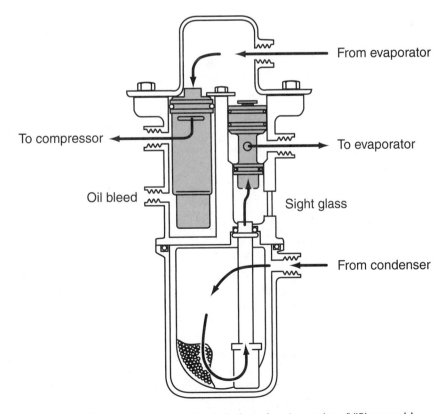

**Figure 7–21** Refrigerant flow through the valves in receiver (VIR) assembly.

*Other A/C systems have an evaporator pressure regulator (EPR) valve mounted in the compressor inlet. The EPR valve performs basically the same function as the STV or POA valve. Some systems with an EPR valve have a compressor inlet service valve and the usual low-side and high-side service valves. An auxiliary gauge in the manifold gauge set may be connected to the compressor inlet service valve. If the refrigerant system and the EPR valve are operating normally, the low-side and high-side gauges should indicate the specified pressures, and the pressure difference between the low-side gauge and the compressor inlet gauge should not exceed 6 psi (41 kPa). The pressure difference in these two gauge readings is the pressure drop across the EPR valve.*

## Task 10    Identify, inspect, and replace A/C system service valves (gauge connections) and valve caps.

31. While discussing the service equipment in Figure 7–22, Technician A says the equipment is designed for refrigerant recovery only.

    Technician B says this equipment is equipped with service valve connections for R-12 refrigerant.

    Who is correct?

    A. A only

    B. B only

    C. Both A and B

    D. Neither A nor B

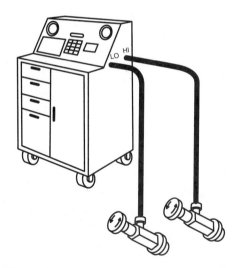

**Figure 7–22** A/C system service equipment.

**Hint**    *R-12 refrigerant systems may have Schrader-type or stem-type service valves. Both types of service valves contain dust caps to keep dirt out of the valves. The Schrader-type service valve contains a valve that is similar to a tire valve. When the manifold gauge set hoses are connected to the service valves, a pin in the hose depresses the Schrader valve (Figure 7–23).*

*Stem-type service valves must be rotated with the proper tool to move the internal valve stem. When the valve stem is front-seated the valve stem completely blocks the refrigerant flow through the valve (view A Figure 7–24). This valve position may be used to isolate the refrigerant system from the compressor for compressor removal. Severe compressor damage may result with the service valves front-seated and the gauge port capped.*

*When the stem-type service valve is back seated, the refrigerant flows through the service valve for normal operation (view B Figure 7–24). In this valve position, refrigerant system pressure is not available at the gauge port.*

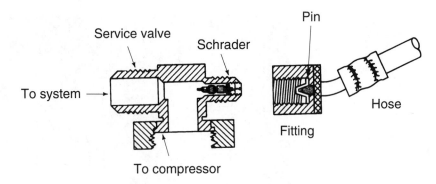

**Figure 7–23** Schrader-type service valve.

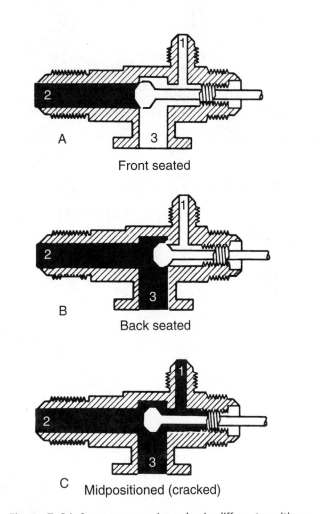

**Figure 7–24** Stem-type service valve in different positions.

*If the stem-type service valve is in the midposition, or cracked position, refrigerant flows normally through the valve, and system pressure is available at the gauge port for diagnostic purposes (view C Figure 7–24).*

*R-134a refrigerant systems have quick-disconnect service valves. An R-12 service valve, (component A Figure 7–25) is compared to an R-134a service valve (component B Figure 7–25). Quick-disconnect hose fittings are located on R-134a manifold gauge sets and recovery/recycling equipment.*

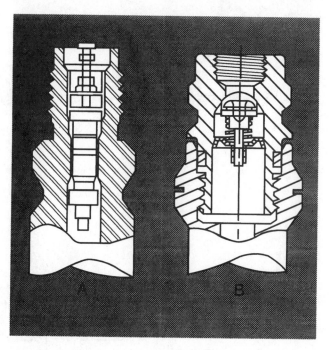

**Figure 7–25**  R-12 service valve (A) compared to an R-134a service valve (B). *(Courtesy of BET, Inc.)*

# Task 11    Inspect and replace A/C system high-pressure relief device.

32. Component 4 in Figure 7–26 discharges refrigerant at approximately
    A. 200 psi (1379 kPa).
    B. 325 psi (2240 kPa).
    C. 375 psi (2585 kPa).
    D. 475 psi (3275 kPa).

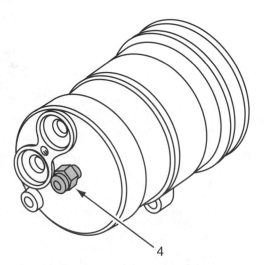

**Figure 7–26** External compressor components.

***Hint***    *If the area around the high-pressure relief valve shows evidence of refrigeration oil, check the refrigeration system pressures. When these pressures are higher than normal, correct the causes of the high pressures such as a refrigerant overcharge, air and moisture in the system, or restricted condenser air passages. If the refrigeration pressures are normal, replace the high-pressure relief valve. The refrigeration system must be discharged before the high-pressure relief valve is removed.*

# Refrigeration System Component Diagnosis and Repair, Heating and Engine Cooling Systems Diagnosis and Repair

## ASE Tasks, Questions, and Related Information

**Task 1**  Diagnose the cause of temperature control problems in the heater/ventilation system; determine needed repairs.

33. The discharge air temperature is higher than specified in an A/C-heater system with a coolant flow control valve with the system operating in the maximum A/C mode.

Technician A says the coolant flow control valve may be stuck open.

Technician B says the coolant flow control valve vacuum hose may have a leak.

Who is correct?

A. A only

B. B only

C. Both A and B

D. Neither A nor B

**Hint**  *If the heater supplies cold air continually, the engine thermostat may be stuck open, or the temperature door may be stuck in the cold position. A coolant flow control valve stuck in the closed position or restricted also causes cold air discharge from the heater.*

*Tape a thermometer to the upper radiator hose and allow the engine to idle for fifteen minutes. The thermometer reading should be within a few degrees or the specified thermostat opening temperature. If the thermometer reading is considerably lower than the thermostat rating, the thermostat is inoperative.*

*Move the A/C-heater controls from full-cold to the maximum heat position and observe the temperature door linkage. If this linkage does not move, the door is sticking or the actuator system is inoperative.*

*When the A/C-heater controls are in the heat position, check the temperature on both sides of the coolant flow control valve. The hose temperature should be the same on both sides of the valve. If this valve is restricted or closed, the temperature at the valve inlet is much higher than the temperature at the outlet.*

*If the A/C-heater system supplies warm or hot air continually, check the refrigeration system pressures. When these pressures are not within specifications, repair the refrigeration system. If the refrigeration pressures are within specification and the air discharge is warm, the temperature blend door may be stuck.*

**Task 2**  Diagnose window fogging problems; determine needed repairs.

34. The inside of the windshield has a sticky film.

Technician A says the engine coolant level should be checked.

Technician B says the heater core may be leaking.

Who is correct?

A. A only

B. B only

C. Both A and B

D. Neither A nor B

**Hint**    *Windshield fogging may be caused by a leaking heater core, or a plugged A/C-heater case drain, which allows water to collect in this case. If the A/C-heater case drain is plugged and water has collected in the case, the water becomes stagnant and provides a very pungent odor in the passenger compartment. If the heater core is leaking, there is a loss of coolant from the cooling system.*

*A leaking evaporator core may cause an oil film on the inside of the windshield. If the evaporator core is leaking, the refrigeration system pressures are lower than specified once the system has lost some refrigerant. Check evaporator leaks with a probe of an electronic leak detector inserted through the resistor assembly opening in the A/C-heater case.*

## Task 3    Perform engine cooling system tests; determine needed repairs.

35. A customer complains about engine coolant loss. The cooling system is pressurized at 15 psi (103 kPa) for 15 minutes. There is no visible sign of coolant leaks in the engine or passenger compartments, but the pressure on the tester gauge decreases to 5 psi (34 kPa). This problem could be caused by any of the following EXCEPT
    A. a leaking heater core.
    B. a leaking transmission cooler.
    C. a leaking head gasket.
    D. a cracked cylinder head.

**Hint**    *A pressure tester may be connected to the radiator filler neck to check for cooling system leaks. Operate the tester pump and apply 15 psi to the cooling system. Inspect the cooling system for external leaks with the system pressurized. If the gauge pressure drops more than specified by the vehicle manufacturer, the cooling system has a leak. If there are no visible external leaks, check the front floor mat for coolant dripping out of the heater core. When there are no external leaks, check the engine for combustion chamber leaks.*

*The radiator pressure cap may be tested with the pressure tester. When the tester pump is operated, the cap should hold the rated pressure. Always relieve the pressure before removing the tester.*

## Task 4    Inspect and replace engine cooling and heater system hoses and pipes.

36. The vacuum valve in the radiator cap is stuck closed. The result of this problem could be
    A. a collapsed upper radiator hose after the engine is shut off.
    B. excessive cooling system pressure at normal engine temperature.
    C. engine overheating when operating under heavy load.
    D. engine overheating during extended idle periods.

**Hint**    *All cooling system hoses should be inspected for soft spots, swelling, hardening, chafing, leaks, and collapsing. If any of these conditions are present, hose replacement is necessary. Hose clamps should be inspected to make sure they are tight. Some radiator hoses contain a wire coil inside them to prevent hose collapse as the coolant temperature decreases. Remember to include heater hoses, and the bypass hose, in the hose inspection. Prior to hose removal, the coolant must be drained from the radiator.*

*Because the friction surfaces are the sides of a V-belt, the belt must be replaced if the sides are worn and the belt is contacting the bottom of the pulley.*

*The belt tension may be checked with the engine shut off, and a belt tension gauge placed over the belt at the center of the longest belt span. A loose, or worn, belt may cause a squealing noise when the engine is accelerated.*

*The belt tension also may be checked by measuring the amount of belt deflection with the engine shut off. Use your thumb to depress the belt at the center of the belt span. If the belt tension is correct, the belt should have 1/2-in. deflection per foot of belt span.*

*Ribbed V-belts usually have a spring-loaded belt tensioner, with a belt wear indicator scale on the tensioner housing. If a power-steering pump belt requires tightening, always pry on the pump ear, not on the housing.*

## Task 5   Inspect, test, and replace radiator, pressure cap, coolant recovery system, and water pump.

37. The coolant level in the coolant recovery container is normal when the engine is cold. This level becomes much higher than normal after the vehicle has been driven for 45 minutes.

    Technician A says some of the radiator tubes may be restricted.

    Technician B says the radiator cap may be the problem.

    Who is correct?

    A. A only

    B. B only

    C. Both A and B

    D. Neither A nor B

**Hint**    *The radiator cap should be inspected for a damaged sealing gasket, or vacuum valve. If the pressure cap sealing gasket, or seat, are damaged, the engine will overheat, and coolant is lost to the coolant recovery system. Under this condition, the coolant recovery container becomes over-filled with coolant.*

*If the cap vacuum valve is sticking, a vacuum may occur in the cooling system after the engine is shut off and the coolant temperature decreases. This vacuum may cause collapsed cooling system hoses. A pressure tester may be used to test the pressure cap, and pressure test the entire cooling system.*

*The coolant level should be at the appropriate mark on the recovery container, depending on engine temperature.*

*With the engine shut off, grasp the fan blades, or the water pump hub, and try to move the blades from side-to-side. This action checks for looseness in the water pump bearing. If there is any side-to-side movement in the bearing, water pump replacement is required.*

*Check for coolant leaks, rust, or residue at the water pump drain hole in the bottom of the pump, and at the inlet hose connected to the pump. When coolant is dripping from the pump drain hole, replace the pump. The water pump may be tested with the pressure tester connected to the radiator filler neck.*

## Task 6   Inspect, test, and replace thermostat, bypass, and housing.

38. A port-fuel-injected engine has an excessively rich air-fuel ratio. This problem could be caused by
    A. engine overheating.
    B. a malfunctioning radiator cap.
    C. the engine thermostat stuck open.
    D. the coolant control valve stuck open.

**Hint**    *The thermostat may be submerged with a thermometer in a container filled with water. Heat the water while observing the thermostat valve and the thermometer. The thermostat valve should begin to open when the temperature on the thermometer is equal to the rated temperature stamped on the thermostat. Replace the thermostat if it does not open at the rated temperature. Many thermostats are marked for installation in the proper direction. Inspect the bypass hose for cracks, deterioration, and restrictions, and replace the hose if these conditions are present.*

## Task 7   Identify, inspect, recover coolant; flush, and refill system with proper coolant.

39. All the following statements about cooling system service are true EXCEPT
    A. When the cooling system pressure is increased, the boiling point is decreased.
    B. If more antifreeze is added to the coolant, the boiling point is increased.
    C. A good quality ethylene glycol antifreeze contains antirust and corrosion inhibitors.
    D. Coolant solutions must be recovered, recycled, or handled as hazardous waste.

**Hint**    *If the radiator tubes and coolant passages in the block and cylinder head are restricted with rust and other contaminants, these components may be flushed. Cooling system flushing equipment is available for this purpose. Always operate the flushing equipment according to the equipment manufacturer's directions, and be sure that your service procedure conforms to pollution laws in your state. Engine coolant must be recycled or handled as a hazardous waste material.*

*Coolant recovery and reconditioning machines are available to remove harmful particles and restore corrosion additives so the coolant can be returned to the cooling system.*

**Task 8    Inspect, test, and replace fan (both electrical and mechanical), fan clutch, fan belts, fan shroud, and air dams.**

40. In the electric cooling fan circuit in Figure 7–27, the low-speed and high-speed fans do not operate unless the air-conditioning is turned on. The cause of this problem could be

A. a blown 10 A, 5C fuse in the instrument panel fuse block.

B. an inoperative engine coolant temperature sensor.

C. an inoperative high-speed coolant fan relay.

D. an inoperative A/C pressure fan switch.

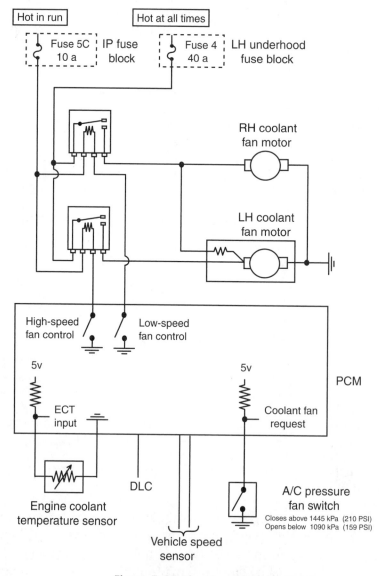

**Figure 7–27** Electric cooling fan circuit.

**Hint**     *If the radiator shroud is loose, improperly positioned, or broken, airflow through the radiator is reduced, and engine overheating may result.*

*The viscous-drive fan clutch should be visually inspected for leaks. If there are oily streaks radiating outward from the hub shaft, the fluid has leaked out of the clutch.*

*With the engine shut off, rotate the cooling fan by hand. If the viscous clutch allows the fan blades to rotate easily both hot and cold, the clutch should be replaced. A slipping viscous clutch results in engine overheating. If there is any looseness between the viscous clutch and the shaft, replace the viscous clutch.*

*If the electric-drive cooling fan does not operate at the coolant temperature specified by the vehicle manufacturer, engine overheating will result, especially at idle and lower speeds when airflow through the radiator is reduced.*

**Task 9**  Inspect, test, and replace heater coolant control valve (manual, vacuum, and electrical types) and auxiliary coolant pump.

41. In Figure 7–28, the adjustment being performed is
    A. the air mix door adjustment.
    B. the manual coolant control valve adjustment.
    C. the ventilation door control rod adjustment.
    D. the defroster door control rod adjustment.

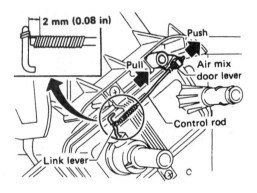

Figure 7–28 A/C-heater system adjustment. *(Used with permission from Nissan North America, Inc.)*

**Hint**     *The heater coolant control valve may be operated manually, or by an electric solenoid. Some coolant control valves are operated by vacuum supplied from the intake manifold through a vacuum solenoid. In many A/C systems, the heater coolant control valve is closed in the maximum A/C mode. This valve remains open in other modes. If the heater coolant control valve is stuck closed or restricted, there is reduced or no heat from the heater core. The hoses on each side of the heater coolant control valve should be the same temperature. When the inlet hose is considerably hotter than the outlet hose, the coolant control valve is restricted. If the coolant control valve is sticking open, cooling may be inadequate in the maximum A/C mode.*

**Task 10**  Inspect, flush, and replace heater core.

42. A gurgling noise is heard inside the passenger compartment. The noise is coming from the A/C-heater case.
    Technician A says the heater core coolant passages may be restricted.
    Technician B says the coolant level in the cooling system may be low.
    Who is correct?
    A. A only
    B. B only
    C. Both A and B
    D. Neither A nor B

*Hint*    *If the heater core is restricted, passenger compartment heating is reduced. Restricted air passages through the core also reduce passenger compartment heating. After the heater hoses are removed, the heater core may be flushed with a water hose. If the heater core is severely restricted, it may be removed and sent to a radiator shop for flushing with a special cleaning solution.*

# Operating Systems and Related Controls Diagnosis and Repair, Electrical

## ASE Tasks, Questions, and Related Information

**Task 1**    Diagnose the cause of failures in the electrical control system of heating, ventilating, and A/C systems; determine needed repairs.

43. An A/C-heater system does not discharge cool air in the A/C mode. The refrigerant system pressures are normal. The temperature door does not move when the A/C-heater controls are moved from maximum A/C to maximum heat (Figure 7–29), but the blower motor operation is normal. All of the following could be the cause of the problem EXCEPT
    A. an open circuit at terminal 6 on the electronic door actuator.
    B. a blown 15A, number 8 fuse in the fuse junction panel.
    C. an open circuit at terminal C2-15 in the automatic temperature control module.
    D. an open circuit at terminal 8 on the electronic door actuator motor.

*Hint*    *Many computer-controlled A/C systems have self-diagnostic capabilities that provide diagnostic trouble codes (DTCs) representing faults in a specific area. The technician usually has to perform some voltmeter or ohmmeter tests to locate the exact cause of the problem. For example, a DTC representing the in-car temperature sensor may be obtained. Thus, the technician has to perform ohmmeter tests on this sensor and the connecting wires to find the exact cause of the problem.*

*If the A/C-heater system does not have self-diagnostic capabilities, the technician has to perform voltmeter and ohmmeter tests to locate the source of the problem.*

**Task 2**    Inspect, test, repair, and replace A/C-heater blower motors, resistors, switches, relay/modules, wiring, and protection devices.

44. A blower motor operates at high speed but does not operate at any other speed (Figure 7–30). The cause of this problem could be
    A. an open circuit between the two lower resistors in the resistor assembly.
    B. an open circuit in the blower motor switch ground connection.
    C. an open circuit at terminal 4 in the blower switch.
    D. an open circuit at terminal 4 in the resistor assembly.

*Hint*    *In many blower motor circuits, the motor switch and a resistor assembly are connected on the ground side of the motor. When the heater or A/C system is turned on, voltage is supplied to one brush in the blower motor. Current flows from the other motor brush through the blower motor switch and resistor assembly to ground. When the blower switch is in the high position, the blower motor brush is ground directly through the switch contacts. If one of the lower speeds is selected, the motor brush is grounded through the proper resistor in the resistor assembly to provide*

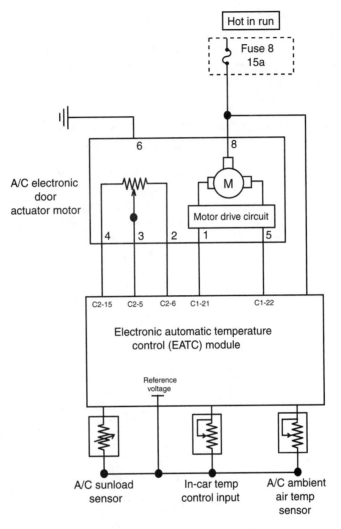

Figure 7–29 A/C-heater electrical system.

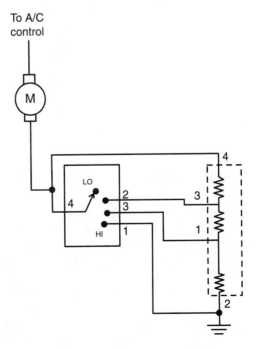

Figure 7–30 Blower motor circuit.

*the blower speed selected. The resistor assembly usually is mounted in the A/C-heater case so airflow past the resistors provides a cooling action.*

*In some computer-controlled A/C systems, the blower motor speed is controlled by the A/C computer and the blower speed control module. The A/C computer commands the blower speed control module to provide the proper blower speed. The blower speed module usually controls the blower speed with a pulse width modulated (PWM) voltage signal.*

## Task 3    Inspect, test, repair, and replace A/C compressor clutch coil, relay/modules, wiring, sensors, switches, diodes, and protection devices.

45. The A/C compressor clutch in Figure 7–31 is de-energized after the vehicle is driven for about 30 minutes, and it does not engage again until the engine is shut off for a period of time. The engine coolant temperature and refrigerant system pressures are normal. The most likely cause of this problem could be

    A. an inoperative engine coolant temperature sensor.

    B. an inoperative A/C pressure cut-off switch.

    C. an inoperative clutch control relay.

    D. an inoperative compressor clutch coil.

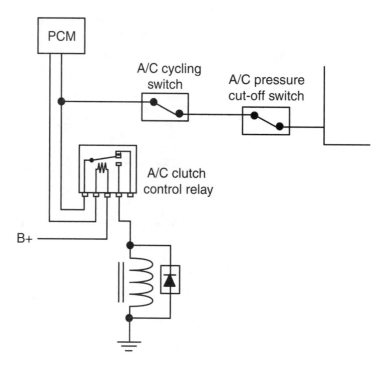

**Figure 7–31** Compressor clutch circuit.

**Hint**    *When the A/C mode is selected in many computer-controlled A/C systems, the A/C computer supplies 12V to the pressure cycling switch. Voltage is supplied through this switch to the A/C pressure cut-off switch and the clutch control relay contacts. A signal is transmitted to the powertrain control module (PCM) from the wire between the A/C pressure switch and the relay contacts. This signal informs the PCM that the A/C mode is selected.*

*The PCM scans the input signals to determine if the compressor clutch operation is appropriate. If the engine coolant temperature sensor indicates very high temperature, or the throttle position sensor indicates wide open throttle, the PCM does not ground the compressor clutch winding. Under this condition, the compressor clutch remains inoperative. The PCM does not energize the clutch control winding for a brief time period after the engine is started. If the PCM detects a low idle speed condition, it does not energize the clutch control relay winding.*

*If the PCM inputs indicate the proper conditions are present for compressor clutch operation, the PCM grounds the clutch control relay winding. This action closes the relay contacts that supply voltage to the compressor clutch coil to engage the clutch.*

*The A/C clutch diode reduces voltage spikes when the compressor clutch is shut off. Some PCMs have a power steering switch input. When this switch input indicates high-power steering pump pressure, the PCM opens the ground circuit from the clutch control relay winding. This action de-energizes the compressor clutch.*

*The A/C pressure cut-off switch opens the compressor clutch circuit if the refrigerant system pressure exceeds 399 to 445 psi (2,751 to 3,068 kPa). This action prevents damage to refrigeration system components.*

## Task 4    Inspect, test, repair, replace, and adjust A/C-related engine control systems.

46. All of the following statements about an electronic fuel injection (EFI) system with an IAC motor are true EXCEPT
    A. If the A/C is turned on, the IAC duty cycle should decrease.
    B. Airflow past the IAC motor pintle increases when the engine is cold.
    C. Airflow past the IAC motor pintle increases when the A/C is turned on.
    D. Airflow past the IAC motor pintle enters the intake manifold below the throttles.

**Hint**    *In many EFI systems, the powertrain control module (PCM) operates the idle air control (IAC) motor to control idle rpm. The IAC motor controls idle rpm by regulating the amount of air flowing past the motor pintle into the intake manifold below the throttle (Figure 7–32). For example, if the engine coolant is cold, the engine coolant temperature (ECT) sensor informs the PCM regarding this condition. In response to this cold ECT signal, the PCM operates the IAC motor so more air flows past the motor pintle into the intake manifold below the throttle. This action increases engine idle rpm to prevent stalling when the engine is cold. If the A/C is turned on, a voltage signal is sent from the A/C system to the PCM. When this signal is received, the PCM operates the IAC motor so more air flows past the motor pintle into the intake manifold below the throttle. This action increases engine rpm slightly to offset the increased load placed on the engine by the A/C compressor. If the duty cycle idle rpm is not increased when the A/C is turned on, engine stalling may occur.*

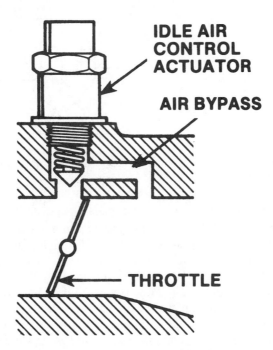

**Figure 7–32** Idle air control (IAC) motor.

*If the IAC duty cycle is too low and the engine idle rpm is too slow with the A/C turned on, the PCM may not be receiving a voltage signal from the A/C system, or the IAC motor may be malfunctioning or sticking. A scan tool may be used to test the IAC motor. On some EFI systems, the scan tool indicates the IAC duty cycle. When the A/C is turned on, the IAC duty cycle should increase.*

## Task 5   Inspect, test, repair, replace, and adjust load sensitive A/C compressor cut-off systems.

47. While discussing compressor clutch control in an A/C system with a pressure transducer (Figure 7–33), Technician A says the pressure transducer is connected in the compressor clutch circuit.

    Technician B says the pressure transducer contains a set of contacts that are normally closed.

    Who is correct?
    A. A only
    B. B only
    C. Both A and B
    D. Neither A nor B

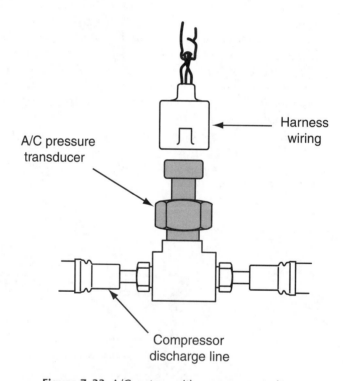

A/C pressure transducer

Harness wiring

Compressor discharge line

**Figure 7–33** A/C system with pressure transducer.

**Hint**   *Some A/C compressor clutch circuits have a high-pressure cut-off switch. This switch opens the compressor clutch circuit and prevents compressor clutch operation if the refrigeration system pressure exceeds 430 psi (2,960 kPa). This high-pressure cut-off switch prevents damage to the compressor or other refrigeration system components.*

*Other A/C systems have an A/C pressure transducer mounted in the compressor discharge line near the compressor. The PCM sends a 5 V signal to the pressure transducer, and the potentiometer in the pressure transducer sends a voltage signal to the PCM in relation to refrigeration system pressure (Figure 7–34). A ground wire is connected from the pressure transducer to the PCM. If the refrigeration system pressure becomes excessively high, the PCM opens the circuit from the compressor clutch relay winding to ground in response to the pressure transducer signal.*

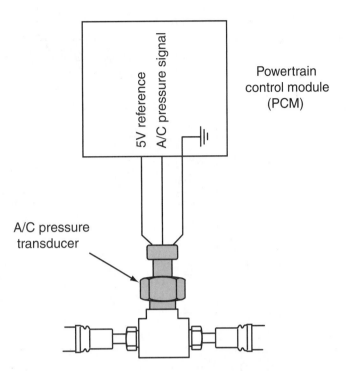

**Figure 7–34** Pressure transducer to PCM wiring connections.

*Under this condition, the compressor clutch relay contacts open, and the compressor clutch is de-energized. The PCM also uses the pressure transducer for cooling fan control.*

*Some compressor clutch circuits contain a thermal limiter switch that senses compressor surface temperature. If the compressor surface temperature becomes excessive, the thermal limiter switch opens the compressor clutch circuit to de-energize the clutch.*

## Task 6  Inspect, test, repair, and replace engine cooling/condenser fan motors, relays/modules, switches, sensors, wiring, and protection devices.

48. The cooling fan system in Figure 7–35 operates normally on high speed, but there is no low-speed fan operation under any operating condition. The cause of this problem could be
    A. an open circuit in the wire between the number 30 terminals in the two fan relays.
    B. a blown 40-amp maxifuse in the LH maxifuse center.
    C. a blown 10-amp cooling fan/TCC fuse in the fuseblock.
    D. an open circuit at terminal B on the A/C head pressure switch.

**Hint**    *Cooling fan circuits vary, depending on the vehicle make and model year. Some cooling fan systems have a low-speed and a high-speed cooling fan. The PCM grounds the low-speed or high-speed relay windings in response to the engine coolant temperature and compressor head pressure switch. When the engine coolant reaches 212°F (100°C), the PCM grounds the low-speed fan relay winding. This action closes these relay contacts, and voltage is supplied through the contacts in the low-speed fan motor.*

*When the engine coolant reaches 226°F (108°C), or the compressor discharge pressure reaches 210 psi (1,448 kPa), the PCM grounds the high-speed relay winding. Under this condition, the relay contacts close and supply voltage to the high-speed fan motor. The compressor head pressure switch normally is closed, and this switch opens at 210 psi (1,448 kPa). When the compressor head pressure switch opens, the PCM grounds the high-speed fan relay winding.*

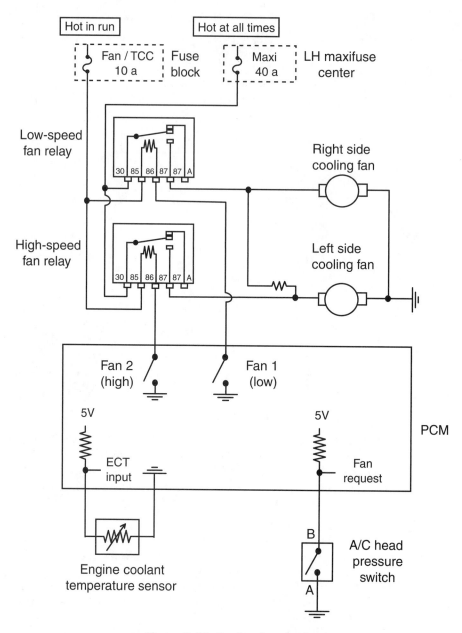

**Figure 7–35** Cooling fan circuit.

## Task 7    Inspect, test, adjust, repair and replace electric actuator motors, relays/modules, switches, sensors, wiring, and protection devices (including dual/multi-zone systems).

49. All of the following statements about computer-controlled A/C system actuator motors are true EXCEPT

   A. Some actuator motors are calibrated automatically in the self-diagnostic mode.

   B. A/C system diagnostic trouble codes represent a fault in a specific component.

   C. The actuator motor control rods must be calibrated manually on some systems.

   D. The actuator motor control rods should only require adjustment after motor replacement or misadjustment.

*Hint*    *Many computer-controlled A/C systems have self-diagnostic capabilities. On Chrysler LH and LHS cars, the self-diagnostic A/C mode is entered by pressing the floor, mix, and defrost buttons simultaneously for a few seconds. The engine must be running with the vehicle stopped and the*

*A/C temperature control set at 75°F (24°C). When the diagnostic mode is entered, the A/C computer calibrates the actuator motors and performs specific system tests. During this time the control head display continues blinking.*

*On some cars, such as a Nissan Maxima, the self-diagnostic mode in the automatic A/C system is entered by starting the engine and pressing the off button in the A/C control head for five seconds. The off button must be pressed within ten seconds after the engine is started, and the fresh vent lever must be in the off position. The self diagnostic mode may be cancelled by pressing the auto button or turning off the ignition switch.*

*The self-diagnostic tests are completed in five steps, and the up arrow for the temperature setting on the A/C control head is pressed to move to the next step. When the down arrow for the temperature setting is pressed, the diagnostic system returns to the previous step. The five steps in the diagnostic tests follow.*

- *Step 1 checks the LEDs and segments in the A/C control head.*
- *Step 2 checks the input sensor signals.*
- *Step 3 checks the mode door position switch.*
- *Step 4 checks the mode door electric actuators.*
- *Step 5 checks the temperature detected by each input sensor.*

*On some systems, the actuator door control rods may be adjusted. These adjustments are necessary after components, such as door actuator motors, are replaced. The procedure for the mode door control rod adjustment follows.*

- *With the mode door installed on the A/C-heater case and the wiring connector attached to the actuator, disconnect the door motor rod from the slide link.*
- *Enter step four in the self-diagnostic mode so 41 is displayed in the A/C control head.*
- *Move the slide link by hand until the mode door is in the vent mode (Figure 7–36).*
- *Connect the motor rod to the slide link.*
- *Continue pressing the DEF button until all six modes have been obtained in step 4, and be sure the door moves to the proper position in each mode.*

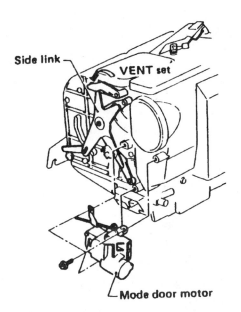

**Figure 7–36** Mode door control rod adjustment. *(Used with permission from Nissan North America, Inc.)*

## Task 8 Inspect, test, service, or replace heating, ventilating, and A/C control panel assemblies.

50. All of the following statements about A/C control panel service are true EXCEPT
    A. The negative battery cable must be removed before A/C control panel removal.
    B. The refrigeration system must be discharged before the A/C control panel is removed.

C. If the vehicle is air bag–equipped, wait the specified time period after negative battery cable removal.

D. Self-diagnostic tests may indicate a defective A/C control panel in a computer-controlled A/C system.

*Hint*    *Self-diagnostic tests should indicate an inoperative A/C control panel. In many systems, the A/C computer is contained in the control panel. Always disconnect the negative battery cable before removing the A/C control panel assembly. If the vehicle is equipped with an air bag, wait for the time specified by the vehicle manufacturer after the negative battery cable is disconnected before starting the A/C control panel removal procedure. Remove the instrument panel molding around the A/C control panel, and then remove the A/C control panel retaining screws. Pull the A/C control panel out of the instrument panel as far as possible. Disconnect the electrical connectors from the A/C control panel and remove the control panel.*

# Operating Systems and Related Controls Diagnosis and Repair, Vacuum/Mechanical

## ASE Tasks, Questions, and Related Information

**Task 1**    Diagnose the cause of failures in the vacuum and mechanical switches and controls of the heating, ventilating, and A/C systems; determine needed repairs.

51. When testing the A/C vacuum system in Figure 7–37, 18 in. Hg is supplied to the vacuum distribution hose connected to the A/C control vacuum valve. Under this condition, there should be 18 in. Hg supplied to
   A. the panel door actuator with the switch in the off position.
   B. the outside recirculate door actuator with the switch in the normal A/C position.
   C. the outside recirculate door actuator with the switch in the floor position.
   D. the panel door actuator with the switch in the maximum A/C position.

*Hint*    *Vacuum hoses and switches may be tested for leaks with a vacuum pump and gauge. When vacuum is supplied to one end of a vacuum hose and the other end of the hose is plugged, the hose should hold 15 to 20 in. Hg without leaking. The same method may be used to test vacuum switches.*

**Task 2**    Inspect, test, service, or replace heating, ventilating, and A/C control panel assemblies.

52. To place some manual A/C control panels in the diagnostic mode, Technician A says the engine should be off and the ignition switch on.
   Technician B says the fan speed control should be in the off position.
   Who is correct?
   A. A only
   B. B only
   C. Both A and B
   D. Neither A nor B

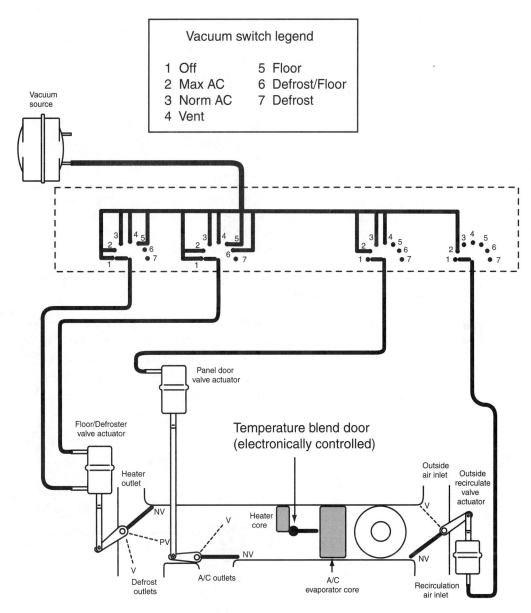

**Figure 7–37** A/C vacuum circuit.

*Hint*    Some manual A/C control panels can be placed in the diagnostic mode with the engine running and the vehicle not moving. The following A/C control panel settings are required to enter the diagnostic mode:

1. *Fully apply the parking brake.*
2. *Start the engine and leave the gear selector in PARK with an automatic transmission, or neutral with a manual transmission.*
3. *Set the fan speed to any speed but OFF.*
4. *Rotate the temperature control fully counterclockwise to full cold.*
5. *Rotate the mode control fully clockwise to the defrost position.*
6. *The A/C button may be on or off.*
7. *Press and hold the electric backlite (EBL) button until the mechanical instrument cluster odometer display indicates AC00 (Figure 7–38). The body control module (BCM) will chime once and the light-emitting diode (LED) near the A/C button will begin flashing.*
8. *Release the EBL button and wait until the LED stops flashing. This indicates the error check and A/C unit door calibration is complete. The diagnostic trouble codes are now displayed in the odometer display one at a time. To display the next DTC, press the A/C*

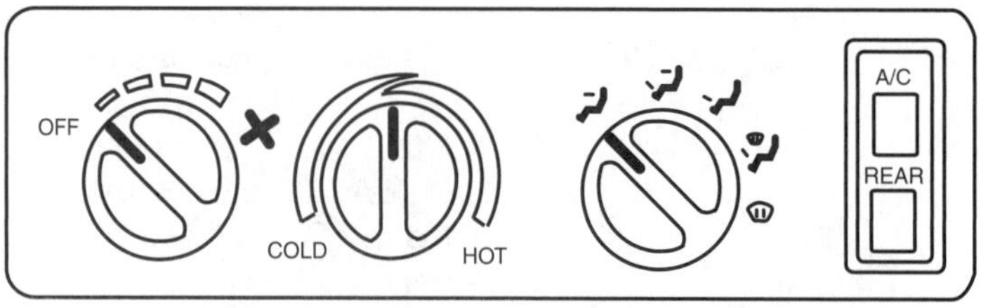

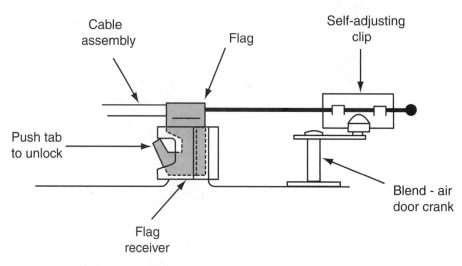

**Figure 7–38** Manual A/C control panel.

*button. Touching any other controls during the diagnostic mode cancels this mode. The odometer returns to a normal display after all the DTCs are shown.*

*Always disconnect the negative battery cable before removing the A/C control panel. Before removing the A/C control panel on an air bag–equipped vehicle, wait for the time period specified by the vehicle manufacturer after the negative battery cable is disconnected.*

## Task 3    Inspect, test, adjust, and replace heating, ventilating, and A/C control cables and linkages.

53. While adjusting the temperature control cable in Figure 7–39
    A. the black cable attaching flag must be removed from the flag receiver.
    B. the self-adjusting clip must be removed from the blend-air door crank.
    C. the blend-air door crank is rotated fully counterclockwise.
    D. the temperature control lever must be in the maximum heat position.

**Figure 7–39** Temperature door cable.

**Hint**    *Many A/C control cables require an adjustment if they have been disconnected or replaced. With the self-adjusting clip installed on the temperature control cable and the blend-air door crank arm, install the black temperature control cable attaching flag. Hold the temperature lever in the maximum cold position and rotate the temperature blend-air door crank arm fully counterclockwise by hand. During this procedure, the self-adjusting clip reaches the proper position on the temperature control cable.*

## Task 4    Inspect, test, and replace heating, ventilating, and A/C vacuum actuators (diaphragms/motors) and hoses.

54. An A/C system has vacuum-operated mode doors except for the temperature door, which is operated by an electric actuator (refer to Figure 7–37). In the A/C mode, air

is discharged from the panel ducts for a short while and then the air discharge switches to the floor ducts. The other modes operate properly. The panel door actuator motor is tested and proven to be satisfactory.

Technician A says the blue vacuum hose from the A/C control switch to the panel door actuator may be leaking.

Technician B says the black vacuum hose connected from the vacuum source to the A/C control switch may be leaking.

Who is correct?

A. A only

B. B only

C. Both A and B

D. Neither A nor B

**Hint**   *Connect the vacuum pump to each vacuum actuator and supply 15 to 20 in. Hg to the actuator (Figure 7–40). Check the vacuum actuator rod to be sure it moves freely. Close the vacuum pump valve and observe the vacuum gauge. The gauge reading should remain steady for at least one minute. If the gauge reading drops slowly, the actuator is leaking.*

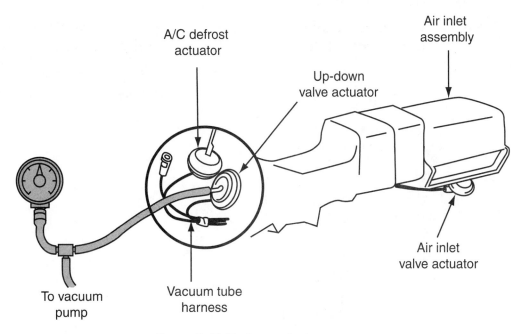

**Figure 7–40** Testing mode door actuator.

## Task 5   Identify, inspect, test, and replace heating, ventilating, and A/C vacuum reservoir, check valve, and restrictors.

55. An A/C system has a vacuum reservoir, check valve, and vacuum-operated mode door actuators including the blend-air door (Figure 7–41). While operating in the A/C mode and climbing a long hill with the throttle nearly wide open, the air discharge temperature gradually becomes warm and the air discharge switches from the panel to the floor ducts. The A/C system operates normally under all other conditions.

The cause of this problem could be

A. a leaking panel door vacuum actuator.

B. an inoperative vacuum reservoir check valve.

C. a leaking blend-air door vacuum actuator.

D. a leaking intake manifold gasket.

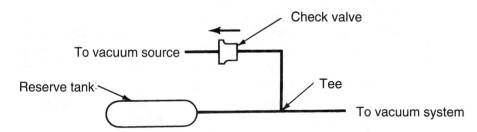

**Figure 7–41** Vacuum reservoir and check valve.

*Hint*    *Vacuum reservoirs may be connected in A/C systems to maintain the vacuum supply during periods of wide open throttle operation when the intake manifold vacuum is very low. A check valve usually is connected between the reservoir and the vacuum source to trap the vacuum in the reservoir when the intake manifold vacuum decreases. Some A/C vacuum systems have a restrictor that delays the vacuum supply to certain components.*

*When vacuum is supplied from a vacuum pump to the reservoir, it should hold 15 to 20 in. Hg without leaking. A check valve should hold vacuum one way, but leak air through it in the opposite direction.*

## Task 6    Inspect, test, adjust, repair, or replace heating, ventilating, and A/C ducts, doors, and outlets (including dual/multi-zone systems).

56.  The outside air-recirculation door is stuck in position A (Figure 7–42).

Technician A says that under this condition outside air is drawn into the A/C-heater case.

Technician B says that under this condition some in-vehicle air leaks past the door.

Who is correct?

A.  A only

B.  B only

C.  Both A and B

D.  Neither A nor B

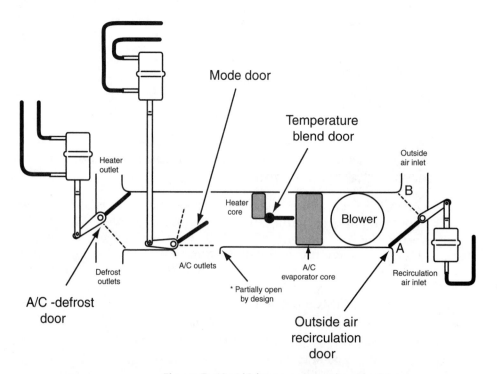

**Figure 7–42** A/C-heater case and mode doors.

*Hint*    *All the mode doors in the A/C-heater case should move freely. Some mode doors are designed to provide a certain amount of air leakage past the door. For example, when the defrost door is directing air to the panel ducts, some air leaks past the door to the defrost ducts. The A/C-heater case and outlet ducts must not have any air leaks.*

# Operating Systems and Related Controls Diagnosis and Repair, Automatic and Semi-Automatic Heating, Ventilating, and A/C Systems

## ASE Tasks, Questions, and Related Information

**Task 1**    Diagnose temperature control system problems; determine needed repairs.

57. In a semiautomatic A/C system, the temperature control is set at 70°F (21°C), and the in-car temperature is 80°F (27°C) after driving the car for one hour. The refrigerant system pressures are normal.

    Technician A says the in-car sensor may be inoperative.

    Technician B says the temperature door may be sticking.

    Who is correct?

    A. A only

    B. B only

    C. Both A and B

    D. Neither A nor B

*Hint*    *When diagnosing improper temperature control, always visually inspect the A/C system for loose or damaged electrical connections and leaking, loose, or deteriorated vacuum hoses. Check the coolant level and engine coolant temperature. Test the refrigerant system pressures. Check the temperature door and related control system.*

*Many automatic and semiautomatic A/C systems have self-diagnostic capabilities that provide diagnostic trouble codes (DTCs). These DTCs represent a fault in a specific area such as a sensor or electric actuator. The technician must perform voltmeter or ohmmeter tests to locate the exact cause of the problem.*

**Task 2**    Diagnose blower system problems; determine needed repairs.

58. The blower motor in Figure 7–43 operates only at low speed. When the fan speed control in the A/C controls is changed from low speed to high speed, the voltage at blower motor terminal A changes from 4 V to 13.5 V. The cause of this problem could be

    A. an inoperative blower motor.

    B. an open blower motor ground.

    C. an inoperative HVAC power module.

    D. an open circuit at programmer terminal D2.

*Hint*    *Blower motor circuits vary depending on the vehicle make and model year. The technician must use the wiring diagram and diagnostic procedure for the system being diagnosed. Check all*

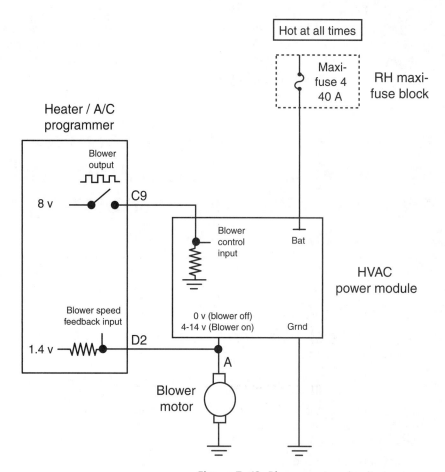

**Figure 7–43** Blower motor circuit.

*the wiring connections and the fuse or fuse link in the blower motor circuit. Be sure the blower motor ground is satisfactory.*

*Check the voltage supplied to the blower motor. If this voltage is less than specified, repair the blower circuit. Supply 12 V and a ground connection to the blower motor terminals with the motor wiring connector disconnected. If the motor operates properly with 12 V supplied, the motor is satisfactory. When the blower speed is slow, replace the motor.*

### Task 3    Diagnose air distribution system problems; determine needed repairs (including dual/multi-zone systems).

59. The computer-controlled A/C system in Figure 7–44 will not go into the recirculation air mode. All the other modes operate normally. The vacuum supplied to the outside air recirculation door vacuum actuator is 1 in. Hg with the engine idling.

    Technician A says to check the check valves, vacuum tank, and vacuum supply to the HVAC programmer.

    Technician B says to check the programmer vacuum switch and hose to the inoperative door actuator.

    Who is correct?

    A.  A only

    B.  B only

    C.  Both A and B

    D.  Neither A nor B

*Hint*    *In some computer-controlled A/C systems, the intake manifold vacuum is supplied through a check valve and vacuum tank to the HVAC programmer. Another check valve may be located in*

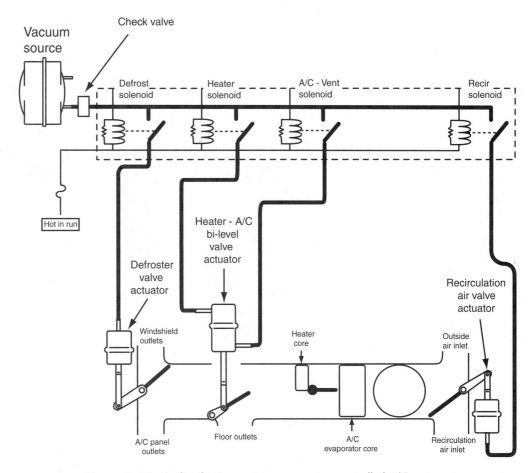

**Figure 7–44** Air distribution system, computer-controlled A/C system.

*the programmer. A vacuum switch supplies vacuum to the appropriate mode door actuator. If none of the mode doors operate properly, check the vacuum source to the programmer. When only one mode door does not operate properly, check the mode door, actuator, vacuum switch, and connecting wires.*

## Task 4  Diagnose compressor clutch control system; determine needed repairs.

60. The compressor clutch is inoperative in the system shown in Figure 7–45. The switches, relay, compressor clutch coil, and connecting wires are tested and proven satisfactory. When the A/C button is pressed in the A/C control panel, voltage is supplied to the clutch control relay contacts.

    Technician A says there may be an open circuit at terminal 69 on the PCM.

    Technician B says the number 18 15A fuse may be open in the fuse junction panel.

    Who is correct?

    A. A only

    B. B only

    C. Both A and B

    D. Neither A nor B

**Hint**    *In some computer-controlled compressor clutch systems, voltage is supplied through the cycling switch, the pressure cut-off switch, and the clutch control relay to the compressor clutch coil. The cycling switch turns the compressor on and off in relation to evaporator suction pressure. The pressure cut-off switch opens the compressor clutch circuit to protect the system if refrigerant pressures become excessive. The clutch control relay is operated by the PCM.*

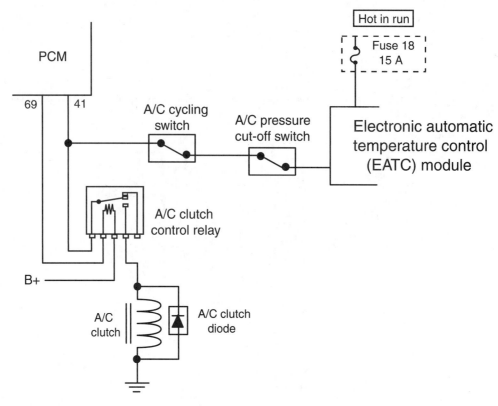

**Figure 7–45** Compressor clutch circuit, computer-controlled A/C system.

## Task 5    Inspect, test, adjust, or replace climate and blower control sensors.

61. When discussing in-vehicle temperature sensor testing with an ohmmeter, Technician A says that as the sensor temperature increases, the sensor resistance should decrease.

   Technician B says that at 50°F (10°C) the sensor should have minimum resistance.

   Who is correct?

   A. A only

   B. B only

   C. Both A and B

   D. Neither A nor B

62. When a cold engine is started, the A/C blower in Figure 7–46 starts immediately with the control in the defrost position. In the lo, auto, hi, or bilevel positions, the blower does not start until the engine is at normal operating temperature.

   Technician A says the engine temperature switch may be inoperative.

   Technician B says there may be an open circuit in the engine temperature switch ground.

   Who is correct?

   A. A only

   B. B only

   C. Both A and B

   D. Neither A nor B

*Hint*    *Many A/C system sensors, such as the in-vehicle sensor and ambient sensor, contain thermistors. The resistance of these sensors increases as sensor temperature decreases. Each sensor must have the resistance specified by the vehicle manufacturer at various temperatures. The sunload sensor*

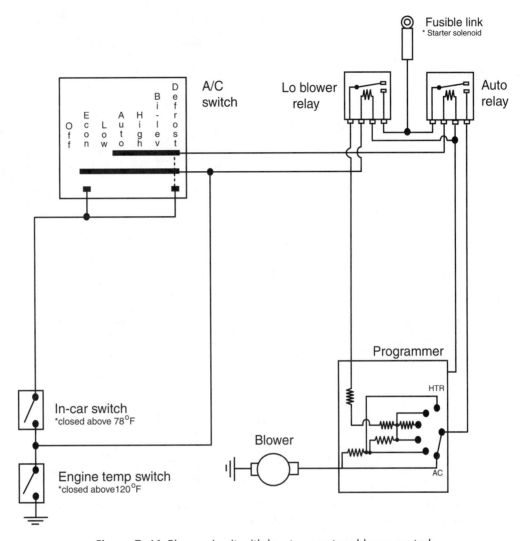

**Figure 7–46** Blower circuit with low temperature blower control.

*usually is mounted on top of the instrument panel. This sensor contains a photovoltaic diode that sends a varying current signal to the A/C computer in relation to the amount of sunlight applied to the sensor.*

*Some blower motor circuits have a low temperature cut-off switch that opens the blower motor circuit below a specific temperature. This temperature switch may sense engine metal or coolant temperature. If the A/C controls are placed in the defrost mode, the low temperature cut-off switch is bypassed and the blower operates normally.*

## Task 6  Inspect, test, adjust, and replace door actuator(s).

63. When diagnosing the power servo system in Figure 7–47, the temperature control is set in the maximum cold position. Solenoid 4 and solenoid 5 are open, and there is vacuum supplied to power servo 2.

   Technician A says this is a normal condition.

   Technician B says there should be no vacuum to power servo 2.

   Who is correct?

   A. A only

   B. B only

   C. Both A and B

   D. Neither A nor B

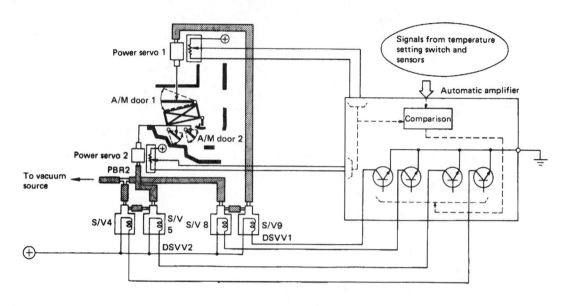

CL: Solenoid valve is closed
OP: Solenoid valve is open

| AIR MIX DOOR CONTROL | | | HOT side | HOLD | COLD side |
|---|---|---|---|---|---|
| Air mix door 1 | Operation of solenoid valve | S/V8 | CL | CL | OP |
| | | S/V9 | OP | CL | OP |
| Air mix door 2 | | S/V4 | CL | CL | OP |
| | | S/V5 | OP | CL | OP |

**Figure 7–47** Computer-controlled A/C system with vacuum solenoids and power servos. *(Used with permission from Nissan North America, Inc.)*

**Hint**   *In some A/C systems, the temperature blend door is controlled by an electric door actuator that is controlled by the A/C computer (Figure 7–48). In other A/C systems, the temperature blend door is controlled by a vacuum actuator that may be called a power servo. A varying vacuum is supplied to this power servo from a solenoid or solenoids controlled by the A/C computer. The computer operates the solenoid to supply the proper vacuum to the servo and position the temperature blend door to provide the temperature selected by the driver.*

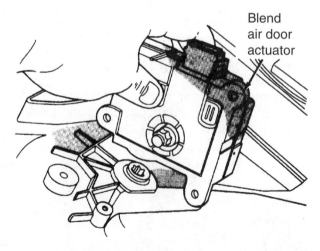

Blend air door actuator

**Figure 7–48** Temperature blend door electric actuator. *(Courtesy of Chrysler Corporation)*

## Task 7  Inspect, test, and replace heater water valve and controls.

64. In Figure 7–49, vacuum is supplied from a hand pump to the water valve solenoid, and battery voltage is supplied to the solenoid terminals, resulting in an audible click from the solenoid. The system holds 16 in. Hg. The water valve does not move.

All of the following may be the cause of the problem EXCEPT

A. a seized water control valve in the heater hose.

B. a plugged vacuum hose between the solenoid and the valve.

C. a jammed linkage from the actuator to the water valve.

D. a seized plunger in the water valve control solenoid.

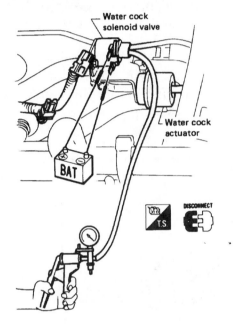

**Figure 7–49** Heater coolant control valve. *(Used with permission from Nissan North America, Inc.)*

*Hint*    *Some A/C systems have a coolant control valve that shuts off the coolant flow through the heater core under certain conditions. This valve may be operated mechanically, electrically, or by engine vacuum. In the vacuum-operated systems, a computer-controlled solenoid keeps the vacuum shut off to the coolant control valve actuator. When the A/C system is placed in the maximum A/C mode, the A/C computer energizes the solenoid, and vacuum is supplied through the solenoid to the coolant valve actuator. Under this condition, the actuator closes the coolant control valve and shuts off the coolant flow through the heater core to maximize the cold air flow through the evaporator.*

## Task 8  Inspect, test, and replace electric and vacuum motors, solenoids, and switches.

65. When diagnosing a computer-controlled A/C system, a diagnostic trouble code (DTC) is obtained indicating a fault in the temperature blend door actuator motor.

Technician A says the first step in the repair procedure is to replace the temperature blend door actuator.

Technician B says the first step in the repair procedure is to check the temperature blend door for a sticking condition.

Who is correct?

A. A only

B. B only

C. Both A and B

D. Neither A nor B

*Hint*   *Most computer-controlled A/C systems have self-diagnostic capabilities. DTCs may be obtained representing various system components. These DTCs represent a fault in a specific area. For example, a DTC representing an electric temperature blend door actuator indicates a fault in this area. The technician must check the door for sticking and test the motor and connecting wires to determine the exact cause of the problem.*

## Task 9   Inspect, test, and replace Automatic Temperature Control (ATC) control panel and/or climate control computer (microprocessor/programmer).

66. A DTC representing the ambient sensor is obtained in the circuit shown in Figure 7–50. The ambient sensor and connector 2 are connected, and connector 7 is disconnected from the ATC control panel. An ohmmeter connected to terminals 9 and 18 in the control panel connector indicates the specified resistance. The most likely cause of this DTC is

   A. an inoperative ambient sensor.

   B. an inoperative A/C control panel.

   C. a loose connection at connector 2 terminal 7.

   D. an open circuit at connector 7 terminal 18.

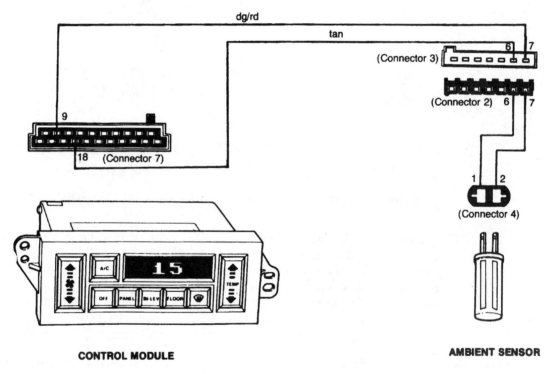

**CONTROL MODULE**                                **AMBIENT SENSOR**

**Figure 7–50** ATC control panel and ambient sensor wiring. *(Courtesy of Chrysler Corporation)*

67. When diagnosing and servicing an automatic A/C system, Technician A says diagnostic trouble codes (DTCs) representing problems in the A/C mechanical system may be displayed on a scan tool.

   Technician B says in an automatic A/C system the microprocessor controls blower speed and compressor clutch operation.

   Who is correct?

   A. A only

   B. B only

   C. Both A and B

   D. Neither A nor B

**Hint**    *In a typical automatic A/C system, the programmer contains a microprocessor which controls the A/C system. The programmer receives input signals from the following sources:*

*The mode selector switch, A/C control, actuators, the sunload sensor, the ambient temperature sensor, the inside temperature sensor, the blower control module, the powertrain control module (PCM), and the heater.*

*The microprocessor in the programmer processes these input signals and controls the following outputs: blower speed, compressor clutch, location of the air inlet, location of the air discharge, and the temperature of the discharge air.*

*When a scan tool is connected to the data link connector, and the A/C diagnostic mode is entered, diagnostic trouble codes (DTCs) representing A/C system electrical/electronic problems may be displayed on the scan tool. The service manual contains detailed diagnostic procedures for each DTC. Before servicing any electrical/electronic component in the A/C system, always disconnect the negative battery cable and wait for the time period specified by the vehicle manufacturer.*

*Always verify the power supply and ground circuits to the A/C programmer before replacing the programmer.*

## Task 10    Check and adjust calibration of Automatic Temperature Control (ATC) system.

68. When measuring the resistance in the A/C computer ground (Figure 7–51), a voltmeter is connected from computer terminal C1-24 to ground. With the ignition switch on, the maximum voltage reading should be

A. .1 V

B. .2 V

C. .5 V

D. .8 V

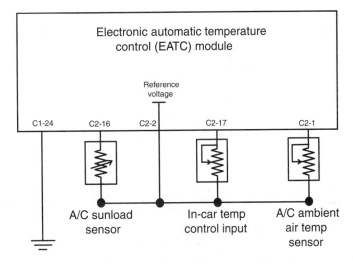

**Figure 7–51** Testing computer ground circuit.

**Hint**    *Before replacing any A/C computer, always be sure the voltage supply wires and ground wire(s) connected to the computer are satisfactory. With the ignition switch on, the maximum voltage drop across computer ground wires should be .1 V. The voltage supply wires should supply 12 V to the computer.*

*When some computer-controlled A/C systems are placed in the diagnostic mode, the A/C computer automatically performs calibration procedures on the electric door actuators. Some other systems require a calibration procedure on the electric door actuators if the actuators or other major components have been replaced. This calibration procedure involves placing the A/C controls in a specific mode with the electric actuator disconnected. The mode door is then moved by hand to a specific position, and the linkage is then connected between the actuator and the door.*

## Task 11  Diagnose data communication issues that affect A/C system operation.

69. When diagnosing an OBD II vehicle, the U1451 diagnostic trouble code (DTC) is displayed on the scan tool.

   Technician A says the defect is located in the body computer system.

   Technician B says the DTC is designated by the Society of Automotive Engineers (SAE).

   Who is correct?

   A. A only

   B. B only

   C. Both A and B

   D. Neither A nor B

**Hint**     *On some vehicles, the A/C control computer is interconnected via data links with other computers such as the PCM. Some sensor signals, like the throttle position sensor (TPS) and engine coolant temperature (ECT), sensor are sent to the PCM and relayed on the data links to the A/C computer. These sensor signals are used by the A/C computer for compressor control. If a defect occurs in the data links, the TPS and ECT sensor signals cannot be relayed to the A/C computer. Under this condition, a diagnostic trouble code(s) (DTCs) are sent in the PCM, and the A/C compressor is not controlled properly. In OBD II systems, vehicles have 5-digit DTCs. For example, a typical DTC is P0351. The prefix indicates the group to which the code belongs. The possible prefixes are P-powertrain, B-body, C-chassis, and U-data link network. The first number in the DTC indicates the group responsible for the code. A "0" indicates the DTC is designated by the SAE, and a "1" indicates the code is designated by the vehicle manufacturer. The second number in the DTC indicates the subgroup to which the code belongs. The possible subgroups are*
   - *air-fuel control*
   - *air-fuel control, injectors*
   - *ignition system, misfire*
   - *auxiliary emission controls*
   - *idle speed control*
   - *PCM input and output*
   - *transmission/transaxle*

   *The last two numbers in the DTC indicate the area in which the defect is located. For example, the P0351 DTC indicates a defect in one of the coil primary circuits.*

# Refrigerant Recovery, Recycling, Handling and Retrofit

## ASE Tasks, Questions, and Related Information

## Task 1  Maintain and verify correct operation of certified equipment.

70. All of the following statements about A/C recovery/recycling equipment are true EXCEPT

   A. The equipment label must indicate UL approval.

   B. The equipment label must indicate SAE J1991 approval.

C. Any size and type of refrigerant storage container over 10 lb. may be used in this equipment.

D. R-12 and R-134a refrigerants or refrigerant oils must not be mixed in the recovery/recycling process.

**Hint**   *Refrigerant recovery/recycling equipment must be used according to the equipment and vehicle manufacturer's recommended procedures (Figure 7–52). Never mix R-12 and R-134a refrigerants or refrigerant oils. This action causes system damage. The recovery/recycling equipment must have a label indicating it meets SAE J1991 standards for this type of equipment. The recovery/recycling equipment must also have an underwriter's laboratory (UL) label. The refrigerant storage tank specified by the A/C recovery/recycling equipment manufacturer must be installed in the recovery/recycling equipment. The tank valve is designed for use with the recovery/recycling equipment, and the unit's overfill limitation system is designed for a specific storage tank.*

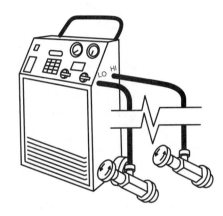

**Figure 7–52** Refrigerant recovery/recycling equipment.

## Task 2   Identify and recover A/C system refrigerant.

71. After the recovery process, the low-side pressure increases above zero after five minutes. This condition indicates
    A. there is still some refrigerant in the system.
    B. there is excessive oil in the refrigerant system.
    C. the refrigerant system is leaking.
    D. there is excessive moisture in the refrigerant system.

**Hint**   *The type of refrigerant in an A/C system may be identified by the underhood refrigerant charge tag. In some R-134a refrigerant systems, the clutch cycling switch is a different color compared to R-12 systems. For example, on some Ford products, the R-134a cycling switch is yellow. Quick-disconnect service fittings are used in R-134a refrigerant systems, whereas R-12 systems have Schrader-type service valves.*

*Cross contamination of an A/C system occurs when the system is partially charged with a refrigerant other than the refrigerant indicated on the system information label. A refrigerant identifier machine is the most accurate way of identifying refrigerants. A cross-contaminated A/C system may have reduced performance, internal damage from chemical reaction, and lubrication problems. If a recovery/recycling machine is connected to a cross-contaminated A/C system, the machine will require major cleaning and the replacement of parts such as filters and dehydrators. If the recovery/recycling machine is connected to other A/C systems before the cross-contamination problem is discovered, the other systems will also be contaminated.*

*Only those refrigerants approved by the Environmental Protection Agency (EPA) may be used in A/C systems. If a cross-contaminated, unknown, flammable, or illegal refrigerant is discovered in an A/C system, it must be disposed of according to EPA and local regulations. A recovery-only machine may be dedicated to recover cross-contaminated, unknown, or illegal refrigerants. Unknown refrigerants must be recovered into a DOT-approved recovery tank that is gray with a yellow top. These tanks must only be filled to 60 percent of their gross weight capacity.*

*The first step in the recovery procedure is to connect the low-side and high-side hoses from the equipment to the service valves. Check the low-side and high-side pressures to be sure there is refrigerant in the system. Do not continue with the recovery process if there is no refrigerant in the system. Open the low-side and high-side valves on the control panel, and open the vapor and liquid valves on the storage tank. Follow the equipment manufacturer's recommended procedure to drain any oil from the separator.*

*Plug in the unit and turn on the main power switch. Press Recover on the key pad. The unit clears itself of refrigerant and displays CL-L on the display. After the refrigerant is cleared, the unit begins the recovery process; this process continues until CPL is displayed. The initial recovery is completed when the refrigerant system is evacuated to 17 in. Hg. When CPL is displayed, the unit flashes the amount of refrigerant and oil recovered. If the low-side gauge pressure increases above zero after five minutes, there is still refrigerant in the system and further recovery is required. If the unit flashes full during the recovery process, the unit tank is full, and it must be replaced with the proper empty tank.*

## Task 3    Recycle or properly dispose of refrigerant.

72. The moisture warning light on the recovery/recycling equipment indicates yellow during the recycling process. This denotes
    A. there is excessive oil in the refrigerant.
    B. the filter/dryer cartridge on the tester must be changed.
    C. there is excessive noncondensable gases in the refrigerant.
    D. the vacuum pump is not producing enough vacuum.

**Hint**    *The controls and key pad vary on different makes of recovery/recycling equipment. Always follow the equipment manufacturer's recommended procedure. Some recovery/recycling equipment automatically recycles the refrigerant during a twenty-minute evacuation process. The Technician can manually select a longer process. When the recovery procedure is completed, press Vacuum to start the vacuum pump. The display shows recycle after five seconds.*

*When the display indicates seventeen minutes remaining in the procedure, press Hold/Continue to stop the vacuum pump. A zero low-side reading indicates a refrigerant system leak. Repair the leak and start the recycling process over again. If the low-side gauge indicates 27 to 30 in. Hg (91 to 101 kPa) close the low-side and high-side valves and wait a few minutes. If this vacuum is not maintained, repair the refrigerant system leak. When the vacuum is maintained, open the low-side and high-side valves and press hold/continue to start the vacuum pump. When the procedure has continued for twenty minutes, the display shows CPL. Some recovery/recycling equipment automatically vents noncondensable gases during the recovery process. This venting provides an audible hissing sound.*

## Task 4    Label and store refrigerant.

73. Technician A says that refrigerant storage containers must be evacuated to 27 in. Hg before the refrigerant is placed in the container.
    Technician B says refrigerant containers may be filled to 90 percent or their gross weight rating.
    Who is correct?
    A. A only
    B. B only
    C. Both A and B
    D. Neither A nor B

**Hint**    *Refrigerant must never be stored in disposable containers. Storage containers for recycled refrigerant must be stamped DOT 4B4 or DOT 4BW. R-12 containers are white, while containers for R-134a are sky blue. Evacuate the storage container to 27 in. Hg before putting* *refrigerant in the container. The refrigerant container must only be filled to 60 percent of its gross weight rating. Refrigerant containers should be stored in a cool location at approximately room temperature.*

## Task 5  Test recycled refrigerant for non-condensable gases.

74. When checking a refrigerant container for noncondensable gases
    A. the container may be stored at 80°F (27°C) for six hours before the test.
    B. the container may be stored near a shop window.
    C. a thermometer should be placed against the container surface.
    D. if the pressure is lower than specified, the refrigerant is ready for use.

**Hint**   *Before a refrigerant container is checked for noncondensable gases, keep the container away from sunlight and store it at 65°F (18°C) for 12 hours. Place a thermometer within 4 inches of the container surface and use a pressure gauge to measure the container pressure. Compare the temperature and pressure readings to the temperature/pressure chart (Figure 7–53). If the pressure of the stored refrigerant is less than specified on the chart, the refrigerant is ready for use.*

*When the pressure of the stored refrigerant is more than indicated on the chart, slowly vent the vapor from the top of the container into the recovery/recycling unit until the pressure is less than specified on the chart.*

| STANDARD TEMPERATURE PRESSURE CHART  FOR 134A | | | | | | | | | |
|---|---|---|---|---|---|---|---|---|---|
| 65 | 69 | 76 | 85 | 87 | 103 | 98 | 125 | 109 | 149 |
| 66 | 70 | 77 | 86 | 88 | 105 | 99 | 127 | 110 | 151 |
| 67 | 71 | 78 | 88 | 89 | 107 | 100 | 129 | 111 | 153 |
| 68 | 73 | 79 | 90 | 90 | 109 | 101 | 131 | 112 | 156 |
| 69 | 74 | 80 | 91 | 91 | 111 | 102 | 133 | 113 | 158 |
| 70 | 76 | 81 | 93 | 92 | 113 | 103 | 135 | 114 | 160 |
| 71 | 77 | 82 | 95 | 93 | 115 | 104 | 137 | 115 | 163 |
| 72 | 79 | 83 | 96 | 94 | 117 | 105 | 139 | 116 | 165 |
| 73 | 80 | 84 | 98 | 95 | 118 | 106 | 142 | 117 | 168 |
| 74 | 82 | 85 | 100 | 96 | 120 | 107 | 144 | 118 | 171 |
| 75 | 83 | 86 | 102 | 97 | 122 | 108 | 146 | 119 | 173 |

**Figure 7–53** Refrigerant temperature/pressure chart.

## Task 6  Follow Federal and local guidelines for retrofit procedures.

75. While discussing retrofits from CFC-12 to HFC-134a refrigerant, Technician A says that on some retrofits the vehicle manufacturer recommends changing the accumulator.
    Technician B says the service valves must be changed during a retrofit.
    Who is correct?
    A. A only
    B. B only
    C. Both A and B
    D. Neither A nor B

**Hint**   *When retrofitting an A/C system from CFC-12 refrigerant to HFC-134a refrigerant, always follow the vehicle manufacturer's recommendations. HFC-134a refrigerant has been the refrigerant of choice for CFC-12 conversions. If you are considering using another refrigerant, be sure it is approved by the EPA and state authorities. Some flammable refrigerants have been introduced to the market, but these refrigerants are illegal in many states because of safety* *concerns. Refrigerant identifying machines are used by many shops to prove the type of refrigerant in the A/C system.*

*Some manufacturers recommend draining the refrigerant oil and placing the specified amount of PAG oil in the system. Others recommend an ester oil in the refrigeration system. Still others do not recommend draining the oil from the system, but instead suggest simply adding the proper amount of PAG oil.*

*Some manufacturers recommend changing replacing the accumulator with an accumulator XH7 or XH9 desiccant. Most receiver/dryers have XH5 desiccant that is not compatible with HFC-134a refrigerant. The receiver/dryer should be changed during a retrofit. Be sure the replacement receiver/dryer is compatible with HFC-134a refrigerant.*

*The service valves must be changed during the retrofit. Adaptors are available to allow the HFC-134a service valves to be installed in the CFC service valve openings.*

*It is not necessary to change the compressor, evaporator, or condenser unless these components are leaking or defective. In a few vehicles, a larger condenser may be required. Be sure there are no openings in the radiator/condenser area that allow air to bypass these components. One vehicle manufacturer recommends installing a fixed orifice tube that has a 0.002 in. smaller orifice, but all other manufacturers do not recommend changing the fixed orifice tube. The thermostatic expansion valve (TXV) does not require replacing unless it is leaking or defective. When replacing the TXV, be sure the replacement valve is recommended for the refrigerant being used. The clutch pressure cycling switch may have to be changed on some retrofits. The clutch pressure cycling switch for HFC-134a is calibrated for slightly higher clutch cycling pressures. An accumulator designed for HFC-134a refrigerant has metric threads in the clutch pressure switch opening, and an accumulator for CFC-12 refrigerant has USC threads. A refrigerant containment device (RCD) pressure switch is installed with many retrofits.*

# Post Test

1. An A/C compressor has a thumping noise in the A/C mode. If the A/C is shut off, the noise disappears.
   Technician A says the refrigerant system is partially blocked.
   Technician B says the refrigerant system pressures may be extremely high.
   Who is correct?
   A. A only
   B. B only
   C. Both A and B
   D. Neither A nor B

2. While discussing A/C systems, Technician A says the fixed orifice tube controls refrigerant flow into the compressor.
   Technician B says the thermostatic expansion valve controls refrigerant flow into the receiver/dryer.
   Who is correct?
   A. A only
   B. B only
   C. Both A and B
   D. Neither A nor B

3. A sight glass contains blue particles. The most likely cause of this problem is
   A. a low refrigerant charge.
   B. dye in the refrigerant.
   C. excessive oil in the refrigerant.
   D. a defective receiver/dryer.

4. When diagnosing an A/C system equipped with a TXV, with the system in the A/C mode
   A. the high-side components should feel cold.
   B. receiver/dryer frosting indicates a restriction in this component.
   C. low-side components should feel warm or hot.
   D. frosting of the compressor suction line indicates that the TXV is stuck closed.

5. While discussing refrigerant system leak testing, Technician A says dye may be added to the refrigerant to locate leaks in the system.
   Technician B says a flame-type leak detector may be used with R-134a refrigerant.
   Who is correct?
   A. A only
   B. B only
   C. Both A and B
   D. Neither A nor B

6. When discharging and recovering refrigerant
   A. R-12 and R-134a refrigerants must not be mixed.
   B. A/C recovery/recycling equipment is limited to use on R-134a systems.
   C. the engine must be running when discharging the refrigerant system.
   D. open only the high-side valve when discharging the refrigerant system.

7. During a refrigerant system evacuation procedure, after five minutes of vacuum pump operation the low-side gauge indicates 20 in. Hg. and the high-side gauge indicates 5 psi.
   Technician A says the condenser may be partially blocked.
   Technician B says the TXV may be stuck open.
   Who is correct?
   A. A only
   B. B only
   C. Both A and B
   D. Neither A nor B

8. While discussing refrigerant system charging, Technician A says high-side charging should be done with the engine running.
   Technician B says if liquid refrigerant enters the compressor, damage to the compressor may result.
   Who is correct?
   A. A only
   B. B only
   C. Both A and B
   D. Neither A nor B

9. When checking the oil level and installing oil in refrigerant systems
   A. excessive oil in a refrigerant system has no effect on cooling efficiency.
   B. compressor damage may result from insufficient oil in the refrigerant system.
   C. mineral oil in R-12 refrigerant systems is nonhydroscopic.
   D. mineral oil in R-12 refrigerant systems has a YN-12 designation.

10. A scraping noise is heard momentarily when the A/C compressor clutch engages. The most likely cause of this problem is
    A. a loose compressor drive belt.
    B. worn internal compressor bearings.
    C. an overcharge of refrigerant.
    D. excessive clearance between the A/C compressor pulley and the armature plate.

11. When replacing an A/C compressor, the refrigerant system is discharged and 2 oz. (59 ml) of oil is recovered from the system. When the compressor is removed, 2 oz. (59 ml) of oil is drained from the compressor. Six ounces (177 ml) of oil is drained from the new compressor. When the new compressor is installed, the amount of oil installed in the compressor should be
    A. 1 oz. (29.5 ml).
    B. 2 oz. (59 ml).
    C. 4 oz. (118 ml).
    D. 6 oz. (177 ml).

12. When testing and replacing A/C compressors
    A. a defective compressor may be indicated by low high-side and high low-side refrigerant system pressures.
    B. if the refrigerant system has stem-type service valves, back-seat these valves to isolate the compressor.
    C. the compressor may be isolated from the refrigerant system with Schrader-type service valves.
    D. oil deposits in the pulley area indicate a defective pulley.

13. All of the following statements about refrigerant lines are true EXCEPT
    A. When the refrigerant system is in operation, high-side lines should feel warm or hot.
    B. Frost formation on the line from the condenser to the receiver/dryer indicates a leak in the line.
    C. Oil accumulation on a refrigerant line or fitting indicates a refrigerant leak.
    D. Some lines are sealed with an O-ring and joined with a spring lock coupling.

14. The low-side and high-side pressures are higher than specified when an A/C system is operating. The condenser air passages and refrigerant passages are not restricted and the sight glass is clear. The most likely cause of this problem is
    A. partially plugged internal radiator passages and engine overheating.
    B. air and moisture in the refrigerant system.
    C. a defective high pressure relief valve.
    D. a compressor clutch cycling switch that is stuck closed.

15. When diagnosing and servicing receiver/driers and accumulators
    A. the receiver/drier is mounted between the compressor and the condenser.
    B. red refrigerant in the sight glass indicates the refrigerant system contains dye.
    C. a large temperature difference between the receiver/drier inlet and outlet temperatures indicates normal operation.
    D. the accumulator is connected between the condenser and the evaporator.

16. A refrigerant system with a TXV has high low-side pressure and there is frost on the suction line between the evaporator and the compressor. The discharge air from the A/C ducts is slightly cool.
    Technician A says the TXV may be stuck open causing evaporator flooding.
    Technician B says there may be moisture in the refrigerant system freezing the TXV.
    Who is correct?
    A. A only
    B. B only
    C. Both A and B
    D. Neither A nor B

17. When leak testing an evaporator, Technician A says to remove the blower resistor assembly and insert an electronic leak detector probe through this opening.

Technician B says to use a flame-type leak detector near the heater/evaporator case.

Who is correct?

A. A only

B. B only

C. Both A and B

D. Neither A nor B

18. The evaporator outlet pipe is frosting and the low-side pressure is lower than specified.

Technician A says the orifice tube is restricted with debris.

Technician B says the evaporator refrigerant passages may be restricted.

Who is correct?

A. A only

B. B only

C. Both A and B

D. Neither A nor B

19. An A/C heater system has a normally open vacuum-operated coolant control valve in one of the heater hoses. During an A/C system performance test, the discharge air temperature is slightly higher than specified, but the refrigerant system pressures are normal.

Technician A says the coolant control valve may be stuck open.

Technician B says there may be a vacuum leak in the hose connected to the coolant control valve.

Who is correct?

A. A only

B. B only

C. Both A and B

D. Neither A nor B

20. While discussing a computer-controlled A/C system diagnosis, Technician A says that on some systems the diagnostic mode is entered by pressing specific buttons simultaneously on the A/C control panel.

Technician B says that on some computer-controlled A/C systems the computer calibrates the actuator motors when the diagnostic mode is entered.

Who is correct?

A. A only

B. B only

C. Both A and B

D. Neither A nor B

# Answers and Analysis

**1.  C**   During an A/C performance test, the temperature at the dash outlets should be 40°F to 50°F. Therefore, C is right, and A, B, and D are wrong.

**2.  C**   Air and moisture in the refrigerant system or a restricted TXV valve cause higher-than-specified system pressures. Therefore, A and D are wrong.

An inoperative compressor may cause low high-side pressure and high low-side pressure, so B is wrong.

A low refrigerant charge causes reduced low-side and high-side pressures, so C is correct.

**3. A**    A restricted TXV valve may cause lower-than-normal low-side pressures, but this condition also results in TXV frosting. Therefore, B is wrong. Air and moisture in the refrigerant system may cause high system pressures, so A is correct.

**4. A**    Clogged condenser air passages cause high low-side and high-side pressure. Thus, A is correct. A slipping A/C compressor belt causes low and varying low-side and high-side pressure. A restricted evaporator core causes low, low-side pressure. An inoperative compressor valve causes low or very low, high-side and low-side pressure. Therefore, B, C, and D are wrong.

**5. A**    Because the pulley is turning with the compressor clutch engaged or disengaged, a damaged pulley bearing provides a growling noise with the clutch disengaged. Therefore, B is wrong.

A low refrigerant charge or an excessive refrigerant charge do not cause a compressor growling noise, so C and D are wrong.

A damaged internal compressor bearing causes a growling noise only with the compressor clutch engaged, because this is the only time the internal compressor components are rotating. Thus, A is right.

**6. A**    A restricted receiver/dryer usually causes frosting of this component. A flooded evaporator causes frosting of the evaporator outlet and compressor suction pipes. A restricted TXV usually causes frosting of this component. Therefore, B, C, and D are wrong. A low refrigerant charge may cause faster than normal clutch cycling without frosting of any components, so A is correct.

**7. B**    An oily residue on refrigerant system components indicates a refrigerant leak in the area of the oily residue. Therefore, A is wrong and B is right.

**8. C**    R-12 refrigerant is sold in white containers, and R-134a is marketed in light blue containers. Both Technicians A and B are right, so C is the correct answer.

**9. A**    Evacuating a refrigerant system removes moisture from the system. Therefore, B, C, and D are wrong, and A is right.

**10. A**    Some manufacturers recommend installing an in-line filter as an alternate to refrigerant system flushing. If this filter contains an orifice tube, the original orifice tube in the system must be removed. Therefore, Technician A is right and Technician B is wrong, making A the right answer.

**11. B**    The high-side charging procedure must be completed when the engine is not running. If liquid refrigerant enters the compressor, this component may be damaged. Therefore, Technician A is wrong and Technician B is correct, making B the right answer.

**12. A**    The oil required with R-134a refrigerant is a polyalkylene glycol (PAG) oil, so B, C, and D are wrong, and A is right.

**13. D**    Restricted evaporator refrigerant passages reduce passenger compartment cooling. Restricted air passages in the heater core reduce air flow in the heat mode, and restricted A/C ducts reduce air flow in the defrost mode. Therefore, A, B, and C are wrong. A restricted passenger compartment filter reduces air flow in heat and A/C modes, so D is correct.

**14. B**    The wait period after the negative battery cable is disconnected allows the backup power supply to power down in the air bag control module. Therefore, Technician A is wrong. If an air bag system is not properly disarmed, accidental air bag deployment may occur. Thus, Technician B is correct, making answer B the right choice.

**15. C**    The purpose of the cycling switch (Item 5 in Figure 7–8) is to cycle the compressor clutch on and off in relation to refrigerant system pressure. Therefore, A, B, and D are wrong, and C is right.

**16. B**    If there is 12V at terminals B, C, and S, the 12V at terminal B indicates there is voltage supplied to the compressor clutch and connecting wire. If the clutch is inoperative, the compressor clutch coil or the wire from the coil to the fuse is open. Therefore, B is right.

**17. A**    A loose A/C compressor belt will likely provide a squealing noise with the compressor clutch engaged. Therefore, B is wrong. Because the air pump does not require much power to turn it, this belt is not likely to provide a squealing noise on acceleration. Thus, C is wrong.

A worn compressor bearing provides a growling noise under all engine operating conditions with the clutch engaged, so D is wrong.

A loose power steering belt may cause a squealing noise on acceleration, so A is right.

**18. A** If the compressor clutch clearance is more than specified, the clutch may slip while the compressor is engaged. This action may cause a scrapping noise, so A is correct.

When the compressor clutch clearance is more than specified, a shim must be removed from behind the armature plate, so B is wrong.

**19. C** If there is 2 oz. of oil recovered from the system and 2 oz. of oil drained from the old compressor, add 4 oz. of oil to the replacement compressor before installation. Thus, C is right.

**20. B** An inoperative compressor may be indicated by high low-side pressure and low high-side pressure. Therefore, B is correct. If both low-side and high-side pressures are high, there may be air and/or moisture in the refrigerant system or the system may have an overcharge of refrigerant, and thus A is wrong.

**21. C** The seal seat O-ring must be installed before the shaft seal, and component 6 is a seal protector. Therefore, both Technicians A and B are correct, and C is the right answer.

**22. A** The tool in Figure 7–14 is used to disconnect spring lock couplings, so B, C, and D are wrong, and A is correct.

**23. C** If the condenser air passages are severely restricted, it may result in high refrigerant system pressures and refrigerant discharge from the high-pressure relief valve. Thus, Technicians A and B are both correct, making C the right answer.

**24. C** When frost is forming on one of the condenser tubes near the bottom of the condenser, the refrigerant passage is restricted at that location in the condenser. Therefore, A, B, and D are wrong, and C is right.

**25. A** If the refrigerant in the sight glass appears red, leak-detecting dye has been added to the refrigerant. This condition does not require any corrective action, so Technician B is wrong. If the receiver/dryer outlet is colder than the inlet, this unit is restricted and should be changed. Thus, Technician A is correct, making A the right answer

**26. D** The symptoms described in the question indicate moisture freezing in the TXV valve. Therefore, D is correct.

**27. C** This question asks us to select the response that is not true. When using the tool to remove an orifice tube, pour a small amount of refrigerant oil on top of the orifice tube, and engage the notch in the tool in the orifice tube. Hold the T-handle and turn the outer sleeve and rotate the outer sleeve to remove the orifice tube. Therefore, statements A, B, and D are correct, but none of these is the requested answer.

Because statement C is not a proper orifice tube removal procedure, this is the requested answer.

**28. A** Restricted evaporator refrigerant passages may cause frosting of the evaporator outlet pipe. Therefore, Technician A is right.

This problem may cause lower-than-specified low-side pressures, so B is wrong, making A the correct answer.

**29. B** A plugged evaporator case drain may cause windshield fogging, but this problem would not result in an oily film on the windshield. Therefore, Technician A is wrong.

An oil film on the windshield may be caused by a refrigerant leak in the evaporator core that may allow some refrigerant oil to escape in the evaporator case. Thus, Technician B is correct, making B the right answer.

**30. A** In some refrigerant systems with an EPR, POA, or STV valve, the compressor runs continually in the A/C mode, so Technician A is right.

Excessive pressure drop across the EPR valve indicates this valve is sticking closed, so Technician B is wrong, making A the right answer.

**31. D** The equipment in Figure 7–22 is designed to recover, recycle, and recharge refrigerant, and this equipment has service valve connections for R-134a refrigerant. Therefore, both Technicians are wrong, and the correct answer is D.

**32.  D**    Component 4 in Figure 7–26 is a high-pressure relief valve that discharges refrigerant at approximately 475 psi (3,275 kPa). Thus, A, B, and C are wrong, and D is right.

**33.  C**    The coolant control valve should be closed in the maximum A/C mode. If this valve is stuck open, heat from the heater core may reduce evaporator cooling to some extent. A leaking vacuum hose connected to the coolant control valve also allows this valve to remain open in the maximum A/C mode. Therefore, both Technicians A and B are right, and C is the correct response.

**34.  C**    A sticky film on the inside of the windshield may be caused by a coolant leak in the heater core, so Technician A is right. Under this condition, the coolant level in the radiator should be checked. Since Technicians A and B are both correct, C is the right answer.

**35.  A**    In this question, we are asked for the problem that would not cause the problem of coolant loss with no visible leaks under the hood or in the passenger compartment. A leaking transmission cooler, head gasket, or cracked cylinder head could result in this problem.

Therefore, B, C, and D are not the requested answer.

A leaking heater core usually causes coolant to leak onto the front floor mat; thus, A is not the cause of the problem, making it the right answer in this scenario.

**36.  A**    A sticking vacuum valve in the radiator cap may cause a collapsed upper radiator hose after the engine is shut off. Therefore, B, C, and D are wrong, and A is the right answer.

**37.  C**    Restricted radiator tubes may cause engine overheating and excessive coolant in the coolant recovery container. Therefore, Technician A is correct.

A defective radiator cap may allow excessive coolant flow into the recovery container, so Technician B also is correct, making C the right answer.

**38.  C**    If the thermostat is stuck open in a fuel-injected engine, the coolant never reaches normal operating temperature. Under this condition, the engine coolant temperature sensor sends a low coolant temperature signal to the PCM. This sensor input results in a rich air-fuel ratio. Therefore, A, B, and D are wrong, and C is correct.

**39.  A**    This question asks for the statement that is not true. When more antifreeze is added to the coolant, the boiling point is increased and coolant solutions must be recovered and recycled.

Most ethylene glycol antifreeze contains antirust and corrosion inhibitors. Therefore, B, C, and D are correct, but these are not the requested answer.

When the cooling system pressure is increased, the boiling point also increases. Therefore, statement A is not true and this is the requested answer.

**40.  B**    A blown 10 A, 5C fuse in the fuse block would prevent both fan motors from operating even with the A/C on. Therefore, A is wrong.

An inoperative high-speed coolant fan relay only prevents high-speed fan operation, so C is wrong.

Because the cooling fans operate with the A/C on, the A/C pressure fan switch must be operating, so D is wrong.

A malfunctioning engine coolant temperature sensor that indicates low coolant temperature may cause the PCM not to ground the fan relay windings and thus prevent the operation of the cooling fans except when the A/C pressure switch closes. Thus, B is correct.

**41.  B**    The manual coolant control valve adjustment is illustrated in Figure 7–28. Therefore, A, C, and D are wrong, and B is right.

**42.  C**    A gurgling noise in the heater core may be caused by a low coolant level in the cooling system or a restricted heater core. Therefore, both Technicians A and B are correct, and C is the right answer.

**43.  C**    This question asks for the response that is not the cause of the problem. An open circuit at terminal 6 or 8 on the electronic door actuator, or a blown 15 A, number 8 fuse, would cause the temperature blend door actuator to be inoperative. Therefore, A, B, and D are correct, but none of these is the requested response.

Terminal C2-15 is connected to the feedback circuit from the electronic temperature door actuator to the automatic temperature control module. An open circuit at this terminal may cause inaccurate temperature blend door position, but the door does move. Therefore, C is correct.

**44.   A**   An open circuit at the blower switch ground or at the terminal causes the blower to be completely inoperative. Therefore, B is wrong.

An open circuit at resistor terminal 4 causes some other blower speeds other than high speed, so D is wrong.

An open circuit at switch terminal 4 prevents any blower speed operated through the switch, but voltage still is available through the circuit to resistor terminal 4 to provide very slow blower speed with the ignition switch on the A/C switch in any position but off.

Therefore, C is wrong.

An open circuit between the two lower resistors prevents current flow through any of the resistors, but high blower speed is available through the high-speed switch contacts directly to ground, so A is correct.

**45.   A**   An inoperative A/C pressure cut-off switch opens the compressor clutch circuit if the refrigerant system pressures are extremely high. However, the question says these pressures are normal, so B is wrong.

An inoperative clutch control relay or clutch coil are not likely to cause the clutch to become de-energized in relation to temperature. Therefore, C and D are wrong.

An inoperative engine coolant temperature sensor may indicate engine overheating to the PCM when the engine is at normal operating temperature. This signal causes the PCM to open the ground circuit on the clutch control relay winding and de-energize the clutch. Thus, A is correct.

**46.   A**   When the engine is cold, airflow past the IAC motor pintle increases. Therefore, B is wrong.

Airflow past the IAC motor pintle increases when the A/C is turned on, and so C is wrong.

Airflow past the IAC motor pintle enters the intake manifold below the throttles. Therefore, D is wrong. If the A/C is turned on, the IAC duty cycle should increase. Therefore, statement A is not true and this is the correct answer.

**47.   D**   The A/C pressure transducer contains a potentiometer that is connected to the PCM.

Therefore, both A and B are wrong and D is the correct answer.

**48.   A**   A blown maxifuse in the LH maxifuse holder, or a blown 10 A cooling fan/TCC fuse in the fuseblock, can cause both cooling fans to be inoperative, so B and C are wrong.

An open circuit at terminal B on the A/C head pressure switch would only prevent cooling fan operation when the A/C pressure is high; thus, D is wrong.

An open circuit between the number 30 terminals on the cooling fan relays prevents voltage supply to the low-speed fan relay contacts, but voltage is still available at the high-speed fan relay contacts. Therefore, A is correct.

**49.   B**   This question asks for the statement that is not true. Some actuator motors are calibrated automatically in the self-diagnostic mode, whereas other actuator motors must be calibrated manually. These actuators should only require calibration after motor replacement or misadjustment. Therefore, statements A, C, and D are correct, but they are not the requested answer.

Diagnostic trouble codes indicate a fault in a certain area not in a specific component.

Thus, statement B is wrong, making it the requested answer.

**50.   B**   This question asks for the statement that is not true. The negative battery cable must be removed and the Technician must wait a specified length of time before A/C control panel removal. Self-diagnostic tests may indicate a malfunctioning A/C control panel. Statements A, C, and D are right, but none of these is the requested answer.

The refrigeration system does not require discharging before A/C panel removal. Therefore, B is wrong and this is the requested answer.

**51.   D**    With 18 in. Hg supplied to the control valve switch assembly, the same vacuum should be applied to the panel door actuator with the switch in the maximum A/C position. Therefore, D is right, and A, B, and C are wrong.

**52.   D**    The engine should be running and the vehicle not moving to enter the A/C panel diagnostic mode. The fan speed control should be in any position but off to enter the A/C panel diagnostic mode. Therefore, both Technician A and Technician B are wrong, and the correct answer is D.

**53.   C**    While adjusting the temperature control cable, the attaching flag and the adjusting clip must be installed, and the temperature control lever must be in the maximum cold position. Thus, A, B, and D are wrong.

During this adjustment, the temperature blend door crank must be rotated fully counter-clockwise, so C is right.

**54.   A**    If the black vacuum hose that supplies vacuum from the intake manifold to the control switch is leaking, none of the mode doors operates properly. Therefore, Technician B is wrong.

A leak in the blue vacuum hose from the control switch to the panel door actuator may cause this actuator to gradually switch from the panel ducts to the floor ducts. Thus, Technician A is correct, making A the right answer.

**55.   B**    A leaking panel door actuator would cause the air discharge to switch from the panel to the floor ducts any time, not just when climbing a hill, so A is wrong.

A leaking temperature blend door actuator causes the system to change temperature any time, not just when climbing a long hill. Thus, C is wrong.

A leaking intake manifold gasket may affect all the vacuum-operated mode doors because this condition causes low-source vacuum. Therefore, D is wrong.

A leaking check valve does not trap the vacuum in the reserve tank when climbing a long hill. This action may cause the temperature blend door to move to the warm air position and the air discharge to switch to the floor ducts, so B is right.

**56.   A**    If the outside recirculation door is stuck in position A, outside air is drawn into the A/C heater case and there is no leakage of in-car air past this door. Therefore, Technician A is right and Technician B is wrong, making A the right answer.

**57.   C**    A sticking temperature blend door or a malfunctioning in-car sensor may cause the in-car temperature to be above the driver-selected temperature. Both A and B are correct and C is the right answer.

**58.   A**    An open blower motor ground would cause this motor to be completely inoperative, so B is wrong. Because the voltage at the blower motor is normal, the HVAC programmer must be satisfactory, thus C is wrong.

Because programmer terminal D2 is only connected to a feedback wire from the blower circuit, an open circuit at this terminal would not cause continual low blower speed, so D is wrong.

An inoperative blower motor may cause continual low blower speed, so A is correct.

**59.   B**    A problem in the check valves, vacuum tank, and vacuum supply affects all the vacuum-operated doors, so A is wrong. The programmer vacuum switch and the hose to the outside/recirculation actuator should be checked first to correct this problem. Thus, B is right.

**60.   A**    If the number 18, 15 A fuse is open, there is no voltage supplied to the compressor clutch, so Technician B is wrong. Because the PCM must ground the clutch control relay winding through PCM terminal 69, an open circuit at this terminal makes it impossible for the PCM to ground this relay winding, so Technician A is correct. Thus, A is the right answer.

**61.   A**    The in-vehicle sensor resistance is at a minimum when the temperature in the vehicle is hot. Therefore, Technician B is wrong. As the temperature of the in-vehicle sensor increases, the sensor resistance should decrease. Therefore, Technician A is correct, making A the right answer.

**62.   A**    If the temperature switch ground is open, the blower motor would not operate except in the defrost mode. Therefore, Technician B is wrong.

The engine temperature switch should close and turn on the blower at 120°F (49°C). A malfunctioning engine temperature switch may delay the blower operation to a higher engine temperature. Thus, Technician A is correct, and A is the right answer.

**63. A**    As indicated in the chart in Figure 7–47, both solenoids 4 and 5 should be open, with the temperature control in the cold position, and vacuum supplied to power servo 2. This is a normal condition, so Technician A is right and Technician B is wrong. Thus, A is the right answer.

**64. D**    This question asks for the choice that is not the cause of the problem. A seized water control valve, a plugged vacuum hose between the solenoid and the valve, or a jammed actuator linkage may cause the water control not to move with 16 in. Hg supplied to the solenoid.

Therefore, A, B, and C may cause the problem, but none of these is the requested answer.

Because the solenoid provides an audible click, the solenoid plunger is not sticking and D is not the cause of the problem. Thus, D is the requested answer.

**65. B**    When a fault code is obtained representing the temperature blend door, the first step in the diagnostic procedure should be to check the temperature blend door for a sticking condition.

So, Technician A is wrong and Technician B is right, making B the right answer,

**66. B**    Because the resistance reading in the ambient sensor and connecting wires is normal, this sensor and connecting wires are satisfactory. Therefore, A, C, and D are wrong. Because the only other component in this circuit is the A/C control panel, this unit must be the problem, so B is correct.

**67. B**    In an automatic A/C system, DTCs represent faults in electrical/electronic systems but not in mechanical systems. Therefore, Technician A is wrong. In an automatic A/C system, the microprocessor controls blower speed and compressor clutch operation. Thus, Technician B is correct.

**68. A**    The specified voltage drop across computer ground wires usually is 0.1 V. Therefore, B, C, and D are wrong, and A is right.

**69. D**    A DTC with a U prefix indicates a defect in the data link network; therefore, A is wrong. In SAE-designated DTCs, the second digit is a zero. Therefore, B is wrong. Because both technicians are wrong, D is the correct answer.

**70. C**    This question asks for the statement that is not true. A/C recover/recycling equipment must have a UL approval, SAE J1991 approval, and refrigerant oils for R-12 and R-134a must not be mixed. Therefore, statements A, B, and D are true, so none of these is the requested answer.

The refrigerant container specified by the recovery/recycling equipment manufacturer must be used in this type of equipment to be sure the container has proper capacity and valving.

Thus, C is right.

**71. A**    After the recovery process, if the low-side gauge rises above 0 psi, there is some refrigerant remaining in the system. Therefore, B, C, and D are wrong, and A is correct.

**72. B**    If the moisture warning light indicates yellow during the recycling process, the refrigerant contains excessive moisture and the filter/dryer cartridge in the recovery/recycling equipment must be changed. Therefore, A, C, and D are wrong, and B is right.

**73. A**    Refrigerant storage containers must be filled to 60 percent of their gross weight rating, so Technician B is wrong. Refrigerant storage containers must be evacuated to 27 in. Hg before refrigerant is placed in the container. Thus, Technician A is correct, making A the right choice.

**74. D**    When checking a refrigerant container for noncondensable gases, the container should be stored away from sunlight at 65°F (18°C) for 12 hours, and a thermometer should be placed 4 inches from the container surface. Therefore, A, B, and C are wrong.

If the container pressure is less than specified, the refrigerant is ready for use. Thus, D is correct.

**75. C**    Some vehicle manufacturers recommend changing the accumulator during a retrofit, so Technician A is right. The service valves must be changed during a retrofit. Thus, Technician B is right. Because Technicians A and B are both right, C is the correct answer.

# Answers to Post Test

**1.  C**    Refrigerant system blockage may cause the A/C compressor to produce a thumping noise. Technician A is right. Extremely high refrigerant system pressures may also cause the compressor to have a thumping noise, so Technician B is also right. Because Technicians A and B are both right, C is the correct answer.

**2.  D**    The fixed orifice tube controls refrigerant flow into the evaporator, so Technician A is wrong. The thermostatic expansion valve controls refrigerant flow in the evaporator, thus Technician B is wrong. Because Technicians A and B are both wrong, D is the correct answer.

**3.  D**    If a sight glass contains blue particles, the receiver/drier is disintegrating, and must be replaced. Therefore, D is correct, and A, B, and C are wrong.

**4.  B**    The high-side components should feel warm or hot; thus, A is wrong. Receiver/drier frosting indicates a restriction in this component, so B is correct. Low-side components should feel cool or cold, so C is wrong. Frosting of the compressor suction line does not indicate the TXV is stuck closed. Thus, D is wrong.

**5.  A**    Dye may be added to the refrigerant to locate system leaks, so Technician A is right. A flame-type leak detector will not work on a system with R-134a refrigerant, so Technician B is wrong. Thus, A is the correct answer.

**6.  A**    R-12 and R-134a refrigerants must not be mixed, so A is correct. A/C recovery/recycling equipment may be used on R-12 and R-134a systems, so B is wrong. The engine must be stopped with discharging the refrigerant system; thus, C is wrong. The low-side and high-side valves should both be open when discharging the refrigerant system, so D is wrong.

**7.  A**    If the high-side gauge indicates a pressure under these conditions, system blockage is indicated. Thus, technician A is correct. A TXV stuck open does not cause this improper high-side pressure during an evacuation. Therefore, Technician B is wrong, making A the correct answer.

**8.  B**    High-side charging should be done with the engine stopped, so Technician A is wrong. If liquid refrigerant enters the compressor, damage to the compressor may result. Thus, Technician B is right, and B is the correct answer.

**9.  B**    Excessive oil in the refrigerant system reduces cooling efficiency, so A is wrong. Compressor damage may result from insufficient oil in the refrigerant system, so B is correct. Mineral oil in an R-12 system is hydroscopic, so C is wrong. Mineral oil in an R-12 system has a YN-9 designation, so D is wrong.

**10.  D**    A loose compressor drive belt causes squealing during acceleration with the compressor clutch engaged, so A is wrong. Worn internal compressor bearings cause a growling noise while the compressor clutch is engaged; thus, B is wrong. A refrigerant overcharge reduces cooling efficiency, so C is wrong. Excessive clearance between the pulley and the armature plate may cause a momentary grinding noise when the compressor clutch engages, making D correct.

**11.  C**    When the new compressor is installed 4 oz. (118 ml) of oil should be installed in the compressor. This amount is equal to the amount recovered from the system and the amount drained from the old compressor, so C is the correct answer.

**12.  A**    High low-side and low high-side pressures indicate a defective compressor, so A is correct. Stem-type service valves must be front-seated to isolate the compressor, so B is wrong. Schrader-type service valves cannot be used to isolate the compressor; thus, C is wrong. Oil deposits in the pulley area indicate a leaking compressor shaft seal, so D is wrong.

**13.  B**    When the refrigerant system is in operation, the high-side lines should feel warm or hot. Thus, A is true, and is not the requested answer. Frost formation on the line from the condenser to the receiver/drier indicates a leak in the line. B is not true, and this is the correct answer. Oil accumulation on lines and fittings indicates a refrigerant leak. Thus, C is true, so it is not the requested answer. Some lines are sealed with an O-ring and joined with a spring coupling. Thus, D is true, and is not the requested answer.

**14.**  **A**    Partially plugged internal radiator passages and engine overheating may cause excessive low-side and high-side refrigerant pressures, so A is correct. Air and moisture in the refrigerant system would also cause bubbles in the sight glass, so B is wrong. A defective pressure relief valve may cause damage to other refrigerant system components, or the premature release of refrigerant, so C is wrong. A compressor clutch cycling switch that is stuck closed causes evaporator freeze-up. Thus, D is wrong.

**15.**  **B**    The receiver/drier is mounted between the condenser and the evaporator, so A is wrong. Red refrigerant in the sight glass indicates dye in the refrigerant, so B is correct. A large temperature difference between the receiver/drier inlet and outlet temperatures indicates a restriction in the receiver/drier, so C is wrong. The accumulator is mounted between the evaporator and the compressor, making answer D wrong.

**16.**  **A**    If the TXV is stuck open, the evaporator becomes flooded with refrigerant, resulting in high low-side pressure and frosting of the suction line between the evaporator and the compressor. Thus, Technician A is correct. If moisture in the refrigerant system freezes in the TXV, the low-side pressure is low and frosting of the TXV occurs. Therefore, Technician B is wrong, and A is the correct answer.

**17.**  **A**    When leak testing an evaporator, an electronic leak detector probe may be inserted through the blower resistor assembly opening, so Technician A is right. The evaporator is not accessible with a flame-type leak detector, and this type of equipment does not work on a system with R-134a refrigerant. Thus, Technician B is wrong, and A is the correct answer.

**18.**  **B**    A restricted orifice tube causes frosting of the orifice tube and a less-than-specified low-side pressure, so Technician A is wrong. Restricted refrigerant passages in the evaporator may cause lower-than-specified low-side pressure and frosting of the evaporator outlet pipe. Thus, Technician B is right, and B is the correct answer.

**19.**  **C**    If the coolant control valve is stuck open, the coolant continues to flow through the heater core during the A/C performance test. This action increases the evaporator and discharge air temperature, so Technician A is right. A leak in the coolant control valve vacuum hose allows this valve to remain open continually. Thus, Technician B is right. Because Technicians A and B are both right, C is the correct answer.

**20.**  **C**    On some computer-controlled A/C systems, the diagnostic mode is entered by pressing specific buttons simultaneously on the A/C control panel, and the computer calibrates the actuator motors when the diagnostic mode is entered. Because Technicians A and B are both right, C is the correct answer.

# 8 Engine Performance

## Pretest

The purpose of this pretest is to determine the amount of review that you may require prior to taking the ASE Engine Performance Test. If you answer all the pretest questions correctly, complete the questions and study the information in this chapter to prepare for the ASE Engine Performance Test. If two or more of your answers to the pretest questions are incorrect, complete a thorough study of the questions and information in this chapter. The pretest answers are located at the end of the pretest, and also in the answer sheets supplied with this book.

1. An engine has a hollow rapping noise during acceleration. This noise is worse when the engine is cold. The cause of this problem could be
   A. worn piston pins.
   B. loose pistons.
   C. worn connecting rod bearings.
   D. loose flywheel bolts.

2. An engine has gray exhaust especially after the engine is restarted after a brief, hot, shutdown.
   Technician A says the head gasket may be leaking.
   Technician B says the piston rings may be worn.
   Who is correct?
   A. only
   B. B only
   C. Both A and B
   D. Neither A nor B

3. During a cylinder leakage test, number 4 cylinder has 45-percent leakage and air is escaping from the PCV valve opening in the rocker arm cover. The cause of this problem could be
   A. a burned exhaust valve.
   B. a bent intake valve.
   C. a broken valve spring.
   D. worn piston rings.

4. A fuel-injected vehicle fails an emission test for high CO and HC emissions.
   Technician A says the fuel pressure may be low.
   Technician B says the $O_2$ sensor signal may be continually low.
   Who is correct?
   A. A only
   B. B only
   C. Both A and B
   D. Neither A nor B

5. When diagnosing an electronic distributor ignition system, a 12 V test light connected from the negative primary coil terminal to ground flutters while cranking the engine. A test spark plug does not fire when it is connected from the coil secondary wire to ground while cranking the engine. The cause of this problem could be
   A. an inoperative pickup coil.
   B. an inoperative ignition module.
   C. an inoperative ignition coil.
   D. a primary circuit.

6. An electronic distributor has a Hall-effect pickup.
   Technician A says this type of pickup may be tested with an ohmmeter.
   Technician B says this type of pickup produces an analog voltage signal.
   Who is correct?
   A. only
   B. B only
   C. Both A and B
   D. Neither A nor B

7. A vehicle with an electronic distributor-less ignition system has a loss of power with no engine misfiring.
   Technician A says the basic ignition timing may be late.
   Technician B says the exhaust system may be restricted.
   Who is correct?
   A. A only
   B. B only
   C. Both A and B
   D. Neither A nor B

8. The malfunction indicator light (MIL) is flashing on a car with an OBD II system. The most likely cause of this problem is
   A. a defective catalytic converter.
   B. a defective upstream heated oxygen sensor ($HO_2$).
   C. a defective exhaust gas recirculation (EGR) valve.
   D. severe engine misfiring.

9. A fuel-injected engine with an electronic distributor ignition system backfires during acceleration, but the engine idles smoothly. The most likely cause of this problem is
   A. malfunctioning injectors.
   B. low fuel-pump pressure.
   C. a cracked distributor cap.
   D. bad spark plugs.

10. A fuel-injected engine has a severe surging problem only at speeds above 55 mph (88 kmh). Engine operation is normal at idle and low speeds.
    Technician A says there may be low voltage at the fuel pump.
    Technician B says the inertia switch may have high resistance.
    Who is correct?
    A. A only
    B. B only
    C. Both A and B
    D. Neither A nor B

11. A port-fuel-injected engine has a rough idle and hard starting problems. The cause of these problems could be
    A. a leaking fuel pump check valve.
    B. dripping injectors.

C. a leaking pressure regulator.

D. low fuel-pump pressure.

12. A vehicle fails an emissions test for high NOx emissions. All of the following could be the cause of the high emissions EXCEPT

A. a plugged EGR exhaust passage.

B. a malfunctioning knock sensor.

C. an open winding in the EGR solenoid.

D. a malfunctioning air charge temperature sensor.

13. When diagnosing an EVAP system with a scan tool, the PCM never provides an on command to the EVAP solenoid at any engine or vehicle speed.

Technician A says to check the ECT sensor signal to the PCM.

Technician B says to check the vacuum hoses from the intake to the EVAP canister.

Who is correct?

A. A only

B. B only

C. Both A and B

D. Neither A nor B

14. When performing a battery capacity test

A. the battery should be discharged at one-third of the cold cranking rating.

B. the battery voltage should be 12.8 V prior to the capacity test.

C. the load should be maintained on the battery for 20 seconds.

D. at 70°F (21°C) the voltage should be above 9.6 V at the end of the test.

# Answers to Pretest

**1. B**   Loose pistons may cause a hollow rapping noise during acceleration. Therefore, A, C, and D are wrong, and B is correct.

**2. A**   Gray exhaust especially after the engine is started indicates a coolant leak into some of the combustion chambers. Therefore, Technician A is right, and Technician B is wrong, making A the correct answer.

**3. D**   If air is leaking from the PCV valve opening during a cylinder leakage test, worn piston rings are indicated. Therefore, A, B, and C are wrong, and D is correct.

**4. B**   Low fuel-pump pressure causes a lean air-fuel ratio and low CO and HC emissions. Thus, Technician A is wrong. A low $O_2$ sensor voltage causes a rich air-fuel ratio and high CO and HC emissions. Therefore, Technician B is correct, making B the right answer.

**5. C**   If a 12 V test light connected from the coil negative terminal flutters when cranking the engine, the pickup coil and module are satisfactory. Therefore, A, B, and D are wrong. If the test spark plug does not fire when connected from the coil secondary wire to ground, the coil is defective. Thus, C is the correct answer.

**6. D**   A Hall-effect pickup is not tested with an ohmmeter, so Technician A is wrong. A Hall-effect pickup produces a digital signal, so Technician B is also wrong. Because Technicians A and B are both wrong, D is the correct answer.

**7. B**   The timing is not adjustable on an electronic distributor-less ignition system, so Technician A is wrong. A restricted exhaust system may cause a loss of power. So, Technician B is right, and B is the correct answer.

**8. D**   If the MIL light is flashing, the PCM has detected a severe engine misfiring problem, so D is correct.

**9. C**   Engine backfiring during acceleration is most likely caused by ignition cross-firing resulting from a cracked distributor cap. Therefore, A, B, and D are wrong, and C is correct.

**10.  C**    Low voltage at the fuel pump reduces fuel pump volume and causes engine surging at higher speeds, so Technician A is right. The inertia switch is in series with the fuel pump, and a resistance problem in this switch causes low voltage at the fuel pump. Thus, Technician B is right. Because Technicians A and B are both right, C is the correct answer.

**11.  B**    A leaking fuel pump check valve causes hard starting, but this defect does not cause rough idle, so A is wrong. Dripping injectors may cause hard starting and rough idle, so B is correct. A leaking pressure regulator may cause hard starting but this should not result in rough idle, so C is wrong. Low fuel-pump pressure may cause engine surging at higher speeds, so D is wrong.

**12.  D**    A plugged EGR exhaust passage may cause high NOx emissions. Thus, A is right, which means it is not the requested answer. A malfunctioning knock sensor may cause reduced spark advance, increased combustion chamber temperatures, and high NOx emissions. Thus, B is right, so it is not the requested answer. An open winding in the EGR solenoid causes an inoperative EGR valve and high NOx emissions. Therefore, C is right, and is not the requested answer. A defective air charge temperature sensor does not cause high NOx emissions. Thus, D is not true, making it the requested answer.

**13.  A**    If the PCM never commands the EVAP solenoid on, the PCM may not be receiving a proper engine coolant temperature (ECT) sensor signal. Thus, Technician A is right. A leak in a vacuum hose from the intake to the canister may cause an inoperative EVAP system, but the PCM should still command the EVAP solenoid on. Therefore, Technician B is wrong, and A is the correct answer.

**14.  D**    When performing a battery load test, the battery should be discharged at one-half the cold cranking rating, the load should be maintained for 15 seconds, and the battery should be above three-quarters charge before the test. Therefore, A, B, and C are wrong. At 70°F, the battery voltage should remain above 9.6 V at the end of the test, so D is correct.

# General Engine Diagnosis

## ASE Tasks, Questions, and Related Information

In this chapter, each task in the Engine Performance category is provided, followed by a question and some information related to the task. If you answer any question incorrectly, study this information very carefully until you understand the correct answer. Question answers and analysis can be found at the end of this chapter and in the answer sheets provided with this book.

**Task 1    Verify driver's complaint, perform visual inspection and/or road test vehicle; determine needed action.**

1. Technician A says one of the first steps in a diagnostic procedure is to identify the problem.

   Technician B says listening to the customer's complaint may help to identify the problem.

   Who is correct?

   A.  A only

   B.  B only

   C.  Both A and B

   D.  Neither A nor B

*Hint*    *The following could be used as a general diagnostic and repair procedure:*
- *Listen to the customer's complaint; be sure the problem is identified; road test the vehicle if necessary.*
- *Check TSB regarding vehicle problems.*
- *Consider possible causes of the problem.*
- *Perform appropriate diagnostic tests.*
- *Fix the problem.*
- *Be sure the problem is eliminated.*

## Task 2   Research applicable vehicle and service information, such as engine management system operation, vehicle service history, service precautions, and technical service bulletins.

2. While discussing applicable vehicle information, Technician A says service bulletin information is available in some scan tool software.

Technician B says vehicle and system-specific wiring diagrams are available via fax from some automotive hot lines.

Who is correct?

A. A only

B. B only

C. Both A and B

D. Neither A nor B

*Hint*    *Technicians must be familiar with the operation of the system being serviced. It is very difficult to diagnose a system if the system operation is unknown. Information regarding the system being diagnosed may be obtained from the vehicle manufacturer's service manual or from other sources such as textbooks or generic manuals. If the vehicle belongs to a regular customer, the vehicle service record may be obtained from the shop computer system. When the vehicle does not belong to a regular customer, service records may be obtained from work order copies supplied by the customer. The vehicle service history indicates the previous work done on the vehicle, and being familiar with this information may save the technician some diagnostic time.*

*Service precautions are highlighted in the manufacturer's service manual, generic service manual, or textbooks. Service bulletins usually inform the technician regarding changes necessary to correct specific drivability or operational problems. The software in many scan tools provides service bulletins related to the vehicle and the problem being diagnosed. Service bulletins are available from the vehicle manufacturer and other publishers. Service manuals and service bulletins are also available on CDs. Technicians must have service bulletin information! Service bulletins often provide information necessary to correct the problem, thereby saving a technician hours of wasted time in diagnosing a problem. Various automotive hot lines supply service information via telephone, and some hot lines also supply vehicle and system-specific wiring diagrams via fax.*

## Task 3   Diagnose noises and/or vibration problems related to engine performance; determine needed action.

3. After sitting overnight, an engine has a heavy thumping noise when it is started. The cause of this problem could be

A. worn camshaft lobes.

B. worn valve lifters.

C. worn main bearings.

D. loose piston pins.

*Hint*    *Worn or sticking valve lifters produce a light clicking noise that is most noticeable at idle and low speeds. A heavy clicking noise at 2,000 rpm may be caused by worn camshaft lobes. Worn piston pins produce a heavy, sharp, rapping noise at idle speed. Loose pistons produce a hollow rapping noise that is loudest during acceleration. Piston ring or ring ridge noise is a sharp metallic rapping noise that is worse during acceleration. Loose connecting rod bearings may cause a lighter rapping noise at vehicle speeds above 35 mph (56 kmh). Loose main bearings cause a heavy thumping noise when the engine is started after being shut off for several hours. A heavy thumping noise with the engine idling may be caused by a loose flywheel or vibration damper.*

## Task 4    Diagnose the cause of unusual exhaust color, odor, and sound; determine needed action.

4. At idle speed, the engine exhaust has a "puff" noise at regular intervals.
   Technician A says this problem may be caused by a burned exhaust valve.
   Technician B says this problem may be caused by excessive coolant temperature.
   Who is correct?
   A. A only
   B. B only
   C. Both A and B
   D. Neither A nor B

**Hint**    *Blue exhaust may be caused by excessive amounts of oil entering the combustion chamber. If the exhaust is black, the air-fuel ratio is too rich. On catalytic-converter-equipped vehicles, excessive sulfur smell indicates a rich air-fuel ratio. A "puff" noise in the exhaust at regular intervals indicates cylinder misfire from a compression, ignition, or fuel system problem.*

## Task 5    Perform engine manifold vacuum or pressure tests; determine needed action.

5. At idle speed, a vacuum gauge connected to the intake manifold indicates a steady 14 in. Hg (Figure 8–1).
   Technician A says this reading is caused by sticking valves.
   Technician B says this problem is caused by a burned valve.
   Who is correct?
   A. A only
   B. B only
   C. Both A and B
   D. Neither A nor B

**Figure 8–1** Vacuum gauge reading. *(Courtesy of Sun Electric Corporation)*

**Hint**    *Normal and abnormal vacuum gauge readings are shown in Figure 8–2.*

## Task 6    Perform cylinder power balance test; determine needed action.

6. During an engine power balance test at idle speed, cylinder number 3 produces a 20-rpm decrease, and all the other cylinders produce a 75-rpm decrease. When the test is repeated at 2,000 rpm, cylinder number 3 has the same rpm drop as the other cylinders.
   This problem could be caused by
   A. an intake manifold vacuum leak.
   B. a bad spark plug.

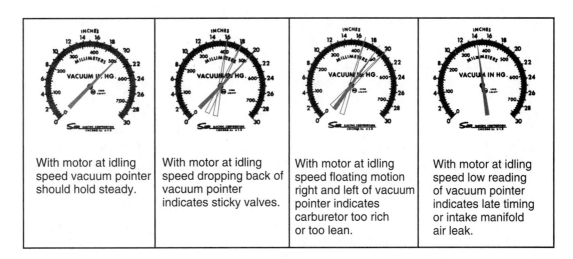

**Figure 8–2** Normal and abnormal vacuum gauge readings. *(Courtesy of Sun Electric Corporation)*

C. an open spark plug wire.

D. a burned exhaust valve.

**Hint**  *During the cylinder power balance test, the tester stops each spark plug from firing for a few seconds and the amount of engine rpm decrease is recorded. If all the cylinders have the specified rpm decrease, the cylinders are contributing equally to engine power. When one cylinder has less rpm decrease, that cylinder is not contributing as much to engine power because of a compression, ignition, or fuel system problem, or an intake manifold vacuum leak. If a vacuum leak is causing the problem, the cylinder rpm drop may be equal to the other cylinders if the test is repeated at a higher engine rpm.*

## Task 7  Perform cylinder cranking compression test; determine needed action.

7. During a compression test, cylinder number 6 has 60 psi and all the other cylinders have 135 psi. When a wet test is preformed on cylinder number 6, the compression increases to 120 psi. The cause of this problem could be

A. a burned exhaust valve.

B. a bent intake valve.

C. worn piston rings.

D. a blown head gasket.

**Hint**  *During a compression test, the ignition and fuel injection system must be disabled. Four compression strokes should be recorded on each cylinder. If some cylinders indicate lower-than-specified compression, a wet test may be performed by squirting about three tablespoons of oil into the cylinder and repeating the test. When the compression reading improves considerably with the oil in the cylinder, worn rings are indicated. If there is very little improvement in the compression reading, the valves or head gasket are probably leaking.*

## Task 8  Perform cylinder leakage/leak-down test; determine needed action.

8. During a leakage test, cylinder number 2 indicates 45-percent leakage and air is escaping from the throttle body.

Technician A says cylinder number 2 may have a bent intake valve.

Technician B says cylinder number 2 may have a bottomed intake valve lifter.

Who is correct?

A. A only

B. B only

C. Both A and B

D. Neither A nor B

**Hint**    *The leakage tester supplies a controlled amount of shop air into each cylinder through the spark plug opening. Both valves in the cylinder must be closed during the leakage test. If cylinder leakage exceeds 20 percent, excessive air is leaking past the rings or valves. Air also could be leaking through the cylinder head or head gasket. When air is escaping through from the crankcase through the PCV valve opening in the rocker cover, the rings and cylinders are worn.*

*Air escaping from the throttle body indicates a damaged intake valve. If air is escaping from the tailpipe, the exhaust valve is burned. Bubbles in the radiator indicate a leaking head gasket or a cracked cylinder head.*

**Task 9**    **Diagnose engine mechanical, electrical, electronic, fuel, and ignition problems with an oscilloscope, engine analyzer, and/or scan tool; determine needed action.**

9. Technician A says some engine analyzers compare test results to specifications and indicate readings that are out of specifications.

Technician B says some engine analyzers provide the option of manual or automatic test modes.

Who is correct?

A. A only

B. B only

C. Both A and B

D. Neither A nor B

**Hint**    *Many engine analyzers have the capability to test ignition, fuel, battery, charging, starting systems, and engine condition. A four- or five-gas emissions analyzer may be contained in the engine analyzer. Some engine analyzers provide the option of manual technician-selected test modes or an automatic test mode. Some engine analyzers contain specifications for various vehicles on a diskette or CD. During the test procedures, the analyzer compares test readings to specifications and identifies test results that are out of specification.*

**Task 10**    **Prepare and inspect the vehicle and analyzer for HC, CO, CO$_2$, and O$_2$ exhaust gas analysis; perform test and interpret exhaust gas readings.**

10. While discussing use of an emissions analyzer to read hydrocarbons (HCs), carbon monoxide (CO), carbon dioxide (CO$_2$), and oxygen (O$_2$) on a catalytic-converter-equipped vehicle, Technician A says a leak in the exhaust system may cause inaccurate emission analyzer readings.

Technician B says that with the tester probe in the tailpipe, the analyzer provides accurate HC and CO readings from the engine.

Who is correct?

A. A only

B. B only

C. Both A and B

D. Neither A nor B

**Hint**    *Emission analyzers usually have a fifteen-minute warm-up and calibration period. Many emission analyzers have an automatic calibration function, but some analyzers with analog meters require manual calibration. Since exhaust leaks cause inaccurate emission readings, the exhaust system on the vehicle must be leak-free. The engine should be at normal operating temperature before the emissions test. Since catalytic converters reduce CO, HC, and oxides of nitrogen (NOx) emissions, if the emissions analyzer probe is in the tailpipe, the analyzer does not indicate the actual emissions coming out of the cylinders. The catalytic converter does not affect CO$_2$ and O$_2$ levels.*

## Task 11  Verify valve adjustment on engines with mechanical or hydraulic lifters.

11. All of the following statements are true about adjusting mechanical valve lifters EXCEPT
    A. If the exhaust valve clearance is less than specified, premature valve burning may occur.
    B. The piston should be at TDC on the exhaust stroke in the cylinder on which the valves are being adjusted.
    C. Excessive intake or exhaust valve clearance causes a clicking noise with the engine idling.
    D. The piston should be at TDC on the compression stroke in the cylinder on which the valves are being adjusted.

*Hint*   *The engine should be at the temperature specified by the vehicle manufacturer before a valve adjustment. When adjusting either type of valve lifter, the piston should be at TDC on the compression stroke in the cylinder on which the valves are being adjusted. When adjusting hydraulic valve lifters, the adjusting nut should be backed off until there is clearance between the rocker arm and valve stem. Rotate the pushrod while tightening the adjusting nut. Continue tightening the adjusting nut until the pushrod turning effort increases slightly. Tighten the adjusting nut the specified amount from this position.*

*The adjusting nut on mechanical lifters is rotated until the specified clearance is available between the rocker arm and valve stem. This clearance is measured with a feeler gauge.*

## Task 12  Verify camshaft timing; determine needed action.

12. When diagnosing valve timing, number 1 piston is positioned at TDC on the exhaust stroke with the timing mark aligned at the 0-degree position on the timing indicator. The valves in number 1 cylinder should be positioned so
    A. the intake valve is beginning to open and the exhaust valve is beginning to close.
    B. the intake valve is completely open and the exhaust valve is closed.
    C. the exhaust valve is completely open and the intake valve is closed.
    D. the intake valve is beginning to close and the exhaust valve is beginning to open.

*Hint*   *The valve timing may be verified with number 1 piston positioned at TDC on the exhaust stroke and the timing marks aligned at the zero mark. Because the exhaust stroke is ending and the intake stroke is beginning, the exhaust valve should be closing and the intake valve should be opening in this crankshaft position. This valve action is called valve overlap. If the valves are in any other position, the valve timing is incorrect.*

## Task 13  Verify engine operating temperature, check coolant level and condition, perform cooling system pressure test; determine needed repairs.

13. All of the following problems may be caused by an engine thermostat that is stuck open EXCEPT
    A. a rich air-fuel ratio and reduced fuel economy.
    B. an inoperative EGR system.
    C. improper cooling fan operation.
    D. an improper air charge temperature sensor signal.

*Hint*   *A thermometer may be taped to the upper radiator hose to check the thermostat operation. When the engine has been idling for 20 minutes, the temperature reading on the thermometer should be close to the specified temperature rating of the thermostat. If the thermostat is sticking closed, the engine overheats. A thermostat that sticks open may cause a rich air-fuel ratio, inoperative emission systems such as EVAP and EGR, improper cooling fan operation, and improper TCC lockup operation.*

*A pressure tester may be installed on the radiator filler neck to check cooling system leaks. Pressurize the cooling system for 15 minutes at the pressure stamped on the radiator cap. If the pressure decreases after 15 minutes, the cooling system is leaking. Inspect the cooling system for external leaks and look for coolant dripping out of the heater case onto the front floor mat. If there are no external leaks, check for combustion chamber leaks or transmission cooler leaks.*

*The radiator cap may be installed on the pressure tester. When the tester pump is operated, the cap should release pressure at or slightly below the cap rating.*

## Task 14   Inspect and test mechanical/electrical fans, fan clutch, fan shroud/ducting, and fan control devices; determine needed repairs.

14. An electric cooling fan is inoperative. With the ignition switch on, there is 12 V supplied to the cooling fan relay contacts and this relay winding (Figure 8–3). When the engine is hot enough to close the temperature switch, the voltage at the switch contacts is 0.2 V. The cause of this problem could be
    A. an inoperative coolant temperature switch.
    B. an open circuit in the main relay winding.
    C. an open circuit between the fan relay and the motor.
    D. a blown 30A fan fuse.

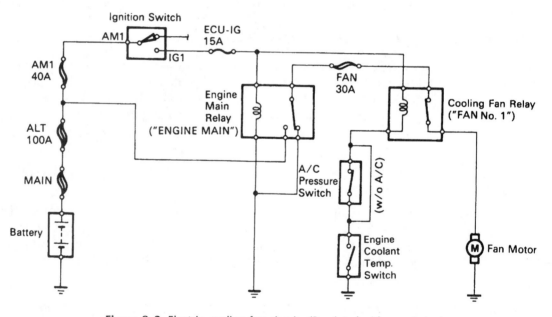

**Figure 8–3**  Electric cooling fan circuit. *(Reprinted with permission)*

*Hint*    *Some electric cooling fan circuits contain a relay and a temperature switch. When the temperature switch closes, the relay winding is grounded through the switch and this action closes the relay contacts. Under this condition, voltage is supplied through the contacts to the fan motor.*

## Task 15   Read and interpret electrical schematic diagrams and symbols.

15. The component D2 in Figure 8–4 is a
    A. transistor.
    B. field effect transistor.
    C. conventional diode.
    D. light-emitting diode.

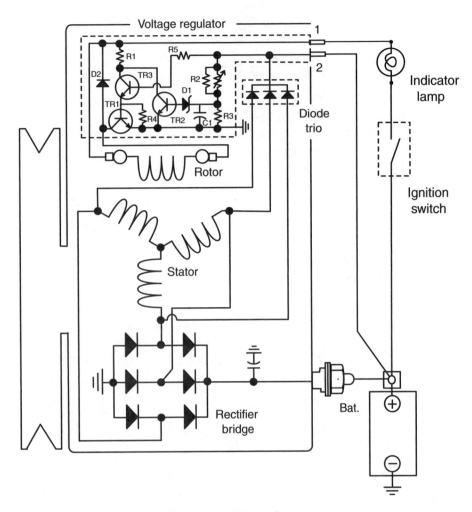

**Figure 8–4** Wiring diagram.

*Hint*    *Electrical and electronic components are represented by symbols on wiring diagrams. Technicians must understand these symbols to properly understand wiring diagrams and electrical/ electronic system operation.*

# Ignition System Diagnosis and Repair

### ASE Tasks, Questions, and Related Information

**Task 1**  **Diagnose ignition system related problems such as no-starting, hard starting, engine misfire, poor drivability, spark knock, power loss, poor mileage, and emissions problems; determine root cause; determine needed repairs.**

16.  When diagnosing a no-start condition on an electronic ignition (EI) system, there is no spark at any of the spark plugs. A 12 V test light does not flutter when connected from the negative primary terminal on each coil to ground with the engine cranking.

The cause of this problem could be

A. an inoperative crankshaft sensor.

B. inoperative ignition coil assembly.

C. damaged spark plug wires.

D. bad spark plugs

17. A vehicle with an EI system has a reduced fuel economy and loss of power complaint (Figure 8–5). The engine runs smoothly, never misfires, and starts easily.

    Technician A says the cause of this problem may be an open circuit in the EST wire from the PCM to the coil module.

    Technician B says the cause of this problem may be a malfunctioning cam sensor.

    Who is correct?

    A. A only

    B. B only

    C. Both A and B

    D. Neither A nor B

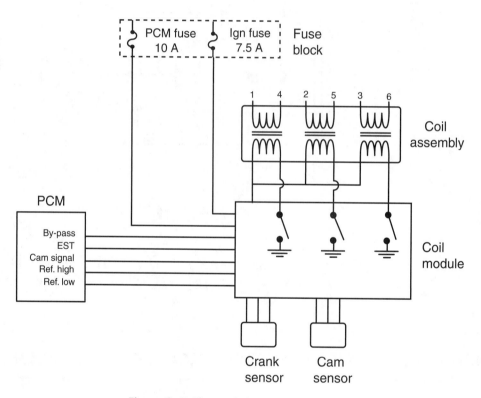

**Figure 8–5** Electronic ignition (EI) system.

18. When diagnosing a no-start condition on an electronic distributor ignition system, a test spark plug fires normally when connected from the coil secondary wire to ground with the engine cranking. The test spark plug fires intermittently when connected from the spark plug wires to ground.

    Technician A says the distributor cap or rotor may have a leakage problem.

    Technician B says the ignition module may be the problem.

    Who is correct?

    A. A only

    B. B only

C. Both A and B

D. Neither A nor B

*Hint*    *When diagnosing a no-start condition on a distributor ignition system, a 12 V test light may be connected from the tach terminal on the ignition coil to ground. If the test light flutters when the engine is cranked, the ignition module is triggering the primary circuit on and off. Therefore, the pickup coil and module are satisfactory. When the test light does not flutter on and off, the pickup coil, pickup coil lead wires, or the module are malfunctioning. The pickup coil and lead wires may be tested with an ohmmeter, and the ignition module may be tested with a module tester. If the module is triggering the primary circuit on and off, a test spark plug may be connected from several spark plug wires to ground. The test spark plug must have the proper voltage requirement for the ignition system being tested. If the test spark plug does not fire when the engine is cranked, the ignition coil is inoperative or the secondary ignition circuit has a leakage problem, which is most likely in the distributor cap or rotor.*

*On electronic (distributor-less) ignition systems, the 12 V test light should be connected to the negative terminal on each ignition coil and ground. If the test light does not flutter on any of the coils when cranking the engine, the crankshaft sensor, camshaft sensor, connecting wires, or the ignition module are malfunctioning. These ignition systems may have a separate module, or the module may be designed into the powertrain control module (PCM). The crankshaft and camshaft sensor signals may be measured with a voltmeter or a lab scope. The test spark plug may be connected from several spark plug wires to ground. Each coil fires two spark plugs. When the engine is cranked and the test spark plug does not fire when connected to the two spark plug wires fired by one of the coils, that coil is likely inoperative, assuming the test light fluttered when connected to the negative coil terminal.*

*Many engines are presently equipped with coil-on-plug or coil-near-plug ignition systems.*

*Because each spark plug has a separate coil, an inoperative coil only causes misfiring on one cylinder. A no-start condition may be caused by a malfunctioning crankshaft sensor, camshaft sensor, or ignition module.*

## Task 2    Interpret ignition system related diagnostic trouble codes (DTC); determine needed repairs.

19. When using a scan tool to diagnose ignition problems, Technician A says if a DTC is obtained representing the crankshaft sensor, this sensor should be replaced.

    Technician B says a malfunctioning crankshaft sensor may cause a no-start condition.

    Who is correct?

    A. A only

    B. B only

    C. Both A and B

    D. Neither A nor B

*Hint*    *Many ignition systems will set DTCs in the powertrain control module (PCM) memory if certain problems occur in ignition system components. For example, DTCs P0335 through P0339 indicate problems in the crankshaft sensor. A malfunctioning camshaft sensor may be indicated by DTCs P0340 through P0349. A scan tool must be connected to the data link connector (DLC) under the instrument panel to retrieve the DTCs. Misfiring on any cylinder will also set a DTC in the PCM. A DTC indicates a problem in a certain area. For instance, a DTC representing the crankshaft sensor indicates a problem in the crankshaft sensor or the connecting wires.*

## Task 3    Inspect, test, repair, or replace ignition primary circuit wiring and components.

20. A pair of ohmmeter leads are connected from one of the distributor pickup leads to ground with the pickup coil connector disconnected from the module. The ohmmeter indicates 14 ohms resistance. This reading indicates

    A. the pickup coil is satisfactory.

    B. the pickup coil winding is shorted.

C. the pickup coil winding is open.

D. the pickup coil winding is grounded.

**Hint**    *The primary ignition coil winding may be tested with an ohmmeter connected to the primary terminals. A reading below the specified value indicates a shorted winding, and a reading higher than specified is caused by a resistance problem or an open circuit. If the ohmmeter leads are connected from one of the primary terminals to ground on the coil container, a low reading indicates a grounded primary winding, whereas an infinity reading proves the primary winding is not grounded. On many coils, the secondary winding may be tested with the ohmmeter leads connected from the coil secondary terminal to one of the primary terminals.*

*The pickup coil may be tested for open and short circuits with the ohmmeter leads connected to the pickup leads when these leads are disconnected from the module. Connect the ohmmeter leads from one of the pickup coil leads to ground to test the pickup coil for a grounded condition.*

## Task 4    Inspect, test, and service distributor.

21. When diagnosing the distributor pickup in Figure 8–6, the pickup may be tested with
    A. an ohmmeter connected across the pickup leads with these leads disconnected.
    B. a test light connected across the pickup leads with these leads disconnected.
    C. a voltmeter connected from the pickup signal wire to ground while cranking the engine.
    D. a 12 V test light connected from the primary positive terminal to ground while cranking the engine.

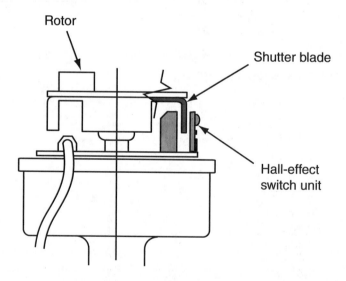

**Figure 8–6**  Hall-effect distributor pick-up.

**Hint**    *A Hall-effect distributor pickup produces a digital voltage signal while cranking the engine. These wires are connected to a Hall-effect pickup. The wires include a ground wire, voltage supply wire, and a signal wire. This type of pickup may be tested with a voltmeter connected from the pickup signal wire to ground while cranking the engine. If the pickup is satisfactory, the voltmeter should indicate the specified low- and high-voltage signals.*

## Task 5    Inspect, test, service, repair or replace ignition system secondary circuit wiring and components.

22. When using an oscilloscope to diagnose the secondary ignition system, the maximum coil voltage on all spark plug wires is 15 kv. The specified maximum coil voltage is 35 kv.

Technician A says there may be a high-resistance problem in the primary ignition circuit.

Technician B says the ignition coil may have a secondary insulation leakage problem. Who is correct?

A. A only

B. B only

C. Both A and B

D. Neither A nor B

**Hint**    *Spark plug wires may be tested with an ohmmeter. If the resistance of any wire exceeds specifications, replace the wire. The normal required secondary coil voltage to fire each spark plug may be checked with an oscilloscope. When the normal required voltage is higher than specified, there is excessive resistance in the secondary circuit, perhaps in spark plug wires or spark plugs.*

*A maximum available coil voltage that is less than specified may be caused by an inoperative coil, a cracked cap or rotor, or high primary circuit resistance causing low primary current.*

## Task 6    Inspect, test, and replace ignition coil(s).

23. In the coil shown in Figure 8–7
    A. 12 V are supplied to the positive side of each primary winding when the ignition switch is turned on.
    B. the negative side of each primary winding is connected to the secondary winding.
    C. one end of each secondary winding is grounded to the coil case.
    D. the secondary windings are interconnected with each other.

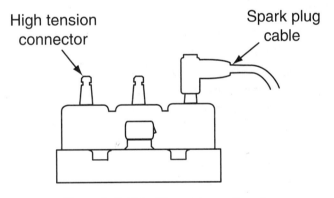

**Figure 8–7** EI ignition system coil pack.

**Hint**    *In a coil used with electronic distributor ignition, the primary winding is connected to the primary terminals. The ends of the secondary winding are connected from the coil tower to one of the primary terminals. On some coils, one end of the secondary winding is connected to ground on the coil frame or case.*

*In a coil pack used with EI systems, 12 V are supplied to the positive side of each primary winding when the ignition switch is turned on. The negative side of each primary winding is connected to the ignition module or PCM. The ends of each secondary winding are connected to the two spark plug wire terminals on each coil.*

## Task 7    Check and adjust, if necessary, ignition system timing and timing advance/retard.

24. Technician A says timing adjustments are required on electronic distributor-less ignition (EI) systems.

Technician B says to set the basic timing on some fuel-injected engines with a distributor, the PCM must be in the limp-in mode.

Who is correct?

A.  A only

B.  B only

C.  Both A and B

D.  Neither A nor B

**Hint**    *When checking the basic ignition timing on some fuel-injected engines with electronic distributor ignition, a set timing connector must be disconnected to prevent the PCM from supplying spark advance during this operation. Some manufacturers recommend disconnecting the engine coolant temperature sensor to place the PCM in the limp-in mode while checking basic ignition timing. Basic timing adjustments are not possible on EI systems.*

## Task 8    Inspect, test, and replace ignition system pick-up sensor or triggering devices.

25.  In the ignition system in Figure 8–8, the ignition switch is on and a digital voltmeter is connected from terminal 7 to 4 on the PCM by back probing the connectors. The voltage indicated on the voltmeter is 0 V. The battery is fully charged, and the voltage supply and ground wires to the PCM are satisfactory.

Technician A says the wire from terminal 7 on the PCM to the crankshaft and camshaft sensors may have an open circuit.

Technician B says the PCM may be malfunctioning.

Who is correct?

A.  A only

B.  B only

C.  Both A and B

D.  Neither A nor B

**Hint**    *In many electronic ignition (EI) systems, the crankshaft and camshaft sensor signals inform the powertrain control module (PCM) regarding crankshaft position and speed and piston position.*

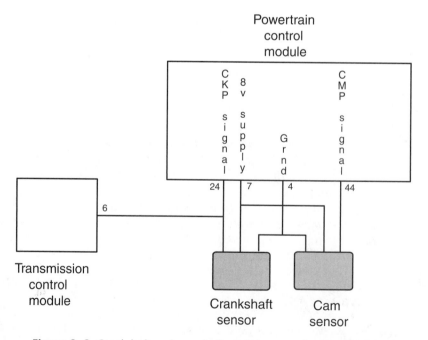

**Figure 8–8** Crankshaft and camshaft sensor connections to the PCM.

*A slotted trigger wheel attached to the crankshaft or flywheel rotates past the crankshaft sensor, and a second trigger wheel attached to the camshaft gear rotates past the camshaft sensor.*

*In some EI systems, the PCM supplies 8 V to the crankshaft and camshaft sensors, and these sensor signals change from 0.3 V to 5.0 V as the slotted trigger wheel rotates past the sensor. A ground wire is connected from each sensor to the PCM. In other systems, the camshaft sensor signal is sent to the PCM and the transmission control module (TCM).*

## Task 9  Inspect, test, and/or replace ignition control module (ICM)/powertrain control module (PCM).

26. When diagnosing a no-start condition, 12 V are available at the positive primary coil terminal with the ignition switch on, and the pickup coil tests are satisfactory. When the voltmeter leads are connected from the negative primary coil terminal to ground, the voltmeter always indicates 12 V while cranking the engine.

    Technician A says the wire from the negative primary coil terminal to the module may have an open circuit.

    Technician B says the ignition module may have an open circuit.

    Who is correct?

    A. A only

    B. B only

    C. Both A and B

    D. Neither A nor B

**Hint**   *When a 12 V test light connected from the negative primary coil terminal to ground does not flutter while cranking the engine, the ignition module or the pickup coil may be the problem. If the ohmmeter tests indicate the pickup is satisfactory, the module is likely inoperative. A variety of ignition module testers are available to test the module individually.*

# Fuel, Air Induction, and Exhaust System Diagnosis and Repair

## ASE Tasks, Questions, and Related Information

## Task 1  Diagnose fuel system related problems, including hot or cold no-starting, hard starting, poor drivability, incorrect idle speed, poor idle, flooding, hesitation, surging, engine misfire, power loss, stalling, poor mileage, and emissions problems; determine root cause; determine needed action.

27. A port-fuel-injected engine has a hard-starting problem after the engine has been shut off for one hour or longer. The engine requires a longer-than-normal cranking time before it starts. All of the following defects may be the cause of the problem EXCEPT

    A. dripping injectors.

    B. a leaking check valve in the fuel pump.

    C. a leaking valve in the fuel pressure regulator.

    D. a leaking injector O-ring.

28. A fuel-injected engine has a higher-than-specified idle speed with the engine at normal operating temperature.

Technician A says the TPS signal voltage may be higher than specified.

Technician B says the engine may have an intake manifold vacuum leak.

Who is correct?

A. A only

B. B only

C. Both A and B

D. Neither A nor B

29. A customer complains about reduced fuel economy on a fuel-injected engine. The engine runs smoothly, but black smoke is emitted from the tailpipe during engine warm-up. The most likely cause of this problem could be

A. inoperative injectors.

B. damaged spark plugs.

C. higher-than-specified fuel pressure.

D. an inoperative fuel pump check valve.

**Hint**   *When diagnosing drivability problems on fuel-injected engines, one of the first tests should be to verify the fuel pressure. If the fuel pressure is lower than specified, the engine may cut out and surge at higher speeds and hesitate on low-speed acceleration. Higher-than-specified fuel pressure causes a rich air-fuel ratio, excessive fuel consumption, and high emissions.*

*Hard starting on fuel-injected engines may be caused by a lean air-fuel ratio. The powertrain control module (PCM) uses the engine coolant temperature (ECT) sensor input to determine the proper amount of air-fuel ratio enrichment when the engine is cold. Therefore, a malfunctioning ECT sensor signal may cause a lean air-fuel ratio on a cold engine, and this results in a hard-starting problem. Hard starting may also be caused by dripping injectors or a leaking check valve in the fuel pump that allows fuel to drain back into the fuel tank when the engine is shut off.*

*Improper idle speed on a fuel-injected engine may be caused by an improper input signal to the PCM. For example, if the throttle position sensor (TPS) voltage signal is lower than specified, the engine idle speed may be too high. If the PCM or the idle air control (IAC) motor have been changed, an idle relearn procedure is required on many PCMs. Some scan tools have idle relearn capabilities. If the idle relearn procedure cannot be performed with a scan tool, an idle relearn procedure without the scan tool is provided in the service manual.*

**Task 2**   **Interpret fuel or induction system related diagnostic trouble codes (DTCs); determine needed repairs.**

30. When diagnosing an on-board diagnostic II (OBD II) vehicle, a scan tool displays a DTC indicating a misfire on the number 2 cylinder.

Technician A says to test the compression on the number 2 cylinder.

Technician B says to test the intake air temperature (IAT) sensor with an ohmmeter.

Who is correct?

A. A only

B. B only

C. Both A and B

D. Neither A nor B

**Hint**   *On fuel-injection systems with a pressure regulator, there are no DTCs that directly indicate improper fuel pressure. On some late model fuel systems, the pressure regulator is discontinued, and a fuel pressure sensor and fuel temperature sensor are located in the fuel rail. In response to these inputs, the PCM pulses the fuel pump on and off to maintain the specified fuel pressure.*

*On these systems, improper fuel pressure causes DTCs to be set in the PCM memory. The IAT sensor is mounted in the air cleaner or air intake system. A malfunctioning IAT sensor may cause a rich air-fuel ratio, decreased fuel economy, and high emissions. If the IAT sensor or connecting wires become inoperative, a DTC is set in the PCM memory. If the IAT sensor becomes coated, it may provide an improper voltage signal.*

*In some fuel injection systems, if an electrical problem occurs in an injector, a DTC is set in the PCM memory. OBD II systems provide cylinder misfire DTCs for each individual cylinder. An inoperative injector may cause a cylinder misfire, but this problem may also be caused by a compression problem, an ignition problem, or a vacuum leak.*

## Task 3 Inspect fuel tank, filler neck, and gas cap; inspect and replace fuel lines, fittings, and hoses; check fuel for contaminants and quality.

31. A fuel-injected engine is hard to start hot or cold and has an acceleration stumble and a loss of power.

    Technician A says there may be an excessive amount of alcohol mixed with the fuel.

    Technician B says the fuel line from the tank to the filter may be restricted.

    Who is correct?

    A. A only

    B. B only

    C. Both A and B

    D. Neither A nor B

**Hint**    *At present most fuel-injected engines will operate satisfactorily on 10-percent alcohol mixed with the gasoline. Excessive amounts of alcohol mixed with the gasoline causes a lean air-fuel ratio, hard starting, loss of power, and a hesitation during acceleration. The fuel systems in flex-fuel vehicles are designed to operate on higher percentages of alcohol. Test equipment is available to test the percentage of alcohol mixed with gasoline. A 100-milliliter (mL) cylinder may be filled with 90 mL of gasoline and 10 mL of water. Place a stopper in the cylinder and shake the contents. Remove the stopper to relieve any pressure and wait five minutes. If the water content is now 15 percent, there was 5 percent alcohol in the fuel.*

## Task 4 Inspect, test, and replace fuel pump and/or fuel pump assembly; inspect, service, and replace fuel filters.

32. A fuel-injected engine will not start and the fuel pump pressure is zero. When the FP and B+ terminals are connected with a jumper wire, the engine starts and runs normally (Figure 8–9). Once the engine starts, the jumper wire may be disconnected and the engine operation is unchanged. The cause of this problem could be

    A. an open circuit in the main relay winding.

    B. a blown 7.5 A IGN fuse.

    C. an open circuit at the B+ fuel pump relay terminal.

    D. a malfunctioning circuit opening relay winding.

**Hint**    *Mechanical fuel pumps should be tested for pressure and volume. A tester is connected at the carburetor inlet fitting to perform these tests. Electric fuel pumps on fuel-injected engines should also be tested for pressure and volume. On a throttle-body-injection system, the fuel pump tester is connected at the throttle body inlet fuel line. In port-fuel-injection systems, a Schrader valve is located on the fuel rail for test equipment connection. On most fuel-injected engines voltage is supplied through a computer-operated relay to the fuel pump. In these systems, the computer shuts off the fuel pump if the ignition switch is on for two seconds and the engine is not cranked or started.*

## Task 5 Inspect and test fuel pump electrical control circuits and components; determine needed repairs.

33. A fuel-injected vehicle has reduced fuel economy. The engine starts easily, runs smoothly, and has normal power.

    Technician A says the fuel return line may be restricted.

    Technician B says the fuel pump check valve may be leaking.

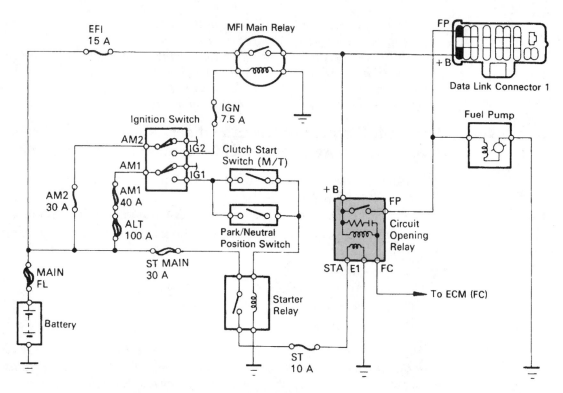

**Figure 8–9** Fuel pump circuit. *(Reprinted with permission)*

Who is correct?
A. A only
B. B only
C. Both A and B
D. Neither A nor B

**Hint**   *Higher-than-specified fuel pressure may be caused by a sticking pressure regulator or a restricted return fuel line. A vacuum leak in the hose connected to the fuel pressure regulator on a port-fuel-injected engine causes higher-than-specified fuel pressure at lower engine speeds. High fuel pressure causes a rich air-fuel ratio and reduced fuel economy.*
*Low fuel-pump pressure causes a loss of power and acceleration stumbles. An inoperative fuel pump, or a restricted in-tank or in-line filter, causes low fuel-pump pressure.*

**Task 6**   **Inspect, test, and repair or replace fuel pressure regulation system and components of fuel injection systems; perform fuel pressure/volume test.**

34. When testing the fuel pressure on a port-injected engine the fuel pressure is higher than specified.
    Technician A says this problem may be caused by a restricted fuel return line.
    Technician B says this problem may be caused by a fuel pressure regulator that is sticking open.
    Who is correct?
    A. A only
    B. B only
    C. Both A and B
    D. Neither A nor B

35. A port-fuel-injected engine with a cold start injector has a rough idle problem and black smoke is emitted from the tailpipe (Figure 8–10). When the electrical connector

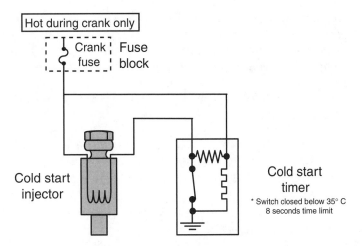

**Figure 8–10** Cold start injector circuit.

is disconnected from the cold start injector, the engine runs normally. The cause of this problem could be

A. an inoperative thermo-time switch.

B. a malfunctioning PCM.

C. an inoperative engine coolant temperature sensor.

D. an inoperative air charge temperature sensor.

36. A port-injected engine with a deceleration fuel reduction feature in the PCM has a reduced fuel economy complaint. When the emissions are tested, this engine has high hydrocarbon (HC) and carbon monoxide (CO) emissions only during deceleration.

Technician A says to listen to the injectors with a stethoscope as the engine is decelerated.

Technician B says to test the fuel pressure at idle and 2,500 rpm.

Who is correct?

A. A only

B. B only

C. Both A and B

D. Neither A nor B

**Hint**      *When performing fuel system pressure and volume tests on a throttle body injection system, the pressure gauge must be connected at the throttle body fuel inlet fitting. On port-injected engines, connect the fuel pressure gauge to the Schrader valve on the fuel rail. Always relieve the fuel system pressure before attempting to connect the pressure gauge. With the pressure gauge connected, cycle the ignition switch several times or start the engine to read the fuel pressure. Low fuel pressure may be caused by an inoperative pump, restricted fuel filter or fuel line, or a fuel pressure regulator that is sticking open. High fuel pressure may be caused by a restricted fuel return line or a fuel pressure regulator that is sticking closed.*

*Some port-fuel-injected engines have a cold start enrichment valve. This valve is operated by a thermo-time switch that senses engine coolant temperature. The thermo-time switch contains a set of contacts and a bimetal switch. When the coolant temperature is below 95°F (35°C), the closed thermo-time switch contacts supply voltage to the cold start enrichment valve. Under this condition, the cold enrichment valve is open and fuel is discharged from this valve into the intake manifold. The bimetal switch action in the thermo-time switch allows the switch contacts to remain closed for a maximum of eight seconds.*

*On many engines with deceleration fuel reduction systems, the PCM stops grounding the injectors in a specific rpm range during deceleration. Place a stethoscope pickup on any injector and accelerate the engine at 3,000 rpm. Decelerate the engine and listen to the injector clicking. The injector should stop clicking momentarily during deceleration if the deceleration fuel reduction system is operating normally. If the deceleration fuel reduction system is not operating, fuel economy will be reduced while HC and CO emissions are increased.*

## Task 7    Inspect, remove, service or replace throttle body assembly; make related adjustments.

37. A throttle body injection system has a rough idle and stalling problem when the engine is at normal operating temperature.
Technician A says the throttle body may require cleaning in the throttle bore area.
Technician B says the PCV valve may be stuck in the open position.
Who is correct?
A. A only
B. B only
C. Both A and B
D. Neither A nor B

**Hint**    *The throttle body may be cleaned with an approved cleaner. Excessive carbon buildup in the throttle bore area may cause rough idle operation and stalling. A PCV valve that is stuck open causes a similar action to an intake manifold vacuum leak. The PCM senses the extra air entering the engine and supplies more fuel to go with the air, which increases idle speed.*

## Task 8    Inspect, test, clean, and replace fuel injectors.

38. During the injector test in Figure 8–11, five of the injectors provide a pressure decrease of 20 psi (138 kPa) and the injector for number 3 cylinder has a 10 psi (69 kPa) decrease. These test results indicate
A. the plunger in the injector for number 3 cylinder is sticking open.
B. the orifice in the injector for number 3 cylinder is restricted.
C. all the injectors are in satisfactory condition.
D. lower fuel pressure is supplied to the number 3 injector.

**Hint**    *Malfunctioning injectors may cause rough idle operation, stalling, or acceleration stumbles.*
*During an injector balance test, a fuel pressure gauge is connected to the Schrader valve in the fuel rail. The injector tester opens each injector for a specific length of time, and the pressure on the fuel gauge should drop the specified amount on each injector. When the pressure drop on an injector is more than specified, the injector plunger is sticking open. If the pressure drop on an injector is less than specified, the injector orifice is restricted.*
*Some injectors may be cleaned with a pressurized container of injector cleaner connected to the Schrader valve on the fuel rail. The fuel pump must be disabled and the fuel return line plugged while cleaning injectors.*

## Task 9    Inspect, service, and repair or replace air filtration system components.

39. An air cleaner element is contaminated with oil.
Technician A says the PCV valve may be stuck open.
Technician B says the PCV clean air hose and fitting may be loose in the rocker arm cover.
Who is correct?
A. A only
B. B only
C. Both A and B
D. Neither A nor B

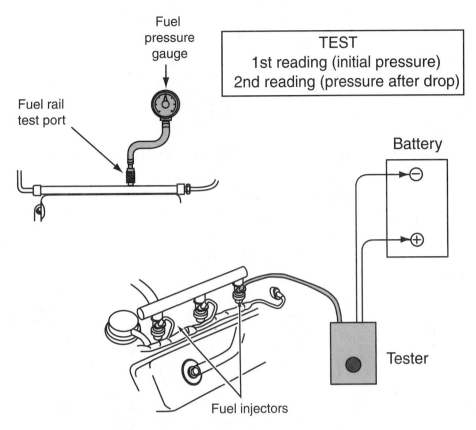

**Figure 8–11** Injector test procedure.

**Hint**     *The air filter element should be inspected for dirt contamination and small holes. A shop light may be placed on the inside of the element. When looking through the element from the outside, the technician can observe the amount of dirt contamination. If the air cleaner element is contaminated with oil, the positive crankcase ventilation (PCV) valve or hose may be restricted.*

*Excessive blowby past the piston rings also results in oil contamination of the air cleaner element. Small holes in the air cleaner element allow dirt particles to enter the engine, which result in rapid cylinder wall, piston, and piston ring wear. The body of the air cleaner must make proper contact with the sealing surfaces on the element to prevent dirt particles from bypassing the element. Shop air pressure may be used to blow dirt out of the air cleaner element. On most air cleaner elements, hold the air nozzle 6 in. (15.24 cm) from the inside of the element. On some late model vehicles, the air flows from the inside to the outside of the air cleaner element. On these elements, hold the air nozzle about 6 in. (15.24 cm) from the outside of the element when blowing dirt out.*

**Task 10**   **Inspect throttle assembly, air induction system, intake manifold and gaskets for vacuum leaks and/or unmetered air.**

40. An engine has a rough idle complaint. A propane cylinder with a precision valve and a hose is used to check for intake vacuum leaks. When propane is charged near the injector in the number 4 cylinder intake port, the engine speed changes and the engine runs smoother.

Technician A says the intake manifold gasket is leaking.

Technician B says the lower injector O-ring is leaking.

Who is correct?

A. A only

B. B only

C.  Both A and B

D.  Neither A nor B

**Hint**   *Intake manifold gaskets and throttle body mounting plates may be checked for leaks with a propane cylinder equipped with a precision metering valve and hose. This equipment is used for the propane enrichment method of adjusting idle mixture screws on carbureted engines. When propane is discharged from the hose on the propane cylinder near an intake manifold vacuum leak, the engine runs smoother and the engine speed changes.*

## Task 11   Check and/or adjust idle speed where applicable.

41.  The throttle body adjustment in Figure 8–12 is

A.  a minimum air rate adjustment.

B.  an idle mixture adjustment.

C.  a fast idle speed adjustment.

D.  a slow idle speed adjustment.

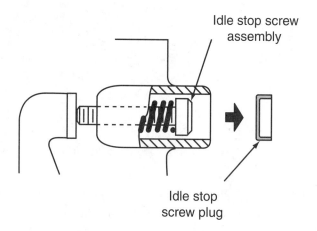

**Figure 8–12** Throttle body adjustment.

**Hint**   *Some throttle body assemblies have a minimum air rate adjustment. This adjustment is only required if the throttle body has been replaced. A metal plug must be removed to access the minimum air rate screw. The idle speed control motor must be in the closed position or the air passage through this motor must be plugged before the minimum air rate adjustment. Adjust the minimum air rate screw to the specified rpm.*

## Task 12   Remove, clean, inspect, test, and repair or replace fuel system vacuum and electrical components and connections.

42.  Technician A says the component in Figure 8–13 allows the throttle to close gradually on deceleration.

Technician B says the component in Figure 8–13 holds the throttle in the specified idle position.

Who is correct?

A.  A only

B.  B only

C.  Both A and B

D.  Neither A nor B

**Hint**   *Some fuel-injected engines have a vacuum-operated throttle opener. This opener allows the throttle to close gradually on deceleration to prevent stalling. With the throttle opener vacuum*

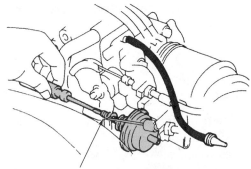

Throttle opener adjusting screw

**Figure 8–13** Vacuum actuator. *(Reprinted with permission)*

*hose removed and plugged, the throttle linkage should strike the opener stem at 1,300 to 1,500 rpm. The opener stem may be adjusted to obtain this rpm.*

**Task 13**  **Inspect, service, and replace exhaust manifold, exhaust pipes, mufflers, resonators, catalytic converters, tail pipes, and heat shields.**

43. A vehicle has a severe loss of power and a top speed of 55 mph (88 kmh). The engine runs smoothly and does not misfire. The cause of this problem could be
    A. a damaged spark plug.
    B. a damaged spark plug wire.
    C. inoperative fuel injectors.
    D. a restricted exhaust pipe.

**Hint**  *A restricted exhaust system, or a restricted air intake system causes a loss of power and reduced maximum vehicle speed. When this problem is present, the engine runs smoothly and does not misfire.*

**Task 14**  **Test for exhaust system restriction; determine needed action.**

44. A port-fuel-injected engine runs smoothly at idle speed but misfires during acceleration. The most likely cause of this problem is
    A. an exhaust gas recirculation (EGR) valve that is stuck open.
    B. a restricted catalytic converter.
    C. dripping fuel injectors.
    D. excessive resistance in a spark plug wire.

**Hint**  *Excessive exhaust backpressure may be caused by a restricted exhaust pipe, catalytic converter, or muffler. If the exhaust backpressure is excessive, engine power and maximum vehicle speed are reduced, but the engine does not misfire. Connect a vacuum gauge to the intake manifold to check for a restricted exhaust system. With the engine idling, the manifold vacuum should be 16 to 21 in. Hg (110.32 to 144.79 kPa). When the engine is accelerated to 2,000 rpm, the vacuum should drop momentarily and then recover to 16 to 21 psi (110.32 to 144 kPa). Hold the engine speed at 2,000 rpm. If the vacuum drops below 16 in. Hg (110.32 kPa) after three minutes, the exhaust system may be restricted.*

**Task 15**  **Inspect, test, clean and repair or replace turbocharger or supercharger and system components.**

45. While testing a turbocharger, the maximum boost pressure is 4 psi (27.5 kPa), and the specified boost pressure is 9 psi (62 kPa).
    Technician A says the engine compression may be lower than specified.

Technician B says the wastegate may be sticking closed.

Who is correct?

A. A only

B. B only

C. Both A and B

D. Neither A nor B

46. The vanes on a turbocharger compressor wheel are severely pitted. The cause of this problem could be

A. a leak in the air intake system.

B. partially seized turbocharger bearings.

C. excessive turbocharger shaft end play.

D. reduced turbocharger coolant circulation.

**Hint**     *During a turbocharger pressure boost test, a vacuum pressure gauge is connected to the intake manifold, and the vehicle is driven at the speed and engine rpm specified by the vehicle manufacturer.*

*If the boost pressure is less than specified, the engine compression should be tested. When the compression is low, the airflow through the engine and boost pressure are reduced. If the wastegate is sticking open, or the turbocharger bearings are damaged, turbocharger boost pressure is reduced.*

*If the turbocharger compressor wheel or turbine wheel housings are scored, the vanes on these wheels have been striking these housings. This problem is caused by excessive turbocharger shaft end play, and the turbocharger or center housing assembly must be replaced. If the compressor wheel vanes are severely pitted, the air intake is leaking.*

*The turbocharger bearings are supplied with oil directly from the main oil gallery. If the engine oil is contaminated because of the lack of oil and filter changes, turbocharger bearing life is shortened. In many turbochargers, engine coolant is circulated through the turbocharger housing to cool the turbocharger bearings. If the turbocharger housing and bearings become too hot because of inadequate coolant circulation, the oil actually burns in the bearings after a hot engine shut down. This action greatly reduces bearing life.*

# Emissions Control Systems Diagnosis and Repair (Including OBD II), Positive Crankcase Ventilation

## ASE Tasks, Questions, and Related Information

**Task 1**   Test and diagnose emissions or drivability problems caused by positive crankcase ventilation (PCV) system.

47. With the PCV valve positioned as shown in Figure 8–14, the engine is

A. operating at wide open throttle.

B. operating at half open throttle.

C. backfiring on acceleration.

D. operating at idle speed.

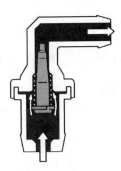

**Figure 8–14** PCV valve. *(Reprinted with permission)*

**Hint**    *During an engine backfire, the PCV valve plunger is seated on the housing, and the passage through the valve is closed. If the engine is idling, the high intake manifold vacuum holds the tapered valve plunger nearly closed. When the throttle is opened and the manifold vacuum decreases, the tapered plunger gradually moves downward to provide more PCV valve opening.*

*If the PCV valve is stuck closed, the air-fuel ratio is richer and HC and CO emissions are higher. A PCV valve stuck in the open position may cause a leaner air-fuel ratio and increased idle rpm.*

**Task 2**    **Inspect, service, and replace positive crankcase ventilation (PCV) filter/ breather cap, valve, tubes, orifice/metering device, and hoses.**

48. The inside of the air cleaner is contaminated with engine oil.
    Technician A says the PCV valve clean air filter in the air cleaner may be plugged.
    Technician B says the hose from the PCV valve to the intake manifold may be severely restricted.
    Who is correct?
    A. A only
    B. B only
    C. Both A and B
    D. Neither A nor B

49. An oil filler cap is loose on the rocker arm cover. The engine is in satisfactory condition with the specified compression. This problem could
    A. allow the PCV system to pull dirt particles into the engine.
    B. reduce the amount of flow through the PCV valve.
    C. increase the amount of flow through the PCV valve.
    D. cause oil to be blown out around the oil cap filter.

**Hint**    *If the inside of the air cleaner is contaminated with engine oil, the engine may have excessive blowby, or the PCV valve and connecting hose may be restricted. When the PCV clean air filter in the air cleaner is plugged, a higher vacuum is built up in the engine. This action may damage engine gaskets and pull dirt particles past some of the engine gaskets, such as rocker cover or oil pan gaskets, into the engine.*

*Some manufacturers recommend removing the PCV valve and shaking it to determine if the valve is faulty. While shaking the valve, the plunger in the valve should rattle if the valve is satisfactory. Other manufacturers recommend blowing through it with a length of hose connected to the valve. Air should pass through the valve easily in one direction, but the valve should block air in the opposite direction.*

# Emissions Control Systems Diagnosis and Repair (Including OBD II), Exhaust Gas Recirculation

## ASE Tasks, Questions, and Related Information

**Task 1**    **Test and diagnose drivability problems caused by the exhaust gas recirculation (EGR) system.**

     50. A vehicle fails an emissions test for NOx emissions. The cause of this emission failure could be

         A. an engine thermostat that is stuck open.

         B. an inoperative engine coolant temperature sensor.

         C. low compression on one cylinder.

         D. restricted exhaust passages under the EGR valve.

**Hint**    *Oxides of nitrogen emissions are caused by oxygen and nitrogen combining at high combustion chamber temperatures. The EGR valve recirculates some exhaust into the intake manifold.*

     *Because there is very little oxygen left in the exhaust, this exhaust does not burn in the combustion chambers, and combustion chamber temperatures are lowered. This action reduces NOx emissions.*

**Task 2**    **Interpret exhaust gas recirculation (EGR) related diagnostic trouble codes (DTCs); determine needed repairs.**

     51. When diagnosing a linear EGR valve, Technician A says the linear EGR valve provides a feedback signal to the PCM.

         Technician B says a scan tool may be used to command the PCM to provide a specific EGR valve opening.

         Who is correct?

         A. A only

         B. B only

         C. Both A and B

         D. Neither A nor B

**Hint**    *In a computer-controlled EGR system with a differential pressure feedback electronic (DPFE) sensor, the PCM pulses the EGR solenoid on and off to supply the precise vacuum to the EGR valve (Figure 8–15). The DPFE sensor senses the EGR flow from the pressure difference across the EGR orifice. This sensor sends an analog voltage signal to the PCM in relation to the EGR flow. The other input sensors inform the PCM regarding the EGR flow required by the engine. If the actual EGR flow does not match the required flow, the PCM changes the solenoid operation to provide the necessary EGR flow. If certain problems occur in this EGR system, a DTC is set in the PCM memory. For example, if the EGR exhaust passages are restricted with carbon, and the EGR flow is reduced, the DPFE sensor informs the PCM regarding the improper EGR flow and a DTC is set in the PCM.*

     *Some EGR systems have a linear EGR valve that is controlled electronically by the PCM without any vacuum controls. The PCM pulses this type of EGR valve on and off to supply the*

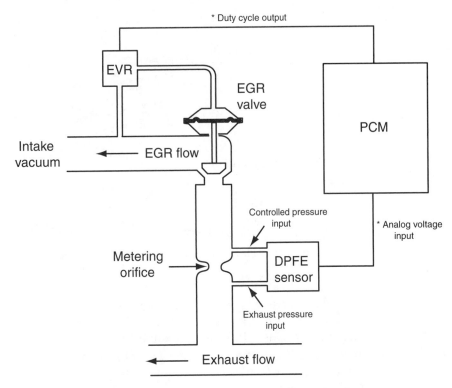

**Figure 8–15** Computer-controlled EGR system with DPFE sensor.

*required EGR valve opening and EGR flow. A linear EGR valve provides a feedback signal that informs the PCM regarding the actual EGR valve position. If an electrical defect occurs in a linear EGR valve, a DTC is set in the PCM. A scan tool may be used to command the PCM to supply a specific EGR valve opening. The feedback signal from the linear EGR valve informs the PCM if this specific EGR valve opening was achieved.*

**Task 3   Inspect, test, service, and replace components of the EGR system, including EGR valve, tubing, exhaust passages, vacuum/pressure controls, filters, hoses, electrical/electronic sensors, controls, solenoids and wiring of exhaust gas recirculation (EGR) systems.**

52. While diagnosing a positive backpressure EGR valve.

    Technician A says that when 18 in. Hg is supplied to the valve with the engine idling, the valve should open.

    Technician B says that when the EGR valve is opened at idle, the engine should slow down at least 150 rpm.

    Who is correct?

    A. A only

    B. B only

    C. Both A and B

    D. Neither A nor B

53. When diagnosing the EGR valve in Figure 8–16

    A. the bleed valve is normally closed with the engine operating at idle speed.

    B. the bleed valve is opened by negative pressure pulses in the exhaust system.

    C. when vacuum is supplied to the valve with the engine stopped, the valve should remain closed.

    D. the bleed valve should be open with the engine operating at 2,000 rpm.

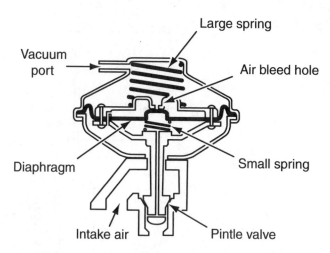

**Figure 8–16** Exhaust gas recirculation (EGR) valve.

54. When diagnosing the computer-controlled EGR valve system in Figure 8–17, Technician A says the EGR valve should be open when the engine coolant is cold and the TPS sensor indicates one-half open throttle.

Technician B says the EGR valve should be open with the engine at normal operating temperature and the throttle wide open.

Who is correct?

A   A only

B.  B only

C.  Both A and B

D.  Neither A nor B

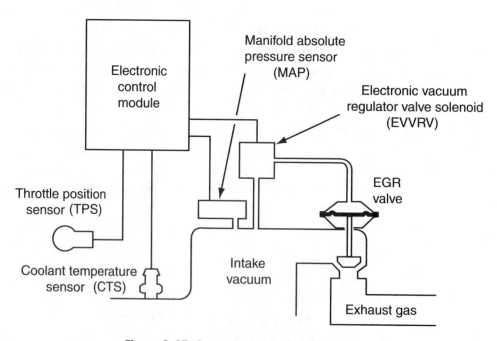

**Figure 8–17** Computer-controlled EGR system.

**Hint**    *When 18 in. Hg are supplied to a conventional EGR valve with the engine idling, the valve should open and the engine should slow down 150 rpm or stall. If this action does not take place, the EGR valve is sticking or the exhaust passages are plugged with carbon.*

*Because a positive backpressure EGR valve contains a vacuum bleed valve, this type of valve does not open when vacuum is supplied to the valve with the engine idling. When the engine speed is increased to 2,000 rpm, exhaust pressure closes the bleed valve. Under this condition, the valve should open when vacuum is supplied to the valve vacuum port.*

*Some engines have a negative backpressure EGR valve in which the bleed valve is normally closed. When the engine is idling or operating at low speed, negative pressure pulses in the exhaust system pull the bleed valve open. Under this condition, vacuum supplied to the valve is bled off, and the valve remains closed. When the engine speed increases, the positive exhaust pressure pulses are closer together and the negative exhaust pressure pulses are reduced. This action allows the bleed valve to close and the vacuum supplied to the EGR valve diaphragm pulls the valve open.*

*In most computer-controlled EGR systems, the PCM operates a solenoid that supplies vacuum to the EGR valve. Since a cold engine does not have high NOx emissions, the PCM does not energize the EGR solenoid when the engine coolant is cold. When the EGR valve is open, the exhaust recirculated into the intake manifold reduces engine power to some extent. Therefore, when the TPS sensor indicates a wide open throttle, the PCM does not energize the EGR solenoid.*

# Emissions Control Systems Diagnosis and Repair (Including OBD II), Secondary Air Injection (AIR) and Catalytic Converter

## ASE Tasks, Questions, and Related Information

**Task 1** Test and diagnose emissions or drivability problems caused by the secondary air injection or catalytic converter systems.

55. A vehicle with a secondary air injection system has high HC and CO emissions and the oxygen sensor voltage is always low (Figure 8–18).

    Technician A says the vacuum hose connected to the air diverter (AIRD) valve may be leaking.

    Technician B says the air pump may be pumping air into the exhaust ports with the engine warmed up.

    Who is correct?
    A. A only
    B. B only
    C. Both A and B
    D. Neither A nor B

**Hint**   *When the engine is started, the PCM does not energize the AIRD or AIRB solenoids and there is no vacuum supplied to the AIRD or AIRB valve. Under this condition, air from the air pump is exhausted to the atmosphere through the AIRB valve.*

*After the engine runs for a brief time, the PCM energizes both the AIRD and AIRB solenoids, and vacuum is supplied through these solenoids to the AIRD and AIRB valves. Under this condition, air from the air pump is directed to the exhaust ports to reduce HC emissions during engine warm-up.*

*When the engine coolant reaches normal operating temperature, the PCM de-energizes the AIRD solenoid. This action shuts off the vacuum supplied to the AIRD valve, and this valve*

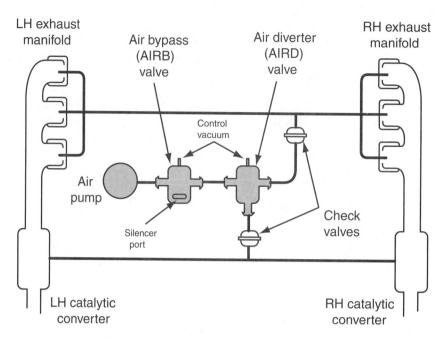

**Figure 8–18** Secondary air injection system.

*plunger moves so air is directed to the catalytic converters. If the air pump continues to pump air into the exhaust ports with a warm engine, the extra oxygen in the exhaust stream causes low $O_2$ sensor voltage. This voltage signal causes the PCM to increase the injector pulse width and provide a rich air-fuel ratio. Under this condition, HC and CO emissions are high.*

## Task 2    Interpret secondary air injection system related diagnostic trouble codes (DTCs); determine needed repairs.

56. When diagnosing an on-board diagnostic II (OBD II) system, Technician A says the upstream and downstream heated oxygen sensors (HO₂S) voltage signals should be similar if the catalytic converter is operating properly.

    Technician B says the PCM illuminates the malfunction indicator light immediately when an inoperative catalytic converter is detected.

    Who is correct?

    A. A only
    B. B only
    C. Both A and B
    D. Neither A nor B

**Hint**    *All cars and light-duty trucks have been equipped with on-board diagnostic II (OBD II) systems since 1996; some 1994 and 1995 cars and light-duty trucks also had these systems. OBD II systems must have the capability to monitor various systems and components, including the secondary air injection system and the catalytic converter. Many OBD II systems have an electric-drive secondary air injection pump that forces air into the exhaust manifolds during engine warm-up to reduce HC emissions. The PCM monitors the secondary air injection system by checking the upstream heated oxygen sensor (HO₂S) voltage when the air pump is turned on. When the air pump is turned on, the airflow into the exhaust manifolds flows past the HO₂S sensor and this immediately lowers the HO₂S voltage signal. If the air pump is turned on and the HO₂S sensor voltage signal does not change a specific amount, the PCM interprets this action as a problem in the secondary air injection system and a pending DTC is set in the PCM. If this problem occurs on two consecutive drive cycles, the PCM illuminates the malfunction indicator light (MIL), and a current DTC is set in the PCM.*

## Task 3 Inspect, test, service, and replace mechanical components and electrical/electronically operated components and circuits of secondary air injection systems.

57. The hose connected from the check valve to AIRD valve is severely burned in a secondary air injection system (see Figure 8–18). The cause of this problem could be
    A. an inoperative check valve.
    B. overheated catalytic converters.
    C. excessive HC emissions.
    D. an inoperative AIRD valve.

58. When the ignition switch is turned on, 12 V are supplied to one terminal on the AIRD and AIRB solenoids. The air is bypassed normally to the atmosphere for a few seconds when the engine is started. When this bypass mode is completed, the airflow is always directed downstream. The voltage on the PCM side of the AIRB solenoid is 0.2 V, and the voltage on the PCM side of the AIRD solenoid is 12 V.
    Technician A says the wire from the AIRD solenoid to the PCM may have an open circuit.
    Technician B says the PCM may not be grounding the circuit from the AIRD solenoid.
    Who is correct?
    A. A only
    B. B only
    C. Both A and B
    D. Neither A nor B

**Hint**    *The check valves in the secondary air injection system prevent exhaust from entering the system hoses. A burned system hose indicates the check valve is allowing exhaust into the secondary air injection system.*

*If the airflow is always directed downstream when the bypass mode is completed, the AIRD valve or control circuit is not operating properly. If there is no vacuum supplied to the AIRD valve, check the vacuum hoses, AIRD solenoid, and connecting wires. In the upstream mode, both solenoids should be energized, and each solenoid should have 12 V at one terminal and a very low voltage at the other terminal. When the voltage is high at the PCM side of the AIRD solenoid, the wire from this solenoid to the PCM is open, or the PCM is not providing a ground for the wire.*

## Task 4 Inspect catalytic converter; interpret catalytic converter related diagnostic trouble codes (DTCs); determine needed repairs.

59. When testing the heated oxygen sensor (HO$_2$S) signals with a digital storage oscilloscope (DSO), the upstream HO$_2$S voltage signal is satisfactory and the downstream HO$_2$S voltage signal is very similar to the upstream HO$_2$S. The most likely cause of this problem is
    A. the downstream HO$_2$S is defective.
    B. there is a leak in the exhaust pipe.
    C. the air-fuel ratio is excessively rich.
    D. the catalytic converter is defective.

**Hint**    *If the catalytic converter rattles when tapped with a soft hammer, the internal components are loose. When this condition is present, the converter should be replaced. A digital pyrometer may be used to check the catalytic converter. If the converter is operating properly, the converter outlet temperature should be 100°F (55°C) hotter compared to the inlet temperature.*

*On-board diagnostic II (OBD II) systems were mandated on 1996 cars. The main difference in an OBD II system is the PCM software that has the capability to monitor various functions controlled by the PCM. A typical OBD II system monitors the catalytic converter, EGR system,* *evaporative system, fuel system, engine misfire, heated oxygen sensor (HO$_2$S), secondary air injection, and thermostat. The PCM also has a comprehensive monitor that includes all the input sensors. Basically, the OBD II monitors are designed to detect any problem in the monitored systems which might increase emission levels to 1.5 times the legal limit for that vehicle year. A drive cycle in an OBD II system is defined as an engine startup and vehicle operation that allows all the monitors to complete their function. Specific driving conditions and times are required to complete the monitors. In many cases, the PCM must sense a defect on two consecutive drive cycles before the PCM illuminates the malfunction indicator light (MIL). If a severe engine misfire occurs, the PCM begins flashing the MIL. Under this condition, continuing to drive the vehicle may damage the catalytic converter.*

*In an OBD II system, an upstream HO$_2$S is located before the catalytic converter and a downstream HO$_2$S is positioned after the catalytic converter. The PCM monitors the catalytic converter by continually checking the signals from the upstream and downstream HO$_2$S. Compared to the upstream HO$_2$S signal, the downstream HO$_2$S signal should have a lower voltage and this voltage should cycle more slowly if the catalytic converter is operating normally. If the upstream and downstream HO$_2$S signals are nearly the same, the catalytic converter is not lowering emissions properly.*

*A significant number of states are using OBD II systems in their emission test facilities. During an emission test, the OBD II system is checked for proper MIL operation, and a scan tool is connected to the DLC to be sure all the monitors have completed their function.*

# Emissions Control Systems Diagnosis and Repair (Including OBD II) Evaporative Emissions Controls

## ASE Tasks, Questions, and Related Information

**Task 1**   **Test and diagnose emissions or drivability problems caused by the evaporative emissions control system.**

60.   A vehicle failed an emissions test for HC and CO at idle speed, and the engine has a rough idle condition (Figure 8–19).
Technician A says the EVAP solenoid plunger may be stuck open.
Technician B says the wire from the EVAP solenoid to the PCM may be open.
Who is correct?
A.  A only
B.  B only
C.  Both A and B
D.  Neither A nor B

**Hint**   *The evaporative (EVAP) emission system prevents fuel tank vapors from escaping to the atmosphere. A hose is connected from the top of the fuel tank to a charcoal canister under the hood. Fuel vapors from the tank flow into the canister where they are absorbed in the charcoal.*

*A tank pressure control or rollover valve usually is connected in this hose. The fuel tank has a filler cap containing pressure and vacuum valves. A purge hose containing a computer-controlled solenoid is connected to the throttle body just above the throttle.*

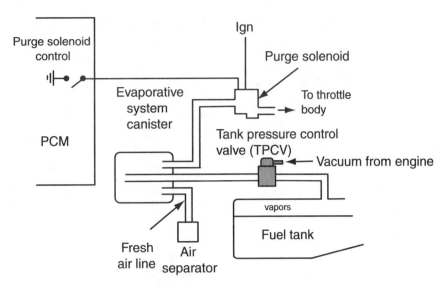

**Figure 8–19** Evaporative control system.

*When the engine is idling below a specific rpm, the purge solenoid remains closed. At a specific engine rpm, the PCM energizes the purge solenoid, and manifold vacuum pulls fuel vapors out of the canister into the intake manifold. Some PCMs energize the purge solenoid at a specific vehicle speed.*

## Task 2   Interpret evaporative emissions related diagnostic trouble codes (DTCs); determine needed repairs.

61. When diagnosing an enhanced evaporative system, Technician A says an improperly tightened fuel tank filler cap may cause the MIL to be illuminated.

    Technician B says a small leak in the hose from the fuel tank to the canister may cause the MIL to be illuminated.

    Who is correct?

    A.  A only

    B.  B only

    C.  Both A and B

    D.  Neither A nor B

**Hint**   *Most vehicles equipped with on-board diagnostic II (OBD II) systems have enhanced evaporative emission systems. A vapor hose is connected from the top of the fuel tank to the canister.*

*The normally closed purge solenoid is connected in the hose from the canister to the intake manifold (Figure 8–20). Another hose is connected from the canister to the normally open vent solenoid.*

*The fuel tank filler cap contains a pressure relief valve and a vacuum valve. A pressure sensor is also mounted in the top of the fuel tank. The PCM begins purging vapors through the evaporative system when the engine coolant temperature is above 122°F (50°C), the engine has been running for three minutes after a cold start or 45 seconds after a warm start, and the PCM is in a closed loop. In response to the fuel tank pressure sensor voltage signal, the PCM pulses the purge and vent solenoids on and off to maintain a specific vacuum in the fuel tank and control vapor flow through the evaporative system into the intake manifold (Figure 8–21). The evaporative system monitor in the PCM monitors the evaporative system for proper operation.*

*If the PCM detects a problem in the system on the second consecutive drive cycle, the PCM sets a current DTC and illuminates the MIL. Any of the following problems may cause an incorrect vacuum in the fuel tank and improper purging of the evaporative system, resulting in a DTC set in the PCM:*

- *A missing, incorrect, malfunctioning, or improperly tightened fuel tank filler cap*
- *A damaged canister*
- *A disconnected or malfunctioning fuel tank pressure sensor*

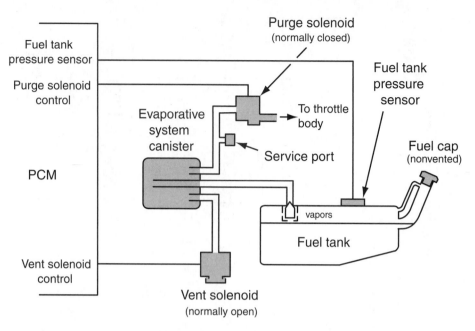

**Figure 8–20** Enhanced evaporative system.

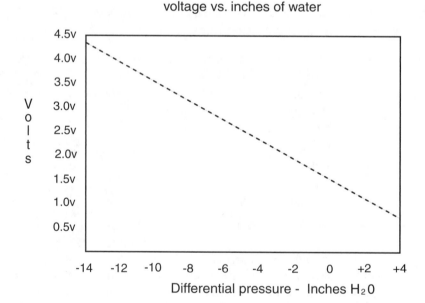

**Figure 8–21** Fuel tank pressure sensor voltage signal in relation to tank pressure.

- *An open circuit in the purge solenoid or vent solenoid windings or connecting wires to the PCM*
- *A disconnected, damaged, pinched, or plugged fuel tank vapor line*
- *A disconnected, damaged, pinched, or plugged purge lines*
- *A disconnected or plugged canister vent hose*

## Task 3    Inspect, test, and replace canister, lines/hoses, mechanical and electrical components of evaporative emissions control system.

62.  While diagnosing an inoperative EVAP system with a scan tool, the tool indicates the purge solenoid is on at the specified rpm, and there are no diagnostic trouble

codes (DTCs) in the PCM memory. There is no vacuum to the canister. This problem may be caused by all of these problems EXCEPT

A. an open winding in the purge solenoid.

B. a purge solenoid plunger stuck in the closed position.

C. a plugged hose from the canister to the intake manifold.

D. a plugged purge port in the intake manifold.

**Hint**   *A scan tool may be used to diagnose the EVAP system. If there is a problem in the purge solenoid winding or connecting wires, a DTC usually is set in the PCM memory. At idle speed, the scan tester should indicate the purge solenoid is off. With the engine at normal operating temperature and running at the specified rpm, the scan tool should indicate the purge solenoid is on.*

*The scan tool only indicates the command from the PCM to the solenoid. The PCM may send an on command to the purge solenoid, but the solenoid may remain inoperative because of such problems as a sticking plunger.*

# Computerized Engine Controls Diagnosis and Repair (Including OBD II)

## ASE Tasks, Questions, and Related Information

**Task 1**  Retrieve and record diagnostic trouble codes (DTCs) and freeze frame data if applicable.

63. Technician A says that on some vehicles the diagnostic trouble codes (DTCs) may be obtained by cycling the ignition switch on and off three times in a ten-second interval. Technician B says that on some vehicles the DTCs are read from the flashes of the malfunction indicator (MIL) light.

Who is correct?

A. A only

B. B only

C. Both A and B

D. Neither A nor B

**Hint**   *On some older vehicles, DTCs are obtained by connecting a jumper wire to the proper terminals in the data link connector (DLC), and then turning on the ignition switch. On other vehicles, the ignition switch is cycled on and off three times in a five-second interval to signal the powertrain control module to enter the diagnostic mode and provide DTCs. On some vehicles, the DTCs may be read by counting the flashes of the MIL in the instrument panel. The MIL flashes each code three times on some vehicles, and codes are in numerical order. The DTCs may also be read on a scan tool connected to the DLC. On OBD II vehicles, a sixteen-terminal data link connector (DLC) is mounted under the dash near the steering column. A scan tool is connected to this DLC to retrieve DTCs, freeze frames, and other data. The data links connected between various on-board computers are also connected to the DLC. Therefore, a scan tool connected to the DLC may be used to obtain DTCs and data from other on-board computers.*

64. A scan tool is connected to the DLC on an OBD II vehicle. The long term fuel trim is +31, and the engine has a rough idle problem.

Technician A says there may be a vacuum leak in the intake manifold.

Technician B says the fuel pressure may be low.

Who is correct?

A. A only

B. B only

C. Both A and B

D. Neither A nor B

**Hint**    *The engine must be at normal operating temperature before attempting to read the DTCs. On some vehicles the DTCs are obtained by connecting a jumper wire to the proper terminals in the DLC, and then turning on the ignition switch. On other vehicles, the ignition switch is cycled on and off three times in a five-second interval to signal the powertrain control module to enter the diagnostic mode and provide DTCs. On some vehicles, the DTCs may be read by counting the flashes of the MIL in the instrument panel. The MIL flashes each code three times on some vehicles, and codes are in numerical order. The DTCs may also be read on a scan tool connected to the DLC. A DTC indicates a problem in a certain area. In many cases, further testing with a voltmeter or ohmmeter is required to locate the exact cause of the problem.*

## Task 2   Diagnose the causes of emissions or drivability problems resulting from failure of computerized engine controls with diagnostic trouble codes (DTCs).

65. A vehicle fails an emissions test for high NOx. The engine has an EGR valve position sensor. It also has a rough idle problem, and a DTC denoting that the EGR valve is stored in the powertrain control module (PCM) memory.

Technician A says the EGR valve may be stuck open.

Technician B says the EGR passages may be restricted with carbon.

Who is correct?

A. A only

B. B only

C. Both A and B

D. Neither A nor B

**Hint**    *The high HC emissions and normal CO and NOx emissions usually indicate a lean misfire condition. If the high HC emissions only occur at idle and low speed, the misfire must be occurring only at this speed. Malfunctioning injectors, low fuel pump pressure, or a malfunctioning ECT sensor may cause a lean air-fuel ratio at idle and 2,500 rpm.*

*If the exhaust gas recirculation (EGR) valve is stuck open, the exhaust flow into the intake manifold dilutes the mixture at idle and low speed, resulting in a lean misfire condition. Because the EGR valve is normally open at 2,500 rpm, it does not cause a lean misfire at this speed.*

66. A vehicle has a poor fuel economy complaint. When tested with a scan tool the $O_2$ sensor voltage is 0.2 V. A DTC representing the $O_2$ sensor is stored in the PCM memory. The injector pulse width is more than specified. When propane is dispersed into the air intake, the $O_2$ sensor voltage and the injector pulse width do not change. The cause of this problem is

A. a malfunctioning PCM.

B. high fuel pressure.

C. an inoperative $O_2$ sensor.

D. an intake manifold vacuum leak.

**Hint**    *If the $O_2$ is satisfactory, the sensor voltage changes when propane is dispersed into the air stream. When the $O_2$ sensor voltage changes but the computer does not change the injector pulse width, the computer may be malfunctioning.*

67. An engine fails an emissions test for high CO and HC. These emissions are high at idle and also at higher speeds, and the engine has rough idle and hard-starting problems.

The fuel pressure is normal, but it decreases after the engine is shut off. The scan tool data indicates an $O_2$ sensor voltage between 0.5 V and 0.8 V. A DTC indicating a rich air-fuel ratio is stored in the PCM memory. The cause of these problems could be

A. a leaking fuel pump check valve.

B. a leaking fuel pressure regulator.

C. dripping injectors.

D. a malfunctioning engine coolant temperature (ECT) sensor.

**Hint**    *Dripping injectors cause a rich air-fuel ratio and rough engine idle. Because fuel also drips out of the injectors after the engine is shut off, the fuel rail is empty when a restart is attempted. This condition causes a longer cranking time before the engine starts.*

68. A fuel-injected engine has a rough idle condition, and high HC emissions only at idle and low speeds. A DTC indicating a lean air-fuel ratio is stored in the PCM memory. Emissions of CO and NOx meet emission standards, and at 2,500 engine rpm, the HC emissions are normal. The most likely cause of this problem is

A. malfunctioning injectors.

B. an EGR valve stuck open.

C. low fuel-pump pressure.

D. a malfunctioning ECT sensor.

**Hint**    *The high HC emissions and normal CO and NOx emissions usually indicate a lean misfire condition. Because the high HC emissions only occur at idle and low speed, the misfire must be occurring only at this speed. Malfunctioning injectors, low fuel-pump pressure, or a malfunctioning ECT sensor may cause a lean air-fuel ratio at idle and 2,500 rpm. If the EGR valve is stuck open, the exhaust flow into the intake manifold dilutes the mixture at idle and low speed, resulting in a lean misfire condition. Because the EGR valve is normally open at 2,500 rpm, it does not cause a lean misfire at this speed.*

69. A vehicle has a secondary air injection system that delivers air to the dual-bed catalytic converter. This vehicle fails an emissions test for HC and CO. Engine operation is normal, but the owner complains about excessive fuel consumption. A DTC is stored in the PCM memory, indicating secondary air injection system airflow is always upstream.

Technician A says the secondary air injection diverter (AIRD) solenoid plunger may be sticking open.

Technician B says the AIRD valve may be sticking.

Who is correct?

A. A only

B. B only

C. Both A and B

D. Neither A nor B

**Hint**    *The secondary air injection system delivers air to the center of many dual-bed catalytic converters. This airflow is necessary to oxidize HC and CO to $O_2$ and $H_2O$ in the rear oxidation catalyst bed.*

## Task 3   Diagnose the causes of emissions or drivability problems resulting from failure of computerized engine controls with no diagnostic trouble codes (DTCs).

70. A vehicle fails an emission test for HC emissions at idle and 2,500 rpm, and the engine has an acceleration stumble. The $O_2$ sensor voltage is 0.1 V to 0.3 V on the scan tool, and the injector pulse width is more than specified. There are no DTCs in the PCM. When propane is directed into the air intake, the $O_2$ sensor voltage immediately increases, and the injector pulse width decreases. The most likely cause of this problem is

A. a malfunctioning $O_2$ sensor.

B. a malfunctioning PCM.

C. low fuel-pump pressure.

D. an open circuit in the $O_2$ sensor signal wire.

*Hint*    *High HC emissions with normal or low CO emissions may be caused by a lean misfire condition. Because the $O_2$ sensor voltage increases when propane is dispersed into the air intake, this sensor is responding. The PCM is supplying more injector pulse width in response to the low $O_2$ sensor voltage, so the PCM is satisfactory. If the $O_2$ sensor signal wire was open, this signal would not change on the scan tool.*

*A lean air-fuel ratio and lean misfire condition plus acceleration stumbles may be caused by low fuel-pump pressure.*

## Task 4    Use a scan tool, digital multimeter (DMM), or digital storage oscilloscope (DSO) to inspect or test computerized engine control system sensors, actuators, circuits, and powertrain control module (PCM); determine needed repairs.

71. A port-injected engine is hard to start after the vehicle is not driven for several hours. When tested, the fuel pressure drops off when the engine is shut off. There are no visible external fuel leaks. With the fuel pressure line and return fuel line plugged, the fuel pressure still drops off when the engine is shut off. The cause of this problem could be

A. dripping injectors.

B. a leaking fuel pump check valve.

C. a leaking fuel pressure regulator.

D. a leaking upper injector O-ring.

*Hint*    *When the ECT sensor is tested with an ohmmeter, the sensor resistance should decrease as the temperature increases. If the resistance of the ECT sensor is lower than specified, the computer thinks the engine is warmer than the actual coolant temperature. Under this condition, the PCM provides a leaner air-fuel ratio than the engine requires. Conversely, if the ECT sensor resistance is higher than specified, the PCM thinks the coolant is colder than the actual coolant temperature.*

*This action causes the PCM to provide a rich air-fuel ratio. The ECT sensor signal affects many computer outputs such as the EGR valve, torque converter clutch, and EVAP purge solenoid. Hard starting after the engine is shut off for several hours may be caused by a leaking fuel-pump check valve or pressure regulator. Dripping injectors or a malfunctioning ECT sensor may cause this problem, but such problems also affect idle operation and emission levels.*

72. When testing MAP sensors, Technician A says many MAP sensors act as a barometric pressure sensor when the ignition switch is turned on and the engine is not running.

Technician B says that on many MAP sensors the signal voltage should increase when the vacuum at the sensor decreases.

Who is correct?

A. A only

B. B only

C. Both A and B

D. Neither A nor B

*Hint*    *Many manifold absolute pressure (MAP) sensors act as a barometric pressure sensor when the ignition is turned on or when the throttle is wide open. Many MAP sensors send an analog voltage signal to the PCM in relation to the intake manifold vacuum. When the ignition switch is turned on, the barometric pressure signal from the MAP sensor to the PCM is usually 4 V to 5 V. If the engine is idling, the MAP sensor signal voltage is approximately 1 V. When the throttle is opened and manifold vacuum decreases, the MAP sensor signal voltage increases.*

*Some MAP sensors provide an AC voltage signal that is measured in hertz (Hz). The signal from these MAP sensors varies from approximately 92 Hz at idle to 162 Hz at wide open throttle. This type of MAP sensor also acts as a barometric pressure sensor with the ignition switch on.*

73. When testing a MAP sensor that produces an analog voltage signal, with 18 in. Hg (40.5 kPa absolute) applied to the sensor, the sensor signal is 1.2 V. When 13 in. Hg (57.3 kPa absolute) is supplied to the sensor, the sensor signal should be
   A. 1.4 V.
   B. 1.6 V.
   C. 2.0 V.
   D. 2.6 V.

## Task 5    Measure and interpret voltage, voltage drop, amperage, and resistance using digital multimeter (DMM) readings.

74. An oxygen ($O_2$) sensor voltage signal should be checked with
   A. a digital ohmmeter.
   B. an analog ohmmeter.
   C. a digital voltmeter.
   D. a milliammeter.

**Hint**    *Some computer inputs, such as the $O_2$ sensor, produce a low voltage signal and a very low current flow. A digital voltmeter must be used to test the $O_2$ sensor voltage. Because an analog voltmeter draws more current compared to a digital voltmeter, $O_2$ sensor damage occurs when this sensor is tested with an analog voltmeter.*

## Task 6    Test, remove, inspect, clean, service, and repair or replace power and ground distribution circuits and connections.

75. In the power distribution circuit (Figure 8–22), with the ignition switch in the accessory (ACCY) position, voltage is supplied to the following fuses:
   A. the $O_2$ heater and HVAC.
   B. the instrument and air BG2.
   C. the wiper and radio.
   D. the instrument and HVAC.

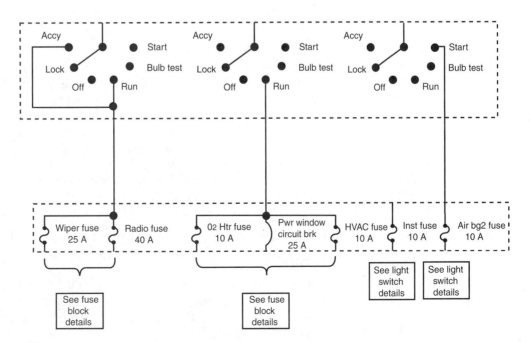

**Figure 8–22** Partial power distribution circuit.

*Hint*   *The power distribution circuit is the power and ground circuits from the battery through the ignition switch and fuses to the individual circuits on the vehicle.*

## Task 7   Practice recommended precautions when handling static sensitive devices and/or replacing the powertrain control module (PCM).

76. All of the following statements are true about installing an EEPROM chip EXCEPT
    A. Do not remove the chip from the protective envelope until you are ready to install it.
    B. Touch a good ground on the vehicle with your finger prior to installing the chip.
    C. Connect a grounding strap from your wrist to a good vehicle ground.
    D. Remove any dust from the chip pins with a shop towel prior to installation.

*Hint*   *Static-sensitive components are shipped in an antistatic envelope. This envelope should not be removed before you are ready to install the chip. Touch a good ground on the vehicle, and connect a grounding strap from your wrist or clothing to a vehicle ground before installing the chip.*
  *Do not handle the chip unnecessarily, and do not move around on the vehicle seat when installing the chip.*

## Task 8   Diagnose drivability and emissions problems resulting from failures of interrelated systems (such as cruise control, security alarms/theft deterrent, torque controls, traction controls, torque management, A/C, non-OEM installed accessories).

77. A torque converter clutch (TCC) is inoperative. The vehicle has a separate PCM and transmission control module (TCM) with interconnected data links. The other transmission functions are normal and the scan tool never indicates a TCC on command from the TCM. Engine drivability is normal.
    Technician A says the data links should be tested with the scan tool.
    Technician B says the ECT sensor signal may not be transmitted from the PCM to the TCM.
    Who is correct?
    A. A only
    B. B only
    C. Both A and B
    D. Neither A nor B

*Hint*   *Some vehicles have separate computers for engine control, transmission control, ABS and traction control, suspension control, and A/C control. On many vehicles, these computers are interconnected by data links. Some input sensor signals are connected to one of these computers and transmitted on the data links to the other computers.*
  *If the data links are inoperative, the necessary inputs will not reach the computer or computers that require these inputs. Damaged data links may cause malfunctioning of some outputs. For example, the ECT sensor signal is sent from the PCM to the TCM for proper TCC control. If the data links are malfunctioning and the ECT signal is not available to the TCM, this computer does not know anything about engine temperature. This condition may cause an inoperative or malfunctioning TCC. On many vehicles, the data links may be tested with a scan tester.*

## Task 9   Diagnose the causes of emissions or drivability problems resulting from computerized spark timing controls; determine needed repairs.

78. An engine has a knock sensor in the right-hand coolant drain plug. With the engine running at 2,000 rpm, the exhaust manifold is tapped above the knock sensor with a small hammer. While observing the timing marks with a timing light, the timing mark does not move.

Technician A says the knock sensor may be the problem.

Technician B says the engine may have a detonation problem.

Who is correct?

A.  A only

B.  B only

C.  Both A and B

D.  Neither A nor B

**Hint**    *Many spark control systems on fuel-injected engines are designed to reduce the spark advance when the engine detonates. If a defective spark control system allows engine detonation, high NOx emissions may occur.*

## Task 10    Verify the repair, and clear diagnostic trouble codes (DTCs).

79. A P1115 is obtained on an OBD II vehicle.

Technician A says this DTC is designated by SAE.

Technician B says this DTC belongs to the idle speed control subgroup.

Who is correct?

A.  A only

B.  B only

C.  Both A and B

D.  Neither A nor B

**Hint**    *Although flash codes could be read on many vehicles in the 1980s, a scan tool must be used to read the DTCs on most vehicles in the 1990s. On OBD II systems, a scan tool must be used to read the DTCs. OBD II DTCs are formatted according to SAE standard J2012. This standard requires five-digit alphanumeric DTC identification. The prefix in each DTC indicates the DTC function: P - powertrain, B - body, C - chassis. The first number in the DTC represents the group responsible for code. If the first number is 0, the code is designated by SAE. When the first number is 1, the vehicle manufacturer is responsible for the DTC. Many DTCs are dictated by SAE, but the vehicle manufacturer may insert some codes. The third digit in the DTC indicates the subgroup to which the code belongs as follows: 0 - total system; 1 - air-fuel control; 2 - air-fuel control; 3 - ignition system misfire; 4 - auxiliary emission controls; 5 - idle speed control; 6 - PCM input and output; 7 - transmission; 8 - transmission or non-EEC powertrain. The fourth and fifth digits in the DTC indicate the area where the problem may be located.*

*For example, in DTC P0155, P indicates it is a powertrain code, 0 indicates it is designated by SAE, 1 indicates the code is in the air-fuel control subgroup, and 55 indicates the code represents a malfunction in the heated oxygen sensor (HO₂S) heater, bank 2, sensor 1. Bank 2 indicates the bank of a V6 or V8 engine that is opposite to the bank where number 1 cylinder is located, and sensor 1 is the upstream HO₂S sensor located ahead of the catalytic converter. An HO₂S sensor designated bank 1 sensor 2 is on the same side of the engine as number 1 cylinder and located downstream from the catalytic converter.*

*A DTC indicates a problem in a certain area. For example, an HO₂S sensor DTC may indicate an inoperative sensor or problems in the wires from the sensor to the PCM. A leak in the exhaust manifold, or a malfunctioning secondary air injection system may also cause an improper HO₂S sensor signal and a DTC representing this sensor. When diagnosing the HO₂S DTC, the Technician must verify the operation of other components or systems that could affect the HO₂S signal. The Technician must also use a digital multimeter (DMM) to test the HO₂S sensor and connecting wires to locate the exact cause of the DTC.*

*DTCs should be erased using the scan tool. Disconnecting battery power from the PCM will erase DTCs, but this procedure can be time-consuming and it also erases some of the PCM memories, which then require relearn procedures.*

# Engines Electrical Systems Diagnosis and Repair, Battery

## ASE Tasks, Questions, and Related Information

**Task 1** **Test and diagnose emissions or drivability problems caused by battery condition, connections, or excessive key-off battery drain; determine needed repairs.**

80. A vehicle has a slow engine cranking condition. The starting motor has been repaired and a battery load test indicates the battery is satisfactory.

   Technician A says to measure the voltage drop across both battery cables with the engine running.

   Technician B says when measuring the voltage drop across the positive battery cable, connect the voltmeter leads from the positive battery post to the other end of the positive cable.

   Who is correct?

   A. A only

   B. B only

   C. Both A and B

   D. Neither A nor B

81. While discussing key-off battery drain testing on a vehicle with port fuel injection and an antilock brake system (ABS), Technician A says when the ignition switch is turned off, the current drain increases after the tester switch is opened.

   Technician B says when the ignition switch is turned off, the key-off drain should be read immediately on the milliameter.

   Who is correct?

   A. A only

   B. B only

   C. Both A and B

   D. Neither A nor B

**Hint**    *Slow engine cranking and hard starting may be caused by high resistance in the battery cables and connections. High resistance in the battery cables and connections may be tested by measuring the voltage drop across the cables with the engine cranking. When measuring the voltage drop across negative battery cable, connect the voltmeter leads from the negative battery post to a ground connection on the engine block. To test the voltage drop across the positive battery cable, connect the voltmeter leads from the positive battery post to the other end of the positive cable, which is usually connected to the starter solenoid. A severely corroded battery cable connection may cause a clicking action from the starter solenoid when the ignition switch is turned to the start position. This problem may be diagnosed by measuring the voltage drop across the cables and connections with the ignition switch in the start position. A discharged battery may also cause a clicking action from the starter solenoid, and the frequency of the clicking action decreases as the battery state of charge decreases. An intermittent open circuit in one of the battery cables may cause an intermittent no-start condition. This problem may also result in charging system voltage spikes that could damage on-board computers.*

   *Computer memories have a very low key-off drain. Other light circuits such as glove compartment lights and trunk lights cause battery drain if they remain on with the ignition key off.*

*Always be sure all electrical systems and components are shut off before testing key-off battery drain. The doors and hood must be closed to be sure the interior lights and underhood light are off.*

*A special tester switch is connected between the negative battery cable and the negative battery terminal. An ammeter with a milliampere scale is connected to the tester switch terminals.*

*The switch is closed to start the engine. When the ignition switch is turned off, the tester switch is opened. Always wait the specified length of time before reading the milliameter. Because computers require a brief time interval to enter the sleep mode, the current drain may be higher than specified when the tester switch is opened initially. If the current drain is excessive, individual circuits and components may be disconnected to locate the exact cause of the drain. Removing fuses and circuit breakers is often the most convenient method of disconnecting individual circuits and components.*

# Engines Electrical Systems Diagnosis and Repair, Starting System

## ASE Tasks, Questions, and Related Information

**Task 1**   **Perform starter current draw test; determine needed action.**

82. The starter draw is more than specified and the cranking rpm and voltage are low. The cause of this problem could be
   A. high resistance in the starter field coils.
   B. worn starter bushings.
   C. high resistance in the battery cables.
   D. high resistance in the solenoid windings.

**Hint**   *High starter current draw with low cranking speed and cranking voltage usually indicate a problem in the starting motor (such as worn bushings), allowing the armature to rub on the field coils. Electrical problems in the armature or field coils also cause this problem.*

*Low cranking speed with low current draw and high cranking voltage usually indicates high resistance in the starter circuit (such as corroded battery cables).*

**Task 2**   **Perform starter circuit voltage drop tests; determine needed action.**

83. With the engine cranking, the voltmeter connected to the starter circuit reads 0.2 V (Figure 8–23). This reading indicates
   A. the positive battery cable has excessive resistance.
   B. the starter ground circuit has excessive resistance.
   C. the starter relay contacts have excessive resistance.
   D. the positive battery cable has normal resistance.

**Hint**   *The voltmeter should be connected from the positive battery terminal to the battery terminal on the starting motor to test voltage drop in the positive battery cable. With the engine cranking, if the voltage drop exceeds 0.2 V, the battery positive cable has excessive resistance.*

*Connect voltmeter leads from the negative battery terminal to the starter case. With the engine cranking, the voltage drop across the battery ground circuit should not exceed 0.2 V.*

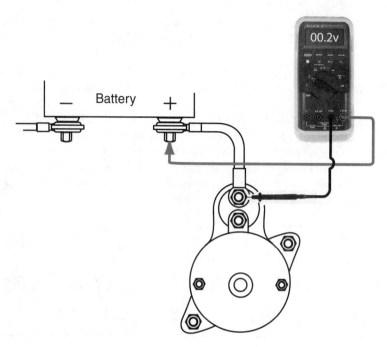

**Figure 8–23** Starter circuit voltage test.

## Task 3   Inspect, test, and repair or replace components and wires in the starter control circuit.

84. The ohmmeter is connected from the solenoid terminal to ground on the solenoid case (Figure 8–24). The ohmmeter indicates an infinity reading.

    Technician A says the solenoid hold-in winding has an open circuit.

    Technician B says the solenoid pull-in winding is shorted to ground.

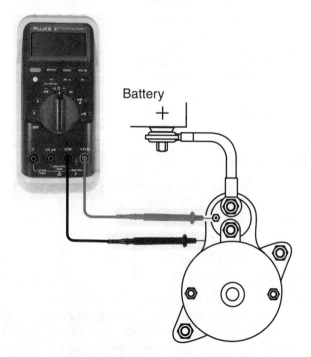

**Figure 8–24** Starter solenoid test.

Who is correct?

A. A only

B. B only

C. Both A and B

D. Neither A nor B

**Hint**    *The ohmmeter leads are connected from the solenoid terminal to ground when checking the hold-in winding. An infinity reading indicates an open hold-in winding and this winding is shorted if the reading is below specifications.*

*Connect the ohmmeter leads from the solenoid terminal to the field coil terminal in the starting motor to test the pull-in winding.*

# Engines Electrical Systems Diagnosis and Repair, Charging System

## ASE Tasks, Questions, and Related Information

**Task 1**    Test and diagnose engine performance problems that cause an undercharge, overcharge, or no-charge condition; determine needed action.

85. The battery gradually becomes undercharged and the vehicle is driven regularly. All of the following may be the cause of the problem EXCEPT

A. a loose alternator belt.

B. a battery drain.

C. a high regulator voltage.

D. a low regulator voltage.

**Hint**    *An undercharged battery may be caused by a loose alternator belt, low regulator voltage, a malfunctioning alternator, or a battery drain. An overcharged battery may be caused by a high regulator voltage. If the alternator has zero output, the field circuit may be open, or the fuse link may be burned out between the battery positive cable and the alternator battery terminal.*

**Task 2**    Inspect, adjust, and replace alternator (generator) drive belts, pulleys, tensioners and fans.

86. Technician A says an alternator V-belt should be replaced if the belt is contacting the bottom of the pulley.

Technician B says many alternator ribbed V-belts have an automatic tensioner with a wear indicator.

Who is correct?

A. A only

B. B only

C. Both A and B

D. Neither A nor B

**Hint**    *A V-belt should be replaced if the belt is oil-soaked, cracked, frayed, or contacting the bottom of the alternator pulley. A belt tension gauge should indicate the specified tension when the gauge is placed in the center of the longest belt span.*

*Many ribbed V-belts have a spring-loaded automatic belt tensioner with a wear indicator.*

## Task 3   Inspect, test, and repair or replace charging circuit connectors and wires.

87. The fuse link is burned out in the wire between the battery positive terminal and the alternator battery terminal. The most likely cause of this problem is
    A. a shorted alternator stator.
    B. reversed battery polarity.
    C. a damaged alternator diode.
    D. high regulator voltage.

**Hint**   *A burned fuse link in the alternator battery wire may be caused by reversed battery polarity when installing the battery or using a booster battery. Replacement fuse links are available, and the old fuse link may be cut out of the wiring harness. Bare the wire in the harness and crimp the replacement fuse link to the wire (Figure 8–25).*

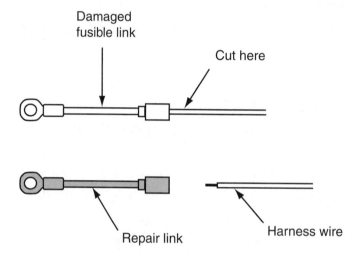

**Figure 8–25** Fuse link replacement.

# Post Test

1. An engine has a sharp rapping noise during idle operation.
   Technician A says one of the connecting rod bearings may be worn.
   Technician B says one of the main bearings is worn.
   Who is correct?
   A. A only
   B. B only
   C. Both A and B
   D. Neither A nor B

2. A cylinder balance test is performed on an in-line six cylinder engine. Cylinders #4 and number 5 have a 20-rpm drop and all the cylinders have an 80-rpm drop. There is no blue smoke in the exhaust, and no indication of blowby from the PCV valve opening. The most likely cause of this problem is
   A. a blown head gasket between cylinders #4 and #5.
   B. worn piston rings in cylinders #4 and #5.
   C. worn valve stem seals and guides in cylinders #4 and #5.
   D. holes in the top of pistons #4 and #5.

3. An oscilloscope indicates an ignition system has a maximum coil voltage of 18 kV. The specified maximum coil voltage is 35 kV.

Technician A says the primary input voltage may be low.

Technician B says the ignition system may have an open spark plug wire.

Who is correct?

A. A only

B. B only

C. Both A and B

D. Neither A nor B

4. An engine misfires only during hard acceleration. It runs smoothly at idle, however. The most likely cause of this problem is

A. a burned exhaust valve.

B. broken piston rings in one cylinder.

C. excessive resistance in a spark plug wire.

D. An intake manifold vacuum leak.

5. An engine experiences spark knock at a steady speed of 55 mph (88.5 kmh) with the engine at normal operating temperature.

Technician A says the EGR valve may be stuck closed.

Technician B says the air-fuel ratio is excessively rich.

Who is correct?

A. A only

B. B only

C. Both A and B

D. Neither A nor B

6. A distributor ignition system experiences repeated ignition module failures.

Technician A says the secondary coil wire may have excessive resistance.

Technician B says the primary coil winding may be shorted.

Who is correct?

A. A only

B. B only

C. Both A and B

D. Neither A nor B

7. When performing an oscilloscope diagnosis on an ignition system, the normal required secondary voltage on cylinders #1 and #4 is 16 kV. The normal required secondary voltage on all the other cylinders is 10 kV.

Technician A says the spark plug wires on cylinders #1 and #4 may have excessive resistance.

Technician B says the spark plugs in cylinders #1 and #4 may be defective.

Who is correct?

A. A only

B. B only

C. Both A and B

D. Neither A nor B

8. An ohmmeter is used to test an electromagnetic-type distributor pickup coil. The specified pickup coil resistance is 650 ohms. The ohmmeter leads are connected from one of the pickup coil leads to ground, and the meter reading is 40 ohms.

Technician A says the turns of wire in the pickup coil are touching each other, resulting in a shorted condition.

Technician B says the pickup coil winding is touching ground on the distributor housing.

Who is correct?

A. A only

B. B only

C. Both A and B

D. Neither A nor B

9. A vehicle with sequential fuel injection has a reduced fuel economy complaint. A fuel pump test indicates the fuel pump pressure is 70 psi (482 kPa), and the specified fuel pressure is 45 psi (310 kPa). The most likely cause of this problem is

A. sticking injector plungers.

B. a pressure regulator valve sticking open.

C. a leak in the pressure regulator vacuum hose.

D. a restricted fuel return line.

10. A pressurized container filled with cleaning solution is connected to the Schrader valve on the rail to clean the injectors in a port-fuel-injected engine with an oil pressure switch in the fuel pump circuit. During this service operation, all of the following actions are necessary EXCEPT

A. disconnecting the wiring connector from the fuel pump relay.

B. disconnecting the wiring connector from the oil pressure switch.

C. installing a new fuel filter.

D. squeezing the fuel return line so it is plugged.

11. With an engine idling, there is no vacuum at the oil filler cap opening, but there is no indication of oil in the air cleaner.

Technician A says the rocker arm cover gasket may be leaking.

Technician B says the PCV valve is plugged.

Who is correct?

A. A only

B. B only

C. Both A and B

D. Neither A nor B

12. All of the following statements about EGR valves are true EXCEPT

A. The EGR valve is closed during hard acceleration.

B. The EGR valve is closed at wide open throttle

C. The EGR valve is open on a cold engine.

D. Vacuum is supplied through an electric solenoid to some EGR valves.

13. While discussing negative backpressure EGR valves, Technician A says that, with the engine shut off, if 18 in. Hg. (40.5 kPa) is supplied to the EGR valve with a vacuum hand pump, the EGR valve should open.

Technician B says that, with the engine idling, if 18 in. Hg. (40.5 kPa) is supplied to the EGR valve with a vacuum hand pump, the EGR valve should remain closed.

Who is correct?

A. A only

B. B only

C. Both A and B

D. Neither A nor B

14. A vehicle with a three-way catalytic converter and a belt-driven air pump system fails an emission test for HC and CO. The engine operation is normal, but the owner has a poor fuel economy complaint. The only stored DTC in the PCM is a lean upstream $HO_2S$ sensor. The most likely cause of this problem is

A. a defective vehicle speed sensor (VSS).

B. a defective mass air flow sensor (MAF).

C. continual upstream air injection.

D. the EGR valve stuck closed.

15. A car fails an enhanced emissions test for high HC and CO emissions. These emissions are high at idle and low vehicle speeds, and the engine has a slight rough idle problem. The EVAP system passed the flow and pressure tests in the enhanced emission test procedure. The most likely cause of this problem is

A. the wire from the EVAP solenoid to the computer is grounded.

B. the EVAP solenoid plunger is stuck closed.

C. one of the EVAP solenoid wires has an open circuit.

D. the hose from the canister to the intake manifold is plugged.

16. A vehicle fails an emission test for high CO and HC. The $HO_2S$ voltage is 0.7 V–0.9 V, and the injector pulse width is less than specified. When a large vacuum leak is created into the intake manifold, the $HO_2S$ voltage is 0.2 V, and the injector pulse width increases. The most likely cause of this problem is

A. a defective $HO_2S$.

B. a defective PCM.

C. high fuel pressure.

D. an open $HO_2S$ ground wire.

17. A vehicle fails an enhanced emissions test for high NOx. The test trace indicates low CO during the entire drive cycle.

Technician A says the air-fuel ratio may be lean.

Technician B says the ignition system may be misfiring.

Who is correct?

A. A only

B. B only

C. Both A and B

D. Neither A nor B

18. When diagnosing a vehicle, a diagnostic trouble code P0156 is obtained on a scan tool. The service manual explains this DTC indicates no activity or voltage from the bank 2 sensor 2 $HO_2S$. This indicates the defective $HO_2S$ is

A. upstream from the catalytic converter and on the same side as number 1 cylinder.

B. upstream from the catalytic converter and on the opposite side from number 1 cylinder.

C. downstream from the catalytic converter and on the same side as number 1 cylinder.

D. downstream from the catalytic converter and on the opposite side from number 1 cylinder.

19. A car has a poor fuel economy complaint and a lack of power especially when driving in hilly terrain. There are no DTCs in the PCM, and all the PCM data is satisfactory except the baro sensor voltage is 2.75 V with the ignition switch on. The baro sensor is built into the MAP sensor, and the specified baro sensor voltage with the ignition switch on is 4.5 V.

Technician A says the MAP vacuum hose may be leaking.

Technician B says to replace the MAP sensor.

Who is correct?

A. A only

B. B only

C. Both A and B

D. Neither A nor B

20. An engine has a hard-starting complaint when cold. At normal engine temperature, the scan tool data is satisfactory. The specified fuel pressure is available, and the fuel

pressure does not decrease after the engine is shut off. The ignition system is always firing. The cause of this problem could be

A. an ECT sensor with less than the specified resistance when cold.

B. a leaking pressure regulator diaphragm.

C. a leaking one-way check valve in the fuel pump.

D. a defective ignition coil.

# Answers and Analysis

**1. C**    One of the first steps in a diagnostic procedure is to identify the problem; listening to the customer's complaint may help to achieve this objective. Both Technicians A and B are correct, so C is the right answer.

**2. C**    Service bulletin information is available in some scan tool software. Therefore, Technician A is correct.

Wiring diagrams are available by fax from some automotive hot lines, and so Technician B is also right, making C the correct answer.

**3. C**    Worn camshaft lobes may cause a heavy clicking noise at higher engine speeds. Loose piston pins may cause a light rapping noise at idle speed, and sticking or worn valve lifters may cause a light clicking noise at idle and low speed. Therefore, A, B, and D are wrong.

The heavy thumping noise when the engine is first started may be caused by worn main bearings, so C is right.

**4. A**    A "puff" noise in the exhaust at regular intervals may be caused by a burned exhaust valve.

Excessive coolant temperature does not result in this symptom. Therefore, Technician A is correct and Technician B is wrong, so A is the right answer.

**5. D**    A low steady vacuum gauge reading may be caused by late ignition timing or an intake manifold vacuum leak. Therefore, both Technicians A and B are wrong, and D is the correct answer.

**6. A**    A lack of rpm decrease during the cylinder power balance test indicates the cylinder is misfiring.

A damaged spark plug, wire, or a burned exhaust valve would cause misfire at low or high speed, so B, C, and D are wrong.

Because the cylinder misfire disappeared at 2,000 rpm, an intake manifold vacuum leak is indicated. Thus, A is right.

**7. C**    There is no increase in the compression reading during the wet test with a burned exhaust valve, bent intake valve, or a blown head gasket. Therefore, A, B, and D are wrong.

Worn piston rings do cause an increase in compression readings during the wet test, so C is correct.

**8. C**    During a cylinder leakage test, excessive leakage and air escaping from the throttle body may be caused by a bent intake valve or a bottomed intake valve lifter. Thus, both Technicians A and B are correct, and C is the right answer.

**9. C**    Some engine analyzers compare test results to specifications and indicated readings that are out of specifications. Manual or automatic test modes may be selected on some engine analyzers. Thus, Technicians A and B are both correct, making C the right answer.

**10. A**    With the emissions analyzer probe in the tailpipe, the analyzer indicates the emissions coming out of the converter, which could be considerably lower than the emissions coming out of the engine. Therefore, Technician B is wrong.

A leak in the exhaust system may cause inaccurate emission readings. Thus, Technician A is correct, making A the right answer.

**11. B** This question asks for the statement that is not true. If the exhaust valve clearance is less than specified, the valve may burn prematurely. Excessive valve clearance causes a clicking noise, especially at idle. The piston should be at TDC on the compression stroke in the cylinder on which the valve is being adjusted. Therefore, A, C, and D are true, so they are not the requested answer.

Both valves are open slightly with the piston at TDC on the exhaust stroke, and the valve clearance must not be adjusted in this piston position. Therefore, B is not true and is the requested answer.

**12. A** With the piston at TDC on the exhaust stroke, the intake valve should be starting to open and the exhaust valve should be starting to close. Therefore, B, C, and D are wrong, and A is right.

**13. D** This question asks for the statement that is not true. If the engine thermostat is stuck open, the air-fuel ratio is rich and fuel economy is reduced because the ECT sensor informs the PCM regarding the lower coolant temperature. This problem also affects the EGR and cooling fan operation. Therefore, A, B, and C are true, so they are not the requested answer.

Because the air charge sensor sends a signal to the PCM in relation to the air intake temperature, this signal is not affected by a thermostat that is stuck open. Therefore, D is not true, making it the requested answer.

**14. C** Because the voltage at the coolant switch contacts is 0.2 V, these contacts must be closed. This indicates the coolant switch is not the cause of the problem, so A is wrong.

If the main relay winding is open, there is no voltage supplied to the cooling fan relay contacts; thus, B is wrong.

A blown 30 A fan fuse also causes 0 V at the cooling fan relay contacts, so D is wrong.

An open circuit from the cooling fan relay to the fan motor causes no cooling fan operation with the voltages mentioned in the question. Thus, C is correct.

**15. C** Component D2 is a conventional diode. Therefore, A, B, and D are wrong, and C is correct.

**16. A** If the ignition system had a defective coil, damaged spark plug wires, or bad spark plugs, the test light still flutters when cranking the engine. Therefore, B, C, and D are wrong. An inoperative crankshaft sensor causes the test light not to flash when cranking the engine. Thus, A is correct.

**17. A** If the cam sensor becomes inoperative with the engine running, the engine continues to run, but the engine does not restart when it is shut off. Therefore, Technician B is wrong.

An open circuit in the EST wire results in no spark advance, which results in reduced fuel economy and performance. Thus, Technician A is correct, making A the right answer.

**18. A** If the ignition module is inoperative, the test spark plug would not fire. Therefore, Technician B is wrong.

A leaking cap and rotor causes the test spark plug to fire normally when connected to the coil secondary wire, but fire intermittently when connected to the spark plug wires. Thus, Technician A is correct, making A the right answer.

**19. B** If a DTC is obtained representing the crankshaft sensor, the wires from this sensor to the module or PCM should be tested before replacing the sensor, and the sensor voltage signal should be tested to prove the sensor condition. Therefore, Technician A is wrong. An inoperative crankshaft sensor may cause a no-start condition. Thus, Technician B is correct, and B is the right answer.

**20. D** An ohmmeter connected from one of the pickup leads to ground should provide an infinity reading. If the ohmmeter provides a low reading, the pickup coil is grounded. Therefore, A, B, and C are wrong, and D is right.

**21. C** The Hall-effect pickup may be tested with a voltmeter connected from the pickup signal wire to ground. Therefore, A, B, and D are wrong, and C is right.

**22. C** A high-resistance problem in the primary circuit reduces primary current flow and magnetic strength, and this action causes low maximum secondary coil voltage. Therefore, A is right.

A coil with a secondary insulation leakage problem also lowers the maximum available secondary coil voltage. Thus, Technicians A and B are both correct, and C is the right answer.

**23.   A**   The negative side of each primary winding is connected to the module or PCM. Therefore, B is wrong.

The two ends of each secondary winding are connected to the spark plug wire terminals on each coil, so C is wrong.

The secondary windings are not interconnected; thus, D is wrong.

When the ignition switch is turned on, 12 V are supplied to the positive side of each primary winding. Thus, A is right.

**24.   B**   Timing adjustments are not possible or required on EI-equipped engines. Therefore, Technician A is wrong.

On some fuel-injected engines with a distributor, the PCM must be placed in the limp-in mode before adjusting basic ignition timing. Thus, Technician B is correct, and B is the right answer.

**25.   B**   The PCM supplies 8 V from terminal 7 to the crankshaft and camshaft sensors. If there is 0 V between this terminal and ground, and the PCM voltage supply and ground wires are satisfactory, the PCM must be malfunctioning. Therefore, Technician A is wrong and Technician B is the correct, making B the right answer.

**26.   C**   If the voltmeter connected from the negative primary coil terminal always indicates 12 V while cranking the engine, the module is open or the wire from the coil negative terminal to the module is open. Thus, Technicians A and B are both correct, and C is the right answer.

**27.   D**   Dripping injectors, a leaking fuel pump check valve, or a leaking pressure regulator valve allow the fuel pressure to slowly decrease after the engine is shut off. This action results in a longer cranking time when restarting the engine. Therefore, A, B, and C are true, but none of these is the requested answer. A leaking injector O-ring may cause an intake manifold vacuum leak and rough engine idle, but this defect does not cause hard starting. Therefore, D is not true, and this is the requested answer.

**28.   B**   A higher-than-specified TPS voltage signal may cause a lower-than-specified idle speed, so Technician A is wrong.

An intake manifold vacuum leak on a fuel-injected engine may cause a faster-than-specified idle speed. Thus, Technician B is right, making B the correct choice.

**29.   C**   Inoperative injectors may cause a rich air-fuel ratio, but this problem also causes rough idle, so A is wrong.

Bad spark plugs cause cylinder misfiring and rough idle, so B is wrong.

An inoperative fuel pump check valve causes fuel drain back into the tank when the engine is shut off, which results in hard starting. Thus, D is wrong.

High fuel-pump pressure causes a rich air-fuel ratio; thus, C is right.

**30.   A**   A malfunctioning IAT sensor may cause a rich air-fuel ratio, reduced fuel economy, and high emissions, but this problem does not result in a cylinder misfire. Therefore, Technician B is wrong.

When diagnosing a cylinder misfire, the cylinder compression must be verified. Thus, Technician A is right, and A is the correct answer.

**31.   A**   A restricted fuel line may cause an acceleration stumble and power loss, but this problem should not result in hard starting, so Technician B is wrong.

Excessive alcohol in the fuel causes a lean air-fuel ratio, hard starting, power loss, and acceleration stumbles, so Technician A is correct, making A the right answer.

**32.   D**   If there is an open in the main relay winding, there is no action from the fuel pump when the FP and B+ terminals are connected. Thus, A is wrong.

When the 7.5 A, IGN fuse is blown, there is no action from the main relay or the fuel pump if the FP and B+ terminals are connected, so B is wrong.

An open circuit at the B+ terminal on the circuit opening relay does not allow the engine to keep running after the jumper wire at the FP and B+ terminals is removed, so C is also wrong.

An open circuit in the winding connected to the STA and E1 terminals in the circuit opening relay causes no action from the fuel pump while cranking the engine. Therefore, D is correct.

**33.  A**  A leaking fuel-pump check valve causes fuel drain back into the tank after the engine is shut off. This results in hard starting, so Technician B is wrong.

A partially plugged return fuel line causes high fuel pressure, a rich air-fuel ratio, and reduced fuel economy. Thus, Technician A is correct, making A the right answer.

**34.  A**  A fuel pressure regulator that is sticking open causes low fuel pressure, so Technician B is wrong. A restricted fuel return line causes high fuel pressure. Thus, Technician A is correct, making A the right answer.

**35.  A**  The PCM, ECT sensor, or ACT sensor do not affect the cold start injector operation. Therefore, B, C, and D are wrong.

Because the cold start injector is controlled by a thermo-time switch, if this component is inoperative, the cold start injector may stay open longer than specified, resulting in a rich air-fuel ratio, rough idle, and black smoke from the tailpipe. A is correct.

**36.  A**  Because the emissions are high only during deceleration and the fuel pressure has no effect on deceleration operation, Technician B is wrong.

An injector deceleration fuel reduction system that does not cut off the injector operation during deceleration may cause high CO and HC emissions during deceleration, so Technician A is correct. Thus, A is the right choice.

**37.  A**  A PCV valve stuck open on a TBI engine acts much like a vacuum leak and causes faster-than-specified idle speed. Thus, Technician B is wrong.

Carbon in the throttle bore area may cause stalling and rough idle operation, so Technician A is correct, making A the right answer.

**38.  B**  An injector plunger sticking in the open position causes excessive pressure drop during the balance test, so A is wrong.

Because the number 3 injector had more than the specified pressure drop, it has a definite problem; thus, C is wrong.

It is impossible to have less pressure at the number 3 injector compared to the other injectors, so D is wrong.

A restricted orifice in the number 3 injector causes less pressure drop on this injector, so B is correct.

**39.  D**  If the PCV valve is stuck open, additional air flows through this valve, and this does not cause oil contamination in the air cleaner element. If the PCV clean air hose and fitting are loose in the rocker arm cover, dirt particles may enter the engine, resulting in worn engine components. Oil contamination in the air cleaner element is caused by a restricted PCV valve or hose. Therefore, both Technicians A and B are wrong, and D is the correct answer.

**40.  B**  Because engine rpm and idle condition both change when the propane hose is placed near the injector, the lower O-ring on the injector must be leaking. Thus, Technician A is wrong and Technician B is correct, making B the right answer.

**41.  A**  The minimum air rate adjustment is shown in the figure. Therefore, B, C, and D are wrong, and A is correct.

**42.  A**  The throttle opener allows the throttle to close gradually on deceleration, so Technician B is wrong and Technician A is right, making A the correct choice.

**43.  D**  A bad spark plug wire, spark plug, or injectors cause cylinder misfiring; thus, A, B, and C are wrong.

A restricted exhaust system causes a severe loss of power and limited maximum vehicle speed, so D is right.

**44.  D**    An EGR valve that is stuck open causes rough idle operation, so A is wrong.

A restricted catalytic converter limits the top vehicle speed, so B is wrong.

Dripping injectors cause rough idle operation, so C is wrong.

Excessive resistance in a spark plug wire causes misfiring during acceleration; thus, D is correct.

**45.  A**    A wastegate sticking closed causes high boost pressure, so Technician B is wrong.

Low engine compression causes reduced air and exhaust flow through the engine, resulting in reduced turbocharger shaft speed. Because this action results in reduced boost pressure, Technician A is correct, and A is the right answer.

**46.  A**    Partially seized bearings, or excessive shaft end play cause reduced turbocharger boost pressure. Thus, B and C are wrong.

Reduced coolant flow through the turbocharger causes damaged turbocharger bearings, so D is wrong.

A leak in the air intake system causes compressor vane pitting, so A is right.

**47.  D**    Wide open throttle operation causes a wide PCV valve opening, so A is wrong.

Operation at half-open throttle results in a medium PCV valve opening; thus, B is wrong.

A backfire causes the PCV to close, so C is wrong.

The high vacuum during idle operation causes a very small PCV valve opening, so D is correct.

**48.  B**    A plugged PCV clean air filter causes excessive vacuum in the engine. This may result in damaged gaskets and dirt particles entering the engine past these gaskets. Thus, Technician A is wrong.

A plugged hose between the PCV valve and the intake manifold causes excessive pressure in the engine. This action causes oil vapors to be forced through the PCV clean air hose into the air cleaner, so Technician B is correct. Thus, B is the right answer.

**49.  A**    A loose oil filler cap would not change the amount of flow through the PCV valve, so B and C are wrong.

Because the engine compression is satisfactory, oil should not be blown out around the loose oil filler cap; thus, D is wrong.

A loose oil filler cap allows the PCV system to pull dirt particles into the engine, so A is right.

**50.  D**    An engine thermostat stuck open, an inoperative engine coolant temperature sensor, or low compression on one cylinder would not increase NOx emissions. Therefore, A, B, and C are wrong.

Plugged exhaust passages under the EGR valve prevent EGR flow and increase NOx emissions. Thus, D is right.

**51.  C**    A linear EGR valve provides a feedback signal to the PCM. Therefore, Technician A is right. A scan tool may be used to command the PCM to provide a specific EGR valve opening, so Technician B is correct. Since Technicians A and B are both right, C is the correct answer.

**52.  B**    Because the bleed valve normally is open in a positive backpressure EGR valve, supplying vacuum to this valve with the engine idling does not open the valve, so Technician A is wrong.

When the EGR valve is opened at idle speed, the engine should slow down 150 rpm, so Technician B is correct, making B the right answer.

**53.  B**    The bleed valve in a negative backpressure EGR valve is pulled open by negative pressure pulses in the exhaust at low engine speeds, so A is wrong.

When vacuum is supplied to this type of valve with the engine stopped, the valve should open because the bleed valve is closed; thus, C is wrong.

If the engine is running at 2,000 rpm, the negative pressure pulses in the exhaust are reduced, and the bleed valve closes. Thus, D is wrong.

The bleed valve is opened by negative pressure pulses in the exhaust at low engine speeds, so B is correct.

**54.  D**   The PCM does not open the EGR valve if the engine coolant is cold or if the throttle is wide open. Therefore, both Technicians A and B are wrong, and D is the correct answer.

**55.  B**   If the vacuum hose to the AIRD valve is leaking, the airflow from the pump is directed through this valve to the catalytic converter; this action does not cause a low $O_2$ sensor signal.

Therefore, A is wrong.

When the airflow from the pump is directed to the exhaust ports with the engine warmed up, the $O_2$ sensor detects the additional oxygen in the exhaust stream. This causes a low $O_2$ sensor voltage signal and the PCM responds to this signal by providing a rich air-fuel ratio, so Technician B is correct. Thus, B is the right answer.

**56.  D**   If the catalytic converter is operating properly, the downstream $HO_2S$ sensor should have a lower and slower cycling voltage compared to the upstream $HO_2S$ sensor. Therefore, Technician A is wrong. The PCM has to detect an inoperative catalytic converter on two consecutive drive cycles to illuminate the MIL. Therefore, Technician B is also wrong, and D is the correct answer.

**57.  A**   If the AIR system hoses are severely burned, one of the check valves is allowing exhaust to enter the system. Therefore, B, C, and D are wrong, and A is right.

**58.  C**   In the upstream mode, the PCM energizes both the AIRB and AIRD solenoids, and vacuum is supplied to both the AIRB and AIRD valves. If the voltage is 12 V on both terminals of the AIRD valve, there is no voltage drop across the winding because there is no current flow through the winding. This condition may be caused by an open wire from this solenoid to the PCM, or the PCM may not be grounding the wire. Therefore, both Technicians A and B are correct, and C is the right answer.

**59.  D**   If the upstream and downstream $HO_2S$ signals are similar, the catalytic converter is defective. A, B, and C are therefore wrong, and D is correct.

**60.  D**   If the wire from the EVAP solenoid to the PCM is open, the EVAP solenoid never opens and the canister is never purged. This may allow HC emissions to escape from the canister to the atmosphere, but it does not affect engine idle. Therefore, B is wrong.

When the EVAP solenoid is stuck open at idle, no purging can occur at idle speed because ported vacuum is supplied to the canister, and there is no ported vacuum at idle. Thus, both Technicians A and B are wrong, and D is the correct answer.

**61.  C**   A loose gasoline tank filler cap changes the pressure in the fuel tank and causes the PCM to illuminate the MIL. Therefore, A is right. A small leak in the hose from the fuel tank to the canister also changes the fuel tank pressure and causes the PCM to illuminate the MIL.

Therefore, Technicians A and B are both right, and C is the correct answer.

**62.  A**   This question asks for the response that is not the cause of the problem. A purge solenoid plunger stuck in the closed position, a plugged hose from the canister to the intake manifold, or a plugged purge port in the intake may cause an inoperative EVAP system without any DTCs in the PCM memory. Therefore, B, C, and D may be the cause of the problem, so none of these is the requested answer.

An open winding in the purge solenoid causes an inoperative EVAP system, but this problem also causes a DTC in the PCM memory. Therefore, A is not the cause of the problem, making it the requested answer.

**63.  B**   On some vehicles, the diagnostic trouble codes may be obtained by cycling the ignition switch on and off three times in a five-second interval, so Technician A is wrong.

On some vehicles, the DTCs are read from the flashes of the malfunction indicator light, so Technician B is correct, making B the right answer.

**64.  A**   A vacuum leak in the intake manifold may cause a rough idle problem. The intake manifold vacuum leak also causes a lean air-fuel ratio, and in response the PCM increases the injector pulse width to try and correct the lean air-fuel ratio. When the PCM continually provides increased injector pulse width, the long term fuel trim is high. So, Technician A is right. Low fuel-pump pressure may cause a lean air-fuel ratio especially at higher speeds, and this could result in increased injector pulse width and long-term fuel trim numbers. However, low fuel-pump pressure is not likely to cause rough idle, so Technician B is wrong. Thus, A is the right answer.

**65.  A**    An EGR valve that is stuck open causes a rough idle problem, high NOx emissions, and a DTC representing the EGR valve when the EGR valve has a valve position sensor. So, Technician A is correct.

High NOx emissions occur if the EGR passages are restricted with carbon, but this condition does not cause rough idle or a DTC representing the EGR valve. Therefore, Technician B is wrong.

**66.  C**    High fuel pressure causes a rich air-fuel ratio and high $O_2$ sensor voltage, so B is wrong.

An intake manifold vacuum leak causes low $O_2$ sensor voltage, but this voltage signal increases when propane is dispersed into the air intake, so D is wrong.

Because the $O_2$ sensor voltage does not change when propane is dispersed into the air intake, this sensor may be the problem, so C is correct.

The PCM does not change the injector pulse width because the $O_2$ sensor signal did not change. Thus, A is wrong.

**67.  C**    A leaking fuel-pump check valve may cause hard starting, but this problem does not cause rough idle operation, so A is wrong.

A leaking fuel pressure regulator may cause hard starting, but this problem does not cause rough idle operation, so B is wrong.

A malfunctioning ECT sensor may cause high CO and HC emissions, but this problem does not cause a decrease in fuel pressure after the engine is shut off. Thus, D is wrong.

Dripping injectors may cause high CO and HC emissions, rough idle, and hard starting after the engine is shut off, and a decrease in fuel pressure after the engine is shut off. So, C is correct.

**68.  B**    Malfunctioning injectors may cause rough idle, but this problem causes high HC and CO emissions at idle and higher speeds. So, A is wrong.

Low fuel-pump pressure or an inoperative ECT sensor cause a lean air-fuel ratio that may result in lean misfiring and high HC emissions at idle and higher speeds. Thus, C and D are wrong.

An EGR valve that is stuck open dilutes the air-fuel ratio at idle and slow speeds, which causes a lean misfire and high HC emissions. Because the EGR valve is normally open at 2,500 rpm, the stuck EGR valve does not affect emissions at this speed. Thus, B is correct.

**69.  C**    If the AIRD solenoid plunger is sticking open, or the AIRD valve is sticking, the airflow from the pump may be continually directed upstream. Because Technicians A and B are both correct, C is the right answer.

**70.  C**    Because the $O_2$ sensor responds to the propane flow into the air intake, this sensor is not the problem, so A is wrong.

When propane flow into the air intake causes a rich signal from the $O_2$ sensor, the PCM responds by reducing the injector pulse width, so the PCM is not the problem, and B is wrong.

If the $O_2$ sensor signal wire is open, there is no signal from this sensor to the PCM, thus D is wrong.

Low fuel-pump pressure causes a lean air-fuel ratio and occasional misfiring. This condition causes high HC emissions at low and high speeds, so C is correct.

**71.  A**    If the fuel line is plugged, a leaking fuel-pump check valve does not cause the fuel pressure to drop off after the engine is shut off. Thus, B is wrong.

If the return fuel line is plugged, a leaking pressure regulator does not cause the fuel pressure to drop off after the engine is shut off. So, C is wrong.

A leaking upper injector O-ring causes a visible leak between the injector and the fuel rail, so D is wrong.

Dripping injectors cause hard starting and fuel pressure drop-off after the engine is shut off with the fuel line and fuel return line plugged. Thus, A is correct.

**72. C** Many MAP sensors act as barometric pressure sensors. On many MAP sensors, the voltage signal increases as the vacuum at the sensor decreases. Therefore, Technicians A and B are both correct, and C is the correct answer.

**73. C** On many MAP sensors, for every 5 in. Hg vacuum change at the sensor, the sensor voltage signal should change 0.7 V to 1 V. If the sensor voltage signal is 1.2 V at 18 in. Hg, this voltage signal should be 2.0 V at 13 in. Hg. Thus, C is correct.

**74. C** An $O_2$ sensor signal must be tested with a digital voltmeter. Therefore, A, B, and D are wrong, and C is right.

**75. C** In Figure 8–22, with the ignition switch in the ACCY position, voltage is supplied only to the wiper and radio fuses; thus, A, B, and D are wrong, and C is right.

**76. D** This question asks for the statement that is not true. When handling an EEPROM chip do not remove the chip from the protective envelope until you are ready to install it. Before touching the chip, touch a good ground, and connect a ground strap from your wrist to a good vehicle ground. Therefore, A, B, and C are true, so they are not the requested answer. The chip pins should not be touched with your fingers or anything else; thus D is not true and is the requested answer.

**77. C** If the TCM is not issuing a TCC on command, the ECT sensor signal may not be received by the TCM. This problem could be caused by the ECT signal not being transmitted from the PCM to the TCM. Under this condition, the data links should be tested with a scan tester. Thus, both Technicians A and B are correct, and C is the right answer.

**78. C** If the timing mark does not move when the exhaust manifold is tapped with a hammer, the knock sensor may be the problem. Under this condition, the engine may detonate. Thus, Technicians A and B are both correct, so C is the right answer.

**79. D** The second digit in the DTC is a 1, indicating the vehicle manufacturer is responsible for this DTC, so Technician A is wrong. The third digit in the DTC is a 1, indicating this DTC belongs to the air-fuel ratio control subgroup; thus, Technician B is also wrong, and D is the correct answer.

**80. B** Voltage drop across battery cables should be measured while the starter is cranking the engine. Thus, Technician A is wrong. When measuring the voltage drop across the positive battery cable, the voltmeter leads should be connected from the positive battery post to the other end of the positive battery cable. Therefore, Technician B is correct.

**81. D** The current drain decreases after the tester switch is opened. The key-off drain should not be read until the computers enter the sleep mode, which takes several minutes. Therefore, both Technicians A and B are wrong, so D is the correct answer.

**82. B** High resistance in the field coils, or battery cables causes low cranking speed, low current draw, and high cranking voltage. Thus, A and C are wrong.
High resistance in the solenoid windings affects the solenoid operation, so D is wrong.
Worn starter bushings allow the armature to rub on the field coils, resulting in high current draw with low cranking speed and low cranking voltage. Thus, B is correct.

**83. D** The voltmeter leads are connected to the ends of the positive battery cable to test the voltage drop across this cable. A voltage drop of 0.2 V across the positive battery cable indicates a normal resistance. Therefore, A, B, and C are wrong, and D is right.

**84. A** The ohmmeter is connected to test the solenoid hold-in winding. An infinity reading indicates an open winding. Therefore, Technician B is wrong and Technician A is correct, meaning A is the right answer.

**85. C** This question asks for the choice that is not the cause of the problem. Battery undercharging may be caused by a loose alternator belt, a battery drain, or a low regulator voltage. Therefore, A, B, and D may be the cause of the problem, so none of these is the requested answer.
High regulator voltage causes battery overcharging, so C is not the cause of the problem, making it the requested answer.

**86.  C**    An alternator V-belt should be replaced if it is touching the bottom of the pulley. Many ribbed V-belts have a spring-loaded automatic tensioner with a belt wear scale. Thus, both Technicians A and B are correct, and C is the right answer.

**87.  B**    A shorted alternator stator or a damaged diode reduce alternator output, but these problems would not blow the fuse link. High regulator voltage causes an overcharged battery. Therefore, A, C, and D are wrong.

Reversed battery polarity causes extremely high current flow through the diodes in the forward direction and also through the alternator battery wire. This high current flow burns out the fuse link in the battery wire. Thus, B is right.

# Answers to Post Test

**1.  D**    A sharp rapping noise at idle is caused by worn piston pins. Therefore, Technicians A and B are both wrong, and D is the correct answer.

**2.  A**    Worn piston rings or worn valve stem seals and guides cause blue smoke in the exhaust. Therefore, B and C are wrong. Holes in the tops of the pistons cause excessive blowby from the PCV valve opening, so D is wrong. A blown head gasket between the #4 and #5 cylinders causes a cylinder rpm drop, so A is correct.

**3.  A**    A low primary input voltage causes low maximum secondary coil voltage. Thus, Technician A is right. An open spark plug wire causes high normal required coil voltage, but this problem does not affect the maximum coil firing voltage. Thus, Technician B is wrong, and A is the correct answer.

**4.  C**    A burned exhaust valve, broken piston rings, or an intake manifold vacuum leak cause rough idle operation. Therefore, A, B, and D are wrong. Excessive resistance in a spark plug wire may cause misfiring only during acceleration, so C is the correct answer.

**5.  A**    If the EGR valve is stuck closed, the lack of EGR flow increases combustion chamber temperatures, and this may result in spark knock. Thus, Technician A is right. A rich air-fuel ratio reduces combustion chamber temperature and this does not result in spark knock. So, Technician B is wrong, and A is the correct answer.

**6.  B**    A secondary coil wire with excessive resistance raises the normal required secondary firing voltage, but this problem does not affect the module. Thus, Technician A is wrong. A shorted primary coil winding increases the primary current flow and this may result in module damage. So, Technician B is right, and B is the correct answer.

**7.  C**    High resistance in the spark plug wires or defective spark plugs may cause higher normal required secondary firing voltage. Technicians A and B are both right, and C is the correct answer.

**8.  B**    When the ohmmeter leads are connected from one pickup coil lead to ground, this coil is being tested for a grounded condition. A satisfactory pickup coil provides an infinite ohmmeter reading and a grounded pickup coil has a low reading. Thus, Technician B is correct; the pickup coil winding is touching the metal on the distributor housing. If the coils of wire in the pickup coil winding are touching together in a shorted condition, the coil has a low resistance when the ohmmeter leads are connected across the pickup coil leads. Therefore, Technician A is wrong, and B is the correct answer.

**9.  D**    Sticking injector plungers do not cause high fuel pressure, so A is wrong. A pressure regulator valve that is stuck open causes low fuel pressure, so B is wrong. A leaking pressure regulator vacuum hose causes only a slightly higher than normal fuel-pump pressure, so C is wrong. A restricted fuel return line causes considerably higher-than-specified fuel pressure and a rich air-fuel ratio that reduces fuel economy. Thus, D is correct.

**10.  C.**    When cleaning fuel injectors, it is necessary to disconnect the wiring connectors from the fuel pump relay and oil pressure switch, and squeeze the fuel return line so it is plugged.

Therefore, A, B, and D are true, but none of these is the requested answer. It is not necessary to install a new fuel filter. Thus, C is not true, making it the requested answer.

**11. A**   If there is no vacuum at the oil filter cap opening with the engine idling, the rocker arm cover gasket may be leaking. Therefore, Technician A is correct. A plugged PCV valve may cause no vacuum at the oil filler cap opening, but there would be an accumulation of oil in the air cleaner. Thus, Technician B is wrong, and A is the correct answer.

**12. C**   The EGR valve is closed during hard acceleration and at wide open throttle. Therefore, A and B are true, so they are not the requested answer. Vacuum is supplied through an electric solenoid to some EGR valves. Thus, C is true, so it also is not the requested answer. The EGR valve is closed on a cold engine. Therefore, D is not true, making it the requested answer.

**13. C**   If 18 in. Hg. (40.5 kPa) is supplied to the EGR valve with the engine shut off, the vacuum bleed valve in the EGR valve should be closed, and the EGR valve should open. Thus, Technician A is right. If 18 in. Hg. (40.5 kPa) is supplied to the EGR valve with the engine idling, the vacuum bleed valve in the EGR valve should be open and the EGR valve should remain closed. So, Technician B is right. Because Technicians A and B are both right, C is the correct answer.

**14. C**   A defective vehicle speed sensor (VSS) or mass air flow (MAF) sensor, or an EGR valve stuck closed, would set DTCs in the PCM. An EGR valve that is stuck closed also causes high NOx emissions. Therefore, A, B, and D are not the cause of the problem. Continual upstream air injections cause a low voltage from the upstream $HO_2S$, and the PCM reacts to this signal by supplying a rich air-fuel ratio. This results in an emission test failure for HC and CO, and a lean upstream $HO_2S$ signal. Thus, C is the correct answer.

**15. A**   If the EVAP solenoid plunger was stuck closed, one of the solenoid wires was open, or the hoses from the canister to the intake manifold were plugged, the EVAP system would fail the flow test. Therefore, B, C, and D are wrong. If the wire from the EVAP solenoid to the computer is grounded, the solenoid remains open continually, allowing purging to occur at idle and low speeds when the solenoid plunger should be closed. This action results in high HC and CO emissions, and rough engine idle. So, A is the correct answer.

**16. C**   Because the $HO_2S$ voltage changes when a lean air-fuel ratio is created, this sensor and related wiring are satisfactory. Thus, A and D are wrong. Because the PCM changes the injector pulse width in response to a lean air-fuel ratio, the PCM is satisfactory, meaning B is wrong. High fuel pressure causes a rich air-fuel ratio and high HC and CO emissions. Thus, C is correct.

**17. A**   A lean air-fuel ratio causes low CO and high NOx, so Technician A is right. Ignition misfiring causes high HC emissions, so Technician B is wrong, and A is the correct answer.

**18. D**   An $HO_2S$ identified as bank 2 sensor 2 is downstream from the catalytic converter and on the opposite side of the engine from number 1 cylinder.

**19. B**   A leaking MAP sensor vacuum hose does not affect the baro sensor reading because it is read with the ignition switch on and there is no vacuum applied to the MAP sensor. Thus, Technician A is wrong, and Technician B is right, making B the correct answer.

**20. A**   An ECT sensor with less than the specified resistance when cold may cause a lean air-fuel ratio during cold starting, resulting in the hard-start problem. Thus, A is correct. A leaking pressure regulator diaphragm may cause a hard-start problem, but it also causes rough idle and a rich air fuel ratio that would be indicated on the scan tool data. Therefore, B is wrong. A leaking one-way check valve in the fuel pump also causes hard starting when the engine and the atmospheric temperature are warm, so C is wrong. The question states that the ignition system is always firing, so D is wrong.

# 9 Advanced Engine Performance

## Pretest

The purpose of this pretest is to determine the amount of review you may require prior to taking the ASE Advanced Engine Performance Specialist Test. If you answer all the pretest questions correctly, complete the questions and study the information in this chapter to prepare for the ASE Advanced Engine Performance Specialist Test. If two or more or your answers to the pretest questions are incorrect, complete a thorough study of the questions and information in this chapter.

The pretest answers are located at the end of the pretest.

1. In an on-board diagnostic II (OBD II) system, the malfunction indicator light (MIL) is illuminated when the emissions exceed
   A. 2.5 times the allowable standard.
   B. 2.0 times the allowable standard.
   C. 1.5 times the allowable standard.
   D. 1.0 times the allowable standard.

2. When discussing five-gas exhaust, analysis Technician A says carbon monoxide (CO) readings are a good indicator of a lean air-fuel ratio.
   Technician B says the high CO readings may be caused by cylinder misfiring.
   Who is correct?
   A. A only
   B. B only
   C. Both A and B
   D. Neither A nor B

3. The LEAST likely cause of engine power loss under load and very low intake manifold vacuum would be
   A. an inoperative EGR valve.
   B. an intake manifold vacuum leak.
   C. a restricted exhaust system.
   D. a lean air-fuel ratio.

4. The composite vehicle is hard to start when the engine is cold, and the engine surges at idle and low speeds. The engine also has an acceleration stumble that is more noticeable when the engine is cold. The scan-tool data in Figure 9–1 was obtained with the ignition switch on and the engine off (KOEO) after the vehicle had been outside overnight. How should the technician proceed?
   A. Replace the oxygen sensor.
   B. Test the ECT sensor and related circuit.
   C. Test the IAT sensor and related circuit.
   D. Test the PCM power and ground circuits.

5. An OBD II vehicle has an enhanced evaporative (EVAP) system (Figure 9–2).
   Technician A says if the vehicle is operated with the gasoline filler cap loose, a DTC is set in the PCM memory.

| SCAN TOOL DATA | | | |
|---|---|---|---|
| Engine Coolant Temperature Sensor (ECT) 0.35 Volts | Intake Air Temperature Sensor (IAT) 4.50 Volts | MAP Sensor    4.5  V<br>MAF Sensor    0.2  V | Throttle Position Sensor (TPS) 0.6 Volts |
| Engine Speed Sensor (RPM) 0 rpm | Heated $O_2$ $HO_2S$ Upstream    0 V<br>Downstream   0 V | Vehicle Speed Sensor (VSS) 0 mph | Battery Voltage (B+) 12.6 Volts |
| Idle Air Control Valve (IAC) 25 percent | Evaporative Emission Canister Solenoid (EVAP) OFF | Torque Converter Clutch Solenoid (TCC) OFF | EGR Valve Control Solenoid  (EGR) 0 percent |
| Malfunction Indicator Lamp (MIL) ON | Diagnostic Trouble Codes NONE | Open/Closed Loop OPEN | Fuel Pump Relay (FP) OFF |
| Fuel Level Sensor 3.5  V | Fuel Tank Pressure Sensor 2.5  V | Transmission Fluid Temperature Sensor 3.5  V | Transmission Turbine Shaft Speed Sensor 0 mph |
| Transmission Range Switch PARK | Transmission Pressure Control Solenoid 0 | Transmission Shift Solenoid 1 ON | Transmission Shift Solenoid 2 OFF |
| Measured Ignition Timing ° BTDC    Base Timing:  10    Actual Timing: | | | |

**Figure 9–1**  OBD II Scan tool data number 1

Technician B says the PCM cycles the purge solenoid and vent solenoid on and off to maintain a vacuum of 2 to 12 in. Hg in the fuel tank.

Who is correct?

A.  A only

B.  B only

C.  Both A and B

D.  Neither A nor B

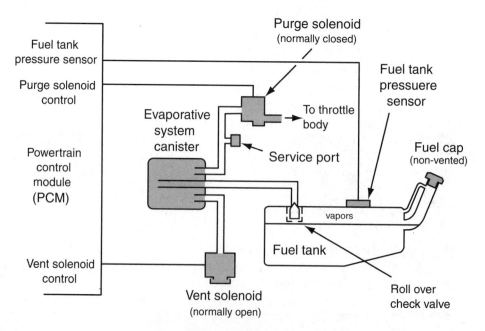

**Figure 9–2** Enhanced EVAP system.

6. Maintaining combustion chamber temperatures below 2,500°F (1372°C) results in
   A. lower CO emissions.
   B. lower $CO_2$ emissions.
   C. lower NOx emissions.
   D. increased $O_2$ emissions.

7. When diagnosing the composite vehicle, Technician A says that comprehensive component monitor faults require only one trip for the PCM to set a DTC and freeze frame data, and to illuminate the MIL.
   Technician B says that "two trip" emissions diagnostic monitors store a temporary DTC in the PCM after a failure occurs on two consecutive trips.
   Who is correct?
   A. A only
   B. B only
   C. Both A and B
   D. Neither A nor B

8. An engine has an idle misfire. A four-gas analyzer connected to the exhaust system indicates 400 ppm HC, 0.2 percent CO, 11 percent $CO_2$, and 6 percent $O_2$. The MOST likely cause of this condition is
   A. a leaking injector.
   B. a clogged air cleaner element.
   C. excessive fuel pressure.
   D. an intake manifold vacuum leak.

9. A customer complains about engine surging at 45 mph (72 kmh), but there are no other drivability problems.
   Technician A says the TFT sensor may be malfunctioning.
   Technician B says there may be a problem in the torque converter clutch (TCC) system.
   Who is correct?
   A. A only
   B. B only
   C. Both A and B
   D. Neither A nor B

10. When diagnosing injector waveforms with a digital storage oscilloscope (DSO) on a V6 engine, the waveform for number 3 injector is 5V above ground potential. The other injector waveforms are close to ground potential. The cause of this problem could be
    A. high resistance in the PCM ground.
    B. high resistance in the voltage supply wire to the injectors.
    C. low resistance in the number 3 injector winding.
    D. high resistance in the wire from the ground side of number 3 injector to the PCM.

11. A vehicle failed an enhanced I/M test with the following readings: HC CO NOx Cutpoints 0.9 gpm 14 gpm 1.9 gpm Actual Readings 1.0 gpm 21 gpm 1.8 gpm. The Technician repaired the cause of the HC and CO failure. When the emission test was repeated, the following test results were obtained: HC CO NOx Cutpoints 0.9 gpm 14 gpm 1.9 gpm Actual readings 0.6 gpm 11 gpm 2.5 gpm. Why are the NOx emissions now higher than the cutpoints?
    A. The heating effect of the lean air-fuel ratio was masking the NOx failure.
    B. The cooling effect of the lean air-fuel ratio was masking the NOx failure.
    C. The heating effect of the rich air-fuel ratio was masking the NOx failure.
    D. The cooling effect of the rich air-fuel ratio was masking the NOx failure.

12. Operating an EI-equipped engine with a spark plug wire removed could result in damage to all the following components EXCEPT
    A. the catalytic converter.
    B. the secondary ignition coil winding.
    C. the primary ignition coil winding.
    D. the spark plug wires.

13. When using a digital storage oscilloscope (DSO) to test an $O_2$ sensor, the correct number of lean/rich transitions from a satisfactory $O_2$ sensor is
    A. 10 to 40 lean/rich transitions in ten seconds.
    B. 1 to 4 lean/rich transitions in ten seconds.
    C. 10/40 lean/rich transitions in sixty seconds.
    D. 1 to 4 lean/rich transitions in sixty seconds.

14. When testing the $HO_2S$ sensors on the composite vehicle with a digital storage oscilloscope (DSO), the upstream $HO_2S$ have satisfactory waveforms and the downstream $HO_2S$ waveform is very similar to the upstream $HO_2S$ with the PCM in closed loop. The MOST likely cause of this condition is
    A. a malfunctioning downstream $HO_2S$.
    B. a damaged ground wire on the downstream $HO_2S$.
    C. high resistance in the signal wire from the downstream $HO_2S$ to the PCM.
    D. an inoperative catalytic converter.

# Answers to Pretest

**1.  C**   The MIL is illuminated when the emissions exceed 1.5 times the allowable standard. Therefore, C is correct, and A, B, and D are wrong.

**2.  D**   Carbon monoxide (CO) readings are high when the air-fuel ratio is rich, but CO readings are not a good indicator of a lean air-fuel ratio. Thus, Technician A is wrong. Cylinder misfiring causes high HC readings but slightly lower CO readings. Therefore, Technician B is wrong. Because Technicians A and B are both wrong, D is the correct answer.

**3.  A**   An inoperative EGR valve has very little effect on engine power, and it does not cause low intake manifold vacuum. So, A is the requested answer. An intake manifold vacuum leak causes a lean air-fuel ratio, power loss, and low intake manifold vacuum. Thus, B is not the requested answer. A restricted exhaust system causes power loss and low intake manifold vacuum at wider throttle openings. Therefore, C is not the requested answer. A lean air-fuel ratio may decrease engine power, and cause lower intake manifold vacuum. Thus, D is not the requested answer.

**4.  B**   The engine coolant temperature (ECT) sensor voltage is much lower than normal. The Technician should test this sensor and the related circuit. Thus, B is correct, and A, C, and D are wrong.

**5.  A**   If the vehicle is operated with the gasoline filler cap loose, the fuel tank pressure is altered and a DTC is set in the PCM memory. Thus, Technician A is right. The PCM operates the purge and vent solenoids to maintain a much lower vacuum in the fuel tank. Therefore, Technician B is wrong, and A is the correct answer.

**6.  C**   Maintaining combustion chamber temperatures below 2,500°F (1372°) lowers NOx emissions.

**7.  A**   Comprehensive component monitor faults on the composite vehicle require only one trip for the PCM to set a DTC and both freeze frame data and illuminate the MIL. Thus, Technician A is right. Two trip emissions diagnostic monitors store a temporary DTC in the PCM when a fault occurs on the first trip. Therefore, Technician B is wrong, and A is the correct answer.

**8.** **D**     The four-gas analyzer readings indicate high HC and low CO. These readings are caused by a lean misfire condition. A leaking injector, clogged air cleaner element, or excessive fuel pressure cause a rich air-fuel ratio, and so A, B, and C are wrong. An intake manifold vacuum leak may cause a lean air-fuel ratio and a lean misfire condition resulting in high HC and low CO. Thus, D is correct.

**9.** **B**     A malfunctioning TFT sensor may affect transmission shifting, but it does not cause surging at 45 mph (72 kmh). Thus, Technician A is wrong. Surging at 45 mph (72 kmh) may be caused by a problem in the TCC system. Therefore, Technician B is right, and B is the correct answer.

**10.** **D**     High resistance in the PCM ground affects all the injectors, so A is wrong. High resistance in the voltage supply wire to the injectors causes low voltage at the injectors, so B is wrong. Low resistance in the number 3 injector winding causes higher current flow in this injector and possible PCM damage. Thus, C is wrong. High resistance in the wire from the ground side of number 3 injector to the PCM may cause the waveform for this injector to be 5V above ground. Therefore, D is correct.

**11.** **D**     The cooling effect of the rich air-fuel ratio was masking the NOx failure. Thus, D is correct, and A, B, and C are wrong.

**12.** **C**     Operating an EI-equipped engine with one spark plug wire removed causes very high HC emissions and maximum secondary coil voltage, and these conditions could damage the catalytic converter, coil secondary winding, and spark plug wires. Therefore, A, B, and D are true, but none of these are the requested answer. High HC emissions and high secondary ignition voltage do not damage the primary coil winding. Thus, C is the correct answer.

**13.** **A**     the correct number of lean/rich transitions is 10 to 40 in ten seconds. So, A is correct, and B, C, and D are wrong.

**14.** **D**     If the upstream and downstream $HO_2S$ waveforms are very similar, the catalytic converter is defective. Thus, D is correct, and A, B, and C are wrong.

# Composite Vehicle

## Introduction

A significant number of questions in the Advance Engine Performance Specialist (L1) test apply to the composite vehicle, and a similar number of questions in this chapter are also based on this vehicle. There are many differences in fuel-injection systems depending on the vehicle make and model year. The panel of experts who developed the L1 test also designed the composite vehicle to provide a typical, generic vehicle and avoid the many individual variations found on different vehicles. Complete information, including a PCM wiring diagram, on the composite vehicle is found in the "composite vehicle Type 3 Preparation/Reference Booklet" supplied to you when you register for the test.

The composite vehicle is equipped with a V6 engine with a firing order of 1, 2, 3, 4, 5, 6. Cylinders 1, 3, and 5 are in bank 1, and cylinders 2, 4, and 6 are in bank 2. This V6 engine has four overhead, chain-driven camshafts, and four valves per cylinder. This vehicle has an on-board diagnostic II (OBD II) system with a standard 16-terminal data link connector (DLC). The composite vehicle has a returnless sequential fuel injection (SFI) system with a pressure regulator mounted in the fuel tank. The regulated fuel pressure is maintained at 50 psi (344 kPa) with a minimum fuel pressure of 45 psi (310 kPa). This vehicle is equipped with an EI ignition system with individual coils mounted on each spark plug. This ignition system has a crankshaft position (CKP) sensor and a camshaft position sensor (CMP) connected to the PCM. Timing is determined by the CKP sensor signal and the PCM. The ignition module is integrated into the PCM, and timing is not adjustable. This engine is equipped with a variable valve timing system that has a hydraulic actuator on each intake camshaft. Engine oil pressure is supplied to each hydraulic actuator through a camshaft position actuator control solenoid operated by the PCM.

The composite vehicle has a four-speed automatic transmission, and transmission shifting is controlled by the transmission control module (TCM). The PCM communicates with the TCM and other on-board computers via the data bus. If the TCM senses an electronic defect, it enters the default mode in which the transmission defaults to second gear, the torque converter clutch is disabled, and maximum line pressure is commanded.

The engine in the composite vehicle has electronic throttle control, and therefore a conventional throttle cable is not required. A throttle actuator control operates the throttle under all engine loads and speeds. Dual accelerator pedal position (APP) sensors provide input from the vehicle operator to the PCM. The APP sensors are operated by the accelerator pedal. Dual throttle position (TP) sensors send inputs to the PCM in relation to the actual throttle angle. If one APP or TP sensor fails, the PCM turns on the malfunction indicator light (MIL) and limits the throttle opening to 33 percent. If any two of the APP or TP sensors fail, the PCM disables the throttle control system and turns on the MIL. If the electronic throttle control system fails, the limp-in mode is entered and the spring-loaded throttle plate returns to a 15-percent opening that provides a no-load fast idle of 1400 to 1500 rpm. After component replacement or a discharged battery, the electronic throttle control system does not require a relearn procedure.

The serial data bus is a high-speed network with a twisted pair of wires. This network allows communication between the PCM, TCM, instrument cluster, immobilizer control module, and a scan tool connected to the data link connector (DLC). The data high circuit switches between 2.5 V in the at-rest state and 3.5 V in the active state. The data low circuit switches between 2.5 V in the at-rest state and 1.5 V in the active state. The serial data bus network uses two 120-ohm terminating resistors. One of these resistors is mounted in the instrument cluster and the second terminating resistor is mounted in the PCM. If either of the data lines short to ground, to each other, or to battery voltage, network DTCs are stored in the PCM and the network is disabled. When the serial data bus is disabled, the PCM continues to provide ignition and fuel control. When the ignition switch is turned on, the immobilizer control module transmits a voltage signal through the antenna around the ignition switch to the transponder in the ignition key. The transponder key responds to this voltage signal with an encrypted key code. The immobilizer control module interprets the key code and compares it to the list of registered keys. When the engine starts, the PCM requests validation of the key code from the immobilizer control module. via the data links. If the key code is valid, the immobilizer control module sends a valid key signal to the PCM and the PCM continues the normal engine operation. When the PCM does not receive a valid key code signal within two seconds after the engine starts, the PCM disables the fuel injectors to shut off the engine. If the ignition switch is cycled off and the engine cranked again, the engine restarts and immediately stalls. Unique encrypted key codes are programmed into the PCM and immobilizer control module during the vehicle manufacturing process. If the PCM or the immobilizer control module are replaced, a scan tool must be used to program the replacement module using the vehicle identification number (VIN) the date, and a factory-assigned PIN number. A maximum of eight keys may be registered in the immobilizer control module.

The composite vehicle has an on-board refueling vapor recovery (ORVR) EVAP system. The ORVR prevents refueling vapors from escaping to the atmosphere by forcing these vapors into the EVAP canister while refueling. The ORVR has a one-inch fill pipe on the fuel tank, and a one-way check valve at the lower end of the fill pipe. A fuel vapor control valve is mounted inside the fuel tank, and a 1/2-in. inside diameter (ID) vent hose is connected from the vapor control valve to the canister. The fuel vapor control valve contains a float that rises to close the vapor hose when the tank is full to prevent liquid fuel from entering the canister. If the vehicle is involved in a roll-over, the float prevents fuel from leaking.

# Input Sensors

## Thermistor-Type Sensors

The composite vehicle has three thermistor-type input sensors, which include the engine coolant temperature (ECT) sensor, intake air temperature (IAT) sensor, and the transmission fluid temperature (TFT) sensor. All three sensor values range from –40°F to 248°F (–40°C to 120°C). At 212°F (100°C) the ECT or TFT sensor voltage signal to the PCM is 0.46 V. At 86°F (30°C), the IAT sensor voltage signal to the PCM is 2.6V (Figure 9–3).

| Temperature °F | Temperature °C | Sensor Voltage |
|---|---|---|
| 248 | 120 | 0.25 |
| 212 | 100 | 0.46 |
| 176 | 80 | 0.84 |
| 150 | 66 | 1.34 |
| 140 | 60 | 1.55 |
| 104 | 40 | 2.27 |
| 86 | 30 | 2.60 |
| 68 | 20 | 2.93 |
| 32 | 0 | 3.59 |
| -4 | -20 | 4.24 |
| -40 | -40 | 4.90 |

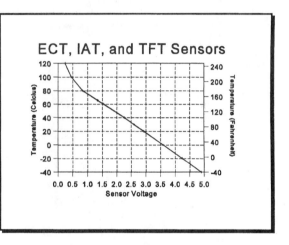

**Figure 9–3** ECT, IAT, and TFT sensor temperature/voltage relationship. *(National Institute of Automotive Service Excellence [ASE])*

## The Mass Air Flow (MAF) Sensor

The MAF sensor sends a voltage signal to the PCM in relation to the amount of air flowing through the air intake system. This voltage signal varies from 0.2 V with the ignition switch on and the engine not running to 4.8 V at maximum airflow (Figure 9–4). With the engine idling at sea level, this voltage signal is 0.7 V. The grams-per-second (gm/sec) airflow through the MAF varies from 0 to 175.

| Mass Airflow (gm/sec) | Sensor Voltage |
|---|---|
| 0 | 0.2 |
| 2 | 0.7 |
| 4 | 1.0 |
| 8 | 1.5 |
| 15 | 2.0 |
| 30 | 2.5 |
| 50 | 3.0 |
| 80 | 3.5 |
| 110 | 4.0 |
| 150 | 4.5 |
| 175 | 4.8 |

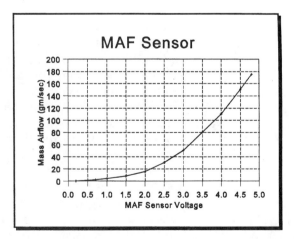

**Figure 9–4** MAF sensor airflow/voltage relationship. *(National Institute for Automotive Service Excellence [ASE])*

# The Manifold Absolute Pressure (MAP) Sensor

The MAP sensor is only used by the PCM for OBD II diagnostics. With the ignition switch on and the engine not running, the MAP sensor voltage signal to the PCM is 4.5 V. With the engine idling at sea level, the MAP sensor voltage signal is 1.5 V (Figure 9–5). The PCM supplies a 5 V reference voltage to the MAP sensor, and a ground wire is connected from this sensor to the PCM.

| Vacuum at sea level (in. Hg.) | Manifold Absolute Pressure (kPa) | Sensor Voltage |
|---|---|---|
| 0 | 101.3 | 4.5 |
| 3 | 91.2 | 4.0 |
| 6 | 81.0 | 3.5 |
| 9 | 70.8 | 3.0 |
| 12 | 60.7 | 2.5 |
| 15 | 50.5 | 2.0 |
| 18 | 40.4 | 1.5 |
| 21 | 30.2 | 1.0 |
| 24 | 20.1 | 0.5 |

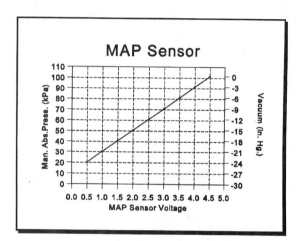

**Figure 9–5**  MAP sensor vacuum/voltage relationship. *(National Institute for Automotive Service Excellence [ASE])*

# The Crankshaft Position (CKP) Sensor

The magnetic-type CKP sensor is mounted on the front engine cover and triggered by a 35-tooth reluctor wheel mounted on the front of the crankshaft behind the balancer pulley. The two terminals on the CKP sensor are connected to the PCM. There is one missing tooth on the CKP reluctor wheel. Each tooth on this wheel is 10 degrees apart, and there is one space for a missing tooth positioned at 60 degrees before top dead center of cylinder 1.

# The Camshaft Position (CMP) Sensor

The CMP sensors are located at the rear of each intake camshaft, and an interrupter ring on each intake camshaft rotates past the sensor. The CMP sensor may be a Hall-effect or optical-type three-wire sensor that sends a voltage signal to the PCM once per intake camshaft revolution. The leading edge of the bank 1 CMP signal occurs when cylinder 1 is on the compression stroke, and the leading edge of the bank 2 CMP signal is sent to the PCM when cylinder 4 is on the compression stroke. If the intake camshafts are fully retarded, the CMP sensor signals occur at top dead center (TDC) of cylinders 1 and 4. Intake camshafts in the fully advanced position cause the CMP signals to occur at 40 crankshaft degrees before TDC. The PCM uses the CMP signals to determine fuel injector, coil sequence, and intake valve timing. If one CMP signal is defective, a diagnostic trouble code is set and intake valve timing remains in the fully retarded position. When neither CMP signal is present while cranking the engine, the PCM disables the fuel injectors, resulting in a no-start condition. injector seqwhen number 1 piston is at TDC on the compression stroke. The PCM uses this signal to provide proper injector and ignition coil sequencing. The CMP is mounted on the front of the bank 1 valve cover, and the interrupter wheel is mounted on the bank 1 exhaust camshaft timing gear. If there is no CMP sensor signal while cranking the engine, the PCM will still operate the injectors and coils and start the engine.

## The Throttle Position (TP) Sensor

The dual, nonadjustable TP sensors are mounted on the throttle body and each sensor sends a voltage signal to the PCM in relation to the throttle opening. The TP sensors are referred to as TP1 and TP2, and each TP sensor contains a potentiometer. The TP1 sensor voltage varies from 4.5 V at closed throttle to 0.5 V at wide open throttle, and the TP2 voltage signal varies from 0.5 V at closed throttle to 4.5 V at wide open throttle (Figure 9–6). If one TP sensor becomes defective, a DTC is set and the PCM limits the maximum throttle opening to 35 percent. If both TP sensors are defective, the throttle actuator control is disabled and the spring-loaded throttle valve is positioned in a 15-percent open position that provides only fast idle operation.

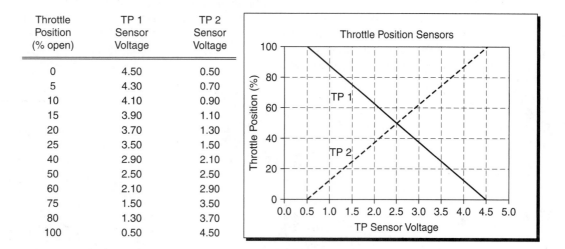

| Throttle Position (% open) | TP 1 Sensor Voltage | TP 2 Sensor Voltage |
|---|---|---|
| 0 | 4.50 | 0.50 |
| 5 | 4.30 | 0.70 |
| 10 | 4.10 | 0.90 |
| 15 | 3.90 | 1.10 |
| 20 | 3.70 | 1.30 |
| 25 | 3.50 | 1.50 |
| 40 | 2.90 | 2.10 |
| 50 | 2.50 | 2.50 |
| 60 | 2.10 | 2.90 |
| 75 | 1.50 | 3.50 |
| 80 | 1.30 | 3.70 |
| 100 | 0.50 | 4.50 |

**Figure 9–6** TPS throttle opening/voltage relationship.*(National Institute for Automotive Service Excellence [ASE])*

## The Vehicle Speed Sensor (VSS)

The vehicle speed sensor (VSS) is a magnetic-type sensor that sends a voltage signal to the PCM in relation to the rotational speed of the final drive. The VSS voltage signal is used by the TCM to control transmission shifting and torque converter clutch (TCC) application. The PCM uses the VSS signal for high-speed fuel cut-off and the instrument cluster uses the VSS for speedometer operation. On a scan tool, the VSS is indicated in miles per hour (mph).

## Heated Oxygen Sensors (HO₂Ss)

The composite vehicle has three electronically heated, zirconia-type $HO_2Ss$. The two upstream $HO_2Ss$ are mounted in the exhaust manifolds and the downstream $HO_2S$ is mounted in the exhaust pipe downstream from the catalytic converter. An $HO_2S$ designated as $HO_2S$ 1/1 is mounted on the same side of the engine as number 1 cylinder and upstream from the catalytic converter. The $HO_2S$ 2/1 is an upstream $HO_2S$ mounted on the opposite side of the engine from number 1 cylinder. The downstream $HO_2S$ is designated as $HO_2S$ 1/2. Each $HO_2S$ contains an electric heater, and voltage is supplied to this heater as long as the ignition switch is turned on. The signal from each $HO_2S$ may vary from 0.0 V to 1.0 V. A rich air-fuel ratio causes upstream $HO_2S$ voltage signals above 0.45 V, and a lean air-fuel ratio results in upstream $HO_2S$ voltage signals below this value. The upstream $HO_2S$ voltage signals are used by the PCM for closed-loop fuel control and OBD II system monitoring. The PCM monitors the catalytic converter operation by comparing the upstream and downstream $HO_2S$ signals. If the catalytic converter is operating properly, the downstream $HO_2S$ voltage should be lower than the upstream $HO_2S$ voltage signals, and the downstream $HO_2S$ voltage signal should cycle more slowly from lean to rich. Typical downstream $HO_2S$ signals are 0.1 to 0.3 V. The PCM provides a ground circuit for both upstream $HO_2S$ as soon as the engine starts, and the PCM provides a ground for the downstream $HO_2S$ after two minutes of engine operation.

# The A/C Pressure Sensor

The A/C pressure sensor is a solid-state three-wire sensor mounted in the high-side vapor line of the A/C system. The voltage signal from this sensor varies from 0.25 V at 25 psi (172 kPa) to 4.5 V at 450 psi (3103 kPa) (Figure 9–7). The PCM uses the A/C pressure sensor signal to control the A/C compressor clutch, cooling fan, and to adjust idle speed. If the A/C system pressure is below 40 psi (276 kPa) or above 420 psi (2,896 kPa), the PCM disables the compressor clutch.

| A/C High Side Pressure (psi) | Sensor Voltage |
|---|---|
| 25 | 0.25 |
| 50 | 0.50 |
| 100 | 1.0 |
| 150 | 1.5 |
| 200 | 2.0 |
| 250 | 2.5 |
| 300 | 3.0 |
| 350 | 3.5 |
| 400 | 4.0 |
| 450 | 4.5 |

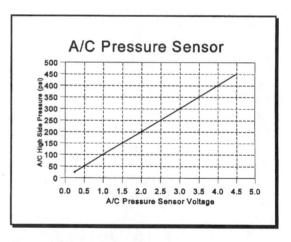

**Figure 9–7** A/C pressure sensor pressure/voltage relationship. *(National Institute for Automotive Service Excellence [ASE])*

# The A/C On/Off Request Switch

The A/C on/off request switch is used by the driver to turn on the A/C system. This switch sends a voltage signal to the PCM to inform the PCM that A/C has been requested. When this signal is received, the PCM increases the airflow through the idle air control (IAC) motor to maintain idle speed and compensate for compressor load on the engine.

# The Brake Pedal Position (BPP) Switch

The brake pedal position (BPP) switch sends a voltage signal to the PCM when the brake pedal is depressed by the driver.

When this signal is received, the PCM releases the TCC.

## The Power-Steering Pressure (PSP) Switch

The power-steering pressure (PSP) switch sends a voltage signal to the PCM when the driver turns the steering wheel and the power-steering pressure increases. This signal is used by the PCM to adjust the airflow through the IAC motor and maintain idle speed to prevent engine stalling.

## The Fuel Level Sensor

The fuel level sensor contains a potentiometer, and this sensor is mounted in the fuel tank. The PCM uses the voltage signal from this sensor to test the evaporative (EVAP) system.
   The fuel level sensor reading varies from 0.5 V/0 percent to 4.5 V/100 percent.

## Fuel Tank Pressure Sensor

The fuel tank pressure sensor sends a signal to the PCM in relation to pressure or vacuum in the fuel tank. With filler cap removed from the fuel tank, the fuel tank pressure sensor voltage signal is 2.5 V. This sensor signal varies from 0.5 V with 0.5 psi (14 in. $H_2O$) vacuum in the fuel tank to 4.5 V with 0.5 psi (14 in. $H_2O$) pressure in the fuel tank (Figure 9–8). The PCM uses the fuel tank pressure sensor signal to diagnose the EVAP system.

| Fuel Tank (EVAP) Pressure | | Sensor |
|---|---|---|
| (in.$H_2O$) | (psi) | Voltage |
| -14.0 | -0.5 | 0.5 |
| -10.5 | -0.375 | 1.0 |
| -7.0 | -0.25 | 1.5 |
| -3.5 | -0.125 | 2.0 |
| 0.0 | 0.0 | 2.5 |
| 3.5 | 0.125 | 3.0 |
| 7.0 | 0.25 | 3.5 |
| 10.5 | 0.375 | 4.0 |
| 14.0 | 0.50 | 4.5 |

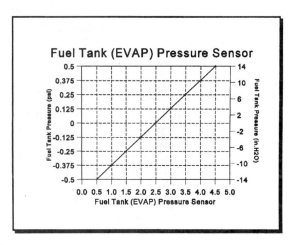

**Figure 9–8** Fuel tank EVAP pressure sensor pressure/voltage relationship. *(National Institute for Automotive Service Excellence [ASE])*

## The Transmission Turbine Shaft Speed (TSS) Sensor

The transmission turbine shaft speed (TSS) sensor sends a voltage signal to the PCM in relation to the transmission input shaft rotational speed. The PCM uses this signal to detect transmission slipping and to control the TCC. The TSS is a magnetic sensor mounted in the transaxle housing.

## The Transmission Range Switch

The transmission range switch sends a voltage signal to the PCM in relation to the gear selector position. This input is used by the PCM to control transmission line pressure and shifting. The transmission range switch is mounted on the transaxle housing.

# The EGR Valve Position Sensor

The EGR valve position sensor is mounted on top of the EGR valve, and senses the pintle position in this valve. The EGR valve position sensor is a three-wire sensor that contains a potentiometer. The voltage signal from the EGR valve position sensor varies from 0.5 V with the EGR valve closed, to 4.5 V with the EGR valve fully open.

# The Knock Sensor

The knock sensor is a two-wire sensor that contains a piezoelectric element. When the engine detonates, a vibration occurs in the metal of the engine, and the knock sensor sends a voltage signal to the PCM in relation to this vibration. The PCM uses the knock sensor signal to reduce spark advance if the engine detonates. When the engine is not detonating, the PCM sends 2.5 V to the knock sensor.

# The Accelerator Pedal Position (APP 1 and APP 2)

The APP 1 and APP 2 sensors are operated by the accelerator pedal. The information regarding these sensors is provided previously in this chapter during the discussion about electronic throttle control.

# Output Actuators

## Fuel Injectors

When the ignition switch is turned on, voltage is supplied to each fuel injector, and each fuel injector is grounded individually by the PCM. The PCM energizes each injector once per camshaft revolution, multiplied by the injector openings in relation to the intake valve openings. The specified ohms resistance in the injector windings is 12 ohms (plus or minus 4 ohms).

## Ignition Coils

The secondary winding in each coil is connected to a spark plug. When the ignition switch is turned on, 12 V are supplied to each coil primary winding. The primary resistance in each coil is 1 ohm (plus or minus 0.5 ohms), and the secondary winding resistance is 10,000 ohms plus or minus 2,000 ohms. The integral ignition module in the PCM controls timing and dwell.

## The Generator Field

The alternator voltage regulator is integral with the PCM. When the ignition switch is turned on, 12 V are supplied to one alternator field terminal. The other field terminal is connected to the PCM. The PCM limits the alternator voltage by controlling the field current with a variable duty cycle principle.

## The Exhaust Gas Recirculation (EGR) Valve Control Solenoid

The PCM operates the EGR control solenoid to supply the right amount of vacuum to the EGR valve and provide the proper EGR valve opening. The PCM controls the EGR solenoid with a duty cycle principle. The PCM begins operating the EGR control solenoid when the coolant temperature reaches 150°F (66°C). If the indicated EGR value is 0 percent, the EGR valve is fully closed, whereas a value of 100 percent is shown when the EGR valve is wide open.

## The Throttle Actuator Control (TAC) Motor

The throttle actuator control (TAC) motor is a bidirectional pulse width modulated DC motor that controls throttle position. A scan-tool data value of 0 percent indicates the PCM has commanded the IAC motor fully closed, and a value of 100 percent means the PCM has commanded the IAC motor to fully open. The PCM controls the TAC motor in relation to the APP 1 and APP 2 sensor signals and other inputs. Any defect in the TAC motor sets a DTC and causes the TAC motor to be disabled. In this mode, the spring-loaded throttle valve returns to the 15-percent open position.

## The A/C Clutch Relay

When the driver selects A/C, the PCM energizes the A/C clutch relay. This action supplies voltage through the relay contacts to the compressor clutch. The resistance of the relay winding is 48 ohms (plus or minus 6 ohms).

## The Fan Control Relay

The PCM energizes the fan control relay when the coolant temperature reaches 210°F (99°C), and the PCM de-energizes this relay when the coolant temperature decreases to 195°F (90°C). The PCM also energizes the fan control relay when the A/C system pressure reaches 300 psi (2068 kPa). When the A/C system pressure drops to 250 psi (1723 kPa), the PCM opens the fan control relay. When the PCM energizes the fan control relay winding, the relay contacts close and voltage is supplied through these contacts to the fan motor. The resistance of the fan control relay winding should be 48 ohms (plus or minus 6 ohms).

## The Fuel Pump Relay

When the ignition switch is turned on, the PCM grounds the fuel pump relay winding. This action closes the relay contacts, and voltage is supplied through these contacts to the fuel pump. If the ignition switch is on for two seconds, and the engine is not cranking or running, the PCM de-energizes the fuel pump relay winding. The resistance of the fuel pump relay winding is 48 ohms (plus or minus 6 ohms).

## The Malfunction Indicator Light (MIL)

When the ignition switch is turned on, the MIL is illuminated to check the bulb. This light should remain off while the engine is running. If an electrical problem occurs in the PCM or TCM system, the PCM illuminates the MIL to inform the driver that a problem exists. When the PCM detects a problem (such as a continual cylinder misfire) that could damage the catalytic converter, the PCM flashes the MIL. When the electrical problem has been repaired and the diagnostic trouble code(s) are erased from the PCM, the PCM turns off the MIL.

# The Evaporative Emission (EVAP) Canister Purge Solenoid

The PCM begins operating the EVAP purge solenoid when the coolant temperature reaches 150°F (66°C). An EVAP purge solenoid duty cycle of 0 percent indicates this solenoid is closed, and a duty cycle of 100 percent indicates this solenoid is fully open to allow maximum purge vapor flow. The PCM operates the EVAP solenoid with a duty cycle principle to control the vacuum in the EVAP system, and thus control the purge flow from the canister into the intake manifold. The PCM also operates the EVAP control solenoid for OBD II testing of the EVAP system. A service port used to pressure test and flow test the EVAP system is located in the hose between the canister and the EVAP control solenoid. The specified resistance of the EVAP solenoid winding is 48 ohms (plus or minus 6 ohms).

# The Evaporative Emission (EVAP) Canister Vent Solenoid

The EVAP canister vent solenoid is normally open, and air flows through this solenoid into the canister. When the PCM monitors the EVAP system operation, the PCM energizes and closes this solenoid to increase the vacuum in the EVAP system. The EVAP vent solenoid should have the same resistance as the EVAP control solenoid.

# Transmission Shift Solenoids: SS1 and SS2

The PCM operates the shift solenoids, SSI and SS2, to control all transmission shifting (Figure 9–9). The transmission shift solenoids control the fluid pressure supplied to the shift valves in the valve body to provide the desired shift. The resistance of each shift solenoid should be 24 ohms (plus or minus 4 ohms).

| Gear | SS 1 | SS 2 |
|---|---|---|
| P, N, R, or 1 | On | Off |
| 2 | Off | Off |
| 3 | Off | On |
| 4 | On | On |

**Figure 9–9** Transmission shift solenoid operation. *(National Institute for Automotive Service Excellence [ASE])*

# The Transmission Pressure Control (PC) Solenoid

The PCM controls the duty cycle of the pressure control (PC) solenoid to supply the appropriate transmission pressure. Fluid pressure from the PC solenoid is supplied to the pressure regulator valve to control transmission fluid pressure. When the PCM supplies a 10-percent duty cycle to the PC solenoid, transmission pressure is highest, while a 90-percent duty cycle provides the lowest transmission pressure. The resistance of the PC solenoid is 6 ohms (plus or minus 1 ohm).

## The Torque Converter Clutch (TCC) Solenoid

The PCM energizes the torque converter clutch (TCC) solenoid and applies the TCC when the coolant temperature increases to 150°F (66°C), the transmission is in third or fourth gear, the brake switch is open, and the vehicle speed is above 40 mph (64 kmh) with a steady throttle opening. If the transmission temperature exceeds 248°F (120°C) the PCM commands TCC lockup.

The PCM varies the TCC solenoid duty cycle from 0 to 100 percent. At 0-percent duty cycle, the solenoid is closed and the TCC is released, while at 100-percent duty cycle the TCC solenoid is open and the TCC is applied. When the TCC solenoid is energized, the fluid passage through the solenoid is open and fluid pressure is supplied from this solenoid to the TCC apply valve in the valve body. The apply valve position directs fluid movement in the torque converter to supply the desired TCC mode. When the brake pedal switch closes, the TCC solenoid duty cycle is cut immediately to 0 percent.

## The Camshaft Position Actuator Control Solenoids

The camshaft position actuator control solenoids control engine oil flow to the camshaft position actuators. The PCM operates the camshaft position actuator solenoids with a duty cycle principle. If the solenoid duty cycle is more than 50 percent, the oil flow through the solenoids causes the camshaft actuators to advance the camshaft timing. When the solenoid duty cycle is less than 50 percent, the oil flow through the solenoids causes the actuators to retard the camshaft position. When the desired camshaft position is achieved, and the throttle opening is steady, the PCM operates these actuator solenoids at a 50-percent duty cycle to ensure the camshaft position is maintained. The camshaft position actuator control solenoids have a resistance of 12 ohms (plus or minus 2 ohms).

# PCM Operating Modes

## The Start Mode

When the driver turns on the ignition switch, the PCM energizes the fuel pump relay, and voltage is supplied through these relay contacts to the fuel pump. The PCM keeps the fuel pump relay energized as long as the engine cranking speed is above 100 rpm or the engine is running. If the engine is not running or cranking above 100 rpm, the PCM de-energizes the fuel pump relay after two seconds.

## The Fuel Cut-Off Mode

If the engine speed exceeds 6,000 rpm, or the vehicle speed is greater than 110 mph (68 kmh), the PCM shuts off the injectors to protect the engine.

## The Acceleration Enrichment Mode

During engine acceleration, the PCM senses the increase in airflow through the MAF. Under this condition, the PCM increases the injector pulse width to provide a richer air-fuel ratio. During wide open throttle operation, the PCM enters closed loop to provide a richer air-fuel ratio.

## The Deceleration Enleanment Mode

The PCM senses engine deceleration from the MAF sensor signal, and the decrease in vehicle speed from the vehicle speed sensor (VSS) signal. During deceleration, the PCM reduces the injector pulse width to provide a leaner air-fuel ratio and reduce exhaust emissions.

## The Clear Flood Mode

The PCM turns off the injectors if the throttle opening is greater than 80 percent and the engine speed is below 400 rpm. This action allows the driver to clear a flooded condition by pressing the accelerator pedal completely and cranking the engine.

## Open Loop and Closed Loop Modes

The PCM operates in open loop until the following conditions are present:
- The coolant temperature is above 68°F (20°C).
- The throttle opening is less than 80 percent.
- Ten seconds have elapsed since the engine was started.
- Both upstream $HO_2S$ and providing valid signals to the PCM.

In open loop, the PCM ignores the $HO_2S$ voltage signals and calculates the injector pulse width from the TPS, ECT, IAT, MAF, and CKP sensors. In addition, in open loop the PCM provides a richer air-fuel ratio. When the preceding conditions are present, the PCM enters closed loop. In this mode, the PCM uses the upstream $HO_2S$ voltage signals to control injector pulse width.

# OBD II Monitors

## Comprehensive Component Monitor

If a problem occurs in an OBD II system that results in emissions being 1.5 times higher than the emission standards for that model year, the PCM illuminates the MIL and stores DTC and freeze frame data. Freeze frame data is placed in the PCM memory when a problem occurs. The comprehensive component monitor includes all the PCM inputs and outputs. The comprehensive component monitor operates continually while the engine is running.

## System Monitors

The OBD II diagnostic system tests several systems for defects while the vehicle is driven. The diagnostic system continuously monitors engine misfire and fuel control. Thermostat operation, variable valve timing, EVAP system leaks, EGR operation, catalytic converter operation, $HO_2S$ heater operation, and $HO_2S$ voltage signals are tested once per trip. A trip may be defined as a key-on cycle in which all enable criteria for a particular diagnostic monitor are met and the monitor is run. The PCM monitors the following specific systems once per trip:
- *Engine misfire.* The PCM detects engine misfire from the CKP signal. If the misfire is severe enough to damage the catalytic converter, the PCM begins flashing the MIL.
- *Fuel control.* The PCM monitors short-term and long-term fuel trim to detect problems in the fuel control system. If the fuel trim numbers are above +30 percent or below –30 percent, the PCM interprets this as a problem.

- *Catalytic converter.* The PCM compares the upstream and downstream HO$_2$S sensors to detect catalytic converter problems. If these two signals are similar, the catalytic converter is not functioning properly.
- *EGR system.* When the PCM energizes the EGR solenoid and opens the EGR valve, the PCM looks for a change in the MAP sensor voltage. If this change does not occur, the PCM interprets this as an EGR system fault.
- *EVAP system.* The PCM energizes the EVAP purge solenoid and the EVAP vent solenoid. This action opens the purge solenoid and closes the vent solenoid to increase the vacuum in the EVAP system. The PCM then turns off the purge solenoid to seal the EVAP system. The PCM monitors the fuel tank pressure sensor signal to check for the proper EVAP system pressure. Reduced vacuum in the EVAP system indicates a system leak. The EVAP system monitor is only run when the engine coolant temperature is below 86°F (30°C) and the fuel tank is between one-quarter and three-quarters full.
- *HO$_2$S.* The PCM monitors the HO$_2$S for proper voltage signals and the correct number of lean to rich transitions.
- *HO$_2$S heaters.* The PCM monitors the time from cold engine start-up to when the HO$_2$S provides a proper voltage signal. If this time is too long, the PCM interprets this as a fault in the HO$_2$S heater.
- *Variable valve timing.* This monitor checks the camshaft position (CMP) sensor signals to compare the desired valve timing and the actual valve timing. If the valve timing is not correct, or requires too long to reach the desired value, a DTC is set in the PCM memory.
- *Engine thermostat.* If the desired engine coolant temperature is not reached in a specific length of time, a DTC is set in the PCM.

## One-Trip Monitors

Some monitors are referred to as one-trip monitors because a failure on one trip sets a DTC, turns on the MIL, and stores freeze frame data. All comprehensive component monitors are one-trip monitors.

## Two-Trip Monitors

Other monitors are designated as two-trip monitors because a problem has to occur on two consecutive trips to set a current DTC, turn on the MIL, and store freeze frame data. When the problem occurs during the first trip, a pending DTC is set in the PCM memory. If the problem occurs on the next trip during which the monitor runs, a confirmed DTC is set in the PCM or TCM and the MIL is illuminated. All the monitors, except the comprehensive component monitor, engine misfire, and fuel control, are two-trip monitors. If a problem does not reoccur on three consecutive trips under similar conditions of engine speed, load, and warm up, the MIL is turned off. However, the DTC and freeze frame data are still stored in the PCM. If the vehicle completes forty warm-up cycles without reoccurrence of the problem, the DTC and the freeze frame data are erased.

Freeze frame data is a "snapshot" of PCM data that is automatically stored in the PCM/TCM memory when a DTC is stored. The freeze frame data may be displayed on a scan tool, and this data is very useful when diagnosing momentary faults in the inputs and outputs.

A warm-up cycle is defined as the period during which the vehicle is started and the coolant temperature increases to at least 40°F (22°C), reaching a minimum of 160°F (71°C). A scan tool may be used to manually erase DTCs and freeze frame data.

## Monitor Readiness Status

The monitor readiness status may be displayed on a scan tool for all the monitors except the comprehensive component, fuel control, and engine misfire monitors. Readiness status indicators are not required for these three monitors because they are run continuously. The oxygen sensors, EVAP system oxygen sensor heaters, variable valve timing, catalytic converter, EGR system, and engine thermostat monitors have readiness status indicators. If a monitor has been run, the readiness status indicator for that monitor indicates "Complete." When a monitor has not run, the readiness status indicator displays "Incomplete."

## Trip and Drive Cycle

A trip may be defined as a key-on cycle during which all the enable criteria for a specific monitor are met and the monitor has run. When the ignition switch is turned off, the trip is completed. A drive cycle is completed when all the monitors on a vehicle have run. The vehicle must be driven under various load and speed conditions to allow all the monitors to run. The following driving conditions allow all the monitors to run.

1. Be sure the fuel tank is between one-quarter and three-quarters full.
2. Start the engine with the coolant temperature below 88°F (30°C) and run the engine until the engine temperature is at least 160°F (71°C).
3. Accelerate the vehicle slowly to 40 to 55 mph (64 to 88 kmh), and maintain this speed for five minutes.
4. Decelerate slowly to 20 mph (32 kmh) or less without using the brakes, and then stop the vehicle. Idle the engine for ten seconds, and then shut the engine off and wait one minute.
5. Start the engine and slowly accelerate to 40–55 mph (64–88 kmh) and maintain this speed for two minutes.
6. Decelerate slowly to 20 mph (32 kmh) or less without using the brakes. Stop the vehicle and allow the engine to idle for ten seconds, then turn the ignition switch off and wait one minute.

System defects prevent a monitor or monitors from running. One system defect (such as a problem in an engine coolant temperature (ECT) sensor) will cause several monitors not to run.

## Exhaust Emissions and Air-Fuel Ratio Relationships

During an I/M 240 emissions test, HC, CO, and NOx are measured in grams per mile (gpm). In this type of emission test, typical cutpoints are HC - 0.8, CO - 15, and NOx - 2.0. Cutpoints may vary in different states, and some states use more lenient cutpoints when I/M testing is initiated. Once the program is operational for a period of time, the cutpoints are reduced. When using an exhaust gas analyzer to test tailpipe emissions in the shop with the secondary air injection system disabled, typical satisfactory readings at idle should be: HC - less than 100 ppm, CO - less than 1 percent, NOx - less than 400 ppm, $CO_2$ - 10 to 12 percent, $O_2$ - less than 1.5 percent.

Typical tailpipe emission readings at 2,000 rpm should be: HC less than 75 ppm, CO - less than 0.8 percent, NOx - less than 600 ppm, $CO_2$ - 12 to 14 percent, $O_2$ - less than 1.5 percent.

The relationship between various air-fuel ratios and tailpipe emissions are provided in Table 9–1.

**AIR/FUEL RATIO**                 **EXHAUST EMISSIONS**

**Very Rich -Below 10:1 at all speeds**

| Engine Speed | HC | CO | $CO_2$ | $O_2$ |
|---|---|---|---|---|
| Idle | 250 ppm | 3% | 7 to 9% | 0.2% |
| Off Idle | 275 ppm | 3% | 7 to 9% | 0.2% |
| Cruise | 300 ppm | 3% | 7 to 9% | 0.2 % |

**Other Symptoms**    Black smoke of sulfur odor, poor fuel ecomony, surge or hesitation, stalling, rough or "lumpy" idle, engine not warming to operating temperature, continuous open-loop operation

**Possible Causes**
- High MAP sensor voltage (vacuum leak or electrical fault)
- Leaking fuel injectors
- High fuel pressure
- High float level of leaking power valve in a carburetor
- Thermostat stuck open or engine otherwise continuously operating at very low temperature

**AIR/FUEL RATIO**                 **EXHAUST EMISSIONS**

**Rich - 10:1 to 12:1 at low speed only**

| Engine Speed | HC | CO | $CO_2$ | $O_2$ |
|---|---|---|---|---|
| Idle | 150 ppm | 1.5% | 7 to 9% | 0.5% |
| Off Idle | 150 ppm | 1.5% | 7 to 9% | 0.5% |
| Cruise | 100 ppm | 1.0% | 11 to 13% | 1.0% |

**Other Symptoms**    Poor fuel economy, surge or hesitation, black smoke and soot-fouled spark plugs, rough idle, vapor canister saturated with fuel or purge valve bad

**Possible Causes**
- High MAP sensor voltage (vacuum leak or electrical fault)
- Leaking fuel injectors
- High fuel pressure
- High float level of leaking power valve in a carburetor
- Engine oil diluted with gasoline
- Thermostat stuck open or engine otherwise continuously operating at very low temperature
- Excessive crankcase blowby

**AIR/FUEL RATIO**                 **EXHAUST EMISSIONS**

**Normal - 13:1 to 15:1 but engine not fully warmed up**

| Engine Speed | HC | CO | $CO_2$ | $O_2$ |
|---|---|---|---|---|
| Idle | 100 ppm | 0.3% | 10 to 12% | 2.5% |
| Off Idle | 80 ppm | 0.3% | 10 to12% | 2.5% |
| Cruise | 50 ppm | 0.3% | 10 to 12% | 2.5% |

**Other Symptoms**    Cold engine emission test failure, Catalytic converter not warmed up

**Table 9–1** Relationship Between Air-Fuel Ratios and Tailpipe Emissions

| AIR/FUEL RATIO | EXHAUST EMISSIONS | | | |
|---|---|---|---|---|
| **Lean - Above 16:1 at high speed** | **Engine Speed** | **HC** | **CO** | **CO$_2$** | **O$_2$** |
| | Idle | 100 ppm | 2.5% | 7 to 9% | 2 to 3% |
| | Off Idle | 80 ppm | 1.0% | 7 to 9% | 2 to 3% |
| | Cruise | 50 ppm | 0.8% | 7 to 9% | 2 to 3% |

**Other Symptoms**    Rough idle, misfire, surging, hesitation

**Possible Causes**
- Intermittent ignition problems causing misfire
- Restricted fuel injectors
- Low fuel pressure
- Vacuum leak
- Low carburetor float level or lean carburetor mixture
- Carburetor heated air intake stuck in cold-air position

| AIR/FUEL RATIO | EXHAUST EMISSIONS | | | |
|---|---|---|---|---|
| **Very Lean - Above 16:1 at all speeds** | **Engine Speed** | **HC** | **CO** | **CO$_2$** | **O$_2$** |
| | Idle | 200 ppm | 0.5% | 7 to 9% | 4 to 5% |
| | Off Idle | 205 ppm | 0.5% | 7 to 9% | 4 to 5% |
| | Cruise | 250 ppm | 1.0% | 7 to 9% | 4 to 5% |

**Other Symptoms**    Rough idle, high-speed misfire, overheating, surging, hesitation, detonation at cruising speeds

**Possible Causes**
- Intermittent ignition problems causing misfire
- Restricted fuel injectors
- Low fuel pressure
- Vacuum leak
- Low carburetor float level or lean carburetor mixture
- Improper ignition timing
- Thermostat stuck closed or engine otherwise operating at very high temperature

**Table 9–1**  Relationship Between Air-Fuel Ratios and Tailpipe Emissions (continued)

# General Powertrain Diagnosis

## ASE Tasks, Questions, and Related Information

In this chapter, each task in the Advanced Engine Performance Specialist category is provided followed by a question and some information related to the task. If you answer any question incorrectly, study this information very carefully until you understand the correct answer. A question-and-answer analysis is provided at the end of this chapter.

## Task 1   Inspect and test for missing, modified, inoperative, or tampered powertrain mechanical components.

1. Installing exhaust headers in place of exhaust manifolds affects the operation of the
   A. catalytic converter.
   B. evaporative emission control system.
   C. coolant temperature sensor.
   D. exhaust gas recirculation (EGR) system.

*Hint*   *When inspecting or servicing a vehicle with electronic fuel injection (EFI), no modifications should be made to the EFI system or powertrain components. According to federal and state laws, it is illegal to modify or tamper with any component that affects emissions. Such action may result in severe penalties for the technician and the shop. If an engine is modified by installing exhaust headers in place of exhaust manifolds, many components in the EFI system are not affected. However, this modification changes the exhaust flow and the exhaust back pressure.*

*This action reduces EGR flow, which may cause increased combustion temperatures, high NOx emissions, and detonation.*

## Task 2   Locate relevant service information.

2. While discussing sources of vehicle information and specifications, Technician A says the vehicle manufacturer's service manual is the most up-to-date source of information and specifications.

   Technician B says the Vehicle Emission Control Information (VECI) label is the most up-to-date source of information and specifications.

   Who is correct?
   A. A only
   B. B only
   C. Both A and B
   D. Neither A nor B

*Hint*   *Vehicle manufacturers write and produce service manuals and this requires a considerable amount of time. Service manuals may be in book form or on CDs. Any changes in adjustments or service procedures are not included in the service manual until the next printing of the manual.*

*Vehicle manufacturers do include changes in adjustments or service procedures required by emission standards or drivability problems on the Vehicle Emission Control Information (VECI) label, and this label may be changed by authorized dealership personnel.*

*Service bulletins are a valuable source of information regarding updated components and/or service procedures to correct specific problems. These bulletins may be in printed form or contained in electronic media such as CDs or scan tool software.*

## Task 3   Research system operation using technical information to determine diagnostic procedure.

3. The injectors have been replaced in an EFI system to correct a rich air-fuel ratio caused by dripping injectors. After the new injectors are installed, the engine idle operation is still rough.

   Technician A says the next step should be to run some combustion chamber cleaner through the air intake with the engine running to remove combustion chamber deposits.

   Technician B says the next step should be to drive the vehicle to allow the PCM adaptive memory to relearn the system.

   Who is correct?
   A. A only
   B. B only

C. Both A and B

D. Neither A nor B

*Hint*    *The operation of an electronic system must be understood from technical information or from previous personal experience. After the system operation is understood and the vehicle problem is clearly identified, the proper diagnostic procedure must be followed to locate the exact cause of the problem.*

## Task 4   Use appropriate diagnostic procedures based on available vehicle data and service information; determine if available information is adequate to proceed with effective diagnosis.

4. All of the following statements about freeze frame and failure records data are true EXCEPT

A. Freeze frame data are stored for emission-related DTCs.

B. The PCM can store up to six freeze frame data records.

C. Failure records are stored for emission-related and other DTCs.

D. The PCM can store data from five failure records.

*Hint*    *The first step in a diagnostic procedure is to listen carefully to the customer's complaint and obtain as much information as possible from the customer. This information should include the vehicle repair history related to the complaint. If necessary, road test the vehicle to identify the complaint. After the complaint is identified on OBD II vehicles, the next diagnostic procedure is a powertrain on-board diagnostic (OBD) system check (Figure 9–10). This diagnostic procedure locates appropriate service bulletins, and uses a scan tool to check for items such as DTCs, freeze frame data, failure records, and PCM data. When DTCs or improper data are displayed, the powertrain on-board diagnostic (OBD) system check directs the technician to the next appropriate diagnostic procedure.*

*Freeze frame data indicating specific vehicle operating conditions are stored in the PCM when certain emission-related history DTCs are stored in the PCM. The PCM only stores one freeze frame record. Freeze frame data are stored for the first failed test that sets the DTC and illuminates the MIL. The PCM does not update freeze frame data if the test failure occurs a second time. Freeze frame data representing fuel trim and cylinder misfire DTCs overwrite other freeze frame data because these data have higher priority compared to data representing other DTCs.*

*Fail record data is similar to freeze frame data, but some PCMs have the capability to store data representing five DTC failure records. Whereas freeze frame data are limited to certain emission-related DTCs, failure record data are not. Failure records are only updated the first time a test fails during each ignition switch cycle. Freeze frame and failure records data help the technician to diagnose specific causes of DTCs.*

## Task 5   Establish relative importance of observed vehicle data.

5. A customer complains about reduced fuel economy on an OBD I vehicle, but the vehicle has no other drivability problems. The data shown in Figure 9–11 are obtained on a scan tool with the engine operating at normal temperature and idle speed. The cause of this problem could be

A. a malfunctioning MAP sensor.

B. an intake manifold vacuum leak.

C. a sticking EGR valve.

D. a malfunctioning IAT sensor.

*Hint*    *When diagnosing vehicle data to determine the cause of a drivability problem, always look at all the data on the scan tool. Some mechanical problems may cause improper data that may be confused with data from malfunctioning input sensors. For example, an intake manifold vacuum*

| | | YES | NO |
|---|---|---|---|
| A | 1. Connect a scan tool to the data link connector.<br><br>2. Turn the key "ON," with the engine "OFF."<br><br>Program the scan tool for the vehicle.<br>Does the scan tool display data? | Go to step C | Go to step B |
| B | Select "Diagnostic Circuit Check,"<br>Class 2 Message Monitor.<br><br>Does the scan tool show other computers as active? | Go to<br>NO Scan<br>Tool Data | Go to<br>Body and<br>Accessories |
| C | Does the engine start and continue to run? | Go to step D | Go to<br>Engine Cranks<br>Does Not Run |
| D | Does the engine continue to run after the vehicle<br>has exceeded 3.2 km/h (2 mph) ? | Go to step E | Go to<br>Body and<br>Accessories |
| E | Are DTCs stored in memory?<br><br>Priority for diagnosis<br><br>     Fuel enable DTCs (P1626, P1631)<br>     PCM malfunction DTCs (P0601, P0602)<br>     System voltage DTCs (P0560, P1635, P1639)<br>     Component level DTCs (switches, sensors, ODMs)<br>     System level DTCs (fuel trim, misfire, EGR flow,<br>             TWC system, EVAP system,<br>             idle control system, and<br>             $HO_2S$ response/transition) | Go to<br>applicable<br>DTC | Go to step F |
| F | Observe the DTC information. If DTC status "Last Test<br>Failed," Test Failed, Test Failed This Ignition, MIL<br>Request, or History DTCs are set, save DTC Freeze<br>Frame and/or Failure Records information.<br><br>Were any DTCs displayed? | Go to the<br>applicable<br>DTC table | Go to step G |
| G | Compare scan data with the values shown in the "Engine<br>Scan Tool Data list."<br><br>Are the values normal or within typical ranges? | Go to<br>Symptoms | Go to<br>Diagnostic Aids<br>and Test<br>Descriptions |

**Figure 9–10** Powertrain on-board diagnositic (OBD) system check.

*leak causes the manifold absolute pressure (MAP) sensor voltage to be higher than specified, and the heated oxygen sensor ($HO_2S$) voltage to be lower than specified.*

## Task 6   Differentiate between powertrain mechanical and electrical/electronic problems, including variable valve timing (VVT) systems.

6. The composite vehicle misfires on two cylinders at idle speed and during acceleration. When an oscilloscope is connected to the engine, cylinders 4 and 5 have very low firing voltages.

## SCAN TOOL DATA

| Engine Coolant Temperature Sensor (ECT)<br>0.42　Volts | Intake Air Temperature Sensor (IAT)<br>2.24　Volts | Manifold Absolute Pressure Sensor (MAP)<br>2.75　Volts | Throttle Position Sensor (TPS)<br>0.6　Volts |
|---|---|---|---|
| Engine Speed Sensor (RPM)<br>750　rpm | Heated Oxygen Sensor ($HO_2S$)<br>0.3–0.7　Volts | Vehicle Speed Sensor (VSS)<br>0　mph | Battery Voltage (B+)<br>14.2　Volts |
| Idle Air Control Valve (IAC)<br>30　percent | Evaporative Emission Canister Solenoid (EVAP)<br>OFF | Torque Converter Clutch Solenoid (TCC)<br>OFF | EGR Valve Control Solenoid (EGR)<br>0　percent |
| Malfunction Indicator Lamp (MIL)<br>OFF | Diagnostic Trouble Codes<br>NONE | Open/Closed Loop<br>CLOSED | Fuel Pump Relay (FP)<br>ON |

Measured Ignition Timing °BTDC　　Base Timing: 10°　Actual Timing: 11°

**Figure 9–11**　OBD I scan tool data number 11.

Technician A says spark plugs 4 and 5 may be fouled.

Technician B says the head gasket may be blown between number 4 and 5 cylinders. Who is correct?

A. A only

B. B only

C. Both A and B

D. Neither A nor B

*Hint*　　*Low ignition firing voltages on the cylinders may be caused by electrical leakage in the distributor cap or spark plug wires on engines with distributor ignition (DI). This problem may also be caused by fouled or bad spark plugs or low cylinder compression. On engines with electronic ignition (EI), low firing voltages may be caused by fouled or bad spark plugs and low cylinder compression. If two cylinders with low firing voltages are adjacent to each other, the low firing voltages may be caused by a blown head gasket between two cylinders.*

## Task 7　Diagnose engine mechanical condition using an exhaust gas analyzer.

7. The composite vehicle has a rough idle problem and experiences reduced fuel economy. The emission readings are shown in Figure 9–12. All of the following may be the cause of the problem EXCEPT

| Engine Speed | Idle | 2000 RPM |
|---|---|---|
| HC (ppm) | 900 | 800 |
| CO (percent) | 0.3 | 0.3 |
| $CO_2$ (percent) | 10 | 10 |
| $O_2$ (percent) | 0.8 | 0.8 |

**Figure 9–12**　Exhaust emission data number 12.

A. bad spark plug wires.

B. bad spark plugs.

C. low cylinder compression.

D. dripping injectors.

*Hint*    *Any problem that causes cylinder misfiring causes high hydrocarbons (HCs), low carbon monoxide (CO), low carbon dioxide ($CO_2$), and high oxygen ($O_2$) emissions. Because CO is a byproduct of combustion, and less combustion occurs during cylinder misfiring, CO does not increase when cylinder misfiring occurs. A rich air-fuel ratio causes high HC, high CO, low $CO_2$, and low $O_2$ emissions.*

## Task 8   Diagnose drivability problems and emission failures caused by cooling system problems.

8. The composite vehicle has an overheating problem and an inoperative cooling fan, but there are no other problems or drivability complaints. When voltage is supplied directly to the cooling fan motor, this motor operates normally.

   Technician A says to test for an open circuit in fuse number 3.

   Technician B says to test for an open circuit between the cooling fan relay and PCM terminal 6.

   Who is correct?

   A. A only

   B. B only

   C. Both A and B

   D. Neither A nor B

*Hint*    *Engine overheating may be caused by a loose water pump drive belt, an engine thermostat sticking closed, plugged radiator tubes, restricted radiator air passages, a blown head gasket, reduced spark advance, or worn water pump impeller blades. Excessive engine operating temperature increases NOx emissions. A thermostat that is sticking open reduces engine temperature, and this affects the engine coolant temperature (ECT) sensor signal and results in a rich air-fuel ratio with high CO and HC emissions. In many cooling systems, the PCM grounds the cooling fan relay winding at a specific engine temperature and closes the relay contacts. Voltage is supplied through these relay contacts to the cooling fan motor.*

## Task 9   Diagnose drivability problems and emission failures caused by engine mechanical problems.

9. An OBD I vehicle has a stalling problem on deceleration with the A/C on. When the problem occurs, snapshot data indicate the knock sensor senses spark knock, and the spark advance drops from 18 degrees to 0. When the engine is running at 1,500 rpm in the shop and the exhaust manifold is tapped with a small hammer, the spark advance decreases and then returns to normal. The MOST likely cause of this problem could be

   A. an inoperative knock sensor.

   B. an open circuit in the knock sensor signal wire.

   C. a malfunctioning PCM.

   D. a broken engine mount.

*Hint*    *Low cylinder compression and misfiring causes high HC and low CO emissions. On OBD II vehicles, freeze frame and failure records data are helpful in diagnosing the cause of DTCs and drivability problems, because the recorded data occurred at the time the DTC was set and the problem occurred. On OBD I vehicles, a scan tool may be used to take snapshot data when the problem occurs to help diagnose the cause of the problem.*

## Task 10   Diagnose drivability problems and emission failures caused by problems or modifications in the transmission and final drive or by incorrect tire size.

10. The transmission in the composite vehicle operates only in second gear and reverse. The MOST likely cause of this problem could be
   A. an open circuit in the wires to shift solenoids 1 and 2.
   B. an open circuit in fuse number 4.
   C. an open circuit in the winding of shift solenoid 1.
   D. a shorted condition in the winding of shift solenoid 2.

**Hint**   *In many PCM-controlled transmissions/transaxles, if both shift solenoids are inoperative, the transmission operates only in second gear and reverse, because second gear is obtained by de-energizing both shift solenoids. Reverse is obtained by the gear selector and manual valve position in the transmission. If the differential ratio is changed, the vehicle speed sensor (VSS) signal is modified, and this changes the torque converter clutch (TCC) lockup time. A modified VSS signal also changes the cruise control operation so the cruise control does not hold the vehicle speed at the selected cruise control speed. On some light-duty trucks, the digital ratio adapter controller (DRAC) module in the instrument panel may be changed or reprogrammed to match various differential ratios or tire sizes.*

## Task 11   Diagnose drivability problems and emission failures caused by exhaust system problems or modifications.

11. The commanded or desired EGR valve opening and the actual EGR valve opening are shown in Figure 9–13. The EGR test results indicated in Figure 9–13 would cause
   A. low combustion chamber temperature.
   B. low NOx emissions.
   C. high NOx emissions.
   D. cylinder misfire.

| Scan Tool Parameter | Units Displayed | Measured Data Value |
|---------------------|-----------------|---------------------|
| Desired EGR         | Percent         | 60%                 |
| Actual EGR          | Percent         | 40%                 |

**Figure 9–13** EGR data.

**Hint**   *If an exhaust system is modified so it has less back pressure, this reduced back pressure forces less exhaust gas through the EGR system into the intake manifold. Reduced EGR flow may cause higher combustion temperatures, engine detonation, and high NOx emissions. Excessive exhaust back pressure reduces airflow through the engine and this reduces engine power. High NOx emissions may be caused by any problem in the EGR system that reduces EGR flow. One of the most common problems in an EGR system is carbon buildup in the EGR passages that reduces EGR flow.*

*On some engines, the PCM supplies a pulse width modulated (PWM) signal to the EGR solenoid, which supplies the proper amount of vacuum to the EGR valve to provide the precise valve opening required by the engine. Other engines have a linear EGR valve that is controlled electronically by the PCM. On some engines with a linear EGR valve, a scan tool may be used to command the PCM to supply various EGR valve openings. The scan tool also indicates the actual EGR valve opening supplied by the PCM. There should be very little difference between the commanded and actual EGR valve openings.*

## Task 12   Determine root cause of failures.

12. An OBD II vehicle with a V6 SFI engine fails an enhanced emission test for HC and CO. The SFI system has a fuel pressure regulator on the fuel rail with a vacuum hose connected to the regulator. The oil level in the crankcase is considerably above the full mark on the dipstick. When tested in the shop the following emission readings were obtained: HC 190 ppm, CO 2.3 percent, $CO_2$ 11.2 percent, $O_2$ 0.2 percent. The "root cause" of the problem could be

A. the crankcase oil is contaminated with gasoline.

B. a bad PCV valve.

C. excessive engine blowby.

D. a leaking fuel pressure regulator diaphragm.

**Hint**   *High HC and CO emissions with low $CO_2$ and $O_2$ emissions indicate a rich air-fuel ratio.*
*Some of the causes of a rich air-fuel ratio on a fuel-injected engine include excessive fuel pressure, dripping injectors, inoperative input sensor signals, a restricted PCV valve, or a damaged pressure regulator diaphragm.*

## Task 13   Determine the root cause of multiple component failures.

13. An OBD II vehicle is equipped with a V6 engine and coil-on-plug ignition with coil driver modules mounted on each ignition coil. The vehicle experiences simultaneous failure of the PCM and two coil driver modules. The charging system voltage is within the manufacturer's specifications. The cause of this problem could be

A. high resistance in the PCM ground.

B. an intermittent open circuit at the alternator battery terminal.

C. excessively wide spark plug gaps.

D. low resistance in some of the ignition coil primary windings.

**Hint**   *Multiple component failures may be caused by high charging system voltage, or high induced voltage peaks. An intermittent open circuit between the alternator battery terminal and the positive battery terminal may cause high voltage peaks in the electrical system. High induced voltage peaks may also come from solenoid windings with damaged voltage suppression diodes. Each time the current flow is shut off in a solenoid winding, an induced voltage occurs in the winding from the magnetic collapse around the winding. The voltage suppression diode normally dissipates these induced voltages. Voltage suppression diodes are typically located in some starter solenoids and A/C compressor clutch windings. If the voltage suppression diode is damaged, the induced voltage may be applied to, and damage, other electronic system components. High induced voltages may also originate from components with a pulsating current such as instrument voltage limiters. These components usually have a suppression choke or capacitor to dissipate induced voltages.*

## Task 14   Determine root cause of repeated component failures.

14. The composite vehicle experiences repeated PCM failure to correct an inoperative cooling fan.
Technician A says the cooling fan relay winding should be tested for a shorted condition.
Technician B says the cooling fan motor should be tested for low resistance.
Who is correct?

A. A only

B. B only

C. Both A and B

D. Neither A nor B

**Hint**   *Repeated component failure of electronic components may be caused by a shorted condition in solenoid or relay windings. A shorted condition in a winding causes less resistance in*

*the winding and increased current flow. This increased current may damage other compo-
nents in the circuit. Repeated failure of input sensors is usually caused by something that
affects the sensor operation. For example, the O₂ sensor may be contaminated by engine
coolant coming from leaking head gaskets, or excessive RTV sealant applied to gaskets during
engine repairs.*

# Computerized Powertrain Controls Diagnosis—Including OBD II

## ASE Tasks, Questions, and Related Information

**Task 1**   Inspect and test for missing, modified, inoperative, or tampered computerized powertrain control components.

15. A customer complains about excessive idle rpm on an OBD II vehicle. The data in Figure 9–14 were obtained on a scan tool with the engine operating at normal temperature and idle speed. The cause of the high idle rpm could be

| SCAN TOOL DATA | | | |
|---|---|---|---|
| Engine Coolant Temperature Sensor (ECT)<br>0.46    Volts | Intake Air Temperature Sensor (IAT)<br>2.24    Volts | MAP Sensor    1.8 V<br>MAF Sensor    1.1 V | Throttle Position Sensor (TPS)<br>0.6    Volts |
| Engine Speed Sensor (RPM)<br>1500    rpm | Heated O₂ HO₂S<br>Upstream    0.2–0.5 V<br>Downstream 0.1–0.3 V | Vehicle Speed Sensor (VSS)<br>0    mph | Battery Voltage (B+)<br>14.2    Volts |
| Idle Air Control Valve (IAC)<br>0    percent | Evaporative Emission Canister Solenoid (EVAP)<br>OFF | Torque Converter Clutch Solenoid (TCC)<br>OFF | EGR Valve Control Solenoid  (EGR)<br>0    percent |
| Malfunction Indicator Lamp (MIL)<br>OFF | Diagnostic Trouble Codes<br>NONE | Open/Closed Loop<br>CLOSED | Fuel Pump Relay (FP)<br>ON |
| Fuel Level Sensor<br>3.5 V | Fuel Tank Pressure Sensor<br>2.0 V | Transmission Fluid Temperature Sensor<br>0.6 V | Transmission Turbine Shaft Speed Sensor<br>0 mph |
| Transmission Range Switch<br>PARK | Transmission Pressure Control Solenoid<br>80% | Transmission Shift Solenoid 1<br>ON | Transmission Shift Solenoid 2<br>OFF |
| Measured Ignition Timing °BTDC    Base Timing:  10    Actual Timing:  20 | | | |

**Figure 9–14** OBD scan tool data number 14.

A. the EGR vacuum hose is plugged with a ball bearing.

B. the PCM has fully seated the IAC motor based on the engine coolant temperature signal.

C. the IAT sensor has excessive resistance or a loose electrical connection.

D. the EVAP solenoid vacuum hose has been disconnected from the intake manifold.

**Hint**   *A vacuum leak in the intake manifold causes a lean air-fuel ratio and a low upstream HO$_2$S signal. The PCM reacts to this signal by increasing the injector pulse width and injecting more fuel, and this action increases idle rpm. Under this condition, the PCM moves the IAC motor toward the seated position to try and decrease the idle rpm.*

## Task 2   Locate relevant service information.

16. While diagnosing PCM systems, Technician A says the DTCs may be used to locate the exact cause of a problem.

    Technician B says diagnostic charts in the vehicle manufacturer's service manual should be followed to locate the exact cause of a problem.

    Who is correct?

    A. A only

    B. B only

    C. Both A and B

    D. Neither A nor B

**Hint**   *When diagnosing electronic systems, DTCs indicate a problem in a certain area. For example, a DTC representing a transmission fluid temperature (TFT) sensor may indicate a problem in this sensor or in the wires connected from the sensor to the PCM. In some cases, a DTC may represent a mechanical problem. DTC P1442 indicates a leak in the EVAP control system on some vehicles. This DTC may be caused by a disconnected or damaged EVAP hose or a loose gasoline tank filler cap. Troubleshooting charts in the vehicle manufacturer's service manual are designed to help the technician locate the exact cause of a problem. Diagnosing a problem with these charts often involves voltmeter or ohmmeter tests to pinpoint the cause of the problem.*

## Task 3   Research system operation using technical information to determine diagnostic procedure.

17. A four-cylinder engine is equipped with multiport fuel injection in which pairs of injectors are connected to the PCM with a common wire on the ground side of the injectors. The engine is misfiring intermittently on two cylinders, but the cylinder compression and ignition system are satisfactory. Noid lights connected to all the injector terminals flash while cranking the engine.

    Technician A says the injector windings may be shorted.

    Technician B says the wire from the ground side of two injectors to the PCM may have a continual open circuit.

    Who is correct?

    A. A only

    B. B only

    C. Both A and B

    D. Neither A nor B

**Hint**   *Refer to the available technical information to become familiar with the system operation. When the system operation is understood, determine the proper diagnostic procedure to locate the root cause of the problem.*

**Task 4**   Use appropriate diagnostic procedures based on available vehicle data and service information; determine if available information is adequate to proceed with effective diagnosis.

18. The composite vehicle starts and runs normally. When the scan tool is connected to the DLC on the composite vehicle, the scan tool is inoperative. The cause of this problem could be

A. an open circuit at DLC terminal 5.

B. a short to ground at PCM terminal 79.

C. an open circuit at DLC terminal 16.

D. a malfunctioning PCM.

**Hint**   *On OBD II vehicles, one terminal on the DLC is supplied with 12 V from the battery. When a scan tool is connected to the DLC, this terminal supplies voltage to the scan tool. The scan tool is completely inoperative if it is not supplied with voltage. A second terminal on the DLC is connected to the serial data terminal on the PCM. The PCM supplies data to this terminal so these data may be displayed on the scan tool.*

**Task 5**   Determine current version of computerized powertrain control system software and updates; perform programming procedures.

19. A customer complains that his or her OBD II vehicle has an acceleration stumble once the engine is warmed up. The engine idles smoothly and has no other drivability complaints. The scan-tool data in Figure 9–15 were obtained during hard acceleration on a road test. The MOST likely cause of the acceleration stumble is

| SCAN TOOL DATA | | | |
|---|---|---|---|
| Engine Coolant Temperature Sensor (ECT)<br>0.44    Volts | Intake Air Temperature Sensor (IAT)<br>2.21    Volts | MAP Sensor    2.0 V<br>MAF Sensor    3.5 V | Throttle Position Sensor (TPS)<br>3.8    Volts |
| Engine Speed Sensor (RPM)<br>2400    rpm | Heated O₂ HO₂S<br>Upstream  0.3–0.8 V<br>Downstream 0.1–0.3 V | Vehicle Speed Sensor (VSS)<br>65    mph | Battery Voltage (B+)<br>14.4    Volts |
| Idle Air Control Valve (IAC)<br>30    percent | Evaporative Emission Canister Solenoid (EVAP)<br>ON | Torque Converter Clutch Solenoid (TCC)<br>ON | EGR Valve Control Solenoid  (EGR)<br>60    percent |
| Malfunction Indicator Lamp (MIL)<br>OFF | Diagnostic Trouble Codes<br>NONE | Open/Closed Loop<br>CLOSED | Fuel Pump Relay (FP)<br>ON |
| Fuel Level Sensor<br>3.5 V | Fuel Tank Pressure Sensor<br>2.1 V | Transmission Fluid Temperature Sensor<br>0.6 V | Transmission Turbine Shaft Speed Sensor<br>65 mph |
| Transmission Range Switch<br>OVERDRIVE | Transmission Pressure Control Solenoid<br>20% | Transmission Shift Solenoid 1<br>OFF | Transmission Shift Solenoid 2<br>ON |
| Measured Ignition Timing °BTDC    Base Timing:  10    Actual Timing:  36 | | | |

**Figure 9–15** OBD scan tool data number 15.

A.  an EGR valve that is sticking open.

B.  an inoperative upstream HO$_2$S sensor.

C.  a partially plugged MAP sensor hose.

D.  an inoperative TPS sensor.

**Hint**   *During hard acceleration, the TPS voltage increases. The MAP sensor voltage signal should increase as the intake manifold vacuum decreases. The injector pulse width should increase during hard acceleration as the PCM provides a richer air-fuel ratio. This richer air-fuel ratio should cause an increase in the upstream HO$_2$S voltage.*

## Task 6   Research OBD II system operation to determine the enable criteria for setting and clearing diagnostic trouble codes (DTCs), and malfunction indicator lamp (MIL) operation.

20.  When diagnosing an OBD II vehicle, Technician A says the PCM has to sense an engine misfire defect on two consecutive trips to set a DTC.

Technician B says a flashing MIL may indicate an engine misfire defect.

Who is correct?

A.  A only

B.  B only

C.  Both A and B

D.  Neither A nor B

**Hint**   *When the problem occurs during the first trip in a two-trip monitor, a pending DTC is set in the PCM memory and freeze frame data is stored. If the problem occurs on the next trip during which the monitor runs, a confirmed DTC is set in the PCM or TCM and the MIL is illuminated. All the monitors, except the comprehensive component monitor, are two-trip monitors. If a problem does not reoccur on three consecutive trips under similar conditions of engine speed, load, and warm up, the MIL is turned off. However, the DTC and freeze frame data are still stored in the PCM. If the vehicle completes forty warm-up cycles without reoccurrence of the problem, the DTC and the freeze frame data are erased.*

## Task 7   Interpret OBD II scan-tool data stream, diagnostic trouble codes (DTCs), freeze frame data, system monitors, monitor readiness indicators, and trip and drive cycle information to determine system condition and verify repair effectiveness.

21.  When diagnosing an OBD II vehicle with an electronically operated EGR valve, the scan tool indicates "Not Complete" on the EGR system monitor. All the other monitors indicate "Complete" and the vehicle was driven under the proper conditions to complete all the monitors.

Technician A says there may be a defect in the EGR valve wiring.

Technician B says there may be a defect in the ECT sensor.

Who is correct?

A.  A only

B.  B only

C.  Both A and B

D.  Neither A nor B

**Hint**   *Specific enable criteria must be met before each monitor will run. The enable criteria for each monitor is available in the vehicle manufacturer's service information. For example, if the ECT sensor is not sending a satisfactory voltage signal to the PCM, many of the monitors will not run. If the vehicle has been driven under the proper conditions to complete the monitors, and a monitor(s) does not run, the technician must check the enable criteria to determine the cause of the problem.*

## Task 8    Establish relative importance of displayed scan-tool data.

22. A customer complains about engine surging at 45 mph (72 kmh). An oscilloscope diagnosis indicates the ignition system is in satisfactory condition.

    Technician A says the TPS may be the problem.

    Technician B says there may be problems in the TCC hydraulic or electronic system.

    Who is correct?

    A. A only

    B. B only

    C. Both A and B

    D. Neither A nor B

**Hint**    *Engine surging just after the torque converter clutch (TCC) locks up may be caused by ignition problems, a lean air-fuel ratio caused by restricted injector orifices or low fuel pressure, an inoperative TPS, or problems in the TCC hydraulic or electronic system.*

## Task 9    Differentiate between electronic powertrain control problems and mechanical problems.

23. A customer complains about engine stalling at idle on his or her OBD I vehicle. The IAC data indicates 3 to 8 percent, but all other data are normal. The customer informed the technician that the PCM had been recently replaced to correct another problem.

    Technician A says to clean and test the IAC motor.

    Technician B says to perform an idle learn procedure.

    Who is correct?

    A. A only

    B. B only

    C. Both A and B

    D. Neither A nor B

**Hint**    *On many OBD I and OBD II vehicles, an idle learn procedure must be performed if components such as the PCM or idle speed control motor have been replaced, or if battery power has been disconnected from the PCM. If this procedure is not performed, the PCM does not control the idle speed properly, and the engine may have a stalling problem at idle and low speeds. Under this condition, the IAC motor data on the scan tool are not within specifications. Always question the customer regarding service work performed previously on the vehicle to determine if the PCM or IAC motor have been replaced or if battery power may have been disconnected from the PCM. An idle learn procedure may be performed with a scan tool or by a manual procedure. If the crankshaft position (CKP) sensor, reluctor wheel, or engine block are replaced on some OBD II vehicles, a CKP system variation learn procedure must be performed. When this procedure is not performed, a false misfire DTC is set in the PCM and the MIL is illuminated with the engine running. On many vehicles, a replacement PCM must be programmed with a scan tool or PC before the engine will start.*

## Task 10   Diagnose no-starting, hard starting, stalling, engine misfire, poor drivability, incorrect idle speed, poor idle, hesitation, surging, spark knock, power loss, poor mileage, illuminated MIL, and emission problems caused by failures of computerized powertrain controls.

24. When diagnosing a vehicle, the technician obtains a long-term fuel trim of +30. The root cause of this problem could be

    A. excessive fuel pressure.

    B. a restricted fuel return line.

C. a continually high upstream HO$_2$S voltage signal.

D. restricted injector orifices.

**Hint**   *Short-term fuel trim and long-term fuel trim numbers displayed on the scan tool are often useful when diagnosing engine performance problems. When the fuel trim numbers are high, the PCM is continually adding more fuel by increasing the injector pulse width. Therefore, the PCM must be receiving a signal that indicates the air-fuel ratio is always lean. If the fuel trim numbers are low, the PCM is continually leaning the air-fuel ratio by decreasing the injector pulse width. Under this condition, the PCM must be receiving an input signal that indicates a continually rich air-fuel ratio.*

## Task 11   Diagnose failures in the data communication bus network; determine needed repairs.

25. When a scan tool is connected to the composite vehicle, a data line DTC is displayed on the scan tool indicating a data link defect.

    Technician A says this defect will prevent the engine from starting.

    Technician B says this defect causes the data link network to be disabled.

    Who is correct?

    A. A only

    B. B only

    C. Both A and B

    D. Neither A nor B

**Hint**   *The data high circuit switches between 2.5V in the at-rest state and 3.5V in the active state. The data low circuit switches between 2.5V in the at-rest state and 1.5V in the active state. The serial data bus network uses two 120-ohm terminating resistors. One of these resistors is mounted in the instrument cluster and the second terminating resistor is mounted in the PCM. If either of the data lines short to ground, to each other, or to battery voltage, network DTCs are stored in the PCM and the network is disabled. When the serial data bus is disabled, the PCM continues to provide ignition and fuel control.*

## Task 12   Diagnose failures in the anti-theft/immobilizer system; determine needed repairs.

26. The composite vehicle starts and immediately stalls. The technician diagnoses the problem as a defective immobilizer control module, and this module is replaced. The technician now must

    A. reprogram the PCM.

    B. reprogram the TCM.

    C. reprogram the data link network.

    D. program the immobilizer control module.

**Hint**   *When the ignition switch is turned on, the immobilizer control module transmits a voltage signal through the antenna around the ignition switch to the transponder in the ignition key. The transponder key responds to this voltage signal with an encrypted key code. The immobilizer control module interprets the key code and compares it to the list of registered keys. When the engine starts, the PCM requests validation of the key code from the immobilizer control module via the data links. If the key code is valid, the immobilizer control module sends a valid key signal to the PCM and the PCM continues the normal engine operation. When the PCM does not receive a valid key code signal within two seconds after the engine starts, the PCM disables the fuel injectors to shut off the engine. If the ignition switch is cycled off and the engine is cranked again, the engine restarts and immediately stalls. Unique encrypted key codes are programmed into the PCM and immobilizer control module during the vehicle manufacturing process. If the PCM or the immobilizer control module are replaced, a scan tool must be used to program the replacement module using the vehicle identification number (VIN), the date,* *and a factory-assigned PIN number. A maximum of eight keys may be registered in the immobilizer control module.*

## Task 13  Perform voltage drop tests on power circuits and ground circuits.

27. When the fuel system injectors in Figure 9–16 are tested with a DSO, all the injector waveforms are 5 V above ground potential when the injectors are turned on.

    Technician A says to test the wires from the injectors to the PCM for high resistance.

    Technician B says to test the PCM ground for high resistance.

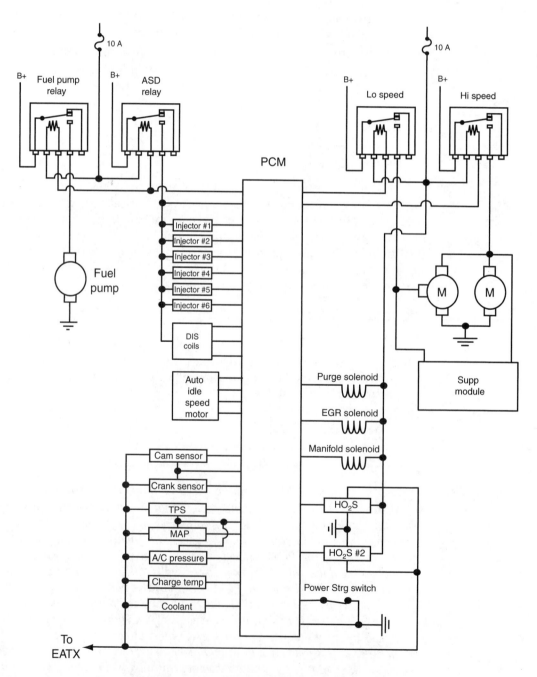

**Figure 9–16** Sequential fuel injection (SFI) system.

Who is correct?

A. A only

B. B only

C. Both A and B

D. Neither A nor B

**Hint**   *When an injector is shut off, the injector waveform on a digital storage oscilloscope (DSO) should be 14 V above ground. If the injector is turned on, the waveform should be very close to ground potential. When the injector waveform is considerably above ground potential with the injector turned on, there is high resistance somewhere from the ground side of the injector through the PCM to ground.*

*Excessive resistance in a PCM ground may cause a variety of problems, including improper operation of the output devices. Extremely high resistance or an open circuit in a PCM ground circuit may cause a no-start condition. The resistance in a PCM ground circuit may be tested by measuring the voltage drop from the PCM ground terminal or terminals to the battery ground with the engine running.*

## Task 14  Perform current flow tests on system circuits.

28. A port-fuel-injected vehicle has a lack of power complaint and the fuel pressure is less than specified. There are no restrictions in the fuel supply line or filter. The voltage supplied to the fuel pump with the engine running is 9.5 V and the voltage drop across the fuel pump ground is 0.1 V. The alternator voltage is 14.2 V, and the fuel pump motor current draw is less than specified. To correct this problem, the technician should:

    A. replace the fuel pump.

    B. repair the resistance problem in the fuel pump ground.

    C. repair the resistance problem in the voltage supply wire to the fuel pump.

    D. perform an alternator output test.

**Hint**   *Low current flow indicates a high resistance or low voltage problem in the circuit. If the current flow is higher than specified, the circuit may have a shorted condition, or the voltage applied to the circuit may be excessive. Current flow readings may be helpful in diagnosing electrical problems. For example, if the voltage supplied to a cooling fan motor is normal and the motor ground circuit resistance is satisfactory, a low current draw indicates high resistance in the motor.*

*High current flow in a primary ignition circuit may indicate a shorted primary coil winding. This condition may cause damage to the ignition module, because the primary current flows through the primary coil winding and the module.*

## Task 15  Perform continuity/resistance tests on system circuits and components.

29. When tested with an ohmmeter, a port fuel injector has a 9-ohms resistance. The specified injector resistance is 15 ohms.

    Technician A says the injector winding is shorted.

    Technician B says this condition allows excessive current flow through the injector.

    Who is correct?

    A. A only

    B. B only

    C. Both A and B

    D. Neither A nor B

**Hint**   *A voltmeter should be used to measure voltage and voltage drop in an electric circuit. Excessive circuit resistance results in too much voltage drop and reduced current flow. An ohmmeter may be used to measure a shorted condition within a coil of wire such as a solenoid winding. If a coil* *is shorted, some of the turns in the coil are touching each other. Under this condition, the coil resistance is reduced, and current flow through the coil increases. When using an ohmmeter, always be sure the battery voltage is disconnected from the circuit. If an ohmmeter is connected to a circuit to which voltage is supplied, the fuse in the meter may be blown or the meter may be damaged.*

## Task 16  Test input sensor/sensor circuit using scan-tool data and/or waveform analysis.

30. An OBD II vehicle fails an enhanced emission test for HC and CO. The data in Figure 9–17 were displayed on a scan tool with the engine running at normal operating temperature and idle speed. The cause of the emission failure could be

| SCAN TOOL DATA | | | |
|---|---|---|---|
| Engine Coolant Temperature Sensor (ECT)<br>0.44  Volts | Intake Air Temperature Sensor (IAT)<br>2.19  Volts | MAP Sensor   1.5 V<br><br>MAF Sensor   0.8 V | Throttle Position Sensor (TPS)<br>0.6  Volts |
| Engine Speed Sensor (RPM)<br>750  rpm | Heated O₂ HO₂S<br>Upstream  0.3–0.8 V<br>Downstream 0.3–0.8 V | Vehicle Speed Sensor (VSS)<br>0  mph | Battery Voltage (B+)<br>14.4  Volts |
| Idle Air Control Valve (IAC)<br>30  percent | Evaporative Emission Canister Solenoid (EVAP)<br>OFF | Torque Converter Clutch Solenoid (TCC)<br>OFF | EGR Valve Control Solenoid  (EGR)<br>0  percent |
| Malfunction Indicator Lamp (MIL)<br>OFF | Diagnostic Trouble Codes<br>NONE | Open/Closed Loop<br>CLOSED | Fuel Pump Relay (FP)<br>ON |
| Fuel Level Sensor<br>3.5 V | Fuel Tank Pressure Sensor<br>2.0 V | Transmission Fluid Temperature Sensor<br>0.6 V | Transmission Turbine Shaft Speed Sensor<br>0 mph |
| Transmission Range Switch<br>PARK | Transmission Pressure Control Solenoid<br>80% | Transmission Shift Solenoid 1<br>ON | Transmission Shift Solenoid 2<br>OFF |
| Measured Ignition Timing °BTDC   Base Timing:  10   Actual Timing:  12 | | | |

**Figure 9–17** OBD II scan tool data number 17.

    A. an inoperative upstream HO₂S sensor.

    B. an inoperative MAP sensor.

    C. an inoperative downstream HO₂S sensor.

    D. an inoperative catalytic converter.

31. When testing a TPS, the DSO voltage trace in Figure 9–18 is obtained. This voltage trace indicates

    A. normal TPS operation throughout the throttle operating range.

    B. an open circuit in the TPS that requires TPS replacement.

    C. a low reference voltage supplied to the TPS from the PCM.

    D. a high resistance in the ground wire between the TPS and the PCM.

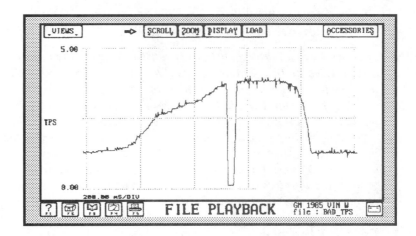

**Figure 9–18** TPS voltage trace.

**Hint**      *Scan-tool data can be very useful when diagnosing PCM system problems. Normal data readings may be obtained from the vehicle manufacturer's service manual. The technician must be familiar with the causes of abnormal data to diagnose electronic systems. A DSO is the most accurate method of testing many input sensors because the DSO reacts much faster than a digital multimeter (DMM) or scan tool. Problems that may not be indicated in scan-tool data or with a DMM will be illustrated on a DSO voltage trace. For example, a momentary dropout in a TPS signal may not appear on a DMM, but this dropout will be shown very clearly on a DSO voltage trace.*

**Task 17**   **Test output actuator/output circuit using scan tool, scan-tool data and/or waveform analysis.**

32. In the injector waveform in Figure 9–19, the purpose of the pulse width modulated (PWM) signals just before the injector shuts off is to
    A. increase the injector duty cycle.
    B. reduce the injector current.

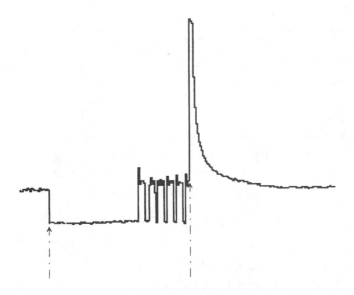

**Figure 9–19** Injector waveform.

C. increase fuel flow.

D. minimize an injector voltage spike.

**Hint**    *Damaged PCM drivers are usually caused by high current flow in the PCM output control circuit operated by the PCM driver. If an output solenoid or relay winding is shorted, the winding has less than the specified resistance and increased current flow. This high current flow may damage the output driver in the PCM. Some PCM drivers have the ability to sense higher-than-normal current flow. When this condition is sensed, the PCM shuts off the driver before the high current flow causes driver damage. Under this condition, the output operated by the driver becomes inoperative.*

*Some PCMs operate the injectors with a peak and hold mode, which allows a higher current to open the injectors and then a lower current to hold the injectors open. After the injector is opened, some PCMs supply a PWM signal to the injector to reduce injector current.*

## Task 18    Confirm the accuracy of observed scan-tool data by directly measuring a system, circuit, or component for actual value.

33. A customer complains that his or her OBD II vehicle stalls during deceleration. The scan-tool data in Figure 9–20 were obtained during deceleration with the engine at normal operating temperature. The cause of the stalling problem could be

| SCAN TOOL DATA | | | |
|---|---|---|---|
| Engine Coolant Temperature Sensor (ECT) 0.42   Volts | Intake Air Temperature Sensor (IAT) 2.26   Volts | MAP Sensor    1.8 V  MAF Sensor    1.0 V | Throttle Position Sensor (TPS) 0.6   Volts |
| Engine Speed Sensor (RPM) 800   rpm | Heated O₂ HO₂S Upstream   0.3–0.7 V Downstream 0.1–0.3 V | Vehicle Speed Sensor (VSS) 15   mph | Battery Voltage (B+) 14.2   Volts |
| Idle Air Control Valve (IAC) 30   percent | Evaporative Emission Canister Solenoid (EVAP) OFF | Torque Converter Clutch Solenoid (TCC) OFF | EGR Valve Control Solenoid  (EGR) 0   percent |
| Malfunction Indicator Lamp (MIL) OFF | Diagnostic Trouble Codes NONE | Open/Closed Loop CLOSED | Fuel Pump Relay (FP) OFF |
| Fuel Level Sensor 3.4 V | Fuel Tank Pressure Sensor 2.0 V | Transmission Fluid Temperature Sensor 0.7 V | Transmission Turbine Shaft Speed Sensor 15 mph |
| Transmission Range Switch OVERDRIVE | Transmission Pressure Control Solenoid 60% | Transmission Shift Solenoid 1 OFF | Transmission Shift Solenoid 2 OFF |
| Measured Ignition Timing °BTDC    Base Timing: 10    Actual Timing:  15 | | | |

**Figure 9–20** OBD II scan tool data number 19.

A. a sticking TCC.

B. an inoperative IAC motor.

C. an inoperative MAP sensor.

D. a VSS sensor.

*Hint*      *When diagnosing injector operation, noid lights may be connected to the injector terminals. During engine cranking, the noid lights flash if the PCM is triggering the injectors on and off. If the noid lights do not flash, the PCM is not turning the injectors on and off. When the noid lights do not flash, be sure the ignition system is firing. On many EFI systems, if the ignition system is not firing, the PCM does not operate the injectors. Be sure the voltage supply to all the injectors is satisfactory, and be sure the wires connected from the injectors to the PCM are satisfactory. Before replacing any PCM, always verify the PCM power supply and ground.*

34. A vehicle is hard to start after it sits overnight with an atmospheric temperature of 60°F (15.5°C). The scan-tool data in Figure 9–21 were obtained with the ignition switch on prior to starting the engine after the vehicle sat overnight. When ECT and IAT sensors were disconnected and tested with an ohmmeter, they had the specified resistance. The cause of the hard-starting problem could be

| SCAN TOOL DATA | | | |
|---|---|---|---|
| Engine Coolant Temperature Sensor (ECT)<br>0        Volts | Intake Air Temperature Sensor (IAT)<br>2.9    Volts | MAP Sensor    4.5 V<br>MAF Sensor    0.2 V | Throttle Position Sensor (TPS)<br>0.6     Volts |
| Engine Speed Sensor (RPM)<br>0          rpm | Heated O$_2$ HO$_2$S<br>Upstream      0 V<br>Downstream   0 V | Vehicle Speed Sensor (VSS)<br>0       mph | Battery Voltage (B+)<br>12.6    Volts |
| Idle Air Control Valve (IAC)<br>30     percent | Evaporative Emission Canister Solenoid (EVAP)<br>OFF | Torque Converter Clutch Solenoid (TCC)<br>OFF | EGR Valve Control Solenoid  (EGR)<br>0     percent |
| Malfunction Indicator Lamp (MIL)<br>ON | Diagnostic Trouble Codes<br>NONE | Open/Closed Loop<br>OPEN | Fuel Pump Relay (FP)<br>OFF |
| Fuel Level Sensor<br>3.3 V | Fuel Tank Pressure Sensor<br>4.5 V | Transmission Fluid Temperature Sensor<br>1.8 V | Transmission Turbine Shaft Speed Sensor<br>0 mph |
| Transmission Range Switch<br>PARK | Transmission Pressure Control Solenoid<br>0 | Transmission Shift Solenoid 1<br>OFF | Transmission Shift Solenoid 2<br>OFF |
| Measured Ignition Timing °BTDC   Base Timing:  10   Actual Timing:  0 | | | |

**Figure 9–21**  OBD II scan tool data number 21.

A. an inoperative ECT sensor.

B. high resistance in the ECT sensor wires.

C. an inoperative IAT sensor.

D. the ECT sensor wires are shorted together.

*Hint*      *Hard engine starting at moderate atmospheric temperatures may be caused by a malfunctioning ECT or IAT sensor signal to the PCM. When these sensor signals are the problem, the PCM does not supply the proper air-fuel ratio and this results in hard starting.*

**Task 19   Test and confirm operation of electrical/electronic circuits not displayed in scan-tool data.**

35. An OBD II vehicle has a fuel economy complaint. A visual inspection does not reveal any problems, and scan-tool data indicate high upstream HO$_2$S sensor voltages and reduced injector pulse width.

Technician A says the next step in diagnosing this problem should be to test fuel pressure.

Technician B says the next step in diagnosing this problem is to check for dripping injectors.

Who is correct?

A. A only

B. B only

C. Both A and B

D. Neither A nor B

36. An OBD II vehicle has a fuel economy complaint. The scan-tool data in Figure 9–22 were obtained while driving the vehicle at 50 mph with the gear selector in the O/D position.

Technician A says the transmission shift linkage may require adjusting.

Technician B says the transmission range switch may require adjusting and testing.

Who is correct?

A. A only

B. B only

C. Both A and B

D. Neither A nor B

## SCAN TOOL DATA

| Engine Coolant Temperature Sensor (ECT) 0.46 Volts | Intake Air Temperature Sensor (IAT) 2.18 Volts | MAP Sensor    2.1  V<br>MAF Sensor    2.5  V | Throttle Position Sensor (TPS) 2.3 Volts |
|---|---|---|---|
| Engine Speed Sensor (RPM) 1600 rpm | Heated O₂ HO₂S<br>Upstream   0.3–0.8 V<br>Downstream 0.1–0.3 V | Vehicle Speed Sensor (VSS) 50 mph | Battery Voltage (B+) 14.2 Volts |
| Idle Air Control Valve (IAC) 30 percent | Evaporative Emission Canister Solenoid (EVAP) ON | Torque Converter Clutch Solenoid (TCC) ON | EGR Valve Control Solenoid (EGR) 80 percent |
| Malfunction Indicator Lamp (MIL) OFF | Diagnostic Trouble Codes NONE | Open/Closed Loop CLOSED | Fuel Pump Relay (FP) ON |
| Fuel Level Sensor 3.5 V | Fuel Tank Pressure Sensor 2.0 V | Transmission Fluid Temperature Sensor 0.6 V | Transmission Turbine Shaft Speed Sensor 50 mph |
| Transmission Range Switch THIRD | Transmission Pressure Control Solenoid 50% | Transmission Shift Solenoid 1 OFF | Transmission Shift Solenoid 2 ON |

Measured Ignition Timing °BTDC    Base Timing: 10    Actual Timing: 27

**Figure 9–22**  OBD II scan tool data number 22.

*Hint*    *On most EFI systems, there are no data to indicate fuel pressure. Therefore, one of the first steps in an EFI diagnosis is a fuel pressure and flow test to verify these items. Low fuel pressure causes a lean air-fuel ratio, acceleration stumbles, and engine cutout at higher speeds. Excessive fuel pressure results in a rich air-fuel ratio and reduced fuel economy.*

*After PCM system components have been replaced, the vehicle should be driven for about ten minutes with the engine at normal operating temperature to allow the PCM adaptive memory to relearn the system. If this action is not completed, engine operation may be adversely affected because the adaptive memory is adapted to the inoperative components.*

## Task 20   Determine root cause of failures.

37. An OBD II vehicle fails an enhanced emission test for HC, CO, and NOx. The scan-tool data in Figure 9–23 were obtained during a road test. The root cause of the problem is
    A. an inoperative EVAP purge solenoid.
    B. an inoperative EGR control solenoid.
    C. an inoperative ECT sensor.
    D. an inoperative MAP sensor.

| SCAN TOOL DATA | | | |
|---|---|---|---|
| Engine Coolant Temperature Sensor (ECT)<br>1.5   Volts | Intake Air Temperature Sensor (IAT)<br>2.1   Volts | MAP Sensor   2.2 V<br>MAF Sensor   3.0 V | Throttle Position Sensor (TPS)<br>2.5   Volts |
| Engine Speed Sensor (RPM)<br>1800   rpm | Heated O₂ HO₂S<br>Upstream   0.3–0.9 V<br>Downstream 0.1–0.3 V | Vehicle Speed Sensor (VSS)<br>55   mph | Battery Voltage (B+)<br>14.2   Volts |
| Idle Air Control Valve (IAC)<br>30   percent | Evaporative Emission Canister Solenoid (EVAP)<br>OFF | Torque Converter Clutch Solenoid (TCC)<br>OFF | EGR Valve Control Solenoid  (EGR)<br>0   percent |
| Malfunction Indicator Lamp (MIL)<br>ON | Diagnostic Trouble Codes<br>PO125 | Open/Closed Loop<br>OPEN | Fuel Pump Relay (FP)<br>ON |
| Fuel Level Sensor<br>3.5 V | Fuel Tank Pressure Sensor<br>4.5 V | Transmission Fluid Temperature Sensor<br>0.6 V | Transmission Turbine Shaft Speed Sensor<br>55 mph |
| Transmission Range Switch<br>OVERDRIVE | Transmission Pressure Control Solenoid<br>40% | Transmission Shift Solenoid 1<br>ON | Transmission Shift Solenoid 2<br>ON |
| Measured Ignition Timing  °BTDC   Base Timing:  10   Actual Timing:  32 | | | |

**Figure 9–23** OBD II scan tool data number 23.

*Hint*      *The TCC, EGR, and EVAP solenoids all require an ECT signal indicating 150°F (65°C) coolant temperature before the PCM operates these solenoids. If these solenoids are inoperative, the root cause of the problem may be in the ECT sensor or related circuit. An inoperative engine thermostat also reduces coolant temperature so this temperature is too low to allow the PCM to operate the TCC, EGR, and EVAP solenoids.*

## Task 21   Determine root cause of multiple component failures.

38. A vehicle has a PCM and a body control module (BCM). The engine has a no-start condition, and diagnostic procedures indicate both the PCM and BCM may be the problem. The fuse link in the alternator battery wire has an open circuit. The MOST likely root cause of this multiple component failure is
    A. battery boosting with reversed polarity.
    B. an open circuit in the alternator field circuit.

C. a short to ground in the alternator battery wire.

D. high resistance in the PCM and BCM ground circuits.

**Hint**    *Computers may be damaged by high voltage spikes caused by damaged suppression diodes in relays or solenoids. Intermittent open circuits may also cause high voltage spikes and damage to electronic components. If the battery polarity is reversed by improper battery cable connections or reversing booster battery cables, severe damage may occur to modules and computers on the vehicle.*

## Task 22  Determine root cause of repeated component failures.

39. The A/C clutch relay has been replaced twice on the composite vehicle. Each time the relay failed, the winding was satisfactory but the relay contacts had an open circuit.

    Technician A says to test the suppression diode in the replacement relay.

    Technician B says to test the A/C clutch coil for a shorted condition.

    Who is correct?

    A. A only

    B. B only

    C. Both A and B

    D. Neither A nor B

**Hint**    *Repeated damage to a PCM driver may be caused by high current flow in the output controlled by the driver. The most likely cause of this high current is a shorted condition. A shorted condition in a solenoid or relay winding may also result in component damage, because this condition causes less resistance and high current flow in the winding.*

## Task 23  Verify effectiveness of repairs.

40. When diagnosing the composite vehicle with a scan tool, a DTC displays representing a defect in the variable valve timing system. This is the only DTC stored in the PCM. The most likely cause of this problem is

    A. a defect in one of the camshaft position (CMP) sensors.

    B. a defect in the CMP sensor wires.

    C. an oil leak in one of the variable valve timing camshaft actuators.

    D. a defective engine coolant temperature (ECT) sensor.

**Hint**    *After an electronic defect on an OBD II vehicle is repaired, the repair may be verified by driving the vehicle through two drive cycles and observing the MIL. If the MIL is not illuminated with the engine running, and a scan tool indicates all the monitors have been run, there are no electrical/electronic defects in the system.*

# Ignition System Diagnosis

## ASE Tasks, Questions, and Related Information

### Task 1  Inspect and test for missing, modified, inoperative, or tampered components.

41. An engine with an EI ignition system has a cylinder misfire condition. When an oscilloscope is connected to the ignition system, number 2 cylinder has a long, low spark line. The cause of this problem is

    A. a spark plug with the wrong heat range.

    B. high resistance in the number 2 spark plug wire.

    C. a fouled spark plug in number 2 cylinder.

    D. a damaged coil connected to the number 2 spark plug.

**Hint**    *A high, short spark line on an oscilloscope pattern indicates high resistance in a spark plug or spark plug wire. A long, low spark line is caused by low cylinder compression, a fouled spark plug, or a grounded spark plug wire.*

## Task 2  Locate relevant service information.

42. Several sets of spark plugs have been replaced in an engine because of electrode burning. This port-fuel-injected engine is in a truck that operates continually under heavy load. The engine temperature, spark advance, and air-fuel ratio are normal.

    Technician A says to check the specified spark plug heat range in the spark plug manufacturer's heat range chart and install hotter range spark plugs.

    Technician B says the fuel pressure may be much higher than specified.

    Who is correct?

    A. A only

    B. B only

    C. Both A and B

    D. Neither A nor B

**Hint**    *Burned spark plugs electrodes may be caused by a lean air-fuel ratio, detonation in the cylinders, engine overheating, or spark plugs with a hotter heat range than specified or required by engine operating conditions. When no other causes are present, spark plugs with a colder heat range may be installed to correct electrode burning.*

## Task 3  Research system operation using technical information to determine diagnostic procedure.

43. The composite vehicle does not start, and there is no voltage supplied to any of the coil primary windings. The fuel pressure is tested and proven to be within specifications.

    Technician A says to test fuse number 4 for an open circuit.

    Technician B says to test for an open circuit in the voltage supply wire to the coil primary windings between coils number 5 and 2.

    Who is correct?

    A. A only

    B. B only

    C. Both A and B

    D. Neither A nor B

**Hint**    *To research an electrical/electronic system, it is necessary to observe a wiring diagram for the system on the year and model of vehicle being diagnosed. After becoming familiar with the details of the electrical/electronic system, a precise diagnostic procedure must be determined and followed.*

## Task 4  Use appropriate diagnostic procedures based on available vehicle data and service information; determine if available information is adequate to proceed with effective diagnosis.

44. A vehicle fails an enhanced emission test. The emission traces in Figure 9–24 were obtained during this test. What should the technician do next to diagnose the emission test failure?

    A. Monitor sensor input data for invalid inputs.

    B. Test the spark plugs and plug wires with an oscilloscope diagnosis.

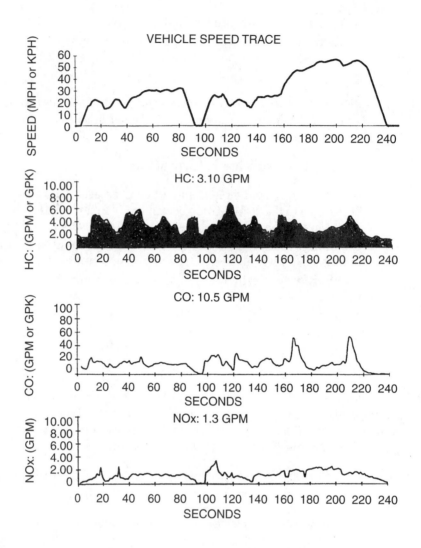

| | CUTPOINTS | |
|---|---|---|
| HC<br>Measured in GPM | CO<br>Measured in GPM | NOx<br>Measured in GPM |
| 0.8 | 15 | 2.0 |

**Figure 9–24** Emission traces.

C. Perform an injector balance test.

D. Test for higher-than-normal fuel pressure.

**Hint**    *Ignition problems that cause cylinder misfiring result in high HC but low CO emissions. A rich air-fuel ratio causes high CO and HC emissions. High NOx emissions are caused by high combustion chamber temperatures that may result from improper EGR valve operation.*

## Task 5    Establish relative importance of displayed scan tool data.

45. An engine has a detonation problem at cruising speed and a loss of power during hard acceleration and at higher speeds. The specified spark advance is 36 degrees at 2,000 rpm, and the actual spark advance is 32 degrees at 2,000 rpm. The scan-tool data indicate the PCM is receiving a knock sensor signal during hard acceleration.

The following readings were obtained on a five-gas analyzer with the engine running at 2,000 rpm on a dynamometer: HC - 40 ppm, CO - 0.3 percent, $CO_2$ - 11 percent, $O_2$ - 4 percent, NOx - 1,200 ppm. The cause of the power loss and detonation could be

A.  an inoperative knock sensor.

B.  lower-than-specified fuel pressure.

C.  a malfunctioning PCM.

D.  dripping injectors.

**Hint**    *A loss of engine power may be caused by a lean air-fuel ratio, a restricted exhaust system, or a lack of spark advance. Engine detonation may be caused by excessive spark advance, an inoperative knock sensor, a lean air-fuel ratio, carbon buildup in the combustion chambers, reduced exhaust valve margins, or an inoperative EGR system.*

## Task 6  Differentiate between ignition electrical/electronic problems and ignition mechanical problems.

46.  A vehicle with a coil-on-spark-plug ignition system stalls intermittently and is difficult to restart. This ignition system has a crankshaft position (CKP) sensor with a reluctor wheel mounted permanently near the rear of the crankshaft. While diagnosing the problem, the technician discovers there is no CKP sensor signal after the engine stalls. The magnetic-type sensor is tested with an ohmmeter and the winding has the specified resistance. The wires connected between the CKP sensor and the PCM are tested and proven to be satisfactory.

Technician A says to check the crankshaft end play.

Technician B says to check the reluctor wheel condition.

Who is correct?

A.  A only

B.  B only

C.  Both A and B

D.  Neither A nor B

**Hint**    *An inoperative CKP sensor signal causes a no-start condition. This voltage signal may be tested with a DSO. If there is no CKP sensor signal, magnet-type sensors may be tested with an ohmmeter.*

*The winding in the CKP sensor must have the specified resistance. The wires connected from the CKP sensor to the PCM may be tested with a voltmeter or ohmmeter. If the CKP sensor and related wires are satisfactory, check the reluctor wheel on the crankshaft and the crankshaft end play. Be sure the sensor is properly installed so it has the specified gap between the sensor and the reluctor wheel.*

## Task 7  Diagnose no-starting, hard starting, stalling, engine misfire, poor drivability, spark knock, power loss, poor mileage, illuminated MIL, and emission problems on vehicles equipped with distributorless electronic ignition (EI) systems; determine needed repairs.

47.  A four-cylinder engine has a waste spark ignition system with two coils that each fire two spark plugs. The ignition module is mounted with the coil assembly, and the PCM signals this module when to turn off the appropriate primary winding and fire the spark plugs at the proper instant. The primary winding in one coil has 0.5-ohms resistance, and the specified resistance is 1 ohm. This condition may cause

A.  damage to the PCM.

B.  damage to the ignition module.

C.  damage to the CKP sensor.

D.  burned spark plug electrodes.

48. In the ignition system in Figure 9–25, Technician A says that the primary windings in coils 1, 2, and 3 are connected in parallel, and spark plugs 1 and 5 are connected in series.

    Technician B says coils 1 and 5 are fired while cylinder 1 is on the compression stroke and cylinder 5 is on the exhaust stroke, or vice versa.

    Who is correct?

    A. A only

    B. B only

    C. Both A and B

    D. Neither A nor B

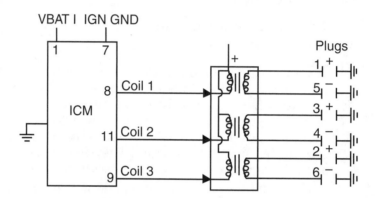

**Figure 9–25** EI ignition system.

49. The EI ignition system in Figure 9–25 misfires only during hard acceleration. All of the following may be the cause of the problem EXCEPT

    A. excessive resistance in the spark plug wires.

    B. low cylinder compression.

    C. a damaged coil with low maximum voltage.

    D. a bad spark plug with an insulation leakage problem.

**Hint**    *Many EI systems are a waste spark system in which the secondary winding of each coil is connected to two spark plugs. The coil fires both spark plugs simultaneously, but one cylinder is on the compression stroke and the other cylinder is on the exhaust stroke when the spark plugs fire.*

*Firing a spark plug while the piston is on the exhaust stroke has no effect on engine operation—thus, the term "waste spark system." Coil-on-spark-plug ignition systems have individual coils connected to each spark plug. In some ignition systems, the coil secondary terminals are connected directly on the spark plugs, whereas other ignition systems have short spark plug wires connected from the coil secondary terminals to the spark plugs.*

*Secondary reserve coil voltage is the difference between the normal required secondary voltage to fire the spark plugs (10,000 V) and the maximum available secondary coil voltage (35,000 V). This voltage reserve is necessary to compensate for wear at spark plug electrodes or in other secondary ignition components that increases the normal required secondary voltage. Secondary voltage reserve is also required to keep firing the spark plugs during hard acceleration when cylinder pressure increases. If the secondary voltage reserve is reduced, the coil(s) run out of voltage during hard acceleration and cylinder misfiring occurs. Secondary voltage reserve may be reduced by high resistance in spark plugs or plug wires, which increases the normal required firing voltage.*

*Secondary voltage reserve may also be reduced by a damaged coil(s) or a leakage problem in the secondary circuit, which reduces the maximum available secondary voltage.*

**Task 8**   Diagnose no-starting, hard starting, stalling, engine misfire, poor drivability, spark knock, power loss, poor mileage, illuminated MIL, and emission problems on vehicles equipped with distributor ignition (DI) systems; determine needed repairs.

50. Which of the following diagnostic tools should be used to check the secondary output of an ignition coil?
   A. A voltmeter
   B. An ohmmeter
   C. An ignition module tester
   D. A test spark plug

51. Which of the following problems would be LEAST likely to cause a continual cylinder misfire condition in a DI ignition system?
   A. Low engine compression
   B. Combustion chamber carbon deposits
   C. An oil-fouled spark plug
   D. Worn distributor bushings.

52. In Figure 9–26, which of the following test results would be considered normal?
   A. Meter 1, 0 ohm
   B. Meter 2, 0 ohm
   C. Meter 1, infinity ohm
   D. Meter 2, infinity ohms

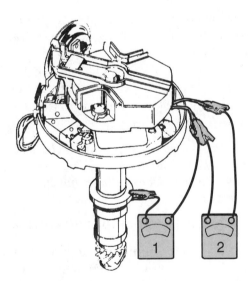

**Figure 9–26** Distributor pickup coil tests.

53. An engine with a DI ignition system does not start. With the engine cranking, a 12 V test light is connected from the ignition coil tachometer terminal to ground, and this light does not flash. The next step in the diagnostic procedure should be to
   A. replace the ignition module.
   B. test the voltage supply to the positive primary coil terminal.
   C. test the secondary coil winding.
   D. test the spark plug wires.

*Hint*    *When diagnosing DI ignition systems, a 12 V test light may be connected from the coil tachometer terminal to ground while cranking the engine. If the test light flashes on and off, the pickup coil and ignition module are triggering the primary ignition circuit on and off. If the 12 V test light does not flash, be sure to verify the voltage supply to the positive primary coil terminal.*

*If this voltage supply is satisfactory and the test light does not flash, the pickup coil, module, or related wiring may be the problem. If the 12 V test light connected from the tachometer terminal to ground flashes while cranking the engine, a test spark plug may be connected from several spark plug wires to ground to test the voltage in the secondary ignition system. If the test spark plug does not fire while cranking the engine, connect the test spark plug from the coil secondary lead to ground. When the test spark plug does not fire in this location, the coil is likely defective. If the test spark plug fires when connected to the coil secondary lead but it does not fire when connected to the spark plug wires, there is an insulation leakage problem in the distributor cap or rotor.*

## Task 9    Test for ignition system failures under various engine load conditions.

54. An engine with a waste spark EI ignition system misfires only during hard acceleration. When an oscilloscope is connected to the ignition system, the normal spark plug firing voltages are 25 kV to 27 kV with the engine idling.

    Technician A says to test the spark plug wires and check the spark plug gaps.

    Technician B says to test the ignition coil windings with an ohmmeter.

    Who is correct?

    A. A only

    B. B only

    C. Both A and B

    D. Neither A nor B

55. A throttle body injected engine with a DI ignition system backfires continually into the intake manifold during hard acceleration. The MOST likely cause of this problem could be

    A. the base timing is less than specified.

    B. the fuel pressure is higher than specified.

    C. the spark plug heat range is hotter than specified.

    D. the distributor cap is cracked.

*Hint*    *Ignition system failures under heavy engine load are usually caused by reduced secondary voltage reserve. Ignition cross firing because of insulation leakage problems in the spark plug wires or distributor cap may cause cylinder misfiring and backfiring into the intake manifold during hard acceleration.*

## Task 10    Test ignition system component operation using waveform analysis.

56. While diagnosing ignition systems by oscilloscope waveforms, Technician A says the normal secondary voltage required to fire the spark plugs is higher at idle speed compared to wide open throttle.

    Technician B says the maximum secondary coil voltage must always exceed the normal required coil voltage.

    Who is correct?

    A. A only

    B. B only

    C. Both A and B

    D. Neither A nor B

*Hint*    *The normal secondary voltage required to fire the spark plugs increases in relation to throttle opening. If the normal required secondary voltage ever exceeds the maximum available secondary coil voltage, misfiring occurs.*

## Task 11  Confirm base ignition timing and/or spark timing control.

57. While diagnosing a coil-on-spark-plug ignition system, Technician A says the CKP sensor gap may be adjusted to set the ignition timing.

Technician B says the timing connector should be disconnected while checking the base timing.

Who is correct?

A. A only

B. B only

C. Both A and B

D. Neither A nor B

**Hint** *On many fuel-injected engines with DI ignition systems, a timing connector must be disconnected while checking or adjusting the base ignition timing. On other DI ignition systems, the ECT sensor must be disconnected to put the PCM in the limp-in mode while checking base timing.*

*Disconnecting the timing connector or the ECT sensor while checking base timing ensures that the PCM is not providing any spark advance. When a timing adjustment is required, the distributor is rotated until the timing mark appears at the specified location on the timing indicator. On waste spark or coil-on-spark-plug EI ignition systems, timing adjustments are not possible. The CKP sensor cannot be rotated vertically to change the ignition timing.*

## Task 12  Determine root cause of failures.

58. A customer complains about poor fuel economy on his or her vehicle. This vehicle has a V8 engine with a knock sensor in the right-hand coolant drain plug in the block. The customer says the problem started after a cooling system flush was performed.

When the scan-tool data were checked, the spark advance was 21 degrees at 2,000 rpm while the specified spark advance is 32 degrees, and the knock sensor data indicated this sensor was sending a signal continually to the PCM. Engine detonation can never be heard during a road test.

Technician A says the knock sensor may have been overtightened and should be replaced.

Technician B says the spark plug heat range may be too hot for the driving conditions, resulting in engine detonation.

Who is correct?

A. A only

B. B only

C. Both A and B

D. Neither A nor B

**Hint** *Engine detonation may be caused by an inoperative knock sensor, excessively thin valve margins, spark plugs with a heat range hotter than specified, a lean air-fuel ratio, or an inoperative EGR system. If the engine has a knock sensor(s), detonation reduces spark advance and fuel economy.*

*If the knock sensor is overtightened during installation, this sensor may become oversensitive and this reduces spark advance and fuel economy. The specified torque on some knock sensors is 15 ft.-lbs.*

## Task 13  Determine root cause of multiple component failures.

59. The spark plugs in an engine have been replaced three times in 20,000 mi (32,000 km) because the electrodes on all the spark plugs are burned. All three sets of spark plugs have the correct heat range. The engine sometimes hesitates during acceleration, but

it idles smoothly when new spark plugs are installed. When the technician checks the scan-tool data, the only abnormal data are the integrator and block learn numbers, which are 170 and 165. The cause of this problem could be

A. low fuel pressure.

B. an intake manifold vacuum leak.

C. a contaminated upstream HO$_2$S sensor.

D. a contaminated IAT sensor.

**Hint**    *Higher-than-normal integrator and block learn numbers indicate the PCM is trying to add more fuel to compensate for a lean air-fuel ratio. A lean air-fuel ratio may be caused by low fuel pressure, an intake manifold vacuum leak, or restricted injector orifices.*

## Task 14  Determine root cause of repeated component failures.

60. An engine with the EI system in Figure 9–27 experiences a loss of power and reduced fuel economy. Each time the problem is diagnosed there is no spark advance. The coil module has been replaced twice to correct this problem, and the problem reoccurs. The cause of this problem could be

A. an intermittent open circuit in the bypass wire.

B. an inoperative CKP sensor.

C. an inoperative camshaft position (CMP) sensor.

D. an intermittent open circuit in the crank sensor signal wire.

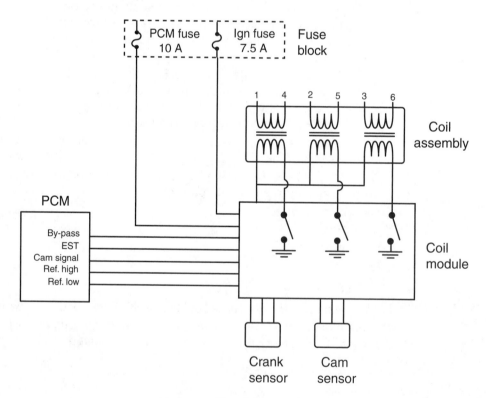

**Figure 9–27** EI ignition system.

**Hint**    *When cranking the engine, the crankshaft position sensor signal goes to the coil module where it is used to inform the module when to open the appropriate primary circuit and fire the spark plugs at the proper instant. When the engine starts, a 5 V signal is sent from the PCM through the bypass wire to the ignition coil module. This signal informs the coil module to enter the* *advance mode. In this mode, the crankshaft sensor signal is sent to the PCM on the reference high wire. The PCM scans the input sensors and sends a signal back to the coil module on the EST wire. This signal informs the module when to open each primary circuit to provide the exact spark advance required by the engine.*

# Fuel Systems and Air Induction Systems Diagnosis

## ASE Tasks, Questions, and Related Information

### Task 1  Inspect and test for missing, modified, inoperative, or tampered components.

61. An open circuit in the ECT signal wire may cause all of the following EXCEPT
    A. hard starting and engine stalling when warm.
    B. improper torque converter clutch operation.
    C. a lean air-fuel ratio.
    D. improper EGR operation.

62. A multiport fuel-injected engine idles at 1,300 rpm when the engine is at normal operating temperature, and the IAC percentage is 0 percent. The MOST likely cause of this problem is
    A. the timing belt has jumped several teeth on the camshaft sprocket.
    B. the ignition timing is retarded.
    C. the fuel pressure is more than specified.
    D. there is a vacuum leak in the intake manifold.

**Hint**   *An open circuit in the ECT sensor or related wiring simulates an extremely cold temperature signal from this sensor, which results in a very rich air-fuel ratio when the engine approaches normal operating temperature.*

*When diagnosing computer-controlled fuel systems, one of the first steps is a visual inspection of the fuel system components. Because fuel pressure above or below specifications affects the air-fuel ratio and engine performance, one of the first steps in diagnosing fuel system components is to verify fuel pressure.*

*If the idle speed is more than specified with the engine warmed up, the intake manifold should be checked for vacuum leaks. An intake manifold vacuum leak allows additional air into the intake manifold, and the PCM supplies more fuel to go with this air. This action increases idle rpm. Several automotive test equipment manufacturers market a vacuum leak tester that blows smoke into the suspected component to check for leaks.*

### Task 2  Locate relevant service information.

63. Which of the following is LEAST useful when inspecting for tampering?
    A. The scan tool
    B. The parts locator book
    C. Service manuals
    D. Service bulletins

**Hint**    *During a tampering inspection, all system components should be inspected for modifications. While doing this inspection, the technician may require a parts locator book, service manual, or service bulletins. Parts locator books may be necessary to locate certain system components, and service manuals or service bulletins may be required to understand the original system design and operation.*

## Task 3  Research system operation using technical information to determine diagnostic procedure.

64. During a road test, the engine in the composite vehicle cuts out and loses power when the engine speed increases above 6,000 rpm while the transaxle downshifts to pass another vehicle.

    Technician A says a restricted fuel line or clogged fuel filter may be the cause of the problem.

    Technician B says some of the ignition coils may be the problem.

    Who is correct?
    A. A only
    B. B only
    C. Both A and B
    D. Neither A nor B

**Hint**    *The difference between normal and abnormal operation of electrical/electronic systems must be clearly understood. If the baseline normal operation of electrical/electronic systems is not clearly defined, it is difficult to determine abnormal operation. It is possible to attempt the diagnosis of symptoms that are actually considered normal operation.*

## Task 4  Evaluate the relationships between fuel trim values, oxygen sensor readings, and other sensor data to determine fuel system control performance.

65. The composite vehicle has a reduced fuel economy complaint. When diagnosed with a scan tool, the long term fuel trim is –28 and the upstream $HO_2S$ both indicate .6 V to 0.8 V, but the engine operation is satisfactory. The other data is satisfactory, and a DTC for long term fuel trim is stored. The most likely cause of this problem is
    A. cylinder misfire.
    B. high fuel pressure.
    C. dripping injectors.
    D. intake manifold vacuum leak.

**Hint**    *If the long-term fuel trim number is high (+15 to +30), the air-fuel ratio is lean and the PCM is increasing the injector pulse width to compensate for the lean mixture. Conversely, if the long-term fuel trim numbers are low (–15 to –30), the air-fuel ratio is rich, and the PCM is decreasing the injector pulse width to bring the mixture back to normal. The technician must observe all the vehicle data and the engine operation to diagnose the root cause of the problem. For example, an intake manifold vacuum leak causes a lean air-fuel ratio, and rough engine idle. This lean air-fuel ratio causes the long-term fuel trim number to be high, and the upstream $HO_2S$ voltages to be low. If the vacuum leak affects one bank in a V6 engine, the $HO_2S$ voltage is only affected on that side of the engine.*

## Task 5  Use appropriate diagnostic procedures based on available vehicle data and service information; determine if available information is adequate to proceed with effective diagnosis.

66. The engine in an OBD II vehicle idles smoothly but experiences a loss of power and engine surging at wide throttle opening. The specified fuel pressure is available with

the engine idling, and when the scan-tool data do not indicate any problems. The most LIKELY cause of this problem could be

A. the injector orifices are partly clogged.

B. the pressure regulator is sticking open.

C. the injectors are dripping.

D. the voltage supplied to the fuel pump is low.

**Hint**     *Low fuel pressure at high vehicle speeds may cause a loss of power and engine surging. When the fuel pressure is normal at idle speed and low at higher speed, always verify the voltage supply to the fuel pump and the voltage drop across the fuel pump ground.*

## Task 6     Establish relative importance of displayed scan-tool data.

67. An OBD I vehicle with an SFI fuel system cuts out and surges above 2,000 rpm. The fuel pressure and volume are satisfactory, and the voltage supplied to the fuel pump is within specifications. The scan-tool data below 2,000 rpm do not indicate any problems. With the engine running at 1,800 rpm, a large vacuum leak is created into the intake manifold, and the $HO_2S$ voltage signal drops immediately to 0.1 V, but the injector pulse width does not change.

Technician A says the PCM may be malfunctioning.

Technician B says the $HO_2S$ signal wire from the sensor to the PCM may be the problem.

Who is correct?

A. A only

B. B only

C. Both A and B

D. Neither A nor B

**Hint**     *If a large vacuum leak is created, the $HO_2S$ voltage signal should immediately decrease, and the PCM should increase the injector pulse width to compensate for the lean air-fuel ratio. When propane is dispersed into the air intake with the engine running, the $HO_2S$ voltage signal should immediately increase, and the PCM should decrease the injector pulse width.*

## Task 7     Differentiate between fuel system mechanical and fuel system electrical/ electronic problems.

68. The composite vehicle cuts out at 55 mph. The engine never misfires or surges. The MIL is not on with the engine running and there are no DTCs in the PCM. All the scan-tool data at idle are normal, as well as the fuel pressure and volume. When the engine cuts out, the fuel pressure, volume, and scan-tool data are normal except the MAF sensor signal, which is 2.1 V and 20 gm/sec. The MAF sensor has been replaced without correcting the problem, and there are no intake manifold leaks or leaks between the MAF sensor and the throttle body.

Technician A says to test for exhaust system restriction.

Technician B says to test for air intake restriction.

Who is correct?

A. A only

B. B only

C. Both A and B

D. Neither A nor B

**Hint**     *Air intake restriction or exhaust system restriction may cause an engine to cut out at a certain speed with no indication of misfiring or surging. If one of these problems exist, the MAF sensor voltage and gm/sec are lower than specified.*

**Task 8**    **Differentiate between air induction system mechanical and air induction system electrical/electronic problems, including electronic throttle actuator control (TAC) systems.**

69. The composite vehicle engine rpm will not exceed 1,500 rpm with the accelerator pedal pushed to the wide open position, and the MIL is illuminated with the engine running.

    Technician A says the throttle actuator control (TAC) motor may be disabled by the PCM because of a TAC motor defect.

    Technician B says these problems may be caused by one defective throttle position (TP) sensor.

    Who is correct?

    A. A only

    B. B only

    C. Both A and B

    D. Neither A nor B

70. An SFI engine has a rough idle problem and an acceleration stumble. This engine has an MAF sensor connected in the air intake hose between the air cleaner and the throttle body. The specified airflow for the MAF is 4 to 7 gm/sec at idle. A scan tool indicates a steady 3 gm/sec at idle. The MOST likely cause of these problems is

    A. an inoperative MAF sensor.

    B. an air leak in the hose between the MAF sensor and the air cleaner.

    C. an air leak in the hose between the MAF sensor and the throttle body.

    D. a loose connection on the MAF sensor electrical connector.

**Hint**    *If a fault occurs in the TAC motor or related circuit, the PCM disables the motor, and a return spring in the TAC motor returns the throttle to the fast idle position. The PCM also disables the TAC motor and illuminates the MIL if a defect occurs in the TP sensors or accelerator pedal position (APP) sensors. If one TP or APP sensor is defective, the throttle position is limited to a 35-percent opening and a DTC is set. When both TP or APP sensors are defective, the throttle position is limited to 15 percent and a DTC is set.*

*On some engines with an MAF sensor, the MAF sensor airflow in gm/sec may be read on a scan tool. This gms/sec reading must be within specifications, and it must increase smoothly as engine speed increases. An erratic MAF gm/sec reading may indicate a loose electrical connection on this sensor or an inoperative sensor. If the gm/second reading is less than specified, there may be an air leak in the hose between the MAF sensor and the throttle body or in the intake manifold.*

**Task 9**    **Diagnose hot or cold no-starting, hard starting, stalling, engine misfire, poor drivability, spark knock, incorrect idle speed, poor idle, flooding, hesitation, surging, power loss, poor mileage, dieseling, illuminated MIL, and emission problems on vehicles equipped with fuel injection fuel systems; determine needed action.**

71. A vehicle passes an enhanced emission test for HC and CO but fails for high NOx emissions. All of the following could be the cause of the problem EXCEPT

    A. excessive carbon buildup on the piston heads.

    B. a plugged EGR exhaust passage.

    C. hot spots in the combustion chamber caused by cooling system deposits.

    D. a rich air-fuel ratio.

72. A vehicle with a DI ignition has a spark knock and poor mileage complaint.

    Technician A says to test the knock sensor using a timing light and a small hammer.

    Technician B says to check the base timing.

Who is correct?

A. A only

B. B only

C. Both A and B

D. Neither A nor B

**Hint**    *High NOx emissions are caused by high combustion chamber temperatures. One of the most common causes of this problem is an inoperative EGR system. High CO and HC readings may be caused by a rich air-fuel ratio. High HC and low CO readings may be caused by cylinder misfire.*

*A leaking cold start injector may cause hard engine starting, rough idle operation, stalling, and reduced fuel economy. Excessive spark advance or an inoperative knock sensor may cause detonation and reduced fuel economy. Remember that base timing is not adjustable on EI ignition systems.*

*On some engines, the PCM shuts off the injectors at a specific engine rpm to prevent engine damage. This action may be confused with problems in the fuel or ignition system.*

## Task 10    Verify fuel quality, fuel system pressure, and fuel system volume.

73. A customer complains that the engine in his or her vehicle requires excessive cranking time before it starts. When the technician tests the fuel pressure, it is within specifications. After the pressure test, the fuel pressure indicated on the test gauge drops to zero in five minutes. When the pressure test is completed and the fuel supply line is manually closed, the fuel pressure drops 3 psi (20.6 kPa) in five minutes. The cause of the long cranking time before starting is

A. a pressure regulator that is stuck open.

B. a leaking fuel pump check valve.

C. leaking injectors.

D. a leak in the fuel return line.

**Hint**    *After the fuel pressure test, the fuel pressure should not decrease more than 5 psi (35 kPa) in ten minutes. If the fuel pressure drops more than this amount, the fuel pump check valve may be leaking, the pressure regulator valve may be leaking, or the injectors may be dripping. When the fuel return line is manually plugged and the fuel pressure no longer drops off excessively, the pressure regulator valve is leaking. If both the fuel supply line and fuel return lines are manually plugged and the fuel pressure still drops off excessively, the injectors are dripping. When the fuel supply line is manually plugged and the pressure no longer drops off excessively, the fuel pump check valve is leaking.*

## Task 11    Evaluate fuel injector and fuel pump performance (mechanical and electrical operation).

74. A vehicle with an SFI engine has a rough idle problem, and the engine shudders at light throttle when the torque converter clutch locks up. The MOST likely cause of this problem is

A. a rich air-fuel ratio caused by an inoperative $HO_2S$ sensor.

B. a lean air-fuel ratio caused by restricted injector orifices.

C. excessive injector pulse width supplied by the PCM.

D. an intake manifold vacuum leak.

75. During an injector balance test on a V6 engine, four injectors provide a 5 psi (35 kPa) pressure decrease. The other two injectors provide a 2 psi (14 kPa) pressure decrease.

Technician A says the injector plungers are sticking open in the two injectors with less pressure decrease.

Technician B says some manufacturers supply self-cleaning injectors, and they do not recommend cleaning these injectors.

Who is correct?

A.  A only

B.  B only

C.  Both A and B

D.  Neither A nor B

**Hint**    *Restricted injector orifices may cause rough idle operation, acceleration stumbles, and shudder after the TCC locks up. Dripping injectors may cause hard starting, reduced fuel economy, and rough idle operation.*

*During an injector balance test, the pressure drop on the injectors must not vary more than specified. If the pressure drop is excessive on some injectors, these injector plungers are sticking open. When the pressure drop is less on some injectors compared to the other injectors, these injector plungers are sticking closed or the injector orifices are partially restricted.*

## Task 12   Determine root cause of failures.

76.  Cylinder number 2 is misfiring on a vehicle, and this misfiring is less noticeable at higher engine speeds. Cylinder compression and the ignition system are tested and proven to be satisfactory. The scan-tool data in Figure 9–28 were obtained with the engine running at normal temperature and idle speed. The MOST likely cause of the cylinder misfire is

A.  a bad number 2 injector.

B.  an intake manifold vacuum leak.

C.  an inoperative $HO_2S$ sensor.

D.  an air leak in the MAP sensor hose.

**Hint**    *Low compression on one cylinder causes continual misfiring on that cylinder. An ignition problem may cause cylinder misfiring only during acceleration. A severe ignition problem may*

| SCAN TOOL DATA | | | |
|---|---|---|---|
| Engine Coolant Temperature Sensor (ECT) 0.44 Volts | Intake Air Temperature Sensor (IAT) 2.21 Volts | MAP Sensor  1.6 V  MAF Sensor  0.8 V | Throttle Position Sensor (TPS) 0.6 Volts |
| Engine Speed Sensor (RPM) 750 rpm | Heated $O_2$ $HO_2S$ Upstream  0.2–0.6 v Downstream 0.1–0.3v | Vehicle Speed Sensor (VSS) 0 mph | Battery Voltage (B+) 14.3 Volts |
| Idle Air Control Valve (IAC) 30 percent | Evaporative Emission Canister Solenoid (EVAP) OFF | Torque Converter Clutch Solenoid (TCC) OFF | EGR Valve Control Solenoid (EGR) 0 percent |
| Malfunction Indicator Lamp (MIL) OFF | Diagnostic Trouble Codes NONE | Open/Closed Loop CLOSED | Fuel Pump Relay (FP) ON |
| Fuel Level Sensor 3.5 V | Fuel Tank Pressure Sensor 1.8 V | Transmission Fluid Temperature Sensor 0.7 V | Transmission Turbine Shaft Speed Sensor 0 mph |
| Transmission Range Switch PARK | Transmission Pressure Control Solenoid 70% | Transmission Shift Solenoid 1 ON | Transmission Shift Solenoid 2 OFF |
| Measured Ignition Timing °BTDC   Base Timing: 10   Actual Timing: 12 | | | |

**Figure 9–28** OBD II scan tool data number 28.

*also cause misfiring at idle. An intake manifold vacuum leak may cause misfiring at idle, but the misfire disappears as the throttle is opened and the vacuum decreases. A bad injector causes continual cylinder misfiring, but this misfiring is most noticeable at idle and low speeds.*

## Task 13   Determine root cause of multiple component failures.

77. The PCM on the composite vehicle has failed twice. Each time this failure occurred, the engine has stopped, and the fuel pump relay driver in the PCM is damaged. The fuse is also blown in the fuel pump relay winding circuit. The root cause of the PCM failure could be

A. a damaged voltage suppression diode in the fuel pump relay.

B. an intermittent short to ground at PCM terminal 4.

C. high resistance at the fuel pump relay contacts.

D. a shorted fuel pump relay winding.

**Hint**    *Drivers in the PCM are usually damaged by high current flow from a shorted winding in one of the solenoids or relays controlled by the PCM. These drivers may also be damaged by inoperative voltage suppression diodes in these solenoids or relays. Fuses are blown by high current flow resulting from a shorted condition in a winding or relay, or a short to ground in a wire. High current flow and blown fuses may also be caused by excessive voltage in the circuit.*

## Task 14   Determine root cause of repeated component failures.

78. A MAP sensor on an OBD I system has been replaced for the third time. Each time the MAP sensor fails, the MIL is on and a MAP sensor DTC is stored in the PCM. The MAP sensor is mounted on the cowl in the engine compartment. The vehicle is operating where the atmospheric temperature is below 32°F (0°C).

Technician A says moisture may be freezing in the MAP sensor hose.

Technician B says the MAP sensor should be mounted on the engine so the sensor has a shorter vacuum hose.

Who is correct?

A. A only

B. B only

C. Both A and B

D. Neither A nor B

**Hint**    *A plugged vacuum hose connected to the MAP sensor causes a MAP sensor DTC to be set in the PCM memory and the MIL to be illuminated.*

# Emission Control Systems Diagnosis

## ASE Tasks, Questions, and Related Information

## Task 1   Inspect and test for missing, modified, inoperative, or tampered components.

79. A customer complains about a rough idle condition on his or her vehicle. A four-gas analyzer indicates the following readings with the engine running at idle speed and

normal operating temperature: HC - 220 ppm, CO - 0.3 percent, $CO_2$ - 11.4 percent, $O_2$ - 4 percent. According to these emission readings, the rough idle problem may be caused by

A. a PCV valve installed backwards.

B. a vacuum leak in the brake booster hose.

C. a disconnected EVAP solenoid electrical connector.

D. a leaking one-way check valve in the secondary air injection system.

*Hint*    *A cylinder misfire caused by a lean air-fuel ratio results in high HC and $O_2$ emissions with low CO and $CO_2$ emissions. A rich air-fuel ratio causes high HC and CO readings with low $O_2$ and $CO_2$ emissions.*

## Task 2  Locate relevant service information.

80. The easiest place to find vehicle-specific emission vacuum hose routing information is

A. in the vehicle manufacturer's service manual.

B. in the vehicle owner's manual.

C. in the vehicle electrical and vacuum hose diagram manual.

D. on the underhood emission label.

*Hint*    *The vehicle manufacturer's service manual contains complete service information on the vehicle. The owner's manual contains basic information regarding vehicle operation. Some vehicle manufacturers publish electrical wiring diagrams and vacuum hose routing information manuals for their vehicles. The quickest place to locate vacuum hose routing information is on the underhood emission label.*

## Task 3  Research system operation using technical information to determine diagnostic procedure.

81. An engine fails an enhanced emissions test for NOx emissions. The engine has a positive backpressure EGR valve with a bleed port integral with the EGR valve stem. A PCM-controlled solenoid supplies vacuum to the EGR valve. The engine has no drivability complaints.

Technician A says the exhaust passage in the EGR valve stem may be plugged with carbon.

Technician B says the EGR solenoid plunger may be stuck closed.

Who is correct?

A. A only

B. B only

C. Both A and B

D. Neither A nor B

*Hint*    *Details regarding emission control system components must be researched and understood to diagnose these components and systems quickly and accurately. For example, there are several different types of EGR valves with different operating characteristics. Positive backpressure EGR valves contain a vacuum bleed valve that is normally opened and closed by exhaust pressure at a certain vehicle speed to allow the EGR valve to open. Negative backpressure EGR valves have an internal vacuum bleed valve that is normally closed and opened by negative pressure pulses in the exhaust system at low speeds. At a certain vehicle speed, these negative pressure pulses become less significant and the bleed valve closes to allow the EGR valve to open.*

## Task 4  Use appropriate diagnostic procedures based on available vehicle data and service information; determine if available information is adequate to proceed with effective diagnosis.

82. Which of the following tests would best isolate a decel valve that is stuck closed?

A. A loaded-mode I/M emission test

B. A four-gas emission test at idle speed

C. A five-gas emission test at 2,000 rpm

D. Compare the MAF and TPS data

*Hint*    *When diagnosing a suspected component, the vehicle must be operated under the conditions when the symptoms are present or when the suspected component is operational.*

## Task 5    Establish relative importance of displayed scan-tool data.

83. An OBD II vehicle with an enhanced EVAP system stalls after it has been driven over 15 mi. (24 km) (Figure 9–29). After the vehicle sits for a few minutes, it does restart. During a road test, scan-tool data indicate the fuel tank pressure sensor voltage is 0.2 V, which is much lower than specified. The MOST likely cause of this problem is

A. the hose from the fuel tank to the canister is plugged.

B. the hose from the canister to the purge solenoid is leaking.

C. the vent solenoid air intake and the vacuum valve in the filler cap are both clogged.

D. the canister is partially clogged.

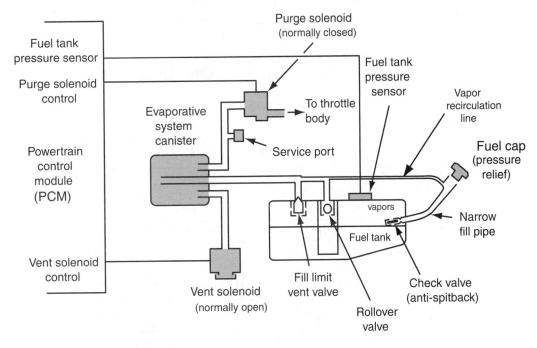

**Figure 9–29** Enhanced EVAP system with on-board refueling vapor recovery (ORVR) system.

*Hint*    *In an enhanced EVAP system, the PCM operates the purge solenoid and the vent solenoid to control fuel tank pressure. The vent solenoid is normally open and the purge solenoid is normally closed. When the PCM runs the EVAP system monitor, the PCM cycles the EVAP vent solenoid to increase the vacuum in the fuel tank to check the EVAP system for leaks. Pressure and vacuum valves in the fuel filler cap prevent excessive high or low pressure in the fuel tank. Some enhanced EVAP systems have an on-board refueling vapor recovery (ORVR) system to prevent the escape of fuel vapors during refueling.*

## Task 6    Differentiate between emission control systems mechanical and electrical/electronic problems.

84. During a road test, scan-tool data on the composite vehicle indicate the EGR solenoid, TCC solenoid, and the EVAP purge solenoid all have a 0-percent duty cycle.

Technician A says to test the voltage supply wires to these three solenoids.

Technician B says to test the engine thermostat.

Who is correct?

A. A only

B. B only

C. Both A and B

D. Neither A nor B

**Hint**   *Engine coolant temperature affects many of the components in the PCM system. For example, improper coolant temperature affects the ECT sensor voltage signal, which in turn causes a rich or lean air-fuel ratio. The PCM does not operate components such as the EGR solenoid, EVAP solenoid, and TCC solenoid until the engine coolant reaches a specific temperature.*

## Task 7   Determine need to diagnose emission control subsystems.

85. A customer complains about reduced fuel economy on his or her OBD I–equipped vehicle with a secondary air injection system that pumps air upstream into the exhaust ports during engine warm-up and downstream into the catalytic converter when the engine is at normal operating temperature. The scan-tool data in Figure 9–30 were obtained with the engine running at normal temperature and idle speed. The cause of the reduced fuel economy could be

A. a restricted fuel return line.

B. dripping injectors.

C. secondary air that is continually upstream.

D. an inoperative MAP sensor.

| SCAN TOOL DATA | | | |
|---|---|---|---|
| Engine Coolant Temperature Sensor (ECT)<br>0.42 Volts | Intake Air Temperature Sensor (IAT)<br>2.20 Volts | Manifold Absolute Pressure Sensor (MAP)<br>1.6 Volts | Throttle Position Sensor (TPS)<br>0.6 Volts |
| Engine Speed Sensor (RPM)<br>750 rpm | Heated Oxygen Sensor (HO$_2$S)<br>0.1–0.3 Volts | Vehicle Speed Sensor (VSS)<br>0 mph | Battery Voltage (B+)<br>14.3 Volts |
| Idle Air Control Valve (IAC)<br>28 percent | Evaporative Emission Canister Solenoid (EVAP)<br>OFF | Torque Converter Clutch Solenoid (TCC)<br>OFF | EGR Valve Control Solenoid (EGR)<br>0 percent |
| Malfunction Indicator Lamp (MIL)<br>OFF | Diagnostic Trouble Codes<br>NONE | Open/Closed Loop<br>CLOSED | Fuel Pump Relay (FP)<br>ON |
| Measured Ignition Timing °BTDC    Base Timing: 10    Actual Timing: 10 | | | |

**Figure 9–30** OBD I scan tool data number 30.

86. The composite vehicle has a restricted catalytic converter. When the converter is replaced, the technician notices the monolith in the old converter is actually melted together.

Technician A says the engine should be checked for high NOx emissions.

Technician B says the engine should be checked for higher-than-normal coolant temperature.

Who is correct?

A. A only

B. B only

C. Both A and B

D. Neither A nor B

**Hint**    *Some secondary air injection systems are designed to pump air into the exhaust ports during engine warm-up. When the engine reaches normal operating temperature, the air injection is directed downstream into the catalytic converter. Many of these systems have two vacuum-operated valves that direct airflow to the exhaust ports or to the catalytic converter. Vacuum is supplied to these valves by two PCM-operated solenoids. If the air system continues to pump air upstream into the exhaust ports with the engine warmed up, the $HO_2S$ sensor senses excessive oxygen in the exhaust stream, and the $HO_2S$ sensor voltage is very low. Under this condition, the PCM supplies a richer air-fuel ratio. If the airflow is directed downstream during engine warmup, HC and CO emissions are high.*

*A lean air-fuel ratio may cause cylinder misfiring at idle, with high HC and $O_2$ emissions. However, if the EGR valve is stuck open allowing exhaust gas to dilute the air-fuel ratio, cylinder misfiring may occur at idle with high HC emissions but low $O_2$ emissions because the exhaust is depleted of oxygen.*

*Melted internal components in a catalytic converter are usually caused by a continually rich air-fuel ratio that causes converter overheating*

## Task 8   Perform functional tests on emission control subsystems; determine needed repairs.

87. An EGR valve should be inoperative under any of these driving conditions EXCEPT
    A. a constant throttle opening at 55 mph (88 kmh).
    B. when the engine coolant is cold.
    C. during wide open throttle operation.
    D. during closed throttle deceleration.

88. A vehicle fails the EVAP flow test during an enhanced emission test. In this EVAP system, the PCM turns the EVAP solenoid either on or off. During a road test, the scan-tool data indicate the EVAP solenoid is on. When a voltmeter is connected from the EVAP solenoid terminal to the PCM, the voltmeter reads 12 V when the scan tool indicates the EVAP solenoid is on. The cause of this problem could be
    A. an open circuit in the EVAP solenoid winding.
    B. an open circuit in the voltage supply wire to the EVAP solenoid.
    C. an open circuit in the wire from the EVAP solenoid to the PCM.
    D. a shorted winding in the EVAP solenoid.

**Hint**    *The PCM opens the EGR valve at moderate vehicle speeds with the engine at normal operating temperature. During other engine operating conditions, the PCM allows the EGR valve to remain closed.*

*When observing output data, the scan tool displays the commands issued by the PCM. For example, the scan tool displays the commanded EVAP solenoid position. However, electrical problems in the EVAP-related circuit, or mechanical problems, can make it impossible for the command to be carried out. For example, the PCM may command the TCC solenoid on, but an open circuit in this solenoid winding makes it impossible for the command to be carried out.*

## Task 9   Determine the effect of exhaust emissions caused by the failure of an emission control component or subsystem.

89. A vehicle fails an emissions test for HC and CO emissions. The engine has a rough idle problem. If the PCV valve is removed from the rocker arm cover, the engine runs smoother and the HC and CO emissions are normal.

Technician A says that the PCV valve is stuck closed.

Technician B says that the engine oil should be checked for contamination.

Who is correct?

A. A only

B. B only

C. Both A and B

D. Neither A nor B

90. When diagnosing catalytic converters, Technician A says the converter inlet temperature should be higher than the outlet temperature.

Technician B says $CO_2$ emissions are lower on a catalytic converter–equipped vehicle compared to a non-converter-equipped vehicle.

Who is correct?

A. A only

B. B only

C. Both A and B

D. Neither A nor B

**Hint**    *If the PCV valve is removed from the rocker arm cover, there should not be much change in engine speed or emission levels. If there is an excessive change in engine speed and emission levels, the engine may have excessive piston ring blowby or the engine oil may be contaminated with gasoline.*

*A positive backpressure EGR valve has a vacuum bleed valve in the center of the EGR valve diaphragm. This bleed valve is normally open. Before vacuum can be supplied to the EGR valve diaphragm to open this valve, a specific exhaust pressure must be supplied up through the EGR valve stem to close the bleed valve. Once this bleed valve is closed and the PCM energizes the EGR solenoid, vacuum is supplied through this solenoid to the EGR valve diaphragm to open the EGR valve.*

*If a catalytic converter is functioning properly, the outlet temperature should be at least 100°F higher compared to the inlet temperature. Catalytic converters create some $CO_2$, and so $CO_2$ emissions are higher on a vehicle with a catalytic converter compared to a non-converter-equipped vehicle.*

## Task 10  Use exhaust gas analyzer readings to diagnose the failure of an emission control component or subsystem.

91. A vehicle fails an I/M emission test for HC at idle. The engine also has a rough idle problem that smoothes out as the engine speed increases. The four-gas analyzer readings at idle and 2,000 rpm are in Figure 9–31. The MOST likely cause of these problems is

A. a shorted fuel injector driver in the PCM.

B. an EGR valve stuck open.

C. an intake manifold vacuum leak.

D. a fouled spark plug.

| Engine Speed | Idle | 2000 RPM |
|---|---|---|
| HC (ppm) | 700 | 20 |
| CO (percent) | 0.45 | 0.20 |
| $CO_2$ (percent) | 12 | 14.5 |
| $O_2$ (percent) | 0.5 | 0.9 |

**Figure 9–31** Exhaust emission readings number 31.

**Hint**   *Exhaust gas analyzer readings can be very helpful in diagnosing engine and emission control system defects. For example, a rich air-fuel ratio causes high HC and CO readings, and low NOx readings. Cylinder misfiring causes very high HC readings and low CO readings.*

**Task 11**   **Diagnose hot or cold no-starting, hard starting, stalling, engine misfire, poor drivability, spark knock, incorrect idle speed, poor idle, flooding, hesitation, surging, power loss, poor mileage, dieseling, illuminated MIL, and emission problems caused by a failure of emission control components or subsystems.**

92. A light-duty truck with an OBD II V8 engine has an acceleration stumble but there are no other drivability complaints. The scan-tool data in Figure 9–32 were obtained during a road test with the engine at normal operating temperature. The fuel pressure was tested and proven to be satisfactory. The engine has a negative backpressure EGR valve. The cause of this acceleration stumble could be

    A. a plugged exhaust passage in the EGR valve stem.

    B. the bleed port stuck open in the EGR valve.

    C. a vacuum leak in the MAP sensor hose.

    D. a vacuum leak in the throttle body to intake gasket.

| SCAN TOOL DATA | | | |
|---|---|---|---|
| Engine Coolant Temperature Sensor (ECT)<br><br>0.42    Volts | Intake Air Temperature Sensor (IAT)<br><br>2.18    Volts | MAP Sensor       3.1 V<br><br>MAF Sensor      3.5 V | Throttle Position Sensor (TPS)<br><br>2.8    Volts |
| Engine Speed Sensor (RPM)<br><br>2400    rpm | Heated O₂ HO₂S<br>Upstream    0.3–0.8 V<br>Downstream 0.1–0.3 V | Vehicle Speed Sensor (VSS)<br><br>70    mph | Battery Voltage (B+)<br><br>14.4    Volts |
| Idle Air Control Valve (IAC)<br><br>30    percent | Evaporative Emission Canister Solenoid (EVAP)<br><br>ON | Torque Converter Clutch Solenoid (TCC)<br><br>ON | EGR Valve Control Solenoid  (EGR)<br>60<br>percent |
| Malfunction Indicator Lamp (MIL)<br>OFF | Diagnostic Trouble Codes<br><br>NONE | Open/Closed Loop<br><br>CLOSED | Fuel Pump Relay (FP)<br><br>ON |
| Fuel Level Sensor<br><br>3.5 V | Fuel Tank Pressure Sensor<br>2.0 V | Transmission Fluid Temperature Sensor<br>0.7 V | Transmission Turbine Shaft Speed Sensor<br>70 mph |
| Transmission Range Switch<br>OVERDRIVE | Transmission Pressure Control Solenoid<br>20% | Transmission Shift Solenoid 1<br>OFF | Transmission Shift Solenoid 2<br>ON |
| Measured Ignition Timing °BTDC   Base Timing: 10   Actual Timing: 37 | | | |

**Figure 9–32** OBD II scan tool data number 32.

93. An OBD I vehicle fails an enhanced emissions test for HC and CO. Scan-tool data indicates the HO₂S sensor voltage is always low. When propane is dispersed into the air intake with the engine idling, the HO₂S sensor voltage does not increase. The cause of this problem could be

    A. the secondary air is always upstream.

    B. the HO₂S signal wire has an open circuit.

    C. the engine has an intake manifold vacuum leak.

    D. one fuel injector is defective.

94. A vehicle hesitates during hard acceleration with the engine at normal operating temperature. This hesitation is less noticeable when the engine is cold.

    The scan-tool data in Figure 9–33 were obtained while accelerating at 75-percent throttle opening during a road test with the engine at normal operating temperature. The cause of this problem could be
    A. an inoperative TPS.
    B. an inoperative MAF sensor.
    C. an inoperative IAT sensor.
    D. an inoperative EVAP solenoid.

| **SCAN TOOL DATA** | | | |
|---|---|---|---|
| Engine Coolant Temperature Sensor (ECT)<br>0.46    Volts | Intake Air Temperature Sensor (IAT)<br>2.25    Volts | MAP Sensor    3.2 V<br>MAF Sensor    3.5 V | Throttle Position Sensor (TPS)<br>2.3    Volts |
| Engine Speed Sensor (RPM)<br>2400    rpm | Heated O₂ HO₂S<br>Upstream    0.3–0.8 V<br>Downstream 0.1–0.3 V | Vehicle Speed Sensor (VSS)<br>70    mph | Battery Voltage (B+)<br>14.4    Volts |
| Idle Air Control Valve (IAC)<br>30    percent | Evaporative Emission Canister Solenoid (EVAP)<br>ON | Torque Converter Clutch Solenoid (TCC)<br>ON | EGR Valve Control Solenoid  (EGR)<br>60    percent |
| Malfunction Indicator Lamp (MIL)<br>ON | Diagnostic Trouble Codes<br>PO122 | Open/Closed Loop<br>CLOSED | Fuel Pump Relay (FP)<br>ON |
| Fuel Level Sensor<br>3.6 V | Fuel Tank Pressure Sensor<br>1.9 V | Transmission Fluid Temperature Sensor<br>0.6 V | Transmission Turbine Shaft Speed Sensor<br>70 mph |
| Transmission Range Switch<br>OVERDRIVE | Transmission Pressure Control Solenoid<br>30% | Transmission Shift Solenoid 1<br>ON | Transmission Shift Solenoid 2<br>ON |
| **Measured Ignition Timing °BTDC   Base Timing: 10   Actual Timing: 35** | | | |

**Figure 9–33** OBD II scan tool data number 33.

95. The composite vehicle has a rough idle problem, but there are no other drivability difficulties. The engine compression and ignition system are tested and proven to be satisfactory. When the vacuum hose is disconnected from the EGR valve, the engine idles smoothly.

    Technician A says to test the wire from the EGR solenoid to the PCM for a grounded condition.

    Technician B says to check for high resistance in the PCM ground.

    Who is correct?
    A. A only
    B. B only
    C. Both A and B
    D. Neither A nor B

*Hint*    *In a negative backpressure EGR valve, exhaust pressure is supplied from the lower end of the EGR valve through the valve stem to a normally closed bleed valve. At lower engine speeds, there are negative pressure pulses in the exhaust between the positive pressure pulses created when each cylinder fires. These negative pressure pulses pull the bleed valve open in the EGR valve, which makes it impossible for vacuum to pull the EGR valve open. To open this type of EGR valve, there must be a specific engine rpm to decrease the negative pressure pulses in the exhaust and close the bleed valve, and vacuum must be supplied through the EGR solenoid to the EGR valve diaphragm.*

*When diagnosing EFI problems, propane may be dispersed into the air intake with the engine idling. When this action is taken, the $HO_2S$ voltage should immediately increase, and the injector pulse width should decrease. If the $HO_2S$ voltage remains the same, this sensor is inoperative or contaminated, or there may be an open circuit in the $HO_2S$ voltage signal wire.*

## Task 12  Determine root cause of failures.

96. A vehicle with a computer-controlled transmission has a poor fuel economy complaint. During a road test, the transmission does shift into all gears, but all the upshifts occur at a higher rpm than specified. The scan-tool data in Figure 9–34 were obtained during a road test with the vehicle operating at 40 mph (64 kmh) with a steady throttle opening. The cause of the problem could be

    A. a damaged shift solenoid.

    B. an inoperative transmission range switch.

    C. an inoperative throttle position sensor.

    D. an inoperative transmission fluid temperature sensor.

**Hint**  *A damaged shift solenoid may cause the transmission to miss one or more shifts. A defective transmission range switch may cause the transmission to be in a gear other than the gear selected by the gear selector position. An inoperative throttle position sensor may affect transmission shifting, but this problem is indicated in the TPS voltage signal. An inoperative transmission fluid temperature (TFT) sensor signal that indicates the transmission fluid is colder than the actual fluid temperature may cause late shifting in all gears.*

| SCAN TOOL DATA | | | |
|---|---|---|---|
| Engine Coolant Temperature Sensor (ECT) 0.44 Volts | Intake Air Temperature Sensor (IAT) 2.18 Volts | MAP Sensor 2.1 V  MAF Sensor 3.0 V | Throttle Position Sensor (TPS) 1.9 Volts |
| Engine Speed Sensor (RPM) 1300 rpm | Heated O₂ HO₂S Upstream 0.3–0.7 V Downstream 0.1–0.3 V | Vehicle Speed Sensor (VSS) 40 mph | Battery Voltage (B+) 14.2 Volts |
| Idle Air Control Valve (IAC) 30 percent | Evaporative Emission Canister Solenoid (EVAP) ON | Torque Converter Clutch Solenoid (TCC) ON | EGR Valve Control Solenoid (EGR) 40 percent |
| Malfunction Indicator Lamp (MIL) OFF | Diagnostic Trouble Codes NONE | Open/Closed Loop CLOSED | Fuel Pump Relay (FP) ON |
| Fuel Level Sensor 3.6 V | Fuel Tank Pressure Sensor 2.0 V | Transmission Fluid Temperature Sensor 4.5 V | Transmission Turbine Shaft Speed Sensor 40 mph |
| Transmission Range Switch OVERDRIVE | Transmission Pressure Control Solenoid 50% | Transmission Shift Solenoid 1 ON | Transmission Shift Solenoid 2 ON |
| Measured Ignition Timing °BTDC   Base Timing: 10   Actual Timing: 28 | | | |

**Figure 9–34** OBD II scan tool data number 34.

## Task 13   Determine root cause of multiple component failures.

97. The composite vehicle has a no-start complaint. Diagnosis proves the PCM is the problem. After the PCM is replaced, the engine starts, but the technician discovers that one headlight, one taillight, and two interior lights are burned out. When the charging circuit voltage is tested, it is satisfactory.

    Technician A says to test the wire from the alternator field circuit to the PCM for an intermittent ground.

    Technician B says to test the wire connected from the alternator battery terminal to the positive battery terminal for an intermittent open circuit.

    Who is correct?

    A. A only

    B. B only

    C. Both A and B

    D. Neither A nor B

**Hint**   *Damage to multiple electrical/electronic components may be caused by high voltage or high-voltage spikes in the electrical system. Continual high voltage may be caused by a damaged voltage regulator. High-voltage spikes may be caused by an intermittent open circuit in the wire connected between the alternator battery terminal and the positive battery terminal. High-voltage spikes may also be caused by damaged suppression diodes in relays or solenoids.*

## Task 14   Determine root cause of repeated component failures.

98. The composite vehicle repeatedly blows fuse number 56. The cause of this problem could be

    A. low resistance in the transmission fluid temperature (TFT) sensor.

    B. the post-catalyst $HO_2S$ heater wire is shorted to ground.

    C. the bank 1 $HO_2S$ wire is shorted to ground at PCM terminal 59.

    D. the power-steering pressure switch contacts are stuck closed.

**Hint**   *Blown fuses are caused by high current flow in the circuit to which the fuse is connected. High current flow may be caused by a wire shorted to ground or a shorted condition in a relay or solenoid winding. Low resistance in a component always results in high current flow.*

## Task 15   Verify effectiveness of repairs

99. An OBD II vehicle has a very intermittent stalling complaint. The customer informs the service writer that the engine idles rough and then stalls. The MIL is illuminated when the problem occurs. The technician connects a scan tool to the DLC, and discovers an EGR valve code. The engine has an electronically operated EGR valve.

    Technician A says the PCM may require reprogramming to install revised EGR control information.

    Technician B says the EGR passage into the intake manifold may be plugged.

    Who is correct?

    A. A only

    B. B only

    C. Both A and B

    D. Neither A nor B

**Hint**   *When verifying the effectiveness of the repairs to correct very intermittent complaints, the customer may be informed that the shop will give him a follow-up phone call after a specific length of time to determine if the repairs are satisfactory.*

# I/M Failure Diagnosis

## ASE Tasks, Questions, and Related Information

**Task 1**  Inspect and test for missing, modified, inoperative, or tampered components.

100. The engine in a vehicle shuts off while driving. A visual inspection of the underhood components does not reveal any problems. The scan-tool data in Figure 9–35 were obtained as the engine shut off while driving at 60 mph (96 kmh) during a road test. The cause of this problem is
    A. the vehicle speed sensor (VSS) is the problem.
    B. the fuel pump relay is faulty.
    C. the IAT sensor.
    D. the EGR valve is stuck open.

*Hint*    *Many PCMs have a fuel cut-off feature that allows the PCM to shut off the injectors to protect the engine if the engine rpm or vehicle speed exceeds a specific limit.*

**Task 2**  Locate relevant service information.

101. A vehicle fails an I/M 240 emission test as indicated by the emission traces and cut-points in Figure 9–36. The cause of this emission failure could be
    A. higher-than-specified fuel pressure.
    B. inoperative upstream $HO_2S$ that provides a low-voltage signal.

| SCAN TOOL DATA | | | |
|---|---|---|---|
| Engine Coolant Temperature Sensor (ECT)<br>0.45  Volts | Intake Air Temperature Sensor (IAT)<br>2.2  Volts | MAP Sensor   2.6 V<br>MAF Sensor   3.0 V | Throttle Position Sensor (TPS)<br>2.8  Volts |
| Engine Speed Sensor (RPM)<br>1900  rpm | Heated $O_2$ $HO_2S$<br>Upstream   0.3–0.8 V<br>Downstream 0.1–0.3 V | Vehicle Speed Sensor (VSS)<br>115  mph | Battery Voltage (B+)<br>14.4  Volts |
| Idle Air Control Valve (IAC)<br>30  percent | Evaporative Emission Canister Solenoid (EVAP)<br>ON | Torque Converter Clutch Solenoid (TCC)<br>ON | EGR Valve Control Solenoid  (EGR)<br>70  percent |
| Malfunction Indicator Lamp (MIL)<br>OFF | Diagnostic Trouble Codes<br>NONE | Open/Closed Loop<br>CLOSED | Fuel Pump Relay (FP)<br>ON |
| Fuel Level Sensor<br>3.5 V | Fuel Tank Pressure Sensor<br>0.7 V | Transmission Fluid Temperature Sensor<br>0.6 V | Transmission Turbine Shaft Speed Sensor<br>60 mph |
| Transmission Range Switch<br>OVERDRIVE | Transmission Pressure Control Solenoid<br>40% | Transmission Shift Solenoid 1<br>ON | Transmission Shift Solenoid 2<br>ON |
| Measured Ignition Timing °BTDC   Base Timing:                Actual Timing: | | | |

**Figure 9–35** OBD II scan tool data number 35.

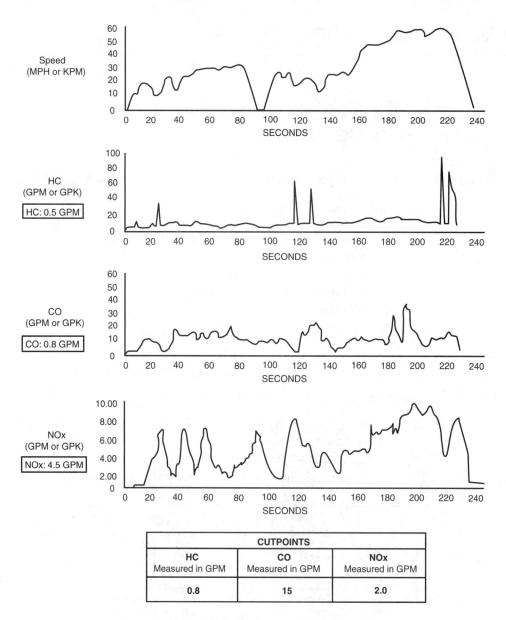

**Figure 9–36** Emission traces, I/M 240 emission test.

C. restricted injector orifices.

D. a partially clogged fuel return line.

**Hint**    *When diagnosing automotive problems, it is extremely important to have recent service bulletin information available. Several companies supply this information on CDs. For example, if the technician is not aware of computer reprogramming updates to correct certain drivability complaints, the technician will waste a lot of valuable time diagnosing the computer system components when the root cause of the problem is in the computer.*

## Task 3  Evaluate emission readings obtained during an I/M test to assist in emission failure diagnosis and repair.

102. An SFI engine fails an emission test for high HC and CO. The engine is equipped with a secondary air injection system that pumps air into the exhaust ports during engine warm-up and into the catalytic converter at normal engine temperature. The vehicle has no drivability complaints and the MIL is not on with the engine running.

Technician A says the ECT sensor voltage signal may be lower than specified.

Technician B says the secondary air injection system may be inoperative.

Who is correct?

A. A only

B. B only

C. Both A and B

D. Neither A nor B

**Hint**    *If a vehicle fails an emission test for high NOx, check the emission traces for indications of a lean air-fuel ratio. A lean air-fuel ratio may cause a lean misfire, which results in high HC and low CO. The high HC may appear during acceleration when misfiring is likely to occur. A lean air-fuel ratio also increases NOx emissions.*

*High HC and CO may be caused by a rich air-fuel ratio. When a secondary air injection system pumps air into the catalytic converter at normal engine temperatures, an inoperative system causes improper operation of the oxidation catalyst in the converter, and this results in high HC and CO emissions.*

## Task 4  Evaluate HC, CO, NOx, $CO_2$, and $O_2$ gas readings; determine the failure relationships.

103. A vehicle fails an enhanced emission test for high NOx and the engine has a detonation problem. The engine has a digital EGR valve. All of the following could be the cause of the problem EXCEPT

A. the EGR vacuum control solenoid is stuck closed.

B. one of the solenoids in the EGR valve has an open circuit.

C. the EGR exhaust passages are plugged.

D. the EGR ground wire has an open circuit.

104. A vehicle fails a non-enhanced emissions test for high HC at idle and 2,000 rpm. A five-gas analyzer indicates the following tailpipe emission readings. Idle: HC - 260 ppm, CO - 0.6 percent, $CO_2$ - 11.5 percent, $O_2$ - 3.8 percent, NOx - 650 ppm. 2,000 rpm: HC - 80 ppm, CO - 0.5 percent, $CO_2$ - 12 percent, $O_2$ - 1.8 percent, NOx - 350 ppm.

Technician A says to check for an intake manifold vacuum leak.

Technician B says to check for high fuel pressure.

Who is correct?

A. A only

B. B only

C. Both A and B

D. Neither A nor B

**Hint**    *High NOx emissions may be caused by an inoperative EGR valve or plugged EGR valve exhaust passages. An inoperative EGR valve increases combustion chamber temperatures, which may cause detonation.*

*A lean air-fuel ratio causes high HC, low CO, low $CO_2$, high $O_2$, and possible high NOx.*

*A rich air-fuel ratio causes high HC, high CO, low $CO_2$, low $O_2$, and low NOx.*

## Task 5  Use test instruments to observe, recognize, and interpret electrical/electronic signals.

105. An OBD II vehicle with a V6 engine has an acceleration stumble when the engine is at normal operating temperature. There are no DTCs in the PCM, and the MIL is not on with the engine running. The technician uses a lab scope to test various input sensors. During an acceleration stumble, the MAP sensor waveform is recorded on the lab scope (Figure 9-37).

Technician A says to test the charging circuit for erratic voltage.

Technician B says to test the voltage supply to the PCM.

Who is correct?

A. A only

B. B only

C. Both A and B

D. Neither A nor B

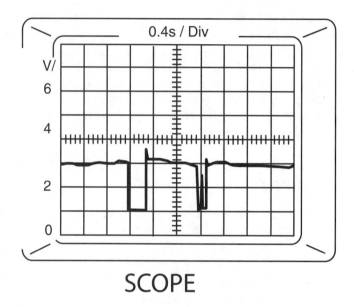

**Figure 9–37** MAP sensor lab scope waveform.

106. The lab scope is used to compare the waveforms on the upstream and downstream HO$_2$S sensors (Figure 9–38). The waveforms in this figure indicate

A. an inoperative downstream HO$_2$S sensor.

B. a rich air-fuel ratio.

C. an intake manifold vacuum leak.

D. an inoperative catalytic converter.

*Hint*    *A lab scope is very useful when diagnosing input sensors because this instrument reacts much faster than a digital multimeter (DMM), and thus the lab scope waveform may indicate minor glitches that do not appear on other instruments.*

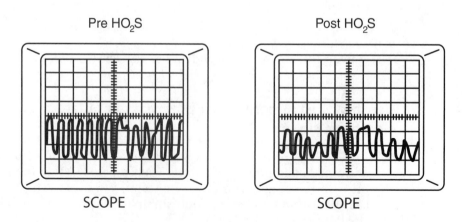

**Figure 9–38** Upstream and downstream HO$_2$S waveforms.

*Sudden voltage dropouts in lab scope waveforms usually indicate a momentary glitch in an input sensor. When comparing upstream and downstream HO₂S sensors, the waveform on the downstream HO₂S should indicate a lower voltage and with slower cycles.*

## Task 6   Analyze HC, CO, NOx, CO₂, and O₂ readings; determine diagnostic test sequence.

107. When using a four-gas analyzer to test the tailpipe emissions on a multiport fuel-injected engine with a three-way catalytic converter, the $O_2$ reading is above 4 percent. The cause of this problem could be
   A. retarded ignition timing.
   B. an inoperative EGR valve.
   C. a lean air-fuel ratio.
   D. an inoperative catalytic converter.

108. A four-gas analyzer is connected to the tailpipe on an OBD II vehicle with an SFI engine. With the engine running at idle speed and normal operating temperature, the following readings are obtained: HC - 500 ppm, CO - 3.8 percent, $CO_2$ - 9 percent, $O_2$ - 0.1 percent. The atmospheric temperature was 98°F (37°C). The scan-tool data obtained at idle speed are shown in Figure 9–39.
   Technician A says the IAT sensor may be the problem.
   Technician B says the MAP sensor vacuum hose may have a leak.
   Who is correct?
   A. A only
   B. B only
   C. Both A and B
   D. Neither A nor B

| SCAN TOOL DATA | | | |
|---|---|---|---|
| Engine Coolant Temperature Sensor (ECT)<br>0.46 Volts | Intake Air Temperature Sensor (IAT)<br>1.5 Volts | MAP Sensor   1.5 V<br>MAF Sensor   0.8 V | Throttle Position Sensor (TPS)<br>0.6 Volts |
| Engine Speed Sensor (RPM)<br>750 rpm | Heated O₂ HO₂S<br>Upstream   0.4–0.9 V<br>Downstream 0.1–0.3 V | Vehicle Speed Sensor (VSS)<br>0 mph | Battery Voltage (B+)<br>14.4 Volts |
| Idle Air Control Valve (IAC)<br>30 percent | Evaporative Emission Canister Solenoid (EVAP)<br>OFF | Torque Converter Clutch Solenoid (TCC)<br>OFF | EGR Valve Control Solenoid (EGR)<br>0 percent |
| Malfunction Indicator Lamp (MIL)<br>OFF | Diagnostic Trouble Codes<br>NONE | Open/Closed Loop<br>CLOSED | Fuel Pump Relay (FP)<br>ON |
| Fuel Level Sensor<br>3.5 V | Fuel Tank Pressure Sensor<br>1.9 V | Transmission Fluid Temperature Sensor<br>0.6 V | Transmission Turbine Shaft Speed Sensor<br>0 mph |
| Transmission Range Switch<br>PARK | Transmission Pressure Control Solenoid<br>75% | Transmission Shift Solenoid 1<br>ON | Transmission Shift Solenoid 2<br>OFF |
| Measured Ignition Timing °BTDC   Base Timing: 10   Actual Timing: 10 | | | |

**Figure 9–39** OBD II scan tool data number 39.

**Hint**    *High $O_2$ emissions are caused by a lean air-fuel ratio. This problem may also cause lean misfiring and high HC emissions and high NOx emissions. High HC and CO emissions with low $CO_2$ and $O_2$ emissions indicate a rich air-fuel ratio, which could be caused by high fuel pressure, dripping injectors, or low $HO_2S$ voltage signals.*

## Task 7  Diagnose the cause of no-load I/M test HC emission failures.

109. An OBD II vehicle with a four-cylinder SFI engine fails an emission test for HC at idle and 2,000 rpm. A four-gas analyzer connected to the tailpipe provided the readings in Figure 9–40.

     Technician A says to test the ignition system for cylinder misfiring.

     Technician B says to check the fuel pressure regulator for a leaking diaphragm and check the crankcase oil for contamination.

     Who is correct?

     A. A only

     B. B only

     C. Both A and B

     D. Neither A nor B

| Engine Speed | Idle | 2000 RPM |
|---|---|---|
| HC (ppm) | 1100 | 900 |
| CO (percent) | 0.3 | 0.3 |
| $CO_2$ (percent) | 12 | 12 |
| $O_2$ (percent) | 1.0 | 1.0 |

**Figure 9–40**  Exhaust emission readings number 40.

**Hint**    *Cylinder misfiring causes very high HC emissions and low CO emissions. A rich air-fuel ratio causes high HC and CO emissions.*

## Task 8  Diagnose the cause of no-load I/M test CO emission failures.

110. An OBD I vehicle with a V6 engine and multiport fuel injection fails an emission test for CO at 2,000 rpm. A four-gas analyzer connected to the tailpipe indicates the emissions shown in Figure 9–41. The cause of the high CO at 2,000 rpm could be

     A. the EVAP canister is saturated with gasoline.

     B. one of the spark plugs is fouled.

| Engine Speed | Idle | 2000 RPM |
|---|---|---|
| HC (ppm) | 50 | 150 |
| CO (percent) | 0.3 | 6 |
| $CO_2$ (percent) | 14 | 12 |
| $O_2$ (percent) | 0.3 | 0.2 |

**Figure 9–41**  Exhaust emission readings number 41.

| SCAN TOOL DATA | | | |
|---|---|---|---|
| Engine Coolant Temperature Sensor (ECT)<br><br>0.45    Volts | Intake Air Temperature Sensor (IAT)<br><br>2.2    Volts | MAP Sensor    1.5 V<br><br>MAF Sensor    2.5 V | Throttle Position Sensor (TPS)<br><br>0.6    Volts |
| Engine Speed Sensor (RPM)<br><br>750    rpm | Heated $O_2$ $HO_2S$<br>Upstream    0.4–0.9 v<br>Downstream 0.1–0.3 v | Vehicle Speed Sensor (VSS)<br><br>0    mph | Battery Voltage (B+)<br><br>14.2    Volts |
| Idle Air Control Valve (IAC)<br><br>30    percent | Evaporative Emission Canister Solenoid (EVAP)<br><br>OFF | Torque Converter Clutch Solenoid (TCC)<br><br>OFF | EGR Valve Control Solenoid  (EGR)<br><br>0    percent |
| Malfunction Indicator Lamp (MIL)<br>OFF | Diagnostic Trouble Codes<br><br>NONE | Open/Closed Loop<br><br>CLOSED | Fuel Pump Relay (FP)<br><br>ON |
| Fuel Level Sensor<br><br>3.6 V | Fuel Tank Pressure Sensor<br><br>1.8 V | Transmission Fluid Temperature Sensor<br><br>0.5 V | Transmission Turbine Shaft Speed Sensor<br><br>0 mph |
| Transmission Range Switch<br>PARK | Transmission Pressure Control Solenoid<br>80% | Transmission Shift Solenoid 1<br>ON | Transmission Shift Solenoid 2<br>OFF |

Measured Ignition Timing °BTDC   Base Timing: 10   Actual Timing:  12

**Figure 9–42** OBD II scan tool data number 42.

    C. the PCV valve is stuck open.

    D. the fuel pressure regulator is sticking open.

111. A vehicle fails an enhanced emission test for high CO and HC. With the engine operating at normal temperature and idle speed, the scan-tool data in Figure 9–42 were obtained. The cause of the emission failure could be

    A. an inoperative upstream $HO_2S$ sensor.

    B. an inoperative MAP sensor.

    C. an inoperative MAF sensor.

    D. an inoperative IAT sensor.

**Hint**    *Problems in the EVAP system may cause a rich air-fuel ratio only at speeds above idle because the EVAP canister is not purged at idle. Malfunctioning input sensors may cause CO and HC emission test failures. For example, if the ECT or IAT sensor is sending a higher-than-specified voltage signal to the PCM, a colder engine coolant temperature or colder air intake temperature is indicated. Under this condition, the PCM provides a richer air-fuel ratio and this causes high CO and HC emissions.*

## Task 9  Diagnose the cause of loaded mode I/M test HC emission failures.

112. The composite vehicle fails an enhanced emission test for high HC. All the other emissions are satisfactory. On the HC emission trace, the high HC emissions only occur when the engine is accelerated.

    Technician A says one of the ignition coils may be the problem and may be causing cylinder misfire under load.

    Technician B says the fuel pressure regulator may be sticking closed.

    Who is correct?

    A. A only

    B. B only

C. Both A and B

D. Neither A nor B

113. An OBD II vehicle fails an enhanced emission test for high HC and CO. The emission traces indicate that high HC and CO emissions occur during deceleration. The scan-tool data did not indicate any problems. When the voltage is measured from the PCM side of the EVAP solenoid to ground with the engine idling, the voltage is 0.2 V. The cause of this emission failure could be

A. an open circuit in the voltage supply wire to the EVAP solenoid.

B. an open circuit in the EVAP solenoid winding.

C. a shorted EVAP solenoid winding.

D. a grounded wire between the EVAP solenoid and the PCM.

**Hint**    *High HC and CO emissions that occur at idle and higher engine speeds may be caused by a rich air-fuel ratio. High HC emissions that occur only during acceleration may be caused by an ignition misfire resulting from insufficient secondary reserve voltage.*

*High HC and CO emissions that occur only during deceleration and idle may be caused by a rich air-fuel ratio resulting from a problem in the EVAP system.*

## Task 10   Diagnose the cause of loaded-mode I/M test CO emission failures.

114. An OBD I vehicle with an SFI four-cylinder engine fails an emission test for high CO and HC. Scan-tool data indicate the $HO_2S$ voltage is 0.6 V to 0.9 V. All of the following could be the cause of the problem EXCEPT

A. an inoperative MAP sensor.

B. an inoperative $HO_2S$ sensor.

C. an inoperative ECT sensor.

D. a partially restricted fuel return line.

115. An OBD II vehicle with an SFI V6 engine fails an enhanced emission test for high CO and HC. The only problem indicated in the scan-tool data is low upstream $HO_2S$ sensor voltage at 0.1 V to 0.4 V. When propane is dispersed into the intake manifold with the engine idling, the upstream $HO_2S$ sensor voltage immediately increases to 0.9 V.

Technician A says to replace the upstream $HO_2S$ sensor.

Technician B says the exhaust manifold may be cracked ahead of the $HO_2S$ sensor.

Who is correct?

A. A only

B. B only

C. Both A and B

D. Neither A nor B

**Hint**    *When diagnosing emission failures with scan-tool data, always look at all the data. Some problems, such as a malfunctioning IAT sensor with higher-than-specified resistance, cause a high-voltage signal from this sensor, which results in a rich air-fuel ratio. Under this condition, the $HO_2S$ sensor voltage signal is higher than specified. The technician must decide if the high $HO_2S$ sensor voltage is the root cause of the problem, or if this signal is just the result of another problem.*

*When a sensor signal is not within specifications, the technician must determine if the sensor is malfunctioning or if some other problem is causing the improper signal.*

## Task 11   Diagnose the cause of loaded-mode I/M test NOx emission failures.

116. An OBD I vehicle with a four-cylinder SFI engine fails an enhanced emission test for high NOx. The engine has a rough idle problem. A five-gas analyzer provides the following emission readings with the engine at normal operating temperature. Idle:

HC - 150 ppm, CO - 0.3 percent, $CO_2$ - 12 percent, $O_2$ - 0.8 percent, NOx - 400 ppm; 2,000 rpm: HC - 200 ppm, CO - 0.3 percent, $CO_2$ - 11.8 percent, $O_2$ - 0.9 percent, NOx - 1,200 ppm. The scan-tool data in Figure 9–43 were obtained with the engine running at normal operating temperature and idle speed. The cause of the high NOx emissions could be

A. restricted injector orifices.

B. an inoperative TPS.

C. an inoperative ECT sensor.

D. a leaking throttle body to intake gasket.

### SCAN TOOL DATA

| Engine Coolant Temperature Sensor (ECT) 0.44 Volts | Intake Air Temperature Sensor (IAT) 2.2 Volts | Manifold Absolute Pressure Sensor (MAP) 1.5 Volts | Throttle Position Sensor (TPS) 0.6 Volts |
|---|---|---|---|
| Engine Speed Sensor (RPM) 800 rpm | Heated Oxygen Sensor ($HO_2S$) 0.2–0.6 Volts | Vehicle Speed Sensor (VSS) 0 mph | Battery Voltage (B+) 14.4 Volts |
| Idle Air Control Valve (IAC) 30 percent | Evaporative Emission Canister Solenoid (EVAP) OFF | Torque Converter Clutch Solenoid (TCC) OFF | EGR Valve Control Solenoid (EGR) 0 percent |
| Malfunction Indicator Lamp (MIL) OFF | Diagnostic Trouble Codes NONE | Open/Closed Loop CLOSED | Fuel Pump Relay (FP) ON |

Measured Ignition Timing °BTDC     Base Timing: 10     Actual Timing: 10

**Figure 9–43** OBD I scan tool data number 43.

117. An OBD II vehicle with a V6 SFI engine fails an enhanced emission test for high NOx. The scan-tool data is satisfactory with the engine operating at idle speed and operating under load. The other emission levels are satisfactory at idle and under load.

Technician A says the fuel pressure may be low.

Technician B says the EGR exhaust passages may be restricted.

Who is correct?

A. A only

B. B only

C. Both A and B

D. Neither A nor B

**Hint**   *High NOx emissions may be caused by an inoperative EGR system or restricted EGR exhaust passages. Excessive carbon buildup in the combustion chambers may cause detonation and high NOx emissions. A malfunctioning knock sensor also causes detonation and high NOx emissions.*

*A lean air-fuel ratio may also contribute to high NOx emissions.*

## Task 12   Evaluate the MIL operation for onboard diagnostic I/M testing.

118. When diagnosing the composite vehicle, a DTC is stored indicating a defect in the variable valve timing system, but the MIL is not illuminated.

Technician A says the fault has occurred on two consecutive trips.

Technician B says the defect has occurred on one trip.

Who is correct?

A. A only

B. B only

C. Both A and B

D. Neither A nor B

*Hint*    *The MIL is located in the instrument cluster. When the ignition switch is turned on, the MIL should be illuminated for fifteen seconds to prove the bulb operation. If the MIL is on with the engine running, the PCM or TCM has sensed a defect in the computer system. A flashing MIL on an OBD II vehicle indicates a severe engine misfire that may result in catalytic converter damage. On some vehicles, such as the composite vehicle, the MIL is illuminated continuously if the instrument cluster cannot communicate via the data links with the PCM and TCM.*

## Task 13   Evaluate monitor readiness status for onboard diagnostic I/M testing.

119. The heated oxygen sensor monitor indicates "Not Completed" on a scan tool. The vehicle is driven under the proper conditions to complete a drive cycle, and this monitor does not run.

Technician A says the fuel level in the fuel tank may be below 15 percent.

Technician B says the battery voltage may be below 10 V.

Who is correct?

A. A only

B. B only

C. Both A and B

D. Neither A nor B

*Hint*    *When using the monitor readiness status in on-board diagnostic I/M testing, all the monitors must indicate "Completed." If a monitor indicates "Not Completed," the vehicle may not have been driven under the proper conditions to allow the monitor to run or there may be a fault in the OBD II system that is preventing the monitor from running. When a defect occurs in an input sensor, such as the engine coolant temperature (ECT) sensor, monitors that are temperature-related (like the EVAP and EGR monitors) will not run. The monitors will not run if the battery voltage supplied to the PCM is less than 10 V. The fuel level in the fuel tank must be above 15 percent to complete the EVAP and HO$_2$S monitors. The technician must check the vehicle manufacturer's service information to determine the enable criteria for each monitor.*

## Task 14   Diagnose communication failures with the vehicle during onboard diagnostic I/M testing.

120. On the composite vehicle, the MIL is illuminated continuously when the ignition switch is turned on. The most likely cause of this problem is a communication failure between the

A. immobilizer control module and the PCM.

B. immobilizer control module and the instrument cluster.

C. PCM and the instrument cluster.

D. A/C control module and the PCM.

*Hint*    *In the OBD II system, the data link network is connected to the TCM, PCM, instrument cluster, and the DLC. When a defect in the data link network prevents communication between the PCM, TCM, and the instrument cluster, the MIL is illuminated continuously.*

## Task 15   Perform functional I/M tests (including fuel cap tests).

121. The enhanced EVAP system is being leak tested on an OBD II vehicle (Figure 9–44). The EVAP system is pressurized to 14 in. $H_2O$. After two minutes, the pressure has dropped to 10 in. $H_2O$.

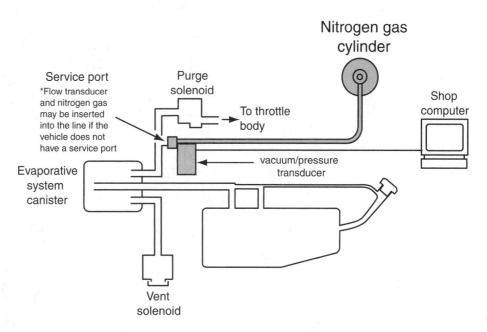

**Figure 9–44** Enhanced EVAP system leak test.

Technician A says the vapor line between the canister and the fuel tank may be leaking.
Technician B says the vacuum management valve (VMV) may be leaking.
Who is correct?
A. A only
B. B only
C. Both A and B
D. Neither A nor B

*Hint*  *The EVAP system pressure test is completed with the engine not running. This test checks for leaks in the EVAP system. On an OBD II vehicle with an enhanced EVAP system, test equipment is used to leak test the EVAP system by pressurizing the system with nitrogen. The EVAP system is pressurized to an average of 14 in. $H_2O$. If the EVAP system pressure remains above 8 in. $H_2O$ after two minutes, the system passes. After the test is completed, there must be pressure in the fuel tank when the filler cap is removed. If the pressure drop was within specifications and there is no pressure in the fuel tank, there is a restriction in the vapor hose preventing pressure from reaching the fuel tank.*

*The EVAP system purge flow test is completed while driving the vehicle on a dynamometer during an enhanced emission test. A flow transducer is connected to the EVAP system service port. If the system does not have a service port, the flow transducer is connected to the purge line between the canister and the intake manifold (Figure 9–45). The flow transducer connected to the EVAP system service port or purge hose senses vacuum in the EVAP system. If this vacuum is between 5 and 50 $H_2O$ for five seconds or more during the enhanced emission test, the EVAP system passes the flow test. When the EVAP system vacuum is within this range for five seconds or more, the EVAP flow during the enhanced emission test is one liter or more.*

## Task 16  Verify effectiveness of repairs.

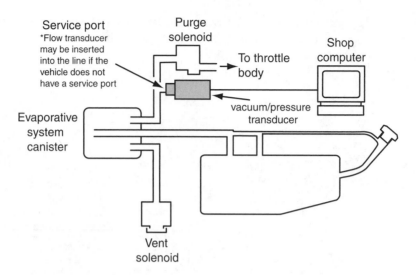

**Figure 9–45** Enhanced EVAP system flow test.

122. An OBD II vehicle with a four-cylinder SFI engine fails an I/M test for high HC. The technician finds and repairs a vacuum leak in the brake booster hose. Before sending the car for an I/M retest, the technician connects a DSO to the upstream HO$_2$S signal wire. The HO$_2$S waveform is shown in Figure 9–46. This waveform indicates

A. the HO$_2$S signal is being used to control the air-fuel ratio and the car may be returned for an I/M retest.

B. the HO$_2$S sensor is contaminated with room temperature vulcanizing (RTV) sealant.

C. the air-fuel ratio is continually rich and the fuel pressure may be too high.

D. a cylinder misfire caused by an ignition problem such as a bad spark plug is occurring.

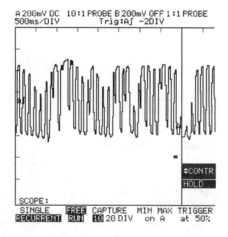

**Figure 9–46** HO$_2$S waveform. *(Reproduced with permission from Fluke Corporation)*

123. An OBD I vehicle fails an emission test for high HC and CO. The technician discovers the MAP sensor is the problem and replaces this sensor. After the replacement, the emission readings in Figure 9–47 were obtained on a four-gas analyzer.

Technician A says the air-fuel ratio is still too rich and further diagnosis and repairs are required.

Technician B says the ignition system or mechanical engine problems must be causing cylinder misfiring.

Who is correct?

A. A only

B. B only

C. Both A and B

D. Neither A nor B

| Engine Speed | Idle | 2000 RPM |
|---|---|---|
| HC (ppm) | 60 | 25 |
| CO (percent) | 0.8 | 0.3 |
| $CO_2$ (percent) | 13 | 13 |
| $O_2$ (percent) | 0.5 | 0.5 |

**Figure 9–47** Exhaust emission readings number 47.

**Hint**    *An automotive repair facility usually does not have the very expensive equipment used in some loaded-mode emission test facilities. Technicians usually have to complete their diagnosis with a four- or five-gas emissions analyzer. The technician must understand what causes each automotive emission to be high or low, and the technician must also have some baseline figures for emission levels on various vehicles. A digital storage oscilloscope (DSO) is very useful in testing input sensors and outputs to determine if their operation is normal or if a problem still exists.*

# Post Test

1. An OBD II vehicle has a loss of power complaint at higher speeds. The engine starts easily at all temperatures, and runs smoothly at idle and lower speeds. An exhaust gas analyzer connected to the tailpipe provides the following readings with the engine at 2,000 rpm and normal operating temperature: HC - 60 ppm, CO - 0.5 percent, $CO_2$ - 9 percent, $O_2$ - 3.5 percent. The most likely cause of this problem is

A. a restricted fuel return line.

B. low fuel pressure.

C. the EGR solenoid is stuck open.

D. the EVAP purge solenoid is stuck open

2. The composite vehicle has a hard-starting problem. The engine cranks properly, but cranking time is much longer than normal. There are no other drivability complaints.

Technician A says the vehicle speed sensor (VSS) may be defective.

Technician B says the camshaft position (CMP) sensor may be defective.

Who is correct?

A. A only

B. B only

C. Both A and B

D. Neither A nor B

3. The PCM fails for the third time on the composite vehicle. When diagnosing the output actuators, the EVAP canister purge solenoid, EVAP canister vent solenoid, and torque converter clutch solenoid all have 14 V applied to them with the engine running. When these solenoids are energized by the PCM, the voltage on the PCM side of the EVAP purge solenoid is 4.5 V and the voltage on the PCM side of the EVAP vent solenoid is 7 V. The root cause of the PCM failures is
   A. high resistance in the wire from the EVAP vent solenoid to the PCM.
   B. high resistance at PCM terminal 83.
   C. a shorted winding in the EVAP vent solenoid causing high current through the PCM.
   D. high resistance in the voltage supply wire to the EVAP vent solenoid.

4. An OBD II vehicle fails an enhanced emissions test for NOx. The engine has no drivability complaints and a scan tool indicates satisfactory data with no DTCs in the PCM memory.
   Technician A says to check the EGR exhaust passages for carbon buildup.
   Technician B says to test the EGR solenoid for a sticking plunger.
   Who is correct?
   A. A only
   B. B only
   C. Both A and B
   D. Neither A nor B

5. When using a digital storage oscilloscope to test the injectors in a V6 engine, the induced voltage spike when the injectors shut off is 325 V on injectors 1, 3, 4, 5, and 6. The induced voltage spike when number 2 injector shuts off is 75 V. The engine has no drivability complaints, but this problem may cause
   A. damage to the PCM.
   B. a rich air-fuel ratio.
   C. engine detonation.
   D. high combustion chamber temperatures

6. An OBD II vehicle has a cylinder misfire problem at idle, and the following readings were obtained on a four-gas analyzer: HC - 1,900 ppm, CO - 0.4 percent, $CO_2$ - 7.5 percent, $O_2$ - 2.5 percent.
   Technician A says the intake manifold has a large vacuum leak.
   Technician B says the one of the injector plungers may be stuck closed.
   Who is correct?
   A. A only
   B. B only
   C. Both A and B
   D. Neither A nor B

7. The composite vehicle has a poor fuel economy complaint, and the following readings were obtained on a four-gas analyzer with the engine operating at idle speed and normal operating temperature: HC - 150 ppm, CO - 1.5 percent, $CO_2$ - 7 percent, $O_2$ - 0.5 percent. All of the following defects could be the cause of the problem EXCEPT
   A. a leaking brake booster vacuum hose.
   B. high fuel pressure.
   C. a partially restricted fuel return line.
   D. a fuel pressure regulator diaphragm that's leaking.

8. An OBD II vehicle has rough idle operation and an acceleration stumble. The scan-tool data is satisfactory except the mass air flow (MAF) sensor reading is erratic with a steady throttle opening.
   Technician A says to clean the terminals in the MAF sensor wiring connector.

Technician B says to clean the hot wire in the MAF with some throttle body cleaner and a Q-tip.

Who is correct?

A. A only

B. B only

C. Both A and B

D. Neither A nor B

9. When diagnosing the composite vehicle, a scan tool indicates the EGR and EVAP monitors have not been run. After the vehicle is driven through the proper drive cycle to cause the monitors to run, the EGR and EVAP monitors still have not run.

Technician A says the engine coolant temperature sensor (ECT) signal may be defective.

Technician B says the fuel level in the fuel tank may be very low.

Who is correct?

A. A only

B. B only

C. Both A and B

D. Neither A nor B

10. The malfunction indicator light (MIL) is blinking on an OBD II vehicle. The most likely cause of this problem is

A. an extremely lean air-fuel ratio.

B. a saturated EVAP canister.

C. a defective fuel tank pressure sensor.

D. a severe engine misfire problem.

11. An OBD vehicle experiences repeated catalytic converter failure. The catalytic converter fails approximately every 10,000 mi (16,000 km). The owner also says the car has never provided the fuel mileage as specified by the vehicle manufacturer, but there are no other drivability complaints. The most likely cause of this problem is

A. dripping fuel injectors.

B. higher-than-specified fuel pressure.

C. a restricted fuel filter.

D. a leak in the fuel return line.

12. The composite vehicle experiences repeated PCM failure. Each time the PCM fails, the PCM driver that operates the fuel pump relay stops functioning and the fuel pump becomes inoperative. The fuel pump relay winding has the specified resistance.

Technician A says the suppression diode in the fuel pump relay may be defective.

Technician B says the fuel pump motor may have shorted windings and high current flow.

Who is correct?

A. A only

B. B only

C. Both A and B

D. Neither A nor B

13. When diagnosing the composite vehicle with a scan tool, the short-term and long-term fuel trim numbers on bank 1 are +30 and +20, respectively. The short-term and long-term fuel trim numbers on bank 2 are +6 and +4, respectively. The cause of this problem may be

A. an intake manifold vacuum leak on the same side of the engine as number 1 cylinder.

B. an exhaust manifold leak on the opposite side of the engine from number 1 cylinder.

    C. dripping injectors on the same side of the engine as number 1 cylinder.

    D. a defective downstream HO$_2$S on the same side as number 1 cylinder.

14. The composite vehicle has an open circuit at PCM terminal 82.

    Technician A says the fuel tank pressure sensor will be inoperative.

    Technician B says the manifold absolute pressure sensor will be inoperative.

    Who is correct?

    A. A only

    B. B only

    C. Both A and B

    D. Neither A nor B

15. An OBD II vehicle stalls frequently during deceleration. During a road test, scan-tool data indicate the PCM does not command the torque converter clutch (TCC) off when the brake pedal is depressed and the vehicle decelerated. The most likely cause of this problem is

    A. a defective TCC solenoid in the transaxle.

    B. a hydraulic problem in the TCC lockup system.

    C. worn friction material on the TCC lockup plate.

    D. a defective brake switch input to the PCM.

16. When testing the engine coolant temperature (ECT) sensor on the composite vehicle, the sensor has a voltage drop of 1V at 80°F (175°C), and a voltage drop of 3 V at 0°F (−20°C).

    Technician A says this sensor reading may cause hard starting.

    Technician B says this sensor reading may cause an excessively rich air-fuel ratio when the engine is cold. Who is correct?

    A. A only

    B. B only

    C. Both A and B

    D. Neither A nor B

17. On an OBD II vehicle, the A/C clutch disengages after the vehicle has been operated for approximately one hour in hot weather.

    Technician A says to use a manifold gauge set to test the A/C system pressures.

    Technician B say to replace the A/C pressure sensor.

    Who is correct?

    A. A only

    B. B only

    C. Both A and B

    D. Neither A nor B

18. The cooling fan is inoperative on the composite vehicle. The voltage inputs to the fan control relay contacts and the relay winding are satisfactory, and the PCM ground is satisfactory. With the ignition switch on and the engine coolant temperature at 220°F (104°C) the voltage at PCM terminal 16 is 12 V. The cause of the inoperative cooling fan is

    A. an open wire from the fan control relay winding to PCM terminal 16.

    B. an open fan control relay winding.

    C. an open suppression diode in the fan control relay.

    D. an open circuit in the PCM.

19. An OBD II vehicle with a coil-on-spark-plug ignition system has a no-start problem. A test spark plug does not fire when it is connected to any of the spark plug wires and the engine is cranked.

    Technician A says to test the fuse in the voltage supply circuit to the primary coil windings.

Technician B says to test the crankshaft sensor.

Who is correct?

A. A only

B. B only

C. Both A and B

D. Neither A nor B

20. The owner of an OBD II vehicle complains about poor fuel economy and a strong "rotten egg" odor coming from the tailpipe. When tested with an exhaust gas analyzer, the following readings were obtained: HC - 150 ppm, CO - 1.5 percent, $CO_2$ - 8 percent, and $O_2$ - 0.5 percent. The fuel pressure is within specifications. The most likely cause of this problem is

A. the engine coolant temperature sensor has less resistance than specified.

B. the fuel pressure regulator diaphragm is leaking.

C. the upstream $HO_2S$ voltage is higher than normal.

D. the MAP sensor voltage is lower than normal.

# Answers and Analysis

**1. D**  A is wrong. Installing exhaust headers in place of exhaust manifolds changes the exhaust flow and reduces exhaust backpressure. This action may not change exhaust temperature, and so the catalytic converter is not affected.

B is wrong. Installing exhaust headers in place of exhaust manifolds does not affect the operation of the evaporative emission system.

C is wrong. Installing exhaust headers in place of exhaust manifolds does not affect the operation of the coolant temperature sensor.

D is correct. Installing exhaust headers in place of exhaust manifolds reduces exhaust backpressure and EGR flow.

**2. B**  A is wrong. Although the service manual is an excellent source of vehicle information and specifications, a corrected VECI label contains the most up-to-date information.

B is correct. Although the service manual is an excellent source of vehicle information and specifications, a corrected VECI label contains the most up-to-date information.

C is wrong because Technician A is wrong and Technician B is correct.

D is wrong because Technician A is wrong and Technician B is correct.

**3. B**  A is wrong because the next step should be to drive the vehicle to allow the PCM adaptive memory to relearn the system.

B is correct because the next step should be to drive the vehicle to allow the PCM adaptive memory to relearn the system.

C is wrong because Technician A is wrong and Technician B is right.

D is wrong because Technician A is wrong and Technician B is right.

**4. B**  In this question, you are asked to select the response that is not true.

A is wrong because freeze frame data are stored for emission-related DTCs.

B is correct because the PCM can only store one freeze frame data record. Therefore, response B is not true and this is the correct answer.

C is wrong because failure records are stored for emission-related, and other, DTCs.

D is wrong because the PCM can store data from five failure records.

**5. A**  A is correct. With the engine operating at idle speed, the MAP sensor voltage should be about 1.2 V. The MAP sensor voltage is 2.75 V, and the scan-tool data do not indicate any other problems. Therefore, the MAP sensor is the problem.

B is wrong because an intake manifold vacuum leak would cause the inoperative MAP sensor voltage, but this problem would also result in low HO$_2$S sensor voltage.

C is wrong because a sticking EGR valve may reduce the intake manifold vacuum and cause the inoperative MAP sensor voltage, but this problem also causes rough idle operation.

D is wrong because the IAT sensor voltage is normal, depending on atmospheric temperature.

**6.  A**    A is correct because fouled spark plugs in cylinders 4 and 5 may cause low firing voltages on both cylinders.

B is wrong because cylinders 4 and 5 are in opposite cylinder banks, and so a blown head gasket could not cause low firing voltages on these cylinders because they are not adjacent to each other.

C is wrong because Technician A is right and Technician B is wrong.

D is wrong because Technician A is right and Technician B is wrong.

**7.  D**    In this question, you are asked to select the response that is not true.

A is wrong, because bad spark plug wires may cause cylinder misfiring, resulting in high HC, low CO, low CO$_2$, and high O$_2$ readings.

B is wrong, because bad spark plugs may cause cylinder misfiring, resulting in high HC, low CO, low CO$_2$ and high O$_2$ readings.

C is wrong, because low cylinder compression may cause cylinder misfiring, resulting in high HC, low CO, low CO$_2$, and high O$_2$ readings.

D is correct because dripping injectors cause a rich air-fuel ratio resulting in high HC, high CO, low CO$_2$, and low O$_2$ emissions. Therefore, response D is not true, and this is the correct answer.

**8.  B**    A is wrong because fuse number 3 also supplies voltage to the fuel pump relay and the A/C clutch relay. If fuse number 3 is open, these components and the cooling fan would all be inoperative.

B is correct because an open circuit between the cooling fan relay and PCM terminal 16 would cause an inoperative cooling fan.

C is wrong because Technician A is wrong and Technician B is right.

D is wrong because Technician A is wrong and Technician B is right.

**9.  D**    A is wrong because the knock sensor responds properly when the exhaust manifold is tapped.

B is wrong because the knock sensor responds properly when the exhaust manifold is tapped.

C is wrong because the PCM decreases the spark advance when the engine is tapped.

D is correct because with a broken engine mount the engine lifts up during acceleration and cruising speed operation. On deceleration, the engine settles down and the broken engine mount allows metal-to-metal contact, which the knock sensor senses as a vibration similar to engine detonation. When the knock sensor and PCM decrease the spark advance to 0 degrees with the A/C on, the engine stalls.

**10.  A**    A is correct because an open circuit in the wires to shift solenoids 1 and 2 causes both solenoids to be inoperative, resulting in transmission operation only in second and reverse.

B is wrong because fuse number 4 supplies voltage to the windings in many of the output control solenoids including the ignition coils.

C is wrong because an open circuit in shift solenoid 1 prevents fourth gear.

D is wrong because a short in shift solenoid 2 may damage this solenoid driver in the PCM.

**11.  C**    A is wrong because if the actual EGR opening is less than the commanded EGR opening, NOx emissions are high.

B is wrong because if the actual EGR opening is less than the commanded EGR opening, NOx emissions are high.

C is correct because if the actual EGR opening is less than the commanded EGR opening, NOx emissions are high.

D is wrong because if the actual EGR opening is less than the commanded EGR opening, NOx emissions are high.

**12. D** A is wrong. The emission readings obtained may be caused by crankcase oil that is diluted with gasoline, but that does not explain how the crankcase oil became diluted with gasoline. A is not the root cause of the problem.

B is wrong because a bad PCV valve may cause high HC and CO emissions, but this problem would not explain the overfilled crankcase.

C is wrong because excessive engine blowby allows oil past the rings, but this would not contaminate the oil with gasoline.

D is correct. A leaking pressure regulator diaphragm allows gasoline to flow through the regulator vacuum hose into the intake manifold. This excessive fuel eventually contaminates the engine oil, and this contamination plus the extra fuel entering the intake causes high HC and CO emissions.

**13. B** A is incorrect because high resistance in the PCM ground may affect the PCM operation, but this problem would not cause failure of the PCM and two ignition coil driver modules.

B is correct because an intermittent open circuit at the alternator battery terminal causes very high voltage peaks in the electrical system, which could damage the PCM and the ignition coil driver modules.

C is wrong because excessively wide spark plug gaps increase the secondary coil voltage required to fire the spark plugs, but this problem would not cause failure of the PCM and the ignition coil driver modules.

D is wrong because low resistance in some of the ignition coil primary windings increases the primary current flow, and this could damage the ignition coil driver modules. Because the ignition coil driver modules turn the primary current on and off, this problem should not damage the PCM.

**14. A** A is correct because a shorted cooling fan relay winding causes high current flow through this winding and the PCM driver. This causes damage to the PCM driver.

B is wrong because low resistance in the cooling fan motor would only increase the current flow through the fuse and the cooling fan relay contacts.

C is wrong because Technician A is right and Technician B is wrong.

D is wrong because Technician A is right and Technician B is wrong.

**15. D** A is wrong because a plugged EGR vacuum hose would not cause a seated IAC motor and 1,500 idle rpm.

B is wrong because the ECT sensor signal is normal.

C is wrong because the IAT sensor voltage signal is normal.

D is correct because a disconnected EVAP solenoid vacuum hose causes a vacuum leak into the intake manifold which results in a vacuum leak and increased idle rpm. The PCM seats the IAC motor to try and reduce the idle rpm.

**16. B** A is wrong because DTCs indicate a problem in a certain area but not in a specific component.

B is correct because diagnostic charts in the vehicle manufacturer's service manual should be followed to locate the exact cause of a problem.

C is wrong because Technician A is wrong and Technician B is right.

D is wrong because Technician A is wrong and Technician B is right.

**17. A** A is correct because shorted injector windings cause high current flow and this may cause the PCM to shut off the driver operating the pair of injectors.

B is wrong because open ground wires from the injectors to the PCM would not allow the noid lights to flash while cranking the engine.

C is wrong because Technician A is right and Technician B is wrong.

D is wrong because Technician A is right and Technician B is wrong.

**18. C** A is wrong because an open circuit at DLC terminal 5 would prevent some data display on the scan tool, but this tool would still be powered from terminal 16 so it would not be completely inoperative.

B is wrong because a short to ground at PCM terminal 79 would prevent data display on the scan tool, but this tool would still be powered from terminal 16 so it would not be completely inoperative.

C is correct because an open circuit at DLC terminal 16 prevents the scan tool from being powered up so it is completely inoperative.

D is wrong because an inoperative PCM would prevent data display on the scan tool, but this tool would still be powered from terminal 16 so it would not be completely inoperative.

**19.  C**   A is wrong because an EGR valve that is sticking open would also cause a rough idle problem.

B is wrong because the upstream $HO_2S$ sensor signal is normal during hard acceleration.

C is correct because the MAP sensor voltage is too low during hard acceleration. A partially plugged MAP sensor hose may cause this problem.

D is wrong because the TPS voltage signal is normal during hard acceleration.

**20.  B**   Technician A is wrong because engine misfire is a continuous monitor. A flashing MIL may indicate an engine misfire defect. Thus, Technician B is correct, making B the right answer.

**21.  A**   If there is a defect in the EGR valve wiring, the EGR monitor may not run, so A is correct. If the ECT sensor is defective, many of the monitors will not run. Therefore, B, C and D are wrong.

**22.  C**   A is wrong. An inoperative TPS that is worn in one position may cause engine surging at a specific speed, but Technician B is also right.

B is wrong. Engine surging after TCC lockup may be caused by problems in the TCC hydraulic or electronic system, but Technician A is also right.

C is correct because both Technicians A and B are right.

D is wrong because both Technicians A and B are right.

**23.  B**   A is wrong because the IAC motor data indicate the PCM is not controlling the IAC motor properly.

B is correct because the PCM has been replaced recently and the IAC data indicate the PCM is not providing proper IAC motor control, but all other data are normal.

C is wrong because Technician A is wrong and Technician B is right.

D is wrong because Technician A is wrong and Technician B is right.

**24.  D**   A and B are wrong because excessive fuel pressure or a restricted fuel return line that causes high fuel pressure result in a rich air-fuel ratio, and this causes the fuel trim numbers to be negative as the PCM reduces the injector pulse width.

C is wrong. Continually high upstream $HO_2S$ voltage also causes the PCM to reduce the injector pulse width.

D is correct. Restricted injector orifices cause a lean air-fuel ratio and the PCM increases the injector pulse width to enrich the air-fuel ratio.

**25.  B**   A data link defect allows the engine to start, but the data link network is disabled. Therefore, A, C, and D are wrong, and B is correct.

**26.  D**   If the immobilizer control module is replaced, it must be programmed. Therefore, A, B, and C are wrong, and D is correct.

**27.  B**   A is wrong because in this sequential fuel injection system, the PCM grounds each injector individually, and it is very unlikely that all eight wires from the injectors to the PCM would have high resistance.

B is correct because high resistance in the PCM ground may cause the injector waveforms to be above ground potential with the injectors on.

C is wrong because Technician A is wrong and Technician B is right.

D is wrong because Technician A is wrong and Technician B is right.

**28.  C**   A is wrong because replacing the fuel pump will not correct the low-voltage problem at the pump.

B is wrong because the resistance in the fuel pump motor ground is normal.

C is correct because the resistance problem in the voltage supply wire to the fuel pump is lowering the voltage at the pump and reducing pump pressure.

D is wrong because the alternator voltage is normal.

**29. C** When the resistance of the injector winding is less than specified, the winding is shorted, and this results in excessive current flow through the injector. Therefore, Technicians A and B are both right, and C is the correct answer.

**30. D** A is wrong because the voltage signal from the upstream $HO_2S$ sensor is normal.

B is wrong because the MAP sensor signal is normal.

C is wrong because the downstream $HO_2S$ sensor signal is satisfactory.

D is correct because the upstream and downstream $HO_2S$ sensor signals are the same, indicating a malfunctioning catalytic converter.

**31. B** A is wrong because the voltage trace indicates an open circuit in the TPS that requires TPS replacement.

B is correct because the voltage trace indicates an open circuit in the TPS that requires TPS replacement.

C is wrong because the voltage trace indicates an open circuit in the TPS that requires TPS replacement.

D is wrong because the voltage trace indicates an open circuit in the TPS that requires TPS replacement.

**32. B** A is wrong because the purpose of the PWM signals is to reduce injector current.

B is correct because the purpose of the PWM signals is to reduce injector current.

C is wrong because the purpose of the PWM signals is to reduce injector current.

D is wrong because the purpose of the PWM signals is to reduce injector current.

**33. A** A is correct because of the TCC sticks on engine stalling during deceleration may occur. The scan-tool data indicate the TCC has been commanded off by the PCM, but the TCC could have mechanical or hydraulic problems causing it to stick on. B is wrong because the IAC data are normal. C is wrong because the MAP sensor signal is normal. D is wrong because the VSS signal is normal.

**34. D** A is wrong because the ECT sensor has the specified resistance.

B is wrong because high resistance in the ECT sensor wires would cause an excessive voltage drop across the wires, but not 0 V as indicated in the scan-tool data.

C is wrong because the IAT sensor signal is normal.

D is correct because a short between the ECT sensor wires would cause a 0 V sensor signal as indicated in the scan-tool data.

**35. A** A is correct because the next step in diagnosing this problem should be to test the fuel pressure. The reduced fuel economy, high $HO_2S$ sensor voltages, and reduced injector pulse width indicate a rich air-fuel ratio, which may be caused by high fuel pressure.

B is wrong because during a diagnostic procedure, the quickest and easiest tests should always be performed first. In this case, the next test should be to test the fuel pressure.

C is wrong because Technician A is right and Technician B is wrong.

D is wrong because Technician A is right and Technician B is wrong.

**36. C** A is wrong. The gear selector is in O/D but the transmission range switch indicates third gear, so the transmission shift linkage may require adjusting and/or testing, but Technician B is also right. C is correct because Technicians A and B are both right.

D is wrong because Technicians A and B are both right.

**37. C** A is wrong because the inoperative ECT sensor is causing the EVAP purge solenoid to be inoperative at 55 mph.

B is wrong because the inoperative ECT sensor is causing the EGR control solenoid to be inoperative at 55 mph.

C is correct because an inoperative ECT sensor is causing a rich air-fuel ratio and inoperative EVAP and EGR solenoids. This is the root cause of the problem.

D is wrong because the MAP sensor data are normal.

**38.  A**     A is correct because battery boosting with reversed polarity may damage the PCM and BCM and blow the fuse link in the alternator battery wire.

B is wrong because an open circuit in the alternator field circuit only causes an inoperative charging circuit.

C is wrong because a short to ground in the alternator battery wire would blow the fuse link in this wire, but this problem is not likely to damage the PCM and BCM.

D is wrong because high resistance in the PCM and BCM ground circuits will affect the operation of these components, but it will not cause component damage.

**39.  B**     A is wrong because a damaged suppression diode would damage the PCM driver, but this condition does not damage the relay contacts.

B is correct because a shorted A/C clutch coil causes high current flow through the relay contacts, which may cause contact damage.

C is wrong because Technician A is wrong and Technician B is right.

D is wrong because Technician A is wrong and Technician B is right.

**40.  A**     A is correct. A defect in one of the CMP sensors may cause a DTC to be stored in the PCM.

B is wrong. A defect in the CMP sensor wires may cancel both CMP sensor signals resulting in a no-start condition.

C is wrong. An oil leak in one of the variable valve timing camshaft actuators may cause the valve timing to be less than fully advanced.

D is wrong. A defective ECT sensor affects several outputs and multiple DTCs to be stored.

**41.  C**     A is wrong because a spark plug with the wrong heat range does not cause a long, low spark line.

B is wrong because high resistance in the number 2 spark plug wire would cause a high, short spark line.

C is correct because a fouled spark plug in number 2 cylinder would cause a long, low spark line.

D is wrong because a bad coil connected to number 2 spark plug would cause misfiring, but it would not cause a long, low spark line.

**42.  D**     A is wrong because hotter range spark plugs would cause more severe electrode burning.

B is wrong because high fuel pressure causes a rich air-fuel ratio, and spark plug electrode burning may be caused by a lean air-fuel ratio.

C is wrong because both Technicians A and B are wrong.

D is correct because both Technicians A and B are wrong.

**43.  D**     Technician A is wrong because if fuse number 4 has an open circuit, the fuel pump would not operate, resulting in zero fuel pressure. The question informs us the fuel pressure is within specifications.

Technician B is wrong because if there was an open circuit in the voltage supply wire to the coil primary windings between coils number 5 and 2, coils 1, 3, and 5 would have voltage supplied to the primary windings. The question says there is zero voltage at all the coil positive primary terminals.

C is wrong because both Technicians A and B are wrong.

D is correct because both Technicians A and B are wrong.

**44.  B**     A is wrong because high HC emissions and low CO emissions are most often caused by cylinder misfiring because of low compression or an ignition problem.

B is correct because high HC emissions and low CO emissions are most often caused by cylinder misfiring because of low compression or an ignition problem.

C is wrong because dripping injectors cause a rich air-fuel ratio, resulting in high HC and CO emissions.

D is wrong because high fuel pressure causes a rich air-fuel ratio, resulting in high HC and CO emissions.

**45. B**   A is wrong because the PCM does receive a knock sensor signal during hard acceleration.

B is correct because lower-than-specified fuel pressure causes a lean air-fuel ratio that results in high NOx and low $CO_2$ emissions with low HC and CO emissions.

C is wrong because the PCM does provide spark advance and there is nothing in the spark advance data or emission readings to indicate a malfunctioning PCM.

D is wrong because dripping injectors cause a rich air-fuel ratio, resulting in high HC and CO emissions with low $CO_2$ and NOx emissions.

**46. C**   A is wrong because excessive crankshaft end play may cause an inoperative CKP sensor signal and this end play should be measured, but Technician B is also right.

B is wrong because the reluctor wheel condition should be checked, but Technician A is also right.

C is correct because both Technicians A and B are right.

D is wrong because both Technicians A and B are right.

**47. B**   A is wrong because the low resistance in the coil primary winding may damage the ignition module.

B is correct because the low resistance in the coil primary winding may damage the ignition module.

C is wrong because the low resistance in the coil primary winding may damage the ignition module.

D is wrong because the low resistance in the coil primary winding may damage the ignition module.

**48. C**   A is wrong because the primary windings in coils 1, 2, and 3 are connected in parallel, and spark plugs 1 and 5 are connected in series, but Technician B is also right.

B is wrong because spark plugs 1 and 5 are fired simultaneously while cylinder 1 is on the compression stroke and cylinder 5 is on the exhaust stroke—or vice versa—but Technician A is also right.

C is correct because both Technicians A and B are right.

D is wrong because both Technicians A and B are right.

**49. B**   In this question, you are asked to select the response that is not the cause of the problem.

A is wrong because excessive resistance in the spark plug wires may cause cylinder misfiring only during hard acceleration.

B is correct because low cylinder compression causes continual cylinder misfiring, and so this problem does not cause misfiring only during hard acceleration.

C is wrong because a bad coil with low maximum voltage may cause cylinder misfiring only during hard acceleration.

D is wrong because a bad spark plug with an insulation leakage problem may cause cylinder misfiring only during hard acceleration.

**50. D**   A is wrong because a voltmeter will not test the secondary output of an ignition coil.

B is wrong because an ohmmeter will not test the secondary output of an ignition coil.

C is wrong because an ignition module tester will not test the secondary output of an ignition coil.

D is correct because a test spark plug may be used to test the secondary output of an ignition coil.

**51. B**   A is wrong because low engine compression causes continual cylinder misfiring.

B is correct because combustion chamber carbon deposits are LEAST likely to cause continual cylinder misfiring.

C is wrong because low engine compression causes continual cylinder misfiring.

D is wrong because worn distributor bushings cause continual cylinder misfiring.

**52. C**    A is wrong because meter 1 should read infinity ohm.

B is wrong because meter 2 should read 650 to 850 ohms.

C is correct because meter 1 should read infinity ohm.

D is wrong because meter 2 should read 650 to 850 ohms.

**53. B**    A is wrong because the voltage supply to the positive primary coil terminal should be tested before replacing the ignition module.

B is correct because the voltage supply to the positive primary coil terminal should be the next test.

C is wrong because the secondary coil winding does not affect the test light connected from the tachometer terminal to ground while cranking the engine.

D is wrong because the spark plug wires do not affect the test light connected from the tachometer terminal to ground while cranking the engine.

**54. A**    A is correct because high resistance in the spark plug wires or spark plug gaps is increasing the normal required coil voltage and reducing the voltage reserve.

B is wrong because bad coil windings would only affect two spark plugs on a waste spark EI system.

C is wrong because Technician A is right and Technician B is wrong.

D is wrong because Technician A is right and Technician B is wrong.

**55. D**    A is wrong because if the base timing is less than specified, it would not cause continual backfiring into the intake manifold during hard acceleration.

B is wrong because if the fuel pressure is higher than specified, it would not cause continual backfiring into the intake manifold during hard acceleration.

C is wrong because if the spark plug heat range is hotter than specified it would not cause continual backfiring into the intake manifold during hard acceleration.

D is correct because a cracked distributor cap may cause continual backfiring into the intake manifold during hard acceleration.

**56. B**    A is wrong because the normal secondary voltage required to fire the spark plugs is lower at idle speed and this voltage increases as the throttle is opened.

B is correct because the maximum secondary coil voltage must always exceed the normal required coil voltage, or misfiring will occur.

C is wrong because Technician A is wrong and Technician B is right.

D is wrong because Technician A is wrong and Technician B is right.

**57. D**    A is wrong because the CKP sensor gap does not affect ignition timing.

B is wrong because timing checks or adjustments are not possible on coil-on-spark-plug ignition systems.

C is wrong because both Technicians A and B are wrong.

D is correct because both Technicians A and B are wrong.

**58. A**    A is correct because an overtightened knock sensor may cause the knock sensor to be over-sensitive so it sends a signal to the PCM indicating detonation even though there is no significant detonation. This results in reduced spark advance and fuel economy.

B is wrong because a higher-than-specified spark plug heat range may cause engine detonation, but the question informs us there is no detonation during a road test.

C is wrong because Technician A is right and Technician B is wrong.

D is wrong because Technician A is right and Technician B is wrong.

**59. A**    A is correct because low fuel pressure causes a lean air-fuel ratio, burned spark plug electrodes, high integrator and block learn numbers, and acceleration stumbles.

B is wrong because an intake manifold vacuum leak would cause rough idle operation, and the question informs us the idle operation is normal.

C is wrong because a contaminated upstream $HO_2S$ sensor causes an improper $HO_2S$ sensor signal and all the data are normal except the integrator and block learn numbers.

D is wrong because a contaminated IAT sensor would cause an improper IAT sensor signal and all the data are normal except the integrator and block learn numbers.

**60.  A**   A is correct because an intermittent open circuit at module terminal B stops the bypass signal from the PCM to the coil module, and this cancels any spark advance.

B is wrong because an inoperative CKP sensor causes a no-start condition.

C is wrong because an inoperative CMP sensor causes a no-start condition once the engine is stopped.

D is wrong because an intermittent open circuit at module terminal G would cause an inoperative CKP sensor and a no-start condition.

**61.  C**   In this question, you are asked to select the response that is not caused by an open circuit in the ECT sensor signal wire.

A is wrong because an open circuit in the ECT sensor signal wire simulates an extremely cold temperature signal from this sensor that could result in hard starting and engine stalling when warm.

B is wrong because an open circuit in the ECT sensor signal wire simulates an extremely cold temperature signal from this sensor that could result in improper torque converter clutch operation.

C is correct because an open circuit in the ECT sensor signal wire simulates an extremely cold temperature signal from this sensor that results in a rich air-fuel ratio.

D is wrong because an open circuit in the ECT sensor signal wire simulates an extremely cold temperature signal from this sensor that could result in improper EGR operation.

**62.  D**   A is wrong because a jumped timing belt would not cause the IAC motor to be at 0 percent.

B is wrong because retarded timing would not cause the IAC motor to be at 0 percent, since this problem would increase the IAC motor opening.

C is wrong because excessive fuel pressure would not cause the IAC motor to be at 0 percent.

D is correct because a vacuum leak in the intake manifold increases idle rpm and causes the PCM to close the IAC motor to compensate for this problem.

**63.  A**   A is correct because a scan tool is NOT used when performing a tampering inspection.

B is wrong because a parts locator book may be used in a tampering inspection.

C is wrong because service manuals may be used in a tampering inspection.

D is wrong because service bulletins may be used in a tampering inspection.

**64.  D**   A is wrong because the PCM normally shuts off the injectors if the engine speed exceeds 6,000 rpm or the vehicle speed increases above 110 mph (70 kmh).

B is wrong because the PCM normally shuts off the injectors if the engine speed exceeds 6,000 rpm or the vehicle speed increases above 110 mph (70 kmh).

C is wrong because both Technicians A and B are wrong.

D is correct because both Technicians A and B are wrong.

**65.  B.**  Cylinder misfire, dripping injectors, or an intake manifold vacuum leak cause rough idle operation, and the only complaint is reduced fuel economy. Therefore, A, C, and D are wrong. High fuel pressure causes a rich air-fuel ratio and high upstream HO$_2$S voltage, and the PCM responds to these signals by decreasing injector pulse width.

**66.  D**   A is wrong because plugged injector orifices cause rough idle operation.

B is wrong because a pressure regulator sticking open causes low pressure at idle speed.

C is wrong because dripping injectors cause rough idle.

D is correct because low voltage at the fuel pump may reduce fuel pressure at high vehicle speeds and cause a loss of power and engine surging.

**67.  A**   A is correct. The PCM may be the problem because the PCM should increase the injector pulse width when the HO$_2$S voltage signal decreases to 0.1 V.

B is wrong because the decreased HO$_2$S voltage signal is displayed on the scan tool, indicating this signal reached the PCM, and so the HO$_2$S signal wire must be satisfactory.

C is wrong because Technician A is right and Technician B is wrong.

D is wrong because Technician A is right and Technician B is wrong.

**68.  C**    A is wrong because the problem may be caused by exhaust system restriction, and the Technician should test for this condition, but Technician B is also right.

B is wrong because the problem may be caused by air intake system restriction, and the technician should test for this condition, but Technician A is also right.

C is correct because both Technicians A and B are right.

D is wrong because both Technicians A and B are right.

**69.  A**    If the TAC motor is defective, the PCM disables the motor and throttle opening is limited to 15 percent. One defective TP sensor causes the PCM to limit throttle opening to 35 percent, which provides more than 1500 rpm. Therefore, A is correct and B is wrong.

**70.  C**    C is correct. When the MAF grams per second is less than specified at idle, air may be entering the intake from a leak in the hose between the MAF sensor and the throttle body.

A defective MAF sensor causes erratic or no grams-per-second reading, so A is wrong. An air leak in the hose between the MAF sensor and the air cleaner does not affect the grams-per-second reading, so B is wrong. A loose connection at the MAF connector causes intermittent MAF operation, so D is wrong.

**71.  D**    In this question, you are asked to select the response that is NOT the cause of the problem.

A is wrong because excessive carbon buildup on the piston heads may cause preignition and high NOx emissions.

B is wrong because a plugged EGR exhaust passage prevents EGR flow and causes high NOx emissions.

C is wrong because hot spots in the combustion chamber caused by cooling system deposits increase combustion temperature and cause high NOx emissions.

D is correct because a rich air-fuel ratio reduces combustion temperatures and NOx emissions.

**72.  C**    A is wrong because the technician should test the knock sensor using a timing light and a small hammer, but Technician B is also right.

B is wrong because the technician should check the base timing, but Technician A is also right.

C is correct because both Technicians A and B are right.

D is wrong because both Technicians A and B are right.

**73.  B**    A is wrong because a pressure regulator that is stuck open causes very low fuel pressure.

B is correct because a leaking fuel pump check valve may cause the fuel to flow through the fuel supply line back into the fuel tank and reduce the pressure after a pressure test. Manually plugging the fuel supply line prevents this action.

C is wrong because leaking injectors would still allow the fuel pressure to decrease after the pressure test with the fuel supply line manually plugged.

D is wrong because a leak in the fuel return line would not cause the fuel pressure to decrease after the pressure test because the pressure regulator is closed.

**74.  B**    A is wrong because a rich air-fuel ratio caused by an inoperative HO$_2$S sensor would not cause a shudder after the TCC locks up.

B is correct because a lean air-fuel ratio caused by restricted injector orifices may cause rough idle and engine shudder after the TCC locks up.

C is wrong because excessive injector pulse width supplied by the PCM does not cause shudder after the TCC locks up.

D is wrong because an intake manifold vacuum leak is less noticeable as the throttle is opened and the vacuum decreases. It is not likely that a vacuum leak would cause engine shudder after the TCC locks up.

**75.  B**    A is wrong because if the pressure drop is less on two injectors, these injector plungers are sticking closed or the injector orifices are partially restricted.

B is correct because some manufacturers supply self-cleaning injectors, and they do not recommend cleaning these injectors.

C is wrong because Technician A is wrong and Technician B is right.

D is wrong because Technician A is wrong and Technician B is right.

**76.  A**  A is correct because an inoperative number 2 injector causes misfiring on number 2 cylinder that is less noticeable at higher engine speeds.

B is wrong because an intake manifold vacuum leak would affect the MAP sensor reading, and the MAP sensor voltage is normal.

C is wrong because an inoperative $HO_2S$ sensor does not cause a cylinder misfire.

D is wrong because an air leak in the MAP sensor hose would cause an improper MAP sensor reading, and the MAP sensor data so far are satisfactory.

**77.  D**  A is wrong because a damaged voltage suppression diode in the fuel pump relay may damage the fuel pump relay driver in the PCM, but this problem would not blow the fuse in the fuel pump relay winding circuit.

B is wrong because a short to ground at PCM terminal 4 would operate the fuel pump continually with the ignition switch on and the engine not running.

C is wrong because high resistance at the fuel pump relay contacts would reduce the voltage at the fuel pump and decrease fuel pressure especially at high speeds.

D is correct because a shorted fuel pump relay winding may damage the fuel pump relay driver in the PCM and blow the fuse in the fuel pump relay winding circuit.

**78.  C**  A is wrong because moisture may be freezing in the MAP sensor hose, but Technician B is also right.

B is wrong because the MAP sensor should be mounted on the engine so the sensor has a shorter vacuum hose, but Technician A is also right.

C is correct because both Technicians A and B are right.

D is wrong because both Technicians A and B are right.

**79.  B**  A is wrong. The high HC and $O_2$ emissions with low CO and $CO_2$ emissions indicate a lean cylinder misfire condition. If the PCV valve is installed backwards, the manifold vacuum pulls this valve to the seated position so it is closed. This action causes a rich air-fuel ratio.

B is correct. The high HC and $O_2$ emissions with low CO and $CO_2$ emissions indicate a lean cylinder misfire condition, and a vacuum leak in the brake booster hose may result in a lean air-fuel ratio.

C is wrong. The high HC and $O_2$ emissions with low CO and $CO_2$ emissions indicate a lean cylinder misfire condition. A disconnected EVAP solenoid electrical connector causes an inoperative solenoid, and this does not affect the idle air-fuel ratio or idle operation.

D is wrong. The high HC and $O_2$ emissions with low CO and $CO_2$ emissions indicate a lean cylinder misfire condition. A leaking one-way check valve in the secondary air injection system causes burned hoses in this system, but this condition does not affect air-fuel ratio.

**80.  D**  A is wrong because the easiest place to locate vacuum hose routing information is on the underhood emission label.

B is wrong because the easiest place to locate vacuum hose routing information is on the underhood emission label.

C is wrong because the easiest place to locate vacuum hose routing information is on the underhood emission label.

D is correct because the easiest place to locate vacuum hose routing information is on the underhood emission label.

**81.  C**  A is wrong because if the exhaust passage in the EGR valve stem is plugged with carbon, the EGR valve is inoperative and this would cause the NOx failure, but Technician B is also right.

B is wrong because if the EGR solenoid plunger is stuck closed, the EGR valve is inoperative and this would cause the NOx failure, but Technician A is also right.

C is correct because both Technicians A and B are right.

D is wrong because both Technicians A and B are right.

**82. A**   A is correct because the decel valve only operates during deceleration, and the loaded-mode I/M test is completed under various engine operating conditions, including decel.

B is wrong because the decel valve only operates during deceleration, and an idle speed emission test would not locate problems in this component.

C is wrong because the decel valve only operates during deceleration, and the emission test at 2,000 rpm would not locate problems in this component.

D is wrong because the decel valve only operates during deceleration, and comparing the MAF and TPS data would not locate problems in this component.

**83. C**   A is wrong because if the hose from the fuel tank to the canister is plugged, no vacuum is supplied to the fuel tank.

B is wrong because if the hose from the canister to the purge solenoid is leaking, fuel tank pressure is zero.

C is correct because if the vent solenoid air intake and the vacuum valve in the filler cap are both clogged, vacuum will increase in the fuel tank.

D is wrong because if the canister is partially clogged, vacuum and purging action in the fuel tank are reduced.

**84. B**   A is wrong because if the voltage supply wire to these solenoids is open, some other components (such as the fuel injectors) are inoperative.

B is correct because the coolant temperature has to reach 150°F (65°C) before the PCM operates the EGR, TCC, and EVAP purge solenoids. With an inoperative thermostat, the coolant temperature may never reach 150°F (65°C).

C is wrong because Technician A is wrong and Technician B is right.

D is wrong because Technician A is wrong and Technician B is right.

**85. C**   A is wrong because a restricted fuel return line causes high fuel pressure, a rich air-fuel ratio and high $HO_2S$ sensor voltage.

B is wrong because dripping injectors cause rough idle, a rich air-fuel ratio, and high $HO_2S$ sensor voltage.

C is correct because if the secondary air is continually upstream, the $HO_2S$ sensor senses excessive oxygen in the exhaust stream, and the $HO_2S$ sensor voltage is low. This causes the PCM to provide a rich air-fuel ratio.

D is wrong because the MAP sensor voltage is normal.

**86. D**   A is wrong because high NOx emissions are caused by high combustion temperature, but this should not melt the monolith in the catalytic converter.

B is wrong because higher-than-normal coolant temperature does not melt the monolith in the catalytic converter. This damage is likely caused by a rich air-fuel ratio.

C is wrong because both Technicians A and B are wrong.

D is correct because both Technicians A and B are wrong.

**87. A**   In this question, you are asked to select the response that indicates when an EGR valve is in operation.

A is correct because at a constant throttle opening at 55 mph (88 kmh), the EGR valve is in operation.

B is wrong because when the engine coolant is cold, the EGR valve is inoperative.

C is wrong because during wide open throttle operation, the EGR valve is inoperative.

D is wrong because during closed throttle deceleration, the EGR valve is inoperative.

**88. C**   A is wrong because an open circuit in the EVAP solenoid winding causes 0 V between the EVAP solenoid and the PCM.

B is wrong because an open circuit in the voltage supply wire to the EVAP solenoid causes 0 V between the EVAP solenoid and the PCM.

C is correct because an open circuit in the wire from the EVAP solenoid to the PCM causes high voltage at the EVAP solenoid terminal connected to the PCM.

D is wrong because a shorted winding in the EVAP solenoid increases the current flow through this winding, which may damage the PCM driver that operates this solenoid.

**89.  B**  A is wrong because if the PCV valve is stuck closed, there is no change in engine operation or emission levels.

B is correct because when the engine oil is contaminated with gasoline, an excessive amount of fuel vapor may be moved through the PCV valve into the intake manifold, resulting in a rich air-fuel ratio and excessive HC and CO emissions with rough idle operation.

C is wrong because Technician A is wrong and Technician B is right.

D is wrong because Technician A is wrong and Technician B is right.

**90.  D**  A is wrong because the converter inlet temperature should be lower than the outlet temperature.

B is wrong because a catalytic converter creates some $CO_2$, and so the $CO_2$ emissions are higher on a catalytic converter–equipped vehicle compared to a non-converter-equipped vehicle.

C is wrong because both Technicians A and B are wrong.

D is correct because both Technicians A and B are wrong.

**91.  B**  A is wrong because a shorted fuel injector driver in the PCM causes an inoperative injector, cylinder misfire, but low HC emissions because fuel is not being injected in one cylinder.

B is correct because an EGR valve stuck open causes rough idle and a lean misfire that causes HC emissions to be higher than normal.

C is wrong because an intake manifold vacuum leak may cause a cylinder misfire at idle, but under this condition the HC and $O_2$ would be high.

D is wrong because a fouled spark plug causes misfiring at idle and very high HC emissions.

**92.  A**  A is correct because a plugged exhaust passage in the EGR valve stem keeps negative pressure pulses in the exhaust from pulling the bleed valve open. This bleed valve is normally closed. If the bleed valve is not pulled open, vacuum supplied through the EGR solenoid can pull the EGR valve open prematurely and cause an acceleration stumble.

B is wrong because if the bleed port is stuck open in the EGR valve, the EGR valve would be completely inoperative.

C is wrong because the MAP sensor voltage signal is normal.

D is wrong because the MAP sensor voltage signal is normal, and the engine has no other drivability complaints such as rough idle.

**93.  B**  A is wrong because if the secondary air is always upstream, the $HO_2S$ voltage should still increase when propane is dispersed into the air intake.

B is correct because if the $HO_2S$ signal wire has an open circuit, the $HO_2S$ voltage signal would be low and it would not increase when propane is dispersed into the air intake.

C is wrong because if the engine has an intake manifold vacuum leak, the $HO_2S$ voltage would still increase when propane is dispersed into the air intake.

D is wrong because one inoperative fuel injector would not cause the $HO_2S$ voltage to remain low when propane is dispersed into the air intake.

**94.  A**  A is correct because the TPS voltage should be 3.5 V at 75-percent throttle opening, and the scan-tool data indicate this voltage is only 2.3 V.

B is wrong because the MAF sensor voltage is satisfactory.

C is wrong because the IAT sensor voltage is satisfactory.

D is wrong because the EVAP solenoid data are satisfactory.

**95.  A**  A is correct because a grounded wire between the EGR solenoid and the PCM energizes the EGR solenoid continually and opens the EGR valve even at idle speed.

B is wrong because high resistance in the PCM ground would not cause the PCM to energize the EGR solenoid continually.

C is wrong because Technician A is right and Technician B is wrong.

D is wrong because Technician A is right and Technician B is wrong.

**96.  D**    A is wrong because an inoperative shift solenoid would not cause all the shifts to be late in relation to engine rpm. This problem would cause the transmission to miss one or more shifts.

B is wrong because a malfunctioning transmission range switch would not cause all the shifts to be late in relation to engine rpm. This problem may cause the transmission to be in another gear besides the one selected by the gear selector position.

C is wrong because an inoperative throttle position sensor may cause all the shifts to be late in relation to engine rpm, but the TPS scan-tool data are satisfactory.

D is correct because an inoperative transmission fluid temperature sensor that indicates low fluid temperature may cause all the shifts to be late.

**97.  C**    A is wrong because an intermittent ground in the alternator field circuit to the PCM bypasses the voltage regulator in the PCM, resulting in very high alternator voltage, but Technician B is also right.

B is wrong because an intermittent open circuit in the wire connected from the alternator battery terminal to the positive battery terminal may cause very high voltage spikes in the electrical system, but Technician A is also right.

C is correct because both Technicians A and B are right.

D is wrong because both Technicians A and B are right.

**98.  B**    A is wrong because the transmission fluid temperature (TFT) sensor is not connected to fuse number 56.

B is correct because a short to ground in the post catalyst $HO_2S$ sensor heater causes high current flow through fuse number 56.

C is wrong. If the bank 1 $HO_2S$ wire is shorted to ground at PCM terminal 59, the $HO_2S$ heater is energized continuously. D is wrong because continually closed power-steering pressure switch contacts do not cause high current in the circuit.

**99.  A**    The rough idle and stalling problem may be caused by the EGR valve sticking open. On some vehicles, new software programming in the PCM causes the PCM to pulsate the EGR valve occasionally to remove carbon from the valve tip, and prevent this problem. Thus, Technician A is correct. If the EGR valve passage is plugged with carbon, NOx emissions are high but this problem should not cause rough idle and stalling. So, Technician B is wrong.

**100.  A**    A is correct because the VSS indicates 115 mph (185 kmh), and the PCM shuts off the injectors if the indicated vehicle speed exceeds 110 mph (177 kmh).

B is wrong because the problem is excessive vehicle speed indicated by the VSS.

C is wrong because the IAT sensor voltage is satisfactory.

D is wrong because the EGR valve data are satisfactory.

**101.  C**    A is wrong because higher-than-specified fuel pressure causes a rich air-fuel ratio, high HC and CO, and low NOx.

B is wrong because an inoperative upstream $HO_2S$ that provides a low-voltage signal causes a rich air-fuel ratio, high HC and CO, and low NOx.

C is correct because restricted injector orifices cause a lean air-fuel ratio, which may cause lean misfiring, intermittent high HC, low CO, and high NOx.

D is wrong because a partially clogged fuel return line causes a rich air-fuel ratio, high HC and CO, and low NOx.

**102.  B**    A is wrong because a lower-than-specified ECT voltage signal causes a lean air-fuel ratio, which results in low CO and high HC and NOx.

B is correct because an inoperative secondary air injection system causes improper operation of the oxidation catalyst in the catalytic converter, and this results in high CO and HC emissions.

C is wrong because Technician A is wrong and Technician B is right.

D is wrong because Technician A is wrong and Technician B is right.

**103.  A**   In this question, you are asked to select the response that is not the cause of the problem.

A is correct because a digital EGR valve does not have a vacuum control solenoid, and so this is not the cause of the high NOx emissions and detonation.

B is wrong because if one of the solenoids in the EGR valve has an open circuit, this could result in high NOx emissions and detonation.

C is wrong because if the EGR exhaust passages are plugged, this could result in high NOx emissions and detonation.

D is wrong because if the EGR ground wire has an open circuit, the EGR valve is inoperative, resulting in high NOx emissions and detonation.

**104.  A**   A is correct because an intake manifold vacuum leak causes a lean air-fuel ratio, which may result in lean misfiring, high HC, low CO, low $CO_2$, high $O_2$, and marginal NOx. The emission readings improve at 2,000 rpm compared to idle because the manifold vacuum decreases as the throttle is opened.

B is wrong because high fuel pressure causes a rich air-fuel ratio and high HC and CO, and low $CO_2$, $O_2$, and NOx.

C is wrong because Technician A is right and Technician B is wrong.

D is wrong because Technician A is right and Technician B is wrong.

**105.  D**   A is wrong because the lab scope waveform indicates a sudden decrease or dropout in the MAP sensor voltage caused by an inoperative MAP sensor or the loss of the 5 V reference voltage supplied to this sensor.

B is wrong because the lab scope waveform indicates a sudden decrease or dropout in the MAP sensor voltage caused by an inoperative MAP sensor, or the loss of the 5 V reference voltage supplied to this sensor.

C is wrong because both Technicians A and B are wrong.

D is correct because both Technicians A and B are wrong.

**106.  D**   A is wrong because the downstream $HO_2S$ sensor waveform is similar to the waveform on the upstream $HO_2S$, indicating an inoperative catalytic converter.

B is wrong because the upstream $HO_2S$ waveform is satisfactory.

C is wrong because the upstream $HO_2S$ waveform is satisfactory.

D is correct because the downstream $HO_2S$ sensor waveform is similar to the waveform on the upstream $HO_2S$, indicating an inoperative catalytic converter.

**107.  C**   A is wrong because retarded ignition timing may cause high HC and CO emissions, but this problem does not cause high $O_2$ emissions.

B is wrong because an inoperative EGR valve causes high NOx emissions, but this problem does not cause high $O_2$ emissions.

C is correct because a lean air-fuel ratio causes high $O_2$ emissions.

D is wrong because the catalytic converter does not affect $O_2$ emissions.

**108.  D**   A is wrong because the IAT sensor voltage signal is satisfactory for the atmospheric temperature.

B is wrong because a leak in the MAP vacuum hose would cause an unsatisfactory MAP sensor voltage signal, but this signal is within specifications.

C is wrong because both Technicians A and B are wrong.

D is correct because both Technicians A and B are wrong.

**109.  A**   A is correct because very high HC emissions and low CO emissions are caused by ignition misfiring.

B is wrong because a leaking fuel pressure regulator diaphragm causes a rich air-fuel ratio with high HC and CO emissions.

C is wrong because Technician A is right and Technician B is wrong.

D is wrong because Technician A is right and Technician B is wrong.

**110.  A**   A is correct because the EVAP canister is not purged at idle. If this canister is saturated with gasoline, the air-fuel ratio becomes richer at 2,000 rpm, resulting in high CO and HC.

B is wrong because one fouled spark plug causes high HC at idle and 2,000 rpm.

C is wrong because an open PCV valve causes a lean air-fuel ratio and low CO.

D is wrong because a fuel pressure regulator that is sticking open causes low fuel pressure and a lean air-fuel ratio, resulting in low CO.

**111.  C**    A is wrong. The upstream $HO_2S$ sensor voltage is high but this is caused by a high MAF signal, causing the PCM to provide a richer air-fuel ratio.

B is wrong because the MAP sensor voltage signal is satisfactory.

C is correct because the MAF sensor voltage is higher than specified, resulting in a rich air-fuel ratio and high CO and HC emissions.

D is wrong because the IAT sensor voltage signal is satisfactory.

**112.  A**    A is correct because one bad coil may cause cylinder misfiring only under load and this results in high HC emissions.

B is wrong because a fuel-pressure regulator that is sticking closed causes continual high fuel pressure and high HC and CO emissions at idle and higher speeds.

C is wrong because Technician A is right and Technician B is wrong.

D is wrong because Technician A is right and Technician B is wrong.

**113.  D**    A is wrong because an open circuit in the voltage supply wire to the EVAP solenoid causes 0 V on either side of the EVAP solenoid.

B is wrong because an open circuit in the EVAP solenoid winding causes 0 V at the PCM side of this solenoid.

C is wrong because a shorted EVAP solenoid winding still provides a very low voltage reading at the PCM side of the solenoid.

D is correct because a grounded wire between the EVAP solenoid and the PCM energizes the EVAP solenoid all the time, causing a rich air-fuel ratio and high HC and CO emissions.

Because there is current flow through the EVAP solenoid winding, most of the system voltage is dropped across the winding, and the voltage at the PCM side of the winding is very low.

**114.  B**    In this question, you are asked to select the response that is not the cause of the problem.

A is wrong because an inoperative MAP sensor could cause a rich air-fuel ratio and high CO and HC emissions.

B is correct because the high $HO_2S$ sensor voltage is the result of a rich air-fuel ratio caused by some other problem. Thus, the high $HO_2S$ sensor voltage is not the cause of the problem.

C is wrong because an inoperative ECT sensor could cause a rich air-fuel ratio and high CO and HC emissions.

D is wrong because a partially restricted fuel return line could cause a rich air-fuel ratio and high CO and HC emissions.

**115.  B**    A is wrong because the $HO_2S$ signal does increase immediately when propane is dispersed into the air intake. This indicates the $HO_2S$ is operational.

B is correct because a cracked exhaust manifold ahead of the $HO_2S$ allows air to be pulled into the exhaust manifold at lower speeds and this causes the low $HO_2S$ voltage, resulting in a rich air-fuel ratio and high HC and CO emissions.

C is wrong because Technician A is wrong and Technician B is right.

D is wrong because Technician A is wrong and Technician B is right.

**116.  A**    A is correct because restricted injector orifices may cause a lean air-fuel ratio, which burns hotter and contributes to high NOx emissions. A lean air-fuel ratio may also cause some lean misfiring and increased HC emissions.

B is wrong because the TPS voltage signal is satisfactory.

C is wrong because the ECT voltage signal is satisfactory.

D is wrong because a leaking throttle body to intake gasket would cause a vacuum leak and higher-than-specified MAP sensor voltage. Because the MAP sensor voltage is satisfactory, this is not the problem.

Answers to Post Test 613

**117. B** A is wrong because low fuel pressure causes a lean air-fuel ratio, resulting in high HC and $O_2$ emissions. This problem also causes a low upstream $HO_2S$ voltage signal. The question states the other emissions levels and scan-tool data are satisfactory.

B is correct because restricted EGR exhaust passages cause only high NOx emissions.

C is wrong because Technician A is wrong and Technician B is right.

D is wrong because Technician A is wrong and Technician B is right.

**118. B** The variable valve timing monitor is a two-trip monitor. If a defect occurs in this system on one trip, a DTC is set, but the MIL is not illuminated. If the defect occurs on two consecutive trips the MIL is illuminated. Therefore, A is wrong.

**119. C** If the fuel level in the tank is below 15 percent or the battery voltage is below 10 V, some monitors will not run. Therefore, Technicians A and B are both right, and C is the correct answer.

**120. C** If a defect in the data link network prevents communication between the PCM and the instrument cluster, the MIL is illuminated continuously. Therefore, A, B, and D are wrong, and C is correct.

**121. D** Technicians A and B are both wrong because the pressure drop obtained is within specifications.

**122. D** A is wrong because the $HO_2S$ signal lean-rich transitions are excessive and this signal is not controlling the air-fuel ratio properly.

B is wrong because the $HO_2S$ signal has excessive lean-rich transitions, but this does not indicate RTV contamination of the sensor.

C is wrong because the $HO_2S$ signal has excessive lean-rich transitions, but this does not indicate a continual rich air-fuel ratio.

D is correct because the $HO_2S$ waveform indicates a cylinder misfire caused by an ignition problem such as a bad spark plug.

**123. D** A is wrong because the four-gas analyzer readings are satisfactory and the car should pass an emissions retest.

B is wrong because the four-gas analyzer readings are satisfactory and the car should pass an emissions retest.

C is wrong because both Technicians A and B are wrong.

D is correct because both Technicians A and B are wrong.

# Answers to Post Test

**1. B** The low $CO_2$ and high $O_2$ readings indicate a lean air-fuel ratio. Low fuel pressure causes a lean air-fuel ratio and a power loss at higher speeds, so B is correct. A restricted fuel return line causes high fuel pressure and a rich air-fuel ratio, so A is wrong. An EGR solenoid stuck open causes the EGR valve to remain open resulting in a rough idle, so C is wrong. An EVAP solenoid stuck open causes canister purging at idle resulting in a rich air-fuel ratio and rough idle, so D is wrong.

**2. B** A defective VSS does not cause hard starting, so Technician A is wrong. A defective CMP sensor causes the PCM to guess at the proper time to open the injectors, and this may cause hard starting. Thus, Technician B is right, and B is the correct answer.

**3. C** High resistance in the wire from the EVAP vent solenoid to the PCM may cause higher voltage at the PCM side of the EVAP vent solenoid, but this problem does not cause PCM damage, so A is wrong. High resistance at the PCM ground terminal 83 would affect all the output actuators, so B is wrong. A shorted EVAP vent solenoid winding has less resistance than specified, and reduced voltage drop. The lower resistance causes higher current through the PCM resulting in PCM damage, so C is correct. High resistance in the voltage supply wire to the EVAP vent solenoid causes low voltage supplied to this solenoid, so D is wrong.

**4.  C**    Carbon buildup in the EGR exhaust passages may cause reduced EGR flow and high combustion temperatures resulting in NOx failure on an emissions test. Thus, Technician A is right. If the EGR solenoid plunger is stuck closed, the EGR valve is inoperative, resulting in a NOx failure on an emissions test. The scan-tool data only indicate the PCM command to the output actuators, and so this problem is not denoted by scan-tool data. Thus, Technician B is right. Because both Technicians A and B are right, C is the correct answer.

**5.  A**    The lower induced voltage in number 2 injector indicates a shorted injector winding that allows higher current flow and may damage the PCM, so A is correct. The shorted injector winding does not cause a rich air-fuel ratio or engine detonation, so B and C are wrong. The shorted injector winding does not cause high combustion chamber temperatures, so D is wrong.

**6.  D**    If the engine had a large intake manifold vacuum leak, the $O_2$ reading would be much higher, so Technician A is wrong. If one injector plunger is stuck closed, that injector would not inject fuel into the cylinder and the HC would be lower. Thus, Technician B is wrong. These readings are likely caused by ignition misfiring. Because both Technicians A and B are wrong, D is the correct answer.

**7.  A**    The four-gas analyzer readings indicate a rich air-fuel ratio. High fuel pressure, a restricted fuel return line, or a leaking pressure regulator diaphragm causes a rich air-fuel ratio. Therefore, B, C, and D are the causes of the problem, meaning they are not the requested answer. A leaking brake booster vacuum hose causes a lean air-fuel ratio. Therefore, A is not a cause of the problem, and thus is the requested answer.

**8.  B**    When the MAF grams-per-second reading is erratic, the hot wire in the MAF may be coated with contaminants. Therefore, Technician A is wrong, and Technician B is right.

**9.  C**    An incorrect ECT sensor voltage signal or a very low fuel level in the tank may cause monitors not to run. Therefore, Technicians A and B are both right, and C is the correct answer.

**10.  D**    A severe engine misfire condition causes the MIL to blink because this condition may cause catalytic converter damage. Thus, D is correct, and A, B, and C are wrong.

**11.  B**    The catalytic converter is being damaged by a continually rich air-fuel ratio. Dripping injectors may cause a rich air-fuel ratio, but they also cause rough idle and hard starting, so A is wrong. High fuel pressure causes a continually rich air-fuel ratio, so B is correct. A restricted fuel filter may cause a lean air-fuel ratio and a loss of power, so C is wrong. A leak in the fuel return line causes a fuel leak, so D is wrong.

**12.  A**    A defective suppression diode in the fuel pump relay may cause the induced voltage in this relay winding to be applied against the PCM driver each time the relay shuts off, so Technician A is correct. High current flow through the fuel pump motor would damage the fuel pump relay contacts, but this defect would not damage the PCM. Thus, Technician B is wrong, and A is the correct answer.

**13.  A**    High fuel trim numbers indicate the PCM is always adding fuel to try and compensate for a lean air-fuel ratio signal to the PCM. Bank 1 is on the same side as number 1 cylinder. A vacuum leak on the same side as number 1 cylinder causes a lean air-fuel ratio signal to the PCM, and the PCM reacts to this signal by adding fuel, resulting in high fuel trim numbers. Thus, A is correct. The fuel trim numbers for bank 2 are normal, so B is wrong. Dripping injectors cause a rich air-fuel ratio, so C is wrong. The downstream $HO_2S$ is used for catalytic converter monitoring, so D is wrong.

**14.  C**    PCM terminal 82 provides a ground for many of the input sensors, including the fuel tank pressure sensor and the MAP sensor. Thus, Technicians A and B are both right, and C is the correct answer.

**15.  D**    If the PCM does not command the TCC off when the brake pedal is depressed during deceleration, the brake switch input to the PCM is the most likely cause of the problem. Thus, D is correct, and A, B, and C are wrong.

**16.  A**    If the ECT sensor has less resistance than specified when it is cold, the engine is hard to start because this input causes the PCM to supply a lean air-fuel ratio. Therefore, Technician A is right, and Technician B is wrong, and A is the correct answer.

**17. A**   An A/C system pressure of 450 psi (3102 kPa) causes the A/C pressure sensor to signal the PCM, and the PCM disengages the A/C compressor clutch. The technician should check the A/C system pressures with a manifold gauge set, because this is the source of the problem. Thus, Technician A is correct. The A/C pressure sensor is probably satisfactory, because it must be sending a signal to the PCM. Therefore, Technician B is wrong, and A is the correct answer.

**18. D**   At 220°F (104°C), the PCM should be grounding the fan control relay winding, and the voltage at terminal 16 should be very low. A 12 V reading at terminal 16 indicates an open circuit in the PCM. Thus, D is correct, and A, B, and C are wrong.

**19. C**   An open fuse in the voltage supply wire to the coil primary windings or a defective crankshaft sensor may cause a no-start condition. Therefore, Technicians A and B are both right, and C is the correct answer.

**20. B**   The exhaust gas analyzer readings of HC, CO, and $O_2$ indicate a rich air-fuel ratio. The vehicle symptoms of poor fuel economy and a strong odor from the tailpipe also indicate a rich air-fuel ratio. If the ECT sensor has less resistance than specified, the air-fuel ratio is leaner than normal, so A is wrong. A leaking fuel pressure regulator diaphragm allows the intake manifold vacuum supplied to the regulator to pull fuel from the regulator into the intake resulting in a rich air-fuel ratio, so B is correct. High $HO_2S$ voltage or lower MAP sensor voltage cause a lean air-fuel ratio, so C and D are wrong.

# Glossary

**Abrasion** The wearing or rubbing away of a part.
*Abrasión* El desgaste o frotamiento de una parte que daña la superficie de una área.

**Acceleration** An increase in velocity or speed.
*Aceleración* Un incremento en la velocidad.

**Access valve** See *Service port* and *Service valve*.
*Válvula de acceso* Ver *Service port* [*Orificio de servicio*] y *Service value* [*Válvula de servicio*].

**Accidental air bag deployment** An unintended air bag deployment caused by improper service procedures.
*Despliegue accidental del Airbag* Despliegue imprevisto del Airbag ocasionado por procedimientos de reparación inadecuados.

**Accumulator** A device used in automatic transmissions to cushion the shock of shifting between gears, providing a smoother feel inside the vehicle.
*Acumulador* Un dispositivo que se usa en las transmisiones automáticas para suavizar el choque de cambios entre las velocidades, así proporcionando una sensación más uniforme en el interior del vehículo.

**A circuit** A generator circuit that uses an external grounded field circuit. In an A circuit, the regulator is on the ground side of the field coil.
*Circuito A* Circuito regulador del generador que utiliza un circuito inductor externo puesto a tierra. En el circuito A, el regulador se encuentra en el lado a tierra de la bobina inductora.

**Actuation test mode** A scan tester mode used to cycle the relays and actuators in a computer system.
*Modo de prueba de activación* Instrumento de pruebas de exploración utilizado para ciclar los relés y los accionadores en una computadora.

**Actuators** Devices that perform the actual work commanded by the computer. They can be in the form of a motor, relay, switch, or solenoid.
*Accionadores* Dispositivos que realizan el trabajo efectivo que ordena la computadora. Dichos dispositivos pueden ser un motor, un relé, un conmutador o un solenoide.

**Adapter** A device or fitting that permits different size parts or components to be fastened or connected to each other.
*Adaptador* Dispositivo o ajuste que permite la sujección o conexión entre sí de piezas de tamaños diferentes.

**Additive** A material added to the engine oil to provide additional properties not originally found in the oil.
*Aditivo* Una materia añadida al aceite de motor para proporcionar unas propiedades adicionales que no se encuentran originalmente en el aceite.

**Adhesion** The property of oils to cling to surfaces.
*Adhesión* La propiedad de los aceites que permite que se adhieren a las superficies

**Adhesives** Chemicals used to hold gaskets in place during the assembly of an engine. They also aid the gasket in maintaining a tight seal by filling in the small irregularities on the surfaces and by preventing the gasket from shifting due to engine vibration.
*Adhesivo* Los productos químicos que se usan para sujetar a los empaques en una posición correcta mientras que se efectúa la asamblea de un motor. También ayudan para que los empaques mantengan un sello impermeable, rellenando a las irregularidades pequeñas en las superficies y previeniendo que se mueva el empaque debido a las vibraciones del motor.

**Adjusting pads, mechanical lifters** Metal discs that are available in various thicknesses, and positioned in the end of the mechanical lifter to adjust valve clearance.
*Cojines de ajuste, elevadores mecánicos* Discos metálicos disponibles en diferentes espesores que se colocan en el extremo del elevador mecánico para ajustar el espacio libre de la válvula.

**Adjustment screen** A display on a computer wheel aligner that allows the technician to see the results of certain suspension adjustments.
*Pantalla para la visualización de ajustes* Representación visual en una computadora para alineación de ruedas que le permite al técnico ver los resultados de ciertos ajustes en la suspensión.

**Advance-type timing light** A timing light that is capable of checking the degrees of spark advance.
*Luz de ensayo de regulación del encendido tipo avance* Luz de ensayo de regulación del encendido capaz de verificar la cantidad del avance de la chispa.

**AERA** Automotive Engine Rebuilder's Association.
*AERA* La Asociación de Reacondicionadores Automotivos.

**Aerobic sealants** Sealants that require the presence of oxygen to cure.
*Sellantes aeróbicos* Los sellantes que requieren la presencia del oxígeno para curarse.

**Aftermarket** A term generally given to a device or accessory that is added to a vehicle by the dealer after the original manufacture, such as an air-conditioning system.
*Postmercado* Término dado generalmente a un dispositivo o accesorio que el distribuidor de automóviles agrega al automóvil después de la fabricación original, como por ejemplo un sistema de acondicionamiento de aire.

**Air bag deployment module** The air bag and deployment canister assembly that is mounted in the steering wheel for the driver's side air bag, or in the dash panel for the passenger's side air bag.
*Unidad de despliegue del Airbag* El conjunto del Airbag y elemento de despliegue montado en el volante de dirección para proteger al conductor, o en el tablero de instrumentos para proteger al pasajero.

**Air bag system** A supplemental restraint that will deploy a bag out of the steering wheel or passenger side dash panel to provide additional protection against head and face injuries during an accident.
*Sistema de Airbag* Resguardo complementario que expulsa una bolsa del volante o del panel de instrumentos del lado del pasajero para proveer protección adicional contra lesiones a la cabeza y a la cara en caso de un accidente.

**Air bleeding** The process of bleeding air from a hydraulic system such as the power-steering system.
*Muestra de aire* Proceso a través del cual se extrae el aire de un sistema hidráulico, como por ejemplo un sistema de dirección hidráulica.

**AIR bypass (AIRB) solenoid** A computer-controlled solenoid that directs air to the atmosphere or to the AIR diverter solenoid.
*Solenoide de paso AIR* Solenoide controlado por computadora que conduce el aire hacia la atmósfera o hacia el solenoide derivador AIR.

**Air charge temperature (ACT) sensor** A sensor that sends a signal to the computer in relation to the intake air temperature.
*Sensor de la temperatura de la carga de aire* Sensor que le envía una señal a la computadora referente a la temperatura del aire aspirado.

**Air-conditioning programmer (ACP)** A control unit that may contain the computer, solenoids, motors, and vacuum diaphragms for air-conditioning control.
Programador del acondicionamiento de aire Unidad de control que puede incluir la computadora, los solenoides, los motores y los diafragmas de vacío para el control del acondicionamiento de aire.

**AIR diverter (AIRD) solenoid** A computer-controlled solenoid in the secondary air injection system that directs air upstream or downstream.
Solenoide derivador AIRD Solenoide controlado por computadora en el sistema de inyección secundaria de aire que conduce el aire hacia arriba o hacia abajo.

**Air gap** The space between two components such as the armature and field coils in a starting motor.
Espacio de aire El espacio entre dos componentes, como por ejemplo el rotor y la armadura de un embrague.

**Air-operated vacuum pump** A vacuum pump operated by air pressure that may be used to pump liquids such as gasoline.
Bomba de vacío accionada hidráulicamente Bomba de vacío accionada por la presión del aire que puede utilizarse para bombear líquidos, como por ejemplo la gasolina.

**Align bore** The process of boring, reaming, or honing the main bearing bores to center all bores onto a true centerline.
Calibrado en serie El proceso de taladrar, escariar o bruñir los taladros de los muñones del cigüeñal para alinearlos todos en una linea recta central.

**Align honing** Machining process of the main bearing journals used to remove in excess of 0.050 inch, but this will result in changing the location of the crankshaft in the block. Align honing restores the original bore size by removing metal from the entire circumference of the bore.
Rectificado en serie Un proceso de rectificar en máquina los muñones del cigüeñal para quitar más de 0.050 de una pulgada, pero ésto resulta en que cambia de lugar el cigüeñal en el monoblock. El rectificado en serie restaura el tamaño original del taladro removiendo el metal de toda su circunferencia.

**Aligning bars** Tools used to determine the proper alignment of the crankshaft saddle bores.
Barras para alinear Las herramientas que sirven para determinar el alineamiento correcto de los taladros del asiento del cigüeñal.

**Alignment** An adjustment to a line or to bring into a line.
Alineación Un ajuste que se efectúa en una linea o alinear.

**Alignment ramp** A metal ramp positioned on the shop floor on which vehicles are placed during wheel alignment procedures.
Rampa de alineación Rampa de metal ubicada en el suelo del taller de reparación de automóviles sobre la que se colocan los vehículos durante los procedimientos de alineación de ruedas.

**Alkaline** A chemical class which is the opposite of acidic, but is just as corrosive to certain materials.
Alcalino Un tipo de químico que es lo contrario del ácido, pero igualmente corrosivo para algunas materiales.

**Alloys** Mixtures of two or more metals. For example, brass is an alloy of copper and zinc.
Aleación La mezcla de dos o más metales. Por ejemplo, el latón es una aleación de cobre con zinc.

**Aluminum hydroxide** Corrosion products that are carried to the radiator and deposited when they cool off. They appear as dark gray when wet, and white when dry.
Hidróxido de aluminio Los productos de corrosión que se llevan al radiador y se depositan al enfriarse. Son de color gris obscuro mojados y se blanquean al secarse.

**Ambient temperature** The temperature of the outside air.
Temperatura ambiente Temperatura del aire ambiente.

**Ambient temperature sensor** Thermistor used to measure the temperature of the air entering the vehicle.
Sensor de temperatura ambiente Termistor utilizado para medir la temperatura del aire que entra al vehículo.

**Ammeter** A test meter used to measure current draw.
Amperímetro Instrumento de prueba utilizado para medir la intensidad de una corriente.

**Amperes** See *Current*.
Amperios Véase *Corriente*.

**Anaerobic sealants** Sealants that will only cure in the absence of oxygen.
Sellantes anaeróbicos Los sellantes que sólo se curan en la ausencia del oxígeno.

**Analog** A voltage signal that is infinitely variable or can be changed within a given range.
Señal analógica Señal continua y variable que debe traducirse a valores numéricos discontinuos para poder ser trataba por una computadora.

**Analog meter** A meter with a movable pointer and a meter scale.
Medidor analógico Medidor provisto de un indicador móvil y una escala métrica.

**Anchor pin** Anchoring pins on which the heel of the brake shoe rotates.
Perno de anclaje Las clavijas de anclaje en las cuales gira el extremo inferior de la zapata de freno.

**Anhydrous calcium grease** A special lubricant used in manual rack and pinion steering gears.
Grasa de calcio anhidro Lubricante especial utilizado en mecanismos de dirección de cremallera y piñón manuales.

**Antidieseling adjustment** An adjustment that prevents dieseling when the ignition switch is turned off.
Ajuste antiautoencendido Ajuste que evita el autoencendido cuando el botón conmutador de encendido está apagado.

**Antifriction bearing** A bearing designed to reduce friction. This type of bearing normally uses ball or roller inserts to reduce the friction.
Cojinetes de antifricción Un cojinete diseñado con el fin de disminuir la fricción. Este tipo de cojinete suele incorporar una pieza inserta esférica o de rodillos para disminuir la fricción.

**Antilock brakes (ABS)** A brake system that automatically pulsates the brakes to prevent wheel lockup under panic stop and poor traction conditions.
Frenos antibloqueo Sistema de frenos que pulsa los frenos automáticamente para impedir el bloqueo de las ruedas en casos de emergencia y de tracción pobre.

**Antirattle spring** A spring that holds parts in clutches and disc brakes together and keeps them from rattling.
Resorte antigolpeteo Un resorte que une las partes en los embragues o en los frenos de disco y los impide de rechinar.

**Antiseize** Thread compound designed to keep threaded connections from damage due to rust or corrosion.
Antiagarrotamiento Un compuesto para filetes diseñado para proteger a las conexiones fileteadas de los daños de la oxidación o la corrosión.

**Anti-theft lug nuts** Locking lug nuts that help prevent wheel theft.
Tuercas de orejetas anti-robo Tuercas de orejetas autobloqueantes que ayudan a evitar el robo de las ruedas.

**Anti-theft wheel covers** Locking wheel covers that help prevent wheel theft.
Cubrerruedas anti-robo Cubrerruedas autobloqueantes que ayudan a evitar el robo de las ruedas.

**A-pillar** The pillar in front of the driver or passenger that supports the windshield.
Soporte A Soporte enfrente del conductor o del pasajero que sostiene el parabrisas.

**Apply devices** Devices that hold or drive members of a planetary gear set. They may be hydraulically or mechanically applied.
**Dispositivos de aplicación** Los dispositivos que sujeten o manejan los miembros de un engranaje planetario. Se pueden aplicar mecánicamente o hidráulicamente.

**Approved power source** A power source that is consistent with the requirements of the equipment so far as voltage, frequency, and ampacity are concerned.
**Fuente aprobada de potencia** Fuente de potencia que cumple con los requisitos del equipo referente a la tensión, frecuencia, y ampacidad.

**Arbor press** A small, hand-operated shop press used when only a light force is required against a bearing, shaft, or other part.
**Prensa para calar** Una prensa de mano pequeña del taller que se puede usar en casos que requieren una fuerza ligera contra un cojinete, una flecha u otra parte.

**Armature** The rotating component in a starting motor.
**Armadura** La parte del embrague que se fija al cigüeñal y se engrana al exitarse el rotor.

**Asbestos** A material that was commonly used as a gasket material in places where temperatures are extreme. This material is being used less frequently today because of health hazards that are inherent to the material.
**Amianto** Una materia que se usaba frecuentemente como materia de empaques en sitios en los cuales las temperaturas son extremas. Esta materia se usa menos actualmente debido a los peligros a la salud que se atribuyen a esta materia.

**ASE blue seal of excellence** An ASE logo displayed by automotive service shops that employ ASE-certified technicians.
**Sello azul de excelencia de la ASE** Logotipo exhibido en talleres de reparación de automóviles donde se emplean mecánicos certificados por la ASE.

**ASE technician certification** ASE provides certification of automotive technicians in eight different areas of expertise.
**Certificación de mecánico de la ASE** Certificación de mecánico de automóviles otorgada por la ASE en ocho áreas diferentes de especialización.

**Aspirator** Tubular device that uses a venturi effect to draw air from the passenger compartment over the in-car sensor. Some manufacturers use a suction motor to draw the air over the sensor.
**Aspirador** Dispositivo tubular que utiliza un efecto venturi para extraer aire del compartimiento del pasajero sobre el sensor dentro del vehículo. Algunos fabricantes utilizan un motor de succión para extraer el aire sobre el sensor.

**ATF** Acronym for Automatic Transmission Fluid.
**ATF** Fluido de Transmisión Automática.

**Atmospheric pressure** The weight of the air.
**Presión atmosférica** El peso del aire.

**Austenitic steel** A corrosion-resistant steel made with carbide or carbon alloys.
**Acero austenítico** Un acero resistente a la corrosión hecho con el carburo o las aleaciones de carbono.

**AUTO** Abbreviation for automatic.
**AUTO** Abreviatura del automático.

**Automatic locking/unlocking hubs** Front-wheel hubs that can engage or disengage themselves from the axles automatically.
**Cubos automáticos de cierre/descerrar** Cubos ubicados en las ruedas delanteras que pueden engranarse o desengranarse de los ejes automáticamente.

**Automatic shutdown (ASD) relay** A computer-operated relay that supplies voltage to the fuel pump, coil primary, and other components on Chrysler fuel-injected engines.

**Relé de parada automática** Relé accionado por computadora que les suministra tensión a la bomba del combustible, al bobinado primario, y a otros componentes en motores de inyección de combustible fabricados por la Chrysler.

**Automatic transmission** A transmission in which gear or ratio changes are self-activated, eliminating the necessity of hand-shifting gears.
**Transmisión automática** Una transmisión en la cual un cambio de engranajes o los cambios en relación son por mando automático, así eliminando la necesidad de cambios de velocidades manual.

**Axial** Parallel to a shaft or bearing bore.
**Axial** Paralelo a una flecha o al taladro del cojinete.

**Axis** The centerline of a rotating part, a symmetrical part, or a circular bore.
**Eje** La línea de quilla de una parte giratoria, una parte simétrica, o un taladro circular.

**Axle** The shaft or shafts of a machine upon which the wheels are mounted.
**Semieje** El eje o los ejes de una máquina sobre los cuales se montan las ruedas.

**Axle carrier assembly** A cast-iron framework that can be removed from the rear axle housing for service and adjustment of the parts.
**Conjunto portador del eje** Armazón de hierro fundido que se puede remover del puente trasero para la reparación y el ajuste de las piezas.

**Axle housing** Designed in the removable carrier or integral carrier types to house the drive pinion, ring gear, differential, and axle shaft assemblies.
**Puente trasero** Diseñado en los tipos de portador desmontables o enterizos para alojar los conjuntos del piñón de mando, de la corona, del diferencial y del árbol motor.

**Axle pullers** A special puller required for axle shaft removal.
**Extractores del eje** Extractor especial requerido para la remoción del árbol motor.

**Axle ratio** The ratio between the rotational speed (rpm) of the drive shaft and that of the driven wheel; gear reduction through the differential, determined by dividing the number of teeth on the ring gear by the number of teeth on the drive pinion.
**Relación del eje** La relación entre la velocidad giratoria (rpm) del árbol propulsor y la de la rueda arrastrada; reducción de los engranajes por medio del diferencial, que se determina por dividir el número de dientes de la corona por el número de los dientes en el piñón de ataque.

**Axle shaft** A shaft on which the road wheels are mounted.
**Flecha del semieje** Una flecha en la cual se monta las ruedas.

**Axle shaft end thrust** A force exerted on the end of an axle shaft that is most pronounced when the vehicle turns corners and curves.
**Golpe en la flecha del semieje** Una fuerza que se aplica en el extremo de la flecha del semieje que se pronuncia más cuando un vehículo da la vuelta.

**Axle shaft tubes** These tubes are attached to the axle housing center section to surround the axle shaft and bearings.
**Tubos del árbol motor** Estos tubos se fijan a la sección central del puente trasero para rodear el árbol motor y los cojinetes.

**Back probe** A term used to mean that a test is being performed on the circuit while the connector is still connected to the component. The test probes are inserted into the back of the wire connector.
**Sonda exploradora de retorno** Término utilizado para expresar que se está llevando a cabo una prueba del circuito mientras el conector sigue conectado al componente. Las sondas de prueba se insertan a la parte posterior del conector de corriente.

**Back seat (service valve)** Turning the valve stem to the left (ccw) as far as possible back-seats the valve. The valve outlet to the system is open and the service port is closed.

**Asentar a la izquierda (válvula de servicio)** El girar el vástago de la válvula al punto más a la izquierda posible asienta a la izquierda la válvula. La salida de la válvula al sistema está abierta y el orificio de servicio está cerrado.

**Backing plate** A plate to which drum braking mechanisms are affixed.
**Placa de respaldo** Una placa a la cual se afija los mecanismos de un freno de tambor.

**Backlash** The clearance between two parts.
**Culateo/juego** La holgura entre dos partes.

**Backpressure** The pressure created within the engine cylinder as a result of a restricted exhaust system. With the creation of backpressure, vacuum cannot be produced as efficiently.
**Contrapresión** La presión creada dentro del cilindro del motor causada por una restricción en el sistema de escape. Al crear esta presión, el vacío no se puede producir eficazmente.

**Backup power supply** A voltage source, usually located in the air bag computer, that is used to deploy an air bag if the battery cables are disconnected in a collision.
**Alimentación de reserva** Fuente de tensión, por lo general localizada en la computadora del Airbag, que se utiliza para desplegar el Airbag si se desconectan los cables de la batería a consecuencia de una colisión.

**Balance** Weight introduced to prevent the vibration of moving parts.
**Equilibrio** Un peso metido para prevenir la vibración de las partes en movimiento.

**Balance valve** A regulating valve that controls a pressure of just the right value to balance other forces acting on the valve.
**Válvula niveladora** Una válvula de reglaje que controla a la presión del valor correcto para mantener el equilibrio contra las otras fuerzas que afectan a la válvula.

**Balanced resistance** A situation where two objects, such as axle shafts, present the same resistance to driving rotation.
**Resistencia equilibrada** Situación en la que dos objetos, tales como los árboles motores, ofrecen la misma resistencia al mando de rotación.

**Balancing** The process of removing vibrations which are caused by the entire engine's reciprocating and rotational mass.
**Equilibrar** El proceso de quitar las vibraciones causadas por la masa recíproca y rotativa del motor.

**Ball-and-trunnion universal joint** A nonconstant velocity universal joint that combines the universal joint and slip joint. It uses a drivable housing connected to a shaft with a ball head through two other balls mounted on trunnions.
**Junta universal de bola y muñequilla** Junta universal de velocidad no constante que combina la junta universal y la junta deslizante. Utiliza un alojamiento accionable conectado a un árbol por medio de un péndulo a través de otras dos bolas montadas sobre las muñequillas.

**Ball bearing** An antifriction bearing consisting of a hardened inner and outer race with hardened steel balls that roll between the two races and supports the load of the shaft.
**Rodamiento de bolas** Un cojinete de antifricción que consiste de una pista endurecida interior y exterior y contiene bolas de acero endurecidas que ruedan entre las dos pistas, y sostiene la carga de la flecha.

**Ball gauge** A common term for a small hole gauge.
**Calibrador de bolas** Término común para calibrador de agujero pequeño.

**Ball joint** A suspension component that attaches the control arm to the steering knuckle and serves as the lower pivot point for the steering knuckle. The ball joint gets its name from its ball-and-socket design. It allows both up-and-down motion as well as rotation. In a MacPherson strut FWD suspension system, the two lower ball joints are nonload carrying.

**Articulación esférica** Un componente de la suspensión que une el brazo de mando a la articulación de la dirección y sirve como un punto pivote inferior de la articulación de la dirección. La articulación esférica derive su nombre de su diseño de bola y casquillo. Permite no sólo el movimiento de arriba y abajo sino también el de rotación. En un sistema de suspensión tipo FWD con poste de MacPherson, las articulaciones esféricas inferiores no soportan el peso.

**Ball joint axial movement** Horizontal movement in a ball joint because of internal joint wear.
**Movimiento axial de la junta esférica** Movimiento vertical en una junta esférica ocasionado por el desgaste interno de la misma.

**Ball joint radial movement** Vertical movement in a ball joint because of internal joint wear.
**Movimiento radial de la junta esférica** Movimiento horizontal en una junta esférica ocasionado por el desgaste interno de la misma.

**Ball joint removal and pressing tools** Special tools required for ball joint removal and replacement.
**Herramientas para la remoción y el ajuste de la junta esférica** Herramientas especiales requeridas para la remoción y el reemplazo de la junta esférica.

**Ball joint unloading** Removing the ball joint tension supplied by the vehicle weight prior to ball joint diagnosis.
**Descarga de la junta esférica** Remoción de la tensión de la junta esférica provista por la carga del vehículo antes de llevarse a cabo la diagnosis de la junta esférica.

**Ball joint wear indicator** A visual method of checking ball joint wear.
**Indicador de desgaste de la junta esférica** Método visual de revisar el desgaste interior de una junta esférica.

**Band** A steel band with an inner lining of friction material. A device used to hold a clutch drum at certain times during transmission operation.
**Banda** Una banda de acero que tiene un forro interior de una materia de fricción. Un dispositivo que retiene al tambor del embrague en algunos momentos durante la operación de la transmisión.

**Barb fitting** A fitting that slips inside a hose and is held in place with a gear-type clamp. Ridges (barbs) on the fitting prevent the hose from slipping off.
**Accesorio arponado** Ajuste que se inserta dentro de una manguera y que se sujeta en su lugar con una abrazadera de tipo engranaje.

**Barometric (Baro) pressure sensor** A sensor that sends a signal to the computer in relation to barometric pressure.
**Sensor de la presión barométrica** Sensor que le envía una señal a la computadora referente a la presión barométrica.

**Battery leakage test** Used to determine if current is discharging across the top of the battery case.
**Prueba de pérdida de corriente de la batería** Prueba utilizada para determinar si se está descargando corriente a través de la parte superior de la caja de la batería.

**Battery terminal test** A test which checks for poor electrical connections between the battery cables and terminals. Use a voltmeter to measure voltage drop across the cables and terminals.
**Prueba de los terminales de la batería** Una prueba que averigua las conexiones eléctricas malas entre los cables y los terminales de la batería. Usa un voltímetro para medir la caída del voltaje entre los cables y los terminales.

**B circuit** A generator regulator circuit that is internally grounded. In the B circuit, the voltage regulator controls the power side of the field circuit.
**Circuito B** Circuito regulador del generador puesto internamente a tierra. En el circuito B, el regulador de tensión controla el lado de potencia del circuito inductor.

**BCM** An abbreviation for Body Control Module.
**BCM** Abreviatura de Módulo regulador del soplador.

**Bead blasters** Parts cleaners which use beads of abrasive media carried by air pressure to clean the parts. The abrasives knock off the contaminates.
**Chorro de perlitas** Una limpiadora de partes que usa una materia abrasiva impulsada por el aire bajo presión para limpiar las partes. Los abrasivos desprenden los contaminantes.

**Bearing** The supporting part that reduces friction between a stationary and rotating part, or between two moving parts.
**Cojinete** La parte portadora que reduce la fricción entre una parte fija y una parte giratoria o entre dos partes que muevan.

**Bearing abrasive roller wear** Fine scratches on the bearing surface.
**Desgaste abrasivo del rodillo del cojinete** Rayados finos en la superficie del cojinete.

**Bearing abrasive step wear** A fine circular wear pattern on the ends of the rollers.
**Desgaste abrasivo escalonado del cojinete** Desgaste circular fino en los extremos de los rodillos.

**Bearing brinelling** Straight line indentations on the races and rollers.
**Acción de Brinnell en un cojinete** Hendiduras en línea recta en los anillos y los rodillos.

**Bearing cage** A spacer that keeps the balls or rollers in a bearing in proper position between the inner and outer races.
**Jaula de cojinete** Espaciador que mantiene las bolas o los rodillos en posición correcta en el cojinete, entre los anillos interior y exterior.

**Bearing caps** In the differential, caps held in place by bolts or nuts which, in turn, hold bearings in place.
**Tapones del cojinete** En un diferencial, las tapas que se sujetan en su lugar por pernos o tuercas, los cuales en su turno, retienen y posicionan a los cojinetes.

**Bearing cone** The inner race, rollers, and cage assembly of a tapered roller bearing. Cones and cups must always be replaced in matched sets.
**Cono del cojinete** La asamblea de la pista interior, los rodillos, y el jaula de un cojinete de rodillos cónico. Se debe siempre reemplazar a ambas partes de un par de conos del cojinete y los anillos exteriores a la vez.

**Bearing crush** The extension of the bearing half beyond the seat, which is crushed into place when the seat is tightened.
**Aplastamiento del cojinete** La parte de extensión del cojinete debajo del asiento que se aplasta en su lugar al apretar el asiento.

**Bearing cup** The outer race of a tapered roller bearing or ball bearing.
**Anillo exterior** La pista exterior de un cojinete cónico de rodillas o de bolas.

**Bearing etching** A loss of material on the bearing rollers and races.
**Corrosión del cojinete** Pérdida de material en los rodillos y los anillos del cojinete.

**Bearing fatigue spalling** Flaking of the surface metal on the rollers and races.
**Escamación y fatiga del cojinete** Condición que ocurre cuando el metal de la superficie de los rodillos y los anillos comienza a escamarse.

**Bearing frettage** A fine, corrosive wear pattern around the races and rollers, with a circular pattern on the races.
**Cinceladura del cojinete** Desgaste corrosivo y fino alrededor de los anillos y los rodillos que se hace evidente a través de figuras circulares en los anillos.

**Bearing galling** Metal smears on the ends of the rollers.
**Desgaste por rozamiento en un cojinete** Ralladuras metálicas en los extremos de los rodillos.

**Bearing heat discoloration** A dark brown or bluish discoloration of the rollers and races caused by excessive heat.
**Descoloramiento del cojinete ocasionado por el calor** Descoloramiento marrón oscuro o azulado de los rodillos y los anillos ocasionado por el calor excesivo.

**Bearing lining** A layer of alloy which is adhered to the bearing back and forms the bearing surface.
**Revestamiento del cojinete** Una capa de aleación que se adhiere a la parte exterior de un cojinete y forma la superficie del cojinete.

**Bearing preload** A tension placed on the bearing rollers and races by an adjustment or assembly procedure.
**Carga previa del cojinete** Tensión aplicada a los rodillos y los anillos del cojinete a través de un procedimiento de ajuste o de montaje.

**Bearing pullers** Special tools designed for bearing removal.
**Extractores de cojinetes** Herramientas especiales diseñadas para la remoción del cojinete.

**Bearing race** The surface upon which the rollers or balls of a bearing rotate. The outer race is the same thing as the cup, and the inner race is the one closest to the axle shaft.
**Pista del cojinete** La superficie sobre la cual ruedan los rodillos o las bolas de un cojinete. La pista exterior es la misma que un anillo exterior, y la pista interior es la más cercana a la flecha del eje.

**Bearing smears** Metal loss from the races and rollers in a circular, blotched pattern.
**Ralladuras en los cojinetes** Pérdida del metal de los anillos y los rodillos que se hace evidente a través de una figura circular oxidada.

**Bearing spread** The distance between the outside parting edges is larger than the diameter of the bore.
**Envergadura del cojinete** La distancia al través de los rebordes exteriores es más grande que el diámetro del taladro.

**Bearing stain discoloration** A light brown or black discoloration of the rollers and races caused by incorrect lubricant or moisture.
**Descoloramiento del cojinete** Descoloramiento marrón claro o negro de los rodillos y los anillos ocasionado por la humedad o la utilización de un lubricante incorrecto.

**Belleville spring** A tempered spring steel cone-shaped plate used to aid the mechanical force in a pressure plate assembly.
**Resorte de tensión** Belleville Un plato de resorte del acero revenido en forma cónica que aumenta a la fuerza mecánica de una asamblea del plato opresor.

**Bellhousing** A housing that fits over the clutch components and connects the engine and the transmission.
**Concha del embrague** Un cárter que encaja a los componentes del embrague y conecta al motor con la transmisión.

**Bellows** Rubber protective covers with accordion-like pleats used to contain lubricants and exclude contaminating dirt or water.
**Fuelles** Cubiertas protectivas de caucho con pliegues en forma de acordeón, utilizadas para contener los lubricantes y evitar la contaminación por polvo o agua.S

**Bellows boots** Accordion-style boots that provide a seal between the tie rods and the housing on a rack and pinion steering gear.
**Botas de fuelles** Botas en forma de acordeón que proveen una junta de estanqueidad entre las barras de acoplamiento y el alojamiento en un mecanismo de dirección de cremallera y piñón.

**Belt** See *V-belt*, *V-groove belt*, and *Serpentine belt*.
**Correa** Ver *V-belt* [*Correa en V*], *V-groove belt* [*Correa ranurada en V*], y *Serpentine belt* [*Correa serpentina*].

**Belt tension** Tightness of a belt or belts.
**Tensión de la correa** Tensión de una correa o correas, medida normalmente en libras-pies (ft.-lbs.) o metros-Newton (Nm).

**Belt tension gauge** A special gauge used to measure drive belt tension.
**Calibrador de tensión de la correa de transmisión** Calibrador especial que se utiliza para medir la tensión de una correa de transmisión.

**Bench test** A term used to indicate that the unit is to be removed from the vehicle and tested.

**Prueba de banco** Término utilizado para indicar que la unidad será removida del vehículo para ser examinada.

**Bevel spur gear** Gear that has teeth with a straight center line cut on a cone.
**Engranaje recto cónico** Engranaje que tiene dientes con un corte de línea central recto, cortado en forma cónica.

**Bias voltage** Voltage applied across a diode.
**Tensión de polarización** El voltaje aplicado atravéz de un diodo.

**Big end** The end of the connecting rod that attaches to the crankshaft.
**Extremo grande** Refiere a la extremidad de la biela que se conecta al cigüeñal.

**Bimetal sensor** A temperature-operated sensor used to control vacuum.
**Sensor bimetal** Sensor accionado por temperatura utilizado para controlar el vacío.

**Bleeder valve** A valve on the master cylinder, caliper, or wheel cylinder that allows air and fluid to be drained from the system.
**Tornillo de purga** Una válvula en el cilindro maestro, en la abrazadera, o en el cilindro de la rueda que permite que el aire y el líquido se vacíen del sistema.

**Bleeding** The process of removing air from a closed system.
**Desangramiento** Proceso a través del cual se remueve el aire de un sistema cerrado.

**Block learn** A chip responsible for fuel control in a General Motors PCM.
**Control de combustible en bloque** Pastilla responsable de controlar el combustible en un módulo del control del tren transmisor de potencia de la General Motors.

**Blowby** The unburned fuel and combustion products that leak past the piston rings and enter the crankcase.
**Soplado** El combustible no consumido y los productos de la combustión que escapen por los anillos de pistón y entran al cárter.

**Blower** See *Squirrel-cage blower.*
**Soplador** Ver *Squirrel-cage blower* [*Soplador con jaula de ardilla*].

**Blower motor** See *Motor.*
**Motor de soplador** Ver *Motor.*

**Blower relay** An electrical device used to control the function or speed of a blower motor.
**Relé del soplador** Dispositivo eléctrico utilizado para regular la función o velocidad de un motor de soplador.

**Blowgun** A device attached to the end of an air hose to control and direct airflow while cleaning components.
**Soplete** Dispositivo fijado en el extremo de una manguera de aire para controlar y conducir el flujo de aire mientras se lleva a cabo la limpieza de los componentes.

**Blueprinting** A technique of building an engine using stricter tolerances than those used by most manufacturers. This results in a smoother running, longer lasting, and higher-output engine.
**Especificado por el plan detallado** Una técnica de construcción del motor usando las tolerancias más exactas de las que usan la mayoría de los fabricantes. Esto resulta en un motor que marcha mejor, dura más, y de mejor rendimiento.

**Bob weights** Devices that are attached to the throws of the crankshaft to simulate the rotating and reciprocating mass of the piston assembly.
**Pesos de contra-balanzón** Los dispositivos que se conectan al brazo excéntrico del cigüeñal para simular la masa giratoria y recíproca del conjunto de los pistones.

**Body sway** Excessive body movement from side to side.
**Oscilación de la carrocería** Movimiento lateral excesivo de la carrocería.

**Boiling point** The temperature at which a liquid changes to a vapor.

**Punto de ebullición** Temperatura a la que un líquido se convierte en vapor.

**Bolt torque** The turning effort required to offset resistance as the bolt is being tightened.
**Torsión del perno** El esfuerzo de torsión que se requiere para compensar la resistencia del perno mientras que esté siendo apretado.

**Boost pressure** The amount of intake manifold pressure created by a turbocharger or supercharger.
**Presión de sobrealimentación** Cantidad de presión en el colector de aspiración producida por un turbocompresor o un compresor.

**Booster** A vacuum or hydraulic device attached to the master cylinder to ease operation and/or increase the effectiveness of a brake system.
**Reforzador** Un dispositivo de vacío o hidráulico conectado al cilindro maestro que facilita la operación y/o hace más eficaz un sistema de frenos.

**Boots** See *Bellows.*
**Botas** Véase *Fuelles.*

**Bore** The diameter of a hole.
**Taladro** El diámetro de un agujero.

**Borescope** A special tool that uses fiber optics to allow the technician to see the internal condition of the engine without having to disassemble it.
**Calibrescopio** Una herramienta especial que usa la tecnología fibro-óptica para permitir que el técnico vea las condiciones internas del motor sin tener que desarmarlo.

**Boring** The process of enlarging a hole.
**Escariar** El proceso de agrandar un agujero.

**Boss** The cast or forged part of a piston that can be machined for balance.
**Resalto** La parte colada o forjada de un pistón que se puede rebajar por máquina para el equilibrio.

**Bottom Dead Center (BDC)** Term used to indicate the piston is at the very bottom of its stroke.
**Punto Muerto Inferior (PMI)** El término que indica que el pistón está en el punto más inferior de su carrera.

**B-pillar** The pillar located over the shoulder of the driver or passenger.
**Soporte B** Soporte ubicado sobre el hombro del conductor o del pasajero.

**Brake anchor** The pivot pin on a brake backing plate upon which the shoe rests.
**Anclaje de freno** Una clavija de pivote en la placa de respaldo sobre la cual descanse la zapata.

**Brake drag** The continuous contact between pad or lining with the brake disc or drum.
**Arrastre de los frenos** Un contacto continuo entre la almohadilla o el forro con el disco o el tambor del freno.

**Brake drum** A bowl-shaped cast iron housing against which the brake shoes press to stop its rotation.
**Tambor** Una caja redonda de fierro colado contra la cual se aprietan las zapatas para detener su rotación.

**Brake drum micrometer** Measures the inside diameter of a brake drum.
**Micrómetro del tambor** Toma la medida del diámetro interior de un tambor de freno.

**Brake fade** Gradual reduction of brake pedal height during a brake application caused by brake drum expansion.
**Amortiguamiento del frenado** La pérdida de la eficiencia de frenar debido al calor excesivo que reduce la fricción entre la almohadilla y el disco o la zapata y el tambor.

**Brake fluid** A hydraulic fluid used to transmit force through brake lines. Brake fluid must be noncorrosive to both the metal and rubber components of the brake system.

**Líquido de freno** Un líquido hidráulico que se usa para transmitir la fuerza por las líneas de freno. El líquido de freno no debe corroer los componentes de metal ni de hule que se encuentran en el sistema de frenos.

**Brake flushing** A procedure for cleaning a brake system by removing fluid and washing out sediment and condensation.
**Purgar los frenos** Un procedimiento para limpiar el sistema de frenos en que se vacia el líquido y se quitan los sedimentos y la condensación.

**Brake grab** A sudden and undesirable increase in braking force.
**Amarro de freno** Un incremento del acción de frenar repentino y no deseado.

**Brake horsepower (bhp)** Power delivered by the engine and available for driving the vehicle; bhp = torque _ rpm/5252.
**Caballo indicado al freno (bhp)** Potencia que provee el motor y que es disponible para el uso del vehículo; bhp = de par mortor _rpm/5252.

**Brake lines** Lines that carry brake fluid from the master cylinder to the wheels.
**Líneas de freno** Las líneas que llevan el líquido de freno del cilíndro maestro hasta las ruedas.

**Brake lining** Heat-resistant friction material that is pressed against the metal drum or disc to achieve braking force in a brake system.
**Forro de frenos** Una material resistente al calor que se oprime contra el tambor o disco de metal para ejecutar una fuerza de frenar en un sistema de frenos.

**Brake pads** The parts of a disc brake system that hold the linings.
**Almohadillas de freno** Las partes en un sistema de frenos de disco que sostienen los forros.

**Brake pedal** The pedal pushed upon to activate the master cylinder.
**Pedal de freno** El pedal que se oprime para accionar el cilíndro maestro.

**Brake pedal jack** A special tool installed between the front seat and the brake pedal to apply the brakes during certain wheel alignment measurements.
**Gato del pedal de freno** Herramienta especial instalada entre el asiento delantero y el pedal de freno que se utiliza para aplicar los frenos durante ciertas medidas de alineación de ruedas.

**Brake shoe** The metal assembly onto which the frictional lining is attached for drum brake systems.
**Zapata de freno** La asamblea metálica a la cual se afija el forro de fricción para el sistema de freno de tambor.

**Break a vacuum** The next step after evacuating a system. The vacuum should be broken with refrigerant or other suitable dry gas, not ambient air or oxygen.
**Romper un vacío** El paso que inmediatamente sigue la evacuación de un sistema. El vacío debe de romperse con refrigerante u otro gas seco apropiado, y no con aire ambiente u oxígeno.

**Breakout box** A terminal box that is designed to be connected in series at Ford PCM terminals to provide access to these terminals for test purposes.
**Caja de desenroscadura** Caja de borne diseñada para conectarse en serie a los bornes del módulo del control del tren transmisor de potencia de la Ford, con el objetivo de facilitar el acceso a dichos bornes para propósitos de prueba.

**Brinnelling** Rough lines worn across a bearing race or shaft due to impact loading, vibration, or inadequate lubrication.
**Efecto brinel** Líneas ásperas que aparecen en las pistas de un cojinete o en las flechas debido al choque de carga, la vibración, o falta de lubricación.

**Broach** To finish the inside surface of a bore by forcing a multiple-edge cutting tool through it.

**Fresar con barrena** Acabar una superficie interior de un taladro al atravesarla con una herramienta que tiene múltiples hojas cortantes.

**Bronze** An alloy of copper and tin.
**Bronce** Una aleación de cobre y hojalata.

**Brushes** Electrically conductive sliding contacts, usually made of copper and carbon.
**Escobillas** Contactos deslizantes de conducción eléctrica, por lo general hechos de cobre y de carbono.

**Bump steer** The tendency of the steering to veer suddenly in one direction when one or both front wheels strike a bump.
**Cambio de dirección ocasionado por promontorios en el terreno** Tendencia de la dirección a cambiar repentinamente de sentido cuando una o ambas ruedas delanteras golpea un promontorio.

**Burn time** The length of the spark line while the spark plug is firing measured in milliseconds.
**Duración del encendido** Espacio de tiempo que la línea de chispas de la bujía permanece encendida, medido en milisegundos.

**Burned valves** Valves that have warped and melted, leaving a groove across the valve head.
**Válvulas quemadas** Las válvulas que se han deformado y fundido, dejándo una ranura a través de la cabeza de la válvula.

**Burnish** To smooth or polish by the use of a sliding tool under pressure.
**Bruñir** Pulir o suavizar por medio de una herramienta deslizando bajo presión.

**Burr** A feather edge of metal left on a part being cut with a file or other cutting tool.
**Rebaba** Una lima espada de metal

**Bus** A common connector used as an information source for the vehicle's various control units.
**Bus** Un conector común que se usa como una fuente de información para los varios aparatos de control del vehículo.

**Bus bar** A common electrical connection to which all of the fuses in the fuse box are attached. The bus bar is connected to battery voltage.
**Barra colectora** Conexión eléctrica común a la que se conectan todos los fusibles de la caja de fusibles. La barra colectora se conecta a la tensión de la batería.

**Bushing** A cylindrical lining used as a bearing assembly made of steel, brass, bronze, nylon, or plastic.
**Buje** Un forro cilíndrico que se usa como una asamblea de cojinete que puede ser hecho del acero, del latón, del bronce, del nylon, o del plástico.

**Bypass** An alternate passage that may be used instead of the main passage.
**Desviación** Pasaje alternativo que puede utilizarse en vez del pasaje principal.

**Bypass hose** A hose that is generally small and is used as an alternate passage to bypass a component or device.
**Manguera desviadora** Manguera que generalmente es pequeña y se utiliza como pasaje alternativo para desviar un componente o dispositivo.

**Bypass valve** A safety feature to prevent engine failure. The valve opens when there is a pressure differential of 5 to 15 psi between the outside and inside of the filter element.
**Válvula de paso** Un componente de seguridad para prevenir los fallos del motor. La válvula se abre cuando hay una diferencial de presión de entre 5 a 15 libras por pulgada cuadrada entre el exterior y el interior del elemento de filtro.

**CAA** Acronym for Clean Air Act.
**CAA** Ley para Aire Limpio.

**Cable reel** A conductive ribbon mounted on top of the steering column to maintain electrical contact between the air bag inflator module and the air bag electrical system. This component may be called a clock-spring electrical connector.

**Bobina de cable** Cinta conductiva montada sobre la columna de dirección que se utiliza para mantener el contacto eléctrico entre la unidad infladora y el sistema eléctrico del Airbag. Dicho componente se conoce también como conector eléctrico de cuerda de reloj.

**Cage** A spacer used to keep the balls or rollers in proper relation to one another. In a constant-velocity joint, the cage is an open metal framework that surrounds the balls to hold them in position.

**Jaula** Un espaciador que mantiene una relación correcta entre los rodillos o las bolas. En una junta de velocidad constante, la jaula es un armazón abierto de metal que rodea a las bolas para mantenerlas en posición.

**Calibrate** To determine the scale of an instrument giving quantitative measurements.

**Calibrar** Determinar la escala de un instrumento dando medidas cuantitativas.

**Calibrator package (CAL-PAK)** A removable chip in some computers that usually contains a fuel backup program.

**Paquete del calibrador** Pastilla desmontable en algunas computadoras; normalmente contiene un programa de reserva para el combustible.

**Caliper** A measuring tool capable of taking readings of inside, outside, depth, and step measurements in 0.001-inch increments.

**Calibre** Una herramienta de medir capaz de tomar las medidas interiores, exteriores, de profundidad y de paso en incrementos de 0.001 de una pulgada.

**Cam ground pistons** Pistons cast or forged into a slight oval or cam shape to allow for expansion. As the piston warms, it will become round.

**Pistones rectificados por leva** Los pistones colados o forjados en una forma lijeramente ovulada o en forma de la leva para permitir la expansión. Al calentarse el pistón se redondea.

**Camber** The amount that the center line of the wheel is titled inward or outward from the true vertical plane of the wheel.

**Combadura** Amplitud de la inclinación, hacia adentro o hacia afuera, de la línea central de una rueda con respecto al verdadero plano vertical de la rueda.

**Camber adjustment** A method of adjusting the inward or outward tilt of a front or rear wheel in relation to the true vertical centerline of the tire.

**Ajuste de la combadura** Método de ajustar la inclinación hacia adentro o hacia afuera de una rueda delantera o trasera con relación a la línea central vertical real del neumático.

**Camber angle** The inward or outward tilt of a line through the center of a front or rear tire in relation to the true vertical centerline of the tire and wheel.

**Ángulo de combadura** Inclinación hacia adentro o hacia afuera de una línea a través del centro de una rueda delantera o trasera con relación a la línea central vertical real del neumático y la rueda.

**Camshaft** The shaft containing lobes to operate the engine valves.

**Arbol de levas** Un eje que tiene lóbulos que operan las válvulas del motor.

**Camshaft degreeing** The altering of the point where the camshaft activates the valves in relation to the crankshaft.

**Graduacion del arbol de levas** Cambiar el punto en donde et arbol de levas acciona las valvulas con respeto al ciguenal.

**Can tap** A device used to pierce, dispense, and seal small cans of refrigerant.

**Macho de roscar para latas** Dispositivo utilizado para perforar, distribuir, y sellar pequeñas latas de refrigerante.

**Can tap valve** A valve found on a can tap that is used to control the flow of refrigerant.

**Válvula de macho de roscar para latas** Válvula que se encuentra en un macho de roscar para latas utilizada para regular el flujo de refrigerante.

**Canceling angles** Opposing operating angles of two universal joints cancel the vibrations developed by the individual universal joint.

**Angulos de supresión** Los ángulos de funcionamiento opuestos de dos juntas universales cancelan las vibraciones producidas por la junta universal individual.

**Canister purge solenoid** A computer-operated solenoid connected in the evaporative emission control system.

**Solenoide de purga de bote** Solenoide accionado por computadora conectado en el sistema de control de emisiones de evaporación.

**Canister-type pressurized injector cleaning container** A container filled with unleaded gasoline and injector cleaner, which is pressurized during the manufacturing process or by the shop air supply.

**Recipiente de limpieza del inyector presionizado tipo bote** Recipiente lleno de gasolina sin plomo y limpiador de inyectores, presionizado durante el proceso de fabricación o mediante el suministro de aire en el taller mecánico.

**Cap** An object that fits over an opening to stop flow.

**Tapón** Un objeto que tapa a una apertura para detener el flujo.

**Cap tube** A tube with a calibrated inside diameter and length used to control the flow of refrigerant. In automotive air-conditioning systems, the tube connecting the remote bulb to the expansion valve or to the thermostat is called the capillary tube.

**Tubo capilar** Tubo de diámetro interior y longitud calibrados; se utiliza para regular el flujo de refrigerante. En sistemas automotrices para el acondicionamiento de aire el tubo que conecta la bombilla a distancia con la válvula de expansión o con el termóstato se llama el tubo capilar.

**Capacity test** A test which checks the battery's ability to perform when loaded.

**Prueba de capacidad** Una prueba que averigua la habilidad de la bateria a funcionar bajo carga.

**Carbon** A nonmetallic element that forms inside of the combustion chamber as a product of burning fuel.

**Carbono** Un elemento no metálico que forma en el interior de la cámara de combustión como un producto del combustible al quemarse.

**Carbon dioxide ($CO_2$)** A gas formed as a byproduct of the combustion process.

**Bióxido de carbono ($CO_2$)** Gas que es un producto derivado del proceso de combustión.

**Carbon monoxide** An odorless, colorless, and toxic gas that is produced as a result of incomplete combustion.

**Monóxido de carbono** Un gas sin olor, sin color y tóxico que se produce como resultado de una combustión incompleta.

**Carbon seal face** A seal face made of a carbon composition rather than from another material such as steel or ceramic.

**Frente de carbono de la junta hermética** Frente de la junta hermética fabricada de un compuesto de carbono en vez de otro material, como por ejemplo el acero o material cerámico.

**Carbonize** The process of carbon formation.

**Carbonizar** El proceso de formar el carbono.

**Carburetor** A fuel delivery device that mixes fuel and air to the proper ratio to produce a combustible gas.

**Carburador** Un dispositivo para entregar el combustible que mezcla el combustible y el aire en proporciones correctas para producir un gas combustible.

**Cardan Universal Joint** A nonconstant velocity universal joint consisting of two yokes with their forked ends joined by a cross. The driven yoke changes speed twice in 360 degrees of rotation.

**Junta Universal Cardan** Una junta universal de velocidad no constante que consiste de dos yugos cuyos extremidades ahorquilladas se unen en cruz. El yugo de arrastre cambia su velocidad dos veces en 360 grados de rotación.

**Carrier** An object that bears, cradles, moves, or transports some other object or objects.
**Portador** Objeto que apoya, acojina, mueve o transporta otro u otros objetos.

**Cartridge fuses** See *Maxi-fuse*.
**Fusibles cartucho** Véase *Maxifusible*.

**Case-harden** To harden the surface of steel. The carburizing method used on low-carbon steel or other alloys to make the case or outer layer of the metal harder than its core.
**Cementar** Endurecer la superficie del acero. El método de carburación que se emplea en el acero de bajo carbono o en otros aleaciones para que el cárter o capa exterior queda más dura que lo que está al interior.

**Case porosity** Leaks caused by tiny holes that are formed by trapped air bubbles during the casting process.
**Porosidad del cárter** Las fugas que se causan por los hoyitos pequeños formados por burbújas de aire entrapados durante el proceso del moldeo.

**Castellate** Formed to resemble a castle battlement, as in a castellated nut.
**Acanalado** De una forma que parece a las almenas de un castillo (véa la palabra en inglés), tal como una tuerca con entallas.

**Castellated nut** A nut with six raised portions or notches through which a cotter pin can be inserted to secure the nut.
**Tuerca con entallas** Una tuerca que tiene seis porciones elevadas o muescas por las cuales se puede insertar un pasador de chaveta para fijar la tuerca.

**Caster** A measurement expressed as an angle of the forward or rearward tilt of the top of the wheel spindle.
**Ángulo de comba de eje** Medida expresada como el ángulo de inclinación hacia adelante o hacia atrás de la parte superior del portamuela.

**Caster adjustment** A method of adjusting the forward or rearward tilt of a line through the center of the upper and lower ball joints, or lower ball joint and upper strut mount, in relation to the true vertical centerline of the tire and wheel viewed from the side.
**Ajuste de comba de eje** Método de ajustar la inclinación hacia adelante o hacia atrás de una línea a través del centro de las juntas esféricas superior e inferior, o la junta esférica inferior y el montaje del montante superior, con relación a la línea central vertical real del neumático y la rueda vista desde la parte lateral.

**Caster angle** A line through the center of the upper and lower ball joints, or lower ball joint and upper strut mount, in relation to the true vertical centerline of the tire and wheel viewed from the side.
**Ángulo de comba de eje** Línea a través del centro de las juntas esféricas superior e inferior, o la junta esférica inferior y el montaje del montante superior, con relación a la línea central vertical real del neumático y la rueda vista desde la parte lateral.

**Catalytic converter vibrator tool** A tool used to remove the pellets from catalytic converters.
**Herramienta vibradora del convertidor catalítico** Herramienta utilizada para remover los granos gordos de los convertidores catalíticos.

**Caustic solutions** Cleaning solutions usually consisting of a mixture of water, sodium hydroxide, and sodium carbonate. This solution is extremely alkaline with a pH rating of 10 or more.
**Soluciones cáusticas** Las soluciones de limpieza normalmente compuestas de una mezcla del agua, el óxido de sodio, y el carbonato sódico. Esta solución es extremadamente alcalino con una clasificación PH de 10 o más.

**Caution** A notice to warn of potential personal injury situations and conditions.
**Precaución** Aviso para advertir situaciones y condiciones que podrían causar heridas personales.

**Cc-ing** A method of measuring the volume of the combustion chamber by measuring the amount of oil the chamber can hold.

**Midiendo en centímetros cúbicos** Un método para medir el volumen de la cámara de combustión al medir la cantidad de aceite que puede contener la cámara.

**C-clip** A C-shaped clip used to retain the drive axles in some rear axle assemblies.
**Grapa de C** Una grapa en forma de C que retiene a las flechas motrices en algunas asambleas de ejes traseras

**CCW** Counterclockwise.
**CCW** Sentido inverso al de las agujas del reloj.

**Celsius** A metric temperature scale using zero as the freezing point of water. The boiling point of water is 100°C (212°F).
**Celsio** Escala de temperatura métrica en la que el cero se utilza como el punto de congelación de agua. El punto de ebullición de agua es 100°C (212°F).

**Center hanger bearing** Ball-type bearing mounted on a vehicle cross member to support the drive shaft and provide a better installation angle to the rear axle.
**Silleta de suspensión central** Cojinete de tipo bola montado sobre la traviesa de un vehículo para apoyar el árbol de mando y proveerle un mejor ángulo de montaje al eje trasero.

**Center section** The middle of the integral axle housing containing the drive pinion, ring gear, and differential assembly.
**Sección central** Centro del puente trasero integral que contiene el piñón de mando, la corona y el conjunto del diferencial.

**Centering joint** Ball socket joint placed between two Cardan universal joints to ensure that the assembly rotates on center.
**Junta centradora** Junta de rótula montada entre dos juntas de cardán para asegurar que el montaje gire en el centro.

**Centimeter (cm)** 0.01 meter or 0.3937 inches.
**Centímetro (cm)** La centésima parte de un metro o 0,3937 pulgadas.

**Centrifugal clutch** A clutch that uses centrifugal force to apply a higher force against the friction disc as the clutch spins faster.
**Embrague centrífugo** Un embrague que emplea a la fuerza centrífuga para aplicar una fuerza mayor contra el disco de fricción mientras que el embrague gira más rapidamente.

**Centrifugal force** A force that tends to move a body away from its center of rotation.
**Fuerza centrífuga** Una fuerza cuya tendencia es de mudar un cuerpo fuera de su centro de rotación.

**Ceramic seal face** A seal face made of a ceramic material instead of steel or carbon.
**Frente cerámica de la junta hermética** Frente de la junta hermética fabricada de un material cerámico en vez del acero o carbono.

**Certified** Having a certificate. A certificate is awarded or issued to those that have demonstrated appropriate competence through testing and/or practical experience.
**Certificado** El poseer un certificado. Se les otorga o emite un certificado a los que han demostrado una cierta capacidad por medio de exámenes y/o experiencia práctica.

**CFC-12** See *Refrigerant-12*.
**CFC-12** Ver *Refrigerante-12*.

**Chamfer** A bevel or taper at the edge of a bore.
**Chaflán** Un bisel o chaflán en el borde de un taladro.

**Chamfer face** A beveled surface on a shaft or part that allows for easier assembly. The ends of FWD drive shafts are often chamfered to make installation of the CV joints easier.
**Cara achaflanada** Una superficie biselada en una flecha o una parte que facilita la asamblea. Los extremos de los árboles de mando de FWD suelen ser achaflandos para facilitar la instalación de las juntas CV.

**Channeling** Local leakage around a valve head caused by extreme temperatures developing at isolated locations on the valve face and head.
Acanalado Una fuga local alrededor de una cabeza de válvula causada por las temperaturas severas que ocurren en puntos aislados en la cara y en la cabeza de la válvula.

**Charge** A specific amount of refrigerant or oil by volume or weight.
Carga Cantidad espécifca de refrigerante o de aceite por volumen o peso.

**Charging system requirement test** A diagnostic test used to determine the total electrical demand of the vehicle's electrical system.
Prueba del requisito del sistema de carga Prueba diagnóstica utilizada para determinar la exigencia eléctrica total del sistema eléctrico del vehículo.

**Chase** To straighten up or repair damaged threads.
Embutir Enderezar o reparar a los filetes dañados.

**Chasing** To clean threads with a tap.
Embutido Limpiar a los filetes con un macho.

**Chassis** The vehicle frame, suspension, and running gear. On FWD cars, it includes the control arms, struts, springs, trailing arms, sway bars, shocks, steering knuckles, and frame. The drive shafts, constantvelocity joints, and transaxle are not part of the chassis or suspension.
Chasis El armazón de un vehículo, la suspensión, y el engranaje de marcha. En los coches de FWD, incluye los brazos de mando, los postes, los resortes (chapas), los brazos traseros, las estabilizadoras, las articulaciones de la dirección y el armazón. Los árboles de mando, las juntas de velocidad constante, y la flecha impulsora no son partes del chasis ni de la suspensión.

**Check valve** A device located in the liquid line or inlet to the dryer. The valve prevents liquid refrigerant from flowing the opposite way when the unit is shut off.
Válvula de retención Dispositivo ubicado en la línea de líquido o en la entrada al secador. Al cerrarse la unidad, la válvula impide que el refrigerante líquido fluya en el sentido contrario.

**Chemical cleaning** The process of using chemical action to remove soil contaminants from the engine components.
Limpieza química El proceso de usar la acción química para desprender los contaminantes sucios de los componentes del motor.

**Cherry picker** A common name for an engine hoist.
Grúa alzavagonetas Término común para montacargas del motor.

**Chuckle** A rattling noise that sounds much like a stick rubbing against the spokes of a bicycle wheel.
Estrépito Ruido muy grande o estruendo parecido al sonido de un palo rozando contra los rayos de la rueda de una bicicleta.

**Circlip** A split steel snap ring that fits into a groove to hold various parts in place. Circlips are often used on the ends of FWD drive shafts to retain the constant-velocity joints.
Grapa circular Un seguro partido circular de acero que se coloca en una ranura para posicionar a varias partes. Las grapas circulares se suelen usar en las extremidades de los árboles de mando en FWD para retener las juntas de velocidad constante.

**Circuit** The path of electron flow consisting of the voltage source, conductors, load component, and return path to the voltage source.
Circuito Trayectoria del flujo de electrones, compuesto de la fuente de tensión, los conductores, el componente de carga y la trayectoria de regreso a la fuente de tensión.

**Clamp bolt** Another name for a pinch bolt.
Perno de aprieto Otro nombre para perno-grapa.

**Clashing** Grinding sound heard when gear and shaft speeds are not the same during a gearshift operation.
Entrechoque Rechinamiento que se escucha cuando las velocidades del engranaje y del árbol no son iguales durante el cambio de velocidades.

**Class A fires** A type of fire in which wood, paper, and other ordinary materials are burning.
Incendio de clase A Incendio en el que se queman la madera, el papel y otros materiales comunes.

**Class B fires** A type of fire involving flammable liquids, such as gasoline, diesel fuel, paint, grease, oil, and other similar liquids.
Incendio de clase B Incendio en el que se queman líquidos inflamables, como por ejemplo, la gasolina, el diesel, la pintura, la grasa, el aceite, y otros líquidos similares.

**Class C fires** Electrical fires.
Incendio de clase C Incendios eléctricos.

**Class D fires** A unique type of fire, for the material burning is a metal. An example of this is a burning "mag" wheel; the magnesium used in the construction of the wheel is a flammable metal and will burn brightly when subjected to high heat.
Incendio de clase D Incendio único, porque la material que se quema es un metal. Ejemplo de esto lo será la quema de una rueda "mag";

**Claw washer** A special locking washer used to retain the tie rods to the rack in some rack and pinion steering gears.
Arandela de garra Arandela de bloqueo especial que se utiliza para sujetar las barras de acoplamiento a la cremallera en algunos mecanismos de dirección de cremallera y piñón. el magnesio con el cual se fabrica la rueda es un metal inflamable que resplandece cuando es sometido a altas temperaturas.

**Clean Air Act** A Title IV amendment signed into law in 1990 that established national policy relative to the reduction and elimination of ozone-depleting substances.
Ley para Aire Limpio Enmienda Título IV firmado y aprobado en 1990 que estableció la política nacional relacionada con la reducción y eliminación de sustancias que agotan el ozono.

**Clearance** The space allowed between two parts.
Holgura El espacio permitido entre dos partes.

**Clearance volume** The volume of the combustion chamber when the piston is at TDC. The size of the clearance volume is a factor in determining compression ratio.
Volumen de la holgura El volumen de la cámara de combustión cuando el pistón está en PMS. El tamaño del volumen de la holgura es un factor en determinar el índice de compresión.

**"C" locks** A thick metal locking device used to lock components in place, such as the rear drive axles.
Retenedores en forma de C Dispositivo de bloqueo metálico grueso que se utiliza para ajustar componentes en su posición, como por ejemplo ejes de mando de tracción trasera.

**Clockspring electrical connector** A conductive ribbon in a plastic container mounted on top of the steering column that maintains electrical contact between the air bag inflator module and the air bag electrical system.
Conector eléctrico de cuerda de reloj Cinta conductiva envuelta en una cubierta plástica montada sobre la parte superior de la columna de dirección, que mantiene el contacto eléctrico entre la unidad infladora y el sistema eléctrico del Airbag.

**Clockwise** A term referring to a clockwise (cw) or left-to-right rotation or motion.
Sentido de las agujas del reloj Término que se refiere a un movimiento en el sentido correcto de las agujas del reloj (cw por sus siglas en inglés), es decir, rotación o movimiento desde la izquierda hacia la derecha.

**Close ratio** A relative term for describing the gear ratios in a transmission. If the gears are numerically close, they are said to be close ratio. This design gives quicker acceleration at the expense of initial acceleration and fuel economy.
Relación próxima Término relativo que describe las relaciones de los engranajes en una transmisión. Si existe una proximidad numérica

entre los engranajes, se dice que su relación es próxima. Este diseño permite una aceleración más rápida a costa de la aceleración inicial y del rendimiento de combustible.

**Closed circuit** A circuit that has no breaks in the path and allows current to flow.
**Circuito cerrado** Circuito de trayectoria ininterrumpida que permite un flujo continuo de corriente.

**Closed loop** A computer operating mode in which the computer uses the oxygen sensor signal to help control the air-fuel ratio.
**Bucle cerrado** Modo de funcionamiento de una computadora en el que se utiliza la señal del sensor de oxígeno para ayudar a controlar la relación de aire y combustible.

**Clunking** A metallic noise most often heard when an automatic transmission is engaged into reverse or drive. It is also often heard when the throttle is applied or released. Clunking is caused by excessive backlash somewhere in the driveline and is felt or heard in the axle.
**Sonido sordo** Sonido metálico escuchado con más frecuencia cuando se engrana una transmisión automática en marcha atrás o en marcha adelante. Se escucha también cuando se hunde o se suelta el acelerador. El sonido metálico sordo se debe al contragolpe excesivo en alguna parte de la línea de transmisión, y se siente o escucha en el eje.

**Cluster assembly** A manual transmission–related term applied to a group of gears of different sizes machined from one steel casting.
**Conjunto desplazable** Término relacionado a la transmisión manual que se aplica a un grupo de engranajes de diferentes tamaños hechos a máquina, de una pieza fundida en acero.

**Clutch** A device for connecting and disconnecting the engine from the transmission, or used for a similar purpose in other units.
**Embrague** Un dispositivo para conectar y desconectar el motor de la transmisión o para tal propósito en otros conjuntos.

**Clutch coil** The electrical part of a clutch assembly. When electrical power is applied to the clutch coil, the clutch is engaged to start and stop the compressor action.
**Bobina del embrague** La parte eléctrica del conjunto del embrague. Cuando se aplica una potencia eléctrica a la bobina del embrague, éste se engrana para arrancar y detener la acción del compresor.

**Clutch control cable** A cable assembly with a flexible outer housing anchored at the upper and lower ends. Moving back and forth inside the flexible housing is a braided wire cable that transfers clutch pedal movement to the clutch release lever.
**Cable de mando del embrague** Conjunto de cable con un alojamiento exterior flexible sujetado a los extremos superior e inferior. Un cable trenzado de alambre que transfiere el movimiento del pedal del embrague a la palanca de desembrague se mueve de atrás para adelante dentro del alojamiento.

**Clutch fork** In the clutch, this is a Y-shaped member into which the throw-out bearing is assembled.
**Horquilla de embrague** En el embrague, una pieza en forma de Y sobre la cual se monta el cojinete de desembrague.

**Clutch housing** A large aluminum or iron casting that surrounds the clutch assembly. Located between the engine and transmission, it is sometimes referred to as bell housing.
**Alojamiento del embrague** Pieza grande fundida en hierro o en aluminio que rodea al conjunto del embrague. Ubicado entre el motor y la transmisión, a veces se le llama alojamiento de campana.

**Clutch linkage** A combination of shafts, levers, or cables that transmits clutch pedal motion to the clutch assembly.
**Articulación de embrague** Combinación de árboles, palancas o cables que transmite el movimiento del pedal del embrague al conjunto del embrague.

**Clutch packs** A series of clutch discs and plates installed alternately in a housing to act as a driving or driven unit.

**Conjuntos de embrague** Una serie de discos y platos de embrague que se han instalado alternativamente en un cárter para funcionar como una unidad de propulsión o arrastre.

**Clutch pedal** A pedal in the driver's compartment that operates the clutch.
**Pedal del embrague** Pedal que hace funcionar el embrague; ubicado en el compartimiento del conductor.

**Clutch pushrod** A solid or hollow rod that transfers linear motion between movable parts—that is, the clutch release bearing and release plate.
**Varilla de empuje del embrague** Varilla sólida o hueca que transfiere un movimiento lineal entre piezas móviles; es decir, el cojinete de desembrague del embrague y la placa de desembrague.

**Clutch safety switch** See *Neutral start switch*.
**Interruptor de seguridad del embrague** Véase *Interruptor de arranque neutro*.

**Clutch shaft** Sometimes known as the transmission input shaft or main drive pinion. The clutch driven disc drives this shaft.
**Árbol de embrague** A veces llamado árbol impulsor de la transmisión o piñón principal de mando. El disco accionado del embrague acciona este árbol.

**Clutch slippage** Engine speed increases but increased torque is not transferred through to the driving wheels because of clutch slippage.
**Resbalado del embrague** La velocidad del motor aumenta pero la torsión aumentada del motor no se transfiere a las ruedas de marcha por el resbalado del embrague. retener a la tuerca.

**Clutch teeth** The locking teeth of a gear.
**Dientes del embrague** Dientes trabador de un engranaje.

**Coefficient of friction** The ratio of the force resisting motion between two surfaces in contact to the force holding the two surfaces in contact.
**Coeficiente de la fricción** La relación entre la fuerza que resiste al movimiento entre dos superficies que tocan y la fuerza que mantiene en contacto a éstas dos superficies.

**Coil preload springs** Coil springs are made of tempered steel rods formed into a spiral that resist compression. They are located in the pressure plate assembly.
**Muelles de embrague** Los muelles espirales son fabricados de varillas de acero revenido y resisten la compresión; se ubican en el conjunto del plato opresor.

**Coil spring** A heavy wirelike steel coil used to support the vehicle weight while allowing for suspension motions. On FWD cars, the front coil springs are mounted around the MacPherson struts. On the rear suspension, they may be mounted to the rear axle, to trailing arms, or around rear struts.
**Muelles de embrague** Un resorte espiral hecho de acero en forma de alambre grueso que soporta el peso del vehículo mientras que permite a los movimientos de la suspensión. En los coches de FWD, los muelles de embrague delanteros se montan alrededor de los postes Macpherson. En la suspensión trasera, pueden montarse en el eje trasero, en los brazos traseros, o alrededor de los postes traseros.

**Coil spring clutch** A clutch using coil springs to hold the pressure plate against the friction disc.
**Embrague de muelle helicoidal** Embrague que emplea muelles helicoidales para mantener la placa de presión contra el disco de fricción.

**Coil spring compressing tool** A special tool required to compress a coil spring prior to removal of the spring from the strut.
**Herramienta para la compresión del muelle helicoidal** Herramienta especial requerida para comprimir un muelle helicoidal antes de remover el muelle del montante.

**Cold cranking amps (CCA)** Rating indicates the battery's ability to deliver a specified amount of current to start an engine at low ambient temperatures.

**Amperios de arranque en frío** Tasa indicativa de la capacidad de la batería para producir una cantidad específica de corriente para arrancar un motor a bajas temperaturas ambiente.

**Collapsible steering column** A steering column that is designed to collapse when impacted by the driver in a collision to help reduce driver injury.
**Columna de dirección plegable** Columna de dirección diseñada para plegarse al ser impactada por el conductor durante una colisión con el propósito de reducir las lesiones que el conductor pueda recibir.

**Color codes** Used to assist in tracing the wires. In most color codes, the first group of letters designates the base color of the insulation and the second group of letters indicates the color of the tracer.
**Códigos de colores** Utilizados para facilitar la identificación de los alambres. Típicamente, el primer alfabeto representa el color base del aislamiento y el segundo representa el color del indicador.

**Combustion** The process of burning.
**Combustión** El proceso de quemar.

**Combustion chamber** The volume of the cylinder above the piston with the piston at TDC.
**Cámara de combustión** El volumen del cilindro arriba del pistón cuando el pistón está en PMS.

**Commutator** A series of conducting segments located around one end of the armature.
**Conmutador** Serie de segmentos conductores ubicados alrededor de un extremo de la armadura.

**Companion flange** A mounting flange that fixedly attaches a drive shaft to another drive train component.
**Brida acompañante** Una brida de montaje que fija un árbol de mando a otro componente del tren de mando.

**Compensating port** An opening in a master cylinder that permits fluid to return to the reservoir.
**Abertura de compensación** Un orificio en el cilindro maestro que permite que el líquido regresa al recipiente.

**Component locator** Service manual used to find where a component is installed in the vehicle. The component locator uses both drawings and text to lead the technician to the desired component.
**Manual para indicar los elementos componentes** Manual de servicio utilizado para localizar dónde se ha instalado un componente en el vehículo. En dicho manual figuran dibujos y texto para guiar al mecánico al componente deseado.

**Composites** Man-made materials using two or more different components tightly bound together. The result is a material consisting of characteristics that neither component possesses on its own.
**Compuestas** Los materiales artificiales que usan dos o más componentes distintos combinados estructuralmente. El resultado es un material que tiene las características que ninguno de los componentes posee individualmente.

**Compound** A mixture of two or more ingredients.
**Compuesta** Una mezcla de dos o más ingredientes.

**Compound gauge** A gauge that registers both pressure and vacuum (above and below atmospheric pressure); used on the low side of the systems.
**Manómetro compuesto** Calibrador que registra tanto la presión como el vacío (a un nivel superior e inferior a la presión atmosférica); utilizado en el lado de baja presión de los sistemas.

**Comprehensive tests** A complete series of battery, starting, charging, ignition, and fuel system tests performed by an engine analyzer.
**Pruebas comprensivas** Serie completa de pruebas realizadas en los sistemas de la batería, del arranque, de la carga, del encendido, y del combustible con un analizador de motores.

**Compression** The reduction in volume of a gas.
**Compresión** La reducción del volumen de un gas.

**Compression fitting** A type of fitting used to connect two or more tubes of the same or different diameter together to form a leakproof joint.
**Ajuste de compresión** Tipo de ajuste utilizado para sujetar dos o más tubos del mismo tamaño o de un tamaño diferente para formar una junta hermética contra fugas.

**Compression gauge** A gauge used to test engine compression.
**Manómetro de compresión** Calibrador utilizado para revisar la compresión de un motor.

**Compression nut** A nut-like device used to seat the compression ring into the compression fitting to ensure a leakproof joint.
**Tuerca de compresión** Dispositivo parecido a una tuerca utilizado para asentar el anillo de compresión dentro del ajuste de compresión para asegurar una junta hermética contra fugas.

**Compression ratio** A comparison between the volume above the piston at BDC and the volume above the piston at TDC.
**Indice de compresión** Una comparación entre el volumen arriba del pistón en el PMI y el volumen arriba del pistón en el PMS.

**Compression rings** The upper rings of the piston designed to hold the compression in the cylinder.
**Anillos (aros) de compresión** Los anillos superiores de un pistón diseñados a mantener la compresión dentro del cilindro.

**Compression testing** A diagnostic test to determine the engine cylinder's ability to seal and to maintain pressure.
**Prueba de compresión** Una prueba diagnóstica que determina la habilidad del cilindro del motor de sellar y mantener la presión.

**Compressor shaft seal** An assembly consisting of springs, snap rings, O-rings, shaft seal, seal sets, and a gasket. The shaft seal is mounted on the compressor crankshaft and permits the shaft to be turned without a loss of refrigerant or oil.
**Junta hermética del árbol del compresor** Conjunto que consiste de muelles, anillos de muelles, juntas tóricas, una junta hermética del árbol, conjuntos de juntas herméticas, y una guarnición. La junta hermética del árbol está montada en el cigüeñal del compresor y permite que el árbol se gire sin una pérdida de refrigerante o aceite.

**Computed timing test** A computer system test mode on Ford products that checks spark advance supplied by the computer.
**Prueba de regulación del avance calculado** Modo de prueba en una computadora de productos fabricados por la Ford que verifica el avance de la chispa suministrado por la computadora.

**Computer** An electronic device that stores and processes data and is capable of operating other devices.
**Computadora** Dispositivo electrónico que almacena y procesa datos y que es capaz de ordenar a otros dispositivos.

**Computer command ride (CCR) system** A computer-controlled suspension system that controls strut or shock absorber firmness in relation to driving and road conditions.
**Sistema de viaje ordenado por computadora** Sistema de suspensión controlado por computadora que controla la firmeza de los montantes o de los amortiguadores de acuerdo a las condiciones de viaje o del camino.

**Computer wheel aligner** A type of wheel aligner that uses a computer to measure wheel alignment angles at the front and rear wheels.
**Computadora para alineación de ruedas** Tipo de alineador de ruedas que utiliza una computadora para medir los ángulos de alineación de las ruedas delanteras y traseras.

**Concentric** Two or more circles having a common center.
**Concéntrico** Dos o más círculos que comparten un centro común.

**Conductor** A substance that is capable of supporting the flow of electricity through it.
**Conductor** Sustancia capaz de conducir corriente eléctrica.

**Cone clutch** The driving and driven parts conically shaped to connect and disconnect power flow. A clutch made from two cones, one fitting inside the other. Friction between the cones forces them to rotate together.

**Embrague cónico** Piezas de accionamiento y accionadas, en forma cónica, empleadas para conectar y desconectar el regulador de fuerza. Un embrague hecho de dos conos, uno se inserta dentro del otro. La fricción entre los conos les hace girar juntos.

**Conformability** The ability of the bearing material to conform itself to slight irregularities of a rotating shaft.

**Conformidad** La habilidad de un material de un cojinete de conformarse a las pequeñas irregularidades de un eje giratorio.

**Connecting rod** The link between the piston and crankshaft.

**Biela** La conexión entre el pistón y el cigüeñal.

**Constant mesh** A manual transmission design that permits the gears to be constantly enmeshed regardless of a vehicle's operating circumstances.

**Engrane constante** Diseño de transmisión manual permite que los engranajes permanezcan siempre engranados a pesar de las condiciones del funcionamiento del vehículo.

**Constant mesh transmission** A transmission in which the gears are engaged at all times and shifts are made by sliding collars, clutches, or other means to connect the gears to the output shaft.

**Transmisión de engrane constante** Transmisión en la que los engranajes están siempre engranados y los cambios se llevan a cabo a través de chavetas deslizantes, embragues u otros medios para conectar los engranajes al árbol de rendimiento.

**Constant-velocity joint** A flexible coupling between two shafts that permits each shaft to maintain the same driving or driven speed regardless of the operating angle, allowing for a smooth transfer of power. The constant-velocity joint (also called CV joint) consists of an inner and outer housing with balls in between, or a tripod and yoke assembly.

**Junta de velocidad constante** Un acoplador flexible entre dos flechas que permite que cada flecha mantenga la velocidad de propulsión o arrastre sin importar el ángulo de operación, efectuando una transferencia lisa del poder. La junta de velocidad constante (también llamada junta CV) consiste en un cárter interior y exterior entre los cuales se encuentran bolas, o de un conjunto de trípode y yugo.

**Contaminated** A term generally used when referring to a refrigerant cylinder or a system that is known to contain foreign substances such as other incompatible or hazardous refrigerants.

**Contaminado** Témino generalmente utilizado al referirse a un cilindro para refrigerante o a un sistema que es reconocido contener sustancias extrañas, como por ejemplo otros refrigerantes incompatibles o peligrosos.

**Continual strut cycling** A test procedure for a computer command ride suspension system that allows continual cycling of the strut armatures.

**Funcionamiento cíclico continuo del montante** Procedimiento de prueba para un sistema de suspensión de viaje ordenado por computadora que permite el funcionamiento cíclico continuo de las armaduras de los montantes.

**Continuity** Refers to the circuit being continuous with no opens.

**Continuidad** Se refiere al circuito ininterrumpido, sin aberturas.

**Continuous self-test** A computer system test mode on Ford products that provides a method of checking defective wiring connections.

**Prueba automática continua** Modo de prueba en una computadora de productos fabricados por la Ford que proporciona un método de verificar conexiones defectuosas del alambrado.

**Contraction** A reduction in mass or dimension; the opposite of expansion.

**Contracción** Una reducción en la masa o en la dimensión; el opuesto de expansión.

**Control arm** A suspension component that links the vehicle frame to the steering knuckle or axle housing and acts as a hinge to allow up-and-down wheel motions. The front control arms are attached to the frame with bushings and bolts and are connected to the steering knuckles with ball joints. The rear control arms attach to the frame with bushings and bolts and are welded or bolted to the rear axle or wheel hubs.

**Brazo de mando** Un componente de la suspensión que une el armazón del vehículo a la articulación de dirección o al cárter del eje y que se porta como una bisagra para permitir a los movimientos verticales de las ruedas. Los brazos de mando delanteros se conectan al armazón por medio de pernos y bujes y se conectan a la articulación de dirección por medio de las articulaciones esféricas. Los brazos de mando traseros se conectan al armazón por medio de pernos y bujes y son soldados o empernados al eje trasero o a los cubos de la rueda.

**Control arm bushing tools** Special tools required for control arm bushing removal and replacement.

**Herramientas para el buje del brazo de mando** Herramientas especiales requeridas para la remoción y el reemplazo del buje del brazo de mando.

**Control valve** The component beneath the master cylinder that contains the hydraulic controls for the brake system.

**Válvula de control** El componente en la parte inferior del cilindro maestro que contiene los controles hidráulicos para el sistema de frenos.

**Convection** A transfer of heat by circulating heated air.

**Convección** Una transferencia del calor por medio de la circulación del aire calentado.

**Coolant** The liquid that is circulated through the engine to absorb heat and transfer it to the atmosphere.

**Fluido refrigerante** El líquido circulado por el motor que absorba el calor y lo transfiere a la atmósfera.

**Coolant hydrometer** A tester designed to measure coolant specific gravity and determine the amount of antifreeze in the coolant.

**Hidrómetro de refrigerante** Instrumento de prueba diseñado para medir la gravedad específica del refrigerante y determinar la cantidad de anticongelante en el refrigerante.

**Cooling fans** Fans used to force airflow through the radiator to help in the transfer of heat from the coolant to the air.

**Ventiladores de enfriamiento** Los ventiladores que sirven para forzar un corriente de aire por el radiador para transferir el calor del fluido refrigerante al aire.

**Cooling system pressure tester** A tester used to test cooling system leaks and radiator pressure caps.

**Instrumento de prueba de la presión del sistema de enfriamiento** Instrumento de prueba utilizado para revisar fugas en el sistema de enfriamiento y en las tapas de presión del radiador.

**Core plugs** Metal plugs screwed or pressed into the block or cylinder head at locations where drilling was required or sand cores were removed during casting.

**Tapones del núcleo** Los tapones metálicos fileteados o prensados en el monoblock o en la cabeza de los cilindros ubicados en donde se había requerido taladrar o quitar los machos de arena durante la fundición.

**Cornering angle** The turning angle of one front wheel in relation to the opposite front wheel during a turn. This angle may be called the turning radius.

**Ángulo de viraje** Ángulo de giro de una rueda delantera con relación a la rueda delantera opuesta durante un viraje. Dicho ángulo se conoce también como radio de giro.

**Corrode** To eat away gradually as if by gnawing, especially by chemical action.

**Corroer** Roído poco a poco, primariamente por acción química.

**Corrosion** Chemical action, usually by an acid, that eats away (decomposes) a metal.

**Corrosión** Una acción química, por lo regular un ácido, que corroe (descompone) un metal.

**Cotter pin** A type of fastener, made from soft steel in the form of a split pin, that can be inserted in a drilled hole. The split ends are spread to lock the pin in position.
**Pasador de chaveta** Un tipo de fijación, hecho de acero blando en forma de una chaveta que se puede insertar en un hueco tallado. Las extremidades partidas se despliequen para asegurar la posición de la chaveta.

**Counterbore** To enlarge a bore to a given depth.
**Ensanchar** Extender un taladro a una profundidad designada.

**Counterclockwise rotation** Rotating in the opposite direction of the hands on a clock.
**Rotación en sentido inverso** Girando en el sentido opuesto de las agujas de un reloj.

**Counter gear assembly** A cluster of gears designed on one casting with short shafts supported by antifriction bearings. Closely related to the cluster assembly.
**Mecanismo contador** Grupo de engranajes diseñados en una sola fundición con árboles cortos apoyados por cojinetes de antifricción. Estrechamente relacionado al conjunto desplazable.

**Countershaft** An intermediate shaft that receives motion from a main shaft and transmits it to a working part, sometimes called a lay shaft.
**Árbol de retorno** Árbol intermedio que recibe movimiento de un árbol primario y lo transmite a una pieza móvil; llamado también árbol secundario.

**Counterweight** Weight cast or forged into the crankshaft to reduce rotational vibration.
**Contrapeso** Un peso colado o forjado en un cigüeñal para reducir las vibraciones giratorias.

**Coupling** A connecting means for transferring movement from one part to another; may be mechanical, hydraulic, or electrical.
**Acoplador** Un método de conexión que transfiere el movimiento de una parte a otra; puede ser mecánico, hidráulico, o eléctrico.

**Coupling phase** A point in the torque converter operation where the turbine speed is 90 percent of the impeller speed and there is no longer any torque multiplication.
**Fase del acoplador** El punto de la operación del convertidor de la torsión en el cual la velocidad de la turbina es el 90 por ciento de la velocidad del impulsor y no queda ningúna multiplicación de la torsión.

**Coupling yoke** A part of the double Cardan universal joint that connects the two universal joint assemblies.
**Yugo de acoplamiento** Pieza de la junta doble de cardán que conecta los dos conjuntos de junta universal.

**Cover plate** A stamped steel cover bolted over the service access to the manual transmission.
**Cubrejuntas** Un cubierto de acero estampado que se emperna en la apertura de servicio de la transmisión manual.

**Cracked position** A mid-seated or open position.
**Posición parcialmente asentada** Posición abierta o media asentada.

**Crankcase** The area of the lower engine block that contains the oil and fumes from the combustion process.
**Cárter** El área inferior del monoblock que contiene el aceite y los vapores del proceso de combustión.

**Cranking vacuum test** Used to compare results with the running vacuum test. Low vacuum during cranking can indicate external leaks such as broken or disconnected vacuum hoses that may not be indicated when running engine vacuum tests.
**Prueba de arranque en vacío** Se efectúa para comparar los resultados con los resultados de la prueba de marcha en vacío. Un nivel bajo de vacío en el arranque puede indicar la presencia de unas fugas externas tal como las mangueras de vacío desconectadas o quebradas que no se pueden descubrir en la prueba de marcha en vacío.

**Crankpins** Another term for connecting rod bearing journals.

**Codo de cigüeñal** Otro término para indicar los muñones de la biela.

**Crankshaft** A mechanical device that converts the reciprocating motions of the pistons into rotary motion.
**Cigüeñal** Un dispositivo mecánico que convierte los movimientos recíprocos de los pistones a un movimiento giratorio.

**Crankshaft end play** The measure of how far the crankshaft can move lengthwise in the block.
**Juego en el extremo del cigüeñal** La medida de cuánto puede moverse el cigüeñal a lo largo del monoblock.

**Crankshaft throw** The measured distance between the centerline of the rod bearing journal and the centerline of the crankshaft.
**Codo de cigüeñal** Una distancia medida entre el centro del muñón del cojinete de biela y el centro del cigüeñal.

**Crimping** The process of bending, or deforming by pinching, a connector so that the wire connection is securely held in place.
**Engarzado** Proceso a través del cual se curva o deforma un conectador mediante un pellizco para que la conexión de alambre se mantenga firme en su lugar.

**Crocus cloth** A very fine polishing paper. It is designed to remove very little metal; therefore, it is safe to use on critical surfaces.
**Tela de óxido férrico** Un papel muy fino para pulir. Fue diseñado para raspar muy poco del metal; por lo tanto, se suele emplear en las superficies críticas.

**Cross member** A steel part of the frame structure that transverses the vehicle body to connect the longitudinal frame rails. Cross members can be welded into place or removed from the vehicle.
**Traviesa** Pieza de acero de la estructura del armazón que atraviesa la carrocería para conectar las barras longitudinales del armazón. Las traviesas se pueden soldar o remover del vehículo.

**Crown** The center area of a bearing half.
**Centrado de la punta** El área central de una mitad del cojinete.

**Crush** The press fit allowance required to maintain bearing position in the bore.
**Aplastamiento** El margen requirido en un ajuste prensado para mantener al cojinete en su posición en el taladro.

**Curb riding height** The distance between the vehicle chassis and the road surface measured at specific locations.
**Altura del cotén del viaje** Distancia entre el chasis del vehículo y la superficie del camino medida en puntos específicos.

**Current** The aggregate flow of electrons through a wire. One ampere represents the movement of 6.25 billion billion electrons (or one coulomb) past one point in a conductor in one second.
**Corriente** Flujo combinado de electrones a través de un alambre. Un amperio representa el movimiento de 6,25 mil millones de mil millones de electrones (o un colombio) que sobrepasa un punto en un conductor en un segundo.

**Current code** A fault code in a computer that is present at all times.
**Código actual** Código de fallo siempre presente en una computadora.

**Current draw test** Diagnostic test used to measure the amount of current that the starter draws when actuated. It determines the electrical and mechanical condition of the starting system.
**Prueba de la intensidad de una corriente** Prueba diagnóstica utilizada para medir la cantidad de corriente que el arrancador tira cuando es accionado. Determina las condiciones eléctricas y mecánicas del sistema de arranque.

**Current output testing** Diagnostic test used to determine the maximum output of the AC generator.
**Prueba de la salida de una corriente** Prueba diagnóstica utilizada para determinar la salida máxima del generador de corriente alterna.

**Custom tests** A series of tests programmed by the technician and performed by an engine analyzer.

**Pruebas de diseño específico** Serie de pruebas programadas por el mecánico y realizadas por un analizador de motores.

**CV joint** See *Constant velocity joint*.
**Junta CV** Véase *Junta de velocidad constante*.

**CW** Abbreviation for clockwise. Also cw.
**CW** Abreviatura del sentido de las agujas del reloj. También cw.

**Cycle** A sequence that is repeated. In the four-stroke engine, four strokes are required to complete one cycle.
**Ciclo** Una secuencia que se repite. En el motor de cuatro carreras, se requieren cuatro carreras para completar un ciclo.

**Cycle clutch time (total)** Time from the moment the clutch engages until it disengages, then reengages. Total time is equal to on time plus off time for one cycle.
**Duración del ciclo del embrague (total)** Espacio de tiempo medido desde el momento en que se engrana el embrague hasta que se desengrane y se engrane de nuevo. El tiempo total es equivalente al trabajo efectivo más el trabajo no efectivo por un ciclo.

**Cycling** A systematic driving method that varies the load on the engine.
**Ciclar** Un método sistemático de conducir que diversifica la carga en el motor.

**Cycling clutch pressure switch** A pressure-actuated electrical switch used to cycle the compressor at a predetermined pressure.
**Autómata manométrico del embrague con funcionamiento cíclico** Interruptor eléctrico accionado a presión utilizado para ciclar el compresor a una presión predeterminada.

**Cycling clutch system** An air-conditioning system in which the air temperature is controlled by starting and stopping the compressor with a thermostat or pressure control.
**Sistema de embrague con funcionamiento cíclico** Sistema de acondicionamiento de aire en el cual la temperatura del aire se regula al arrancarse y detenerse el compresor con un termóstato o regulador de presión.

**Cylinder** A hole bored into the engine block which the piston travels in, and which makes up part of the combustion chamber where the air-fuel mixture is compressed.
**Cilindro** Un agujero taladrado en el monoblock en el cual viaja el pistón, y que forma parte de la cámara de combustión en donde se comprime la mezcla de aire/combustible.

**Cylinder block** The main structure of the engine. Most of the other engine components are attached to the block.
**Monoblock** La estructura principal del motor. La mayoría de los otros componentes del motor se conectan al monoblock.

**Cylinder bore dial gauge** An instrument used to measure the cylinder bore for wear, taper, and out-of-round.
**Calibre carátula del taladro del cilindro** Un instrumento de medida que sirve para averiguar si el taladro del cilindro está gastado, cónico, u ovalado.

**Cylinder end stopper** A circular bushing in the end of a rack and pinion steering gear housing.
**Tapador en el extremo del cilindro** Buje circular en el extremo del alojamiento de un mecanismo de dirección de cremallera y piñón.

**Cylinder head** On most engines, the cylinder head contains the valves, valve seats, valve guides, valve springs, and the upper portion of the combustion chamber.
**Cabeza del cilindro** En la mayoría de los motores la cabeza del cilindro contiene las válvulas con sus asientos, sus guías, sus resortes y la parte superior de la cámara de combustión.

**Cylinder leakage test** A test which determines the condition of the piston ring, intake or exhaust valve, and the head gasket. It uses a controlled amount of air pressure to determine the amount of leakage.

**Prueba de fugas en el cilindro** Una prueba para determinar la condición del anillo de pistón, la válvula de entrada o escape, y la junta de la cabeza. Se emplea una cantidad controlada de presión de aire para determinar la cantidad de la fuga.

**Cylinder leakage tester** A tester designed to measure the amount of air leaking from the combustion chamber past the piston rings or valves.
**Instrumento de prueba de la fuga del cilindro** Instrumento de prueba diseñado para medir la cantidad de aire que se escapa desde la cámara de combustión y que sobrepasa los anillos de pistón o las válvulas.

**Cylinder ridge** An area of no wear resulting from the piston ring not travelling the full height of the cylinder.
**Cilindro con reborde** Una área sin desgaste que resulta cuando el anillo no suba la altura completa del cilindro.

**Data link connector (DLC)** An electrical connector for computer system diagnosis mounted under the instrument panel or in the engine compartment.
**Conector de enlace de datos** Conector eléctrico montado debajo del tablero de instrumentos o en el compartimiento del motor, que se utiliza para la diagnosis del sistema informático.

**Datum line** A straight reference line such as the top of a dedicated bench system.
**Línea de datos** Línea recta de referencia, como por ejemplo la parte superior de un sistema de banco dedicado.

**Dead axle** An axle that only supports the vehicle and does not transmit power.
**Eje portante** Eje que sirve sólo para apoyar el vehículo y que no transmite fuerza motriz.

**Decal** A label that is designed to stick fast when transferred. A decal affixed under the hood of a vehicle is used to identify the type of refrigerant used in a system.
**Calcomanía** Etiqueta diseñada para pegarse fuertemente al ser transferido. Una calcomanía pegada debajo de la capota se utiliza para identificar el tipo de refrigerante utilizado en un sistema.

**Deceleration** A reduction of speed.
**Deceleración** Una reducción de la velocidad.

**Deck** The top of the engine block where the cylinder head is attached.
**Cubierta** La parte superior del monoblock en donde se conecta la cabeza del cilindro.

**Deck clearance** A measure of the distance that the top of the piston is below, or above, the deck of the engine block (when the piston is at TDC).
**Holgura de la cubierta** La distancia que queda la parte superior del pistón abajo, o arriba, de la cubierta del monoblock (cuando el pistón está en PMS).

**Dedicated bench system** A heavy steel bed with special fixtures for aligning unitized bodies.
**Sistema de banco dedicado** Asiento pesado de acero con aparatos especiales que se utiliza para la alineación de carrocerías unitarias.

**Deep cycling** Discharging the battery completely before recharging it.
**Operación cíclica completa** La descarga completa de la batería previo al recargo

**Deflection** The bending or movement away from normal due to loading.
**Desviación** El abarquillamiento o movimiento fuera de lo normal debido a la carga.

**Deglazing** The process of roughening the cylinder wall without changing its diameter.
**Lijar** El proceso de poner áspero el muro del cilindro sin cambiar su diámetro.

**Degree** A unit of measurement equal to 1/360th of a circle.
**Grado** Una unidad de medida que iguala a la 1/360 parte de un círculo.

**Density** The compactness or relative mass of matter in a given volume.
Densidad  Lo compacto o la masa relativa de la materia en un volumen designado.

**Department of Transportation** The United States Department of Transportation is a federal agency charged with regulation and control of the shipment of all hazardous materials.
Departamento de Transportes  El Departamento de Transportes de los Estados Unidos de América es una agencia federal que tiene a su cargo la regulación y control del transporte de todos los materiales peligrosos.

**Dependability** Reliability; trustworthiness.
Carácter responsable  Digno de confianza; integridad.

**Depressing pin** A pin located in the end of a service hose to press (open) a Schrader-type valve.
Pasador depresor  Pasador ubicado en el extremo de una manguera de servicio para forzar que se abra una válvula de tipo Schrader.

**Depth gauge** Based on a micrometer, this gauge is used to measure the depth of holes, slots, and keyways.
Galga de profundidades  Principalmente como un micrómetro, esta galga se usa para medir la profundidad de los hoyos, las ranuras y las mortajas.

**Depth micrometers** An instrument designed to measure the depth of a bore.
Micrómetro de profundidad  Una herramienta diseñada para medir la profundidad de un taladro.

**Detent** A small depression in a shaft, rail, or rod into which a pawl or ball drops when the shaft, rail, or rod is moved. This provides a locking effect.
Detención  Un pequeño hueco en una flecha, una barra o una varilla en el cual cae una bola o un linguete al moverse la flecha, la barra o la varilla. Esto provee un efecto de enclavamiento.

**Detent mechanism** A shifting control designed to hold the manual transmission in the gear range selected.
Aparato de detención  Un control de desplazamiento diseñado a sujetar a la transmisión manual en la velocidad seleccionada.

**Detonation** A defect which occurs if the air-fuel mixture in the cylinder is burned too fast.
Detonación  Un defecto que ocurre si la mezcla aire/combustible en el cilindro se quema demasiado rápido.

**Diagnosis** The use of instruments, the service manual, and experience to determine the action of parts or systems to ascertain the cause of the failure.
Diagnosis  El uso de los instrumentos, el manual de servicio, y la experiencia para determinar la acción de los partes o los sistemas y descubrir la causa de un fallo.

**Diagnostic drawing and text screen** A display on a computer wheel aligner that provides illustrations and written instructions for the technicians.
Pantalla para la visualización de textos y diagramas diagnósticos  Representación visual en una computadora para alineación de ruedas que le provee a los mecánicos ilustraciones e instrucciones escritas.

**Diagnostic trouble code (DTC)** A code retained in a computer memory representing a fault in a specific area of the computer system.
Códigos indicadores de fallas para propósitos diagnósticos  Código almacenado en la memoria de una computadora que representa una falla en un área específica de la computadora.

**Dial bore gauge** A commonly used name for a bore gauge.
Verificador de calibrador con indicador  Nombre comúnmente utilizado para calibrador de diámetro interior.

**Dial calipers** Vernier-style calipers with a dial gauge installed to make reading of the vernier scale easier and faster.

Calibres de carátula  Los calibres tipo vernier equipados con una carátula para facilitar y acelerar la lectura de una escala vernier.

**Dial indicator** A measuring instrument consisting of a dial face with a needle. The dial is usually calibrated in 0.001-inch increments. A spring-loaded plunger or toggle lever transfers movement to the dial needle.
Indicador de carátula  Un instrumento de medida que consiste de una carátula con una aguja. La carátula suele ser calibrada en incrementos de 0.001 de una pulgada. Un pistón tubular cargado de resorte o por una palanca acodada transfiere el movimiento a la aguja de la carátula.

**Diamond-shaped chassis** A vehicle chassis that is shaped like a diamond from collision damage.
Chasis en forma de diamante  Chasis que ha adquirido la forma de un diamante a causa del impacto recibido durante una colisión.

**Diaphragm spring** A circular disc shaped like a cone, with spring tension that allows it to flex forward or backward. Often referred to as a Belleville spring.
Muelle de diafragma  Disco circular en forma de cono, con tensión en el muelle que le permite moverse hacia adelante o hacia atrás. Conocido también como muelle de Belleville.

**Diaphragm spring clutch** A clutch in which a diaphragm spring, rather than a coil spring, applies pressure against the friction disc.
Embrague de muelle de diafragma  Embrague en el que un muelle de diafragma, en vez de un muelle helicoidal, ejerce presión contra el disco de fricción.

**Die** A tool used to repair or cut new external threads.
Matriz  Una herramienta para reparar o cortar las roscas exteriores.

**Diesel particulates** Carbon particles emitted in diesel engine exhaust.
Partículas de diesel  Partículas de carbón presentes en el escape de un motor diesel.

**Differential** A mechanism between drive axles that permits one wheel to run at a different speed than the other while turning.
Diferencial  Un mecanismo entre dos semiejes que permite que una rueda gira a una velocidad distinta que la otra en una curva.

**Differential action** An operational situation where one driving wheel rotates at a slower speed than the opposite driving wheel.
Acción del diferencial  Una situación durante la operación en la cual una rueda propulsora gira con una velocidad más lenta que la rueda propulsora opuesta.

**Differential case** The metal unit that encases the differential side gears and pinion gears, and to which the ring gear is attached.
Caja de satélites  La unidad metálica que encaja a los engranajes planetarios (laterales) y a los satélites del diferencial, y a la cual se conecta la corona.

**Differential case spread** Another name for preload.
Extensión de la caja del diferencial  Otro nombre para carga previa.

**Differential drive gear** A large circular helical gear that is driven by the transaxle pinion gear and shaft and drives the differential assembly.
Corona  Un engranaje helicoidal grande circular que es arrastrado por el piñón de la flecha de transmisión y la flecha y propela al conjunto del diferencial.

**Differential housing** A cast-iron assembly that houses the differential unit and the drive axles. Also called the rear axle housing.
Cárter del diferencial  Una asamblea de acero vaciado que encaja a la unidad del diferencial y los semiejes. También se llama el cárter del eje trasero.

**Differential pinion gears** Small beveled gears located on the differential pinion shaft.
Satélites  Engranajes pequeños biselados que se ubican en la flecha del piñón del diferencial.

**Differential pinion shaft** A short shaft locked to the differential case. This shaft supports the differential pinion gears.
Flecha del piñón del diferencial Una flecha corta clavada en la caja de satélites. Esta flecha sostiene a los satélites.

**Differential ring gear** A large circular hypoid-type gear enmeshed with the hypoid drive pinion gear.
Corona Un engranaje helicoidal grande circular endentado con el piñón de ataque hipoide.

**Differential side gears** The gears inside the differential case that are internally splined to the axle shafts, and which are driven by the differential pinion gears.
Planetarios (laterales) Los engranajes adentro de la caja de satélites que son acanalados a los semiejes desde el interior, y que se arrastran por los satélites.

**Digital** A voltage signal is either on-off, yes-no, or high-low.
Digital Una señal de tensión está Encendida-Apagada, es Sí-No o Alta- Baja.

**Digital EGR valve** An EGR valve that contains a computer-operated solenoid or solenoids.
Válvula EGR digital Una válvula EGR que contiene un solenoide o solenoides accionados por computadora.

**Digital meter** A meter with a digital display.
Medidor digital Medidor con lectura digital.

**Dimmer switch** A switch in the headlight circuit that provides the means for the driver to select either high beam or low beam operation, and to switch between the two. The dimmer switch is connected in series within the headlight circuit and controls the current path for high and low beams.
Conmutador reductor Conmutador en el circuito para faros delanteros que le permite al conductor elegir la luz larga o la luz corta, y conmutar entre las dos. El conmutador reductor se conecta en serie dentro del circuito para faros delanteros y controla la trayectoria de la corriente para la luz larga y la luz corta.

**Dimpling** Brinelling, or the presence of indentations in a normally smooth surface.
Abolladura Acción de Brinell o la formación de hendiduras en una superficie normalmente lisa.

**Diode** An electrical one-way check valve that allows current to flow in one direction only.
Diodo Válvula eléctrica de retención, de una vía, que permite que la corriente fluya en una sola dirección.

**Diode rectifier bridge** A series of diodes that are used to provide a reasonably constant DC voltage to the vehicle's electrical system and battery.
Puente rectificador de diodo Serie de diodos utilizados para proveerles una tensión de corriente continua bastante constante al sistema eléctrico y a la batería del vehículo.

**Diode trio** Used by some manufacturers to rectify the stator of an AC generator current so that it can be used to create the magnetic field in the field coil of the rotor.
Trío de diodos Utilizado por algunos fabricantes para rectificar el estátor de la corriente de un generador de corriente alterna y poder así utilizarlo para crear el campo magnético en la bobina inductora del rotor.

**Dipstick** A metal rod used to measure the fluid in an engine or transmission.
Varilla de medida Una varilla de metal que se usa para medir el nivel de flúido en un motor o en una transmisión.

**Direct drive** One turn of the input driving member compared to one complete turn of the driven member, such as when there is direct engagement between the engine and drive shaft where the engine crankshaft and the drive shaft turn at the same rpm.
Mando directo Una vuelta del miembro de ataque o de propulsión que se compara a una vuelta completa del miembro de arrastre, tal como cuando hay un enganchamiento directo entre el motor y el árbol de transmisión cuando el cigüeñal y el árbol de transmisión giran al mismo rpm.

**Directional stability** The tendency of a vehicle steering to remain in the straight-ahead position when driven straight ahead on a smooth, level road surface.
Estabilidad direccional Tendencia de la dirección del vehículo a permanecer en línea recta al ser así conducido en un camino cuya superficie es lisa y nivelada.

**Disarm** To turn off; to disable a device or circuit.
Desarmar Apagar; incapacitar un dispositivo o circuito.

**Disc brake** A brake design in which the member attached to the wheel is a metal disc, and braking force is applied by two brake pads that are squeezed against the disc by the caliper.
Frenos de disco Un diseño de frenos en el cual el miembro conectado a la rueda es un disco de metal y la fuerza del frenado se aplica por medio de la mordaza que aprieta dos almohadillas de freno contra el disco.

**Disc runout** A measurement of how much a brake disc wobbles from side to side as it rotates.
Corrimiento del disco Una medida del movimiento oscilatorio de un disco de un lado al otro mientras que gira.

**Disco accionado** Pieza del conjunto del embrague que recibe su fuerza motriz del conjunto del volante y del conjunto de la placa de presión.

**Disengage** When the operator moves the clutch pedal toward the floor to disconnect the driven clutch disc from the driving flywheel and pressure plate assembly.
Desembragar Cuando el operador mueva el pedal de embrague hacia el piso para desconectar el disco de embrague del volante impulsor y del conjunto del plato opresor.

**Displacement** A measure of engine volume. The larger the displacement, the greater the power output.
Desplazamiento Una medida del volumen del motor. Lo más grande el desplazamiento, lo mejor la producción de potencia.

**Distortion** A warpage or change in form from the original.
Distorción Un abarquillamiento o un cambio en la forma de la original.

**Distributor** The mechanism within the ignition system that controls the primary circuit and directs the secondary voltage to the correct spark plug.
Distribuidor El mecanismo dentro del sistema de arranque que controla al circuito primario y dirige el voltaje secundario a la bujía correcta.

**Distributor ignition (DI) system** SAE J1930 terminology for any ignition system with a distributor.
Sistema de encendido con distribuidor Término utilizado por la SAE J1930 para referirse a cualquier sistema de encendido que tenga un distribuidor.

**Dog tooth** A series of gear teeth that are part of the dog clutching action in a transmission synchronizer operation. The locking teeth of a gear.
Diente de sierra Serie de dientes del engranaje que forman parte de la acción del embrague de garras durante una sincronización de transmisión. Dientes de cierre de un engranaje.

**Double cardan universal joint** A near constant velocity universal joint that consists of two Cardan universal joints connected by a coupling yoke.
Junta doble de cardán Junta universal de velocidad casi constante compuesta de dos juntas de cardán conectadas por un yugo de acoplamiento.

**Double-offset constant velocity joint** Another name for the type of plunging, inner CV joint found on many GM, Ford, and Japanese FWD cars.
Junta de velocidad constante de desviación doble Otro nombre para el tipo de junta de velocidad constante interior de pistón tubular, instalada en muchos automóviles de tracción delantera de la GM, de la Ford y japoneses.

**Double reduction axle** A drive axle construction in which two sets of reduction gears are used for extreme reduction of the gear ratio.
**Eje de reducción doble** Eje de mando en el que se utilizan dos juegos de reductores para lograr una mayor reducción de la relación de engranajes.

**Dowel** A metal pin attached to one object which, when inserted into a hole in another object, ensures proper alignment.
**Espiga** Una clavija de metal que se fija a un objeto, que al insertarla en el hoyo de otro objeto, asegura una alineación correcta.

**Dowel pin** A pin inserted in matching holes in two parts to maintain those parts in fixed relation one to another.
**Clavija de espiga** Una clavija que se inserta en los hoyos alineados en dos partes para mantener esos dos partes en una relación fijada el uno al otro.

**Downshift** To shift a transmission into a lower gear.
**Cambio descendente** Cambiar la velocidad de una transmisión a una velocidad más baja.

**Downstream air** Air injected into the catalytic converter.
**Aire conducido hacia abajo** Aire inyectado dentro del convertidor catalítico.

**Drag link** A connecting rod or link between the steering gear, Pitman arm, and the steering linkage.
**Varilla de arrastre** Biela o unión entre el mecanismo de dirección, el brazo pitman y el cuadrilátero de la dirección.

**Drive cycle diagnosis** A test procedure when diagnosing computer-controlled suspension.
**Diagnosis del ciclo propulsor** Procedimiento de prueba llevado a cabo durante la diagnosis de una suspensión controlada por computadora.

**Drive pinion flange** A rim used to connect the rear of the drive shaft to the rear axle drive pinion.
**Brida de piñón de mando** Corona utilizada para conectar la parte trasera del árbol de mando al piñón de mando del eje trasero.

**Drive pinion gear** One of the two main driving gears located within the transaxle or rear driving axle housing. Together, the two gears multiply engine torque.
**Engranaje de piñón de ataque** Uno de dos engranajes de ataque principales que se ubican adentro de la flecha de transmisión o en el cárter del eje de propulsión. Los dos engranajes trabajan juntos para multiplicar la potencia.

**Drive shaft** An assembly of one or two universal joints connected to a shaft or tube; used to transmit power from the transmission to the differential. Also called the propeller shaft.
**Árbol de mando** Una asamblea de una o dos uniones universales que se conectan a un árbol o un tubo; se usa para transferir la potencia desde la transmisión al diferencial. También se le refiere como el árbol de propulsión.

**Drive shaft installation angle** The angle that the drive shaft is mounted off the true horizontal line measured in degrees.
**Ángulo de montaje del árbol de mando** El ángulo al que se monta el árbol de mando fuera de la línea horizontal verdadera, medido en grados.

**Driveline** The universal joints, drive shaft, and other parts connecting the transmission with the driving axles.
**Línea de transmisión** Las juntas universales, el árbol de mando y otras piezas que conectan la transmisión a los ejes motores.

**Driveline torque** Relates to rear-wheel driveline and is the transfer of torque between the transmission and the driving axle assembly.
**Potencia de la flecha motríz** Se relaciona a la flecha motríz de las ruedas traseras y transfiere la potencia de la torsión entre la transmisión y el conjunto del eje trasero.

**Driveline wrap-up** A condition where axles, gears, U-joints, and other components can bind or fail if the 4WD mode is used on pavement where 2WD is more suitable.

**Falla de la línea de transmisión** Condición que ocurre cuando los ejes, los engranajes, las juntas universales, y otros componentes se traban o fallan si se emplea la tracción a las cuatro ruedas sobre un pavimiento donde la tracción a las dos ruedas es más adecuada.

**Driven disc** The part of the clutch assembly that receives driving motion from the flywheel and pressure plate assemblies.

**Driven gear** The gear meshed directly with the driving gear to provide torque multiplication, reduction, or a change of direction.
**Engranaje de arrastre** El engranaje endentado directamente al engranaje de ataque para proporcionar la multiplicación, la reducción, o los cambios de dirección de la potencia.

**Driving axle** A term related collectively to the rear driving axle assembly where the drive pinion, ring gear, and differential assembly are located within the driving axle housing.
**Eje motor** Término relacionado colectivamente al conjunto del eje motor trasero donde el piñón de mando, la corona, y el conjunto del diferencial están ubicados dentro del puente trasero.

**Drop forging** A piece of steel shaped between dies while hot.
**Estampado** Un pedazo de acero que se forma entre bloques mientras que esté caliente.

**Drum brake** A brake design in which the component attached to the wheel is shaped like a drum. Shoes press against the inside of the drum to provide braking action.
**Frenos de tambor** Un diseño de frenos en el cual el componente conectado a la rueda tiene la forma de un tambor. Las zapatas oprimen contra la superficie interior para efectuar la acción de frenar.

**Dry compression testing** A compression test performed with no additional oil added to the cylinders.
**Prueba de compresión en seco** Una prueba de compresión que se efectúa sin añadir aceite adicional a los cilindros.

**Dry-disc clutch** A clutch in which the friction faces of the friction disc are dry, as opposed to a wet-disc clutch, which runs submerged in oil. The conventional type of automobile clutch.
**Embrague de disco seco** Embrague en el que las placas de fricción del disco de fricción están secas, lo opuesto de un disco mojado, que funciona sumergido en aceite. Tipo convencional de embrague de automóviles.

**Dry friction** The friction between two dry solids.
**Fricción seca** Fricción entre dos sólidos secos.

**Dry nitrogen** The element nitrogen (N) which has been processed to ensure that it is free of moisture.
**Nitrógeno seco** El elemento nitrógeno (N) que ha sido procesado para asegurar que esté libre de humedad.

**Dry sleeves** Replacement cylinder sleeves that do not come into contact with engine coolant. They are surrounded by the cylinder bore.
**Camisas secas** Las camisas de repuesto del cilindro que no se ponen en contacto con el fluido refrigerante del motor. Se ajusten en el taladro del cilindro.

**Dual** Two.
**Doble** Dos.

**Dual master cylinder** A master cylinder consisting of two separate sections, one for the rear brakes and the other for the front brakes.
**Cilindro maestro dual** Un cilindro maestro que consiste de dos secciones distintas, una para los frenos traseros y la otra para los frenos delanteros.

**Dual reduction axle** A drive axle construction with two sets of pinions and gears, either of which can be used.
**Eje de reducción doble** Conjunto de eje de mando que tiene dos juegos de piñones y engranajes. Se puede utilizar cualquiera de los dos.

**Dual-servo-action brakes** Both brake shoes are self-energizing—that is, they tend to multiply braking forces as they are applied.

**Freno duo-servo** Ambas zapatas son autoenergéticas; quiere decir, suelen multiplicar la fuerza de frenar al aplicarlas.

**Dual system** Two systems; usually refers to two evaporators in an air-conditioning system; one in the front and one in the rear of the vehicle, driven off a single compressor and condenser system.
**Sistema doble** Dos sistemas; se refiere normalmente a dos evaporadores en un sistema de acondicionamiento de aire; uno en la parte delantera y el otro en la parte trasera del vehículo; los dos son accionados por un solo sistema compresor condensador.

**Duct** A tube or passage used to provide a means to transfer air or liquid from one point or place to another.
**Conducto** Tubo o pasaje utilizado para proveer un medio para transferir aire o líquido desde un punto o lugar a otro.

**Dummy shaft** A shaft, shorter than the countershaft, used during disassembly and reassembly in place of the countershaft.
**Árbol falso** Árbol, más corto que el árbol de retorno, empleado durante el desmontaje y el remonte en vez del árbol de retorno.

**Duration** The length of time, expressed in degrees of crankshaft rotation, that the valve is open.
**Duración** La cantidad del tiempo, representado por los grados de rotación del cigüeñal, que está abierta la válvula.

**Duty-cycle** The percentage of on time to total cycle time.
**Ciclo de trabajo** Porcentaje del trabajo efectivo a tiempo total del ciclo.

**Dynamic** In motion.
**Dinámico** En movimiento.

**Dynamic balance** The balance of an object when it is in motion—for example, the dynamic balance of a rotating drive shaft.
**Balance dinámico** El balance de un objeto mientras que esté en movimiento: por ejemplo, el balance dinámico de un árbol de mando giratorio.

**Dynamometer** Test equipment that places a load on the engine, causing the engine to work. It measures the amount of rotating force placed against the load.
**Dinamómetro** El equipo de prueba que aplica una carga en el motor, causando que el motor trabaja. Mide la cantidad de la fuerza giratoria puesta contra la carga.

**EATC** Electronic Automatic Temperature Control.
**EATC** Regulador Automático y Electrónico de Temperatura.

**ECC** Electronic Climate Control.
**ECC** Regulador Electrónico de Clima.

**Eccentric** One circle within another circle with neither having the same center.

**Eccentric camber bolt** A bolt with an out-of-round metal cam on the bolt head that may be used to adjust camber.
**Perno de combadura excéntrica** Perno con una leva metálica con defecto de circularidad en su cabeza, que puede utilizarse para ajustar la combadura.

**Eccentric cams or bushings** Out-of-round metal cams mounted on a retaining bolt with the shoulder of the cam positioned against a component. When the cam is rotated, the component position is changed.
**Bujes o levas excéntricas** Levas metálicas con defecto de circularidad montadas sobre un perno de retenida; el apoyo de la leva está colocado contra un componente. Al girar la leva, cambia la posición del componente.

**Eccentric washer** A normal looking washer whose hole is offset from its center.
**Arandela excéntrica** Arandela de apariencia normal pero cuyo agujero se encuentra fuera del centro. El agujero se desvía del centro.

**Eccentricity** The physical characteristics designed into some bearings calling for an inside assembled vertical diameter that is slightly smaller than the horizontal diameter.

**Excentricidad** Las características físicas diseñadas en algunos cojinetes que requieren que el diámetro interior asemblado sea un poco más pequeño que el diámetro horizontal.

**Eddy currents** Small induced currents.
**Corriente de Foucault** Pequeñas corrientes inducidas.

**Efficiency** A ratio of the amount of energy put into an engine as compared to the amount of energy produced by the engine.
**Rendimiento** Un índice de la cantidad de energía introducido en el motor comparado a la cantidad de energía que produce el motor.

**EGR pressure transducer (EPT)** A vacuum switching device operated by exhaust pressure that opens and closes the vacuum passage to the EGR valve.
**Transconductor de presión EGR** Dispositivo de conmutación de vacío accionado por la presión del escape que abre y cierra el paso del vacío a la válvula EGR.

**EGR vacuum regulator (EVR) solenoid** A solenoid that is cycled by the computer to provide a specific vacuum to the EGR valve.
**Solenoide regulador de vacío EGR** Solenoide ciclado por la computadora para proporcionarle un vacío específico a la válvula EGR.

**Elastomer** Any rubber-like plastic or synthetic material used to make bellows, bushings, and seals.
**Elastómero** Cualquiera materia plástica parecida al hule o una materia sintética que se utiliza para fabricar a los fuelles, los bujes y las juntas.

**Electric throttle kicker** An electric solenoid that controls throttle opening at idle speed and prevents dieseling.
**Nivelador eléctrico de la mariposa** Solenoide eléctrico que controla la apertura de la mariposa a una velocidad de marcha lenta y evita el autoencendido.

**Electrical load** The working device of the circuit.
**Carga eléctrica** Dispositivo de trabajo del circuito.

**Electrochemical** The chemical action of two dissimilar materials in a chemical solution.
**Electroquímico** Acción química de dos materiales distintos en una solución química.

**Electrolysis** The result of two different metals in contact with each other. The lesser of the two metals is eaten away.
**Electrólisis** El resultado de dos metales distintos que se ponen en contacto. El menor de los dos metales se consume.

**Electrolyte** A solution of 64-percent water and 36-percent sulfuric acid.
**Electrolito** Solucion de un 64% de agua y un 36% de ácido sulfúrico.

**Electromagnetic gauge** Gauge that produces needle movement by magnetic forces.
**Calibrador electromagnético** Calibrador que genera el movimiento de la aguja mediante fuerzas magnéticas.

**Electromagnetic induction** The production of voltage and current within a conductor as a result of relative motion within a magnetic field.
**Inducción electromagnética** Producción de tensión y de corriente dentro de un conductor como resultado del movimiento relativo dentro de un campo magnético.

**Electromagnetic interference (EMI)** An undesirable creation of electromagnetism whenever current is switched on and off.
**Interferencia electromagnética** Fenómeno de electromagnetismo no deseable que resulta cuando se conecta y se desconecta la corriente.

**Electromagnetism** A form of magnetism that occurs when current flows through a conductor.
**Electromagnetismo** Forma de magnetismo que ocurre cuando la corriente fluye a través de un conductor.

**Electromotive force (EMF)** See *Voltage*.
**Fuerza electromotriz** Véase *Tensión*.

**Electronic fuel injection (EFI)** A generic term applied to various types of fuel injection systems.
Inyección electrónica de combustible  Término general aplicado a varios sistemas de inyección de combustible.

**Electronic ignition (EI) system**  SAE J1930 terminology for any ignition system without a distributor.
Sistema de encendido electrónico  Término utilizado por la SAE J1930 para referirse a cualquier sistema de encendido que no tenga distribuidor.

**Electronic neutral check**  A check while servicing an electronically controlled four-wheel steering system.
Revisión electrónica neutra  Revisión llevada a cabo durante la reparación de un sistema de dirección en las cuatro ruedas controlado electrónicamente.

**Electronically erasable programmable read-only memory (EEPROM)**  A chip in a computer that may be erased easily with special equipment.
Memoria de solo lectura borrable y programable electrónicamente (EEPROM)  Pastilla en una computadora que puede borrarse fácilmente con equipo especial.

**Emission system**  Helps to reduce the harmful emissions resulting from the combustion process.
Sistema de emisión  Ayuda en reducir las emisiones nocivos que resultan del proceso de combustión.

**End clearance**  Distance between a set of gears and their cover, commonly measured on oil pumps.
Holgura del extremo  La distancia entre un conjunto de engranajes y su placa de recubrimiento, suele medirse en las bombas de aceite.

**End play**  The amount of axial or end-to-end movement in a shaft due to clearance in the bearings.
Juego de las extremidades  La cantidad del movimiento axial o del movimiento de extremidad a extremidad en una flecha debido a la holgura que se deja en los cojinetes.

**Energy**  The ability to do work.
Energía  La habilidad de hacer un trabajo.

**Engage**  When the vehicle operator moves the clutch pedal up from the floor, this engages the driving flywheel and pressure plate to rotate and drive the driven disc.
Accionar  Cuando el operador del vehículo deja subir el pedal del embrague del piso, ésto acciona la volante de ataque y el plato opresor para impulsar al disco de arrastre.

**Engagement chatter**  A shaking, shuddering action that takes place as the driven disc makes contact with the driving members. Chatter is caused by a rapid grip and slip action.
Chasquido de enganchamiento  Un movimiento de sacudo o temblor que resulta cuando el disco de ataque viene en contacto con los miembros de propulsión. El chasquido se causa por una acción rápida de agarrar y deslizar.

**Engine**  The power plant that propels the vehicle.
Motor  El central de energía que propulsa el vehículo.

**Engine analyzer**  A tester designed to test engine systems such as battery, starter, charging, ignition, and fuel, plus engine condition.
Analizador de motores  Instrumento de prueba diseñado para revisar sistemas de motores, como por ejemplo los de la batería, del arranque, de la carga, del encendido, y del combustible, además de la condición del motor.

**Engine block**  The main structure of the engine that houses the pistons and crankshaft. Most other engine components attach to the engine block.
Monoblock  La estructura principal del motor que contiene los pistones y el cigüeñal. La mayoría de los otros componentes del motor se conectan al monoblock.

**Engine coolant temperature (ECT) sensor**  A sensor that sends a voltage signal to the computer in relation to coolant temperature.

Sensor de la temperatura del refrigerante del motor  Sensor que le envía una señal de tensión a la computadora referente a la temperatura del refrigerante.

**Engine hoist**  A special lifting tool designed to remove the engine through the hood opening. Most are portable or fold-up for easy storage. The lifting of the boom is performed by a special long reach hydraulic jack.
Grúa  Una herramienta especial de izar diseñada para remover el motor por la apertura del cofre. Suelen ser portátiles o se doblan fácilmente para guardarse. El brazo de la grúa se levanta con un gato hidráulico de larga extensión.

**Engine lift**  A hydraulically operated piece of equipment used to lift the engine from the chassis.
Elevador de motores  Equipo accionado hidráulicamente que se utiliza para levantar el motor del chasis.

**Engine stand**  A special holding fixture that attaches to the back of the engine, supporting it at a comfortable working height. In addition, most stands allow the engine to be rotated for easier disassembly and assembly.
Bancada para motor  Un accesorio de apoyo especial que se conecta a la parte trasera del motor. Permite que se soporta el motor en una altura ideal para trabajar. Además, la mayoría de las bancadas permiten la rotación del motor para facilitar el desmontaje y montaje.

**Engine torque**  A turning or twisting action developed by the engine, measured in foot-pounds or newton-meters.
Torsión del motor  Una acción de girar o torcer que crea el motor, ésta se mide en libras-pie o kilos-metros.

**English fastener**  Any type of fastener with English size designations, numbers, decimals, or fractions of an inch.
Asegurador inglés  Cualquier tipo de asegurador provisto de indicaciones, números, decimales, o fracciones de una pulgada del sistema inglés.

**English measuring system**  The USCS measuring system.
Sistema Imperial Británico  Sistema de medida de USCS [Sistema Usual U.S.].S

**Environmental Protection Agency (EPA)**  An agency of the U.S. government charged with the responsibility of protecting the environment and enforcing the Clean Air Act (CAA) of 1990.
Agencia para la Protección del Medio Ambiente (EPA)  Agencia del gobierno estadounidense que tiene a su cargo la responsabilidad de proteger el medio ambiente y ejecutar la Ley para Aire Limpio (CAA por sus siglas en inglés) de 1990.

**EPA**  Environmental Protection Agency.
EPA  Agencia para la Protección del Medio Ambiente.

**Epoxies**  Synthetic resins that produce the strongest adhesives in current use, as well as plastics and corrosion coatings. Epoxy adhesives are thermosetting; that is, after initial hardening, they cannot be remelted by heat. They have excellent resistance to solvents and weathering agents, and have high electrical and temperature resistance.
Resinas epósicas  Las resinas sintéticas que producen los adhesivos más fuertes hasta la fecha, así como los recubrimientos de plástico y anticorrosivos. Los adhesivos epósicos son termoendurecible; o sea, después de curarse inicialmente, no se derriten al exponerse al calor. Poseen una resistencia excelente a los disolventes y los agentes destructivos del clima, y una resistencia muy fuerte contra la electricidad y la temperatura.

**Equivalent series load (equivalent resistance)**  The total resistance of a parallel circuit. It is equivalent to the resistance of a single load in series with the voltage source.
Carga en serie equivalente (resistencia equivalente)  Resistencia total de un circuito en paralelo, equivalente a la resistencia de una sola carga en serie con la fuente de tensión.

**Essential tool kit**  A set of special tools designed for a particular model of car or truck.

**Estuche de herramientas principales** Un conjunto de herramientas especiales diseñadas para un modelo específico de coche o camión.

**Etching** A discoloration or removal of some material caused by corrosion or some other chemical reaction.
**Grabado por ácido** Una descoloración o remueva de una materia que se efectua por medio de la corrosión u otra reacción química.

**Evacuate** To create a vacuum within a system to remove all traces of air and moisture.
**Evacuar** El dejar un vacío dentro de un sistema para remover completamente todo aire y humedad.

**Evacuation** See *Evacuate*.
**Evacuación** Ver *Evacuate* [*Evacuar*].

**Evaporative (EVAP) emission canister purge solenoid** A computer-operated solenoid in the EVAP system that controls the amount of fuel vapor purged from the canister into the intake manifold.
**Solenoide de purga de bote de emisiones de evaporación** Solenoide accionado por computadora en el sistema de evaporación que controla la cantidad de los vapores del combustible removidos desde el bote al colector de aspiración.

**Evaporative (EVAP) emission canister vent solenoid** A computer-operated solenoid in the EVAP system that controls the fuel tank pressure on OBD II systems.
**Solenoide de salida del sistema de bote** Solenoide accionado por computadora en el sistema de evaporación que controla la presión del tanque del combustible en los sistemas OBD II.

**Evaporative (EVAP) system** A system that collects fuel vapors from the fuel tank and directs them into the intake manifold rather than allowing them to escape to the atmosphere.
**Sistema de evaporación** Sistema que acumula los vapores del combustible que escapan del tanque del combustible y los conduce hacia el colector de aspiración en vez de permitir que los mismos se escapen hacia la atmósfera.

**Evaporator core** The tube and fin assembly located inside the evaporator housing. The refrigerant fluid picks up heat in the evaporator core when it changes into a vapor.
**Núcleo del evaporador** El conjunto de tubo y aletas ubicado dentro del alojamiento de evaporador. El refrigerante acumula calor en el núcleo del evaporador cuando se convierte en vapor.

**Excitation current** Current that magnetically excites the field circuit of the AC generator.
**Corriente de excitación** Corriente que excita magnéticamente al circuito inductor del generador de corriente alterna.

**Exhaust gas analyzer** A tester that measures carbon monoxide, carbon dioxide, hydrocarbons, and oxygen in the engine exhaust.
**Analizador del gas del escape** Instrumento de prueba que mide el monóxido de carbono, el bióxido de carbono, los hidrocarburos, y el oxígeno en el escape del motor.

**Exhaust gas recirculation (EGR) valve** A valve that circulates a specific amount of exhaust gas into the intake manifold to reduce NOx emissions.
**Válvula de recirculación del gas del escape** Válvula que hace circular una cantidad específica del gas del escape hacia el colector de aspiración para disminuir emisiones de óxidos de nitrógeno.

**Exhaust gas recirculation valve position (EVP) sensor** A sensor that sends a voltage signal to the computer in relation to the EGR valve position.
**Sensor de la posición de la válvula de recirculación del gas del escape** Sensor que le envía una señal de tensión a la computadora referente a la posición de la válvula EGR.

**Exhaust gas temperature sensor** A sensor that sends a voltage signal to the computer in relation to exhaust temperature.
**Sensor de la temperatura del gas del escape** Sensor que le envía una señal de tensión a la computadora referente a la temperatura del escape.

**Exhaust manifold** A component which collects and then directs engine exhaust gases from the cylinders.
**Múltiple de escape** Un componente que colecciona y luego dirige los gases de escape del motor desde los cilindros.

**Exhaust manifold gasket** A part which seals the connection between the cylinder head and the exhaust manifold.
**Junta del múltiple de escape** Una parte que sella la conexión entre la cabeza del cilindro y el múltiple de escape.

**Exhaust system** A system which removes the by-products of the combustion process from the cylinders.
**Sistema de escape** Un sistema que remueva los subproductos del proceso de combustión de los cilindros.

**Exhaust valve** An engine part which controls the expulsion of spent gases and emissions out of the cylinder.
**Válvula de escape** Una parte del motor que controla la expulsión de los gases consumidos y las emisiones del cilindro.

**Expansion** An increase in size.
**Dilatación** Un incremento en el tamaño.

**Expansion tank** An auxiliary tank that is usually connected to the inlet tank or a radiator, and which provides additional storage space for heated coolant. Often called a coolant recovery tank.
**Tanque de expansión** Tanque auxiliar que normalmente se conecta al tanque de entrada o a un radiador y que provee almacenaje adicional del enfriante calentado. Llamado con frecuencia tanque para la recuperación del enfriante.

**Extension housing** An aluminum or iron casting of various lengths that encloses the transmission output shaft and supporting bearings.
**Cubierta de extensión** Una pieza moldeada de aluminio o acero que puede ser de varias longitudes que encierre a la flecha de salida de la transmisión y a los cojinetes de soporte.

**External** On the outside.
**Externo** Al exterior.

**External cone clutch** The external surface of one part has a tapered surface to mate with an internally tapered surface to form a cone clutch.
**Embrague cónico externo** La superficie externa de una pieza tiene una superficie cónica para hacer juego con una superficie internamente cónica y así formar un embrague de cono.

**External gear** A gear with teeth across the outside surface.
**Engranaje exterior** Un engranaje cuyos dientes están en la superficie exterior.

**External snap ring** A snap ring found on the outside of a part such as a shaft.
**Anillo de muelle exterior** Anillo de muelle que se encuentra en el exterior de una pieza, como por ejemplo un árbol.

**Externally balanced engine** Engines balanced by using counterweights on the flywheel and vibration dampener in conjunction with the crankshaft counterweights.
**Motor de equilibración externo** Los motores que se equilibran empleando los contrapesos en el volante y un amortiguador de vibraciones junto con los contrapesos del cigüeñal.

**Externally tabbed clutch plates** Clutch plates that are designed with tabs around the outside periphery to fit into grooves in a housing or drum.
**Placas de embrague de orejas externas** Las placas de embrague que se diseñan de un modo para que las orejas periféricas de la superficie se acomoden en una ranura alrededor de un cárter o un tambor.

**Extreme-pressure lubricant** A special lubricant for use in hypoid-gear differentials; needed because of the heavy wiping loads imposed on the gear teeth.
**Lubricante de presión extrema** Un lubricante especial que se usa en las diferenciales de tipo engranaje hipóide; se requiere por la carga de transmisión de materia pesada que se imponen en los dientes del engranaje.

**Face** The front surface of an object.
**Cara** La superficie delantera de un objeto.

**Face shield** A clear plastic shield that protects the entire face.
**Careta** Un escudo transparente de plástico que proteja la cara entera.

**Fail-safe function** A mode entered by a computer if the computer detects a fault in the system.
**Función de autoprotección** Modo adaptado por una computadora si la misma detecta un fallo en el sistema.

**Failsoft** Computer substitution of a fixed input value if a sensor circuit should fail. This provides for system operation, but at a limited function.
**Falla activa** Sustitución por la computadora de un valor fijo de entrada en caso de que ocurra una falla en el circuito de un sensor. Esto asegura el funcionamiento del sistema, pero a una capacidad limitada.

**False guides** Devices which are similar to inserts.
**Guías falsas** Los dispositivos muy parecidos a las piezas insertas.

**Fan relay** A relay for the cooling and/or auxiliary fan motors.
**Relé del ventilador** Relé para los motores de enfriamiento y/o los auxiliares.

**Fast burn combustion chamber** A chamber designed to increase the speed of combustion by creating a turbulence as the air-fuel mixture enters the chamber.
**Cámara de combustión rápido** Una cámara diseñada para aumentar la rapidez de la combustión creando una turbulencia mientras que la mezcla de aire/combustible entra a la cámara.

**Fast charging** Battery charging using a high amperage for a short period of time.
**Carga rápida** Carga de la batería que utiliza un amperaje máximo por un corto espacio de tiempo.

**Fatigue** The deterioration of metal under excessive loads.
**Fatiga** El deterioro de un metal bajo cargas excesivas.

**Fault codes** Numeric codes stored in a computer memory representing specific computer system faults.
**Códigos de fallo** Códigos numéricos almacenados en la memoria de una computadora que representan fallos específicos en el sistema informático.

**Federal Clean Air Act** See *Clean Air Act.*
**Ley Federal para Aire Limpio** Ver *Clean Air Act* [*Ley para Aire Limpio*].

**Feedback** Data concerning the effects of the computer's commands are fed back to the computer as an input signal. Used to determine if the desired result has been achieved. 2. A condition that can occur when electricity seeks a path of lower resistance, but the alternate path operates another component than that intended. Feedback can be classified as a short.
**Realimentación** Datos referentes a los efectos de las órdenes de la computadora se suministran a la misma como señal de entrada. La realimentación se utiliza para determinar si se ha logrado el resultado deseado. 2. Condición que puede ocurrir cuando la electricidad busca una trayectoria de menos resistencia, pero la trayectoria alterna opera otro componente que aquel deseado. La realimentación puede clasificarse como un cortocircuito.

**Feeler gauge** A measuring tool consisting of a series of metal strips cut to precise thicknesses. A feeler gauge pack usually consists of blades between 0.002 to 0.025 inch in 0.001- or 0.002-inch increments.
**Galga calibrada** Una serie de hojas metálicas cortadas a los espesores precisos. Un conjunto de galgas calibradas suele tener las hojas entre el 0.002 a 0.025 de una pulgada en incrementos de .001 o .002 de una pulgada.

**Ferrous metals** Metals which contain iron. Cast iron and steel are examples of ferrous metals. Magnets are attracted to ferrous metals.

**Metales férreos** Los metales que contienen el hierro. El hierro colado y el acero son ejemplos de los metales férreos. Estos metales atraerán un imán.

**Fiber composites** A mixture of metallic threads along with a resin form a composite offering weight and cost reduction, long-term durability, and fatigue life. Fiberglass is a fiber composite.
**Compuestos de fibra** Una mezcla de hilos metálicos y resina forman un compuesto que ofrece una disminución de peso y costo, mayor durabilidad y resistencia a la fatiga. La fibra de vidrio es un compuesto de fibra.

**Fiber optics** A medium of transmitting for the transmission of light through polymethylmethacrylate plastic that keeps the light rays parallel even if there are extreme bends in the plastic.
**Transmisión por fibra óptica** Técnica de transmisión de luz por medio de un plástico de polimetacrilato de metilo que mantiene los rayos de luz paralelos aunque el plástico esté sumamente torcido.

**Field current draw test** Diagnostic test that determines if there is current available to the field windings.
**Prueba de la intensidad de una corriente inductora** Prueba diagnóstica que determina si se está generando corriente a los devanados inductores.

**Field service mode** A computer diagnostic mode that indicates whether the computer is in open or closed loop on General Motors PCMs.
**Modo de servicio de campo** Modo diagnóstico de una computadora que indica si la misma se encuentra en bucle abierto o cerrado en módulos del control del tren transmisor de potencia de la General Motors.

**Fill neck** The part of the radiator on which the pressure cap is attached. Most radiators, however, are filled via the recovery tank.
**Cuello de relleno** La parte del radiador a la que se fija la tapadera de presión. Sin embargo, la mayoría de radiadores se llena por medio del tanque de recuperación.

**Fillets** Small, rounded corners machined on the edges of journals to increase strength.
**Cantos redondeados** Las esquinas pequeñas, redondeadas labradas por máquina en los bordes de un muñón para reinforzarla.

**Filter** A device used with the dryer or as a separate unit to remove foreign material from the refrigerant.
**Filtro** Dispositivo utilizado con el secador o como unidad separada para extraer material extraño del refrigerante.

**Filter/dryer** A device that has a filter to remove foreign material from the refrigerant and a desiccant to remove moisture from the refrigerant.
**Secador del filtro** Dispositivo provisto de un filtro para remover el material extraño del refrigerante y un desecante para remover la humedad del refrigerante.

**Final drive gears** Main driving gears located in the axle area of the transaxle housing.
**Engranajes de la transmisión final** Mecanismos de accionamiento principales ubicados en la región del eje del alojamiento del transeje.

**Final drive ratio** The ratio between the drive pinion and ring gear.
**Relación del mando final** La relación entre el piñón de ataque y la corona.

**Fire extinguisher** A portable apparatus that contains chemicals, water, foam, or special gas that can be discharged to extinguish a small fire.
**Extintor de encendios** Un aparato portátil que contiene los químicos, el agua, la espuma o un gas especial que se puede discargar en un fuego pequeño para apagarlo.

**First gear** A small diameter driving helical- or spur-type gear located on the cluster gear assembly. First gear provides torque multiplication to get the vehicle moving.
**Engranaje de primera velocidad** Mecanismo de mando helicoide o recto de diámetro pequeño ubicado en el conjunto del tren desplazable. El engranaje de primera velocidad inicia la multiplicación de par de torsión para impulsar el vehículo.

**Fit** The contact between two machined surfaces.
**Ajuste** El contacto entre dos superficies maquinadas.

**Fixed caliper** A disc brake caliper that has pistons on both sides of the rotor. It is rigidly fixed to the suspension.
**Mordaza fija** Una mordaza de freno de disco que tiene pistones en ambos lados del rotor. Se fija rígidamente a la suspensión.

**Fixed-type constant-velocity joint** A joint that cannot telescope or plunge to compensate for suspension travel. Fixed joints are always found on the outer ends of the drive shafts of FWD cars. A fixed joint may be of either Rzeppa or tripod type.
**Junta tipo fijo de velocidad constante** Una junta que no tiene la capacidad de los movimientos telescópicos o repentinos que sirven para compensar en los viajes de suspensión. Las juntas fijas siempre se ubican en las extremidades exteriores de los árboles de mando en los coches de FWD. Una junta tipo fijo puede ser de un tipo Rzeppa o de trípode.

**Flange** A projecting rim or collar on an object for keeping it in place.
**Reborde** Una orilla o un collar sobresaliente de un objeto cuya función es de mantenerlo en lugar.

**Flange yoke** The part of the rear universal joint attached to the drive pinion.
**Yugo de brida** Pieza de la junta universal trasera fijada al piñón de mando.

**Flap** A strip of emery cloth, or other abrasive material, wound around a slotted mandrel. The loose end of the flap is allowed to slap against the surface being machined.
**Aleta** Un tiro de tela de esmeril, u otro material abrasivo, envuelto alrededor de un mandrino hendido. Se permite que la extremidad suelta de la aleta golpea la superficie que se está trabajando por máquina.

**Flare** A flange or cone-shaped end applied to a piece of tubing to provide a means of fastening it to a fitting.
**Abocinado** Brida o extremo en forma cónica aplicado a una pieza de tubería para proveer un medio de asegurarse a un ajuste.

**Flash code diagnosis** Reading computer system diagnostic trouble codes (DTCs) from the flashes of the malfunction indicator light (MIL).
**Diagnosis con código de destello** La lectura de códigos indicativos de fallas para propósitos diagnósticos de una computadora mediante los destellos de la luz indicadora de funcionamiento defectuoso.

**Flex fans** Fans designed to widen their pitch at slow engine speeds when more airflow is required through the radiator fins. At higher engine speeds, the pitch is decreased to reduce the horsepower required to turn the fan.
**Ventiladores ajustables** Un diseño de los ventiladores que amplían su paso en las velocidades más bajas que requieren más flujo del aire por los devanados del radiador. En las velocidades más altas el paso se disminuye, reduciendo los requerimientos del par de dar las vueltas al ventilador.

**Flex plate** A stamped steel coupler bolted to the rear of the crankshaft. The flex plate provides a mounting for the torque converter.
**Placa flexible** Un acoplador de acero embutido empernado a la parte trasera del cigüeñal. La placa flexible provee el asiento del convertidor de par.

**Floating caliper** A disc brake caliper that has one piston. The caliper is free to move on pins in response to the piston pressing against the rotor.
**Mordaza flotante** Una mordaza de freno de disco que tiene un pistón. La mordaza mueve libremente sobre clavijas respondiendo al pistón oprimiendo contra el rotor.

**Floating piston pin** A piston pin that is not locked in the connecting rod or the piston, allowing it to turn or oscillate in both the connecting rod and piston.
**Perno flotante del pistón** Un perno del pistón que no está clavado en la biela o en el pistón, permitiéndolo girar u osilar en la biela y el pistón.

**Floor jack** A portable hydraulic tool used to raise and lower a vehicle.

**Gato** Una herramienta portátil hidráulica que sirve para levantar y bajar un vehículo.

**Flow control valve** A special valve that controls fluid movement in relation to system demands.
**Válvula de control de flujo** Válvula especial que controla el movimiento del fluido de acuerdo a las exigencias del sistema.

**Fluid coupling** A device in the powertrain consisting of two rotating members; transmits power from the engine, through a fluid, to the transmission.
**Acoplamiento de fluido** Un dispositivo en el tren de potencia que consiste de dos miembros rotativos; transmite la potencia del motor, por medio de un fluido, a la transmisión.

**Fluid drive** A drive in which there is no mechanical connection between the input and output shafts, and power is transmitted by moving oil.
**Dirección fluido** Una dirección en la cual no hay conexiones mecánicas entre las flechas de entrada o salida, y la potencia se transmite por medio del aceite en movimiento. de fuerza del motor; también sirve como parte del embrague y del sistema de arranque.

**Flywheel** A heavy circular component located on the rear of the crankshaft that keeps the crankshaft rotating during nonproductive strokes.
**Volante** Un componente pesado circular ubicado en la parte trasera del cigüeñal que procure que el cigüeñal sigue girando durante las carreras no productivas.

**Flywheel ring gear** A gear, fitted around the flywheel, that is engaged by teeth on the starting-motor drive to crank the engine.
**Engranaje anular del volante** Un engranaje, colocado alrededor del volante que se acciona por los dientes en el propulsor del motor de arranque y arranca al motor.

**Followers** Similar to rocker arms, followers are used on many OHC engines. Followers run directly off of the camshaft.
**Seguidor** Parecidos a los balancines, los seguidores se emplean en muchos motores OHC. Los seguidores se accionan directamente por el árbol de levas.

**Foot-pound (ft.-lb.)** A measurement for torque in the USC or British system.
**Pie-libra (pies-lb)** Una medida de la cantidad de energía o fuerza que requiere mover una libra la distancia de un pie.

**Force** Any push or pull exerted on an object; measured in pounds and ounces, or in newtons (N) in the metric system.
**Fuerza** Cualquier acción empujado o jalado que se efectua en un objeto; se mide en pies y onzas, o en newtones (N) en el sistema métrico.

**Forced air** Air that is moved mechanically, such as by a fan or blower.
**Aire forzado** Aire que se mueve mecánicamente, como por ejemplo por un ventilador o soplador.

**Forward-bias** A positive voltage that is applied to the P-type material, and negative voltage to the N-type material, of a semiconductor.
**Polarización directa** Tensión positiva aplicada al material P y tensión negativa aplicada al material N de un semiconductor.

**Forward coast side** The side of the ring gear tooth the drive pinion contacts when the vehicle is decelerating.
**Cara de cabotaje delantera** Cara del diente de la corona con el cual el piñón de mando entra en contacto mientras el vehículo deacelera.

**Forward drive side** The side of the ring gear tooth that the drive pinion contacts when accelerating or on the drive.
**Cara de mando delantera** Cara del diente de la corona con el cual el piñón de mando entra en contacto mientras acelera o está en marcha.

**Four gas engine analyzer** A device that measures the exhaust of the engine. It enables the technician to look at the effects of the combustion process by measuring hydrocarbons, carbon monoxide, carbon dioxide, and oxygen levels in the exhaust.

**Analizador de cuatro gases del motor** Un dispositivo que mide los vapores de escape del motor. Permite que el técnico vea los efectos del proceso de combustión midiendo los niveles de los hidrocarburos, el monóxido de carbono, el bióxido de carbono y el oxígeno en los vapores del escape.

**Four-wheel drive** A vehicle that has driving axles at both the front and rear so that all four wheels can be driven.
**Tracción a cuatro ruedas** En un vehículo, se trata de los ejes de dirección fronteras y traseras, para que cada una de las ruedas puede impulsar.

**Four wheel high** A transfer case shift position where both front and rear drive shafts receive power and rotate at the speed of the transmission output shaft.
**Alto de cuatro ruedas** Posición en la caja de cambios donde los árboles de mando delantero y trasero reciben fuerza y giran a la misma velocidad que el árbol de rendimiento de la transmisión.

**Frame** The main understructure of the vehicle to which everything else is attached. Most FWD cars have only a subframe for the front suspension and drivetrain. The body serves as the frame for the rear suspension.
**Armazón** La estructura principal del vehículo al cual todo se conecta. La mayoría de los coches FWD sólo tiene un bastidor auxiliar para la suspensión delantera y el tren de propulsión. El carrocería del coche sirve de chassis para la suspensión trasera.

**Frame flange** The upper or lower horizontal edge on a vehicle frame.
**Brida del armazón** Borde horizontal superior o inferior en el armazón del vehículo.

**Frame web** The vertical side of a vehicle frame.
**Malla del armazón** Lado vertical del armazón del vehículo.

**Free speed test** Diagnostic test that determines the free rotational speed of the armature. This test is also referred to as the no-load test.
**Prueba de velocidad libre** Prueba diagnóstica que determina la velocidad giratoria libre de la armadura. A dicha prueba se le llama prueba sin carga.

**Free-wheel** To turn freely and not transmit power.
**Volante libre** Da vueltas libremente sin transferir la potencia.

**Free-wheeling clutch** A mechanical device that will engage the driving member to impart motion to a driven member in one direction but not the other. Also known as an *overrunning clutch*.
**Embrague de volante libre** Un dispositivo mecánico que acciona el miembro de tracción y da movimiento al miembro de tracción en una dirección pero no en la otra. También se conoce bajo el nombre de un "embrague de sobremarcha".

**Free-wheeling engine** An engine in which valve lift and angle prevent valve-to-piston contact if the timing belt or chain breaks.
**Motor de piñón libre** Un motor en el cual el levantamiento y el ángulo de las válvulas previenen el contacto entre la válvula y el pistón si se quiebra la correa o la cadena de sincronización.

**Freeze frame data** Computer system data that is recorded in the powertrain control module (PCM) memory when an electronic fault is sensed in the PCM inputs or outputs.
**Imagen congelada de datos** Datos de la computadora registrados en la memoria de la unidad de control del tren transmisor de potencia cuando un fallo electrónico está sentido en las entradas y salidas de la unidad de control del tren transmisor de potencia.

**Freon** A term commonly used for R-12.
**Freón** Térmimo comúnmente utilizado para R-12.

**Friction** The resistance to motion between two bodies in contact with each other.
**Fricción** La resistencia al movimiento entre dos cuerpos que están en contacto.

**Friction bearing** A bearing in which there is sliding contact between the moving surfaces. Sleeve bearings, such as those used in connecting rods, are friction bearings.

**Rodamientos de fricción** Un cojinete en el cual hay un contacto deslizante entre las superficies en movimiento. Los rodamientos de manguitos, como los que se usan en las bielas, son rodamientos de fricción.

**Friction disc** In the clutch, a flat disc, faced on both sides with friction material and splined to the clutch shaft. It is positioned between the clutch pressure plate and the engine flywheel. Also called the clutch disc or driven disc.
**Disco de fricción** En el embrague, un disco plano al cual se ha cubierto ambos lados con una materia de fricción y que ha sido estriado a la flecha del embrague. Se posiciona entre el plato opresor del embrague y el volante del motor. También se llama el disco del embrague o el disco de arrastre.

**Friction facings** A hard-molded or woven asbestos or paper material that is riveted or bonded to the clutch driven disc.
**Superficie de fricción** Un recubrimiento remachado o aglomerado al disco de arrastre del embrague que puede ser hecho del amianto moldeado o tejido o de una materia de papel.

**Front and rear wheel alignment angle screen** A display on a computer wheel aligner that provides readings of the front and rear wheel alignment angles.
**Pantalla para la visualización del ángulo de alineación de las ruedas delanteras y traseras** Representación visual en una computadora para alineación de ruedas que provee lecturas de los ángulos de alineación de las ruedas delanteras y traseras.

**Front bearing retainer** An iron or aluminum circular casting fastened to the front of a transmission housing to retain the front transmission bearing assembly.
**Retenedor del cojinete de rueda delantero** Pieza circular fundida en hierro o en aluminio fijada a la parte delantera de un alojamiento de la transmisión para sujetar el conjunto del cojinete de transmisión delantero.

**Front differential/axle assembly** Like a conventional rear axle but having steerable wheels.
**Conjunto de diferencial/eje delantero** Igual que el puente trasero convencional, pero con ruedas orientales.

**Front main steering angle sensor** An input sensor mounted in the front steering gear in an electronically controlled four-wheel steering system.
**Sensor principal del ángulo de la dirección delantera** Sensor de entrada montado en el mecanismo de dirección delantera en un sistema de dirección en las cuatro ruedas controlado electrónicamente.

**Front pump** Pump located at the front of the transmission. It is driven by the engine through two dogs on the torque converter housing. It supplies fluid whenever the engine is running.
**Bomba delantera** Una bomba ubicada en la parte delantera de la transmisión. Se arrastre por el motor a través de dos álabes en el cárter del convertidor de la torsión. Provee el fluido mientras que funciona el motor.

**Front seat** Closing off the line leaving the compressor open to the service port fitting. This allows service to the compressor without purging the entire system. Never operate the system with the valves front-seated.
**Asentar a la derecha** El cerrar la línea dejando abierto el compresor al ajuste del orificio de servicio, lo cual permite prestar servicio al compresor sin purgar todo el sistema. Nunca haga funcionar el sistema con las válvulas asentadas a la derecha.

**Front substeering angle sensor** An input sensor mounted in the front steering gear in an electronically controlled four-wheel steering system.
**Sensor auxiliar del ángulo de la dirección delantera** Sensor de entrada montado en el mecanismo de dirección delantera en un sistema de dirección en las cuatro ruedas controlado electrónicamente.

**Front-wheel drive (FWD)** System in which the vehicle has all drivetrain components located at the front.
**Tracción de las ruedas delanteras (FWD)** El vehículo tiene todos los componentes del tren de propulsión en la parte delantera.

**Fuel cut rpm** The rpm range in which the computer stops operating the injectors during deceleration.
**Detención de combustible según las rpm** Margen de revoluciones al que la computadora detiene el funcionamiento de los inyectores durante la desaceleración.

**Fuel pressure test port** A threaded port on the fuel rail to which a pressure gauge may be connected to test fuel pressure.
**Lumbrera de prueba de la presión del combustible** Lumbrera fileteada que se encuentra en el carril del combustible a la que puede conectársele un calibrador de presión para revisar la presión del combustible.

**Fuel pump volume** The amount of fuel the pump delivers in a specific time period.
**Volumen de la bomba del combustible** Cantidad de combustible que la bomba envía dentro de un espacio de tiempo específico.

**Fuel system** A system that includes the intake system which brings air into the engine and the components that deliver the fuel to the engine.
**Sistema de combustible** Un sistema que incluye al sistema de admisión que introduce el aire dentro del motor y a los componentes que entregan el combustible al motor.

**Fuel tank pressure sensor** A pressure sensor mounted in the top of the fuel tank on OBD II systems that sends a voltage signal to the powertrain control module (PCM) in relation to the vacuum or pressure in the fuel tank.
**Sensor de presión del tanque del combustible** Sensor de presión montado sobre el tanque del combustible de los sistemas de diagnósticos OBD II que le envía una señal de tensión a la unidad de control del tren transmisor de potencia referente al vacío o a la presión en el tanque del combustible.

**Fuel tank purging** Removing fuel vapors and foreign material from the fuel tank.
**Purga del tanque del combustible** La remoción de vapores de combustible y de material extraño del tanque del combustible.

**Fulcrum rings** A circular ring over which the pressure plate diaphragm spring pivots.
**Anillos de fulcro** Anillo circular sobre el cual gira el muelle del diafragma de la placa de presión.

**Full field** Field windings that are constantly energized with full battery current. Full fielding will produce maximum AC generator output.
**Campo completo** Devanados inductores que se excitan constantemente con corriente total de la batería. EL campo completo producirá la salida máxima de un generador de corriente alterna.

**Full-field test** Diagnostic test used to isolate if the detected problem lies in the AC generator or the regulator.
**Prueba de campo completo** Prueba diagnóstica utilizada para determinar si el problema descubierto se encuentra en el generador de corriente alterna o en el regulador.

**Full-floating rear axle** An axle that only transmits driving force to the rear wheels. The weight of the vehicle (including payload) is supported by the axle housing.
**Eje trasero enteramente flotante** Eje que solamente transmite la fuerza motriz a las ruedas traseras. El puente trasero soporta el peso del vehículo (incluyendo la carga útil).

**Fully synchronized** In a manual transmission, the synchronizer assembly operates to improve the shift quality in all forward gears.
**Enteramente sincronizado** En una transmisión manual, el conjunto sincronizador funciona para mejorar la calidad del desplazamiento en todos los engranajes delanteros. Se transfiere el metal de una superficie a la otra, lo cual deja un aspecto corroído o raspado.

**Fuse** A replaceable circuit protection device that will melt should the current passing through it exceed its rating.
**Fusible** Dispositivo reemplazable de protección del circuito que se fundirá si la corriente que fluye por el mismo excede su valor determinado.

**Fuse box** A term used that indicates the central location of the fuses contained in a single holding fixture.
**Caja de fusibles** Término utilizado para indicar la ubicación central de los fusibles contenidos en un solo elemento permanente.

**Fusible link** A wire made of meltable material with a special heat-resistant insulation. When there is an overload in the circuit, the link melts and opens the circuit.
**Cartucho de fusible** Alambre hecho de material fusible con aislamiento especial resistente al calor. Cuando ocurre una sobrecarga en el circuito, el cartucho se funde y abre el circuito.

**FWD** Abbreviation for front-wheel drive.
**FWD** Abreviación de tracción de las ruedas delanteras.

**Gallery plugs** Metal plugs used to cap the drilled oil passages in the cylinder head or engine block. They can be threaded or pressed into the hole.
**Tapones de la canalización de aceite** Los tapones metálicos que sirven para estancar los pasajes taladrados del aceite en la cabeza del cilindro o en el monoblock. Pueden ser fileteados o prensados en el agujero.

**Galling** A displacement of metal.
**Raspar** El desplazamiento del metal.

**Galvanic corrosion** A type of corrosion that occurs when two dissimilar metals, such as magnesium and steel, are in contact with each other.
**Corrosión galvánica** Tipo de corrosión producida cuando dos metales distintos, como por ejemplo, el magnesio y el acero, entran en contacto el uno con el otro.

**Gasket** A rubber, felt, cork, or metallic material used to seal surfaces of stationary parts.
**Empaque** Un material de caucho, fieltro, corcho o metal que sirve para sellar las superficies de las partes fijas.

**Gasket cement** A liquid adhesive material, or sealer, used to install gaskets.
**Mastique para empaques** Una substancia líquida adhesiva, o una substancia impermeable, que se usa para instalar a los empaques.

**Gauge** A device that displays the measurement of a monitored system by the use of a needle or pointer that moves along a calibrated scale. 2. The number that is assigned to a wire to indicate its size. The larger the number, the smaller the diameter of the conductor.
**Calibrador** 1. Dispositivo que muestra la medida de un sistema regulado por medio de una aguja o indicador que se mueve a través de una escala calibrada. 2. El número asignado a un alambre indica su tamaño. Mientras mayor sea el número, más pequeño será el diámetro del conductor.

**Gauss gauge** A meter that is sensitive to the magnetic field surrounding a wire conducting current. The gauge needle will fluctuate over the portion of the circuit that has current flowing through it. Once the ground has been passed, the needle will stop fluctuating.
**Calibrador gauss** Instrumento sensible al campo magnético que rodea un alambre conductor de corriente. La aguja del calibrador se moverá sobre la parte del circuito a través del cual fluye la corriente. Una vez se pasa a tierra, la aguja dejará de moverse.

**Gear** A wheel with external or internal teeth that serves to transmit or change motion.
**Engranaje** Una rueda que tiene dientes interiores o exteriores que sirve para transferir o cambiar el movimiento.

**Gear backlash, or lash** Movement between gear teeth that are meshed with each other.
**Contragolpe o juego en los engranajes** Movimiento entre los dientes del engranaje que están endentados entre sí.

**Gear clash** The noise that results when two gears are traveling at different speeds and are forced together.
**Choque de engranajes** Ruido producido cuando dos engranajes giran a velocidades diferentes y se unen por fuerza.

**Gear lubricant** A type of grease or oil blended especially to lubricate gears.
**Lubricante para engranaje** Un tipo de grasa o aceite que ha sido mezclado específicamente para la lubricación de los engranajes.

**Gear noise** The howling or whining of the ring gear and pinion due to an improperly set gear pattern, gear damage, or improper bearing preload.
**Ruido del engranaje** Aullido o silbido de la corona y del piñón debido al montaje incorrecto de los engranajes, a averías en los engranajes o a carga previa incorrecta del cojinete.

**Gear ratio** The number of revolutions of a driving gear required to turn a driven gear through one complete revolution. For a pair of gears, the ratio is found by dividing the number of teeth on the driven gear by the number of teeth on the driving gear.
**Relación de los engranajes** El número de las revoluciones requeridas del engranaje de propulsión para dar una vuelta completa al engranaje arrastrado. En una pareja de engranajes, la relación se calcula al dividir el número de los dientes en el engranaje de arrastre por el número de los dientes en el engranaje de propulsión.

**Gear rattle** A repetitive metallic impact or rapping noise that occurs when the vehicle is lugging in gear. The intensity of the noise increases with operating temperature and engine torque, and decreases with increasing vehicle speed.
**Estruendo del engranaje** Impacto metálico repetitivo o golpeteo que ocurre cuando el vehículo arrastra los engranajes durante la marcha. La intensidad del ruido aumenta con alta temperatura de funcionamiento y del par de torsión del motor, y disminuye al aumentar la velocidad del vehículo.

**Gear reduction** When a small gear drives a large gear, there is an output speed reduction and a torque increase, which results in a gear reduction.
**Velocidad descendente** Cuando un engranaje pequeño impulsa a un engranaje grande, hay una reducción en la velocidad de salida y un incremento en la torsión que resulta en un cambio descendente de las velocidades.

**Gear whine** A high-pitched sound developed by some types of meshing gears.
**Ruido del engranaje** Un sonido agudo que proviene de algunos tipos de engranajes endentados.

**Gearshift** A linkage-type mechanism by which the gears in an automobile transmission are engaged and disengaged.
**Varillaje de cambios** Un mecanismo tipo eslabón que acciona y desembraga a los engranajes de la transmisión.

**Geometric centerline** An imaginary line through the exact center of the front and rear wheels.
**Línea central geométrica** Línea imaginaria a través del centro exacto de las ruedas delanteras y traseras.

**Glaze** Polishing of the cylinder wall resulting from piston ring travel in the cylinder in conjunction with combustion heat and engine oil.
**Porcelana** El pulido de la pared del cilindro que resulta del viaje del anillo del pistón en el cilindro en combinación con el calor de combustión y el aceite del motor.

**Governor pressure** The transmission's hydraulic pressure that is directly related to output shaft speed. It is used to control shift points.
**Regulador de presión** La presión hidráulica de una transmisión se relaciona directamente a la velocidad de la flecha de salida. Se usa para controlar los puntos de cambios de velocidad.

**Governor valve** A device used to sense vehicle speed. The governor valve is attached to the output shaft.
**Válvula reguladora** Un dispositivo que se usa para determinar la velocidad de un vehículo. La válvula reguladora se monta en la flecha de salida.

**Grade marks** Radial lines on the bolt head that indicate the strength of the bolt.
**Marcos de grado** Las líneas radiales en la cabeza del perno que indican su fuerza.

**Graduated container** A measure such as a beaker or measuring cup that has a graduated scale for the measure of a liquid.
**Recipiente graduado** Una medida, como por ejemplo un cubilete o una taza de medir, provista de una escala graduada para la medición de un líquido.

**Graphite** Very fine carbon dust with a slippery texture used as a lubricant.
**Grafito** Un polvo de carbón muy fino con una calidad grasosa que se usa para lubricar.

**Graphite oil** An oil with a graphite base that may be used for special lubricating requirements such as door locks.
**Aceite de grafito** Aceite con una base de grafito que puede utilizarse para necesidades de lubrificación especiales, como por ejemplo en cerraduras de puertas.

**Grind** To finish or polish a surface by means of an abrasive wheel.
**Esmerilar** Acabar o pulir una superficie con una rueda abrasiva.

**Grinding** A machining process of removing metal from the crankshaft journals by use of a special machine and stones.
**Rectificado a esmeril** Un proceso maquinario de rebajar el metal de los muñones del cigüeñal usando una máquina especial y las piedras de afilar.

**Grit** Cleaning sand that is angular in shape and is used for aggressive cleaning.
**Grano** La arena de limpieza cuyos partículos son de forma angular y que se emplea en la limpieza agresiva.

**Gross weight** The weight of a substance or matter that includes the weight of its container.
**Peso bruto** Peso de una sustancia o materia que incluye el peso de su recipiente.

**Ground** The common negative connection of the electrical system that is the point of lowest voltage.
**Tierra** Conexión negativa común del sistema eléctrico. Es el punto de tensión más baja.

**Ground circuit test** A diagnostic test performed to measure the voltage drop in the ground side of the circuit.
**Prueba del circuito a tierra** Prueba diagnóstica llevada a cabo para medir la caída de tensión en el lado a tierra del circuito.

**Ground side** The portion of the circuit that is from the load component to the negative side of the source.
**Lado a tierra** Parte del circuito que va del componente de carga al lado negativo de la fuente.

**Grounded** An intentional or unintentional connection of a wire, positive (+) or negative (−), to the ground.
**Puesto a tierra** Una conexión prevista o imprevista de un alambre, positiva (+) o negativa (−), a la tierra. Se dice que un cortocircuito es puesto a tierra.

**Grounded circuit** An electrical defect that allows current to return to ground before it has reached the intended load component.
**Circuito puesto a tierra** Falla eléctrica que permite el regreso de la corriente a tierra antes de alcanzar el componente de carga deseado.

**Growler** Test equipment used to test starter armatures for shorts and grounds. It produces a very strong magnetic field that is capable of inducing a current flow and magnetism in a conductor.
**Indicador de cortocircuitos** Equipo de prueba utilizado para localizar cortociruitos y tierra en armaduras de arranque. Genera un campo magnético sumamente fuerte, capaz de inducir flujo de corriente y magnetismo en un conductor.

**Half-shaft** Either of the two drive shafts that connect the transaxle to the wheel hubs in FWD cars. Half-shafts have constant velocity joints attached to each end to allow for suspension motions and steering. The shafts may be of solid or tubular steel and may be of different lengths.

**Semieje** Cualquiera de los dos árboles de mando que conecta el transeje a los cubos de rueda en automóviles de tracción delantera. Los semiejes tienen juntas de velocidad constante fijadas a cada extremo para permitir el movimiento de suspensión y la dirección. Los semiejes pueden ser de acero sólido o tubular y sus longitudes pueden variar.

**Hall-effect switch** A sensor that operates on the principle that if a current is allowed to flow through thin conducting material being exposed to a magnetic field, another voltage is produced.

**Conmutador de efecto Hall** Sensor que funciona basado en el principio de que si se permite el flujo de corriente a través de un material conductor delgado que ha sido expuesto a un campo magnético, se produce otra tensión.

**Halogen** The term used to identify a group of chemically related nonmetallic elements. These elements include chlorine, fluorine, and iodine.

**Halógeno** Término utilizado para identificar un grupo de elementos no metálicos relacionados químicamente. Dichos elementos incluyen el cloro, el flúor y el yodo.

**Hand-held digital pyrometer** A tester for measuring component temperature.

**Pirómetro digital de mano** Instrumento de prueba para medir la temperatura de un componente.

**Hand press** A hand-operated device for pressing precision-fit components.

**Prensa de mano** Dispositivo accionado manualmente para prensar componentes con un fuerte ajuste de precisión.

**Hand tools** Tools that use only the force generated from the body to operate. They multiply the force through leverage to accomplish the work.

**Herramienta de mano** Las herramientas que sólo emplean la fuerza proporcionado por el cuerpo. Utilizan la acción de palanca para multiplicar la fuerza recibida a efectuar el trabajo.

**Hard fault code** A fault code in a computer that represents a fault that is present at all times.

**Código de fallo duro** Código de fallo en una computadora que representa un fallo que está siempre presente.

**Hard pedal** A loss in braking efficiency so that an excessive amount of pressure is needed to actuate the brakes.

**Pedal de freno duro** Una pérdida de la eficiencia del frenado en que se requiere una cantidad excesiva de presión para activar los frenos.

**Hard-shell connector** An electrical connector that has a hard plastic shell that holds the connecting terminals of separate wires.

**Conectador de casco duro** Conectador eléctrico con casco duro de plástico que sostiene separados los bornes conectadores de alambres individuales.

**Harmonic Balancer (vibration dampener)** A component attached to the front of the crankshaft used to reduce the torsional or twisting vibration that occurs along the length of the crankshaft.

**Equilibrador armónico (amortiguador de vibraciones)** Un componente conectado a la parte delantera de un cigüeñal que sirve para reducir las vibraciones torsionales que ocurren por la longitud del cigüeñal.

**Harshness** A bumpy ride caused by a stiff suspension. Can be often cured by installing softer springs or shock absorbers.

**Aspereza** Viaje de muchas sacudidas ocasionadas por una suspensión rígida. Puede remediarse con la instalación de muelles más flexibles o amortiguadores.

**Hazardous material** A material that could cause injury or death to a person, or could damage or pollute land, air, or water.

**Materiales peligrosos** Un material que podría causar daños o la muerte a una persona o podría causar los daños o la contaminación de la tierra, el aire o el agua.

**Hazardous Materials Inventory Roster** A required poster that lists all hazardous materials the employee may come into contact with. It also lists the area of the workplace where the material is used.

**Escalafón de inventario de los materiales peligrosos** Un cartel mandatorio que indica todos los materiales peligrosos con los cuales un empleado puede ponerse en contacto. Tambien indica donde en el área del espacio de trabajo se usa dicho material.

**HC** The abbreviation for hydrocarbons. Hydrocarbons are present in particles of unburned gasoline.

**HC** La abreviación de hidrocarburos. Los hidrocarburos se encuentran en los partículos no consumidos de la gasolina.

**HCFC** Hydrochlorofluorocarbon refrigerant.

**HCFC** Refrigerante de hidroclorofluorocarbono.

**Head gasket** Gasket used to prevent compression pressures, gases, and fluids from leaking. It is located on the connection between the cylinder head and engine block.

**Junta de la cabeza** Una junta que se emplea en prevenir que se escapen las presiones, los gases y los fluidos de compresión. Se ubica en la conexión entre la cabeza de los cilindros y el monoblock.

**Header tanks** The top and bottom tanks (downflow) or side tanks (crossflow) of a radiator. The tanks in which coolant is accumulated or received.

**Tanques para alimentación por gravedad** Los tanques superiores e inferiores (flujo descendente) o los tanques laterales (flujo transversal) de un radiador. Tanques en los cuales el enfriador se acumula o se recibe.

**Heat exchanger** An apparatus in which heat is transferred from one medium to another on the principle that heat moves to an object with less heat.

**Intercambiador de calor** Aparato en el que se transfiere el calor de un medio a otro, lo cual se basa en el principio que el calor se atrae a un objeto que tiene menos calor.

**Heat treated** Metal hardened by a process of heating it to a high temperature, then quenching it in a cool bath.

**Tratado térmico (cementado)** El metal endurecido por el proceso de calentándolo a una temperatura muy elevada, luego amortiguándolo en un baño frio.

**Heated oxygen sensor (HO$_2$S)** A sensor mounted in the exhaust manifold that sends a voltage signal to the engine computer in relation to the amount of oxygen in the exhaust.

**Sensor térmico de oxígeno (HO$_2$S)** Sensor montado en el múltiple de escape que le envía una señal de tensión a la computadora del motor referente a la cantidad de oxígeno en el escape.

**Heated resistor-type MAF sensor** An MAF sensor that uses a heated resistor to sense air intake volume and temperature, and sends a voltage signal to the computer in relation to the total volume of intake air.

**Sensor MAF tipo resistor térmico** Sensor MAF que utiliza un resistor térmico para advertir el volumen y la temperatura del aire aspirado, y que le envía una señal de tensión a la computadora referente al volumen total de aire aspirado.

**Heater core** A radiator-like heat exchanger located in the case/duct system through which coolant flows to provide heat to the vehicle interior.

**Núcleo del calentador** Intercambiador de calor parecido a un radiador y ubicado en el sistema de caja/conducto a través del cual fluye el enfriador para proveer calor al interior del vehículo.

**Heat-shrink tubing** A hollow insulation material that shrinks to an airtight fit over a connection when exposed to heat.

**Tubería contraída térmicamente** Material aislante hueco que se contrae para acomodarse herméticamente sobre una conexión cuando se encuentra expuesto al calor.

**Heavy spots** A spot in a tire casing that is heavier than the rest of the tire.

**Puntos pesados** Punto en la cubierta del neumático más pesado que el resto del neumático.

**Heel** The outside, larger half of the gear tooth.

**Talón** Mitad exterior más grande del diente de engranaje.

**Helical** Shapes like a coil spring or a screw thread.
Helicoidal Formas parecidas a un muelle helicoidal o a un filete de tornillo.

**Helical gear** Gears with the teeth cut at an angle to the axis of the gear.
Engranaje helicoidal Engranajes que tienen los dientes cortados a un ángulo del pivote del engranaje.

**Heli-coil** A spiral thread used to restore damaged threads to original size.
Heli-coil Un hilo espiral que se emplea para restaurar las roscas dañadas a su tamaño original.

**Herringbone gear** A pair of helical gears designed to operate together. The angle of the pair of gears forms a V.
Engranaje bihelocoidal Par de engranajes helicoidales diseñados para funcionar juntos. El ángulo del par de engranajes forma una V.

**HI** The designation for high, as in blower speed or system mode.
HI Indicación para indicar marcha rápida, como por ejemplo la velocidad de un soplador o el modo de un sistema.

**High pedal** A clutch pedal that has an excessive amount of pedal travel.
Pedal alto Pedal del embrague con exceso de avance.

**High-side gauge** The right-side gauge on the manifold used to read refrigerant pressure in the high side of the system.
Calibrador del lado de alta presión El calibrador del lado derecho del múltiple utilizado para medir la presión del refrigerante en el lado de alta presión del sistema.

**High-side hand valve** The high-side valve on the manifold set used to control flow between the high side and service ports.
Válvula de mano del lado de alta presión Válvula del lado de alta presión que se encuentra en el conjunto del múltiple, utilizada para regular el flujo entre el lado de alta presión y los orificios de servicio.

**High-side service valve** A device located on the discharge side of the compressor; this valve permits the service technician to check the high-side pressures and perform other necessary operations.
Válvula de servicio del lado de alta presión Dispositivo ubicado en el lado de descarga del compresor; dicha válvula permite que el mecánico verifique las presiones en el lado de alta presión y lleve a cabo otras funciones necesarias.

**High-side switch** See *Pressure switch*.
Autómata manométrico del lado de alta presión Ver *Pressure switch* [*Autómata manométrico*].

**High-torque clutch** A heavy-duty clutch assembly used on some vehicles known to operate with higher-than-average head pressure.
Embrague de alto par de torsión Conjunto de embrague para servicio pesado utilizado en algunos vehículos que funcionan con una altura piezométrica más alta que la normal.

**History code** A fault code in a computer that represents an intermittent defect.
Código histórico Código de fallo en una computadora que representa un defecto intermitente.

**Hoist** A lift used to raise the entire vehicle.
Elevador Una grúa que se emplea en levantar el vehículo completo.

**Honing** A process whereby an abrasive material is used to smoothe a surface. Honing is a common operation in preparing cylinder walls for the installation of pistons.
Rectificación Un proceso por el cual una material abrasiva se usa para tersar una superficie. Rectificar es una operación común en preparar los interiores de los cilindros para la instalación de los pistones.

**Horsepower** A measure of mechanical power, or the rate at which work is done. One horsepower equals 33,000 ft.-lbs. (foot-pounds) of work per minute. It is the power necessary to raise 33,000 pounds a distance of one foot in one minute.

Caballo de fuerza Una medida de fuerza mecánica, o el régimen en el cual se efectúa el trabajo. Un caballo de fuerza iguala a 33,000 lb.p. (libras pie) de trabajo por minuto. Es la fuerza requerida para transportar a 33,000 libras una distancia de 1 pie en 1 minuto.

**Hot** A term given the positive (+) side of an energized electrical system. Also refers to an object that is heated.
Cargado/caliente Término utilizado para referirse al lado positivo (+) de un sistema eléctrico excitado. Se refiere también a un objeto que es calentado.

**Hot knife** A knife-like tool that has a heated blade. Used for separating objects—for example, evaporator cases.
Cuchillo en caliente Herramienta parecida a un cuchillo provista de una hoja calentada. Utilizada para separar objetos; p.e. las cajas de evaporadores.

**Hot spots** The small areas on a friction surface that are a different color, normally blue, or are harder than the rest of the surface.
Zonas de calor Zonas pequeñas sobre una superficie de rozamiento de un color diferente, normalmente azul, o más duras que el resto de la superficie.

**Hot spray tank** Cleaning tank that sprays caustic solutions onto the components and also soaks them.
Tanque de rocío caliente Un tanque de limpieza que rocía las soluciones cáusticas sobre los componentes mientras que éstos remojan.

**Hot wire–type MAF sensor** An MAF sensor that uses a heated wire to sense air intake volume and temperature, and sends a voltage signal to the computer in relation to the total volume of intake air.
Sensor MAF tipo térmico Sensor MAF que utiliza un alambre térmico para advertir el volumen y la temperatura del aire aspirado, y que le envía una señal de tensión a la computadora referente al volumen total de aire aspirado.

**Hotchkiss drive** A type of rear suspension in which leaf springs absorb the rear axle housing torque.
Transmisión Hotchkiss Tipo de suspensión trasera en la cual muelles de láminas absorben el par de torsión del puente trasero.

**Hub** The center part of a wheel to which the wheel is attached.
Cubo La parte central de una rueda a la cual se monta la rueda.

**Hydraulic clutch** A clutch that is actuated by hydraulic pressure; used in cars and trucks when the engine is some distance from the driver's compartment so that it would be difficult to use mechanical linkages.
Embrague hidráulico Embrague accionado por presión hidráulica; utilizado en automóviles y camiones cuando el motor está lejos del compartimiento del conductor para dificultar la utilización de bielas motrices mecánicas.

**Hydraulic fluid reservoir** A part of a master cylinder assembly that holds reserve fluid.
Depósito de fluido hidráulico Parte del conjunto del cilindro primario que contiene el fluido de reserva.

**Hydraulic lifters** Lifters that use oil to absorb the resultant shock of valve train operation.
Elevadores hidráulicas Los elevadores que usan el aceite para absorber los golpes causados por la operación del tren de válvulas.

**Hydraulic press** A piece of shop equipment that develops a heavy force by use of a hydraulic piston-and-jack assembly.
Prensa hidráulica Una herramienta del taller que provee una fuerza grande por medio de una asamblea de gato con un pistón hidráulico.

**Hydraulic pressure** Pressure exerted through the medium of a liquid.
Presión hidráulica La presión esforzada por medio de un líquido.

**Hydraulic valve lifters** Round, cylindrical, metal components used to open the valves. These components are operated by oil pressure to maintain zero clearance between the valve stem and rocker arm.
Desmontaválvulas hidráulicas Componentes metálicos, cilíndricos y redondos utilizados para abrir las válvulas. Estos componentes se

accionan mediante la presión del aceite para mantener cero espacio libre entre el vástago de válvula y el balancín.

**Hydrocarbon** A chemical composition made up of hydrogen and carbon.
*Hidrocarburo* Una composición quimica compuesta del hidrógeno y el carbono.

**Hydrometer** A test instrument used to check the specific gravity of the electrolyte to determine the battery's state of charge.
*Hidrómetro* Instrumento de prueba utilizado para verificar la gravedad específica del electrolito y así determinar el estado de carga de la batería.

**Hydroscopic** The property of a fluid in which it has a tendency to absorb moisture from the atmosphere.
*Hidroscópico* Cualidad que posee un fluido para absorber la humedad de la atmósfera.

**Hydrostatic lock** The result of attempting to compress a liquid in the cylinder. Since liquid is not compressible, the piston is not able to travel in the cylinder.
*Bloqueo hidrostático* El resultado de un intento de comprimir un líquido en el cilindro. Como no se puede comprimir un líquido, el pistón no puede moverse en el cilindro.

**Hygiene** A system of rules and principles intended to promote and preserve health.
*Higiene* Sistema de normas y principios cuyo propósito es promover y preservar la salud.

**Hypoid gear** A gear that is similar in appearance to a spiral bevel gear, but the teeth are cut so that the gears match in a position where the shaft center lines do not meet; cut in a spiral form to allow the pinion to be set below the center line of the ring gear so that the car floor can be lower.
*Engranaje hipoide* Engranaje parecido a un engranaje cónico con dentado espiral, pero en el cual los dientes se cortan para que los engranajes se engranen en una posición donde las líneas centrales del árbol no se crucen; cortado en forma de espiral para permitir que el piñon sea colocado debajo de la línea central de la corona y que así el piso del vehículo sea más bajo.

**Hypoid gear lubricant** An extreme pressure lubricant designed for the severe operation of hypoid gears.
*Lubrificante del engranaje hipoide* Lubrificante de extrema presión diseñado para el funcionamiento riguroso de los engranajes hipoides.

**ID** Inside diameter.
*DI* Diámetro interior.

**Idle** Engine speed when the accelerator pedal is fully released and there is no load on the engine.
*Marcha lenta* La velocidad del motor cuando el pedal accelerador está completamente desembragada y no hay carga en el motor.

**Idle air control bypass air (IAC BPA) motor** An IAC motor that controls idle speed by regulating the amount of air bypassing the throttle.
*Motor para el control de la marcha lenta con el paso de aire* Un motor IAC que controla la velocidad de la marcha lenta regulando la cantidad de aire que se desvía de la mariposa.

**Idle air control bypass air (IAC BPA) valve** A valve operated by the IAC BPA motor that regulates the air bypassing the throttle to control idle speed.
*Válvula para el control de la marcha lenta con el paso de aire* Válvula accionada por el motor IAC BPA que regula el aire que se desvía de la mariposa para controlar la velocidad de la marcha lenta.

**Idle air control (IAC) motor** A computer-controlled motor that controls idle speed under all conditions.
*Motor para el control de la marcha lenta con aire* Motor controlado por computadora que controla la velocidad de la marcha lenta bajo cualquier condición del funcionamiento del motor.

**Idle contact switch** A switch in the IAC motor stem that informs the computer when the throttle is in the idle position.
*Conmutador de contacto de la marcha lenta* Conmutador en el vástago del motor IAC que le advierte a la computadora cuándo la mariposa está en la posición de marcha lenta.

**Idle stop solenoid** An electric solenoid that maintains the throttle in the proper idle speed position and prevents dieseling when the ignition switch is turned off.
*Solenoide de detención de la marcha lenta* Solenoide eléctrico que mantiene la mariposa en la posición de velocidad de marcha lenta adecuada y evita el autoencendido al desconectarse el botón conmutador de encendido.

**Idler** A pulley device that keeps the belt whip out of the drive belt of an automotive air conditioner. The idler is used as a means of tightening the belt.
*Polea loca* Polea que mantiene la vibración de la correa fuera de la correa de transmisión de un acondicionador de aire automotriz. Se utiliza la polea loca para proveerle tensión a la correa.

**Idler pulley** A pulley used to tension or torque the belt(s).
*Polea tensora* Polea utilizada para proveer tensión o par de torsión a la(s) correa(s).

**Ignition** The process of igniting the air-fuel mixture in the combustion chamber.
*Encendido* El proceso de encender la mezcla de aire/combustible en la cámara de combustión.

**Ignition crossfiring** Ignition firing between distributor cap terminals or spark plug wires.
*Encendido por inducción* Encendido entre los bornes de la tapa del distribuidor o los alambres de las bujías.

**Ignition module tester** An electronic tester designed to test ignition modules.
*Instrumento de prueba del módulo del encendido* Instrumento de prueba electrónico diseñado para revisar módulos del encendido.

**Ignition system** System that delivers the spark used to ignite the compressed air-fuel mixture.
*Sistema de encendido* Un sistema que entrega la chispa que sirve para encender la mezcla comprimida de aire/combustible.

**Impedance** The combined opposition to current created by the resistance, capacitance, and inductance of a test meter or circuit.
*Impedancia* Oposición combinada a la corriente generada por la resistencia, la capacitancia y la inductancia de un instrumento de prueba o de un circuito.

**Impeller** The pump or driving member in a torque converter.
*Impulsor* La bomba o el miembro impulsor en un convertidor de torsión.

**Inboard constant velocity joint** The inner constant velocity joint, or the one closest to the transaxle. The inboard joint is usually a plunging-type joint that telescopes to compensate for suspension motions.
*Junta de velocidad constante del interior* Junta de velocidad constante interior, o la que está más cerca del transeje. La junta del interior es normalmente una junta de tipo sumergible que se extiende para compensar el movimiento de la suspensión.

**In-car temperature sensor** A thermistor used in automatic temperature control units for sensing the in-car temperature. Also see *Thermistor*.
*Sensor de temperatura del interior del vehículo* Termistor utilizado en unidades de regulación automática de temperatura para sentir la temperatura del interior del vehículo. Ver también *Thermistor* [*Termistor*].

**Inch** 25.4 mm, 2.54 cm, or 0.0254 meters.
*Pulgada* 25,4 milímetros, 2,54 centímetros, o 0,0254 metros.

**Inclinometer** Device designed with a spirit level and graduated scale to measure the inclination of a driveline assembly. The inclinometer connects to the drive shaft magnetically.

**Inclinómetro** Instrumento diseñado con nivel de burbuja de aire y escala graduada para medir la inclinación de un conjunto de la línea de transmisión. El inclinómetro se conecta magnéticamente al árbol de mando.

**Increments** Series of regular additions from small to large.
**Incrementos** Una serie de incrementos regulares que va de pequeño a grande.

**Independent rear suspension (IRS)** The vehicle's rear wheels move up and down independently of each other.
**Suspensión trasera independiente** Las ruedas traseras del vehículo realizan un movimiento de ascenso y descenso de manera independiente la una de la otra.

**Independent suspension** A suspension system that allows one wheel to move up and down without affecting the opposite wheel. Provides superior handling and a smoother ride.
**Suspensión independiente** Sistema de suspensión que permite que una rueda realice un movimiento de ascenso y descenso sin afectar la rueda opuesta. Permite un manejo excelente y un viaje mucho más cómodo.

**Index** To orient two parts by marking them. During reassembly the parts are arranged so the index marks are next to each other. Used to preserve the orientation between balanced parts.
**Índice** Orientar a dos partes marcándolas. Al montarlas, las partes se colocan para que las marcas de índice estén alineadas. Se usan los índices para preservar la orientación de las partes balanceadas.

**Indexing** The process of offsetting the crankshaft so the rod journals are centered in the grinding machine. The offset from the main bearing journal is the same distance as from the center of the main journal to the center of the crank throw.
**Rectificación graduada** El proceso de descentrar el cigüeñal para que los muñones de las bielas sean centrales en la rectificadora. La desviación del muñón del cojinete principal queda la misma distancia que la distancia del centro del muñón del cojinete principal al centro del codo del cigüeñal.

**Indicated horsepower** The amount of horsepower the engine can theoretically produce.
**Potencia disponible** La cantidad del caballo de fuerzas que puede producir un motor según la teoría.

**Induction hardening** A process which uses an electromagnet to heat the seat through induction. The seat is heated to a temperature of about 1700°F (930°C). The hardened depth is about 0.060 inch (1.5 mm).
**Endurecimiento por inducción** Un proceso que usa un electroimán para calentar el asiento por medio de la inducción. El asiento se calienta a una temperatura de aproximadamente 1700°F (930°C). La profundidad del endurecimiento es aproximadamente de un 0.060 pulgada (1.5 mm).

**Inertia** The tendency of objects in motion to remain in motion and of objects at rest to remain at rest.
**Inercia** La tendencia de que los objetos en movimiento queden en movimiento y los objetos en reposo permanezcan en reposo.

**Injector balance tester** A tester designed to test port injectors.
**Instrumento de prueba del equilibro del inyector** Instrumento de prueba diseñado para revisar inyectores de lumbreras.

**Inlet check valve** Keeps the oil filter filled at all times so when the engine is started an instantaneous supply of oil is available.
**Válvula de retención** Mantiene siempre lleno el filtro de aceite para que cuando se arranque el motor dispone de un suministro instantáneo de aceite.

**Inner bearing race** The inner part of a bearing assembly on which the rolling elements, ball or roller, rotate.
**Anillo de cojinete interior** Parte interior de un conjunto de cojinete sobre la cual los elementos rodantes, o sea, las bolas o los rodillos, giran.

**Input shaft** The shaft carrying the driving gear by which the power is applied, as to the transmission.

**Flecha de entrada** La flecha que porta el engranaje propulsor por el cual se aplica la potencia, como a la transmisión.

**Insert bearings** An interchangeable type of bearing. The bearing is a self-contained part that is inserted into the bearing housing.
**Cojinetes de inserción** Un cojinete de tipo intercambiable. El cojinete es de una parte entera que se puede insertar en la cubierta del cojinete.

**Insert fitting** A fitting that is designed to fit inside, such as a barb fitting that fits inside a hose.
**Ajuste inserto** Ajuste diseñado para insertarse dentro de un objeto, como por ejemplo un ajuste arponado que se inserta dentro de una manguera.

**Insert guides** Removable valve guides that are pressed into the cylinder head.
**Guías para inserción** Las guías desmontables de las válvulas que han sido prensadas en la cabeza de los cilindros.

**Insert springs** Round wire springs that hold the inserts or shift plates in contact with the synchronizer sleeve. Located around the synchronizer hub.
**Muelles insertos** Muelles redondos de alambre que mantienen el contacto entre las piezas insertas o placas de cambio de velocidades y el manguito sincronizador. Ubicados alrededor del cubo sincronizador.

**Inserts** One of several terms that could apply to the shift plates found in a synchronizer assembly.
**Piezas insertas** Uno de los varios téminos que puede aplicarse a las placas de cambio de velocidades encontradas en un conjunto sincronizador.

**Inside micrometer** Precision instrument designed to measure the inside diameter of a hole.
**Micrómetro interior** Una herramienta de medir diseñada para medir el diámetro interior de un agujero.

**Inspection cover** A removable cover that permits entrance for inspection and service work.
**Cubierta de inspección** Cubierta desmontable que permite la entrada para la inspección y la reparación.

**Inspection, maintenance (I/M) testing** Emission inspection and maintenance programs that are usually administered by various states.
**Pruebas de inspección y mantenimiento** Programas de inspección y mantenimiento de emisiones que normalmente administran diferentes estados.

**Instrument panel cluster (IPC)** A display in the instrument panel that may include gauges, indicator lights, or digital displays to indicate system functions.
**Instrumentos agrupados en el tablero de instrumentos** Representación visual en el tablero de instrumentos que puede incluir calibradores, luces indicadoras, o visualizaciones digitales para indicar las funciones del sistema.

**Instrument voltage regulator (IVR)** Provides a constant voltage to the gauge regardless of the voltage output of the charging system.
**Instrumento regulador de tensión** Le provee tensión constante al calibrador, sin importar cual sea la salida de tensión del sistema de carga.

**Insulated circuit resistance test** A voltage drop test that is used to locate high resistance in the starter circuit.
**Prueba de la resistencia de un circuito aislado** Prueba de la caída de tensión utilizada para localizar alta resistencia en el circuito de arranque.

**Insulated side** The portion of the circuit from the positive side of the source to the load component.
**Lado aislado** Parte del circuito que va del lado positivo de la fuente al componente de carga.

**Insulator** A substance that is not capable of supporting the flow of electricity.

**Aislador** Sustancia que no es capaz de soportar el flujo de electricidad.

**Intake manifold** Component that delivers the air or air-fuel mixture to each engine cylinder.
**Múltiple de admisión** Un componente que entrega el aire o la mezcla del aire/combustible a cada cilindro del motor.

**Intake manifold gasket** Gasket which fits between the manifold and cylinder head to seal the air-fuel mixture or intake air.
**Junta del múltiple de admisión** Una junta que queda entre el múltiple y la cabeza del cilindro para sellar la mezcla de aire/combustible o el aire de admisión.

**Intake valve** The control passage of the air-fuel mixture entering the cylinder.
**Válvula de entrada** El pasaje de control para la mezcla de aire/combustible entrando al cilindro.

**Integral** Built into, as part of the whole.
**Integral** Incorporado, una parte de la totalidad.

**Integral axle housing** A rear axle housing-type where the parts are serviced through an inspection cover and adjusted within, and relative to, the axle housing.
**Puente trasero integral** Tipo de puente trasero, donde se reparan las piezas a través de una cubierta de inspección y se ajustan dentro del y con relación al puente trasero.

**Integral guides** Valve guides that are manufactured and machined as part of the cylinder head.
**Guías integrales** Las guías de las válvulas fabricadas e incorporadas a máquina como parte de la cabeza del cilindro.

**Integral reservoir** A reservoir that is joined with another component such as a power-steering pump.
**Tanque integral** Tanque unido a otro componente, como por ejemplo una bomba de la dirección hidráulica.

**Integrated circuit (IC chip)** A complex circuit of thousands of transistors, diodes, resistors, capacitors, and other electronic devices that are formed onto a small silicon chip. As many as 30,000 transistors can be placed on a chip that is 1/4 inch (6.35 mm) square.
**Circuito integrado (Fragmento CI)** Circuito complejo de miles de transistores, diodos, resistores, condensadores, y otros dispositivos electrónicos formados en un fragmento pequeño de silicio. En un fragmento de 1/4 de pulgada (6,35 mm) cuadrada, pueden colocarse hasta 30.000 transistores.

**Integrator** A chip responsible for fuel control in a General Motors PCM.
**Integrador** Pastilla responsable de controlar el combustible en un módulo del control del tren transmisor de potencia de la General Motors.

**Interference angle** Valve design where the seat angle is 1 degree greater than the valve face angle to provide a more positive seal.
**Angulo de interferencia** Un diseño de la válvula en el cual el ángulo del asiento de la válvula es un grado además del ángulo de la cara de la válvula para proveer un sello más positivo.

**Interference engine** An engine design in which, if the timing belt or chain breaks or is out of phase, the valves will contact the pistons. In addition, in multivalve interference engines, valve-to-valve contact is possible if the belt or chain is out of phase.
**Interferencia del motor** Un diseño del motor en el cual, si se quiebra la correa o la cadena de sincronización, o si está fuera de fase, las válvulas se pondrán en contacto con los pistones. Además, en los motores de interferencia de múltiples válvulas, es posible el contacto de válvula a válvula si la correa o la cadena está fuera de fase.

**Interference fit** A press fit—for example, the inside diameter of a bore is 0.001 inch smaller than the outside diameter of a shaft, so when the shaft is fitted into the bore it must be pressed in to overcome the 0.001-inch interference fit.
**Ajuste a interferencia** Ajuste en prensa; por ejemplo, el diámetro interior de un calibre es un 0,001 de pulgada más pequeño que el diámetro exterior de un árbol, así que, cuando el árbol se inserta dentro del calibre, tiene que ser comprimido para compensar el ajuste a interferencia de un 0,001 de pulgada.

**Interlock mechanism** A mechanism in the transmission shift linkage that prevents the selection of two gears at one time.
**Mecanismo de enganche** Mecanismo en la biela motriz del cambio de velocidades de la transmisión que impide la selección de dos velocidades a la vez.

**Intermediate bearing plate** Another name for the center support plate of a transmission.
**Placa intermedia del cojinete** Otro nombre para la placa central de soporte de una transmisión.

**Intermediate drive shaft** Located between the left and right drive shafts, it equalizes drive shaft length.
**Árbol de mando intermedio** Ubicado entre los árboles de mando izquierdo y derecho, compensa la longitud del árbol de mando.

**Intermediate plate** A mechanism in the transmission shift linkage that prevents the selection of two gears at one time.
**Placa intermedia** Mecanismo en la biela motriz del cambio de velocidades de la transmisión que impide la selección de dos velocidades a la vez.

**Intermittent fault code** A fault code in a computer memory that is caused by a defect that is not always present.
**Código de fallo intermitente** Código de fallo en la memoria de una computadora ocasionado por un defecto que no está siempre presente.

**Internal** Inside; within.
**Interno** Al interior, dentro de una cosa.

**Internal combustion engines** An engine that burns its fuels within the engine.
**Motores de combustión interna** Un motor que quema su combustible dentro del motor.

**Internal gear** A gear with teeth pointing inward, toward the hollow center of the gear.
**Engranaje internal** Un engranaje cuyos dientes apuntan hacia el interior, al hueco central del engranaje.

**Internal snap ring** A snap ring used to hold a component or part inside a cavity or case.
**Anillo de muelle interno** Anillo de muelle utilizado para sujetar un componente o una pieza dentro de una cavidad o caja.

**Internally balanced engine** Engines balanced by the counterweights only.
**Motor de equilibración interno** Los motores que se equilibran solamente por medio de los contrapesos.

**International system (SI)** A system of weights and measures.
**Sistema internacional** Sistema de pesos y medidas.

**Jack stands (safety stands)** Support devices used to hold the vehicle off the floor after it has been raised by the floor jack.
**Torres (soportes de seguridad)** Los dispositivos de soporte que se emplean para mantener el vehículo levantado del piso despues de que haya sido levantado del piso por un gato hidráulico.

**Jam nut** A second nut tightened against a primary nut to prevent it from working loose. Used on inner and outer tie-rod adjustment nuts and on many pinion-bearing adjustment nuts.
**Contra tuerca** Una tuerca secundaria que se aprieta contra una tuerca primaria para prevenir que ésta se afloja. Se emplean en las tuercas de ajustes interiores y exteriores para las barras de acoplamiento y también en muchas de las tuercas de ajuste de portapiñones.

**Jet valves** Valves used by some manufacturers to direct a stream of oil to the underside of the piston head. Jet valves are common on turbocharged engines to keep the piston cool to prevent detonation and piston damage.
**Válvulas de chorro** Las válvulas empleadaos por algunos fabricantes para dirigir un chorro de aceite a la parte inferior de la cabeza del

pistón. Las válvulas de chorro suelen usarse en los motores turbo-cargados para mantener frío al pistón y para prevenir la detonación y los daños al pistón.

**Joint angle** The angle formed by the input and output shafts of constant velocity joints. Outer joints can typically operate at angles up to 45 degrees, whereas inner joints have more restricted angles.
**Ángulo de las juntas** Ángulo formado por el árbol impulsor y el árbol de rendimiento de las juntas de velocidad constante. Las juntas exteriores pueden funcionar típicamente en ángulos de hasta 45 grados, mientras que las juntas interiores funcionan dentro de ángulos más limitados.

**Journal** An inner bearing operated by a shaft.
**Muñón** Un cojinete interior operado por una flecha.

**Jumper** A wire used to temporarily bypass a device or component for the purpose of testing.
**Barreta** Alambre utilizado para desviar un dispositivo o componente de manera temporal para llevar a cabo una prueba.

**Key** A small block inserted between the shaft and hub to prevent circumferential movement.
**Chaveta** Un tope pequeño que se meta entre la flecha y el cubo para prevenir un movimiento circunferencial.

**Key on engine off (KOEO) test** A computer system test mode on Ford products that displays diagnostic trouble codes (DTCs) with the key on and the engine stopped.
**Prueba con la llave en la posición de encendido y el motor apagado** Modo de prueba en una computadora de productos fabricados por la Ford que muestra códigos indicadores de fallas para propósitos diagnósticos cuando la llave está en la posición de encendido y el motor está apagado.

**Key on engine running (KOER) test** A computer system test mode on Ford products that displays diagnostic trouble codes (DTCs) with the engine running.
**Prueba con la llave en la posición de encendido y el motor encendido** Modo de prueba en una computadora de productos fabricados por la Ford que muestra códigos indicadores de fallas para propósitos diagnósticos cuando el motor está encendido.

**Keyway** A groove or slot cut to permit the insertion of a key.
**Ranura de chaveta** Un corte de ranura o mortaja que permite insertar una chaveta.

**Kilogram** A unit of measure in the metric system. One kilogram is equal to 2.205 pounds in the English system.
**Kilogramo** Unidad de medida en el sistema métrico. Un kilogramo equivale a 2,205 libras en el sistema inglés.

**Kilopascal** A unit of measure in the metric system. One kilopascal (kPa) is equal to 0.145 pound per square inch (psi) in the English system.
**Kilopascal** Unidad de medida en el sistema métrico. Un kilopascal (kPa) equivale a 0,145 libras por pulgada cuadrada en el sistema inglés.

**Kilovolts (kV)** Thousands of volts.
**Kilovoltios (kV)** Miles de voltios.

**Kinetic energy** Energy that is working.
**Energía cinética** La energía trabajando.

**King pin** A metal rod or pin on which steering knuckles turn.
**Clavija maestra** Varilla o chaveta de metal sobre la cual giran los muñones de dirección.

**Knock** A descriptive term used to identify various noises occurring in an engine.
**Golpe** Un término descriptivo que sirve para identificar los varios ruidos que ocurren en un motor.

**Knock sensor** A sensor that sends a voltage signal to the computer in relation to engine detonation.

**Sensor de golpeteo** Sensor que le envía una señal de tensión a la computadora referente a la detonación del motor.

**Knock sensor module** An electronic module that changes the analog knock sensor signal to a digital signal and sends it to the PCM.
**Módulo del sensor de golpeteo** Módulo electrónico que convierte la señal analógica del sensor de golpeteo en una señal digital y se la envía al módulo del control del tren transmisor de potencia.

**Knuckle** The part of the suspension that supports the wheel hub and serves as the steering pivot. The bottom of the knuckle is attached to the lower control arm with a ball joint, and the upper portion is usually bolted to the strut.
**Muñón** Parte de la suspensión que apoya al cubo de la rueda y que sirve como pivote de dirección. La parte inferior del muñón se fija al brazo de mando inferior por medio de una junta esférica, y la superior normalmente se emperna al montante.

**Knurl** A special bit that rolls a thread into the guide and causes the metal to rise.
**Moleta** Una broca especial que enreda un hilo en la guía y causa que se levanta el metal.

**Knurling** A machining process that decreases the size of a bore by forcing a bit that swells the metal much like a tap does when it cuts threads.
**Moletear** Un proceso de rebajar a máquina el tamaño de un taladro forzando una broca que hincha al metal de una manera muy parecida de un macho cortando las roscas.

**kPa** kilopascal.
**kPa** kilopascal.

**Lamp** A device that produces light as a result of current flow through a filament. The filament is enclosed within a glass envelope and is a type of resistance wire that is generally made from tungsten.
**Lámpara** Dispositivo que produce luz como resultado del flujo de corriente a través de un filamento. El filamento es un tipo de alambre de resistencia hecho por lo general de tungsteno, que es encerrado dentro de una bombilla.

**Lapping** The process of fitting one surface to another by rubbing them together with an abrasive material between the surfaces.
**Pulido** El proceso de ajustar una superficie a otra rozándolas juntas con un material adhesivo entre las superficies.

**Lash** The amount of free motion in a gear train, between gears, or in a mechanical assembly, such as the lash in a valve train.
**Juego** La cantidad del movimiento libre en un tren de engranajes, entre los engranajes o en una asamblea mecánica, tal como el juego en un tren de válvulas.

**Lateral movement** Movement from side to side.
**Movimiento lateral** Movimiento de un lado a otro.

**Lateral runout** The variation in side-to-side movement.
**Desviación lateral** Variación del movimiento de un lado a otro.

**Leaf spring** A spring made up of a single flat steel plate or of several plates of graduated lengths assembled one on top of another; used on vehicles to absorb road shocks by bending or flexing.
**Muelle de láminas** Muelle compuesto de una sola placa de acero plana o de varias placas de longitudes graduadas montadas la una sobre la otra; utilizado en vehículos para absorber la aspereza de la carretera a través de la flexión.

**Leakdown** The relative movement of the lifter's plunger in respect to the lifter body.
**Tiempo de fuga** El movimiento relativo del émbolo del levantaválvulas con relación al cuerpo del levantaválvulas.

**Leakdown testing** Testing which determines the lifter's ability to hold hydraulic pressure and maintain zero lash.
**Prueba de fuga** La prueba que verifica la habilidad del levantaválvulas de mantener la presión y un juego cero.

**Lean-burn miss** The result of incomplete combustion due to the lack of oxygen.
Marcha irregular de mezcla pobre El resultado de la combustión incompleta debido a la falta de oxígeno.

**Lift** A device used to raise a vehicle.
Elevador Dispositivo utilizado para levantar un vehículo.

**Lifters** Mechanical (solid) or hydraulic connections between the camshaft and the valves. Lifters follow the contour of the camshaft lobes to lift the valve off its seat.
Levantaválvulas Las conexiones mecánicas (sólidas) o hidráulicas entre el árbol de levas y las válvulas. Los levantaválvulas siguen el contorno de los lóbulos del árbol de levas para levantar la válvula de su asiento.

**Light-emitting diode (LED)** A special diode that emits light when current flows through the diode.
Diodo emisor de luz (LED) Diodo especial que emite luz cuando una corriente fluye a través del mismo.

**Limited-slip differential** A differential designed so that when one wheel is slipping, a major portion of the drive torque is supplied to the wheel with the better traction. Also called a nonslip differential.
Diferencial de deslizamiento limitado Diferencial diseñado para que cuando una rueda se deslice, una mayor parte del par de torsión de mando llegue a la rueda que tiene mejor tracción. Llamado también diferencial antideslizante.

**Line boring** A machining process of the main bearing journals that restores the original bore size by cutting metal using cutting bits.
Taladrar en serie Un proceso de rectificación a máquina de los muñones del cigüeñal que restaura el tamaño original del taladro cortando el metal con brocas para cortar.

**Line wrench** A common term for a flare nut wrench.
Llave de tuerca de línea Término común para llave de tuerca abocinada.

**Linear EGR valve** An EGR valve containing an electric solenoid that is pulsed on and off by the computer to provide a precise EGR flow.
Válvula EGR lineal Válvula EGR con un solenoide eléctrico que la computadora enciende y apaga para proporcionar un flujo exacto de EGR.

**Liners** Thin tubes placed between two parts.
Manguitos Los tubos delgados puestos entre dos partes.

**Linkage** Any series of rods, yokes, levers, and so on, used to transmit motion from one unit to another.
Biela Cualquiera serie de barras, yugos, palancas, y todo lo demás, que se usa para transferir los movimientos de una unidad a otra.

**Lip seals** Molded synthetic rubber seal with a slight raise (lip) that is the actual sealing point. The lip provides a positive seal while allowing for some lateral movement of the shaft.
Junta con reborde Las juntas de caucho sintético moldeado que tienen un reborde ligero (un labio) que es el punto efectivo del sello. El reborde provee un sello positivo mientras que permite algún movimiento lateral de la flecha.

**Liquid** A state of matter; a column of fluid without solids or gas pockets.
Líquido Estado de materia; columna de fluido sin sólidos ni bolsillos de gas.

**Liquid crystal display (LCD)** A display that sandwiches electrodes and polarized fluid between layers of glass. When voltage is applied to the electrodes, the light slots of the fluid are rearranged to allow light to pass through.
Visualizador de cristal líquido Visualizador digital que consta de dos láminas de vidrio selladas, entre las cuales se encuentran los electrodos y el fluido polarizado. Cuando se aplica tensión a los electrodos, se rompe la disposición de las moléculas para permitir la formación de carácteres visibles.

**Lithium-based grease** A special lubricant that is used on the rack bearing in manual rack and pinion steering gears.

Grasa con base de litio Lubricante especial utilizado en el cojinete de la cremallera en mecanismos de dirección de cremallera y piñón manuales.

**Live axle** A shaft that transmits power from the differential to the wheels.
Eje motor Árbol que transmite fuerza motriz del diferencial a las ruedas.

**Lobe** The part of the camshaft that raises the lifter.
Lóbulo La parte del árbol de levas que alza al levantaválvulas.

**Lock pin** Used in some ball sockets (inner tie-rod end) to keep the connecting nuts from working loose. Also used on some lower ball joints to hold the tapered stud in the steering knuckle.
Clavija de cerrojo Se usan en algunas rótulas (las extremidades interiores de la barra de acoplamiento) para prevenir que se aflojen las tuercas de conexión. También se emplean en algunas juntas esféricas inferiores para retener el perno cónico en la articulación de dirección.

**Locked differential** A differential with the side and pinion gears locked together.
Diferencial trabado Diferencial en el que el engranaje lateral y el de piñón están sujetos entre sí.

**Locking** A condition of a bearing caused by large particles of dirt that become trapped between a bearing and its race.
Cierre Condición de un cojinete causada por partículas de polvo que se atrapan entre un cojinete y su anillo.

**Locknut** A second nut turned down on a holding nut to prevent loosening.
Contra tuerca Una tuerca secundaria apretada contra una tuerca de sostén para prevenir que ésta se afloja.

**Lockplates** Metal tabs bent around nuts or bolt heads.
Placa de cerrojo Chavetas de metal que se doblan alrededor de las tuercas o las cabezas de los pernos.

**Lockwasher** A type of washer which, when placed under the head of a bolt or nut, prevents the bolt or nut from working loose.
Arandela de freno Un tipo de arandela que, al colocarse bajo la cabeza de un perno, previene que el perno o la tuerca se aflojan.

**Low-refrigerant switch** A switch that senses low pressure due to a loss of refrigerant and stops compressor action. Some alert the operator and/or set a trouble code.
Interruptor para advertir un nivel bajo de refrigerante Interruptor que siente una presión baja debido a una pérdida de refrigerante y que detiene la acción del compresor. Algunos interruptores advierten al operador y/o fijan un código indicador de fallas.

**Low-side gauge** The left-side gauge on the manifold used to read refrigerant pressure in the low side of the system.
Calibrador del lado de baja presión El calibrador en el lado izquierdo del múltiple utilizado para medir la presión del refrigerante del lado de baja presión del sistema.

**Low-side hand valve** The manifold valve used to control flow between the low side and service ports of the manifold.
Válvula de mano del lado de baja presión Válvula de distribución utilizada para regular el flujo entre el lado de baja presión y los orificios de servicio del colector.

**Low-side service valve** A device located on the suction side of the compressor, which allows the service technician to check low-side pressures and perform other necessary service operations.
Válvula de servicio del lado de baja presión Dispositivo ubicado en el lado de succión del compresor; dicha válvula permite que el mecánico verifique las presiones del lado de baja presión y lleve a cabo otras funciones necesarias de servicio.

**Low speed** The gearing that produces the highest torque and lowest speed of the wheels.
Velocidad baja La velocidad que produce la torsión más alta y la velocidad más baja a las ruedas.

**Lower end** Refers to the main bearing journals of the cylinder block.
**Extremo inferior** Refiere a los muñones del cigüeñal del monoblock.

**Lubricant** Any material, usually a petroleum product such as grease or oil, that is placed between two moving parts to reduce friction.
**Lubricante** Cualquier substancia, normalmente un producto de petróleo como la grasa o el aciete, que se coloca entre dos partes en movimiento para reducir la fricción.

**Lubrication system** System which supplies oil to high friction and wear locations.
**Sistema de lubricación** Un sistema que proporciona el aceite a las áreas de alta fricción y desgaste.

**Lug nut** The nuts that fasten the wheels to the axle hub or brake rotor. Missing lug nuts should always be replaced. Overtightening can cause warpage of the brake rotor in some cases.
**Tuerca de orejetas** Tuercas que sujetan las ruedas al cubo del eje o el rotor de freno. Se deben reemplazar siempre las tuercas de orejetas que se hayan perdido. En algunos casos el apretar demasiado puede causar el torcimiento del rotor de freno.

**Lugging** A term used to describe an operating condition in which the engine is operating at too low of an engine speed for the selected gear.
**Arrastrar** Término utilizado para describir una condición en la que el motor funciona a una velocidad demasiado baja para la marcha elegida.

**Machinist's rule** A multiple scale ruler used to measure distances or components that do not require precise measurement.
**Regla de acero** Una regla con graduaciones múltiples para medir las distancias o los componentes que no requieren una medida precisa.

**Magnetic-base thermometer** A thermometer that may be retained to metal components with a magnetic base.
**Termómetro con base magnética** Termómetro que puede sujetarse a componentes metálicos por medio de una base magnética.

**Magnetic probe-type digital tachometer** A digital tachometer that reads engine rpm and uses a magnetic probe pickup.
**Tacómetro digital tipo sonda magnética** Tacómetro digital que lee las rpm del motor y utiliza una captación de sonda magnética.

**Magnetic probe-type digital timing meter** A digital reading that displays crankshaft degrees and uses a magnetic-type pickup probe mounted in the magnetic timing probe receptacle.
**Medidor de regulación digital tipo sonda magnética** Lectura digital que muestra los grados del cigüeñal y utiliza una sonda de captación tipo magnético montada en el receptáculo de la sonda de regulación magnética.

**Magnetic pulse generator** Sensor that uses the principle of magnetic induction to produce a voltage signal. Magnetic pulse generators are commonly used to send data concerning the speed of the monitored component to the computer.
**Generador de impulsos magnéticos** Sensor que funciona según el principio de inducción magnética para producir una señal de tensión. Los generadores de impulsos magnéticos se utilizan comúnmente para transmitir datos a la computadora relacionados a la velocidad del componente regulado.

**Magnetic sensor** A sensor that produces a voltage signal from a rotating element near a winding and a permanent magnet. This voltage signal is often used for ignition triggering.
**Sensor magnético** Sensor que produce una señal de tensión desde un elemento giratorio cerca de un devanado y de un imán permanente. Esta señal de tensión se utiliza con frecuencia para arrancar el motor.

**Magnetic timing offset** An adjustment to compensate for the position of the magnetic receptacle opening in relation to the TDC mark on the crankshaft pulley.
**Desviación de regulación magnética** Ajuste para compensar la posición de la abertura del receptáculo magnético de acuerdo a la marca TDC en la roldana del cigüeñal.

**Magnetic timing probe receptacle** An opening in which the magnetic timing probe is installed to check basic timing.
**Receptáculo de la sonda de regulación magnética** Abertura en la que está montada la sonda de regulación magnética para verificar la regulación básica.

**Magnetic wheel alignment gauge** A wheel alignment gauge that is held on each front hub with a magnet.
**Calibrador magnético para la alineación de ruedas** Calibrador para la alineación de ruedas que se fija a cada uno de los cubos delanteros con un imán.

**Magnetism** An energy form resulting from atoms aligning within certain materials, giving the materials the ability to attract other metals.
**Magnetismo** Forma de energía que resulta de la alineación de átomos dentro de ciertos materiales y que le da a éstos la capacidad de atraer otros metales.

**Main bearing** A bearing used as a crankshaft support.
**Cojinete del cigüeñal** Un cojinete que se emplea como un soporte para el cigüeñal.

**Main bearing clearance** The distance between the main bearing journal and the main bearings.
**Holgura del cojinete principal** La distancia entre el muñón del cigüeñal y los cojinetes principales del cigüeñal.

**Main bearing journal** The crankshaft journal that is supported by the main bearing.
**Muñón del cojinete principal** El muñón del cigüeñal apoyado por el cojinete del cigüeñal.

**Main bearing saddle bore** The housing that is machined to receive a main bearing.
**Taladro del asiento del cojinete principal** Un cárter se ha labrado a máquina para aceptar un cojinete de cigüeñal.

**Main menu** A display on a computer wheel aligner from which the technician selects various test procedures.
**Menú principal** Representación visual en una computadora para alineación de ruedas de la que el mecánico puede elegir varios procedimientos de prueba.

**Main oil pressure regulator valve** Regulates the line pressure in a transmission.
**Válvula reguladora de la línea de presión** Regula la presión en la línea de una transmisión.

**Mainline pressure** The hydraulic pressure that operates apply devices and is the source of all other pressures in an automatic transmission. It is developed by pump pressure and regulated by the pressure regulator.
**Línea de presión** La presión hidráulica que opera a los dispositivos de aplicación y es el orígen de todas las presiones en la transmisión automática. Proviene de la bomba de presión y es regulada por el regulador de presión.

**Major thrust surface** The side of the piston skirt that pushes against the cylinder wall during the power stroke.
**Superficie de empuje principal** El lado de la faldilla que empuja contra el muro del cilindro durante la carrera de fuerza es la superficie de empuje principal.

**Malfunction indicator light (MIL)** A light in the instrument panel that is illuminated by the PCM if certain defects occur in the computer system.
**Luz indicadora de funcionamiento defectuoso** Luz en el panel de instrumentos que el módulo del control del tren transmisor de potencia ilumina si ocurren ciertas fallas en la computadora.

**Manifold** A device equipped with a hand shutoff valve. Gauges are connected to the manifold for use in system testing and servicing.
**Múltiple** Dispositivo provisto de una válvula de cierre accionada a mano. Calibradores se conectan al múltiple para ser utilizados para llevar a cabo pruebas del sistema y para servicio.

**Manifold absolute pressure (MAP) sensor** An input sensor that sends a signal to the computer in relation to intake manifold vacuum.
**Sensor de la presión absoluta del colector** Sensor de entrada que le envía una señal a la computadora referente al vacío del colector de aspiración.

**Manifold and gauge set** A manifold complete with gauges and charging hoses.
**Conjunto del múltiple y calibrador** Múltiple provisto de calibradores y mangueras de carga.

**Manifold hand valve** Valves used to open and close passages through the manifold set.
**Válvula de distribución accionada a mano** Válvulas utilizadas para abrir y cerrar conductos a través del conjunto del múltiple.

**Manual control valve** A valve used to manually select the operating mode of the transmission. It is moved by the gearshift linkage.
**Válvula de control manual** Una válvula que se usa para escoger a una velocidad de la transmisión por mano. Se mueva por la biela de velocidades.

**Manufacturer** A person or company whose business is to produce a product or components for a product.
**Fabricante** Persona o empresa cuyo propósito es fabricar un producto o componentes para un producto.

**Manufacturer's procedures** Specific step-by-step instructions provided by the manufacturer for the assembly, disassembly, installation, replacement, and/or repair of a particular product manufactured by them.
**Procedimientos del fabricante** Instrucciones específicas a seguir paso por paso; dichas instrucciones son suministradas por el fabricante para montar, desmontar, instalar, reemplazar, y/o reparar un producto específico fabricado por él.

**Master cylinder** The liquid-filled cylinder in the hydraulic brake system or clutch, where hydraulic pressure is developed when the driver depresses a foot pedal.
**Cilindro primario** Cilindro lleno de líquido en el sistema de freno hidráulico o en el embrague, donde se produce presión hidráulica cuando el conductor oprime el pedal.

**Matched gear set code** Identification marks on two gears that indicate they are matched. They should not be mismatched with another gear set and placed into operation.
**Código del juego de engranaje emparejado** Señales de identificación en dos engranajes que indican que ambos hacen pareja. No deben emparejarse con otro juego de engranaje y ponerse en funcionamiento.

**Material Safety Data Sheet (MSDS)** A sheet which contains detailed information concerning hazardous materials. The MSDS must be maintained by the employer.
**Hoja de Dato de Seguridad de los Materiales (MSDS)** Una hoja que contiene la información detallada referente a los materiales peligrosos. El patrón debe conservar el MSDS.

**MAX** A mode, maximum, for heating or cooling. Selecting MAX generally overrides all other conditions that may have been programmed.
**MAX (Máximo)** Modo máximo para calentamiento o enfriamiento. El seleccionar MAX generalmente anula todas las otras condiciones que pueden haber sido programadas.

**Maxi-fuse** A circuit protection device that looks similar to blade-type fuses except they are larger and have a higher amperage capacity. Maxi-fuses are used because they are less likely to cause an underhood fire when there is an overload in the circuit. If the fusible link burns in two, it is possible that the "hot" side of the fuse could come into contact with the vehicle frame and the wire could catch on fire.
**Maxifusible** Dispositivo de protección del circuito parecido a un fusible de tipo de cuchilla, pero más grande y con mayor capacidad de amperaje. Se utilizan maxifusibles porque existen menos probabilidades de que ocasionen un incendio debajo de la capota cuando ocurra una sobrecarga en el circuito. Si el cartucho de fusible se quemase en dos partes, es posible que el lado "cargado" del fusible entre en contacto con el armazón del vehículo y que el alambre se encienda.

**Mechanical efficiency** A comparison of the power actually delivered by the crankshaft to the power developed within the cylinders at the same rpm.
**Rendimiento mecánica** Una comparación entre la energía que actualmente entrega el cigüeñal y la energía desarrollado dentro de los cilindros en la misma rpm.

**Mechanical valve lifters, or solid tappets** Round, cylindrical, metal components mounted between the camshaft lobes and the pushrods to open the valves.
**Desmontaválvulas mecánicas, o alzaválvulas sólidas** Componentes metálicos, cilíndricos y redondos montados entre los lóbulos del árbol de levas y las varillas de empuje para abrir las válvulas.

**Memory calibrator (MEM-CAL)** A removable chip in some computers that replaces the PROM and CAL-PAK chips.
**Calibrador de memoria** Pastilla desmontable en algunas computadoras que reemplaza las pastillas PROM y CAL-PAK.

**Memory steer** The tendency of the vehicle steering not to return to the straight-ahead position after a turn, but to keep steering in the direction of the turn.
**Dirección de memoria** Tendencia de la dirección del vehículo a no regresar a la posición de línea recta después de un viraje, sino a continuar girando en el sentido del viraje.

**Meshing** The mating, or engaging, of the teeth of two gears.
**Engrane** Embragar o endentar a los dientes de dos engranajes.

**Meter** 1/10,000,000 of the distance from the North Pole to the Equator, or 39.37 inches.
**Metro** Un 1/10,000,000 de la distancia del polo del norte al ecuador.

**Meter impedance** The total internal electrical resistance in a meter.
**Impedancia de un medidor** La resistencia eléctrica interna total en un medidor.

**Metering valve** A component that momentarily delays the application of front disc brakes until the rear drum brakes begin to move. Helps to provide balanced braking.
**Válvula de medición** Un componente que retrasa momentáneamente la aplicación de los frenos de disco delanteras hasta que comienzan a moverse los frenos de tambor traseros. Ayuda en proporcionar el enfrenado más equilibrado.

**Metric fastener** Any type fastener with metric size designations, numbers or millimeters.
**Asegurador métrico** Cualquier asegurador provisto de indicaciones, números o milimetros.

**Microinch** One millionth of an inch. The microinch is the standard measurement for surface finish in the American customary system.
**Micropulgada** La millonésima parte de una pulgada. Una micropulgada es una unidad común para el acabado de la superficie en el sistema de medida americana.

**Micrometer** One millionth of a meter. The micrometer is the standard measurement for surface finish in the metric system.
**Micrómetro** Una millonésima parte de un metro. La medida común para el acabado de la superficie en el sistema métrica.

**Micrometer** Precision measuring instrument designed to measure outside, inside, or depth measurements.
**Micrómetro** Un instrumento de medidas precisas diseñado para tomar medidas exteriores, interiores o de profundidad.

**Micron** A thousandth of a millimeter or about .0008 inch.
**Micrón** La milésima parte de un milímetro o aproximadamente un .0008 de una pulgada.

**Mid-positioned** The position of a stem-type service valve where all fluid passages are interconnected. Also referred to as *cracked*.
**Ubicación central** Posición de una válvula de servicio de tipo vástago donde todos los pasajes que conducen fluidos se interconectan.

**Millimeter (mm)** 0.001 meter or 0.03937 inches.
**Milímetro (mm)** 0,001 de un metro o 0,03937 de una pulgada.

**Minor thrust side** The side opposite the side of the rod that is stressed during the power stroke.
**Lado de empuje menor** El lado opuesto del lado cuyo biela recibe el esfuerzo durante la carrera de potencia.

**Minor thrust surface** The area of the piston skirt that pushes against the cylinder wall during the compression stroke.
**Superficie de empuje menor** El faldón del pistón que empuja contra el muro del cilindro durante la carrera de compresión.

**Misalignment** When bearings are not on the same centerline.
**Desalineamiento** Cuando los cojinetes no comparten la misma línea central.

**Modulator** A vacuum-diaphragm device connected to a source of engine vacuum. It provides an engine load signal to the transmission.
**Modulador** Un dispositivo de diafragma de vacío que se conecta a un orígen de vacío en el motor. Provee un señal de carga del motor a la transmisión.

**Molded connector** An electrical connector that usually has one to four wires that are molded into a one-piece component.
**Conectador moldeado** Conectador eléctrico que por lo general tiene hasta un máximo cuatro alambres que se moldean en un componente de una sola pieza.

**Molybdenum disulphide lithium-based grease** A special lubricant containing molybdenum disulphide and lithium that may be used on some steering components.
**Grasa de bisulfuro de molibdeno con base de litio** Lubricante especial compuesto de bisulfuro de molibdeno y litio que puede utilizarse en algunos componentes de la dirección.

**Motor** An electrical device that produces a continuous turning motion. A motor is used to propel a fan blade or a blower wheel.
**Motor** Dispositivo eléctrico que produce un movimiento giratorio continuo. Se utiliza un motor para impeler las aletas del ventilador o la rueda del soplador.

**Mounts** Made of rubber to insulate vibrations and noise while they support a powertrain part, such as engine or transmission mounts.
**Monturas** Hecho de hule para insular a las vibraciones y a los ruidos mientras que sujetan una parte del tren de propulsión, tal como las monturas del motor o las monturas de la transmisión.S

**Mounting boss** See *Flange*.
**Protuberancia de montaje** Ver *Flange* [*Brida*].

**Mounting flange** See *Flange*.
**Brida de montaje** Ver *Flange* [*Brida*].

**MSDS** Material Safety Data Sheet.
**MSDS** Hojas de información sobre la seguridad de un material.

**Muffler chisel** A chisel that is designed for cutting muffler inlet and outlet pipes.
**Cincel para silenciadores** Cincel diseñado para cortar los tubos de entrada y salida del silenciador.

**Multiple disc** A clutch with a number of driving and driven discs as compared to a single plate clutch.
**Discos múltiples** Un embrague que tiene varios discos de propulsión o de arraste al contraste con un embrague de un sólo plato.

**Multiport fuel injection (MFI)** A fuel injection system in which the injectors are grounded in the computer in pairs or groups of three or four.
**Inyección de combustible de paso múltiple** Sistema de inyección de combustible en el que los inyectores se ponen a tierra en la computadora en pares o en grupos de tres o cuatro.

**Multipull unitized body straightening system** A hydraulically operated system that pulls in more than one location when straightening unitized bodies.
**Sistema de tiro múltiple para enderezar la carrocería unitaria** Sistema activado hidráulicamente que tira hacia más de una dirección al enderezar carrocerías unitarias.

**Mushroomed** A condition caused by pounding of a punch or a chisel, producing a mushroom-shaped end which should be ground off to ensure maximum safety.
**Hinchado** Condición ocasionada por el golpeo de un punzón o cincel, lo cual hace que el extremo vuelva en forma de un hongo y que debe ser afilado para asegurar máxima seguridad.

**National Institute for Automotive Service Excellence (ASE)** An organization that provides voluntary automotive technician certification in eight areas of expertise.
**Instituto Nacional para la excelencia en la reparación de automóviles** Organización que provee una certificación voluntaria de mecánico de automóviles en ocho áreas diferentes de especialización.

**Necking** A valve stem defect where the stem narrows near the head.
**Vástago ahusado** Un defecto del vástago de válvula en el cual el vástago se adelgaza cerca de la cabeza del cilindro.

**Needle bearing** An antifriction bearing using a great number of long, small-diameter rollers. Also known as a quill bearing.
**Rodamiento de agujas** Un rodamiento (cojinete) antifricativo que emplea una gran cantidad de rodillos largos y de diámetro muy pequeños.

**Needle deflection** Distance of travel from zero of the needle on a dial gauge.
**Desviación de la aguja** La distancia del cero que viaja una aguja de un indicador.

**Negative back-pressure EGR valve** An EGR valve containing a vacuum bleed valve that is operated by negative pulses in the exhaust.
**Válvula EGR de contrapresión negativa** Válvula EGR que contiene una válvula de descarga de vacío accionada por impulsos negativos en el escape.

**Neoprene** A synthetic rubber that is not affected by the various chemicals that are harmful to natural rubber.
**Neoprene** Un hule sintético que no se afecta por los varios productos químicos que pueden dañar al hule natural.

**Net valve lift** The actual amount a valve lifts off its seat. It is found by subtracting the lash specification and amount of component deflection from the gross valve lift.
**Producto neto del levantamiento de la válvula** La cantidad actual que se levanta una válvula de su asiento. Para determinarlo se substrae la especificación del juego de las válvulas más la cantidad de desviación del componente del total bruto de la cantidad del levantamiento de la válvula.

**Net weight** The weight of a product only; container and packaging not included.
**Peso neto** Peso de sólo el producto mismo; no incluye el recipiente y encajonamiento.

**Neutral** In a transmission, the setting in which all gears are disengaged and the input shaft is disconnected from the drive wheels.
**Neutral** En una transmisión, la velocidad en la cual todos los engranajes están desembragados y el árbol de salida está desconectada de las ruedas de propulsión.

**Neutral/drive switch (NDS)** A switch that sends a signal to the computer in relation to a gear selector position.
**Conmutador de mando neutral** Conmutador que le envía una señal a la computadora referente a la posición del selector de velocidades.

**Neutral-start switch** A switch wired into the ignition switch to prevent engine cranking unless the transmission shift lever is in neutral or the clutch pedal is depressed.

**Interruptor de arranque en neutral** Un interruptor eléctrico instalado en el interruptor de encendido que previene el arranque del motor al menos de que la palanca de cambio de velocidad esté en una posición neutral o que se pisa en el embrague.

**Newton-meter (Nm)** Metric measurement of torque or twisting force.
**Metro newton (Nm)** Una medida métrica de la fuerza de torsión.

**No-crank** A term used to mean that when the ignition switch is placed in the START position, the starter does not turn the engine.
**Sin arranque** Término utilizado para expresar que cuando el botón conmutador de encendido está en la posición START, el arrancador no enciende el motor.

**No-crank test** Diagnostic test performed to locate any opens in the starter or control circuits.
**Prueba sin arranque** Prueba diagnóstica llevada a cabo para localizar aberturas en los circuitos de arranque o de mando.

**Noise** Any unwanted or annoying sound.
**Ruido** Cualquier sonido perturbador o molesto.

**Nominal shim** A shim with a designated thickness.
**Laminilla fina** Una cuña de un espesor especificado.

**Noncycling clutch** An electro-mechanical compressor clutch that does not cycle on and off as a means of temperature control. It is used to turn the system on when cooling is desired, and off when cooling is not desired.
**Embrague sin funcionamiento cíclico** Embrague electromecánico del compresor que no se enciende y se apaga como medio de regular la temperatura; se utiliza para arrancar el sistema cuando se desea enfriamiento y para detener el sistema cuando no se desea enfriamiento.

**Nonhardening** A gasket sealer that never hardens.
**Sinfragua** Un cemento de empaque que no endurece.

**Normally closed (NC) switch** A switch designation denoting that the contacts are closed until acted upon by an outside force.
**Conmutador normalmente cerrado** Nombre aplicado a un conmutador cuyos contactos permanecerán cerrados hasta que sean accionados por una fuerza exterior.

**Normally open (NO) switch** A switch designation denoting that the contacts are open until acted upon by an outside force.
**Conmutador normalmente abierto** Nombre aplicado a un conmutador cuyos contactos permanecerán abiertos hasta que sean accionados por una fuerza exterior.

**Nose switch** A switch in the IAC motor stem that informs the computer when the throttle is in the idle position.
**Conmutador de contacto de la marcha lenta** Conmutador en el vástago del motor IAC que le advierte a la computadora cuándo la mariposa está en la posición de marcha lenta.

**Nucleate boiling** The process of maintaining the overall temperature of a coolant to a level below its boiling point, but allowing the portions of the coolant actually contacting the surfaces (the nuclei) to boil into a gas.
**Ebullición nucleido** El proceso de mantener un nivel de temperatura general de un fluido refrigerante menos de la de su punto de ebullición, pero permitiendo que las porciones del fluido refrigerante que actualmente están en contacto con las superficies (el nucleido) hiervan para formar un gas.

**Nut** A removable fastener used with a bolt to lock pieces together; made by threading a hole through the center of a piece of metal that has been shaped to a standard size.
**Tuerca** Un retén removible que se usa con un perno o tuerca para unir a dos piezas; se fabrica al filetear un hoyo taladrado en un pedazo de metal que se ha formado a un tamaño especificado.

**OBD II** See *On-board diagnostics II*.
**OBD II** Véase Diagnósticos a bordo II.

**Observe** To see and note; to perceive; notice.
**Observar** Ver y anotar; percibir; fijarse en algo.

**Occupational safety glasses** Eye protection device designed with special high-impact lenses and frames and side protection.
**Lentes de seguridad** Un dispositivo de protección para los ojos diseñado con los cristales y el armazón resistentes a los impactos fuertes, y que provee protección a los lados de los ojos.

**OD** Outside diameter.
**DE** Diámetro exterior.

**Odometer** A mechanical counter in the speedometer unit that indicates total miles accumulated on the vehicle.
**Odómetro** Aparato mecánico en la unidad del velocímetro con el que se cuentan las millas totales recorridas por el vehículo.

**OEM** Original Equipment Manufacturer.
**OEM** Fabricante Original del Equipo.

**Off-square** When the seat and valve stem are not properly aligned, they are off-square. This condition causes the valve stem to flex as the valve face is forced into the seat by spring and combustion pressures.
**Fuera de escuadra** Refiere a que el asiento y el vástago de la válvula no estén alineadas correctamente. Esta condición causa que el vástago de la válvula se dobla cuando la cara de la válvula se asienta bajo la fuerza de las presiones del resorte y la combustión.

**Off-the-road** Generally refers to vehicles that are not licensed for road use, such as harvesters, bulldozers, and so on.
**Fuera de carretera** Generalmente se refiere a vehículos que no son permitidos operar en la carretera, como por ejemplo cosechadoras, rasadoras, etcétera.

**Ohm** Unit of measure for resistance. One ohm is the resistance of a conductor such that a constant current of one ampere in it produces a voltage of one volt between its ends.
**Ohmio** Unidad de resistencia eléctrica. Un ohmio es la resistencia de un conductor si una corriente constante de 1 amperio en el conductor produce una tensión de 1 voltio entre los dos extremos.

**Ohmmeter** A test meter used to measure resistance and continuity in a circuit.
**Ohmiómetro** Instrumento de prueba utilizado para medir la resistencia y la continuidad en un circuito.

**Ohm's law** Defines the relationship between current, voltage, and resistance.
**Ley de Ohm** Define la relación entre la corriente, la tensión y la resistencia.

**Oil breakdown** Condition of oil which results from high temperatures for extended periods of time. The oil will combine with oxygen and can cause carbon deposits in the engine.
**Deterioro del aceite** Un resultado de las temperaturas elevadas por largos periodos de tiempo. El aceite combinará con el oxígeno y puede causar los depósitos del carbono en el motor.

**Oil clearance** The difference between the inside bearing diameter and the journal diameter.
**Holgura de aceite** La diferencia entre el diámetro interior de un cojinete y el diámetro del muñón.

**Oil gallery** The main oil supply line in the engine block.
**Canalización de aceite** La línea principal del suministro de aceite en el monoblock.

**Oil pan gaskets** Gaskets used to prevent leakage from the crankcase at the connection between the oil pan and the engine block.
**Empaque de la tapa del cárter** Los empaques que se emplean para prevenir las fugas del cárter en la conexión entre el colector de aceite y el monoblock.

**Oil pressure gauge** A gauge used to test engine oil pressure.
**Manómetro de la presión del aceite** Calibrador utilizado para revisar la presión del aceite del motor.

**Oil pressure test** Test to determine the condition of the bearings and other internal engine components.
Prueba de presión de aceite Una prueba que se efectua para determinar la condición de los cojinetes u otros componentes interiores del motor.

**Oil pump** A rotor- or gear-type positive displacement pump used to take oil from the sump and deliver it to the oil galleries. The oil galleries direct the oil to the high wear areas of the engine.
Bomba de aceite Un rotor o una bomba de tipo desplazamiento positivo de engrenajes que sirve para tomar el aceite de un suministro y entregarlo a las canalizaciones de aceite. Las canalizaciones de aceite dirigen el aceite a las áreas del motor que imponen gastos severos.

**Oil seal** A seal placed around a rotating shaft or other moving part to prevent leakage of oil.
Empaque de aceite Un empaque que se coloca alrededor de una flecha giratoria para prevenir el goteo de aceite.

**On-board diagnostics II (OBD II)** An engine computer system mandated on all vehicles beginning in 1996. The powertrain control module (PCM) must be able to monitor specific functions and systems such as engine misfire, the catalytic converter, the exhaust gas recirculation (EGR) system, the evaporative (EVAP) system, fuel control, heated oxygen sensors (HO$_2$S), and HO$_2$S heaters.
Diagnósticos a bordo II Sistema de computadora del motor mandado para todos vehículos desde el 1996. La unidad de control del tren transmisor de potencia debe poder monitorizar funciones y sistemas específicos tales como fallos del motor, el convertidor catalítico, el sistema de recirculación del gas del escape, el sistema de evaporación, el control del combustible y los sensores térmicos de oxígeno y de los calentadores.

**One-piece valves** A valve stem design where the head is not welded to the stem. In a two-piece valve, the head is welded to the stem.
Válvulas de una pieza Un diseño del vástago de válvula que no tiene una cabeza soldada al vástago. Una válvula de dos piezas tiene una cabeza soldada al vástago.

**One-way clutch** See *Sprag clutch*.
Embrague de una vía Vea *Sprag clutch*.

**Open circuit** A term used to indicate that current flow is stopped. By opening the circuit, the path for electron flow is broken.
Circuito abierto Término utilizado para indicar que el flujo de corriente ha sido detenido. Al abrirse el circuito, se interrumpe la trayectoria para el flujo de electrones.

**Open circuit voltage test** Test to determine the battery's state of charge. It is used when a hydrometer is not available or cannot be used.
Prueba de voltaje de circuito abierto Una prueba para determinar el estado de carga de la batería. Se usa cuando no es disponible o no se puede usar un hidrómetro.

**Open loop** A computer operating mode in which the computer controls the air-fuel ratio and ignores the oxygen sensor signal.
Bucle abierto Modo de funcionamiento de una computadora en el que se controla la relación de aire y combustible y se pasa por alto la señal del sensor de oxígeno.

**Open pressure** Spring tension when the valve spring is compressed and the valve is fully open.
Presión abierta La tensión de un resorte de la válvula al estar comprimido con la válvula completamente abierta.

**Operating angle** The difference between the drive shaft and transmission installation angles is the operating angle.
Ángulo de funcionamiento El ángulo de funcionamiento es la diferencia entre los ángulos del montaje del árbol de mando y de la transmisión.

**Opposed cylinder engine** Engine block design in which the cylinders are across from each other. Also referred to as horizontally opposed or "pancake" engines.

Motor de cilindros opuestos Un diseño de monoblock que coloca los cilindros en lados opuestos. Tambien conocido con el nombre motores de cilindros opuestos horizontalmente o motores achatados.

**Optical toe gauge** A piece of equipment that uses a light beam to measure front wheel toe.
Calibrador óptico del tope Equipo que utiliza un rayo de luz para medir el tope de la rueda delantera.

**Optical-type pickup** A pickup that contains a photo diode and a lightemitting diode with a slotted plate between these components.
Captación tipo óptico Captación que contiene un fotodiodo y un diodo emisor de luz entre los cuales está colocada una placa ranurada.

**Orifice** A small hole. A calibrated opening in a tube or pipe to regulate the flow of a fluid or liquid.
Orificio Agujero pequeño. Apertura calibrada en un tubo o cañería para regular el flujo de un fluido o de un líquido.

**O-ring** A type of sealing ring, usually made of rubber or a rubber-like material. In use, the O-ring is compressed into a groove to provide the sealing action.
Anillo en O Un tipo de sello anular, suele ser hecho de hule o de una materia parecida al hule. Al usarse, el anillo en O se comprime en una ranura para proveer un sello.

**Oscillate** To swing back and forth like a pendulum.
Oscilar Moverse alternativamente en dos sentidos contrarios como un péndulo.

**Oscilloscope** A cathode ray tube (CRT) that displays voltage waveforms from the ignition system.
Osciloscopio Tubo de rayos catódicos que muestra formas de onda de tensión provenientes del sistema de encendido.

**OSHA** Occupational Safety and Health Administration.
OSHA Dirección para la Seguridad y Salud Industrial.

**Outboard constant velocity joint** The outer constant velocity joint, or the one closest to the wheels. The outer joint is a fixed joint.
Junta de velocidad constante fuera de borde Junta de velocidad constante exterior, o la que está más cerca de las ruedas. La junta exterior es una junta fija.

**Outer bearing race** The outer part of a bearing assembly on which the balls or rollers rotate.
Pista exterior de un cojinete La parte exterior de una asamblea de cojinetes en la cual ruedan las bolas o los rodillos.

**Out-of-round** The condition when measurements of a diameter differ at different locations. The term out-of-round applies to inside or outside diameters.
Ovulado Una condición en la cual las medidas de un diámetro varían en lugares distintos. El término ovulado aplica a los diámetros interiores o exteriores.

**Out-of-round gauges** Instruments used to measure the concentricity of connecting rod bores.
Calibradores de ovulado Los instrumentos para medir la concentricidad de los taladros de las bielas.

**Output shaft** The shaft or gear that delivers the power from a device, such as a transmission.
Flecha de salida La flecha o la velocidad que transmite la potencia de un dispositivo, tal como una transmisión.

**Output state test** A computer system test mode on Ford products that turns the relays and actuators on and off.
Prueba del estado de producción Modo de prueba en una computadora de productos fabricados por la Ford que enciende y apaga los relés y los accionadores.

**Outside micrometers** Tool designed to measure the outside diameter or thickness of a component.

**Micrómetros exteriores** Una herramienta diseñada para medir los diámetros exteriores o el espesor de un componente.

**Outside temperature sensor** See *Ambient sensor*.
**Sensor de la temperatura ambiente** Ver *Ambient sensor* [*Sensor ambiente*].

**Overall ratio** The product of the transmission gear ratio multiplied by the final drive or rear axle ratio.
**Relación global** El producto de multiplicar la relación de los engranajes de la transmisión por la relación del impulso final o por la relación del eje trasero.

**Overcenter spring** A heavy coil spring arrangement in the clutch linkage to assist the driver with disengaging the clutch and returning the clutch linkage to the full engagement position.
**Muelle sobrecentro** Distribución de muelles helicoidales gruesos en la biela motriz del embrague para ayudar al conductor a desengranar el embrague y devolver la biela motriz del embrague a la posición de enganche total.

**Overcharge** Indicates that too much refrigerant or refrigeration oil is added to the system.
**Sobrecarga** Indica que una cantidad excesiva de refrigerante o aceite de refrigeración ha sido agregada al sistema.

**Overdrive** Any arrangement of gearing that produces more revolutions of the driven shaft than of the driving shaft.
**Sobremultiplicación** Un arreglo de los engranajes que produce más revoluciones de la flecha de arrastre que los de la flecha de propulsión.

**Overdrive ratio** Identified by the decimal point indicating less than one driving input revolution compared to one output revolution of a shaft.
**Relación del sobremultiplicación** Se identifica por el punto decimal que indica menos de una revolución del motor comparado a una revolución de una flecha de salida.

**Overhang** The area of the face between the seat contact and the margin.
**Sobresaliente** El área de la cara entre el contacto del asiento y el margen.

**Overhead hoist** A lifting tool that uses a chain fall or electric motor to hoist the engine out of the hood opening. The hoist can be attached to a movable A-frame or on an I-beam across the shop ceiling.
**Grúa (montacarga) en alto** Una herramienta de izar que utiliza una caída de cadena o un motor eléctrico para levantar el motor por la apertura del capó. El aparato de izar puede conectarse a un armazón en forma de A o a un hierro en T a través del techo del taller.

**Overhead valve engine** An engine with the camshaft located in the engine block and the valves in the cylinder head.
**Motor con válvulas en cabeza** Un motor que tiene el árbol de levas ubicado en el monoblock y las válvulas en la cabeza del cilindro.

**Overload** Excess current flow in a circuit.
**Sobrecarga** Flujo de corriente superior a la que tiene asignada un circuito.

**Overrun coupling** A free-wheeling device to permit rotation in one direction but not in the other.
**Acoplamiento de sobremarcha** Un dispositivo de marcha de rueda libre que permite las giraciones en una dirección, pero no en la otra dirección.

**Overrunning clutch** A device consisting of a shaft or housing linked together by rollers or sprags operating between movable and fixed races. As the shaft rotates, the rollers or sprags jam between the movable and fixed races. This jamming action locks together the shaft and housing. If the fixed race should be driven at a speed greater than the movable race, the rollers or sprags will disconnect the shaft.
**Embrague de sobremarcha** Un dispositivo que consiste de una flecha o un cárter eslabonados por medio de rodillos o palancas de detención que operan entre pistas fijas y movibles. Al girar la flecha, los rodillos o palancas de detención se aprietan entre las pistas fijas y movibles. Este acción de apretarse enclava el cárter con la flecha. Si la pista fija se arrastra en una velocidad más alta que la pista movible, los rodillos o palancas de detención desconectarán a la flecha.

**Oversize bearings** Bearings that are thicker than standard to increase the outside diameter of the bearing to fit an oversize bearing bore. The inside diameter is the same as standard bearings.
**Cojinete de medidas superiores** Los cojinetes que son de un espesor más grueso de lo normal para aumentar el diámetro exterior del cojinete para quedarse en un taladro que rebasa la medida. El diámetro interior es lo mismo del cojinete normal.

**Oversteer** The tendency of a vehicle to turn sharper than the turn selected by the driver.
**Sobreviraje** Tendencia de un vehículo a girar más de lo que el conductor desea.

**Oxidation** Burning or combustion; the combining of a material with oxygen. Rusting is slow oxidation, and combustion is rapid oxidation.
**Oxidación** Quemando o la combustión; la combinación de una materia con el oxígeno. El orín es una oxidación lenta, la combustión es la oxidación rápida.

**Oxygen ($O_2$)** A gaseous element that is present in air.
**Oxígeno ($O_2$)** Elemento gaseoso presente en el aire.

**Oxygen ($O_2$) sensor** A sensor mounted in the exhaust system that sends a voltage signal to the computer in relation to the amount of oxygen in the exhaust stream.
**Sensor de oxígeno ($O_2$)** Sensor montado en el sistema de escape que le envía una señal de tensión a la computadora referente a la cantidad de oxígeno en el caudal del escape.

**Ozone friendly** Any product that does not pose a hazard or danger to the ozone.
**Sustancia no dañina al ozono** Cualquier producto que no es peligrosa o amenaza al ozono.

**Parallel** The quality of two items being the same distance from each other at all points; usually applied to lines and, in automotive work, to machined surfaces.
**Paralelo** La calidad de dos artículos que mantienen la misma distancia el uno del otro en cada punto; suele aplicarse a las líneas y, en el trabajo automotívo, a las superficies acabadas a máquina.

**Parallel circuit** A circuit that provides two or more paths for electricity to flow.
**Circuito en paralelo** Circuito que provee dos o más trayectorias para que circule la electricidad.

**Parasitic loads** Electrical loads that are still present when the ignition switch is in the OFF position.
**Cargas parásitas** Cargas eléctricas que todavía se encuentran presente cuando el botón conmutador de encendido está en la posición OFF.

**Park** Generally refers to a component or mechanism that is at rest.
**Reposo** Generalmente se refiere a un componente o mecanismo que no está funcionando.

**Park/neutral switch** A switch connected in the starter solenoid circuit that prevents starter operation except in park or neutral.
**Conmutador PARK/neutral** Conmutador conectado en el circuito del solenoide del arranque que evita el funcionamiento del arranque si el selector de velocidades no se encuentra en las posiciones PARK o NEUTRAL.

**Park switch** Contact points located inside the wiper motor assembly that supply current to the motor after the wiper control switch has been turned to the PARK position. This allows the motor to continue operating until the wipers have reached their PARK position.
**Conmutador PARK** Puntos de contacto ubicados dentro del conjunto del motor del frotador que le suministran corriente al motor después de que el conmutador para el control de los frotadores haya sido colocado en la posición PARK. Esto permite que el motor continue su funcionamiento hasta que los frotadores hayan alcanzado la posición original.

**Parking brake** A mechanically operated brake that's used to hold the vehicle when it is parked.
Freno de estacionamiento (emergencia) Un freno que se opera a mano para sostener al vehículo al estacionarse.

**Parking brake strut** A bar between the brake shoes. When the parking lever is actuated, the parking brake strut pushes the leading brake shoe into the drum.
Poste del freno de estacionamiento Una barra entre las zapatas de freno. Al actuarse la palanca de estacionamiento, el poste del freno de estacionamiento empuje la zapata delantera contra el tambor.

**Parts per million (ppm)** The volume of a gas such as hydrocarbons in ppm in relation to one million parts of the total volume of exhaust gas.
Partes por millón (ppm) Volumen de un gas, como por ejemplo los hidrocarburos, en partes por millón de acuerdo a un millón de partes del volumen total del gas del escape.

**Parts washers** Parts washers generally use a mild solvent to soak the components. Some provide for agitation and spraying of the solvent.
Lavadora de partes Las lavadoras de partes generalmente remojan los componentes en un solvente débil. Algunos proveen la agitación y rocían el solvente.

**Pascal's law** The law of fluid motion.
Ley de pascal La ley del movimiento del fluido.

**Passive seat belt system** Seat belt operation that automatically puts the shoulder and/or lap belt around the driver or occupant. The automatic seat belt is moved by DC motors that move the belts by means of carriers on tracks.
Sistema pasivo de cinturones de seguridad Función de los cinturones de seguridad que automáticamente coloca el cinturón superior y/o inferior sobre el conductor o pasajero. Motores de corriente continua accionar los cinturones automáticos mediante el uso de portadores en pistas.

**Pawl** A lever that pivots on a shaft. When lifted, it swings freely and when lowered, it locates in a detent or notch to hold a mechanism stationary.
Trinquete Una palanca que gira en una flecha. Levantado, mueve sín restricción, bajado, se coloca en una endentación o una muesca para mantener sín movimiento a un mecanismo.

**PCM** Power train control module.
PCM Módulo regulador del transmisor de potencia.

**Pedal play** The distance the clutch pedal and release bearing assembly move from the fully engaged position to the point where the release bearing contacts the pressure plate release levers.
Holgura del pedal Distancia a la que el pedal del embrague y el conjunto del cojinete de desembrague se mueven de una posición enteramente engranada a un punto donde el cojinete de desembrague entra en contacto con las palancas de desembrague de la placa de presión.

**Peen** To stretch or clinch over by pounding.
Martillazo Estirar o remachar por machacado.

**Peening** The process of removing stress in a metal by striking it.
Martillar El proceso de quitar la fatiga de un metal golpéandolo.

**Performance test** Readings of the temperature and pressure under controlled conditions to determine if an air-conditioning system is operating at full efficiency.
Prueba de rendimiento Lecturas de la temperatura y presión bajo condiciones controladas para determinar si un sistema de acondicionamiento de aire funciona a un rendimiento completo.

**pH** A value expressing acidity or basicity in terms of the relative amounts of hydrogen ions (H+) and hydroxide ions (OH−) present in a solution.
pH Un valor que expresa la acidez o lo básico en términos de las cantidades relativas de los iones de hidrógeno (H+) y los iones hidróxidos (OH−) presentes.

**Phasing** Rotational position of the universal joints on the drive shaft.
Fasaje Posición de rotación de las juntas universales sobre el árbol de mando.

**Phosgene gas** The gas formed when R-12 comes in contact with heat. It is poisonous and will make you sick or fatally ill.
Gas fosgeno Gas formado por la mezcla del R-12 y el calor. Sustancia venenosa que produce graves trastornos o hasta la muerte.

**Photocell** A variable resistor that uses light to change resistance.
Fotocélula Resistor variable que utiliza luz para cambiar la resistencia.

**Photoelectric tachometer** A tachometer that contains an internal light source and a photoelectric cell. This meter senses rpm from reflective tape attached to a rotating component.
Tacómetro fotoeléctrico Tacómetro que contiene una fuente interna de luz y una célula fotoeléctrica. Este medidor advierte las rpm mediante una cinta reflectora adherida a un componente giratorio.

**Phototransistor** A transistor that is sensitive to light.
Fototransistor Transistor sensible a la luz.

**Photovoltaic diodes** Diodes capable of producing a voltage when exposed to radiant energy.
Diodos fotovoltaicos Diodos capaces de generar una tensión cuando se encuentran expuestos a la energía de radiación.

**Pick-up coil** The stationary component of the magnetic pulse generator consisting of a weak permanent magnet that is wound around by fine wire. As the timing disc rotates in front of it, the changes of magnetic lines of force generate a small voltage signal in the coil.
Bobina captadora Componente fijo del generador de impulsos magnéticos compuesta de un imán permanente débil devanado con alambre fino. Mientras gira el disco sincronizador enfrente de él, los cambios de las líneas de fuerza magnética generan una pequeña señal de tensión en la bobina.

**Pick-up tube** A tube used by the oil pump to deliver oil from the bottom of the oil pan. The bottom of the tube has a screen to filter larger contaminates.
Tubo de captación Un tubo empleado por la bomba de aceite para entregar el aceite del fondo del colector de aceite. La parte inferior del tubo tiene una rejilla para filtrar los contaminantes más grandes.

**Piercing pin** The part of a saddle valve that is used to pierce a hole in the tubing.
Pasador perforador Parte de la válvula de silleta utilizada para perforar un agujero en la tubería.

**Piezoresistive sensor** A sensor that is sensitive to pressure changes.
Sensor piezoresistivo Sensor susceptible a los cambios de presión.

**Pilot bearing** A small bearing, such as in the center of the flywheel end of the crankshaft, which carries the forward end of the clutch shaft.
Cojinete piloto Cojinete pequeño, como por ejemplo, el ubicado en el centro del extremo del volante del cigüeñal, que soporta el extremo delantero del árbol del embrague.

**Pilot bushing** A plain bearing fitted in the end of a crankshaft. The primary purpose is to support the input shaft of the transmission.
Buje piloto Cojinete sencillo insertado en el extremo de un cigüeñal. El propósito principal es apoyar el árbol impulsor de la transmisión.

**Pilot shaft** A shaft used to align parts and that is removed before final installation of the parts; a dummy shaft.
Árbol piloto Árbol que se utiliza para alinear piezas y que se remueve antes del montaje final de las mismas; árbol falso.

**Pin boss** A bore machined into the piston that accepts the piston pin to attach the piston to the connecting rod.
Mamelón Un taladro tallado a máquina en el pistón que acomoda la espiga del pistón para conectarlo a la biela.

**Pinging noise** A shop term for engine detonation that sounds like a rattling noise in the engine cylinders.

**Sonido agudo** Término utilizado en el taller mecánico para referirse a la detonación del motor cuyo ruido se asemeja a un estrépito en los cilindros de un motor.

**Pinion and bearing assembly** The pinion gear and supporting bearings in a rack and pinion steering gear.
**Conjunto del piñón y cojinete** Engranaje del piñón y cojinetes de soporte en un mecanismo de dirección de cremallera y piñón.

**Pinion carrier** The mounting or bracket that retains the bearings supporting a pinion shaft.
**Portador de piñón** Montaje o soporte que sujeta los cojinetes que apoyan un árbol del piñón.

**Pinion gear** The smaller of two meshing gears.
**Engranaje de piñón** El más pequeño de los dos engranajes de engrane.

**Pinning** A cold crack repair process that avoids altering the characteristics of the metal. This process uses a series of tapered iron pins threaded into overlapping holes along the length of the crack.
**Chavetear** Un proceso de reparar sin calor a las grietas que no cambia las características del metal. Este proceso usa una serie de espárragos cónicos fileteados en los agujeros extendidos uno sobre otro por toda la longitud de la grieta.

**Pinpoint tests** Specific test procedures to locate the exact cause of a fault code in a computer system.
**Pruebas llevadas a cabo con precisión** Procedimientos de prueba específicos que se llevan a cabo para localizar la causa exacta de un código de fallo en un sistema informático.

**Pin-type connector** A single or multiple electrical connector that is round- or pin-shaped and fits inside a matching connector.
**Conectador de tipo pasador** Conectador eléctrico único o múltiple en forma redonda o en forma de pasador que se inserta dentro de un conectador emparejado.

**Pipe expander** A tool designed to expand exhaust system pipes.
**Expansor de tubo** Herramienta diseñada para expandir los tubos del sistema de escape.

**Piston** An engine component in the form of a hollow cylinder that is enclosed at the top and open at the bottom. Combustion forces are applied to the top of the piston to force it down. The piston, when assembled to the connecting rod, is designed to transmit the power produced in the combustion chamber to the crankshaft.
**Pistón** Un componente del motor que consiste de un cilindro hueco cerrado en la parte de arriba y abierto en la parte de abajo. Las fuerzas de combustión se aplican en la parte superior del pistón para forzarlo hacia abajo. El pistón, al conectarse a la biela, es diseñado para transmitir la fuerza producida en la cámara de combustión al cigüeñal.

**Piston balance pads** Some manufacturers provide balance pads just below the pin boss. The piston can be balanced by removing material from this area.
**Placa (pastilla) de equilibración del pistón** Algunos fabricantes proveen las placas para equilibrar ubicadas justo abajo del mamelón. Se puede equilibrar al pistón quitando el material de esta área.

**Piston collapse** A condition describing a collapse or reduction in the diameter of the piston skirt due to heat or stress.
**Caída del pistón** Una condición que describe el fallo o la reducción en el diámetro de la falda del pistón debido al calor o la fatiga.

**Piston dwell time** The length of time in crankshaft degrees the piston remains at top dead center without moving.
**Angulo de cierre del pistón** La cantidad del tiempo en los grados del ángulo del cigüeñal que el pistón se queda en la posición de punto muerto superior sin moverse.

**Piston head (crown)** The top of the piston that forms the bottom of the combustion camber.
**Cabeza de piston** La parte superior del pistón que forma la parte inferior de la cámara de combustión.

**Piston land** Area used to confine and support the piston rings in their grooves.
**Meseta de la pared del pistón** El área de un pistón que sirve para restringir y sostener los anillos del pistón en sus muescas.

**Piston offset** A piston design that offsets the pin bore to provide more effective downward force onto the crankshaft by increasing the leverage applied to the crankshaft.
**Desviación del pistón** Un diseño del piston que desvía el taladro del eje para proveer una fuerza descendente en el cigüeñal más eficáz aumentando la acción de palanca que se aplica en el cigüeñal.

**Piston pin** Component which connects the piston to the connecting rod. There are three basic designs used: a piston pin anchored to the piston and floating in the connecting rod, a piston pin anchored to the connecting rod and floating in the piston, and a piston pin full floating in the piston and connecting rod.
**Eje del pistón** Un componente que conecta el pistón a la biela. Se usan tres diseños básicos: un eje de pistón fijo al pistón y libre en la biela, un eje de pistón fijo a la biela y libre en el pistón, y un eje de pistón completamente libre en el pistón y la biela.

**Piston pin knock** A noise caused by a worn piston pin or bushing, worn piston pin boss, and worn bearings.
**Golpe del eje del pistón** Un ruido causado por un eje o manguito de pistón desgastado, un mamelón desgastado, y los cojinetes desgastados.

**Piston rings** Components which seal the compression and expansion gases, and prevent oil from entering the combustion chamber.
**Anillos (aros) del pistón** Un componente que sella los gases de compresión y expansión, y previene que entra el aceite en la cámara de combustión.

**Piston seal** The seal fitted to the disc brake pistons. It provides the return motion to the piston as well as sealing in the brake fluid.
**Sello del pistón** El sello que se ajusta a los pistones de los frenos de disco. Provee el movimiento de regreso al pistón mientras que previene una fuga del líquido de freno.

**Piston skirt** A component which forms a bearing area in contact with the cylinder wall and helps to prevent the piston from rocking in the cylinder.
**Faldilla del pistón** Un componente que forma una área de apoyo en contacto con el muro del cilindro y ayuda en prevenir que el pistón oscila en el cilindro.

**Piston slap** A sound which results from the piston hitting the side of the cylinder wall.
**Golpeteo del pistón** Un ruido resultando del pistón golpeando contra el muro del cilindro.

**Piston stroke** The distance the piston travels from TDC to BDC.
**Carrera del pistón** La distancia que viaja el piston del PMS al PMI.

**Pitch** The number of threads per inch on any threaded part.
**Paso** El número de filetes por pulgada de cualquier parte fileteada.

**Pitman arm** A short arm in the steering linkage that connects the steering gear to other steering components.
**Brazo pitman** Brazo corto en el cuadrilátero de la dirección que articula el mecanismo de dirección a los otros componentes de dirección.

**Pitman arm puller** A special puller required to remove a pitman arm.
**Extractor del brazo pitman** Extractor especial requerido para la remoción de un brazo pitman.

**Pivot** A pin or shaft upon which another part rests or turns.
**Pivote** Una chaveta o una flecha que soporta a otra parte o sirve como un punto para girar.

**Planet carrier** The carrier or bracket in a planetary gear system that contains the shafts upon which the pinions or planet gears turn.
**Perno de arrastre planetario** El soporte o la abrazadera que contiene las flechas en las cuales giran los engranajes planetarios o los piñones.

**Planet gears** The gears in a planetary gear set that connect the sun gear to the ring gear.
Engranajes planetarios Los engranajes en un conjunto de engranajes planetario que conectan al engranaje propulsor interior (el engranaje sol) con la corona.

**Planet pinions** In a planetary gear system, the gears that mesh with, and revolve about, the sun gear; they also mesh with the ring gear.
Piñones planetarios En un sistema de engranajes planetarios, los engranajes que se endentan con, y giran alrededor, el engranaje propulsor (sol); también se endentan con la corona.

**Planetary gear set** A system of gearing that is modeled after the solar system. A pinion is surrounded by an internal ring gear and planet gears are in mesh between the ring gear and pinion around which all revolve.
Conjunto de engranajes planetarios Un sistema de engranaje cuyo patrón es el sistema solar. Un engranaje propulsor (la corona interior) rodea al piñón de ataque y los engranajes satélites y planetas se endentan entre la corona y el piñón alrededor del cual todo gira.

**Plasma** A material containing positive ions and unbound electrons in which the total number of positive and negative charges are almost equal. The properties of plasma are sufficiently different from those of solids, liquids, and gases for it to be considered a fourth state of matter.
Plasma Un material comprendido de los iones positivos y los electrones libres en el cual el número total de cargas positivas y negativas son casi iguales. Las propriedades del plasma son bastante diferentes de las de los sólidos, los líquidos y los gases para que se considera un cuarto estado de materia.

**Plastigage** A string-like plastic that is available in different diameters used to measure the clearance between two components. The diameter of the plastic gage is exact, thus any crush of the gage material will provide an accurate measurement of oil clearance.
Plastigage Un plástico en forma de hilo disponible en diámetros distintos que se emplea en medir la holgura entre dos componentes. El diámetro del calibre plástico es preciso, así cualquier aplastamiento del material del calibre provee una medida precisa de la holgura del aceite.

**Plate loading** Force developed by the pressure plate assembly to hold the driven disc against the flywheel.
Carga de placa Fuerza producida por el conjunto de la placa de presión para sujetar el disco accionado contra el volante.

**Plateau honing** A cylinder honing process that uses a coarse stone to produce the finished cylinder size and a very fine stone to remove an immeasurable amount and plateau the cut.
Esmerilado de nivelación Un proceso de esmerilar el cilindro usando una piedra tosca para producir el tamaño del cilindro acabado y una piedra muy fina para quitar una cantidad inmensurable y nivelar el corte.

**Plug** Anything that will fit into an opening to stop fluid or airflow.
Tapón Cualquier cosa que se ajuste en una apertura para prevenir el goteo o el escape de un corriente del aire.

**Plumb bob** A weight with a sharp, tapered point that is suspended and centered on a string.
Plomada Balanza con un extremo cónico puntiagudo, suspendida y centrada en una hilera.

**Plunge grinding** A crankshaft grinding method that uses a dressed stone the exact shape of the new journal surface. The stone is fed straight into the journal.
Rectificación empujado Un método de afilar el cigüeñal usando una piedra de afilar en la forma exacta de la superficie del muñón nuevo. La piedra de afilar se empuja directamente dentro del muñón.

**Plunging action** Telescoping action of an inner front-wheel-drive universal joint.
Acción sumergible Acción telescópica de una junta universal interior de tracción delantera.

**Plunging constant velocity joint** Usually the inner constant velocity joint. The joint is designed so that it can telescope slightly to compensate for suspension motions.
Junta de velocidad constante sumergible Normalmente junta de velocidad constante interior. La junta está diseñada para que pueda extenderse ligeramente y así compensar el movimiento de la suspensión.

**Ply separation** A parting of the plies in a tire casing.
Separación de estrias División de las estrias en la cubierta de un neumático.

**Pneumatic tools** Tools powered by compressed air.
Herramientas neumáticas Las herramientas que derivan su poder del aire bajo presión.

**Polishing** The process of removing light roughness from the journals by using a fine emery cloth. Polishing can be used to remove minor scoring of journals which do not require grinding.
Bruñido El proceso de quitar la aspereza ligera de los muñones usando una tela abrasiva muy fina. El bruñido puede emplearse en quitar las rayas superficiales de los muñones que no requieren rectificación.

**Poly belt** See *Serpentine belt*.
Correa poli Ver *Serpentine belt* [*Correa serpentina*].

**Polyol ester (ESTER)** A synthetic oil-like lubricant that is occasionally recommended for use in an HFC-134a system. This lubricant is compatible with both HFC-134a and CFC-12.
Polioléster Lubrificante sintético parecido a aceite que se recomienda de vez en cuando para usar en un sistema HFC 134a. Dicho lubrificante es compatible tanto con HFC 134a como CFC 12.

**Polyurethane air cleaner cover** A circular polyurethane ring mounted over the air cleaner element to improve cleaning capabilities.
Cubierta de poliuretano del filtro de aire Anillo circular de poliuretano montado sobre el elemento del filtro de aire para facilitar la limpieza.

**Poppet valve** A valve design consisting of a circular head with a stem attached in the center. Poppet valves are used to control the opening or closing of a passage by linear movement.
Válvula champiñón Un diseño de una válvula que consiste de una cabeza redonda con un vástago conectado en el centro. Las válvulas champiñones controlan la apertura o cerradura de un pasaje por medio de un movimiento linear.

**Porosity** A statement of how porous or permeable to liquids a material is.
Porosidad Una expresión de lo poroso o permeable a los líquidos es una materia.

**Port EGR valve** An EGR valve operated by ported vacuum from above the throttle.
Válvula EGR lumbrera Válvula EGR accionada por un vacío con lumbreras desde la parte superior de la mariposa.

**Port fuel injection (PFI)** A fuel injection system with an injector positioned in each intake port.
Inyección de combustible de lumbrera Sistema de inyección de combustible que tiene un inyector colocado en cada una de las lumbreras de aspiración.

**Positive back-pressure EGR valve** An EGR valve with a vacuum bleed valve that is operated by positive pressure pulses in the exhaust.
Válvula EGR de contrapresión positiva Válvula EGR con una válvula de descarga de vacío accionada por impulsos de presión positiva en el escape.

**Positive crankcase ventilation (PCV) system** An emission control system that routes blowby gases and unburned oil/fuel vapors to the intake manifold to be added to the combustion process.
Sistema (PCV) de ventilación positiva de la caja del cigüeñal Un sistema de emisión que lleva los gases soplados y los vapores del aceite/combustible no quemados al múltiple de entrada para que se pueden añadir al proceso de combustión.

**Positive crankcase ventilation (PCV) valve** A valve that delivers crankcase vapors into the intake manifold rather than allowing them to escape to the atmosphere.
**Válvula de ventilación positiva del cárter** Válvula que conduce los vapores del cárter hacia el colector de aspiración en vez de permitir que los mismos se escapen hacia la atmósfera.

**Positive displacement pumps** Pumps that deliver the same amount of oil with every revolution, regardless of speed.
**Bombas de desplazamiento positivo** Las bombas que entregan la misma cantidad del aceite con cada revolución, sin que importa la velocidad.

**Positive pressure** Any pressure above atmospheric.
**Presión positiva** Cualquier presión sobre la de la atmosférica.

**Potentiometer** A variable resistor that acts as a circuit divider, providing accurate voltage drop readings proportional to movement.
**Potenciómetro** Resistor variable que actúa como un divisor de circuito para obtener lecturas de perdidas de tensión precisas en proporción con el movimiento.

**Pound** A weight measure; 16 ounces. A term often used when referring to a small can of refrigerant, although the can does not necessarily contain 16 ounces.
**Libra** Medida de peso, 16 onzas. Término utilizada con frecuencia al referirse a una lata pequeña de refrigerante, aunque es posible que la lata contenga menos de 16 onzas.

**Pound of refrigerant** A term used by some technicians when referring to a small can of refrigerant that actually contains less than 16 ounces.
**Libra de refrigerante** Término utilizado por algunos mecánicos al referirse a una lata pequeña de refrigerante que en realidad contiene menos de 16 onzas.

**Power balance test** Test used to determine if all cylinders are producing the same amount of power output. In an ideal situation, all cylinders would produce the exact same amount of power.
**Prueba del equilibrio de fuerza** Una prueba para determinar si todos los cilindros producen la misma cantidad de potencia de salida. En una situación ideal, todos los cilindros producirían exactamente la misma cantidad de poder.

**Power balance tester** A tester designed to stop each cylinder from firing for a brief time and record the rpm decrease.
**Instrumento de prueba del equilibro de la potencia** Instrumento de prueba diseñado para detener el encendido de cada cilindro por un breve espacio de tiempo y registrar el descenso de las rpm.

**Power booster** Used to increase pedal pressure applied to a brake master cylinder.
**Reforzador** Se usa para incrementar la presión del pedal aplicada al cilindro maestro del freno.

**Power brakes** A brake system that employs vacuum or hydraulics to assist the driver in producing braking force.
**Frenos de potencia** Un sistema de frenos que emplea un vacío o las hidráulicas para asistir al conductor en efectuar una fuerza de enfrenado.

**Power module** Controls the operation of the blower motor in an automatic temperature control system.
**Transmisor de potencia** Regula el funcionamiento del motor del soplador en un sistema de control automático de temperatura.

**Power-steering pump pressure gauge** A gauge and valve with connecting hoses for checking the power-steering pump pressure.
**Calibrador de presión de la bomba de la dirección hidráulica** Calibrador y válvula con mangueras de conexión utilizadas para verificar la presión de la bomba de la dirección hidráulica.

**Power tools** Tools that use other forces than that generated from the body. They can use compressed air, electricity, or hydraulic pressure to generate and multiply force.
**Herramientas de motor** Las herramientas que usan otras fuerzas que las producidas por el cuerpo. Pueden usar el aire bajo presión, la electricidad, o la presión hidráulica para engendrar y multiplicar la fuerza.

**Powertrain** The mechanisms that carry the power from the engine crankshaft to the drive wheels; these include the clutch, transmission, driveline, differential, and axles.
**Tren impulsor** Los mecanismos que transfieren la potencia desde el cigüeñal del motor a las ruedas de propulsión; éstos incluyen el embrague, la transmisión, la flecha motríz, el diferencial y los semiejes.

**Powertrain control module (PCM)** A computer that controls engine and possibly transmission functions.
**Unidad de control del tren transmisor de potencia** Computadora que controla las funciones del motor y posiblemente las de la transmisión.

**Prealignment inspection** A check of steering and suspension components prior to a wheel alignment.
**Inspección antes de una alineación** Verificación de los componentes de la dirección y de la suspensión antes de llevarse a cabo una alineación de ruedas.

**Predetermined** A set of fixed values or parameters that have been programmed or otherwise fixed into an operating system.
**Predeterminado** Valores fijos o parámetros que han sido programados o de otra manera fijados en un sistema de funcionamiento.

**Pre-ignition** Defect that is the result of a spark occurring too soon.
**Autoencendido** Un defecto que resulta de una chispa que ocurre demasiado temprano.

**Preliminary inspection screen** A display on a computer wheel aligner that allows the technician to check and record the condition of many suspension and steering components.
**Pantalla para la visualización durante una inspección preliminary** Representación visual en una computadora para alineación de ruedas que le permite al mecánico revisar y anotar la condición de un gran número de componentes de la suspensión y de la dirección.

**Preload** A load applied to a part during assembly so as to maintain critical tolerances when the operating load is applied later.
**Carga previa** Una carga aplicada a una parte durante la asamblea para asegurar sus tolerancias críticas antes de que se le aplica la carga de la operación.

**Prelubrication** Lubrication of components such as turbocharger bearings prior to starting the engine.
**Prelubrificación** Lubrificación de componentes, como por ejemplo los cojinetes del turbocompresor, antes del arranque del motor.

**Press fit** Forcing a part into an opening that is slightly smaller than the part itself to make a solid fit.
**Ajustamiento a presión** Forzar a una parte en una apertura que es de un tamaño más pequeño de la parte para asegurar un ajustamiento sólido.

**Pressure** Force per unit area, or force divided by area. Usually measured in pounds per square inch (psi) or in kilopascals (kPa) in the metric system.
**Presión** La fuerza por unidad de una area, o la fuerza dividida por el area. Suele medirse en libras por pulgada cuadrada (lb/pulg2) o en kilopascales (kPa) en el sistema métrico.

**Pressure bleeding** Pressure bleeding uses air to pressurize brake fluid in order to force air out of the hydraulic brake system.
**Purga con presión** En purgar con presión uno usa el aire para sobre-comprimir el líquido de freno así forzando el aire fuera del sistema hidráulico de frenos.

**Pressure differential valve** Used in dual brake systems to sense unequal hydraulic pressure between the front and rear brakes.
**Válvula del diferencial de presión** Usado en los sistemas de frenos dobles para sentir una presión desigual entre los frenos delanteros y traseros.

**Pressure gauge** A calibrated instrument for measuring pressure.
**Manómetro** Instrumento calibrado para medir la presión.

**Pressure plate** That part of the clutch which exerts force against the friction disc; it is mounted on, and rotates with, the flywheel.
Plato opresor  Una parte del embrague que aplica la fuerza en el disco de fricción; se monta sobre el volante, y gira con éste.

**Pressure relief valve** A valve designed to limit pump pressure, such as a power-steering pump.
Válvula de alivio de presión  Válvula diseñada para limitar la presión de una bomba, como por ejemplo una bomba de la dirección hidráulica.

**Pressure switch** An electrical switch that is activated by a predetermined low or high pressure. A high-pressure switch is generally used for system protection; a low-pressure switch may be used for temperature control or system protection.
Autómata manométrico  Interruptor eléctrico accionado por una baja o alta presión predeterminada. Generalmente se utiliza un autómata manométrico de alta presión para la protección del sistema; puede utilizarse uno de baja presión para la regulación de temperatura o protección del sistema.

**Pressurized injector cleaning container** A small, pressurized container filled with unleaded gasoline and injector cleaner for cleaning injectors with the engine running.
Recipiente presionizado para la limpieza del inyector  Pequeño recipiente presionizado lleno de gasolina sin plomo y limpiador de inyectores para limpiar los inyectores cuando el motor está encendido.

**Primary shoe** When the car is moving forward, the shoe facing the front of the car is the leading or primary shoe.
Zapata primaria  Al moverse hacia frente el coche, la zapata en la dirección hacia la parte delantera del coche es la zapata de guía o primaria.

**Primary wiring** Conductors that carry low voltage and low current. The insulation of primary wires is usually thin.
Hilos primarios  Hilos conductores de tensión y corriente bajas. El aislamiento de hilos primarios es normalmente delgado.

**Profilometer** A tool capable of electrically sensing the distances between peaks to determine the finish of a cut.
Perfilómetro  Una herramienta capaz de detectar electrónicamente las distancias entre dos puntos altos para determinar cuando terminar un corte.

**Program** A set of instructions that the computer must follow to achieve desired results.
Programa  Conjunto de instrucciones que la computadora debe seguir para lograr los resultados deseados.

**Programmed ride control (PRC) system** A computer-controlled suspension system in which the computer operates an actuator in each strut to control strut firmness.
Sistema de control programado del viaje  Sistema de suspensión controlado por computadora en el que la computadora opera un accionador en cada uno de los montantes para controlar la firmeza de los mismos.

**PROM (programmable read-only memory)** Memory chip that contains specific data that pertains to the exact vehicle that the computer is installed in. This information may be used to inform the CPU of the accessories that are equipped on the vehicle.
PROM (memoria de sólo lectura programable)  Fragmento de memoria que contiene datos específicos referentes al vehículo particular en el que se instala la computadora. Esta información puede utilizarse para informar a la UCP sobre los accesorios de los cuales el vehículo está dotado.

**Propane** A combustible liquified petroleum gas ($C_3H_8$) that becomes liquid when compressed.
Propano  Gas inflamable utilizado como propulsor para el detector de fugas de halogenuro.

**Propeller shaft** See *Drive shaft*.
Flecha de Propulsión  Vea *Flecha motríz*.

**Proportioning valve** This valve regulates the hydraulic pressure in the rear brake system. It is located between the inlet and outlet ports of the rear system in the control valve. It allows equal pressure to be applied to both the front and rear brakes until a particular pressure is obtained.
Válvula dosificadora  Esta válvula regula la presión hidráulica en el sistema de frenos trasero. Se ubica entre las aberturas de entrada y salida del sistema trasero en la válvula de control. Permite que una presión equilibrada se aplica a ambos los frenos delanteros y traseros hasta que se obtiene una presión específica.

**Protection device** Circuit protector that is designed to "turn off" the system that it protects. This is done by creating an open to prevent a complete circuit.
Dispositivo de protección  Protector de circuito diseñado para "desconectar" el sistema al que provee protección. Esto se hace abriendo el circuito para impedir un circuito completo.

**Prove-out circuit** A function of the ignition switch that completes the warning light circuit to ground through the ignition switch when it is in the START position. The warning light is on during engine cranking to indicate to the driver that the bulb is working properly.
Circuito de prueba  Función del botón conmutador de encendido que completa el circuito de la luz de aviso para que se ponga a tierra a través del botón conmutador de encendido cuando éste se encuentra en la posición START. La luz de aviso se encenderá durante el arranque del motor para avisarle al conductor que la bombilla funciona correctamente.

**Prussian blue** A blue pigment; in solution, useful in determining the area of contact between two surfaces.
Azul de Prusia  Un pigmento azul; en forma líquida, ayuda en determinar el área de contacto entre dos superficies.

**PSI** Abbreviation for pounds per square inch; a measurement of pressure.
Lb/pulg2  Una abreviación de libras por pulgada cuadrada, una medida de la presión.

**Psig** Pounds per square inch gauge.
Psig  Calibrador de libras por pulgada cuadrada.

**Puller** Generally, a shop tool used to separate two closely fitted parts without damage. Often contains a screw, or several screws, which can be turned to apply a gradual force.
Extractor  Generalmente, una herramienta del taller que sirve para separar a dos partes apretadas sin incurrir daños. Suele tener una tuerca o varias tuercas, que se pueden girar para aplicar la fuerza gradualmente.

**Pulsation** To move or beat with rhythmic impulses.
Pulsación  Moverse o batir con impulsos rítmicos.

**Pulse width** The length of time in milliseconds that an actuator is energized.
Duración de impulsos  Espacio de tiempo en milisegundos en el que se excita un accionador.

**Pulse width modulation** On/off cycling of a component. The period of time for each cycle does not change, only the amount of on time in each cycle changes.
Modulación de duración de impulsos  Modulación de impulsos de un componente. El espacio de tiempo de cada ciclo no varía; lo que varía es la cantidad de trabajo efectivo de cada ciclo.

**Pulsed secondary air injection system** A system that uses negative pressure pulses in the exhaust to move air into the exhaust system.
Sistema de inyección secundaria de aire por impulsos  Sistema que utiliza impulsos de la presión negativa en el escape para conducir el aire hacia el sistema de escape.

**Pulsing pedal** A condition where the brake pedal moves up and down when it is applied. Normally due to an unparallel brake rotor.
Pedal pulsante  Una condición en la cual el pedal de freno se mueve hacia arriba y abajo al aplicarse. Normalmente se debe a un rotor de freno que esta fuera de paralelo.

**Purge** To remove moisture and/or air from a system or a component by flushing with a dry gas such as nitrogen (N) to remove all refrigerant from the system.
**Purgar** Remover humedad y/o aire de un sistema o un componente al descargarlo con un gas seco, como por ejemplo el nitrógeno (N), para remover todo el refrigerante del sistema.

**Purity test** A static test that may be performed to compare the suspect refrigerant pressure to an appropriate temperature chart to determine its purity.
**Prueba de pureza** Prueba estática que puede llevarse a cabo para comparar la presión del refrigerante con un gráfico de temperatura apropiado para determinar la pureza del mismo.

**Pushrod** A connecting link between the lifter and rocker arm. Engines designed with the camshaft located in the block use pushrods to transfer motion from the lifters to the rocker arms.
**Varilla de presión** Una conexión entre la levantaválvulas y el balancín empuja válvulas. Los motores diseñados con el árbol de levas en el bloque usan las varillas de presión para transferir el movimiento de la levantaválvulas a los balancines.

**Quad driver** A group of transistors in a computer that controls specific outputs.
**Excitador cuádruple** Grupo de transistores en una computadora que controla salidas específicas.

**Quadrant** A section of a gear. A term sometimes used to identify the shift lever selector mounted on the steering column.
**Cuadrante** Sección de un engranaje. Término utilizado en algunas ocasiones para identificar el selector de la palanca de cambio de velocidades montado sobre la columna de dirección.

**Quick-disconnect fuel line fittings** Fuel line fittings that may be disconnected without using a wrench.
**Conexiones de la línea del combustible de desmontaje rápido** Conexiones de la línea del combustible que se pueden desmontar sin la utilización de una llave de tuerca.

**Quill shaft** The term used by some manufacturers to refer to the protruding hollow shaft of the transmission's front bearing retainer.
**Árbol de manguito** Término utilizado por algunos fabricantes para referirse al árbol hueco proyectado del retenedor del cojinete delantero de la transmisión.

**R-12** The type of refrigerant used on most cars prior to the early 1990s; is commonly referred to as freon.
**R-12** Tipo de refrigerante utilizado en la mayoría de los automóviles; comúnmente llamado freón.

**Race** A channel in the inner or outer ring of an antifriction bearing in which the balls or rollers roll.
**Pista** Un canal en el anillo interior o exterior de un cojinete antifricción en el cual ruedan las bolas o los rodillos.

**Raceway** A groove or track designed into the races of a bearing or universal joint housing to guide and control the action of the balls or trunnions.
**Anillo de rodadura** Ranura o canal construido en el interior de los anillos de un cojinete o de un alojamiento de junta universal para guiar y controlar el movimiento de las bolas o de las muñequillas.

**Rack bushing** A bushing that supports the rack in the rack and pinion steering gear housing.
**Buje de la cremallera** Buje que apoya la cremallera en el alojamiento del mecanismo de dirección de cremallera y piñón.

**Radial** The direction moving straight out from the center of a circle. Perpendicular to the shaft or bearing bore.
**Radial** La dirección al moverse directamente del centro de un círculo. Perpendicular a la flecha o al taladro del cojinete.

**Radial clearance** Clearance within the bearing and between balls and races perpendicular to the shaft. Also called radial displacement.

**Holgura radial** La holgura en un cojinete entre las bolas y las pistas que son perpendiculares a la flecha. También se llama un desplazamiento radial.

**Radial load** A force perpendicular to the axis of rotation.
**Carga radial** Una fuerza perpendicular al centro de rotación.

**Radial runout** The variations in diameter of a round object such as a tire.
**Desviación radial** Variaciones en el diámetro de un objeto circular, como por ejemplo un neumático.

**Radiation** The transfer of heat without heating the medium through which it is transmitted.
**Radiación** La transferencia de calor sin calentar el medio por el cual se transmite.

**Radiator** A component consisting of a series of tubes and fins that transfer the heat from the coolant to the air.
**Radiador** Un componente que consiste de una serie de tubos y aletas que transfieren el calor del fluido refrigerante al aire.

**Radiator shroud** A circular component positioned around the cooling fan to concentrate the airflow through the radiator.
**Bóveda del radiador** Componente circular que rodea el ventilador de enfriamiento para concentrar el flujo de aire a través del radiador.

**Ram air** Air that is forced through the radiator and condenser coils by the movement of the vehicle or the action of the fan.
**Aire admitido en sentido de la marcha** Aire forzado a través de las bobinas del radiador y del condensador por medio del movimiento del vehículo o la acción del ventilador.

**Ratcheting mechanism** Uses a pawl and gear arrangement to transmit motion or to lock a particular mechanism by having the pawl drop between gear teeth.
**Mecanismo de trinquete** Utiliza un conjunto de retén y engranaje para transmitir movimiento o para bloquear un mecanismo específico haciendo que el retén caiga entre los dientes del engranaje.

**Ratio** The relation or proportion that one number bears to another.
**Relación** La correlación o proporción de un número con respecto a otro.

**Ravigneaux** Designer of a planetary gear system with small and large sun gears, long and short planetary pinions, planetary carriers, and ring gear.
**Ravigneaux** Fabricante de un sistema de engranaje planetario que consiste de engranajes principales pequeños y grandes, piñones planetarios largos y cortos, portadores planetarios y corona.

**Real-time damping (RTD)** The real-time damping module controls the road-sensing suspension system.
**Amortiguamiento en tiempo real** La unidad del amortiguamiento en tiempo real controla el sistema de suspensión con equipo sensor.

**Ream** Process of accurately finishing a hole with a rotating fluted tool.
**Escariar** El proceso de acabar un taladro precisamente con una herramiente acanalada giratoria.

**Reamer** A round metal-cutting tool with a series of sharp cutting edges; enlarges a hole when turned inside it.
**Escariador** Una herramienta redonda para cortar a los metales que tiene una seria de rebordes mordaces agudos; al girarse en un agujero lo agranda.

**Rear axle offset** A condition in which the complete rear axle assembly has turned so one rear wheel has moved forward and the opposite rear wheel has moved rearward.
**Desviación del eje trasero** Condición que ocurre cuando todo el conjunto del eje trasero ha girado de manera que una de las ruedas traseras se ha movido hacia adelante y la opuesta se ha movido hacia atrás.

**Rear axle sideset** A condition in which the rear axle assembly has moved sideways from its original position.
**Resbalamiento lateral del eje trasero** Condición que ocurre cuando el conjunto del eje trasero se ha movido lateralmente desde su posición original.

**Rear axle torque** The torque received and multiplied by the rear driving axle assembly.
**Torsión del eje trasero** Par de torsión recibido y multiplicado por el conjunto del eje motor trasero.

**Rear main steering angle sensor** An input sensor in the rear steering actuator of an electronically controlled four-wheel steering system.
**Sensor principal del ángulo de la dirección trasera** Sensor de entrada en el accionador de la dirección trasera de un sistema de dirección en las cuatro ruedas controlado electrónicamente.

**Rear steering center lock pin** A special pin used to lock the rear steering during specific service procedures in an electronically controlled four-wheel steering system.
**Pasador de cierre central de la dirección trasera** Pasador especial que se utiliza para bloquear la dirección trasera durante procedimientos de reparación específicos en un sistema de dirección en las cuatro ruedas controlado electrónicamente.

**Rear substeering angle sensor** An input sensor in the rear steering actuator of a four-wheel steering system.
**Sensor auxiliar del ángulo de la dirección trasera** Sensor de entrada en el accionador de la dirección trasera de un sistema de dirección en las cuatro ruedas.

**Rear-wheel drive** A term associated with a vehicle where the engine is mounted at the front and the driving axle and driving wheels are at the rear of the vehicle.
**Tracción trasera** Un término que se asocia con un vehículo en el cual el motor se ubica en la parte delantera y el eje propulsor y las ruedas propulsores se encuentran en la parte trasera del vehículo.

**Rear wheel tracking** Refers to the position of the rear wheels in relation to the front wheels.
**Encarrilamiento de las ruedas traseras** Se refiere a la posición de las ruedas traseras con relación a las ruedas delanteras.

**Rebuilt** To build after having been disassembled, inspected, and worn and damaged parts and components are replaced.
**Reconstruído** Fabricar después de haber sido desmontado y revisado, y luego reemplazar las piezas desgastadas y averiadas.

**Receiver/dryer** A tank-like vessel having a desiccant and used for the storage of refrigerant.
**Receptor/secador** Recipiente parecido a un tanque provisto de un desecante y utilizado para el almacenaje de refrigerante.

**Reciprocating** An up-and-down or back-and-forth motion.
**Alternativo** Un movimiento oscilante de arriba a abajo o de un lado a otro.

**RECIRC** An abbreviation for the recirculate mode, as with air.
**RECIRC** Abreviatura del modo recirculatorio, como por ejemplo con aire.

**Recovery system** A term often used to refer to the circuit inside the recovery unit used to recycle and/or transfer refrigerant from the air-conditioning system to the recovery cylinder.
**Sistema de recuperación** Término utilizado con frecuencia para referirse al circuito dentro de la unidad de recuperación interior utilizada para reciclar y/o transferir el refrigerante del sistema de acondicionamiento de aire al cilindro de recuperación.

**Recovery tank** An auxiliary tank, usually connected to the inlet tank of a radiator, which provides additional storage space for heated coolant.
**Tanque de recuperación** Tanque auxiliar que normalmente se conecta al tanque de entrada de un radiador, lo cual provee almacenaje adicional para el enfriante calentado.

**Rectification** The converting of AC current to DC current.
**Rectificación** Proceso a través del cual la corriente alterna es transformada en una corriente continua.

**Reed valve** A one-way check valve. The reed opens to allow the air-fuel mixture to enter from one direction, while closing to prevent movement in the other direction.

**Válvula de lengüeta** Una válvula de una vía. La lengüeta se abre para permitir entrar la mezcla de aire/combustible de una dirección, mientras que se cierre para prevenir el movimiento de la otra dirección.

**Reference pickup** A pickup assembly that is often used for ignition triggering.
**Captación de referencia** Conjunto de captación que se utiliza con frecuencia para el arranque del encendido.

**Reference voltage** A constant voltage supplied from the computer to some of the input sensors.
**Tensión de referencia** Tensión constante que le suministra la computadora a algunos de los sensores de entrada.

**Refrigerant-12** The refrigerant used in automotive air conditioners, as well as other air-conditioning and refrigeration systems. The chemical name of refrigerant-12 is dichlorodifluoromethane. The chemical symbol is $CCl_2F_2$.
**Refrigerante 12** Refrigerante utilizado tanto en acondicionadores de aire automotrices como en otros sistemas de acondicionamiento de aire y refrigeración. El nombre químico del refrigerante 12 es diclorodi-florometano, y el símbolo químico es CCl2F2.

**Relay** An electrical switch device that is activated by a low-current source and controls a high-current device.
**Relé** Interruptor eléctrico que es accionado por una fuente de corriente baja y regula un dispositivo de corriente alta.

**Release bearing** A ball-type bearing moved by the clutch pedal linkage to contact the pressure plate release levers to either engage or disengage the driven disc with the clutch driving members.
**Cojinete de desembrague** Cojinete de tipo bola accionado por la biela motriz del pedal del embrague para entrar en contacto con las palancas de desembrague de la placa de presión o para engranar o desengranar el disco accionado con los mecanismos de accionamiento del embrague.

**Release levers** In the clutch, levers that are moved by throwout-bearing movement, causing clutch spring force to be relieved so that the clutch is disengaged, or uncoupled from the flywheel.
**Palancas de desembrague** En el embrague, las palancas accionadas por el movimiento del cojinete de desembrague, que hacen disminuir la fuerza del muelle del embrague para que el embrague se desengrane, o se desacople del volante.

**Release plate** Plate designed to release the clutch pressure plate's loading on the clutch driven disc.
**Placa de desembrague** Placa diseñada para desembragar la carga de la placa de presión del embrague en el disco accionado del embrague.

**Relief valve** Valve used to prevent excessive oil pressure. Since the oil pump is positive displacement, pressures could increase to a hazardous level at higher engine speeds. The relief valve opens to return oil to the sump and drop pressure in the system.
**Válvula de rebose** Una válvula que previene una presión excesiva de aceite. Como la bomba de aceite es de desplazamiento positivo, las presiones podrían aumentar a un nivel peligroso en las velocidades más altas. La válvula abre para regresar el aceite al resumidero y bajar la presión del sistema.

**Remote reservoir** A container containing fluid that is mounted separately from the fluid pump.
**Tanque remoto** Recipiente lleno de líquido que se monta separado de la bomba del fluido.

**Removable carrier housing** A type of rear axle housing from which the axle carrier assembly can be removed for parts service and adjustment.
**Alojamiento portador desmontable** Tipo de puente trasero del cual se puede desmontar el conjunto del portador del eje para la reparación y el ajuste de las piezas.

**Reserve-capacity rating** An indicator, in minutes, of how long the vehicle can be driven with the headlights on, if the charging system should fail. The reserve-capacity rating is determined by the length of time, in minutes, that a fully charged battery can be discharged at 25 amperes before battery cell voltage drops below 1.75 volts per cell.

**Clasificación de capacidad en reserva** Indicación, en minutos, de cuánto tiempo un vehículo puede continuar siendo conducido, con los faros delanteros encendidos, en caso de que ocurriese una falla en el sistema de carga. La clasificación de capacidad en reserva se determina por el espacio de tiempo, en minutos, en el que una batería completamente cargada puede descargarse a 25 amperios antes de que la tensión del acumulador de la batería disminuya a un nivel inferior de 1,75 amperios por acumulador.

**Reserve tank** A storage vessel for excess fluid. See *Recovery tank*, *Receiver/dryer*, and *Accumulator*.

**Tanque de reserva** Recipiente de almacenaje para un exceso de fluido. Ver *Recovery tank* [*Tanque de recuperación*], *Receiver/dryer* [*Receptor/secador*] y *Accumulator* [*Acumulador*].

**Resistance** Opposition to current flow.

**Resistencia** Oposición que presenta un conductor al paso de la corriente eléctrica.

**Resistive shorts** Shorts to ground that pass through a form of resistance first.

**Cortocircuitos resistivos** Cortocircuitos a tierra que primero pasan por una forma de resistencia.

**Resistor** A voltage-dropping device that is usually wire wound and provides a means of controlling fan speeds.

**Resistor** Dispositivo de caída de tensión que normalmente es devando con alambre y provee un medio de regular la velocidad del ventilador.

**Resistor block** A series of resistors with different values.

**Bloque resistor** Serie de resistores que tienen valores diferentes.

**Resource Conservation and Recovery Act (RCRA)** Law that makes users of hazardous materials responsible for the material from the time it becomes a waste until disposal is complete.

**Acta de Conservación y Recobro de Recursos (RCRA)** Una ley que hace responsable a los que usan los materiales peligrosos desde el tiempo que se convierte en un producto residual hasta que se haya completado su disposición.

**Respirator** A mask or face shield worn in a hazardous environment to provide clean fresh air and/or oxygen.

**Mascarilla** Máscara o protector de cara que se lleva puesto en un ambiente peligroso para proveer aire limpio y puro y/o oxígeno.

**Responsibility** Being reliable and trustworthy.

**Responsabilidad** Ser confiable y fidedigno.

**Restricted** Having limitations. Keeping within limits, confines, or boundaries.

**Restringido** Que tiene limitaciones. Mantenerse dentro de límites, confines, o fronteras.

**Restrictor** An insert fitting or device used to control the flow of refrigerant or refrigeration oil.

**Limitador** Pieza inserta o dispositivo utilizado para regular el flujo de refrigerante o aceite de refrigeración.

**Retaining ring** A removable fastener used as a shoulder to retain and position a round bearing in a hole.

**Anillo de retén** Un seguro removible que sirve de collarín para sujetar y posicionar a un cojinete en un agujero.

**Retractor clips** Spring steel clips that connect the diaphragm's flexing action to the pressure plate.

**Grapas retractoras** Grapas de acero para muelles que conectan el movimiento flexible del diafragma a la placa de presión.

**Reverse idler gear** In a transmission, an additional gear that must be meshed to obtain reverse gear; a gear used only in reverse that does not transmit power when the transmission is in any other position.

**Piñón de marcha atrás** En una transmisión, engranaje adicional que debe engranarse para obtener un engranaje de marcha atrás; engranaje utilizado solamente durante la inversión de marcha, que no transmite fuerza cuando la transmisión se encuentra en cualquier otra posición.

**Reversed-bias** A positive voltage is applied to the N-type material and negative voltage is applied to the P-type material of a semiconductor.

**Polarización inversa** Tensión positiva aplicada al material N y tensión negativa aplicada al material P de un semiconductor.

**Revolutions per minute (rpm) drop** The amount of rpm decrease when a cylinder stops firing for a brief time.

**Descenso de las revoluciones por minuto (rpm)** Cantidad que descienden las rpm cuando un cilindro detiene el encendido por un breve espacio de tiempo.

**Rheostat** A two-terminal variable resistor used to regulate the strength of an electrical current.

**Reóstato** Resistor variable de dos bornes utilizado para regular la resistencia de una corriente eléctrica.

**Ribbed V-belt** A belt containing a series of small "V" grooves on the underside of the belt.

**Correa nervada en V** Correa con una serie de ranuras pequeñas en forma de "V" en la superficie inferior de la misma.

**Ride height screen** A display on a computer wheel aligner that illustrates the locations for ride height measurement.

**Pantalla para la visualización de la altura del viaje** Representación visual en una computadora para alineación de ruedas que muestra los puntos donde se mide la altura del viaje.

**Ridge reamer** A cutting tool used to remove the ridge at the top of the cylinder.

**Escariador de reborde** Una herramienta de cortar que sirve para quitar el reborde en la parte superior del cilindro.

**Rim clamps** Special clamps designed to clamp on the wheel rims and allow the attachment of alignment equipment such as the wheel units on a computer wheel aligner.

**Abrazaderas de la llanta** Abrazaderas especiales diseñadas para sujetar las llantas de las ruedas y así permitir la fijación del equipo de alineación, como por ejemplo las unidades de la rueda en una computadora para alineación de ruedas.

**Ring gear** A gear that surrounds or rings the sun and planet gears in a planetary system. Also, the name given to the spiral bevel gear in a differential.

**Corona** Engranaje que rodea los engranajes planetario y el principal en un sistema planetario. También es el nombre que se le da al engranaje cónico con dentado espiral en un diferencial.

**Ring noise** A noise caused by worn rings or cylinders. Other causes include broken piston ring lands and too little tension of the ring against the cylinder wall.

**Ruido de los anillos** Un ruido causado por los anillos o cilindros desgastados. Otra causas incluyen las mesetas de pistones rotas o una tensión insuficiente del anillo contra el muro del cilindro.

**Ring ridge** A ridge near the top of the cylinder created by wear in the ring travel area of the cylinder.

**Reborde del anillo** Reborde cerca de la parte superior del cilindro ocasionado por un desgaste en el área de carrera del anillo del cilindro.

**Ring seating** The process of lapping the rings against the cylinder wall accomplished by the movement of the piston in the cylinder.

**Asiento del anillo** El proceso de asentar los anillos a pulso con el muro del cilindro que se lleva acabo por el movimiento del pistón dentro del cilindro.

**Rivet** A headed pin used for uniting two or more pieces by passing the shank through a hole in each piece and securing it by forming a head on the opposite end.

**Remache** Una clavija con cabeza que sirve para unir a dos piezas o más al pasar el vástago por un hoyo en cada pieza y asegurarlo por formar una cabeza en el extremo opuesto.

**Road feel** A feeling experienced by a driver during a turn when the driver has a positive feeling that the front wheels are turning in the intended direction.
**Sensación del camino** Sensación experimentada por un conductor durante un viraje cuando está completamente seguro de que las ruedas delanteras están girando en la dirección correcta.

**Road sensing suspension (RSS)** A computer-controlled suspension system that senses road conditions and adjusts suspension firmness to match these conditions in a few milliseconds.
**Suspensión con equipo sensor** Sistema de suspensión controlado por computadora que advierte las condiciones actuales del camino y ajusta la firmeza de la suspensión para equiparar dichas condiciones en un período de milisegundos.

**Rocker arm** Pivots that transfer the motion of the pushrods or followers to the valve stem.
**Balancín** Un punto pivote que transfiere el movimiento de las levantaválvulas o de los seguidores al vástago de la válvula.

**Rocker arm ratio** A mathematical comparison of rocker arm dimensions. The rocker arm ratio compares the center-to-valve-stem measurement against the center-to-pushrod measurement.
**Indice del balancín** Una comparación de las dimensiones del balancín. El índice del balancín compara las dimensiones del balancín del centro- al-vástago de la válvula con las dimensiones del centro-al-levantaválvulas.

**Rod beaming** The process of polishing the beams of the rods to prevent stress risers. This is done by blending the casting seam on the sides of the rods.
**Pulido de las varillas** El proceso de pulir los resaltos de las varillas para prevenir la deformación de colada. Esto se lleva acabo puliendo la mazarota en los lados de las varillas.

**Rod length to stroke ratio** A mathematical comparison between the length of the connecting rod and the length of the engine's stroke. It is determined by dividing the connecting rod length by the stroke.
**Indice de longitud de la biela a la carrera** Una comparación matemática entre la longitud de la biela y la longitud de la carrera del motor. Se determina dividiendo la longitud de la biela por la carrera.

**Roller bearing** An inner and outer race upon which hardened steel rollers operate.
**Cojinete de rodillos** Una pista interior y exterior en la cual operan los rodillos hecho de acero endurecido.

**Rollers** Round steel bearings that can be used as the locking element in an overrunning clutch or as the rolling element in an antifriction bearing.
**Rodillos** Articulaciones redondas de acero que pueden servir como un elemento de enclavamiento en un embrague de sobremarcha o como el elemento que rueda en un cojinete antifricción.

**Room temperature vulcanizing (RTV) sealant** A type of sealant that may be used to replace gaskets, or to help to seal gaskets, in some applications.
**Compuesto obturador vulcanizador a temperatura ambiente** Tipo de compuesto obturador que puede utilizarse para reemplazar guarniciones, o para ayudar a sellarlas, en algunas aplicaciones.

**Rotary flow** A fluid force generated in the torque converter that is related to vortex flow. The vortex flow leaving the impeller is not only flowing out of the impeller at high speed but is also rotating faster than the turbine. The rotating fluid striking the slower turning turbine exerts a force against the turbine, which is defined as rotary flow.
**Flujo rotativo** Una fuerza fluida producida en el convertidor de torsión que se relaciona al flujo torbellino. El flujo torbellino saliendo del rotor no sólo viaja en una alta velocidad sino también gira más rápidamente que el turbino. El fluido rotativo chocando contra el turbino que gira más lentamente, impone una fuerza contra el turbino que se define como flujo rotativo.

**Rotary valve** A valve that rotates to cover and uncover the intake port. A rotary valve is usually designed as a flat disc that is driven from the crankshaft.
**Válvula rotativa** Una válvula que gira para cubrir y descubrir la puerta de admisión. Suele ser diseñada como un disco plano impulsado por el cigüeñal.

**Rotor** The component of the AC generator that is rotated by the drive belt and creates the rotating magnetic field of the AC generator.
**Rotor** Parte rotativa del generador de corriente alterna accionada por la correa de transmisión y que produce el campo magnético rotativo del generador de corriente alterna.

**RPM** Abbreviation for revolutions per minute, a measure of rotational speed.
**RPM** Abreviación de revoluciones por minuto, una medida de la velocidad rotativa.

**RTV sealer** Room-temperature vulcanizing gasket material, which cures at room temperature; a plastic paste squeezed from a tube to form a gasket of any shape.
**Sellador RTV** Una materia vulcanizante de empaque que cura en temperaturas del ambiente; una pasta plástica exprimida de un tubo para formar un empaque de cualquiera forma.

**Rubber coupling** Rubber-based disc used as a universal joint between the driving and driven shafts.
**Acoplamiento de caucho** Disco con base de caucho; utilizado como junta universal entre el árbol de accionamiento y el árbol accionado.

**Running design change** A design change made during a current model/year production.
**Cambio al diseño corriente** Un cambio al diseño hecho durante la fabricación del modelo/año actual.

**Runout** Deviation of the specified normal travel of an object. The amount of deviation or wobble a shaft or wheel has as it rotates. Runout is measured with a dial indicator.
**Corrimiento** Una desviación de la carrera normal y especificada de un objeto. La cantidad de desviación o vacilación de una flecha o una rueda mientras que gira. El corrimiento se mide con un indicador de carátula.

**RV** Recreational vehicle.
**RV** Vehículo para el recreo.

**RWD** Abbreviation for rear-wheel drive.
**RWD** Abreviación de tracción trasera.

**Rzeppa constant velocity joint** The name given to the ball-type constant velocity joint (as opposed to the tripod-type constant velocity joint). Rzeppa joints are usually the outer joints on most FWD cars. Named after its inventor, Alfred Rzeppa, a Ford engineer.
**Junta de velocidad constante Rzeppa** Nombre que se le da a la junta de velocidad constante de tipo bola (en contraste con la junta de velocidad constante de tipo trípode). Las juntas Rzeppa normalmente son las juntas exteriores en la mayoría de automóviles de tracción delantera. Nombrada por su creador, Alfred Rzeppa, ingeniero de la Ford.

**Saddle** The portion of the crankcase bore that holds the bearing half in place.
**Asiento del cojinete (silleta)** La parte del taladro del cigüeñal que mantiene en su lugar a la mitad con el cojinete.

**Saddle valve** A two-part accessory valve that may be clamped around the metal part of a system hose to provide access to the air-conditioning system for service.
**Válvula de silleta** Válvula accesoria de dos partes que puede fijarse con una abrazadera a la parte metálica de una manguera del sistema para proveer acceso al sistema de acondicionamiento de aire para llevar a cabo servicio.

**SAE** Society of Automotive Engineers.
**SAE** Asociación de Ingenieros Automotrices.

**Safety** Freedom from danger or injury; the state of being safe.
**Seguridad** Libre de peligro o daño; calidad o estado de seguro.

**Safety goggles** Safety devices which provide eye protection from all sides. Goggles fit against the face and forehead to seal off the eyes from outside elements.
**Gafas de seguridad** Proporciona la protección a los ojos de todos lados siendo que quedan apretados contra la cara y la frente para formar un sello para los ojos contra los elementos exteriores.

**Safety stands** Commonly called jack stands and are used to support a vehicle when it is raised by a jack or hoist.
**Soportes de seguridad** Comúnmente llamados soportes de gato. Son utilizados para sostener un vehículo cuando es levantado por un gato o montacargas.

**Scale** The distance of the marks from each other on a measuring tool.
**Escala** La distancia entre las marcas en una herramienta de medir.

**Scan tool** A digital computer system tester used to read trouble codes and perform other diagnostic functions.
**Verificador de exploración** Verificador digital del sistema informático que se utiliza para leer códigos indicativos de problemas y llevar a cabo otras funciones diagnósticas.

**Schrader valve** A spring-loaded valve similar to a tire valve. The Schrader valve is located inside the service valve fitting and is used on some control devices to hold refrigerant in the system. Special adapters must be used with the gauge hose to allow access to the system.
**Válvula Schrader** Válvula con cierre automático parecida al vástago del neumático. La válvula Schrader está ubicada dentro del ajuste de la válvula de servicio y se utiliza en algunos dispositivos de regulación para guardar refrigerante dentro del sistema. Deben utilizarse adaptadores especiales con una manguera calibrador para permitir acceso al sistema.

**Score** A scratch, ridge, or groove marring a finish surface.
**Raya** Un rasguño, una arruga, o una muesca que echa a perder una superficie.

**Scuffing** Scraping and heavy wear between two surfaces.
**Erosión** El rozamiento y desgaste fuerte entre dos superficies.

**Seal** Component used to seal between a stationary part and a moving one.
**Junta** Un componente que se emplea para sellar entre una parte fija y una que mueva.

**Seal drivers** Designed to maintain even contact with a seal case and prevent seal damage during installation.
**Empujadores para sellar** Diseñados para mantener un contacto uniforme con el revestimiento de la junta de estanqueidad y evitar el daño de la misma durante el montaje.

**Seal seat** The part of a compressor shaft seal assembly that is stationary and matches the rotating part, known as the seal face or shaft seal.
**Asiento de la junta hermética** Parte del conjunto de la junta hermética del árbol del compresor que es inmóvil y que se empareja a la parte rotativa; conocido como la frente de junta hermética o la junta hermética del árbol.

**Sealant** A special liquid material commonly used to fill irregularities between the gasket and its mating surface. Some sealants are designed to be used in place of a gasket.
**Compuesto obturador** Un material líquido especial que normalmente se emplea para rellenar las irregularidades entre un empaque y su superficie de contacto. Algunos compuestos son diseñados de uso sin empaque.

**Sealed-beam headlight** A self-contained glass unit that consists of a filament, an inner reflector, and an outer glass lens.
**Faro delantero sellado** Unidad de vidrio que contiene un filamento, un reflector interior y una lente exterior de vidrio.

**Sealer** A thick, tacky compound, usually spread with a brush, which may be used as a gasket or sealant to seal small openings or surface irregularities.

**Sellador** Un compuesto pegajoso y espeso, comúnmente aplicado con una brocha, que puede usarse como un empaque o un obturador para sellar a las aperturas pequeñas o a las irregularidades de la superficie.

**Seat** A surface, usually machined, upon which another part rests or seats—for example, the surface upon which a valve face rests.
**Asiento** Una superficie, comúnmente maquinada, sobre la cual yace o se asienta otra parte; por ejemplo, la superficie sobre la cual yace la cara de la válvula.

**Seat pressure** Term which indicates spring tension with the spring at installed height and the valve closed.
**Presión del asiento** Un término que indica la tensión del resorte en su altura de instalación con la válvula cerrada.

**Seat runout (concentricity)** A measure of how circular the valve seat is in relation to the valve guide.
**Excentricidad del asiento (concentricidad)** Una medida de lo circular del asiento de la válvula con relación a la guía de la válvula.

**Second order vibration** Vibration which occurs twice per revolution.
**Vibración de segunda orden** Una vibración que ocurre dos veces por revolución.

**Secondary air injection (AIR) system** A system that injects air into the exhaust system from a belt-driven pump.
**Sistema de inyección secundaria de aire** Sistema que inyecta aire dentro del sistema de escape desde una bomba accionada por correa.

**Secondary shoe** When the car is moving forward, the shoe facing the rear is the trailing or secondary shoe.
**Zapata secundaria** Al moverse hacia frente el coche, la zapata en la dirección hacia la parte trasera del coche es la zapata seguidora o secundaria.

**Secondary wiring** Conductors, such as battery cables and ignition spark plug wires, that are used to carry high voltage or high current. Secondary wires have extra thick insulation.
**Hilos secundarios** Conductores, tales como cables de batería e hilos de bujías del encendido, utilizados para transmitir tensión o corriente alta. Los hilos secundarios poseen un aislamiento sumamente grueso.

**Section modulus** The measurement of a frame's strength based on height, width, thickness, and the shape of the side rails.
**Coeficiente de sección** Medida de la resistencia de un armazón, basada en la altura, el ancho, el espesor y la forma de las vigas laterales.

**Seize** When one surface moving upon another causes scratches. The metal transfer can become severe enough to cause the moving component to stop.
**Rayar** Cuando los movimientos de una superficie sobre otra causan las rayas. La transferencia del metal puede ser tan severa que para al componente en movimiento.

**Selective shim** Shims of different thicknesses that are used to provide an adjustment such as bearing preload.
**Laminillas selectivas** Laminillas de diferentes espesores que se utilizan para proveer un ajuste, como por ejemplo la carga previa del cojinete.

**Self-adjusting clutch linkage** Monitors clutch pedal play through a clutch control cable and ratcheting mechanism to automatically adjust clutch pedal play.
**Biela motriz del embrague de ajuste automático** Controla el juego del pedal del embrague mediante un cable de mando del embrague y un mecanismo de trinquete para ajustar automáticamente el juego del pedal del embrague.

**Self-energizing** The increase in friction contact between the toe of the brake shoe caused by the drum rotation tending to pull the shoe into the drum.
**Autoenergético** El incremento del contacto frotativo entre la parte superior del freno producido por la rotación del tambor que tiene una tendencia a jalar a la zapata hacia el tambor.

**Self-locking nut** A special nut designed so it will not loosen because of vibration.
Tuerca de cierre automático Tuerca con un diseño especial que evita su aflojamiento a causa de la vibración.

**Self-powered test light** A test light powered by an internal battery.
Luz de prueba propulsada automáticamente Luz de prueba propulsada por una batería interna.

**Self-test input wire** A diagnostic wire located near the data link connector (DLC) on Ford vehicles.
Alambre de entrada de prueba automática Alambre diagnóstico ubicado cerca del conector de enlace diagnóstico en vehículos fabricados por la Ford.

**Semicentrifugal pressure plate** The release levers of this pressure plate are weighted to take advantage of centrifugal force to increase plate loading, resulting in reduced driven disc slip.
Placa de presión semicentrífuga A las palancas de desembrague de esta placa de presión se les añade peso para aprovechar la fuerza centrífuga y hacer que ésta aumente la carga de la placa. El resultado será un deslizamiento menor del disco accionado.

**Semiconductors** An element that is neither a conductor nor an insulator. Semiconductors are materials that conduct electric current under certain conditions, yet will not conduct under other conditions.
Semiconductores Elemento que no es ni conductor ni aislante. Los semiconductores son materiales que transmiten corriente eléctrica bajo ciertas circunstancias, pero no la transmiten bajo otras.

**Semifloating rear axle** An axle that supports the weight of the vehicle on the axle shaft in addition to transmitting driving forces to the rear wheels.
Eje trasero semi-flotante Eje que apoya el peso del vehículo sobre el árbol motor además de transmitir las fuerzas motrices a las ruedas traseras.

**Sender unit** The sensor for the gauge. It is a variable resistor that changes resistance values with changing monitored conditions.
Unidad emisora Sensor para el calibrador. Es un resistor variable que cambia los valores de resistencia según cambian las condiciones reguladas.

**Sensor** Any device that provides an input to the computer.
Sensor Cualquier dispositivo que le transmite información a la computadora.

**Separators** A component in an antifriction bearing that keeps the rolling components apart.
Separadores Componente en un cojinete de antifricción que mantiene los componentes de rodamiento separados.

**Sequential fuel injection (SFI)** A fuel injection system in which the injectors are individually grounded into the computer.
Inyección de combustible en ordenamiento Sistema de inyección de combustible en el que los inyectores se ponen individualmente a tierra en la computadora.

**Series circuit** A circuit that provides a single path for current flow from the electrical source through all the circuit's components, and back to the source.
Circuito en serie Circuito que provee una trayectoria única para el flujo de corriente de la fuente eléctrica a través de todos los componentes del circuito, y de nuevo hacia la fuente.

**Series-parallel circuit** A circuit that has some loads in series and some in parallel.
Circuito en series paralelas Circuito que tiene unas cargas en serie y otras en paralelo.

**Serpentine belt** A ribbed V-belt drive system in which all the belt-driven components are on the same vertical plane.
Correa serpentina Sistema de transmisión con correa nervada en V en el que todos los componentes accionados por una correa se encuentran sobre el mismo plano vertical.

**Service bay diagnostics** A diagnostic system supplied by Ford Motor Company to their dealers that diagnoses vehicles and provides communication between the dealer and manufacturer.
Diagnóstico para el puesto de reparación Sistema diagnóstico que les suministra la Ford a sus distribuidores de automóviles; dicho sistema diagnostica vehículos y permite una buena comunicación entre el distribuidor y el fabricante.

**Service check connector** A diagnostic connector used to diagnose computer systems. A scan tester may be connected to this connector.
Conector para la revisión de reparaciones Conector diagnóstico que se utiliza para diagnosticar sistemas informáticos. A dicho conector se le puede conectar un verificador de exploración.

**Service manual** One of the most important tools for today's technician, the service manual provides information concerning engine identification, service procedures, and specifications. In addition, the service manual provides information regarding wiring harness connections and routing, component location, and fluid capacities. Service manuals may be obtained from the vehicle manufacturer or through aftermarket suppliers.
Manual de servicio Una de las herramientas más importantes del técnico moderno. Proporciona la información sobre la identificación del motor, los procedimientos del servicio, y las especificaciones. Además, el manual de servicio provee la información sobre las conexiones y los rumbos del mazo de alambres, ubicación de los componentes y las capacidades de los fluidos. Los manuales se pueden obtener del fabricante del vehículo o por los proveedores de repuestas.

**Service port** A fitting found on the service valves and some control devices; the manifold set hoses are connected to this fitting.
Orificio de servicio Ajuste ubicado en las válvulas de servicio y en algunos dispositivos de regulación; las mangueras del conjunto del colector se conectan a este ajuste.

**Service procedure** A suggested routine for the step-by-step act of troubleshooting, diagnosing, and/or repairs.
Procedimiento de servico Rutina sugerida para la acción a seguir paso a paso para detectar fallas, diagnosticar, y/o reparar.

**Service sleeve (speedy-sleeve)** A metal sleeve that is pressed over a damaged sealing area to provide a new, smooth surface for the seal lip.
Camisa de servicio (manguito rápido) Una camisa de metal que se coloca sobre una área de sello dañada que provee una superficie nueva y lisa para el borde del sello.

**Service valve** See *High-side* (*Low-side*) *service valve.*
Válvula de servicio Ver *High-side* (*Low-side*) *service valve* [*Válvula de servicio del lado de alta presión (baja presión)*].

**Servo** A device that converts hydraulic pressure into mechanical movement, often multiplying it. Used to apply the bands of a transmission.
Servo Un dispositivo que convierte la presión hidráulica al movimiento mecánico, frequentemente multiplicándola. Se usa en la aplicación de las bandas de una transmisión.
Materia de laminillas Las hojas de metal cuyo espesor se conoce precisamente que pueden cortarse en tiras y usarse para medir o corregir a las holguras.

**Servomotor** An electrical motor that produces rotation of less than a full turn. A feedback mechanism is used to position itself to the exact degree of rotation required.
Servomotor Motor eléctrico que genera rotación de menos de una revolución completa. Utiliza un mecanismo de realimentación para ubicarse al grado exacto de la rotación requerida.

**Setback** Occurs when one front wheel is driven rearward in relation to the opposite front wheel.
Retroceso Condición que ocurre cuando una de las ruedas delanteras se mueve hacia atrás con relación a la rueda delantera opuesta.

**Shaft key** A soft metal key that secures a member on a shaft to prevent it from slipping.

**Chaveta del árbol** Chaveta de metal blando que fija una pieza a un árbol para evitar su deslizamiento.

**Shaft seal** See *Compressor shaft seal.*
**Junta hermética del árbol** Ver *Compressor shaft seal* [*Junta hermética del árbol del compresor*].

**Sheared injected plastic** Is inserted into steering column shafts and gearshift tubes to allow these components to collapse if the driver is thrown against the steering column in a collision.
**Plástico cortado inyectado** Insertado en los árboles de la columna de dirección y en los tubos del cambio de engranajes de velocidad para permitir que estos componentes se pleguen si la columna de dirección es impactada por el conductor durante una colisión.

**Shift forks** Mechanisms attached to shift rails that fit into the synchronizer hub for change of gears.
**Horquillas de cambio de velocidades** Las ranuras en el anillo sincronizador del embrague cónico deben ser afiladas para lograr la sincronización.

**Shift lever** The lever used to change gears in a transmission. Also the lever on the starting motor that moves the drive pinion into or out of mesh with the flywheel teeth.
**Palanca del cambiador** La palanca que sirve para cambiar a las velocidades de una transmisión. También es la palanca del motor de arranque que mueva al piñón de ataque para engranarse o desengranarse con los dientes del volante.

**Shift rails** Rods placed within the transmission housing that are a part of the transmission gearshift linkage.
**Barras de cambio de velocidades** Varillas ubicadas dentro del alojamiento de transmisión que forman parte de la biela motriz del cambio de velocidades de la transmisión.

**Shift valve** A valve that controls the shifting of the gears in an automatic transmission.
**Válvula de cambios** Una válvula que controla a los cambios de las velocidades en una transmisión automática.

**Shim** Thin sheets used as spacers between two parts, such as the two halves of a journal bearing.
**Laminilla de relleno** Hojas delgadas que sirven de espaciadores entre dos partes, tal como las dos partes de un muñón.

**Shim select screen** A display on a computer wheel aligner that shows the technician the thickness and location of the proper shims for rear wheel alignment.
**Pantalla para la visualización durante la selección de laminillas** Representación visual en una computadora para alineación de ruedas que le muestra al mecánico el espesor y la ubicación de las laminillas correctas para la alineación de las ruedas traseras.

**Shim stock** Sheets of metal of accurately known thickness which can be cut into strips and used to measure or correct clearances.
**Materia de laminillas** Las hojas de metal cuyo espesor se conoce precisamente que pueden cortarse en tiras y usarse para medir o corregir a las holguras.

**Shock absorber manual test** A test in which the lower end of the shock absorber is disconnected and the shock absorber is operated manually to determine its condition.
**Prueba manual del amortiguador** Prueba en la que se desconecta el extremo inferior del amortiguador con el fin de poder operarlo manualmente y lograr determinar su condición.

**Shock absorber or strut bounce test** A test in which the vehicle is bounced by leaning on the bumper to determine the shock absorber condition.
**Prueba de rebote del montante o del amortiguador** Prueba en la que se ejerce presión sobre el parachoques con el fin de hacer rebotar el vehículo y lograr determinar la condición del amortiguador.

**Shop layout** The location of all shop facilities including service bays, equipment, safety equipment, and offices.

**Arreglo del taller de reparación** Ubicación de todas las instalaciones del taller, incluyendo los puestos de reparación, el equipo, el equipo de seguridad, y las oficinas.

**Short** An electrical defect that allows electrical current to bypass its normal path.
**Cortocircuito** Un defecto eléctrico que permite que el corriente eléctrico sobrepasa su rumbo normal.

**Shot** Round beads which are used for cleaning when etching of the metal is not desired.
**Granalla** Las bolitas que se emplean en la limpieza cuando el grabado del metal no es deseado.

**Shot-peening** A tempering process that uses shot under pressure to tighten the outside surface of the metal. Shot peening is used to help strengthen the component and reduce the chances of stress or surface cracks.
**Chorreo con granalla** Un proceso de templado que usa la granalla bajo presión para estrechar la superficie exterior del metal. El chorreo con granalla fortalece al componente y disminuye la apariencia del fatiga o de grietas en la superficie.

**Shudder** A shake or shiver movement.
**Estremecimiento** Sacudida o temblor.

**Shut-off valve** A valve that provides positive shut-off of a fluid or vapor passage.
**Válvula de cierre** Válvula que provee el cierre positivo del pasaje de un fluido o un vapor.

**Shutter wheel** A metal wheel consisting of a series of alternating windows and vanes. It creates a magnetic shunt that changes the strength of the magnetic field from the permanent magnet of the Hall-effect switch or magnetic pulse generator.
**Rueda obturadora** Rueda metálica compuesta de una serie de ventanas y aspas alternas. Genera una derivación magnética que cambia la potencia del campo magnético, del imán permanente del conmutador de efecto Hall o del generador de impulsos magnéticos.

**SI measuring system** The metric measuring system.
**Sistema de medida SI** Sistema métrico de medida.

**Side clearance** The clearance between the sides of moving parts when the sides do not serve as load-carrying surfaces.
**Holgura lateral** La holgura entre los lados de las partes en movimiento mientras que los lados no funcionan como las superficies de carga.

**Side gears** Gears that are meshed with the differential pinions and splined to the axle shafts (RWD) or drive shafts (FWD).
**Engranajes laterales** Engranajes que se engranan con los piñones del diferencial y son ranurados a los árboles motores en vehículos de tracción trasera o a los árboles de mando en vehículos de tracción delantera.

**Side thrust** Longitudinal movement of two gears.
**Empuje lateral** Movimiento longitudinal de dos engranajes.

**Silicon** A nonmetallic element that can be doped to provide good lubrication properties. The melting point of silicon is 2,570°F (1,410°C).
**Silicio** Un elemento no metálico que se puede agregar para proporcionar las propriedades buenas de la lubricación. El silicio se funde hacia 2,570°F (1,410°C).

**Silicone grease** A heat-dissipating grease placed on components such as ignition modules.
**Grasa de silicón** Grasa para disipar el calor utilizada en componentes, como por ejemplo módulos del encendido.

**Single-pull unitized body straightening system** Hydraulically operated equipment that pulls in one location while straightening unitized bodies.
**Sistema enderezador de tiro único para la carrocería unitaria** Sistema activado hidráulicamente que tira hacia una dirección al enderezar carrocerías unitarias.

**Sizing point** The location the manufacturer designates for measuring the diameter of the piston to determine clearance.
Punto de calibración El lugar indicado por el fabricante para efectuar las medidas del diámetro del pistón para determinar la holgura.

**Slave cylinder** Located at a lower part of the clutch housing. Receives fluid pressure from the master cylinder to engage or disengage the clutch.
Cilindro secundario Ubicado en la parte inferior del alojamiento del embrague. Recibe presión de fluido del cilindro primario para engranar o desengranar el embrague.

**Sleeving** The process of boring the cylinder to accept a sleeve.
Preparar para camisa El proceso de taladrar un cilindro para aceptar una camisa.

**Sliding-fit** Where sufficient clearance has been allowed between the shaft and journal to allow free-running without overheating.
Ajuste corredera Donde se ha dejado una holgura suficiente entre la flecha y el muñón para permitir una marcha libre sin sobrecalentamiento.

**Sliding gear transmission** A transmission in which gears are moved on their shafts to change gear ratios.
Transmisión por engranaje desplazable Transmisión en la cual los engranajes se mueven sobre sus árboles para cambiar la relación de los engranajes.

**Sliding yoke** Slides on internal and external splines to compensate for driveline length changes.
Yugo deslizante Se desliza sobre las lengüetas internas y externas para compensar los cambios de longitud de la línea de transmisión.

**Slip fit** Running or sliding fit.
Ajuste corredizo Ajuste deslizante o de marcha.

**Slip joint** In the powertrain, a variable-length connection that permits the drive shaft to change its effective length.
Junta corrediza En el tren transmisor de potencia, una conexión de longitud variable que le permite al árbol de mando cambiar su longitud eficaz.

**Slipper skirt** A piston skirt ground to provide additional clearance between the piston and the counterweights of the crankshaft. Without this recessed area, the piston would contact the crankshaft when shorter connecting rods are used.
Faldilla deslizante Una faldilla del pistón rectificada para proveer una holgura adicional entre el pistón y los contrapesos del cigüeñal. Sin esta área rebajada, el pistón podría rozar contra el cigüeñal cuando se emplean las bielas más cortas.

**Slitting tool** A special chisel designed for slitting exhaust system pipes.
Herramienta de hender Cincel especial diseñado para hendir los tubos del sistema de escape.

**Slow charging** Battery charging rate between 3 and 15 amps for a long period of time.
Carga lenta Indice de carga de la batería de entre 3 y 15 amperios por un largo espacio de tiempo.

**Slow cranking** A defect that occurs when the starter drive engages the ring gear, but the engine turns at too slow of a speed to start. Some manufacturers provide specifications for engine cranking speed.
Arranque lento Un defecto que ocurre cuando el acoplamiento del motor de arranque impulsa a la corona del volante, pero el motor gira con una velocidad demasiado lento para arrancar. Algunos fabricantes proveen las especificaciones para la velocidad del arranque del motor.

**Small end** Term that refers to the end of the connecting rod that accepts the piston pin.
Extremo pequeño Un término que refiere a la extremidad de la biela que acepta el eje del pistón.

**Small-hole gauge** An instrument used to measure holes or bores that are smaller than a telescoping gauge can measure.

Calibre de taladros chicos Un instrumento que se emplea para medir los agujeros o taladros que son demasiado pequeños para medirse con un calibrador telescópico.

**Snap ring** Split spring-type ring located in an internal or external groove to retain a part.
Anillo de seguridad Un anillo partido tipo resorte que se coloca en una muesca interior o exterior para retener a una parte.

**Snap shot testing** The process of freezing computer data into the scan tester memory during a road test and then reading this data later.
Prueba instantánea Proceso de capturar datos de la computadora en la memoria del instrumento de pruebas de exploración durante una prueba en carretera y leer dichos datos más tarde.

**Soak tanks** Cleaning tanks equipped with a large basket which holds the parts while they are submerged into a caustic solution or detergent. Some soak tanks are equipped with an agitation system.
Tanques (cubos) de remojo Los tanques equipados con un capacho para sostener las partes que se sumergen en una solución cáustica o en el detergente. Algunos tanques tienen un sistema de agitación.

**Society of Automotive Engineers** A professional organization of the automotive industry. Founded in 1905 as the Society of Automobile Engineers, the SAE is dedicated to providing technical information and standards to the automotive industry. Present goals are to assure a skilled engineering and technical workforce for the year 2000 and beyond. The goal, known as VISION 2000, encompasses all of SAE's educational programs, including student competitions, scholarships, teacher recognition, and more.
Sociedad de Ingenieros Automotrices Organización profesional de la industria automotriz. Establecido en 1905 como la Sociedad de Ingenieros de Automóviles (SAE por sus siglas en inglés), dicha sociedad se dedica a proveerle información técnica y normas a la industria automotriz. Sus metas actuales son asegurar una fuerza laboral capacitada en la ingeniería y en el campo técnico para el año 2000 y después. La meta, conocida como VISION 2000, abarca todos los programas educativos de la SAE, e incluye concursos entre estudiantes, el otorgar becas, el reconocer al profesor, y más.

**Soft codes** Codes are those that have occurred in the past, but were not present during the last BCM test of the circuit.
Códigos suaves Códigos que han ocurrido en el pasado, pero que no estaban presentes durante la última prueba BCM del circuito.

**Soft-jaw vise** A vise equipped with soft metal, such as copper, in the jaws.
Tornillo con tenacilla maleable Tornillo equipado con un metal blando, como por ejemplo cobre, en las tenacillas.

**Soldering** The process of using heat and solder (a mixture of lead and tin) to make a splice or connection.
Soldadura Proceso a través del cual se utiliza calor y soldadura (una mezcla de plomo y de estaño) para hacer un empalme o una conexión.

**Solderless connectors** Hollow metal tubes that are covered with insulating plastic. They can be butt connectors or terminal ends.
Conectadores sin soldadura Tubos huecos de metal cubiertos de plástico aislante. Pueden ser extremos de conectadores o de bornes.

**Solenoid** An electromagnetic device that uses movement of a plunger to exert a pulling or holding force.
Solenoide Dispositivo electromagnético que utiliza el movimiento de un pulsador para ejercer una fuerza de arrastre o de retención.

**Solenoid circuit resistance test** Diagnostic test used to determine the electrical condition of the solenoid and the control circuit of the starting system.
Prueba de la resistencia de un circuito solenoide Prueba diagnóstica utilizada para determinar la condición eléctrica del solenoide y del circuito de mando del sistema de arranque.

**Solenoid valve** An electromagnetic valve controlled remotely by electrically energizing and deenergizing a coil.

**Válvula de solenoide** Válvula electromagnética regulada a distancia por una bobina al excitar y deexcitar una bobina electrónicamente.

**Solid axle** A rear axle design that places the final drive, axles, bearings, and hubs into one housing.
**Eje sólido** Diseño del eje trasero que coloca la transmisión final, los ejes, los cojinetes y los cubos dentro de un solo alojamiento.

**Solid lifters (mechanical lifters)** Components which provide a rigid connection between the camshaft and the valves.
**Levantaválvulas macizas (o mecánicas)** Los componentes que proveen una conexión rígida entre el árbol de levas y las válvulas.

**Solid state** Referring to electronics consisting of semiconductor devices and other related nonmechanical components.
**Estado sólido** Se refiere a componentes electrónicos que consisten en dispositivos semiconductores y otros componentes relacionados no mecánicos.

**Spade-type connector** A single or multiple electrical connector that has flat spade-like mating provisions.
**Conectador de tipo azadón** Conectador único o múltiple provisto de dispositivos planos de tipo azadón para emparejarse.

**Spalling** A condition where the material of a bearing surface breaks away from the base metal.
**Escamación** Una condición en la cual una materia de la superficie de un rodamiento se separa del metal base.

**Span gauge** A commonly used term for a telescoping gauge.
**Indicador de extensión** Término común para indicador telescópico.

**Specific gravity** A unit measurement for determining the sulfuric acid content of an electrolyte.
**Gravedad específica** Una unidad de medida para determinar el contenido del ácido sulfúrico en el electrolito.

**Specifications** Design characteristics of a component or assembly noted by the manufacturer. Specifications for a vehicle include fluid capacities, weights, and other pertinent maintenance information.
**Especificaciones** Características de diseño de un componente o conjunto indicadas por el fabricante. Las especificaciones para un vehículo incluyen capacidades del fluido, pesos, y otra información pertinente para mantenimiento del vehículo.

**Specifications menu** A display on a computer wheel aligner that allows the technician to select, enter, alter, and display vehicle specifications.
**Menú de especificaciones** Representación visual en una computadora para alineación de ruedas que le permite al mecánico seleccionar, introducir datos, cambiar e indicar especificaciones referentes al vehículo.

**Speed gears** Driven gears located on the transmission output shaft. This term differentiates between the gears of the counter gear and cluster assemblies and gears on the transmission output shaft.
**Engranajes de velocidades** Engranajes accionados ubicados en el árbol de rendimiento de la transmisión. Este término distingue entre los engranajes de los conjuntos del mecanismo contador y de los engranajes desplazables y los engranajes sobre el árbol de rendimiento de la transmisión.

**Speedometer** An instrument panel gauge that indicates the speed of the vehicle.
**Velocímetro** Calibrador en el panel de instrumentos que marca la velocidad del vehículo.

**Spike** In our application, an electrical spike. An unwanted momentary high-energy electrical surge.
**Impulso afilado** En nuestro campo, un impulso afilado eléctrico. Una elevación repentina eléctrica de alta energía no deseada.

**Spindle** The shaft on which the wheels and wheel bearings mount.
**Husillo** La flecha en la cual se montan las ruedas y el conjunto del cojinete de las ruedas.

**Spiral bevel gear** A ring gear and pinion wherein the mating teeth are curved and placed at an angle with the pinion shaft.
**Engranaje cónico con dentado espiral** Corona y piñón cuyos dientes emparejados son curvos y están montados en ángulo con el árbol de piñón.

**Spiral cable** A conductive ribbon mounted in a plastic container on top of the steering column that maintains electrical contact between the air bag inflator module and the air bag electrical system. A spiral cable may be called a clock spring electrical connector.
**Cable espiral** Cinta conductiva montada en un recipiente plástico sobre la columna de dirección que mantiene el contacto eléctrico entre la unidad infladora y el sistema eléctrico del Airbag. El cable espiral se conoce también como conector eléctrico de cuerda de reloj.

**Spiral gear** A gear with teeth cut according to a mathematical curve on a cone. Spiral bevel gears that are not parallel have center lines that intersect.
**Engranaje helocoidal** Engranaje con dientes cortados en un cono según una curva matemática. Los engranajes cónicos con dentado espiral que no son paralelos tienen líneas centrales que se cruzan.

**Splice** The joining of single wire ends or the joining of two or more electrical conductors at a single point.
**Empalme** La unión de los extremos de un alambre o la unión de dos o más conductores eléctricos en un solo punto.

**Splice clip** A special connector used along with solder to ensure a good connection. The splice clip is different from solderless connectors in that it does not have insulation.
**Grapa para empalme** Conectador especial utilizado junto con la soldadura para garantizar una conexión perfecta. La grapa para empalme se diferencia de los conectadores sin soldadura porque no está provista de aislamiento.

**Spline** Slot or groove cut in a shaft or bore; a splined shaft onto which a hub, wheel, gear, and so on, with matching splines in its bore is assembled so that the two must turn together.
**Acanaladura (espárrago)** Una muesca o ranura cortada en una flecha o en un taladro; una flecha acanalada en la cual se montan un cubo, una rueda, un engranaje, y todo lo demás que tiene una acanaladura pareja en el taladro de manera de que las dos deben girar juntos.

**Splined hub** Several keys placed radially around the inside diameter of a circular part, such as a wheel or driven disc.
**Cubo ranurado** Varias chavetas ubicadas de manera radial alrededor del diámetro interior de una pieza circular, como por ejemplo, una rueda o un disco accionado.

**Split lip seal** Typically, a rope seal sometimes used to denote any two-part oil seal.
**Sello hendido** Típicamente, un sello de cuerda que se usa a veces para demarcar cualquier sello de aceite de dos partes.

**Split pin** A round split spring steel tubular pin used for locking purposes—for example, locking a gear to a shaft.
**Chaveta hendida** Una chaveta partida redonda y tubular hecho de acero para resorte que sirve para el enclavamiento; por ejemplo, para enclavar un engranaje a una flecha.

**Spongy pedal** A condition where the brake pedal does not give firm resistance to foot pressure. Normally caused by air in the hydraulic system.
**Resistencia esponjosa** Una condición en la cual el pedal de freno no ofrece una resistencia firme a la presión del pie. Suele ser causado por la presencia del aire en el sistema hidráulico.

**Spontaneous combustion** A fire that occurs spontaneously. For example, heat is slowly generated from oxidation of oil on rags; the heat continues to increase until the flash point is reached and the rags suddenly begin to burn.
**Combustión espontánea** Un fuego que ocurre espontáneamente. Por ejemplo, el calor se produce lentamente por la oxidación del aceite

en los trapos; el calor continua a aumentarse hasta que llega a la temperatura de inflamabilidad y los trapos comienzan a quemarse.

**Spool valve** A cylindrically shaped valve with two or more valleys between the lands. Spool valves are used to direct fluid flow.
**Válvula de carrete** Una válvula de forma cilíndrica que tiene dos acanaladuras de cañón o más entre las partes planas. Las válvulas de carrete sirven para dirigir el flujo del fluido.

**Sprag clutch** A member of the overrunning clutch family using a sprag to jam between the inner and outer races used for holding or driving action.
**Embrague de puntal** Un miembro de la familia de embragues de sobremarcha que usa a una palanca de detención trabada entre las pistas interiores y exteriores para realizar una acción de asir o marchar.

**Spread** Descriptive term applied when the diameter at the outside parting edges of a bearing shell exceeds the inside diameter of the mating housing bore.
**Aplastamiento** Un término descriptivo que se aplica cuando el diámetro de los bordes exteriores del casquillo de un cojinete es más grande que el diámetro interior del taladro en el superficie de contacto del cárter.

**Spring** A device that changes shape when it is stretched or compressed but returns to its original shape when the force is removed; the component of the automotive suspension system that absorbs road shocks by flexing and twisting.
**Resorte** Un dispositivo que cambia de forma al ser estirado o comprimido, pero que recupera su forma original al levantarse la fuerza; es un componente del sistema de suspensión automotivo que absorba los choques del camino al doblarse y torcerse.

**Spring fill diagnostics** A diagnostic procedure in a computer-controlled air suspension system.
**Diagnóstico con relleno para muelles** Procedimiento diagnóstico en un sistema de suspensión de aire controlado por computadora.

**Spring free length** The height the spring stands when not loaded.
**Longitud libre del resorte** La altura del resorte cuando no tiene carga.

**Spring insulators, or silencers** Rings usually made from plastic and positioned on each end of a coil spring to reduce noise and vibration transfer to the chassis.
**Aisladores de muelles, o silenciadores** Anillos, por lo general fabricados de plástico y colocados a ambos extremos de un muelle helicoidal, que se utilizan para disminuir la transferencia de ruido y de vibración al chasis.

**Spring lock fitting** A special fitting using a spring to lock the mating parts together forming a leakproof joint.
**Ajuste de cierre automático** Ajuste especial utilizando un resorte para cerrar piezas emparejadas para formar así una junta hermética contra fugas.

**Spring retainer** A steel plate designed to hold a coil or several coil springs in place.
**Retén de resorte** Una chapa de acero diseñado a sostener en su posición a un resorte helicoidal o más.

**Spring sag** Occurs when a spring becomes weak, and the curb riding height is reduced compared to the original height.
**Agotamiento de los muelles** Condición que ocurre cuando el muelle se debilita; la altura del cotén del viaje disminuye si se la compara con la altura orginal.

**Spring shims** Components used to correct installed height of the valve spring. Spring shims are used to correct for machining tolerance.
**Chapas de relleno** Los componentes que se emplean para ajustar la altura de instalación del resorte de la válvula. El sobreespesor del maquinado se puede ajustar por medio de estas chapas.

**Spring squareness** Refers to how true to vertical the entire spring is.
**Escuadrado del resorte** Refiere a si la posición del resorte entero está en línea recta al vertical.

**Spur gear** Gears cut on a cylinder with teeth that are straight and parallel to the axis.
**Engranaje recto** Engranajes cortados en un cilindro que tienen dientes rectos y paralelos al pivote.

**Squeak** A high-pitched noise of short duration.
**Chillido** Un ruido agudo de poca duración.

**Squeal** A continuous high-pitched noise.
**Alarido** Un ruido agudo continuo.

**Squirrel-cage blower** A blower wheel designed to provide a large volume of air with a minimum of noise. The blower is more compact than the fan and air can be directed more efficiently.
**Soplador con jaula de ardilla** Rueda de soplador diseñada para proveer un gran caudal de aire con un mínimo de ruido. El soplador es más compacto que el ventilador y el aire puede dirigirse con un mayor rendimiento.

**Squish area** The area of the combustion chamber where the piston is very close to the cylinder head. The air-fuel mixture is rapidly pushed out of this area as the piston approaches TDC, causing turbulence and forcing the mixture toward the spark plug. The squish area can also double as the quench area.
**Area de compresión** El área de la cámara de combustión en donde el pistón está muy cerca a la cabeza del cilindro. La mezcla del aire/combustible se expulsa rápidamente de esta área al aproximarse el pistón al PMS, causando una turbulencia y empujando la mezcla hacia la bujía. El área de compresión tambien puede servir de área de extinción.

**Stabilize** To make steady.
**Estabilizar** Quedarse detenida una cosa.

**Stabilizer bar** Also called a sway bar. It prevents the vehicle's body from diving into turns.
**Barra estabilizadora** Llamada también barra de oscilación lateral. Impide que la carrocería del vehículo se desestabilice durante los virajes.

**Staking punch** A chisel-like punch used to create a large dimple in metal. This dimple prevents the nut from self-adjusting.
**Punzón de estacas** Punzón parecido a un cincel, utilizado para hacer un agujero grande en un metal. Esta abolladura impide que la tuerca se ajuste automáticamente.

**Stall** A condition where the engine is operating and the transmission is in gear, but the drive wheels are not turning because the turbine of the torque converter is not moving.
**Paro** Una condición en la cual opera el motor y la transmisión está embragada pero las ruedas de impulso no giran porque no mueva el turbino del convertidor de la torsión.

**Stall test** A test of the one-way clutch in a torque converter.
**Prueba de paro** Una prueba del embrague de una vía en un convertidor de la torsión.

**Star-adjuster** Star-shaped rotor used as an adjustment device in drum brakes.
**Ajustador de estrella y tornillo** Un rotor en forma de estrella que se usa como dispositivo de ajuste en los frenos de tambor.

**Starter drive** The part of the starter motor that engages the armature to the engine flywheel ring gear.
**Transmisión de arranque** Parte del motor de arranque que engrana la armadura a la corona del volante de la máquina.

**Starting air valve** A vacuum-operated valve that supplies more air into the intake manifold when starting the engine.
**Válvula de aire para el arranque** Válvula accionada por vacío que le suministra mayor cantidad de aire al colector de aspiración durante el arranque del motor.

**State of charge** The condition of a battery's electrolyte and plate materials at any given time.
**Estado de carga** Condición del electrolito y de los materiales de la placa de una batería en cualquier momento dado.

**Static** A form of electricity caused by friction.
**Estático** Una forma de la electricidad causada por la fricción.

**Static imbalance** Refers to the imbalance of a wheel and tire at rest.
**Desequilibrio estático** Se refiere al desequilibrio de una rueda y de un neumático cuando se ha detenido la marcha del vehículo.

**Stator** The stationary coil of the AC generator in which current is produced.
**Estátor** Bobina fija del generador de corriente alterna donde se genera corriente.

**Steam cleaners** A pressure washer that uses a soap solution, heated under pressure to a temperature higher than its normal boiling point. The super-heated solution boils once it leaves the nozzle as it shoots against the object being cleaned.
**Limpiadoras de vapor** Una limpiadora a presión que utiliza una solución de jabón, calentado bajo presión a una temperatura más elevada de su punto de ebullición. La solución sobrecalentada hierve al salir de la boquilla proyectada hacia el objeto para limpiar.

**Steering effort** The amount of effort required by the driver to turn the steering wheel.
**Esfuerzo de dirección** Amplitud de esfuerzo requerido por parte del conductor para girar el volante de dirección.

**Steering pull** The tendency of the steering to pull to the right or left when the vehicle is driven straight ahead on a smooth, straight road surface.
**Tiro de la dirección** Tendencia de la dirección a desviarse hacia la derecha o hacia la izquierda mientras se conduce el vehículo en línea recta en un camino cuya superficie es lisa y nivelada.

**Steering wheel free play** The amount of steering wheel movement before the front wheels begin to turn.
**Juego libre del volante de dirección** Amplitud de movimiento del volante de dirección antes de que las ruedas delanteras comiencen a girar.

**Steering wheel locking tool** A special tool used to lock the steering wheel during certain wheel alignment procedures.
**Herramienta para el cierre del volante de dirección** Herramienta especial que se utiliza para bloquear el volante de dirección durante ciertos procedimientos de alineación de ruedas.

**Stellite** A hard facing material made from a cobalt-based material with a high chromium content.
**Estelita** Un material de recarga compuesto de un material de base cobáltico con un contenido muy alto del cromo.

**Stepped resistor** A resistor that has two or more fixed resistor values.
**Resistor de secciones escalonadas** Resistor que tiene dos o más valores de resistencia fija.

**Stepper motor** An electrical motor that contains a permanent magnet armature with two or four field coils. Can be used to move the controlled device to whatever location is desired. By applying voltage pulses to selected coils of the motor, the armature will turn a specific number of degrees. When the same voltage pulses are applied to the opposite coils, the armature will rotate the same number of degrees in the opposite direction.
**Motor paso a paso** Motor eléctrico que contiene una armadura magnética fija con dos o cuatro bobinas inductoras. Puede utilizarse para mover el dispositivo regulado a cualquier lugar deseado. Al aplicárseles impulsos de tensión a ciertas bobinas del motor, la armadura girará un número específico de grados. Cuando estos mismos impulsos de tensión se aplican a las bobinas opuestas, la armadura girará el mismo número de grados en la dirección opuesta.

**Stethoscope** A special tool that amplifies sound to help diagnose noise location.
**Estetoscopio** Herramienta especial que amplifica el sonido para ayudar a diagnosticar la procedencia de los ruidos.

**Still timing** The process of adjusting base ignition timing without the engine running.

**Regulación sin marcha** El proceso de ajustar el avance del encendido fundamental sin que esté en marcha el motor.

**Stone dressing** Dressing a stone refers to using a diamond tool to clean and restore the stone's surface.
**Reacondicionar la muela** Para reacondicionar una muela se emplea una herramienta de diamante para limpiar y rectificar la agudeza de la superficie de la muela.

**Stratify** Arrange or form into layers. To fully blend.
**Estratificar** Arreglar o formar en capas. Mezclar completamente.

**Stress** The force to which a material, mechanism, or component is subjected.
**Esfuerzo** La fuerza a la cual se somete a una materia, un mecanísmo o un componente.

**Stress risers** Defects in the component resulting in a weakness of the metal. The defect tends to decrease the tensile strength of the metal, and stress applied to the area of the defect tends to cause breakage.
**Deformación de colada** Los defectos en el componente que causan una debilidad del metal. El defecto suele disminuir la resistencia a la tracción del metal y al aplicar una carga en el área del defecto muchas veces causa la quebradura.

**Strikeout, or rebound, bumper** A rubber block that prevents the control arm from striking the chassis when the wheel hits a large road irregularity.
**Parachoques de rebote** Bloque de caucho que evita que el brazo de mando choque contra el chasis cuando la rueda golpea una irregularidad en el camino.

**Stroke** The distance traveled by the piston from TDC to BDC.
**Carrera** La distancia que viaja el pistón del PMS al PMI.

**Strut assembly** Refers to all the strut components, including the strut tube, shock absorber, coil spring, and upper bearing assembly.
**Conjunto de montante** Se refiere a todas las piezas del montante, inclusive al tubo de montante, al amortiguador, al muelle helicoidal, y al conjunto del cojinete superior.

**Strut cartridge** The inner components in a strut that may be replaced rather than replacing the complete strut.
**Cartucho del montante** Componentes internos de un montante que pueden ser reemplazados en vez de tener que reemplazarse todo el montante.

**Strut chatter** A chattering noise as the steering wheel is turned often caused by a binding upper strut mount.
**Vibración del montante** Rechinamiento producido mientras se gira el volante de dirección. A menudo ocasionado por el trabamiento del montaje del montante superior.

**Strut rod, or radius rod** A rod connected from the lower control arm to the chassis to prevent forward and rearward control arm movement.
**Varilla del montante o varilla radial** Varilla conectada del brazo de mando inferior al chasis que se utiliza para evitar el movimiento hacia adelante o hacia atrás del brazo de mando.

**Strut tower** A circular, raised, reinforced area inboard of the front fenders that supports the upper strut mount and strut assembly.
**Torre del montante** Área circular, elevada y reinforzada en la parte interior de los guardafangos delanteros que apoya el conjunto del montante y montaje del montante superior.

**Stub shaft** A very short shaft.
**Árbol corto** Árbol sumamente corto.

**Subsystem** A system within a system.
**Subsistema** Sistema dentro de un sistema.

**Sulfation** A chemical action within a battery that interferes with the ability of the cells to deliver current and accept a charge.

**Sulfatación** Una reacción química dentro de la batería que estorba la habilidad de las celulas de entregar el corriente y aceptar una carga.

**Sulfuric acid** A very corrosive acid used in automotive batteries.
**Ácido sulfúrico** Ácido sumamente corrosivo utilizado en las baterías de automóviles.

**Sun gear** The central gear in a planetary gear system around which the rest of the gears rotate. The innermost gear of the planetary gear set.
**Engranaje principal (sol)** El engranaje central en un sistema de engranajes planetarios alrededor del cual giran los otros engranajes.

**Sun load** Heat intensity and/or light intensity produced by the sun.
**Carga del sol** Intensidad calorífica y/o de la luz generada por el sol.

**Sunburst frame cracks** Radiate outward from an opening in the vehicle frame.
**Grietas del armazón en forma de rayos de sol** Grietas que se proyectan hacia afuera desde una abertura en el armazón del vehículo.

**Superheat switch** An electrical switch activated by an abnormal temperature-pressure condition (a superheated vapor); used for system protection.
**Interruptor de vapor sobrecalentado** Interruptor eléctrico accionado por una condición anormal de presión y temperatura (vapor sobrecalentado); utilizado para la protección del sistema.

**Supplemental inflatable restraint (SIR)** An air bag system.
**Sistema de seguridad inflable suplementario** Sistema del Airbag.

**Surface-to-volume ratio** A mathematical comparison between the surface area of the combustion chamber and the volume of the combustion chamber. The greater the surface area, the more area the mixture can cling to, and mixture which clings to the metal will not burn completely because the metal cools it. Typical surface-to-volume ratio is 7.5:1.
**Índice superficie-volumen** Una comparación matemática entre el área de la superficie de la cámara de combustión y el volumen de la cámara de combustión. Lo mayor el área de la superficie, lo mayor el área en donde puede pegarse la mezcla y la mezcla pegada al metal no se quemará completamente puesto que la enfría el metal. El índice típico del superficie-volumen es el 7.5:1.

**Sway bar** Also called a stabilizer bar. It prevents the vehicle's body from diving into turns.
**Barra de oscilación lateral** Llamada también barra estabilizadora. Impide que la carrocería del vehículo se desestabilice durante los virajes.

**Sweep grinding** A crankshaft grinding method using a stone that is swept back and forth across the journal surface.
**Rectificado de barrido** Un método de rectificación del cigüeñal empleando una muela que mueve de un lado al otro encima de la superficie de un muñón.

**Switch test** A computer system test mode that tests the switch input signals to the computer.
**Prueba de conmutación** Modo de prueba de una computadora que revisa las señales de entrada de conmutación hechas a la computadora.

**Synchromesh transmission** Transmission gearing that aids the meshing of two gears or shift collars by matching their speed before engaging them.
**Transmisión de engranaje sincronizado** Engranaje transmisor que facilita el engrane de dos engranajes o collares de cambio de velocidades al igualar la velocidad de éstos antes de engranarlos.

**Synchronization timing** Timing design used on vehicles with computer-controlled fuel injection systems.
**Regulación (tiempo) sincronizado** Un diseño de regulación empleado en los vehículos equipados con sistemas de inyección de combustible de control computarizado. Un captador en el distribuidor sirve para sincronizar la posición del cigüeñal con la posición del árbol de levas.

**Synchronize** To cause two events to occur at the same time—for example, to bring two gears to the same speed before they are meshed to prevent gear clash.
**Sincronizar** Hacer que dos sucesos ocurran al mismo tiempo; por ejemplo, hacer que dos engranajes giren a la misma velocidad antes de que se engranen para evitar el choque de engranajes.

**Synchronizer assemblies** Device that uses cone clutches to bring two parts rotating at two speeds to the same speed. A synchronizer assembly operates between two gears; first and second gear, third and fourth gear.
**Conjuntos sincronizadores** Mecanismo que utiliza embragues cónicos para hacer que dos piezas que giran a dos velocidades giren a una misma velocidad. Un conjunto sincronizador funciona entre dos engranajes; engranajes de primera y segunda velocidad, y engranajes de tercera y cuarta velocidad.

**Synchronizer blocker ring** Usually a brass ring that acts as a clutch and causes driving and driven units to turn at the same speed before final engagement.
**Anillo de bloque sincronizador** Normalmente un anillo de latón que sirve de embrague y hace que las piezas de accionamiento y las accionadas giren a la misma velocidad antes del acoplamiento final.

**Synchronizer hub** Center part of the synchronizer assembly that is splined to the synchronizer sleeve and transmission output shaft.
**Cubo sincronizador** Pieza central del conjunto sincronizador ranurada al manguito sincronizador y al árbol de rendimiento de la transmisión.

**Synchronizer (SYNC) pickup** A pickup assembly that produces a voltage signal for ignition triggering or injector sequencing.
**Captación sincronizadora** Conjunto de captación que produce una señal de tensión para el arranque del encendido o para el ordenamiento del inyector.

**Synchronizer sleeve** The sliding sleeve that fits over the complete synchronizer assembly.
**Manguito sincronizador** Manguito deslizante que cubre todo el conjunto sincronizador.

**System** All of the components and lines that make up an air-conditioning system.
**Sistema** Todos los componentes y líneas que componen un sistema de acondicionamiento de aire.

**Tach-dwellmeter** A meter that reads engine rpm and ignition dwell.
**Tacómetro y medidor de retraso** Medidor que lee las rpm del motor y el retraso del encendido.

**Tachometer** An instrument that measures the speed of the engine in revolutions per minute (rpm).
**Tacómetro** Instrumento que mide la velocidad del motor en revoluciones por minuto (rpm).

**Tachometer (TACH) terminal** The negative primary coil terminal.
**Borne del tacómetro** Borne negativo de la bobina primaria.

**Tail shaft** A commonly used term for a transmission's extension housing.
**Extremo del árbol** Término comúnmente utilizado para el alojamiento de extensión de una transmisión.

**Tank** See *Header tanks* and *Expansion tank*.
**Tanque** Ver *Header tanks* [*Tanques para alimentación por gravedad*] y *Expansion tank* [*Tanque de expansión*].

**Tap** A tool used to repair or cut new internal threads.
**Macho** Una herramienta que sirve para reparar o cortar las roscas interiores nuevas.

**Tapered-head bolts** A bolt with a taper on the underside of the head.

**Pernos de cabeza cónica** Perno con una cabeza cuya superficie inferior es cónica.

**Tare weight** The weight of the packaging material. See *Net weight* and *Gross weight*.
**Taraje** Peso del material de encajonamiento. Ver *Net weight* [*Peso neto*] y *Gross weight* [*Peso bruto*].

**Teardown** A term often used to describe the process of disassembling a transmission.
**Desmontaje** Un término común que describe el proceso de desarmar a una transmisión.

**Teflon ring compressing tool** A special tool used to compress the Teflon rings on the pinion prior to installation in a rack and pinion steering gear.
**Herramienta para la compresión de anillos de teflón** Herramienta especial que se utiliza para comprimir los anillos de teflón en el piñón antes de ser instalados en un mecanismo de dirección de cremallera y piñón.

**Teflon ring expander** A special tool required to expand the Teflon rings prior to installation on the pinion in a rack and pinion steering gear.
**Expansor para anillos de teflón** Herramienta especial requerida para expandir los anillos de teflón antes de ser instalados en un mecanismo de dirección de cremallera y piñón.

**Telescoping gauge** A precision tool used in conjunction with outside micrometers to measure the inside diameter of a hole. Telescoping gauges are sometimes called snap gauges.
**Calibrador telescópico** Una herramienta de precisión que se emplea junta con los micrómetros exteriores para medir los diámetros interiores de un agujero. Tambien se llaman calibradores de brocha.

**Temper** To change the physical characteristics of a metal by applying heat.
**Templar** Cambiar las características físicas de un metal mediante una aplicación del calor.

**Temperature door** A door within the case/duct stem to direct air through the heater and/or evaporator core.
**Puerta de temperatura** Puerta ubicada dentro del vástago de caja/conducto para conducir el aire a través del núcleo del calentador y/o del evaporador.

**Temperature switch** A switch actuated by a change in temperature at a predetermined point.
**Interruptor de temperatura** Interruptor accionado por un cambio de temperatura a un punto predeterminado.

**Tensile strength** The metal's resistance to be pulled apart.
**Resistencia a la tracción** La resistencia de un metal a ser estirado.

**Tension** Effort that elongates or "stretches" a material.
**Tensión** Un esfuerzo que alarga o "estira" a una materia.

**Tension gauge** A tool for measuring the tension of a belt.
**Manómetro para tensión** Herramienta para medir la tensión de una correa.

**Test spark plug** A spark plug with the electrodes removed so it requires a much higher firing voltage for testing such components as the ignition coil.
**Bujía de prueba** Bujía que a consecuencia de habérsele removido los electrodos necesitará mayor tensión de encendido para revisar componentes, como por ejemplo la bobina del encendido.

**Thermal cleaners (pyrolytic ovens)** A parts cleaner that uses high temperatures to bake the grease and grime into ash.
**Limpiadores térmicos (hornos pirolíticos)** Una limpiadora de partes que usa las temperaturas elevadas para convertir la grasa y el lodo en cenizas.

**Thermal efficiency** A measurement comparing the amount of energy present in a fuel and the actual energy output of the engine.

**Rendimiento térmico** Una medida que compara la cantidad de la energía presente en un combustible y la potencia actual producido por el motor.

**Thermal vacuum switch (TVS)** A vacuum switching device operated by heat applied to a thermo-wax element.
**Conmutador de vacío térmico** Dispositivo de conmutación de vacío accionado por el calor aplicado a un elemento de termocera.

**Thermal vacuum valve (TVV)** A valve that is opened and closed by a thermo-wax element mounted in the cooling system.
**Válvula térmica de vacío** Válvula de vacío que un elemento de termocera montado en el sistema de enfriamiento abre y cierra.

**Thermistor** A solid-state variable resistor made from a semiconductor material that changes resistance in relation to temperature changes.
**Termistor** Resistor variable de estado sólido hecho de un material semiconductor que cambia su resistencia en relación con los cambios de temperatura.

**Thermodynamics** The study of the relationship between heat energy and mechanical energy.
**Termodinámica** El estudio de la relación entre la energía del calor y la energía mecánica.

**Thermostat** A control device that allows the engine to reach normal operating temperatures quickly and maintains the desired temperatures.
**Termostato** Un dispositivo de control que permite que el motor llegue rápidamente a las temperaturas de funcionamiento normales y mantenga las temperaturas deseadas.

**Thermostat tester** A tester designed to measure thermostat opening temperature.
**Instrumento de prueba del termostato** Instrumento de prueba diseñado para medir la temperatura inicial del termostato.

**Thickness gauge** Strips of metal made to an exact thickness, used to measure clearances between parts.
**Calibre de espesores** Las tiras del metal que se han fabricado a un espesor exacto, sirven para medir las holguras entre las partes.

**Thread chaser (thread restorer)** A tool designed to roll the threads back into shape, not to cut new threads.
**Peine de roscar a mano** Una herramienta diseñada a repujar las roscas dañadas hacia su estado original, no a cortar las roscas nuevas.

**Thread depth** The height of the thread of a bolt from its base to the top of its peak.
**Profundidad de la rosca** La longitud de la rosca de un perno desde su fondo a la parte superior.

**Thread insert (heli-coil)** A device that allows for major thread repairs while keeping the same size fastener.
**Inserto preroscado (heli-coil)** Un dispositivo que permite las reparaciones completas de las roscas manteniendo los retenes del mismo tamaño.

**Three-coil gauge** A gauge design that uses the interaction of three electromagnets and the total field effect upon a permanent magnet to cause needle movement.
**Calibrador de tres bobinas** Calibrador diseñado para utilizar la interacción de tres electroimanes y el efecto inductor total sobre un imán permanente para producir el movimiento de la aguja.

**Three-minute charge test** A reasonably accurate method for diagnosing a sulfated battery on conventional batteries.
**Prueba de carga de tres minutos** Un método bastante preciso de diagnosticar una batería sulfatada con las baterías ordinarias.

**Three-quarter floating axle** The axle housing carries the weight of the vehicle while the bearings support the wheels on the outer ends of the axle housing tubes.
**Eje flotante de tres cuartos** Puente trasero que soporta el peso del vehículo mientras los cojinetes soportan las ruedas en los extremos exteriores de los tubos del puente trasero.

**Throating** Machining process using a 60-degree stone to narrow the contact surface.
**Rectificación** Un proceso de rebajar a máquina empleando una muela de 60 grados para disminuir la superficie de contacto.

**Throttle body injection (TBI)** A fuel injection system with the injector or injectors mounted above the throttle.
**Inyección del cuerpo de la mariposa** Sistema de inyección de combustible en el que el inyector, o los inyectores, están montados sobre la mariposa.

**Throttle position sensor (TPS)** A sensor mounted on the throttle shaft that sends a voltage signal to the computer in relation to throttle opening.
**Sensor de la posición de la mariposa** Sensor montado sobre el árbol de la mariposa que le envía una señal de tensión a la computadora referente a la apertura de la mariposa.

**Throttle position switch** A switch that informs the computer whether the throttle is in the idle position. This switch is usually part of the TPS.
**Conmutador de la posición de la mariposa** Conmutador que le advierte a la computadora si la mariposa se encuentra en la posición de marcha lenta. Este conmutador normalmente forma parte del sensor de la posición de la mariposa.

**Throw** The distance from the center of the crankshaft main bearing to the center of the connecting rod journal.
**Codo del cigüeñal** La distancia del centro del muñón principal del cigüeñal al centro del muñón de la biela.

**Throw-off** The quantity of oil that escapes at the end of the bearings and lubricates adjacent engine parts while the engine is running.
**Expulsión** La cantidad del aceite que escapa de las extremidades de los cojinetes y lubrifica las partes contiguas del motor al estar en marcha el motor.

**Throwout bearing** In the clutch, the bearing that can be moved inward to the release levers by clutch-pedal action to cause declutching, which disengages the engine crankshaft from the transmission.
**Cojinete de desembrague** En el embrague, cojinete que puede moverse hacia adentro hasta las palancas de desembrague, por medio de la acción del pedal del embrague, para lograr el desembrague. Esta acción desengrana el cigüeñal del motor de la transmisión.

**Thrust bearing** A bearing designed to resist or contain side or end motion as well as reduce friction.
**Cojinete de empuje** Un cojinete diseñado a detener o reprimir a los movimientos laterales o de las extremidades y también reducir la fricción.

**Thrust line** A line positioned at a 90-degree angle to the rear axle and projected toward the front of the vehicle.
**Línea de empuje** Línea colocada a un ángulo de 90 grados con relación al eje trasero y proyectada hacia la parte frontal del vehículo.

**Thrust load** A load that pushes or reacts through the bearing in a direction parallel to the shaft.
**Carga de empuje** Una carga que empuja o reacciona por el cojinete en una dirección paralelo a la flecha.

**Thrust washer** A washer designed to take up end thrust and prevent excessive end play.
**Arandela de empuje** Una arandela diseñada para rellenar a la holgura de la extremidad y prevenir demasiado juego en la extremidad.

**Tie rod** The linkage between the steering rack and the steering knuckle arm. The tie rod is threaded into a tie-rod end or has a threaded split member for making toe adjustments.
**Barra de acoplamiento** Biela mortiz entre la cremallera y el brazo del muñón de dirección. La barra de acoplamiento se filetea en un extremo de la barra de acoplamiento o tiene una pieza fileteada hendida para ajustar el tope.

**Tie-rod end** The fittings on the ends of the tie rods. The outer tie-rod end connects to the steering arm, and the inner one connects to the steering rack. Both ends include ball sockets to allow pivotal action, as well as up and down flexing.
**Extremo de la barra de acoplamiento** Conexiones en los extremos de las barras de acoplamiento. El extremo exterior de la barra de acoplamiento se conecta al brazo de dirección, y el extremo interior a la cremallera. Ambos extremos incluyen juntas de rótula para permitir tanto el movimiento giratorio como el de ascenso y descenso.

**Tie rod end and ball joint puller** A special puller designed for removing tie rod ends and ball joints.
**Extractor para la junta esférica y el extremo de la barra de acoplamiento** Extractor especial diseñado para la remoción de las juntas esféricas y los extremos de barras de acoplamiento.

**Tie rod sleeve adjusting tool** A special tool required to rotate tie rod sleeves without damaging the sleeve.
**Herramienta para el ajuste del manguito de la barra de acoplamiento** Herramienta especial requerida para girar los manguitos de las barras de acoplamiento sin estropearlos.

**Timing** The process of identifying when an event is to occur.
**Sincronización** El proceso de identificar cuando debe ocurrir un evento.

**Timing chain** The chain that drives the camshaft off of the crankshaft.
**Cadena de sincronización** La cadena procedente del cigüeñal que acciona al árbol de levas.

**Timing connector** A wiring connector that must be disconnected while checking basic ignition timing on fuel-injected engines.
**Conector de regulación** Conector del alambrado que debe desconectarse mientras se verifica la regulación básica del encendido en motores de inyección de combustible.

**Tire changer** Equipment used to demount and mount tires on wheel rims.
**Cambiador de neumáticos** Equipo que se utiliza para desmontar y montar los neumáticos en las llantas de las ruedas.

**Tire condition screen** A display on a computer wheel aligner that allows the technician to check and enter the tire condition.
**Pantalla para la visualización de la condición del neumático** Representación visual en una computadora para alineación de ruedas que le permite al mecánico revisar e introducir datos referentes a la condición del neumático.

**Tire conicity** Occurs when the tire belt is wound off center in the manufacturing process creating a cone-shaped belt, which results in steering pull.
**Conicidad del neumático** Condición que ocurre cuando la correa del neumático se devana fuera del centro durante el proceso de fabricación creando así una correa en forma cónica y trayendo como resultado el tiro de la dirección.

**Tire rotation** Involves moving each wheel and tire to a different location on the vehicle to increase tire life.
**Cambio de posición de los neumáticos** Se refiere a la rotación de cada rueda y neumático a una posición diferente en el vehículo para aumentar la vida útil del neumático.

**Tire thump** A pounding noise as the tire and wheel rotate; usually caused by improper wheel balance.
**Ruido sordo del neumático** Ruido similar al de un golpe pesado que se produce mientras el neumático y la rueda están girando. Por lo general este ruido lo ocasiona el desequilibrio de la rueda.

**Tire tread depth gauge** A special tool required to measure tire tread depth.
**Calibrador de la profundidad de la huella del neumático** Herramienta especial requerida para medir la profundidad de la huella del neumático.

**Tire vibration** Vertical or sideways tire oscillations.
**Vibración del neumático** Oscilaciones verticales o laterales del neumático.

**Toe**  A suspension dimension that reflects the difference in the distance between the extreme front and extreme rear of the tires.
**Tope**  Dimensión de la suspensión que refleja la diferencia de la distancia entre los extremos delantero y trasero de las ruedas.

**Toe gauge**  A special tool used to measure front or rear wheel toe.
**Calibrador del tope**  Herramienta especial que se utiliza para medir el tope de las ruedas delanteras o traseras.

**Toe-in**  A condition in which the distance between the front edges of the tires is less than the distance between the rear edges of the tires.
**Convergencia**  Condición que ocurre cuando la distancia entre los bordes frontales de los neumáticos es menor que la distancia entre los bordes traseros de los mismos.

**Toe-out**  A condition in which the distance between the front edges of the tires is more than the distance between the rear edges of the tires.
**Divergencia**  Condición que ocurre cuando la distancia entre los bordes delanteros de los neumáticos es mayor que la distancia entre los bordes traseros de los mismos.

**Toe-out on turns**  A steering system feature that allows the front wheels to be at different angles when the vehicle is turning a corner.
**Divergencia**  Característica del sistema de dirección que permite que las ruedas delanteras se inclinen a ángulos diferentes cuando el vehículo hace un viraje.

**Tolerance**  A permissible variation between the two extremes of a specification or dimension.
**Tolerancia**  Una variación que se permite entre los dos extremos de una especificación o una dimensión.

**Top dead center (TDC)**  Term used to indicate the piston is at the very top of its stroke.
**Punto muerto superior (PMS)**  El término que indica que el pistón está en la cima de su carrera.

**Topping**  Refers to the use of the 30-degree stone to lower the contact surface on the valve face.
**Despunte**  Refiere al uso de una muela de 30 grados para rebajar la superficie de contacto en la cara de la válvula.

**Torque**  A rotating force around a pivot point. For example, the twisting force applied to a bolt or shaft is called torque.
**Torsión**  Una fuerza que gira alrededor de un punto de pivote. Por ejemplo, la fuerza giratoria que se aplica en un perno o en un eje se llama torsión (par).

**Torque capacity**  The ability of a converter clutch to hold torque.
**Capacidad de la torsión**  La habilidad de un convertidor de embrague a sostener a la torsión.

**Torque converter**  A series of components that work together to reduce slip and multiply torque. It provides a fluid coupling between the engine and the automatic transmission.
**Convertidor del par**  Una serie de componentes que trabajan juntos para disminuir el deslizamiento y multiplicar el par. Provee un acoplamiento flúido entre el motor y la transmisión automática.

**Torque curve**  A line plotted on a chart to illustrate the torque personality of an engine. When the engine operates on its torque curve, it is producing the most torque for the quantity of fuel being burned.
**Curva de la torsión**  Una línea delineada en una carta para ilustrar las características de la torsión del motor. Al operar un motor en su curva de la torsión, produce la torsión óptima para la cantidad del combustible que se consuma.

**Torque multiplication**  The result of meshing a small driving gear and a large driven gear to reduce speed and increase output torque.
**Multiplicación de la torsión**  El resultado de engranar a un engranaje pequeño de ataque con un engranaje más grande arrastrado para reducir la velocidad e incrementar la torsión de salida.

**Torque plates**  Metal blocks about 2 inches thick that are bolted to the cylinder block at the cylinder head mating surface to prevent twisting during honing and boring operations.
**Chapas de torsión**  Los bloques de metal midiendo unas 2 pulgadas de grueso empernados al monoblock en la superficie de contacto de la cabeza del cilindro que previenen que se retuerce durante las operaciones de esmerilado y taladreo.

**Torque steer**  Steering pull felt in the steering wheel as the result of increased torque during hand acceleration.
**Dirección la torsión**  Una acción que se nota en el volante de dirección como resultado de un aumento de la torsión.

**Torque tube**  A fixed tube over the drive shaft on some cars. It helps locate the rear axle and takes torque reaction loads from the drive axle so the drive shaft will not sense them.
**Tubo de eje cardán**  Tubo fijo sobre el árbol de mando en algunos automóviles. Ayuda a colocar el eje trasero y remueve las cargas de reacción de torsión del eje de mando para que el árbol de mando no las reciba.

**Torque wrench**  Wrench that measures the amount of twisting force applied to a fastener.
**Llave de torsión**  Una llave que sirve para medir la cantidad de la fuerza de torsión que se aplica a una fijación.

**Torsional springs**  Round, stiff coil springs placed in the driven disc to absorb the torsional disturbances between the driving flywheel and pressure plate and the driven transmission input shaft.
**Muelles de torsión**  Muelles helicoidales redondos y rígidos ubicados en el disco accionado para absorber las alteraciones de torsión entre el volante motor y la placa de presión, y el árbol impulsor accionado de la transmisión.

**Total engine displacement**  The sum of displacements for all cylinders in an engine.
**Cilindrada total del motor**  La suma de todos los desplazamientos de los cilindros de un motor.

**Total pedal travel**  The total amount the pedal moves from no free play to complete clutch disengagement.
**Avance total del pedal**  Distancia total a la que el pedal se mueve de cero juego al desengrane total del embrague.

**Total runout**  The sum of the maximum readings below and above the zero line on the indicator.
**Desviación total**  Suma de las lecturas máximas bajo y sobre la línea del cero en el indicador de cuadrante.

**Total toe**  The sum of the toe angles on both wheels.
**Tope total**  Suma de los ángulos del tope en ambas ruedas.

**Total travel**  The distance the clutch pedal and release bearing move from the fully engaged position until the clutch is fully disengaged.
**Avance total**  Distancia a la que el pedal del embrague y el cojinete de desembrague se mueven de la posición enteramente engranada hasta que el embrague se desengrane por completo.

**Track gauge**  A long straight bar with adjustable pointers used to measure rear-wheel tracking in relation to the front wheels.
**Calibrador del encarrilamiento**  Barra larga y recta con agujas ajustables que se utiliza para medir el encarrilamiento de las ruedas traseras con relación a las ruedas delanteras.

**Traction**  The gripping action between the tire tread and the road's surface.
**Tracción**  La acción de agarrar entre la cara de la rueda y la superficie del camino.

**Traction control system (TCS)**  A computer-controlled system that prevents one drive wheel from spinning more than the opposite drive wheel on acceleration.

**Sistema para el control de la tracción** Sistema de suspensión controlado por computadora que evita que una rueda motriz gire más que la opuesta durante la aceleración del vehículo.

**Tram gauge** A long, straight bar with adjustable pointers used to measure unitized bodies.
**Calibrador del tram** Barra larga y recta con agujas ajustables que se utiliza para medir carrocerías unitarias.

**Transaxle** Type of construction in which the transmission and differential are combined in one unit.
**Flecha de transmisión** Un tipo de construcción en el cual la transmisión y el diferencial se combinan en una unidad.

**Transaxle assembly** A compact housing most often used in front-wheel-drive vehicles that houses the manual transmission, final drive gears, and differential assembly.
**Conjunto del transeje** Alojamiento compacto que normalmente se utiliza en vehículos de tracción delantera y que aloja a la transmisión manual, a los engranajes de transmisión final y al conjunto del diferencial.

**Transfer case** An auxiliary transmission mounted behind the main transmission. Used to divide engine power and transfer it to both front and rear differentials, either full-time or part-time.
**Cárter de la transferencia** Una transmisión auxiliar montada detrás de la transmisión principal. Sirve para dividir la potencia del motor y transferirla a ambos diferenciales delanteros y traseros todo el tiempo o la mitad del tiempo.

**Transmission** The device in the powertrain that provides different gear ratios between the engine and drive wheels as well as reverse.
**Transmisión** El dispositivo en el trén de potencia que provee las relaciones diferentes de engranaje entre el motor y las ruedas de impulso y también la marcha de reversa.

**Transmission case** An aluminum or iron casting that encloses the manual transmission parts.
**Caja de transmisión** Pieza fundida en aluminio o en hierro que encubre las piezas de la transmisión manual.

**Transmission pressure control solenoid** A computer-operated solenoid in the automatic transmission that controls transmission fluid pressure.
**Solenoide de regulador de la presión de la transmisión** Solenoide accionado por computadora en la transmisión automática que controla la presión del fluido de la transmisión.

**Transmission shift solenoids** Computer-controlled solenoids that control transmission shifting.
**Solenoides de control de los cambios de las velocidades de la transmisión** Solenoides accionados por computadora que controlan los cambios de las velocidades de la transmisión.

**Transverse mounted engine** Engine placement where the block faces from side to side instead of front to back within the vehicle.
**Motor de montaje transversal** La ubicación del motor en la cual la longitud del monoblock queda de un lado a otro dentro del vehículo en vez de estar colocado de frente a atrás.

**Tread wear indicators** Raised portions near the bottom of the tire tread that are exposed at a specific tread wear.
**Indicadores del desgaste del neumático** Secciones elevadas cerca de la parte inferior de la huella del neumático que quedan expuestas cuando la huella alcanza una cantidad de desgaste específica.

**Trim height** The normal chassis riding height on a computer-controlled air suspension system.
**Altura equilibrada** Altura normal de viaje del chasis en un sistema de suspensión controlado por computadora.

**Triple evacuation** A process of evacuation that involves three pumpdowns and two system purges with an inert gas such as dry nitrogen (N).
**Evacuación triple** Proceso de evacuación que involucra tres envíos con bomba y dos purgas del sistema con un gas inerte, como por ejemplo el nitrógeno seco (N).

**Tripod (also called tripot)** A three-prong bearing that is the major component in tripod constant velocity joints. It has three arms (or trunnions) with needle bearings and rollers that ride in the grooves or yokes of a tulip assembly.
**Trípode** Cojinete de tres puntas; componente principal en juntas trípode de velocidad constante. Tiene tres brazos (o muñequillas) con cojinetes de agujas y rodillos que van montados sobre las ranuras o los yugos de un conjunto tulipán.

**Tripod universal joints** Universal joint consisting of a hub with three arm and roller assemblies that fit inside a casting called a tulip.
**Juntas universales de trípode** Juntas universales que consisten de un cubo con conjuntos de tres brazos y de rodillos que se insertan dentro de una pieza fundida llamada un tulipán.

**Trouble codes** Output of the self-diagnostics program in the form of a numbered code that indicates faulty circuits or components. Trouble codes are two- or three-digit characters that are displayed in the diagnostic display if the testing and failure requirements are both met.
**Códigos indicadores de fallas** Datos del programa autodiagnóstico en forma de código numerado que indica los circuitos o los componentes defectuosos. Dichos códigos se componen de dos o tres carácteres digitales que se muestran en el visualizador diagnóstico si se llenan los requisitos de prueba y de falla.

**Troubleshooting** The diagnostic procedure of locating and identifying the cause of the fault. It is a step-by-step process of elimination by use of cause-and-effect.
**Detección de fallas** Procedimiento diagnóstico a través del cual se localiza e identifica la falla. Es un proceso de eliminación que se lleva a cabo paso a paso por medio de causa y efecto.

**Trunnion** One of the projecting arms on a tripod or on the cross of a four-point universal joint. Each trunnion has a bearing surface that allows it to pivot within a joint or slide within a tulip assembly.
**Muñequilla** Uno de los brazos salientes en un trípode o en la cruz de una junta universal de cuatro puntas. La superficie del cojinete de cada muñequilla permite que la muñequilla gire dentro de una junta o se deslice dentro de un conjunto tulipán.

**Tulip assembly** The outer housing containing grooves or yokes in which trunnion bearings move in a tripod constant velocity joint.
**Conjunto tulipán** Alojamiento exterior que contiene ranuras o yugos en los cuales se mueven los cojinetes de muñequilla en una junta trípode de velocidad constante.

**Turn tables** Heavy steel plates placed under the front wheels during a wheel alignment that allow front wheel turning.
**Plataformas giratorias** Placas pesadas de acero que se colocan debajo de las ruedas delanteras durante una alineación de ruedas para permitir el giro de las mismas.

**Turning angle screen** A display on a computer wheel aligner that displays the turning angle.
**Pantalla para la visualización del ángulo de giro** Representación visual en una computadora para alineación de ruedas que muestra el ángulo de giro.

**Turning imbalance** A condition in which more effort is required to turn the steering wheel in one direction compared to the opposite direction.
**Desequilibrio de giro** Condición que ocurre cuando se requiere más esfuerzo para girar el volante en una dirección que en la dirección opuesta.

**Turning radius** The turning angle of the front wheel on the inside of a turn compared to the front wheel turning angle on the outside of the turn.
**Radio de giro, o círculo de giro** Ángulo de giro de la rueda delantera en el interior de un viraje con relación al ángulo de giro de la rueda delantera en el exterior del viraje.

**Turning radius gauge** A gauge with a degree scale mounted on turn tables under the front wheels.
**Calibrador del radio de giro** Calibrador con una escala de grados montado sobre plataformas giratorias debajo de las ruedas delanteras.

**Two-coil gauge** A gauge design that uses the interaction of two electromagnets and the total field effect upon an armature to cause needle movement.
**Calibrador de dos bobinas** Calibrador diseñado para utilizar la interacción de dos electroimanes y el efecto inductor total sobre una armadura para generar el movimiento de la aguja.

**Two-disk clutch** A clutch with two friction discs for additional holding power; used in heavy-duty equipment.
**Embrague de dos discos** Embrague con dos discos de fricción para proporcionar más fuerza de retención; utilizado en equipo de gran potencia.

**Two-gas emissions analyzer** An analyzer designed to measure hydrocarbons and carbon monoxide in the exhaust.
**Analizador de dos tipos de emisiones** Analizador diseñado para medir los hidrocarburos y el monóxido de carbono en el escape.

**Two-speed rear axle** See *Double-reduction differential.*
**Eje trasero de dos velocidades** Véase diferencial de reducción doble.

**TXV** Thermostatic expansion valve.
**TXV** Válvula de expansión termostática.

**U-bolt** An iron rod with threads on both ends, bent into the shape of a U and fitted with a nut at each end.
**Perno en U** Varilla de hierro con filetes de tornillo en los dos extremos, acodada en forma de U y provista de una tuerca a cada extremo.

**Undersize bearing** Bearing with the same outside diameter as standard bearings but constructed of thicker bearing material in order to fit an undersize crankshaft journal.
**Cojinete de dimensión inferior** Un cojinete que tiene el diámetro exterior del mismo tamaño que los cojinetes normales pero que se ha fabricado de un material más grueso para que puede quedar en un muñón de cigüeñal más pequeño.

**Understeer** The tendency of a vehicle not to turn as much as desired by the driver.
**Dirección pobre** Tendencia de un vehículo a no girar tanto como lo desea el conductor.

**United States Customary (USC)** A system of weights and measures patterned after the British system.
**Sistema usual estadounidense (USC)** Sistema de pesos y medidas desarrollado según el modelo del Sistema Imperial Británico.

**Universal joint** A mechanical device that transmits rotary motion from one shaft to another shaft at varying angles.
**Junta universal** Dispositivo mecánico que transmite movimiento giratorio de un árbol a otro a ángulos cambiantes.

**Universal joint operating angle** The difference in degrees between the drive shaft and transmission installation angles.
**Ángulo de funcionamiento de la junta universal** Diferencia en grados entre los ángulos del árbol de mando y del montaje de la transmisión.

**U-joint** A four-point cross connected to two U-shaped yokes that serves as a flexible coupling between shafts.
**Junta de U** Una cruceta de cuatro puntos que se conecta a dos yugos en forma de U que sirven de acoplamientos flexibles entre las flechas.

**Ultraviolet (UV)** The part of the electromagnetic spectrum emitted by the sun that lies between visible violet light and X-rays.
**Ultravioleta** Parte del espectro electromagnético generado por el sol que se encuentra entre la luz violeta visible y los rayos X.

**Unsprung weight** The weight of the tires, wheels, axles, control arms, and springs.
**Peso no suspendido** Peso de las ruedas, llantas, ejes, brazos de mando, y muelles.

**Upshift** To shift a transmission into a higher gear.
**Cambio ascendente** Cambiar a la velocidad de una transmisión a una más alta.

**Upstream air** Air injected into the exhaust ports.
**Aire conducido hacia arriba** Aire inyectado dentro de las lumbreras del escape.

**USCS measuring system** United States Customary System is also known as the English system.
**Sistema de medidade USCS** Sistema Usual U.S.; conocido también como Sistema Imperial Británico.

**Vacuum** A pressure lower than atmospheric pressure. Vacuum in the engine is created when the volume of the cylinder above the piston is increased. This results in the atmospheric pressure pushing the air-fuel mixture into the area of lowered pressure above the piston.
**Vacío** Una presión más baja que la presión atmosférica. El vacío en un motor se crea al aumentar el volumen del cilindro arriba del pistón. Esto resulta en que la presión atmosférica empuja la mezcla aire/combustible dentro del área arriba del pistón.

**Vacuum brake booster** A diaphragm-type booster that uses manifold vacuum and atmospheric pressure for its power.
**Reforzador de vacío** Un reforzador de tipo de diafragma que usa el vacío del colector y la presión atmosférica para operar.

**Vacuum delay valve** A vacuum valve with a restrictive port that delays a vacuum increase through the valve.
**Válvula de retardo de vacío** Válvula de vacío con una lumbrera restrictiva que retarda el aumento de vacío a través de la válvula.

**Vacuum distribution valve** A valve used in vacuum-controlled concealed headlight systems. It controls the direction of the vacuum to various vacuum motors or to the vent.
**Válvula de distribución al vacío** Válvula utilizada en el sistema de faros delanteros ocultos controlado al vacío. Regula la dirección del vacío a varios motores al vacío o sirve para dar salida del sistema.

**Vacuum fluorescent display (VFD)** A display type that uses anode segments coated with phosphor and bombarded with tungsten electrons to cause the segments to glow.
**Visualización de fluorescencia al vacío** Tipo de visualización que utiliza segmentos ánodos cubiertos de fósforo y bombardeados de electrones de tungsteno para producir la luminiscencia de los segmentos.

**Vacuum gauge** A gauge used to measure below atmospheric pressure.
**Vacuómetro** Calibrador utilizado para medir a una presión inferior a la de la atmósfera.

**Vacuum hand pump** A mechanical pump with a vacuum gauge and hose used for testing vacuum-operated components.
**Bomba de vacío manual** Bomba mecánica con un calibrador de vacío y una manguera que se utiliza para probar componentes a depresión.

**Vacuum motor** A device designed to provide mechanical control through the use of a vacuum.
**Motor de vacío** Dispositivo diseñado para proveer regulación mecánica mediante un vacío.

**Vacuum-operated decel valve** A valve that allows more air into the intake manifold during deceleration to improve emission levels.
**Válvula de desaceleración accionada por vacío** Válvula accionada por vacío que admite una mayor cantidad de aire en el colector de aspiración durante una desaceleración a fin de reducir los niveles de emisiones.

**Vacuum pressure gauge** A gauge designed to measure vacuum and pressure.
**Manómetro de la presión del vacío** Calibrador diseñado para medir el vacío y la presión.

**Vacuum pump** A mechanical device used to evacuate the refrigeration system in order to rid it of excess moisture and air.
**Bomba de vacío** Dispositivo mecánico utilizado para evacuar el sistema de refrigeración para purgarlo de un exceso de humedad y aire.

**Vacuum signal** The presence of a vacuum.
Señal de vacío Presencia de un vacío.

**Vacuum testing** Testing that determines the engine's ability to provide sufficient pressure differentials to allow the induction of the air-fuel mixture into the cylinder. Results of vacuum testing indicate internal engine condition, fuel delivery abilities, ignition system condition, and valve timing.
Prueba del vacío Una prueba que determina la habilidad del motor en proveer las diferenciales de presión adecuadas que permiten la inducción de la mezcla aire/combustible al cilindro. Los resultados indicarán la condición interna del motor, las habilidades de entregar el combustible, la condición del sistema de encendido y la sincronización de las válvulas.

**Vacuum throttle kicker** A vacuum diaphragm with a stem that provides more throttle opening under certain conditions, such as deceleration or A/C on.
Nivelador de la mariposa de vacío Diafragma de vacío con un vástago que proporciona una apertura más extensa de la mariposa bajo ciertas condiciones, como por ejemplo una desaceleración o cuando el aire acondicionado está encendido.

**Valley pan** A pan located under the intake manifold in the valley of a "V" type engine used to prevent the formation of deposits on the underside of the intake manifold.
Cárter en V Un cárter ubicado en la parte inferior del múltiple de admisión en la parte más baja de un motor tipo "V" que previene la formación de los depósitos en la parte inferior del múltiple de admisión.

**Valve** Device that controls the flow of gases into and out of the engine cylinder.
Válvula Un dispositivo que controla el flujo de los gases entrando y saliendo del cilindro del motor.

**Valve body** Main hydraulic control assembly of a transmission containing the components necessary to control the distribution of pressurized transmission fluid throughout the transmission.
Cuerpo de la válvula Asamblea principal del control hidráulico de una transmisión que contiene los componentes necesarios para controlar a la distribución del fluido de la transmisión bajo presión por toda la transmisión.

**Valve cover gaskets** Components that seal the connection between the valve cover and cylinder head. The valve cover gasket is not subject to pressures, but must be able to seal hot, thinning oil.
Juntas de la tapa de válvula Los componentes que sellan la conexión entre la tapa de las válvulas y la cabeza del cilindro. La junta de la tapa de válvula no se sujeta a la presión, pero si tiene que sellar el aceite caliente muy fluido.

**Valve cupping** A deformation of the valve head caused by heat and combustion pressures.
Embutición de la válvula Una deformación de la cabeza de la válvula causada por el calor y las presiones de la combustión.

**Valve float** A condition that allows the valve to remain open longer than it is intended. Valve float is the effect of inertia on the valve.
Flotación de la válvula Una condición que permite que queda abierta la válvula por más tiempo de lo designado. Es el efecto que tiene la inercia en la válvula.

**Valve guide** A part of the cylinder head that supports and guides the valve stem.
Guía de válvula Una parte de la cabeza del cilindro que apoya y guía al vástago de la válvula.

**Valve guide bore gauge** Instrument that provides a quick measurement of the valve guide. It can also be used to measure taper and out-of-round.
Verificador del calibrador para la guía de válvula Un instrumento que provee una medida rápida de la guía de la válvula. Tambien puede servir para medir lo cónico y lo ovulado.

**Valve guide clearance** Measurement of the difference between the valve stem diameter and the guide bore diameter.

Holgura de la guía de válvula La medida de la diferencia entre el diámetro del vástago de la válvula y el diámetro del taladro de la guía.

**Valve overhang** The area of the face between the seat contact and the margin.
Desborde de la válvula El área de la cara entre el contacto del asiento y el margen.

**Valve overlap** The length of time, measured in degrees of crankshaft revolution, which the intake and exhaust valves of the same combustion chamber are open simultaneously.
Períodos de abertura de las válvulas El período del tiempo, que se mide en los grados de las revoluciones del cigüeñal, en el cual las válvulas de admisión y de escape de la misma cámara de combustión están abiertas simultáneamente.

**Valve seat** Machined surface of the cylinder head that provides the mating surface for the valve face. The valve seat can be either machined into the cylinder head or a separate component that is pressed into the cylinder head.
Asiento de la válvula La superficie acabada a máquina de la cabeza del cilíndro que provee una superficie de contacto para la cara de la válvula. El asiento puede ser labrado a máquina en la cabeza del cilindro o puede ser un componente aparte para prensar en la cabeza de la válvula.

**Valve seat grinding** The process of restoring the surface of the valve seat by removing small amounts of material through the use of special grinding stones.
Rectificado de los asientos de las válvulas El proceso de restaurar la superficie al quitar las cantidades pequeñas de material por medio de las muelas especiales.

**Valve seat recession** The loss of metal from the valve seat which causes the seat to recede into the cylinder head.
Encastre del asiento de la válvula La pérdida del metal del asiento de la válvula, lo que causa que el asiento retroceda dentro de la cabeza del cilindro.

**Valve seat runout gauge** Instrument that provides a quick measurement of the valve seat concentricity.
Calibrador de la excentricidad del asiento de la válvula Un instrumento que provee una medida rápida de la concentricidad del asiento de la válvula.

**Valve spring** A coil of specially constructed metal used to force the valve closed, providing a positive seal between the valve face and seat.
Resorte de válvula Un rollo de metal de construcción especial que provee la fuerza para mantener cerrada la válvula, así proveyendo un sello positivo entre la cara de la válvula y el asiento.

**Valve spring tension tester** Device used to measure the open and closed valve spring pressures.
Probador de tensión del resorte de válvula Un dispositivo que se emplea en medir las presiones de los resortes de las válvulas en la posición abierta o cerrada.

**Valve stem installed height** The distance between the top of the valve retainer and the valve spring seat on the cylinder head.
Altura instalada del vástago de la válvula Distancia entre la parte superior del retenedor de la válvula y el asiento del resorte de la válvula en la culata del cilindro.

**Valve train** The series of components that work together to open and close the valves.
Tren de válvulas Una serie de componentes que trabajan juntos para abrir y cerrar las válvulas.

**Vane-type MAF sensor** A MAF sensor containing a pivoted vane that moves a pointer on a variable resistor. This resistor sends a voltage signal to the computer in relation to the total volume of intake air.
Sensor MAF tipo paleta Sensor MAF con una paleta articulada que mueve un indicador en un resistor variable. Este resistor le envía una

señal de tensión a la computadora referente al volumen total de aire aspirado.

**Variable resistor** A resistor that provides for an infinite number of resistance values within a range.
**Resistor variable** Resistor que provee un número infinito de valores de resistencia dentro de un margen.

**V-belt** A drive belt with a V-shape.
**Correa en V** Correa de transmisión en forma de V.

**Vehicle Identification Number (VIN)** An alphanumeric code consisting of seventeen characters used to properly identify the vehicle and its major components.
**Número de Identificación del Vehículo (VIN)** Un código alfa- numérico que consiste de diez y siete caracteres que sirve para identificar correctamente al vehículo y sus componentes principales.

**Vehicle lift** A hydraulically or air-operated mechanism for lifting vehicles.
**Levantamiento del vehículo** Mecanismo activado hidráulicamente o por aire que se utiliza para levantar vehículos.

**Vehicle lift points** The areas that the manufacturer recommends for safe vehicle lifting. Lift points are structurally strong enough to sustain the stress of lifting. They are usually illustrated in the service manual or by the jack manufacturer. If in doubt, ask your instructor.
**Puntos para izar el vehículo** Las áreas recomendadas por el fabricante en donde se puede levantar al vehículo con seguridad. Son áreas que tienen la fuerza estructural para sostener la carga de izar. Los puntos para izar suelen ser ilustrados en el manual de servicio o por el fabricante del gato. Si hay dudas, consulte su instructor.

**Vehicle speed sensor (VSS)** A sensor that is usually mounted in the transmission and sends a voltage signal to the computer in relation to vehicle speed.
**Sensor de la velocidad del vehículo** Sensor que normalmente se monta en la transmisión y que le envía una señal de tensión a la computadora referente a la velocidad del motor.

**Vehicle wander** The tendency of the steering to pull to the right or left when the vehicle is driven straight ahead on a straight road.
**Desviación de la marcha del vehículo** Tendencia de la dirección a desviarse hacia la derecha o hacia la izquierda cuando se conduce el vehículo en línea recta en un camino cuya superficie es lisa.

**Ventilation** The act of supplying fresh air to an enclosed space, such as the inside of an automobile.
**Ventilación** Proceso de suministrar el aire fresco a un espacio cerrado, como por ejemplo al interior de un automóvil.

**Vernier calipers** Calipers that use a vernier scale which allows for measurement to a precision of 10 to 25 times as fine as the base scale.
**Pie de rey** Los calibres que emplean una escala vernier permitiendo una medida precisa de 10 a 25 veces más finas que la escala fundamental.

**V-groove belt** See V-belt.
**Correa con ranuras en V** Ver V-belt [Correa en V].

**Vibration** A quivering trembling motion felt in the vehicle at different speed ranges.
**Vibratcion** Estremecimento y temblor que se adviete en el vehiulo a diferentes gamas de velocidades.

**Vibration dampener** See Harmonic balancer.
**Amortiguador de vibraciones** Vea Equilibrador armónico.

**Viscosity** The measure of oil thickness.
**Viscosidad** La medida de lo espeso de un aceite.

**Viscous** Thick; tending to resist flowing.
**Viscoso** Espeso; que tiende a resistir el movimiento.

**Viscous-drive fan clutch** A cooling fan drive clutch that drives the fan at a higher speed when the temperature increases.
**Embrague de mando viscoso del ventilador** Embrague de mando del ventilador de enfriamiento que acciona el ventilador para que gire más rápido a temperaturas más altas.

**Viscous friction** The friction between layers of a liquid.
**Fricción viscosa** Fricción entre las capas de un fluido.

**Volt** The unit used to measure the amount of electrical force.
**Voltio** Unidad práctica de tensión para medir la cantidad de fuerza eléctrica.

**Voltage** The difference or potential that indicates an excess of electrons at the end of the circuit the farthest from the electromotive force. It is the electrical pressure that causes electrons to move through a circuit. One volt is the amount of pressure required to move one amp of current through one ohm of resistance.
**Tensión** Diferencia o potencial que indica un exceso de electrones al punto del circuito que se encuentra más alejado de la fuerza electromotriz. La presión eléctrica genera el movimiento de electrones a través de un circuito. Un voltio equivale a la cantidad de presión requerida para mover un amperio de corriente a través de un ohmio de resistencia.

**Voltage drop** A resistance in the circuit that reduces the electrical pressure available after the resistance. The resistance can be either the load component, the conductors, any connections, or unwanted resistance.
**Caída de tensión** Resistencia en el circuito que disminuye la presión eléctrica disponible después de la resistencia. La resistencia puede ser el componente de carga, los conductores, cualquier conexión o resistencia no deseada.

**Voltage regulator** Used to control the output voltage of the AC generator, based on charging system demands, by controlling field current.
**Regulador de tensión** Dispositivo cuya función es mantener la tensión de salida del generador de corriente alterna, de acuerdo a las variaciones en la corriente de carga, controlando la corriente inductora.

**Volt-amp tester** A tester designed to test volts and amps in such circuits as battery, starter, and charging.
**Instrumento de prueba de voltios y amperios** Instrumento de prueba diseñado para revisar los voltios y los amperios en circuitos como por ejemplo, de la batería, del arranque y de la carga.

**Voltmeter** A test meter used to read the pressure behind the flow of electrons.
**Voltímetro** Instrumento de prueba utilizado para medir la presión del flujo de electrones.

**Volumetric efficiency** A measurement of the amount of air-fuel mixture that actually enters the combustion chamber compared to the amount that could be drawn in.
**Rendimiento volumétrico** Una medida de la cantidad de la mezcla del aire/combustible que actualmente entra en la cámara de combustión comparada con la cantidad que podría entrar.

**Vortex** Path of fluid flow in a torque converter. The vortex may be high, low, or zero, depending on the relative speed between the pump and turbine.
**Vórtice** La vía del flujo de los fluidos en un convertido de torsión. El vórtice puede ser alto, bajo, o cero, depende de la velocidad relativa entre la bomba y la turbina.

**Vortex flow** Recirculating flow between the converter impeller and turbine that causes torque multiplication.
**Flujo del vórtice** El fluyo recirculante entre el impulsor del convertidor y la turbina que causa la multiplicación de la torsión.

**Warning label** A label that is supplied by the manufacturer of the hazardous material. Each warning label lists the chemical name, applicable hazard warnings, hazardous ingredients, and the manufacturer's name and address.
**Marbete de aviso** Un marbete proveido por el fabricante de los materiales peligrosos. Cada marbete especifica el nombre químico,

los avisos pertinentes del peligro, y el nombre y la dirección del fabricante.

**Warning light** A lamp that is illuminated to warn the driver of a possible problem or hazardous condition.
Luz de aviso Lámpara que se enciende para avisarle al conductor sobre posibles problemas o condiciones peligrosas.

**Wastegate stroke** The amount of turbocharger wastegate diaphragm and rod movement.
Carrera de la compuerta de desagüe Cantidad de movimiento del diafragma y de la varilla de la compuerta de desagüe del turbocompresor.

**Watt** The unit of measure of electrical power, which is the equivalent of horsepower. One horsepower is equal to 746 watts.
Watio Unidad de potencia eléctrica, equivalente a un caballo de vapor. 746 watios equivalen a un caballo de vapor (CV).

**Wattage** A measure of the total electrical work being performed per unit of time.
Vataje Medida del trabajo eléctrico total realizado por unidad de tiempo.

**Wear-in** The required time needed for the ring to conform to the shape of the cylinder bore.
Tiempo de estreno El tiempo que se requiere para que el anillo se conforme a la forma del taladro del cilindro.

**Wear indicator** A device that visually indicates ball joint wear.
Indicador de desgaste Dispositivo que indica de manera visual el desgaste de una junta esférica.

**Weather-pack connector** An electrical connector that has rubber seals on the terminal ends and on the covers of the connector half to protect the circuit from corrosion.
Conectador resistente a la intemperie Conectador que tiene sellos de caucho en los extremos de los bornes y en las cubiertas de la parte del conectador para proteger el circuito contra la corrosión.

**Wet compression testing** Compression test done after adding a small amount of oil to the cylinder. Wet compression testing is performed if the cylinder fails the dry compression test.
Prueba de compresión húmedo La prueba de compresión que se lleva acabo añadiendo una pequeña cantidad del aceite al cilindro. Se efectúa si el cilindro reprueba la prueba de compresión en seco.

**Wet-disc clutch** A clutch in which the friction disc (or discs) is operated in a bath of oil.
Embrague de disco flotante Un embrague en el cual el disco (o los discos) de fricción opera en un baño de aceite.

**Wet sleeves** Replacement cylinder sleeves that are surrounded by engine coolant.
Camisas húmedas Las camisas de repuesta del cilindro que se rodean por el fluido refrigerante del motor.

**Wheel** A disc or spokes with a hub at the center that revolves around an axle, and a rim around the outside for mounting the tire on.
Rueda Un disco o rayo que tiene en su centro un cubo que gira alrededor de un eje, y tiene un rim alrededor de su exterior en la cual se monta el neumático.

**Wheel balancer** Equipment for wheel balancing that is usually computer controlled.
Equilibrador de ruedas Equipo para la equilibración de ruedas que por lo general es controlado por computadora.

**Wheel cylinder** A mechanism located at each wheel in a drum brake system. It uses hydraulic pressure to force the brake shoes against the drum to stop the wheel from turning.

Cilindro de la rueda Un mecanismo ubicado en cada rueda de un sistema de frenos de tambor. Utiliza la presión hidráulica para forzar las zapatas contra el tambor así previniendo que gira la rueda.

**Wheel offset** The amount of the wheel assembly that is to the side of the wheel's mounting hub.
Desviación de la rueda La porción del conjunto de la rueda ubicada al lado del cubo de montaje de la rueda.

**Wheel runout compensation screen** A display on a computer wheel aligner that is shown during the wheel runout compensation procedure.
Pantalla para la visualización de la compensación de desviación de la rueda Representación visual en una computadora para alineación de ruedas que aparece durante el procedimiento de compensación de desviación de la rueda.

**Wheel shimmy** The wobble of a tire.
Baileteo de la rueda Movimiento de zigzag de una rueda.

**Wiring diagram** An electrical schematic that shows a representation of actual electrical or electronic components and the wiring of the vehicle's electrical systems.
Esquema de conexiones Esquema en el que se muestran las conexiones internas de los componentes eléctricos o electronicos reales y las de los sistemas eléctricos del vehículo.

**Wiring harness** A group of wires enclosed in a conduit and routed to specific areas of the vehicle.
Cableado preformado Conjunto de alambres envueltos en un conducto y dirigidos hacia áreas específicas del vehículo.

**Woodruff key** A half-moon shaped metal key used to retain a component, such as a pulley, on a shaft.
Chaveta woodruff Chaveta de metal en forma de media luna que se utiliza para sujetar un componente, como por ejemplo una roldana, en un árbol.

**Worm gear** A gear with teeth that resemble a thread on a bolt. It is meshed with a gear that has teeth similar to a helical tooth except that it is dished to allow more contact.
Engranaje sinfín Engranaje con dientes parecidos a los filetes de tornillo en un perno. Se engrana con un engranaje cuyos dientes son parecidos a un diente helicoidal, pero se comba para permitir un mejor contacto.

**Yield strength** A measurement of the material strength from which a frame is manufactured.
Límite para fluencia Medida de la resistencia del material del que está fabricado un armazón.

**Yoke** In a universal joint, the drivable torque-and-motion input and output member, attached to a shaft or tube.
Yugo En una junta universal, el miembro de la entrada y la salida que transfiere a la torsión y al movimiento, que se conecta a una flecha o a un tubo.

**Yoke bearing** A U-shaped, spring-loaded bearing in the rack-and-pinion steering assembly that presses the pinion gear against the rack.
Cojinete de yugo En el conjunto de dirección de cremallera y piñón, cojinete en forma de U, con cierre automático, que sujeta el piñón contra la cremallera.

**Zerk fitting** A common name for grease fittings. A very small check valve that allows grease to be injected into a component part, but keeps the grease from squirting out again.
Conexión Zerk Término común para conexiones de engrase. Válvula de retención sumamente pequeña que permite inyectar la grasa en un componente, y que a la vez impide que esa grasa se derrame nuevamente.